Bulk Metallic Glasses, Second Edition

Bulk Metallic Glasses, Second Edition

C. Suryanarayana and A. Inoue

CRC Press
Taylor & Francis Group
Boca Raton London New York

CRC Press is an imprint of the
Taylor & Francis Group, an **informa** business

CRC Press
Taylor & Francis Group
6000 Broken Sound Parkway NW, Suite 300
Boca Raton, FL 33487-2742

First issued in paperback 2020

ISBN-13: 978-1-4987-6367-7 (hbk)
ISBN-13: 978-0-367-65750-5 (pbk)

This book is fondly

dedicated

to the memory of

Professor T.R. Anantharaman

(November 25, 1927–June 18, 2009)

who pioneered research on metallic glasses in India

Contents

Foreword

On September 3, 1960, there appeared in print a short article, barely over one column long, that shifted a paradigm: Klement, Willens, and Duwez reported that the alloy $Au_{75}Si_{25}$ (composition in atomic %) could be obtained in glassy form by rapid solidification of the liquid [*Nature* 187, 869 (1960)]. It was already known that metallic alloys could be obtained in unstable noncrystalline form by vapor deposition onto cold substrates, but the realization of a metallic *glass* (formed when a liquid solidifies without crystallization) was a surprise. At first, the very possibility of a metallic glassy state was disputed, but the active research stimulated by that 1960 paper has gone on to show not only that metallic alloys can form true glasses, but also that such glasses can be formed in bulk without the need for rapid solidification, that they have distinctive, sometimes record-breaking, properties, and that they have and will find a wide range of applications.

It is time for a golden jubilee celebration! Thankfully, we find one well executed in the form of this book that comprehensively covers the modern field of bulk metallic glasses. Better still, the book's authors have prominent and complementary profiles in the field. With publication records spanning some 80 man-years, the authors can put metallic glasses into context in the broad fields of advanced metallic materials and nonequilibrium processing in which they have both been so active.

Over the last 50 years, interest has waxed and waned, but surveying the field at this point, we can see not only that metallic glasses are an important new class of material, but also that their appearance has transformed fundamental understanding in a wide range of areas. For example, studies of the structure of metallic glasses have stimulated much work on equilibrium and supercooled metallic liquids. The structure of such liquids was first described by Bernal as dense random packing. While a high packing density is a characteristic of metallic liquids and glasses, it is now better appreciated that high densities are not associated with randomness, but rather with high degrees of order, albeit noncrystalline. The kinetics of crystallization of liquids is often discussed in terms of time–temperature–transformation diagrams. Whereas such diagrams were once purely schematic, glass-forming metallic systems allow them to be determined experimentally. Transmission electron microscopy of metallic glasses has permitted atomic-level investigations of crystal nucleation, relevant for standard solidification yet difficult or impossible to conduct on the liquids themselves. A comparison of metallic glasses with other longer-established members of the broad family of glasses has improved understanding of the glass transition, and has revealed remarkable correlations of thermal, elastic, and mechanical properties that hint at fundamentals of the glassy state spanning different bonding types.

Metallic glasses are of interest in establishing the fundamental basis of physical, mechanical, and chemical properties. In their early days, for example, they were important in showing something not universally expected at the time—that crystallinity is not a prerequisite for ferromagnetism. And at the present time, the distinctive shear-band deformation mode of metallic glasses is stimulating the development of theory on the ultimate strength of materials.

Such contributions to the broader understanding of condensed matter can only support interest in the metallic glasses themselves as new materials. This book covers their remarkable development since 1960: production methods, understanding of structure and glass-forming ability, composition ranges available, and properties. Above all, it is the availability of many metallic glasses in bulk form (with minimum dimension exceeding 1 mm to 1 cm) that has stimulated the current intense research on these materials.

Among the reasons for studying the crystallization of metallic glasses is that attractive microstructures can be obtained. The metallic glasses can act as precursors, rather as oxide glasses do in the production of conventional glass ceramics. There are, indeed, several ways in which composites based on metallic glasses are of interest. Among those covered in this book are dispersions of a primary ductile crystalline phase in a metallic–glass matrix, composites that show remarkable combinations of strength and toughness. Also covered is the emerging topic of metallic glass foams.

Very early in studies of metallic glasses, it was recognized that they would have properties attractive for applications. Their excellent soft magnetic properties were the first to be exploited on a large commercial scale. As made clear in this book, their physical and mechanical properties, and their corrosion resistance, can also be highly attractive. So much of our technology is limited by materials performance. The current challenges in clean power generation and other aspects of sustainability may yet be materials science's finest hour, when solutions emerge in the form of new materials and new processing routes: bulk metallic glasses will surely have a prominent role. They may not be used on the largest scales associated with structural steels, but on the scales associated with personal items, such as sports equipment down to micro- and nano-devices, they have much to offer.

In this important reference work for the field, the authors provide comprehensive yet critical coverage of all major aspects of bulk metallic glasses. They recognize that it is important to understand the drawbacks as well as the advantages of these materials. They also show that while there have been great advances in understanding, many challenges remain. Bulk metallic glasses are at the heart of an exciting field with much still to deliver.

A. Lindsay Greer
Cambridge, UK

The future of metallurgy lies in mastering disorder. In fact, one can argue that the trajectory of our field has followed a path of increasing topological disorder ever since the realization that pure metals are crystalline. Although crystallinity provides symmetry and order, it is the regions where ideal crystalline packing is disrupted that capture our attention. Metallurgists focus on "defects," including dislocations and grain boundaries, recognizing that these positions of broken periodicity, even when present in vanishingly small volume fractions, utterly dominate the most important properties. Recently, the interest in so-called nanocrystalline or nanostructured metals has driven attention to a regime where more atoms than not lie in disordered regions, evoking a variety of new mechanisms and properties of immediate technological interest.

In this landscape, amorphous metals are a limit—they are completely noncrystalline, with a negligible fraction of atoms sitting in a symmetric, crystalline environment. They are a mystery—their structure is often more definable in terms of what they are not than what they are. When first discovered in the 1950s, they were rare and unusual; today they are pervasive, and can be produced by a wide array of synthesis routes. Their properties are not limited by the presence of lattice defects; they promise a suite of genuinely exceptional properties. Extreme values of strength, fracture toughness, magnetic properties, corrosion resistance, and other properties have been recorded in amorphous metals. In some cases, these come not individually, but in combinations unparalleled by any other material known to humankind.

If our future is to master disorder, then amorphous metals will certainly point the way. In particular, metallic glasses (i.e., amorphous metals produced from the melt) are formed by intentionally stabilizing the disordered liquid structure, and the rules for doing so become better understood with each passing year. And, although we may glibly describe them as "disordered," metallic glasses are structured in rich and complex ways that are just beginning to be unraveled. Their properties emerge in enigmatic ways from their disordered structure, and the means of optimizing those properties remain tantalizingly out of reach.

Metallic glasses have been actively studied for decades, and with particular fervor in the one just past. With many thousands of scholarly articles on the subject, and as many more expected in the coming years, a research text that overviews the entire field is overdue. Although there have been conference proceedings, edited collections, and review articles on metallic glasses, this book is unique in that its authors have designed a text that systematically covers each of the important aspects of the field, from processing, to structure, to properties. The authors have been at the center of this field through its most crucial period, have extensive experience in writing reviews, and bring complementary views to bear in the book. Each of the chapters presents a balance of breadth and depth: the coverage of the field is extensive with copious citations to the research literature and extensive compilation of data on glass properties. Attention is paid to critical nuances that highlight the complexity of glasses, such as the difference between "bulk" glass-forming alloys formed by casting (which is the principal focus of the book), and similar alloys with subtly different structures produced by rapid solidification or solid-state routes.

With all of these qualities, this book is sure to be a welcome source of reflection for those already immersed in the field, and an inspiration to those entering it. I expect that this book will be an important milestone on the path to mastering disorder.

Christopher A. Schuh
MIT, Cambridge, Massachusetts

Metallic glasses with amorphous structures, first discovered in 1960, are of both scientific and technological interest. The successful synthesis of bulk metallic glasses during the late 1980s stimulated great enthusiasm in the study of this class of metallic materials. Substantial progress has been made in understanding the physical, chemical, and mechanical properties of bulk metallic glasses since the beginning of the 1990s. Characterization on the atomic scale indicates the formation of tightly bonded atomic clusters and loosely bonded free-volume zones in the amorphous state. Various polyhedral packings have been observed in these clusters, and have also been confirmed by atomic simulations. Unlike crystalline materials with periodic lattice structures, the plastic deformation of bulk metallic glasses proceeds via localized shear bands, resulting in inhomogeneous deformation at ambient temperatures. Bulk metallic glasses show extremely high strength, close to the theoretical strength of solids in the glassy state, and superplastic behavior in the supercooled liquid state. Experimental studies and theoretical analyses both reveal shear transformation zones, instead of dislocations, as the basic unit for plastic deformation in bulk metallic glasses. Furthermore, recent studies indicate that bulk metallic glasses possess attractive physical and mechanical properties for high-tech applications in micro/-nanosystems. These interesting results, as well as all fundamental and applied topics, are systematically described in 11 chapters in this book, authored by Professor Suryanarayana from the University of Central Florida, Orlando, Florida, and Professor Inoue from Tohoku University, Sendai, Japan.

Professor Inoue has conducted pioneering work on the synthesis and development of bulk metallic glasses over the past 20 years. He is familiar with all aspects of the properties and behaviors of these materials. There is no doubt that Professor Inoue is the right person to lead the effort in writing this book. In contrast to the edited book published previously, this book covers all topics in a coherent manner. Bulk metallic glasses are presently emerging as a new class of metallic materials with unique physical and mechanical properties for structural and functional uses. This book is suitable for young researchers in materials science and applied physics who are interested in learning about bulk metallic glasses, and are looking for a guidebook to help launch their research into this exciting materials field. This book is also perfect for use as a textbook for students in graduate schools.

C.T. Liu
Hong Kong Polytechnic University
Hong Kong
Auburn University
Auburn, Alabama

Preface to the Second Edition

We are extremely happy with the reception that the first edition of our book on bulk metallic glasses received from senior and experienced researchers, junior researchers, and fresh graduate students. There have been several positive comments about the style adopted in the book to explain the concepts and the present status of this exciting new field of materials science and engineering.

The fast pace of research in this field has been maintained from the beginning, and new findings are continuously being reported in both archival journals and conference proceedings. Therefore, we have decided to bring out a new edition.

In this second edition, we have decided to maintain the easy readability of the material without sacrificing accuracy. Additionally, we have incorporated all the new information that has been published in the period since the publication of the first edition until the end of 2016. The new information generated in the last few years has essentially been in the areas of glass-forming ability, corrosion behavior, mechanical properties, and applications of bulk metallic glasses. We have tried to include all the new data and to provide up-to-date assessments in all the sections. We have also updated all the references in every chapter of the book.

We have been very fortunate in receiving constructive comments and criticism from the readers. Particular mention must be made of Professors S. Ranganathan of the Indian Institute of Science in Bangalore, India, and Dmitri Louizguine-Luzgin of the WPI-Advanced Materials Research Institute of Tohoku University in Sendai, Japan. We are very thankful to them and all others who provided feedback. We are also thankful to Dr. F.L. Kong of the International Institute of Green Materials of the Josai International University in Togane, Japan, and Dr. Zhi Wang of the South China University of Technology, Guangzhou, China, for help in different aspects of the preparation of the draft.

Despite taking great care to present the wealth of new information carefully and accurately, it is possible that some errors may have crept in. If the readers notice any errors in the data or presentation, we would be extremely grateful if they could please contact us at Surya@ucf.edu or ainouebmg@yahoo.co.jp.

C. Suryanarayana
Orlando, Florida

A. Inoue
Togane, Japan

Acknowledgments

It is with great pleasure that we acknowledge the invaluable assistance we have received from many colleagues and friends who have contributed, in different ways, to the successful completion of this book.

Both of us have been involved with the synthesis, characterization, and applications of rapidly solidified materials in general, and metallic glasses in particular, for over three decades. During this long journey, we have had the good fortune of interacting with a number of leaders in this field, from whom we have learned immensely and benefited greatly in understanding the complexities of metallic glasses. In alphabetical order, they are T.R. Anantharaman, R.W. Cahn, H.S. Chen, P. Duwez, T. Egami, H. Fujimori, A.L. Greer, K. Hashimoto, C.T. Liu, T. Masumoto, J.H. Perepezko, S. Ranganathan, K. Suzuki, and A.R. Yavari. We are pleased to have had their guidance and advice at different stages in our professional careers.

Some chapters of the book were read by colleagues. We are grateful to Professor Dmitri Louzguine-Luzgin and Dr. Nobuyuki Nishiyama of Tohoku University, Sendai, Japan, and Professor R. Vaidyanathan of the University of Central Florida, Orlando, Florida, for their critical comments and constructive suggestions. The incorporation of these suggestions has certainly improved the clarity and readability of the book.

Many of the figures in this book have been prepared by friends and students. We are thankful to Dr. Akira Takeuchi, Dr. Ichiro Seki, C.L. Qin, and Dr. U.M.R. Seelam for their hard work and patience in preparing the several iterations of the figures. The aesthetics of the figures are mostly due to their skills. Professors Dmitri Louzguine-Luzgin and Yoshihiko Yokoyama have also provided some of the figures used in this book. We are also grateful to the authors and publishers from whom we have borrowed several figures.

Parts of this book were written while one of the authors (C.S.) was a visiting professor at the International Frontier Center for Advanced Materials at Tohoku University in Sendai, Japan. C.S. is grateful to the Tohoku University for the award of this professorship. A number of colleagues have helped us during this stay. Particular mention should be made of Professor Mingwei Chen, Dr. Hisamichi Kimura, Professor Dmitri Louzguine-Luzgin, Dr. Nobuyuki Nishiyama, Dr. Tokujiro Yamamoto, and Dr. Yoshihiko Yokoyama, with whom we had stimulating discussions. C.S. is also grateful to Professor Kazuhiro Hono of the National Institute for Materials Science, Tsukuba, Japan, for several useful discussions on different aspects of bulk metallic glasses. The quality of the material presented in the book has vastly improved as a result of these discussions.

C.S. would also like to thank Professors Ranganathan Kumar and David Nicholson, successive chairmen of the Department of Mechanical, Materials, and Aerospace Engineering at the University of Central Florida for providing a conducive environment in which to complete the book.

We wish to thank the staff of Taylor & Francis/CRC Press for their high level of cooperation and interest in successfully producing a high-quality and aesthetically pleasing book. We are particularly thankful to Allison Shatkin for her patience in waiting for the delivery of the final manuscript.

Last, but by no means least, we owe a huge debt of gratitude to our wives Meena and Mariko, who encouraged and supported us with love, understanding, and patience throughout this endeavor.

C. Suryanarayana
Orlando, Florida

A. Inoue
Sendai, Japan

Authors

C. Suryanarayana, PhD, is a professor in the Department of Mechanical and Aerospace Engineering at the University of Central Florida in Orlando, Florida. He has conducted research investigations in the areas of rapid solidification processing, mechanical alloying, innovative synthesis/processing techniques, metallic glasses, superconductivity, quasicrystals, and nanostructured materials for almost 50 years and has concentrated his research efforts on bulk metallic glasses and mechanical alloying and milling for the past 20 years. He has published more than 380 technical papers in archival journals and authored/edited 9 books and 12 conference proceedings. Earlier, he was a professor of metallurgy at Banaras Hindu University, in Varanasi, India. He held visiting assignments at the University of Oxford, Oxford, United Kingdom; the Atomic Energy Center in Mol, Belgium; the Wright-Patterson Air Force Base in Dayton, Ohio; the University of Idaho in Moscow, Idaho; the Colorado School of Mines in Golden, Colorado; the GKSS Research Center in Geesthacht, Germany; the Helmut Schmidt University in Hamburg, Germany; Tohoku University in Sendai, Japan; the National Institute for Materials Science in Tsukuba, Japan; Chungnam National University in Taejon, Korea; Shenyang National Laboratory for Materials Science in Shenyang, China; and Northeastern University in Shenyang, China. Most recently, he was selected as a Jefferson Science Fellow by the U.S. National Academies and served as a senior science adviser in the U.S. Department of State in Washington, DC.

Professor Suryanarayana is on the editorial boards of several prestigious materials science journals. He has received several awards for his research contributions to nonequilibrium processing of materials, including the Science Academy Medal for young scientists of the Indian National Science Academy, the Pandya Silver Medal of the Indian Institute of Metals, the National Metallurgists Day Award of the Government of India, the Distinguished Alumnus Award of Banaras Hindu University, and the Lee Hsun Research Award from the Chinese Academy of Sciences. He was also awarded the Lifetime Achievement Award in Engineering by the Central Florida Engineers, and the Lifetime Achievement Award in Electron Microscopy by the Electron Microscope Society of India. In 2011, Thomson Reuters announced that Professor Suryanarayana was one of the world's top 40 researchers in the field of materials science, based on those achieving the highest citation impact scores for their papers published since January 2000. Most recently, TMS, the premier materials society of the United States, awarded him the 2016 Educator Award. He is a fellow of ASM International, the Institute of Materials, Minerals and Mining, London, United Kingdom, and the Electron Microscope Society of India. He earned his BE in metallurgy from the Indian Institute of Science, Bangalore, and his MS and PhD in physical metallurgy from Banaras Hindu University, Varanasi, India.

A. Inoue , PhD, is an Emeritus Professor of Tohoku University and has conducted research investigations on rapidly solidified glassy alloys; nanocrystalline alloys for engineering applications; iron and steel metallurgy; metallic glasses; superconductivity; and synthesis, characterization, and commercialization of metallic glasses in general, and bulk metallic glasses (BMGs) in particular. He has pioneered the field of BMGs through systematic synthesis, characterization, and applications for the past 25 years. He has published more than 2000 papers in archival journals and edited several conference proceedings and books. He also holds more than 200 patents and has been a member of the editorial boards of several archival journals. Professor Inoue was president of Tohoku University in Sendai, Japan, from 2006 to 2012. Currently, he is a professor and special adviser to the chancellor of Josai University, Tokyo, Japan, and director of the International Institute of Green Materials, Josai International University, Togane, Japan. Since 2013, he has been the 1000 Talented Invited Professor at the School of Materials Science and Engineering, Tianjin University, Tianjin, China, and since the beginning of 2017, he has been the special invited professor and director of the Institute of Massive Amorphous Metal Science, Chinese University of Mining Technology, Suzhou, China.

Professor Inoue has held visiting assignments at the Royal Institute of Technology, Stockholm, Sweden; the Swedish Institute of Metals Research; AT&T Bell Laboratories, Murray Hill, New Jersey; and the Institute für Forschung Werkstoff (IFW), Dresden, Germany. He was awarded honorary doctorate degrees from the Swedish Royal Institute of Technology, Stockholm, Sweden; from Dong-Eui University, Busan, Korea; and from the Shanghai Jiao Tong University, Shanghai, China. He was appointed a fellow of Churchill College, University of Cambridge, Cambridge, United Kingdom. He holds honorary professor positions in several prestigious universities.

Professor Inoue has received several awards and recognitions for his research contributions. He is a member of the Japan Academy and a foreign member of the U.S. National Academy of Engineering. Some of the most important awards he has received include the Japan Academy Prize (in recognition of his outstanding scholarly contributions to the pioneering development of bulk metallic glasses [BMGs]), the Japan Prime Minister's Prize (in recognition of his outstanding Industry–University–Government Cooperation Achievement of BMGs), and the James C. McGroddy Prize for New Materials (in recognition of the development of slow cooling methods for the fabrication of BMGs with remarkable mechanical properties and the characterization and applications of these materials) from the American Physical Society. He was most recently awarded the 2010 Acta Materialia, Inc. Gold Medal. He has also delivered the Kelly Lecture (University of Cambridge, Cambridge, United Kingdom) and the Dr. Morris Traverse Lecture (Indian Institute of Science, Bangalore, India). The Institute for Scientific Information has selected him as one of the most cited researchers in the field of materials science and engineering. He earned his BS in metallurgical engineering from the Himeji Institute of Technology, and his MS and PhD in materials science and engineering from Tohoku University, Sendai, Japan.

1

Introduction

1.1 Motivation

The search for new and advanced materials has been the major preoccupation of materials scientists during the past several years. Recent investigations have focused on the improvement of the properties and performance of existing materials and/or synthesis and development of completely novel materials. Significant improvements have been achieved in the mechanical, chemical, and physical properties of materials by the addition of alloying elements, microstructural modification, and by subjecting the materials to thermal, mechanical, or thermomechanical processing methods. Completely new materials, unheard of earlier, have also been synthesized. These include metallic glasses [1–3], quasicrystals [4,5], nanocrystalline materials [6–10], and high-temperature superconductors [11], to name a few. The high-technology industries have provided the opportunity and fillip to develop these novel materials. Along with the development of the synthesis methods for these materials, there has also been the development of newer and improved techniques to characterize these materials with better resolution capabilities to determine the crystal structure and microstructure at different levels (nanometer, micrometer, and mesoscale levels), phase identification, and composition of phases of ever-decreasing dimensions and with higher and higher resolutions, down to the atomic level.

The rapid progress of technology during the last 50–60 years and the current demands in the twenty-first century have been putting tremendous pressure on materials scientists to develop ever-newer materials that have further improved properties, such as higher strength or improved stiffness, and materials that could be used at much higher temperatures and in more aggressive environments than is possible with traditional and commercially available materials. These efforts have resulted in the design and development of advanced materials that are "stronger, stiffer, and lighter" and also those that could be used at much higher temperatures ("hotter") than the existing materials. The synthesis and development of such materials have been facilitated by exploring the interrelationship between the processing, structure, properties, and performance of materials; the basic underlying theme of materials science and engineering.

1.2 Advanced Materials

Advanced materials have been defined as those where first consideration is given to the systematic synthesis and control of crystal structure and microstructure to provide a precisely tailored set of properties for demanding applications [12]. Thus, the attraction of advanced materials is that they could be designed and synthesized with improved and well-defined properties for specific applications.

Naturally available materials have excellent property combinations in some cases. They are even being imitated to develop newer materials, and a new area of the biomimetic synthesis of materials has emerged. But, in order to achieve a combination of properties and performance better than that of the existing materials, it is well recognized that materials need to be processed under far from equilibrium or nonequilibrium conditions [13]. This realization has led to the development of a number of nonequilibrium processing techniques during the second half of the twentieth century. Among these, special mention may be made of rapid solidification processing (RSP) [1–3,14–17], mechanical alloying (MA) [18–22], plasma processing [23], vapor deposition [24], and spray deposition [25]. Considerable research is being done in these different areas to develop materials for a variety of applications, and this is clearly evident in the large number of publications every year and also the number of conferences devoted to these topics.

The basis of nonequilibrium processing is to "energize and quench" a material, as proposed by Turnbull [26]. Processes such as solid-state quenching, rapid solidification from the liquid state, irradiation, and condensation from the vapor phase were considered to evaluate the departures from equilibrium. However, there are several other methods of nonequilibrium processing that do not involve quenching. These include, among others, the static undercooling of liquid droplets, electrodeposition of alloys, MA, severe plastic deformation, application of high pressures, and absorption of hydrogen. Therefore, instead of calculating the quench rate, it may be useful to evaluate the maximum departure from equilibrium in each of these processing methods.

The process of "energize and quench" to synthesize metastable phases has been described earlier [27]. In brief, the process of energization involves bringing the equilibrium crystalline material into a highly energetic condition by some external dynamic forcing, for example, through an increase in temperature (melting or evaporation), irradiation, the application of pressure, or the storing of mechanical energy by plastic deformation. Such energized materials were referred to as *driven materials* by Martin and Bellon [28]. The process of energization may also involve a possible change of state from solid to liquid (melting) or gas (evaporation). For example, during RSP the starting solid material is melted, and during vapor deposition the material is vaporized. The energized material is then "quenched" into a configurationally frozen state by methods such as RSP so that the resulting material is in a highly metastable condition. This "quenched" phase can then be used as a precursor to obtain the desired chemical constitution (other less metastable phases) and/or microstructures (e.g., nanocrystalline material) by subsequent heat treatment/processing. It has been shown that materials processed in this way possess improved physical, chemical, and mechanical characteristics in comparison to their conventional ingot (solidification) processed materials. These metastable phases can be used either as they are or subsequently transformed to other less metastable or near-equilibrium or equilibrium crystalline phases to achieve the desired microstructural features and properties by annealing methods.

It has long been known that the properties of materials can be changed by altering the crystal structure (through polymorphic changes) or, more significantly, through microstructural modifications (by introducing crystal defects such as dislocations and grain boundaries), or a combination of both. Gleiter [10] has recently suggested that, apart from nanocrystalline materials, one could also consider nanoglasses to improve the properties of materials further. Nanoglasses are similar to nanocrystalline materials,

TABLE 1.1

Departure from Equilibrium Achieved in Different Nonequilibrium Processing Methods

Technique	Effective Quench Rate (K s⁻¹) Ref. [26]	Maximum Departure from Equilibrium (kJ mol⁻¹)	
		Ref. [29]	Refs. [30,31]
Solid-state quench	10^3	0.5–1.0	16
Rapid solidification processing	10^5–10^8	2–3	24
Mechanical alloying	—	10	30
Mechanical cold work	—	—	1
Irradiation/ion implantation	10^{12}	—	30
Condensation from vapor	10^{12}	100	160

except that the "grains," instead of being crystalline as in nanocrystalline materials, are glassy.

The ability of the different processing methods to synthesize nonequilibrium structures may be conveniently evaluated by measuring or estimating the achieved departure from equilibrium, that is, the maximum energy that could be stored in excess of the equilibrium/stable value. This has been done by different groups of researchers for different nonequilibrium processing methods [26,29–32]. While the excess energy is expressed in kilojoule per mole (kJ mol^{-1}) in Refs. [29–31], Turnbull [26] expressed this as an "effective quenching rate." The way the departure is calculated is also different in these different approaches (the reader should refer to the original papers for details of calculations), and therefore the results do not correspond exactly in all cases. But, they at least provide a means to compare the efficiencies of the different techniques to achieve the metastable effects. Table 1.1 summarizes the departures calculated for some of the different nonequilibrium processing methods mentioned here. It is clear from Table 1.1 that RSP introduces large departure from equilibrium, even though higher departures have been noted when the material is processed either by MA, vapor deposition, or by ion implantation.

A large variety of techniques are available to process materials under far from equilibrium conditions. But, we will briefly describe the technique of RSP, since it is the variation of this technique that led to the synthesis and development of bulk metallic glasses, the subject matter of this book. MA is another technique that has also been found to be very useful in synthesizing nonequilibrium materials, including metallic glasses. Therefore, a brief description of this process is also presented.

1.3 Rapid Solidification Processing

The RSP of metallic melts was first conducted by Pol Duwez and his colleagues at the California Institute of Technology (Caltech) in Pasadena, California, during the 1959–1960 time period [1]. A historical description of the discovery of metallic glasses has been provided by Duwez himself [33]. In this method, a molten metal or alloy is solidified very rapidly at typical rates of about 10^6 K s^{-1}, but at least 10^4 K s^{-1}. Such high solidification rates have been achieved traditionally by any of the following three variants.

1. *Droplet methods*: In this group of methods, a molten metal is atomized into small droplets, and these are allowed to solidify either in the form of splats (on good thermally conducting substrates, e.g., as in "gun" quenching) or by impinging a cold stream of air or an inert gas against the molten droplets (as, for example, in atomization solidification).
2. *Jet methods*: In these methods, a flowing molten stream of metal is stabilized so that it solidifies as a continuous filament, ribbon, or sheet in contact with a moving chilled surface (e.g., chill block melt spinning and its variants).
3. *Surface melting technologies*: These methods involve rapid melting at the surface of a bulk metal followed by high rates of solidification achieved through rapid heat extraction into the unmelted block (as in laser surface treatments).

A number of techniques based on these three basic categories have been developed over the years, and these have been reviewed and summarized earlier [34,35]. Therefore, we will not describe any of these techniques here. The interested reader can refer to the reviews or original publications for the details.

The technique of RSP has revolutionized many traditional concepts of metallurgy and materials science. The most dramatic of these, for example, is that metallic materials can exist, in addition to their normal crystalline state, either in a glassy (noncrystalline or amorphous) state [1–3] or in a quasicrystalline state (in which the traditionally forbidden rotational crystal symmetries could be observed) [4,5]. Furthermore, it has been possible to synthesize a variety of other metastable phases, such as supersaturated solid solutions, and nonequilibrium intermediate phases. Rapidly solidified alloys have been finding a multitude of applications, including a range of soft (for transformer core laminations) and hard magnetic materials, wear-resistant light alloys, materials with enhanced catalytic performance and for fuel-cell applications, powder metallurgy tool steels and superalloys, and new alloys for medical implants and dental amalgams [36].

One of the requirements to achieve high solidification rates during RSP and, consequently, the metastable effects in materials, is that heat must be extracted very rapidly from the melt. The achievement of solidification rates of the order of 10^6 K s^{-1} was possible only when at least one of the dimensions of the specimen was small, usually of the order of about 20–50 μm. As a result, the products of RSP were ribbons, wires, or powders. Finding applications for such thin materials was not easy unless these powders, ribbons (after pulverization), and other forms were consolidated to full density as bulk materials. On the other hand, if a glassy alloy could be obtained at slow solidification rates, for example, by water quenching where the solidification rate was only about 10^2 K s^{-1}, the section thickness of the glassy phase would be expected to be much larger. Thus, materials scientists have been on the lookout for materials and/or processes to produce metallic glasses in thicker sections at slower solidification rates, very similar to what is commonly done for silicate or oxide glasses. These attempts were successful in the late 1980s, mostly due to the efforts of Professors Akihisa Inoue and Tsuyoshi Masumoto at the Tohoku University in Sendai, Japan. They produced 1.2-mm diameter rods of $La_{55}Al_{25}Ni_{20}$ alloy in a fully glassy condition by water quenching [37] and 3-mm diameter rods and 2.3×5 mm square sections of the same La–Al–Ni–alloy with glassy phase by metallic mold casting [38]. Even though millimeter-sized glassy alloys were produced earlier [39,40], this was the first time that glass formation was demonstrated in such sizes in alloys without the presence of a noble metal and by copper mold casting. This glassy alloy had a wide supercooled liquid region ΔT_x (=$T_x - T_g$, where T_x and T_g represent the crystallization and glass-transition temperatures, respectively). Such glasses with large section thicknesses are now referred to as *bulk metallic glasses*. Later, Professor Bill Johnson from Caltech and his group also produced a number of Zr-based bulk metallic glasses with thicknesses in the centimeter range [41]. Currently, activity in this research area of materials science is global, and it is estimated that more than a thousand research papers are published annually.

1.4 Mechanical Alloying

Mechanical alloying (MA) is a powder processing method that was developed in the mid-1960s by John Benjamin at INCO International [42] to produce nickel-based oxide-dispersion strengthened (ODS) superalloys for gas turbine applications [20,21]. In this process, involving repeated cold welding, fracturing, and rewelding of powder particles in a high-energy ball mill, the blended elemental powder particles and grinding medium (usually stainless steel or tungsten carbide balls) are loaded into a vial and agitated at a high speed for the desired length of time. The soft powder particles of each metal or constituent get crushed and become flat like pancakes. These flat particles of different metals form layered structures. This process gets repeated a few hundreds to thousands of times, resulting in the convolution of powders. After milling the powders for some time, very thin layered structures consisting of the individual components are formed and the layer thickness is very small. Because of the heavy plastic deformation experienced by the powders, crystal defects such as dislocations, grain boundaries, vacancies, and others are introduced into the powder particles. Simultaneously, there is also a small rise in the temperature of the powder particles. Due to the combined effects of thin lamellae (and therefore reduced diffusion distances), increased diffusivity (due to the presence of a high concentration of crystal defects), and the slight increase in powder temperature, diffusion is easily facilitated, and this allows the blended elemental particles to alloy with each other at room or near-room temperature.

Mechanically alloyed powders also have been shown to display a variety of constitutional and microstructural changes. Using this technique, it has been possible to produce equilibrium alloys starting from blended elemental powders at temperatures significantly lower than by conventional methods. It has also been shown that it is possible to synthesize a variety of nonequilibrium phases, such as supersaturated solid solutions, metastable intermediate phases, quasicrystalline alloys, nanostructured materials, and metallic glasses [20,21]. In fact, all the nonequilibrium effects achieved by RSP of metallic melts have also been achieved in mechanically alloyed powders.

The mechanically alloyed powders can be consolidated to full density by conventional or advanced methods, such as vacuum hot pressing, hot extrusion, hot isostatic pressing, shock consolidation, spark plasma sintering, or combinations of these, to obtain bulk samples. In fact, this technique of MA is a very viable alternative to produce bulk glassy alloys of any size, even if there is a limit to the section thickness obtained by solidification methods. This is because the amorphous powders obtained by MA can be consolidated in the supercooled liquid region (the temperature interval between the glass transition and crystallization temperatures) to any size, without the amorphous phase crystallizing. Such examples are available in the literature (see, for example, Ref. [43]).

1.5 Outline of the Book

In Chapter 2, we will introduce the basic concepts of metallic glasses and differentiate between crystalline and glassy materials, with special emphasis on the differences between noncrystalline, glassy, and amorphous solids. We will then describe the general background to glass formation by different processing methods and then lead the reader to the concepts of bulk metallic glass formation. The chapter will end with the various literature resources that are available to researchers (both new and experienced), including details of the dedicated conferences on the subject of bulk metallic glasses.

Glass formation requires that the alloy satisfies certain basic criteria. Furthermore, a number of different criteria have been developed during the last several years to understand which alloy systems show a high glass-forming ability. These different criteria will be reviewed and critically discussed and compared in Chapter 3.

Chapter 4 will then describe the different experimental methods available to synthesize bulk metallic glasses in different sizes and shapes. Emphasis here will be on the production of bulk metallic glasses starting from the liquid state. But, other methods, for example, MA, to achieve similar results, will also be briefly described. The relative advantages and disadvantages of the different methods will be discussed.

Metallic glasses, including bulk metallic glasses, are metastable in nature. Therefore, on annealing them at increasingly higher temperatures, the metallic glass will go through structural relaxation to annihilate the excess quenched-in free volume, then transform into the supercooled liquid region, and eventually crystallization occurs. That is, given sufficient time, they will transform into the equilibrium crystalline phases. The kinetics of these transformations (frequently referred to as *crystallization* or *devitrification*) and the mechanisms of transformations will be described in Chapter 5.

The next few chapters will discuss the different properties of bulk metallic glasses, including physical, chemical, mechanical, and magnetic properties. Chapter 6 will discuss the physical properties of bulk metallic glasses and cover density, diffusivity, thermal expansion, electrical resistivity, specific heat, and viscosity. It will be shown how an evaluation of these properties will aid in understanding the structural relaxation and crystallization behavior of bulk metallic glasses.

Chapter 7 will then focus on the corrosion behavior of bulk metallic glasses. Extensive literature has been generated on this aspect, especially on Cu-, Fe-, and Zr-based bulk metallic glasses. The available results will be reviewed and the methodology to improve the corrosion resistance of bulk metallic glasses, a very important aspect for industrial applications, will be highlighted.

There is plenty of literature on the mechanical properties of bulk metallic glasses, since it is expected that the bulk metallic glasses could potentially be structural materials. Chapter 8 will therefore discuss the very large amount of literature on the mechanical behavior of metallic glasses. The inhomogeneous and homogeneous deformation behaviors of bulk metallic glasses, their general mechanical properties, and reasons for their high strength will be described. The chapter will then proceed to describe fatigue and fracture behavior and then discuss the different ways in which plasticity of bulk metallic glasses can be enhanced.

Chapter 9 will review the different magnetic properties of bulk metallic glasses and what potential applications these glasses may have on the basis of their magnetic properties.

Even though bulk metallic glasses are relatively new (they have just turned 25!), they have an interesting combination of properties, as described in the different chapters.

The bulk metallic glasses have already found a number of uses and these materials are being exploited for more applications. The existing and potential applications of bulk metallic glasses are described in Chapter 10.

The last chapter (Chapter 11) presents concluding remarks and expectations and prospects for these novel materials in the future.

The first edition of the book has been thoroughly revised and a significant amount of new material is added in almost every chapter. As such, we believe that the current edition is most up to date.

References

1. Duwez, P. (1967). Structure and properties of alloys rapidly quenched from the liquid state. *Trans. ASM Q.* 60: 607–633.
2. Anantharaman, T.R., ed. (1984). *Metallic Glasses: Production, Properties, and Applications*. Zürich, Switzerland: Trans Tech Publications.
3. Liebermann, H.H., ed. (1993). *Rapidly Solidified Alloys: Processes, Structures, Properties, Applications*. New York: Marcel Dekker.
4. Shechtman, D., I. Blech, D. Gratias, and J.W. Cahn (1984). Metallic phase with long-range orientational order and no translational symmetry. *Phys. Rev. Lett.* 53: 1951–1953.
5. Suryanarayana, C. and H. Jones (1988). Formation and characteristics of quasicrystalline phases: A review. *Int. J. Rapid Solidif.* 3: 253–293.
6. Gleiter, H. (1989). Nanocrystalline materials. *Prog. Mater. Sci.* 33: 223–315.
7. Suryanarayana, C. (1995). Nanocrystalline materials. *Int. Mater. Rev.* 40: 41–64.
8. Gleiter, H. (2000). Nanostructured materials: Basic concepts and microstructure. *Acta Mater.* 48: 1–29.
9. Suryanarayana, C. (2005). Recent developments in nanostructured materials. *Adv. Eng. Mater.* 7: 983–992.
10. Gleiter, H. (2008). Our thoughts are ours, their ends none of our own: Are there ways to synthesize materials beyond the limitations of today? *Acta Mater.* 56: 5875–5893.
11. Krabbes, G., G. Fuchs, W.-R. Canders, H. May, and R. Palka (2006). *High Temperature Superconductor Bulk Materials: Fundamentals—Processing—Properties Control—Application Aspects*. Weinheim, Germany: Wiley-VCH.
12. Bloor, D., R.J. Brook, M.C. Flemings, and S. Mahajan, eds. (1994). *The Encyclopedia of Advanced Materials*. Oxford, UK: Pergamon.
13. Suryanarayana, C., ed. (1999). *Non-Equilibrium Processing of Materials*. Oxford, UK: Pergamon.
14. Anantharaman, T.R. and C. Suryanarayana (1971). Review: A decade of quenching from the melt. *J. Mater. Sci.* 6: 1111–1135.
15. Suryanarayana, C. (1980). *Rapidly Quenched Metals: A Bibliography 1973–1979*. New York: IFI Plenum.
16. Jones, H. (1982). *Rapid Solidification of Metals and Alloys*. London, UK: Institution of Metallurgists.
17. Anantharaman, T.R. and C. Suryanarayana (1987). *Rapidly Solidified Metals: A Technological Overview*. Zürich, Switzerland: Trans Tech Publications.
18. Koch, C.C. (1991). Mechanical milling and alloying. In *Processing of Metals and Alloys*, ed. R.W. Cahn. Vol. 15 of *Materials Science* and *Technology—A Comprehensive Treatment*, pp. 193–245. Weinheim, Germany: VCH.
19. Suryanarayana, C. (1995). *Bibliography on Mechanical Alloying and Milling*. Cambridge, UK: Cambridge International Science Publishing.
20. Suryanarayana, C. (2001). Mechanical alloying and milling. *Prog. Mater. Sci.* 46: 1–184.
21. Suryanarayana, C. (2004). *Mechanical Alloying and Milling*. New York: Marcel Dekker.

22. Murty, B.S. and S. Ranganathan (1998). Novel materials synthesis by mechanical alloying/milling. *Int. Mater. Rev.* 43: 101–141.
23. Lieberman, M.A. and A.J. Lichtenberg (2005). *Principles of Plasma Discharges and Materials Processing*, 2nd edn. Hoboken, NJ: Wiley Interscience.
24. Sree Harsha, K.S. (2006). *Principles of Vapor Deposition of Thin Films*. Oxford, UK: Elsevier.
25. Lavernia, E.J. and Y. Wu (1996). *Spray Atomization and Deposition*. Chichester, UK: Wiley.
26. Turnbull, D. (1981). Metastable structures in metallurgy. *Metall. Trans. A* 12: 695–708.
27. Suryanarayana, C. (2004). Introduction. In *Mechanical Alloying and Milling*, ed. C. Suryanarayana, pp. 5–7. New York: Marcel Dekker.
28. Martin, G. and P. Bellon (1997). Driven alloys. *Solid State Phys.* 50: 189–331.
29. Shingu, P.H. (1993). Thermodynamic principles of metastable phase formation. In *Processing Materials for Properties*, eds. H. Henein and T. Oki, pp. 1275–1280. Warrendale, PA: TMS. See also the paper presented at the RQ 15 International Conference in Shanghai, China, August 24–28, 2014.
30. Froes, F.H., C. Suryanarayana, K.C. Russell, and C.M. Ward-Close (1995). Far from equilibrium processing of light metals. In *Novel Techniques in Synthesis and Processing of Advanced Materials*, eds. J. Singh and S.M. Copley, pp. 1–21. Warrendale, PA: TMS.
31. Froes, F.H., C. Suryanarayana, K.C. Russell, and C.G. Li (1995). Synthesis of intermetallics by mechanical alloying. *Mater. Sci. Eng. A* 192/193: 612–623.
32. Klassen, T., M. Oehring, and R. Bormann (1997). Microscopic mechanisms of metastable phase formation during ball milling of intermetallic TiAl phases. *Acta Mater.* 45: 3935–3948.
33. Duwez, P. (1981). Metallic glasses—Historical background. In *Glassy Metals I*, eds. H.-J. Güntherodt and H. Beck, pp. 19–23. Berlin, Germany: Springer-Verlag.
34. Suryanarayana, C. (1991). Rapid solidification. In *Processing of Metals and Alloys*, ed. R.W. Cahn. Vol. 15 of *Materials Science* and *Technology—A Comprehensive Treatment*, pp. 57–110. Weinheim, Germany: VCH.
35. Jones, H. (2001). A perspective on the development of rapid solidification and nonequilibrium processing and its future. *Mater. Sci. Eng. A* 304–306: 11–19.
36. Suryanarayana, C. (2002). Rapid solidification processing. In *Encyclopedia of Materials: Science and Technology—Updates*, eds. K.H.J. Buschow, R.W. Cahn, M.C. Flemings, E.J. Kramer, and S. Mahajan, pp. 1–10. Oxford, UK: Pergamon.
37. Inoue, A., K. Kita, T. Zhang, and T. Masumoto (1989). An amorphous $La_{55}Al_{25}Ni_{20}$ alloy prepared by water quenching. *Mater. Trans., JIM* 30: 722–725.
38. Inoue, A., T. Zhang, and T. Masumoto (1990). Production of amorphous cylinder and sheet of $La_{55}Al_{25}Ni_{20}$ alloy by a metallic mold casting method. *Mater. Trans., JIM* 31: 425–428.
39. Chen, H.S. (1974). Thermodynamic considerations on the formation and stability of metallic glasses. *Acta Metall.* 22: 1505–1511.
40. Kui, H.W., A.L. Greer, and D. Turnbull (1984). Formation of bulk metallic glass by fluxing. *Appl. Phys. Lett.* 45: 615–616.
41. Peker, A. and W.L. Johnson (1993). A highly processable metallic glass: $Zr_{41.2}Ti_{13.8}Cu_{12.5}Ni_{10.0}Be_{22.5}$. *Appl. Phys. Lett.* 63: 2342–2344.
42. Benjamin, J.S. (1970). Dispersion strengthened superalloys by mechanical alloying. *Metall. Trans.* 1: 2943–2951.
43. Sherif El-Eskandarany, M. and A. Inoue (2006). Synthesis of new bulk metallic glassy $Ti_{60}Al_{15}Cu_{10}W_{10}Ni_5$ alloy by hot-pressing the mechanically alloyed powders at the supercooled liquid region. *Metall. Mater. Trans. A* 37A: 2231–2238.

2

Metallic Glasses

2.1 Introduction

Metallic materials are traditionally considered crystalline in nature, possessing translational symmetry; that is, their constituent atoms are arranged in a regular and periodic manner in three dimensions. However, a revolution in the concept of metals was brought about in 1960 when Pol Duwez, at the California Institute of Technology in Pasadena, California, synthesized an Au–25 at.% Si alloy in the glassy state by rapidly solidifying the liquid at rates approaching a million degrees per second [1]. These high solidification rates were achieved by propelling a small droplet of the liquid alloy tangentially onto a highly conducting substrate, such as copper, to enable the liquid to be spread out in the form of a thin foil on the substrate surface. The good thermal contact between the substrate and the molten metal film ensured that heat was extracted rapidly by the large substrate through the small thickness of the foil. In this technique, known as the *gun technique* of liquid quenching, the solidification rates were estimated to vary from as low as 10^4 to as high as 10^{10} K s^{-1}, depending on the thickness of the foil, the nature of the substrate, the type of material solidified, and how good the thermal contact was between the foil and the substrate. A typical solidification rate for a foil of 50 μm thickness is about 10^6 K s^{-1}.

The Au–Si alloy rapidly solidified by Duwez did not show any crystalline peaks in its x-ray diffraction (XRD) pattern. Instead, the microphotometer trace of the Debye–Scherrer diffraction pattern from this alloy showed a couple of somewhat broad and diffuse peaks and Professor Duwez was not convinced (but, the two graduate students were!) that this was really amorphous. The authors had, however, interpreted this as indicative of the presence of a noncrystalline structure in the material. This ambiguity in the interpretation was probably due to the low thermal stability of this alloy, since the rapidly solidified Au–25 at.% Si alloy had completely transformed into a nonequilibrium crystalline state after 24 h at room temperature. A detailed account of this important observation and the historical development leading to this discovery has been narrated by Professor Pol Duwez [2].

Investigations on other alloys rapidly solidified from the liquid state at rates of about 10^5–10^6 K s^{-1} have unambiguously confirmed that these alloys are truly amorphous and lack crystallinity, which is typical of metallic materials. The amorphous nature of the quenched material was confirmed by transmission electron microscopy and electron diffraction techniques, in addition to the XRD method. The presence of a broad diffuse halo in the electron diffraction pattern and the absence of diffraction contrast in the electron micrographs have proven, beyond doubt, that the material is truly amorphous. Several other advanced techniques have also been brought to bear on this aspect later. Since these amorphous materials are based on metals, these were referred to as *glassy metals* or *metallic glasses*. Since the first discovery of a metallic glass in 1960, thousands of alloys of different compositions have been prepared as metallic glasses. An early and somewhat detailed compilation of the available literature on metallic glass formation by rapid solidification processing (RSP) methods is available in Refs. [3,4].

A large variety of metallic glasses have been developed during the last 50-plus years. These could be broadly classified into metal–metalloid or metal–metal types, even though other classifications also exist. In a typical metal–metalloid-type glass, the metal atoms constitute about 80 at.% and the metalloid atoms (typically B, C, P, and Si) about 20 at.%. The metal atoms may be of one type or a combination of different metals, but the total amount of the metal atoms is about 80 at.%. Similarly, the metalloid component may be of one type or a combination of different metalloid atoms, and again the total amount of the metalloid atoms is about 20 at.%. Some well-investigated compositions in this category include $Pd_{80}Si_{20}$, $Pd_{77}Cu_6Si_{17}$, $Fe_{80}B_{20}$, $Fe_{40}Ni_{40}B_{20}$, $Ni_{75}Si_8B_{17}$, $Fe_{40}Ni_{40}P_{14}B_6$, $Fe_{70}Cr_{10}P_{13}C_7$, $Ni_{49}Fe_{29}B_6P_{14}Si_2$, and some exotic compositions such as $W_{35}Mo_{20}Cr_{15}Fe_5Ni_5P_6B_6C_5Si_3$. (The subscripts represent the atomic percentages of the elements in the alloy.)

In the metal–metal type of metallic glasses, only metal-type, and no metalloid-type, atoms are involved. Some of the alloys that have been well investigated in this category include $Ni_{60}Nb_{40}$, $Cu_{57}Zr_{43}$, $Mg_{70}Zn_{30}$, $La_{80}Au_{20}$, and $Fe_{90}Zr_{10}$. One can immediately notice an important difference between the compositions of the metal–metalloid-type and metal–metal-type metallic glasses. Whereas the metalloid content is usually around 20 at.% in the metal–metalloid-type glasses and the rest is metallic, there is no such compositional restriction in the case of metal–metal-type metallic glasses. The second metal component can be as small as 9–10 at.% or as large as nearly 50 at.%.

2.2 Distinction between Crystals and Glasses

Metals and alloys are traditionally considered crystalline in nature. That is, their constituent atoms are arranged in a periodic manner in three dimensions, with the caveat that it is not necessary that the periodicity is the same in the three directions. In other words, if we know the coordinates of one atom in one unit cell of the crystal, and the size and shape of the unit cell (i.e., periodicity in the three directions), then it is possible to predict the position of other atoms. Further, the concept of space lattice requires that every atom in the crystal has identical surroundings. That is, the nearest neighbor distances and the coordination number (CN) (the number of nearest neighbors) for any atom is the same irrespective of where the atom position is considered. Thus, for a face-centered cubic (fcc) structure, for example, the first nearest neighbor distance is $a\sqrt{2}/2$, where a is the lattice parameter of the crystal. Further, every atom has 12 nearest neighbors in the crystal, that is, the CN for the fcc structure is 12. These nearest neighbor distances and CN are different for different structures and details about these and how they can be calculated can be found in standard textbooks on crystallography [5–7]. Because of the constraints of symmetry on the different ways in which the constituent atoms can be arranged, only a limited number of arrangements are possible in the three-dimensional space. Thus, we have only 7 crystal systems, 14 Bravais lattices, 230 space groups, and so on. Different metals with the same Bravais lattice can have different lattice parameters and/or inter-axial angles, and therefore new structures could be generated. Additionally, by placing different number of atoms at each lattice point in a given Bravais lattice, it is again possible to generate a new crystal structure. Thus, the actual number of crystal structures is, of course, unlimited.

But, all of these theoretically infinite numbers of arrangements can be described in terms of one of the 14 Bravais lattices.

In this description, we have assumed that the crystal is perfect and that there are no crystal defects such as vacancies, dislocations, or grain boundaries in it. The presence of such defects introduces "imperfection" into the lattice, and the atomic positions are displaced with respect to their ideal positions, that is, the crystal regions are distorted. Depending on whether the severely distorted region extends many interatomic distances in zero, one, or two dimensions, crystal defects are classified into point (vacancies, interstitials), line (dislocations), or planar (grain boundaries, stacking faults) types. Real crystals can thus be considered as perfect crystals containing a number of "defects."

Any solid in which the regular arrangement of atoms, that is, periodicity, is absent is considered "noncrystalline" in character. That is, these materials do not possess any crystallinity. *Amorphous* and *glassy* are the other terms normally used to describe such an arrangement of atoms. Since the atomic arrangement is random (i.e., there is no periodicity) in these noncrystalline materials, it will be difficult to define either the nearest neighbor distances or the CN. Each atom in the noncrystalline solid will have different nearest neighbors and CNs. But, it can be safely stated that the nearest neighbor distances are longer and the CNs smaller in a noncrystalline solid in comparison to its crystal counterpart, assuming that a solid can exist in both the crystalline and noncrystalline states. This suggests that there is no unique description of the "structure" of a noncrystalline solid and that there is no limit to the possible atomic arrangements in these materials. This means that it is possible to have an infinite number of atomic arrangements, and this poses a problem in describing the atomic structure of noncrystalline materials accurately.

Generally speaking, solid materials may be considered to be either crystalline (possessing long-range translational periodicity) or noncrystalline (without any long-range order present). However, in 1984, Shechtman et al. [8] reported the formation of a new type of solid that did not possess long-range translational symmetry (so it was not a crystal) but did possess long-range order (so it was not noncrystalline). This discovery was awarded the Nobel Prize in Chemistry in 2011. These materials, now referred to as *quasicrystals* [9,10], exhibit a structure in which rotational symmetry, but no translational symmetry, exists over large distances. These rotational symmetries are exemplified by the so-called forbidden symmetries (fivefold, sevenfold, tenfold, and so on). The quasicrystals have been designated with specific names, depending on the type of symmetry they exhibit. For example, materials possessing the fivefold symmetry are known as *icosahedral crystals*, those possessing tenfold symmetry decagonal crystals, and so on. A new ordered phase showing the apparent fivefold symmetry was observed by Sastry et al. [11] in 1978 in a rapidly solidified Al–Pd alloy, but was interpreted to arise from a microstructure consisting of a series of fine twins. This was later shown to be a two-dimensional (or decagonal) quasicrystal with one periodic axis normal to two nonperiodic axes.

2.3 Differences between Amorphous Alloys and Metallic Glasses

The terms *noncrystalline*, *amorphous*, and *glassy* refer to similar (random) atomic arrangements in solid materials, and therefore these terms have been used interchangeably (quite understandably so) in the literature, leading to some confusion. Further, some researchers have been preferentially using one or other of these terms. In addition, some researchers refer to the thin ribbon glassy materials as *amorphous* and to the bulk glassy alloys only as *glasses*. Therefore, several different terms have been used to describe these noncrystalline materials. To avoid ambiguity and confusion, we would like to define/explain each of these terms for use in this book. It has been generally agreed upon that *noncrystalline* is a generic term used to describe any solid material that does not possess crystallinity. Researchers have tried to distinguish between glassy and amorphous materials. A noncrystalline solid formed by continuous cooling from the liquid state is known as a *glass*. (People were able to produce glasses only from the liquid state in the early years, and hence this usage.) On the other hand, a noncrystalline material, obtained by any other process, for example, vapor deposition or solid-state processing methods such as mechanical alloying, but not directly from the liquid state, is referred to as an *amorphous* material. But, it should be remembered that both glasses and amorphous solids are noncrystalline. Since we will be dealing in this book with noncrystalline solids produced mostly by continuous cooling from the liquid state, we will be using the term *glass* to describe these materials. However, we will try to continue to maintain the distinction between "glassy" and "amorphous" solids to refer to the different materials throughout the book. Thus, to recapitulate,

> *Glass* is any noncrystalline solid obtained by continuous cooling from the liquid state, and *amorphous solid* is any noncrystalline material obtained by any other method, except by continuous cooling from the liquid state.

It has been occasionally emphasized that the presence of a glass transition temperature, T_g, which we will define in Section 2.4, is the hallmark of a true glass. In the case of metallic glass ribbons obtained by RSP methods, T_g is not commonly observed. Therefore, only when Chen and Turnbull [12,13] reported the presence of T_g in the differential scanning calorimeter (DSC) plot of a rapidly solidified $Au_{77}Ge_{13.6}Si_{9.4}$ alloy, were researchers convinced that the noncrystalline alloys produced by RSP were true glasses. The glass transition was manifested thermally by an abrupt rise of 5.5 cal g-atom^{-1} K^{-1} in specific heat, C_p, as the alloy was heated from 285 to 297 K. Just above 297 K, C_p started to fall with increasing temperature on a reasonable extrapolation of the high-temperature C_p–T relation of the liquid alloy. This thermal behavior could be repeated in the same sample after it had been carried through the transition and quenched again. Further, the viscosity of the glassy alloy increased rapidly with decreasing temperature from 0.9×10^8 Pa s at 305 K to 1.4×10^{12} Pa s at 285 K. The T_g value increased only by 1–3 K as the heating rate was changed by a factor of 16. These variations of C_p and viscosity with temperature, at this temperature, are as expected for a glass → liquid transition, and therefore, this temperature was designated as T_g for this alloy.

Let us consider the case when the material continuously solidified from the liquid state (above the critical cooling rate required to form the noncrystalline structure) did not show any diffraction contrast in the transmission electron micrographs and the x-ray and electron diffraction patterns showed the presence of only a broad and diffuse halo. Let us also note that even the high-resolution transmission electron micrographs showed only a salt-and-pepper contrast. But, the DSC curves did not show the clear presence of a glass transition temperature.

According to the listed attributes, this material should be referred to as a *glass*, since structurally all the requirements for it to be called a glass are satisfied. But, some researchers question the designation of such a material as glass, since T_g is not observed. There could be different reasons for the nonobservance of T_g in noncrystalline materials. It is possible that the glass transition temperature, T_g, and the crystallization temperature, T_x, for this material are so close to each other that the presence of T_g is completely masked by the strong exothermic peak representing T_x. If these two temperatures are sufficiently well separated, then the presence of T_g can be easily recognized. In the case of bulk metallic glasses (BMGs), there is usually a large supercooled liquid region, $\Delta T_x = T_x - T_g$, and in such cases it is relatively easy to locate T_g. But, even in the case of BMGs, there are materials that exhibit a very large critical diameter, but the presence of T_g cannot be clearly identified. For example, an $Nd_{70}Fe_{20}Al_{10}$ ternary alloy melt can be cast into a 7 mm diameter glassy rod, but the DSC curves do not indicate the presence of T_g [14]. Further, it may not be easy to recognize the presence of T_g if the glassy alloy is reheated at a slow rate; instead it may be possible to locate T_g if the alloy is heated rapidly, say at a rate of 200 K s^{-1}. Thus, there could be many different reasons for not detecting T_g in an alloy. Therefore, if a material is structurally proven noncrystalline, and it is obtained by continuous cooling from the liquid state, it should be designated a glass, even if T_g is not detected. In fact, Angell [15] mentions that the presence of T_g is not essential for a material to be called a glass!

It may also be relevant in this context to mention that amorphous alloys produced by non-solidification methods may not show the presence of T_g, since it is expected to be observed only during solidification from the liquid state.

2.4 Concepts of Glass Formation

As described earlier, a glass is a noncrystalline solid formed by continuous cooling from the liquid state. The metallic glasses are basically no different from those of silicate or oxide glasses or organic polymers. Figure 2.1 shows the variation of specific volume (volume per unit mass) as a function of temperature. When the temperature of a liquid metal is reduced, its volume decreases up to the freezing/melting point, T_m. At the freezing temperature, there is a precipitous drop in the specific volume of the metal until it reaches the

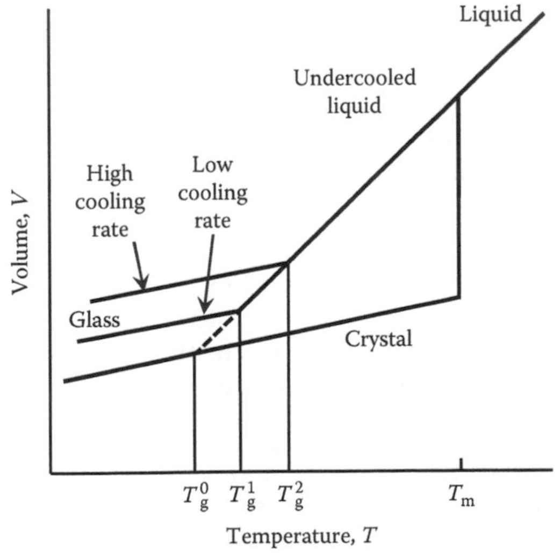

FIGURE 2.1
Variation of specific volume with temperature for a normal and a glass-forming material.

value characteristic of the solid crystalline metal. Further decrease in temperature below T_m results in a slow decrease of the volume of the metal, depending on its coefficient of thermal expansion.

Even though it has been mentioned that the liquid exhibits a sudden decrease in its specific volume at the freezing temperature to transform into the crystal, a liquid normally undercools (or supercools, i.e., the liquid state can be maintained without solidification occurring at temperatures well below the melting temperature) before crystallization can begin. This is because an activation energy barrier needs to be overcome before solid nuclei can form in the melt, and this activation barrier is smaller the larger the value of undercooling. The amount of undercooling achieved depends on several factors, including the initial viscosity of the liquid, the rate of increase of viscosity with decreasing temperature, the interfacial energy between the melt and the crystal, the temperature dependence of the free energy difference between the undercooled melt and the crystal phases, the imposed cooling rate, and the efficiency of heterogeneous nucleating agents. The actual value of undercooling is different for different metals, but in general practice the value is at best only a few tens of degrees. But, if special efforts are made to remove the heterogeneous nucleating sites, for example, by fluxing methods, then the undercooling achieved can be a few hundred degrees. Another way of increasing the amount of undercooling achieved is to increase the imposed cooling rate; the higher the cooling rate, the larger is the amount of undercooling.

The situation in the case of a glass-forming liquid is different. The volume of most of the materials decreases with decreasing temperature from the liquid state. But, the difference between a normal metal and a glass-forming metal is that, in the case of the glass-forming liquid, the liquid can be significantly undercooled, either due to the imposition of a high cooling rate, or removal of heterogeneous nucleating sites, or other reasons. The volume decreases even in the undercooled region, and its viscosity continues to increase. At some temperature, normally well below T_m, the viscosity becomes so high that the liquid gets "frozen-in," and this frozen-in liquid (it is like a solid for all practical purposes) is referred to as *glass*. The temperature at which the viscosity of the undercooled liquid reaches a value of 10^{12} Pa s is traditionally designated T_g. But, in reality, there is no sharply defined temperature at which this occurs. Rather, there is a temperature interval in which the liquid becomes a glass, and Kauzmann [16, p. 227] preferred to call this the *glass-transformation interval*.

Since viscosity plays a very important role in glass formation, let us look at this in some detail. Viscosity indicates the resistance to flow of a system and is a measure of its internal friction. The International System unit of viscosity is the pascal-second (1 Pa s = 1 kg m^{-1} s^{-1}). An older unit is Poise, P, with the relationship:

$$1\,\mathrm{P} = 0.1\,\mathrm{Pa\,s} \tag{2.1}$$

As a reference point, water at 20°C has a viscosity of 1 centiPoise, cP (10^{-2} Poise). The viscosities of some substances of common use are

Water at 20°C	1.002 cP (1.002 × 10^{-3} Pa s)
Mercury at 20°C	1.554 cP (1.554 × 10^{-3} Pa s)
Pancake syrup at 20°C	2500 cP (2.5 Pa s)
Peanut butter at 20°C	250,000 cP (250 Pa s)
Soda glass at 575°C	1 × 10^{15} cP (1 × 10^{12} Pa s)

As previously defined, the glass transition temperature, T_g, is the temperature at which the supercooled liquid becomes solid glass. To be more accurate, this should be called the *thermal* or *calorimetric* glass transition. It is also important to realize that this "transition" is not a true thermodynamic phase transition, but its origin is strictly kinetic, since the value of T_g depends on the cooling rate and, more generally, on the way the glass is prepared.

The glassy state below T_g is often referred to as the *thermodynamic state* of a vitrified (glassy) substance. It is true that the properties of the glassy solid do not show any time dependence for short observation times and/or well below T_g. Therefore, it appears that the glassy solid is in a properly defined thermodynamic state. However, even in this case, time will always play a fundamental role in the formation and description of the glass, the important reason simply being that it is not an equilibrium state. Glassy substances that look like a solid on experimental timescales of seconds or even years may look like a liquid on geological timescales. The non-static nature of glass is best seen near the glass transition.

The phenomenon of glass formation, as described herein, is often referred to as *vitrification* for nonmetallic materials and organic polymers. Similarly, the formation of the crystalline phase(s) on subsequent heating of the glassy phase to higher temperatures is referred to as *crystallization* (in the case of metallic materials) and *devitrification* for nonmetallic materials. We will use the terms *glass formation* and *crystallization* in the present book to refer to the formation of the glassy phase on cooling from the liquid state and its transformation to the crystalline phase(s) on reheating of the glassy alloy, respectively.

It may be realized from this description that the T_g value is dependent upon several factors, such as the imposed cooling rate and the magnitude of undercooling. In other words, there is no unique T_g value for a given material (unlike the melting or freezing temperature, which is thermodynamically defined and so it is fixed), but it is a kinetic parameter. Thus, the T_g and the structure of the glass are cooling rate dependent. The faster the alloy is solidified from the liquid state, the higher the T_g value (Figure 2.1). The structure of the glass also depends on the extent to which structural relaxation has occurred during subsequent heating to higher temperatures, but below the crystallization temperature, T_x. The higher the structural relaxation, the closer the alloy moves toward a "true" glass.

The variations of specific heat, C_p, and viscosity, η, with temperature are shown in Figure 2.2. The C_p of the undercooled liquid increases with decreasing temperature and the difference between the C_p of the undercooled liquid and the glass continues to increase

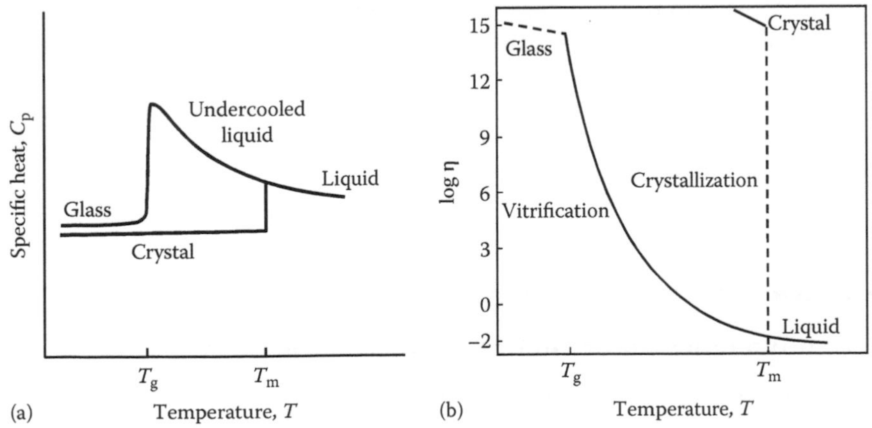

FIGURE 2.2
Variation of (a) specific heat and (b) viscosity with temperature for crystal and glass formation.

till T_g. At T_g, there is a sudden drop in the C_p value of the undercooled liquid, a manifestation of the fewer degrees of freedom as a result of freezing of the liquid. Once the glass is formed, there is very little difference in the C_p between the crystal and the glass.

During cooling from the liquid state, the viscosity of a metallic liquid increases slowly with a decrease in temperature. But, at the freezing temperature, the viscosity of the material increases suddenly by about 15 orders of magnitude. The variation of viscosity in a glass-forming liquid is, however, different. The viscosity increases gradually in the liquid state with decreasing temperature. But, this trend continues below the freezing point, even though the rate of increase is more rapid with a further decrease in temperature in the supercooled liquid. But, at the glass transition temperature, T_g, the viscosity is so high that there is no more flow of the liquid and the material is a solid, for all practical purposes. That is, its viscosity remains practically constant.

But, it is most important to remember one point about glasses. There is no unique "structure" for a given glass as there is for a crystal. The temperature at which departure from configurational equilibrium occurs determines the structure. And this temperature is a function of the cooling rate. Further, at any given temperature, there is a temporal variation in the structure and the relaxation times are different at different temperatures.

Metallic glasses often exhibit a reversible glass ↔ liquid transition at the glass transition temperature, T_g, which is manifested in significant changes in the specific heat, C_p, or viscosity, η. During reheating of glassy samples, there is a sudden increase in C_p and simultaneously there is a decrease in the viscosity at T_g. These observed reversible changes suggest that, similar to other types of glasses, metallic glasses can revert to the supercooled liquid state without crystallization. In the scientific community working with BMGs, this state is normally referred to as the *supercooled liquid region*. This reversibility also suggests that the structure (atomic arrangement) of the glass is closely related to the atomic arrangements present in the liquid state. At T_x, which is higher than T_g, the supercooled liquid transforms into the crystalline phase(s). As mentioned earlier, the temperature interval between T_x and T_g is referred to as the *width of the supercooled liquid region*, that is, $\Delta T_x = T_x - T_g$. The value of ΔT_x is different for different glasses, and is usually taken as an indication of the thermal stability of the glass produced. In the case of BMGs, this temperature interval is usually quite large, and values of over 120 K have been reported; the highest reported to date is 131 K in a $Pd_{43}Cu_{27}Ni_{10}P_{20}$ BMG alloy [17]. In the case of rapidly solidified thin ribbon glasses, and marginal glass formers, the value of ΔT_x is very small, if observed at all.

There is a difference in the transformation behavior of ordinary (crystalline) solids and glasses. On heating, while a crystal transforms into the liquid state at the melting temperature, T_m (unless some allotropic changes take place), a metallic glass, on the other hand, first transforms into the supercooled liquid state at the glass transition temperature, T_g, and then into the crystalline state at the crystallization temperature, T_x. The crystalline solid will eventually melt at T_m.

2.5 Thermodynamics and Kinetics of Glass Formation

It is important to realize that any glass (this is even more true for metallic glasses) is not in a thermodynamically stable (equilibrium) state. From a physics point of view, glasses are in an excited state, and at any given temperature, provided sufficient time (maybe a few minutes to thousands of years, depending on the type of glass and the way it is made) is provided, they will relax and eventually transform into the crystalline ground state. So, the question now is whether the principles of thermodynamics, in which the functions are defined for equilibrium states, are applicable to systems that are far from equilibrium.

Johnson [18] and Leuzzi and Nieuwenhuizen [19] have dealt with this aspect in some detail based on the concepts of undercooling developed by Turnbull [20]. It is true that the thermodynamic principles are, strictly speaking, applicable only to a system that is in equilibrium. But, they can also be used when the system under consideration is an undercooled liquid.

It is well known that metallic liquids can be significantly undercooled for extended periods of time without crystallization occurring [20]. This is because the critical size of the nucleus to form the solid crystalline phase, at the freezing temperature, T_m, is infinitely large. Put differently, the timescale for the nucleation of the crystalline phase is a function of the degree of undercooling experienced by the melt. It is exceedingly long at the freezing temperature and decreases with increasing amounts of undercooling. When the timescale for nucleation is sufficiently long so that the liquid phase can still explore the different possible energy states in which it could exist, one can still define the entropy and other thermodynamic functions of the liquid. That is, the liquid is in a metastable state for which entropy, free energy, and other thermodynamic parameters can be defined. In this sense, one can use the thermodynamic principles that are applicable to equilibrium systems in these situations also.

2.5.1 Thermodynamic Stability

The thermodynamic stability of a system at constant temperature and pressure is determined by its Gibbs free energy, *G*, defined as

$$G = H - TS \tag{2.2}$$

where:

- H is the enthalpy
- T is the absolute temperature
- S is the entropy

Thermodynamically, a system will be in stable equilibrium, that is, it will not transform into any other phase(s) under the given conditions of temperature and pressure, if it has attained the lowest possible value of the Gibbs free energy. Equation 2.2 predicts that a system at any temperature can be made more stable either by increasing the entropy or decreasing the enthalpy or both. Metallic crystalline solids have the strongest atomic bonding and therefore the lowest enthalpy, *H*. Consequently, solids are the most stable phases at low temperatures. On the other hand, the atomic vibration frequency increases with increasing temperature and consequently, the entropy, *S*, is high at elevated temperatures. As a result, the product of temperature and entropy increases and, therefore, the value of the $-TS$ term dominates at higher temperatures. Therefore, phases with more freedom of atomic movement, that is, liquids and gases, become more stable at elevated temperatures [21,22].

Using these concepts, it may be stated that a glass becomes more "stable" when the free energy of the glassy phase is lower than that of the competing crystalline phase(s). In other words, the change in free energy, ΔG (= $G_{glass} - G_{crystal}$) becomes negative. Mathematically expressed,

$$\Delta G = \Delta H_f - T\Delta S_f \tag{2.3}$$

where:

Δ	represents the change in these quantities between the final and initial states
H_f and S_f	represent the enthalpy of fusion and entropy of fusion, respectively

The system becomes stable when the value of G is at its lowest, or ΔG is negative. A negative value of ΔG can be obtained either by decreasing the value of ΔH_f or increasing the value of ΔS_f, or both. Since entropy is nothing but a measure of the different ways in which the constituent atoms can be arranged (microscopic states), this value will increase with increasing number of components in the alloy system. Thus, even if ΔH_f were to remain constant, the free energy will be lower because of the increased entropy when the alloy system consists of a large number of components. (This is one reason why even a small amount of impurity lowers the free energy of a metal, and consequently, it is impossible to find a 100% pure metal! For the same reason, it is also true that the thermodynamic stability of multicomponent alloys is much higher than that of an alloy with fewer components.) But, the value of ΔH_f will not remain constant because of the chemical interaction among the different constituent elements.

The free energy of the system can also be decreased, at a constant temperature, in cases of low chemical potential due to low enthalpy, and large interfacial energy between the liquid and solid phases. Since it will be difficult to intentionally control these parameters in an alloy system, the easiest way to decrease the free energy would be to increase ΔS_f by having a large number of components in the alloy system. This is why it has been easier to synthesize glassy phases in ternary and higher-order alloy systems than in binary alloy systems. Obviously, it will be much easier to produce the glassy phases in alloys containing a number of components, that is, in multicomponent alloy systems. BMGs, which can be produced in the glassy state at very slow solidification rates, are typically multicomponent alloy systems.

Increases in ΔS_f also result in an increase in the degree of dense random packing of atoms, which leads to a decrease in ΔH_f and, consequently, an increase in the solid/liquid interfacial energy.

2.5.2 Kinetics of Glass Formation

According to the free-volume model [23] or the entropy model [24] of the liquid state, it is expected that every liquid will undergo a transition to the glassy state, provided that crystallization can be bypassed or avoided. Thus, the problem of glass formation turns out to be purely kinetic in nature. Therefore, if a liquid could be cooled sufficiently rapidly to prevent the formation of detectable amount of a crystalline phase, glass formation could be achieved. Hence, whether a glass forms or not is related to the rapidity with which the liquid can be cooled and also to the kinetic constants.

The kinetics of crystallization were first treated by Turnbull [25]. The following assumptions were made:

1. The composition of the crystals forming is the same as that of the liquid.
2. Nucleation transients are unimportant.
3. Bulk free energy change associated with the transformation of the undercooled liquid to the crystal phase, ΔG_v, is given by the linear approximation, $\Delta G_v = \Delta H_f \cdot \Delta T_r$, where H_f is the molar enthalpy of fusion and ΔT_r is the reduced undercooling ($\Delta T_r = (T_l - T)/T_l$, where T_l is the melting [liquidus] temperature).

The homogeneous nucleation rate, I, for the formation of crystalline nuclei from a supercooled melt (in a liquid free of nuclei or heterogeneous, i.e., preferred nucleation sites) can be expressed [25] as

$$I = \frac{k_n}{\eta(T)} \exp\left[-\frac{b\alpha^3\beta}{T_r(\Delta T_r)^2}\right] \tag{2.4}$$

where:

b	is a shape factor (=$16\pi/3$ for a spherical nucleus)
k_n	is a kinetic constant
$\eta(T)$	is the shear viscosity of the liquid at temperature T
T_r	is the reduced temperature ($T_r = T/T_l$)
ΔT_r	is the reduced supercooling ($\Delta T_r = 1 - T_r$)
α and β are	dimensionless parameters related, respectively, to the liquid/solid interfacial energy (σ) and to the molar entropy of fusion, ΔS_f

Thus,

$$\alpha = \frac{\left(N_A \bar{V}^2\right)^{1/3} \sigma}{\Delta H_f} \tag{2.5}$$

and

$$\beta = \frac{\Delta S_f}{R} \tag{2.6}$$

where:

N_A	is Avogadro's number
$\bar{V}$	is the molar volume of the crystal
R	is the universal gas constant

It is clear from Equation 2.4 for I that for a given temperature and η, as $\alpha^3\beta$ increases, the nucleation rate decreases very steeply. Increasing α and β means an increase in σ and ΔS_f and/or a decrease in ΔH_f, all consistent with the thermodynamic approach of increased glass-forming ability (GFA) explained previously.

It is also interesting to note that η is closely related to the reduced glass transition temperature, T_{rg} (=T_g/T_l), and $\alpha^3\beta$ determines the thermal stability of the supercooled liquid. In agreement with experimental results, the value of $\alpha\beta^{1/3}$ for metallic melts has been estimated to be about 0.5. The importance of $\alpha^3\beta$ can be appreciated from the following two examples. When $\alpha\beta^{1/3} > 0.9$, unseeded liquid will not crystallize at any cooling rate by homogeneous nucleation. In other words, the glass will continue to be stable, unless crystal nucleation takes place at heterogeneous sites. On the other hand, when $\alpha\beta^{1/3} \leq 0.25$, it would be impossible to suppress crystallization. Thus, the higher the $\alpha\beta^{1/3}$ value, the easier it is to suppress crystallization and achieve glass formation.

The equation for the growth rate, U, of a crystal from an undercooled liquid can be expressed as

$$U = \frac{10^2 f}{\eta}\left[1 - \exp\left(-\frac{\Delta T_r \Delta H_f}{RT}\right)\right] \tag{2.7}$$

where f represents the fraction of sites at the crystal surfaces where atomic attachment can occur (=1 for close-packed crystals and 0.2 ΔT_r for faceted crystals). Here also we can see that U decreases as η increases, and will thus contribute to increased glass formability.

Since both I and U vary, at any given temperature, as $1/\eta$, both the glass-forming tendency and the stability of the glass should increase with reduced glass transition temperature, T_{rg}, and increasing values of α and β. Reducing the value of f through atomic rearrangements such as local ordering or segregation would also lower the growth rate. Since the value of $\alpha\beta^{1/3}$ is approximately 0.5 for metallic alloys, it may be easily shown that liquids for which $T_{rg} > 2/3$ may readily be quenched into a glassy state, whereas if $T_{rg} = 0.5$, a cooling rate of about 10^6 K s^{-1} is required for the melt to be quenched into the glassy state.

Based on the treatment of Uhlmann [26], Davies [27] combined the values of I and U (calculated using Equations 2.4 and 2.7, respectively) with the Johnson–Mehl–Avrami treatment of transformation kinetics, and calculated the fraction of transformed phase x in time t, for small x, as

$$x = \tfrac{1}{3}\pi I U^3 t^4 \tag{2.8}$$

Substituting the values of I and U in Equation 2.8, the time needed to achieve a small fraction of crystals from the melt was calculated as

$$t \approx \frac{9.3\eta a_o^2 x}{kTf^3\overline{N_v}}\left[\frac{\exp\left(\dfrac{1.07}{\Delta T_r^2 T_r^3}\right)}{\left\{1-\exp\left(-\dfrac{\Delta H_f \Delta T_r}{RT}\right)\right\}^3}\right]^{1/4} \tag{2.9}$$

where:

a_o is the mean atomic diameter

$\overline{N_v}$ is the average volume concentration of atoms, and all the other parameters have the same meaning, as described earlier

A time–temperature–transformation (*T–T–T*) curve was then computed by calculating the time, t, as a function of T_r, to transform to a barely detectable fraction of crystal, which was arbitrarily taken to be $x = 10^{-6}$. Figure 2.3 shows the *T–T–T* curves calculated in this manner for pure metal Ni, reasonably good glass formers such as $Au_{78}Ge_{14}Si_8$ and $Pd_{82}Si_{18}$ alloys, and a good glass former, such as $Pd_{78}Cu_6Si_{16}$. The critical cooling rate, R_c, to obtain the glassy phase was then obtained as

$$R_c = \frac{\Delta T}{t_n} \tag{2.10}$$

where:

ΔT is the undercooling

t_n is the time at the nose of the *T–T–T* curve

The calculated critical cooling rates ranged from ~10^{10} K s^{-1} for nickel to ~10^6 K s^{-1} for $Au_{78}Ge_{14}Si_8$, ~3000 K s^{-1} for $Pd_{82}Si_{18}$, and ~35 K s^{-1} for $Pd_{78}Cu_6Si_{16}$ alloys. For the three alloy glasses, the T_{rg} values were 1/4, 1/2, and 2/3, respectively, confirming that the GFA increases with increasing values of T_{rg}.

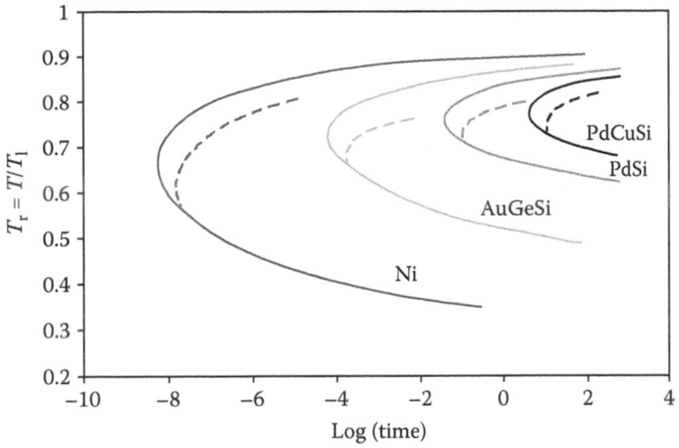

FIGURE 2.3
Time–temperature–transformation (*T–T–T*) curves (solid lines) and the corresponding continuous cooling transformation curves (dashed lines) for the formation of a small volume fraction for pure metal Ni, and $Au_{78}Ge_{14}Si_8$, $Pd_{82}Si_{18}$, and $Pd_{78}Cu_6Si_{16}$ alloys.

2.6 Methods to Synthesize Metallic Glasses

Noncrystalline materials have been prepared by a variety of methods, starting from the vapor, liquid, or solid states. We will now briefly describe the different techniques available to synthesize noncrystalline (or glassy) materials. Cahn [28] and Johnson [18] have succinctly summarized the different available techniques to produce amorphous and glassy alloys, and described some of the results. The full details of the preparation of BMGs will be described in Chapter 4.

2.6.1 Vapor-State Processes

The earliest report of the formation of an amorphous phase in a metal is by vapor deposition methods. The physicists produced amorphous metals by the vapor deposition route to study their superconducting properties. According to the available reports, the German physicist Kramer [29,30] synthesized amorphous Sb by first evaporating the Sb metal and allowing it to condense on a cold substrate. Subsequently, Buckel and Hilsch [31,32] synthesized thin films of Bi, Ga, and Sn–Bi alloys by the vapor deposition process. Even though they believed that their films were ultrafine-grained from *in situ* electron diffraction experiments, it became clear later that the films were actually amorphous. There is still debate in the literature as to whether the vapor-deposited pure metals Sb and Bi were truly amorphous or microcrystalline. Irrespective of whether they were amorphous or not, this technique is presently used to synthesize amorphous materials in small quantities.

Several variants of the vapor deposition method are used nowadays to synthesize amorphous alloys and other metastable phases [33]. These include, for example, sputtering, chemical vapor deposition, and electron beam evaporation. One of the most popular among these is the sputtering method, in which atoms from the surface of the material are knocked off using high-energy ions, and the atoms so removed from the surface are subsequently allowed to deposit onto a cold substrate. Dahlgren [34] has used this method extensively. Bickerdike et al. [35,36] used high-rate thermal evaporation methods to deposit Al alloy vapor onto cold substrates to make very large objects. But, these techniques seem to produce mostly crystalline, not amorphous, alloys. Further, Dahlgren's approach appears to be expensive. Therefore, these vapor deposition methods are used mainly to produce small quantities for mostly scientific investigations and for electronic and magnetic applications.

Another technique that has been extensively used to prepare amorphous alloys, again mostly in thin film form, is to implant high-energy solute ions into metallic surfaces. This technique has come to be known as *ion implantation*. In a related technique, referred to as *ion mixing*, successively vapor-deposited layers of different elements are mixed together and alloyed by means of the thermal energy of injected rare-gas ions. Several amorphous and metastable crystalline phases have been produced by these methods. The details of the techniques and the results obtained using these techniques may be found in Refs. [37–40].

2.6.2 Liquid-State Processes

Metallic glasses were usually prepared by rapidly solidifying their metallic melts at solidification rates of about 10^6 K s^{-1}, using techniques such as melt spinning and its variants. The product of RSP is in the form of thin ribbons, even though wires, powders, and other shapes have also been produced. The formation of a glassy structure in metallic materials requires that the melt is cooled at very high solidification rates to prevent crystal nuclei from forming. Consequently, the minimum cooling rate required to form the glass (critical cooling rate) is about 10^5–10^6 K s^{-1}, since mostly binary and ternary alloys are used. An achievement of such high solidification rates is possible only when the heat is removed from the melt rapidly, and as a result, the section thickness is limited to a few tens of micrometers.

Starting with the Au–Si alloy in 1960, a number of other binary, ternary, quaternary, and higher-order alloy systems have been solidified into the glassy state by RSP methods. It may be noted that it is very difficult to produce a pure metal in the glassy state, since the critical cooling rate required to form the glassy phase has been estimated to be of the order of 10^{12}–10^{13} K s^{-1} [41]. Recently, however, Zhong et al. [42] have generated very high cooling rates of about 10^{14} K s^{-1} and produced the pure metals Ta, V, W, and Mo in the glassy state. Rapidly solidified crystalline alloys have been shown to exhibit a variety of constitutional and microstructural features that are different from their equilibrium counterparts. The constitutional effects of RSP include formation of supersaturated solid solutions, metastable intermediate phases, quasicrystalline phases, and metallic glasses, while the microstructural changes include refinement of grains, second phases, and segregation patterns.

Through the proper choice of the alloy system and its composition, it is possible to reduce the critical cooling rate for glass formation. Consequently, the section thickness increases from a few micrometers (typically in the range 20–50 μm) in the RSP alloys to a few millimeters or even a few centimeters. Such glasses are referred to as BMGs, the subject matter of this book. The preparation of BMGs starting from the liquid state has been the most popular method. All the different techniques available to prepare BMGs will be described fully in Chapter 4.

Apart from the RSP methods, electrodeposition methods have also been used to synthesize amorphous alloys. It was in 1950, for the first time, that Brenner et al. [43] reported the formation of an amorphous phase in electrodeposited Ni–P alloys containing >10 at.% P. The amorphous nature of the deposit was inferred from the presence of only one broad diffuse peak in the XRD pattern. These alloys have a very high hardness, and consequently these are used as wear- and corrosion-resistant coatings [44]. It is also possible to deposit some amorphous alloys by electroless deposition, particularly for magnetic investigations

[45]. The limited size of the product formed in one experiment puts a limitation on the usefulness of this technique to produce large quantities of metallic glasses.

2.6.3 Solid-State Processes

The preparation of amorphous phases starting from the solid state has been another technique, second only to RSP methods in popularity. A number of different methods in this major category are available, and the most commonly used method among these is mechanical alloying/milling. Details about this method will also be provided in Chapter 4, since this method has been used to produce large quantities of metallic glassy alloys. *Mechanical alloying* refers to the process when one starts with a mixture of elemental powders and alloying occurs due to interdiffusion in the constituent elements, resulting in the formation of a variety of phases. On the other hand, one starts with prealloyed homogeneous powders in *mechanical milling*. That is, while material transfer is involved in mechanical alloying, no such transfer occurs during mechanical milling. In brief, the process of mechanical alloying/milling involves putting the required amounts of the elemental or prealloyed powders in a hardened steel container along with hardened steel balls and agitating the whole mixture for the required length of time. Alloying occurs between the metal powders and can result in the formation of supersaturated solid solutions, intermetallic phases, quasi-crystalline phases, or amorphous alloys, depending on the powder composition and milling conditions. The choice of the phase formed depends on the thermodynamic stability of the different competing phases. The milled powder particles experience heavy plastic deformation, leading to the generation of a variety of crystal defects, such as dislocations, grain boundaries, and stacking faults. These defects raise the free energy of the crystalline system to a level higher than that of a hypothetical amorphous phase, and consequently, the crystalline phase becomes destabilized and an amorphous phase forms [46,47].

The first report of the formation of an amorphous phase by mechanical milling was in a Y–Co intermetallic compound [48] and that by mechanical alloying in a Ni–Nb powder blend [49]. Subsequently, amorphous phases have been formed in a large number of binary, ternary, and higher-order systems by this method. Unlike in RSP methods, the conditions for the formation of an amorphous phase by mechanical alloying/milling seem to be quite different. For example, amorphous phases are formed, not necessarily near eutectic compositions, but in a much wider composition range. The nature and transformation behavior of the amorphous phases formed by mechanical alloying/milling also appear to be different from those formed by RSP methods.

A number of alloy compositions that have been cast into BMGs by solidification methods have also been obtained in the amorphous state by mechanical alloying/milling methods. The details of these alloys and the process parameters will be discussed at length in Chapter 4.

There are several other techniques to synthesize amorphous phases in the solid state. All these methods could be grouped under the generic name *solid-state amorphization reactions*.

2.6.3.1 Hydrogen-Induced Amorphization

Yeh et al. [50] reported the formation of an amorphous metal hydride $Zr_3RhH_{5.5}$ by a reaction of hydrogen with a metastable crystalline Zr_3Rh compound at sufficiently low temperatures (less than about 200°C). A similar amorphous hydride phase was also reported to form when the amorphous Zr_3Rh alloy obtained in a glassy state was reacted with hydrogen. No difference was found between these two types of amorphous phases. Based on the experimental results, the authors mentioned that an amorphous phase could form when (1) at least two components were present, (2) a large asymmetry existed in the diffusion rates of the two species of atoms, and (3) a corresponding crystalline polymorph for the amorphous phase did not exist. Aoki [51] has reported that hydrogen-induced amorphization is possible in many binary metal compounds.

2.6.3.2 Multilayer Amorphization

An amorphous phase was found to form when thin metal films (10–50 nm in thickness) of La and Au were allowed to interdiffuse at relatively low temperatures (50°C–100°C). Similar to the case of hydrogen-induced amorphization, here also the large asymmetry in the diffusion coefficients of the two elements was found to be responsible for the formation of the amorphous phase [52]. Since then, a large number of cases where multilayer amorphization occurs have been reported.

In a somewhat related process, Atzmon et al. [53] reported that an amorphous phase was also produced by cold-rolling of thin foils of Ni and Zr, and then annealing them at a low temperature. Dinda et al. [54] extended this to achieving amorphization by cold-rolling in Cu–Zr, Cu–Ti–Zr, and Cu–Ti–Zr–Ni multilayers.

2.6.3.3 Pressure-Induced Amorphization

Alloys in the systems Cd–Sb, Zn–Sb, and Al–Ge were subjected to high pressures when they formed unstable crystalline phases and then decayed in a short time to amorphous phases [55]. An amorphous phase was also found to form in Cu–12–17 at.% Sn alloys when they were heated to high temperatures under pressure, and the power was suddenly turned off. In a related process, designated *sonochemical synthesis*, Suslick et al. [56] reported that they were able to produce amorphous iron by irradiation of liquid iron pentacarbonyl.

2.6.3.4 Amorphization by Irradiation

A number of intermetallics, for example, NiTi, Zr_3Al, Cu_4Ti_3, $CuTi_2$, FeTi, and so on, have been amorphized by irradiation with high-energy electrons, heavy ions, or fission fragments [18]. It is only intermetallics, and not solid solutions, that have been amorphized this way. Several different criteria have been proposed for this purpose, which include low temperature, high doses, high dose rates, intermetallics with an extremely narrow homogeneity range, permanently ordered compounds (i.e., intermetallics that are ordered up to the melting temperature and do not show an order–disorder transformation) that also have high-ordering energies, destruction of long-range order by irradiation, and so on. Fecht and Johnson [57] have analyzed the fundamental issues involved in the amorphization of intermetallics by irradiation methods.

2.6.3.5 Severe Plastic Deformation

Intense plastic deformation at low temperatures has been shown to refine the microstructural features of metallic specimens. With a view to obtain such ultrafine microstructures in massive billets, Segal developed the equal channel angular pressing (ECAP, sometimes also referred to as *equal channel angular extrusion* [ECAE], although the former acronym is more popular) method in the early 1980s [58]. The process is based on the observation that metalworking by simple shear provides close to an ideal method for the formation of microstructure and texture. In this method, a solid die containing a channel that is bent into an L-shaped configuration is used. A sample that just fits into the channel is pressed through the die using a plunger. The billets usually have either a square or circular cross section, not exceeding about 20 mm. Thus, the sample undergoes shear deformation but comes out without a change in the cross-sectional dimensions, suggesting that repetitive pressings can be conducted to obtain very high total strains. The strain introduced into the sample is primarily dependent on the angle subtended by the two parts of the channel. This method has been used since the 1990s to produce ultrafine-grained materials [59,60]. The development of the high-pressure torsion (HPT) method has extended the capabilities of these methods further with the formation of amorphous phases in select alloy systems. All these methods involving heavy deformation are now grouped together under the category of severe plastic deformation (SPD) methods.

The SPD methods have some advantages over other methods in synthesizing ultrafine-grained materials. Firstly, ultrafine-grained materials with high-angle grain boundaries can be synthesized. Secondly, the samples are dense and there is no porosity in them since they have not been produced by the consolidation of powders. Thirdly, the grain size is fairly uniform throughout the structure. Lastly, the technique may be applied directly to commercial cast metals. However, the limitations of this technique are that the minimum grain size is not very small (typically it is about 200–300 nm), and that amorphous phases or other metastable phases have not been synthesized frequently. But, these methods are very useful in producing bulk ultrafine-grained materials.

In the HPT process, a small disk-like specimen (typically 10 mm in diameter and 0.3 mm in thickness) is deformed by torsion under high pressure. The extent of plastic deformation induced in the specimen can be very high by simply increasing the number of rotations. A very large true strain is achieved by increasing the number of rotations to, say, 20. Amorphous phases have been produced in binary Cu–Zr [61,62] and ternary Cu–Zr–Al alloy samples by this method [62].

Figure 2.4 shows schematics of the ECAP and HPT processing methods. The interested reader should refer to the original papers for further details of processing.

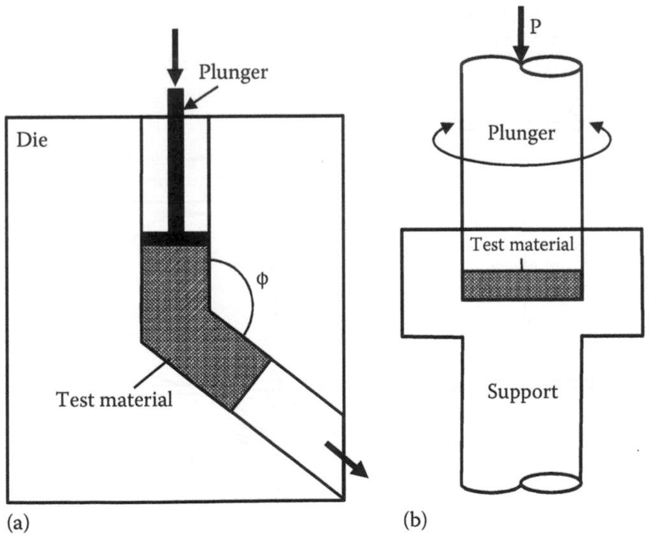

FIGURE 2.4
Schematic diagrams showing the (a) equal channel angular pressing (ECAP) and (b) high-pressure torsion (HPT) processes.

2.6.3.6 Accumulative Roll Bonding

The methods of SPD previously described have two drawbacks. Firstly, forming machines with large load capacities and expensive dies are required. Secondly, the productivity is relatively low. To overcome these two limitations, Saito et al. [63] developed the accumulative roll-bonding (ARB) process, shown schematically in Figure 2.5. In this process, the stacking of materials and conventional roll bonding are repeated. First, a strip is neatly placed on the top of another strip. The surfaces of the strips are treated, if necessary, to enhance the bond strength. The two layers of the material are joined together by rolling as in a conventional roll-bonding process. Then the length of the rolled material is sectioned into two halves. The sectioned strips are again surface treated, stacked to the initial dimensions, and roll bonded. The whole process is repeated several times.

The ARB process is conducted at elevated temperatures, but below the recrystallization temperature, since recrystallization annihilates the accumulated strain. Very low temperatures would result in insufficient ductility and bond strength. It has been shown that if the homologous temperature of roll bonding is <0.5, a sound joining can be achieved for reductions >50%. Very high-strength materials were obtained by this process in different alloy systems. Tsuji [64] recently presented a brief overview of the developments in this process.

It has been recently reported that amorphization occurs in materials processed by ARB at the interfacial regions of the constituent metals [65]. Cu/Zr multilayered sheets, processed at room temperature by ARB, showed a nanolamellar structure. But, when this specimen was annealed at 400°C, most of the specimen transformed into an amorphous state [66]. It may be useful to subject the samples to an HPT treatment after processing the multi-stacks by ARB. In such a situation, it may be possible to obtain an amorphous phase fully in the whole sample. One difficulty that may arise in such a situation is that the ARB-processed material may be highly brittle, and so HPT may not be a feasible method to process it further.

From this description, it is clear that a large number of techniques are available to synthesize amorphous alloys and metallic glasses. Among all the methods discussed, it is

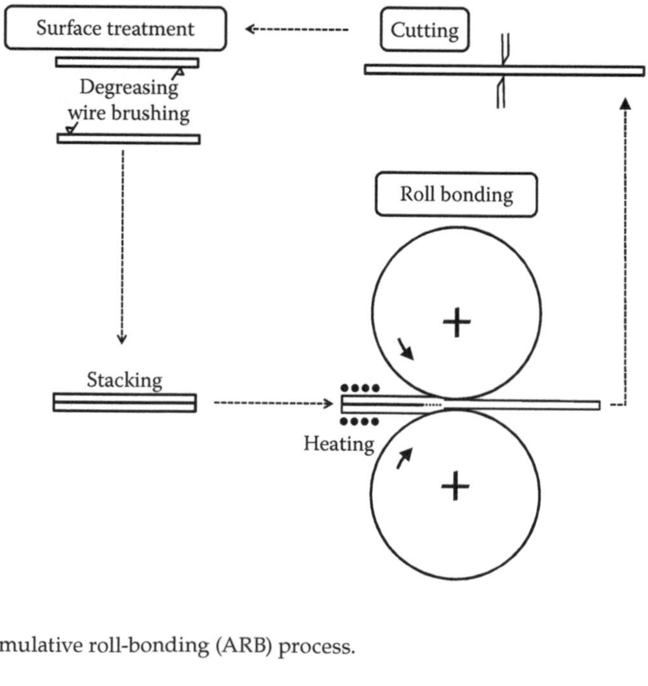

FIGURE 2.5
Schematic of the accumulative roll-bonding (ARB) process.

easy to realize that the processing of amorphous alloys from the vapor route is expensive and slow, and therefore it is difficult and not economically viable to obtain large quantities. This technique has its niche in producing small quantities of amorphous alloys for electronic and magnetic applications.

Starting from the solid state is another important route, and it is possible to obtain large quantities of amorphous alloys. There are two different versions here—single-stage and two-stage processes. In the single-stage process, the desired product is obtained directly in one stage, whereas in the two-stage process, the final product is obtained after an intermediate processing step. For example, one can obtain the final product in a single stage using processes such as SPD. But, one may not easily obtain amorphous phases by using these processing methods. On the other hand, using techniques like mechanical alloying or milling, it is easy to obtain the amorphous phases. But, this powder needs to be consolidated into bulk shapes by methods such as vacuum hot pressing, hot isostatic pressing, spark plasma sintering, and so on. Thus, this is a two-stage process, wherein the powder is produced first and is then consolidated into bulk shape to obtain the final product. It is possible to introduce contamination and impurities during these different processing stages, and therefore, purity of the final product is of concern. The other solid-state processing methods, for example, irradiation or ion mixing of multilayers, also produce only small quantities of materials, and therefore they may also turn out to be expensive.

Liquid-state processing is the ideal way to obtain metallic glasses, especially the bulk variety. Traditionally, most of the engineering materials for structural applications were produced starting from the liquid state, producing an ingot and then obtaining the final product through secondary processing stages. Therefore, an added advantage of using the solidification-processing approach is that one could use the traditional casting equipment and others for subsequent thermomechanical processing, if desired and necessary. Therefore, we will be concentrating on the solidification processing of BMGs, but make oblique references to other methods to compare the properties of the products obtained by the different methods.

2.7 Bulk Metallic Glasses

BMGs are those noncrystalline solids obtained by continuous cooling from the liquid state, which have a section thickness of at least a few millimeters. More commonly, metallic glasses with at least a diameter or section thickness of 1 mm are considered "bulk." (Nowadays, researchers tend to consider 10 mm as the minimum diameter or section thickness at which a glass can be designated as bulk.)

2.7.1 Characteristics of Bulk Metallic Glasses

As will become clear in the subsequent chapters, BMGs have the following four important characteristics:

- The alloy systems have a minimum of three components; more commonly the number is much larger, and that is why they are frequently referred to as *multicomponent alloy systems*. There have also been reports of binary BMGs. But, the maximum diameter of the rod that could be obtained in a fully glassy condition, in a binary alloy, is usually reported to be 1 or 2 mm. And even in these sections, a small volume fraction of nanocrystalline precipitates have been frequently observed to be dispersed in the glassy matrix.

- They can be produced at slow solidification rates, typically 10^3 K s^{-1} or less. The lowest solidification rate at which BMGs have been obtained was reported as 0.067 K s^{-1}, that is, 4 K min^{-1} [67]; a really slow solidification rate indeed!
- BMGs exhibit large section thicknesses or diameters, a minimum of about 1 mm. The largest diameter of a BMG rod produced to date is 80 mm in a $Pd_{42.5}Cu_{30}Ni_{7.5}P_{20}$ alloy [68].
- They exhibit a large supercooled liquid region. The difference between the glass transition temperature, T_g, and the crystallization temperature, T_x, that is, $\Delta T_x = T_x - T_g$, is large, usually a few tens of degrees, and the highest reported value so far is 131 K in a $Pd_{43}Ni_{10}Cu_{27}P_{20}$ alloy [17].

There has been some confusion in the literature regarding which noncrystalline alloys may be designated as BMGs. Ideally, a BMG satisfies all four criteria. But, some researchers have used the term *bulk* to denote that they are referring to a multicomponent system, even though they had produced the glassy phase by melt spinning in the form of thin ribbons, only about 50 μm in thickness. Even though it should be easy to produce such an alloy composition in bulk form at slower cooling rates, this designation acquires credibility only when it has been shown that the same alloy composition can be produced in large diameters. Secondly, the literature contains references to results showing that a melt-spun metallic glass ribbon has a large ΔT_x value, and therefore this alloy has been referred to as a BMG. This result must again be corroborated with a large section thickness. In other words, the most important criterion for a BMG is that it should be possible to produce it in *a large section thickness*.

Another important point to be remembered is that a metallic glass, or for that matter any noncrystalline alloy, should not contain any crystalline phases, for it to be called a *noncrystalline* alloy. If it contains a crystalline phase in addition to the glassy phase, it should more appropriately be referred to as a *composite*. BMG composites also have interesting properties, but they should not simply be referred to as BMGs alone. When the as-cast alloy is characterized by XRD techniques, the presence of a broad and diffuse peak is often taken to be evidence for the presence of a glassy phase. This is normally true. But, it should be realized that the technique of XRD is not very sensitive to the presence of a small volume fraction of a crystalline phase in a glassy phase, especially when the crystals are in nanocrystalline condition. Therefore, even if the XRD pattern shows a broad halo, the material may contain a small volume fraction of a crystalline phase dispersed in the glassy matrix. Further, a structure consisting of a glassy phase, or extremely fine grains, or a nanocrystalline material with a small grain size of less than about 10 nm, will exhibit a broad and diffuse halo. Thus, it is always desirable to confirm the lack of crystallinity in the material by conducting (high-resolution) transmission electron microscopy investigations.

2.7.2 Origins of Bulk Metallic Glasses

The origins of BMGs go back to 1974, when Chen [69] reported that by water quenching ternary alloys of different compositions in the $(Pd_{1-x}M_x)_{0.835}Si_{0.165}$ (with M = Fe, Co, Ni, Cu, Ag, and Au), Pt–Ni–P, and Pd–Ni–P systems, he was able to produce long glassy rods, 1–3 mm in diameter and several centimeters in length. The quenching rate was estimated to be less than 10^3 K s^{-1}. Also, 2 mm diameter rods of water-quenched $Pd_{77.5}Cu_6Si_{16.5}$ glasses were produced later [70]. Subsequently, Drehman et al. [71] reported in 1982 that by eliminating surface heterogeneities through etching, they were able to produce a 6 mm diameter glass in the $Pd_{40}Ni_{40}P_{20}$ alloy system by slowly cooling from the liquid state at a rate of 1.4 K s^{-1}. Later, Kui et al. [72] used the fluxing technique in 1984, by which, through adding B_2O_3 flux to the $Pd_{40}Ni_{40}P_{20}$ alloy melt, they were able to completely eliminate all the heterogeneous nucleation sites and consequently produce a 10 mm diameter glass.

Starting in the late 1980s, the RSP group at Tohoku University in Sendai, Japan, systematically investigated the GFA of different alloy systems and was able to produce bulk glassy alloys in some of the systems at solidification rates of 10^3 K s^{-1} or lower. A brief account of the development of BMGs at Tohoku University follows.

Al–La–Ni [73] and Mg–Ni–La [74] glassy alloys were produced in a wide composition range of 3–83 at.% La and 0–60 at.% Ni in the La–Al–Ni system and 10–40 at.% Ni and 5–30 at.% La in the Mg–Ni–La system. But, more important than just producing the glassy phases in these systems, these glasses showed clear glass transition temperatures, T_g, and very wide supercooled liquid regions, ΔT_x. The ΔT_x values reported were 69 K for the $La_{55}Al_{25}Ni_{20}$ glass and 58 K for the $Mg_{50}Ni_{30}La_{20}$ glass. These ΔT_x values were much larger than those reported for the noble-metal-based glasses, Pd–Ni–P and $Pt_{60}Ni_{15}P_{25}$ (35–40 K), reported earlier.

Since the large ΔT_x value suggests a high thermal stability of the supercooled liquid against nucleation of a crystalline phase, it was argued that the GFA of these alloys should be high enough to produce them in a glassy state at relatively slow cooling rates, as obtained by water quenching. Indeed, a fully glassy phase was reported to form in water-quenched $La_{55}Al_{25}Ni_{20}$ alloy melt [75]. Figure 2.6 shows the XRD pattern recorded from the water-quenched alloy rod with a diameter of 0.8 mm, which shows only a broad diffuse diffraction peak and no other sharp peaks corresponding to any crystalline phase. The XRD pattern from a 20 μm thick melt-spun ribbon of the same alloy composition is also presented in Figure 2.6 for comparison. It may be clearly seen that there is no difference in the diffraction patterns of the two samples. The broad peak is present at the same diffraction angle in both cases, suggesting that the glassy phase is similar in both. The maximum diameter of the rod that could be produced in a fully glassy state at this composition was reported to be 1.2 mm. A transverse cross section of the etched glassy rod did not show the presence of any crystalline phase.

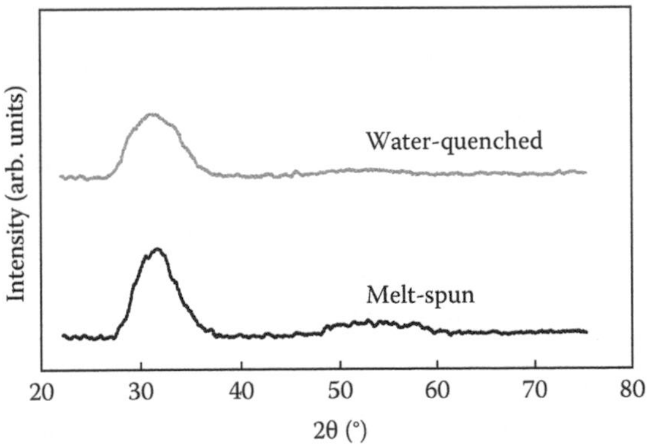

FIGURE 2.6
X-ray diffraction patterns of $La_{55}Al_{25}Ni_{20}$ glassy samples in the water-quenched rod (0.8 mm diameter) and melt-spun ribbon (20 μm thick) conditions. Note that in both cases, the diffraction pattern shows only a broad peak, and sharp peaks indicative of any crystalline phase are absent. Furthermore, the position of the broad peak is the same in both samples, suggesting that the glassy phase produced is the same in both cases. (Reprinted from Inoue, A. et al., *Mater. Trans., JIM*, 30, 722, 1989. With permission.)

This early publication also brought out the following very important points regarding the thermal characteristics of BMGs in comparison with the melt-spun ribbons of the same composition:

- The glass transition temperature, T_g, and the crystallization temperature, T_x, for both the rod and ribbon samples were identical at $T_g = 476$ K and $T_x = 545$ K.
- The heat of crystallization was the same for both the melt-spun ribbon and slowly cooled rod samples.
- However, the heat of irreversible structural relaxation (below T_g), ΔH_{relax}, was found to be significantly different. While the melt-spun ribbon had a value of $\Delta H_{relax} = 598$ J mol^{-1}, the bulk rod sample had a value of about half this, *viz.*, $\Delta H_{relax} = 287$ J mol^{-1}. The lower value for the water-quenched rod sample suggests that this slowly cooled sample had a more relaxed atomic configuration than the rapidly solidified melt-spun ribbon sample.

These observations have subsequently been confirmed in a number of other alloy systems and by several researchers. Subsequent to this initial discovery, Peker and Johnson [76] produced a 14 mm diameter fully glassy rod in the composition $Zr_{41.2}Ti_{13.8}Cu_{12.5}Ni_{10.0}Be_{22.5}$ (most commonly referred to as *Vitreloy 1* or *Vit 1*, and now redesignated as liquid metal or LM-001), and since then there has been an explosion in research activity in this area all over the world.

Starting with the initial discovery of the formation of BMGs in the La–Al–Ni system, the Tohoku group has produced a very large number of BMGs in different alloy systems based on Mg, Zr, Ti, Pd, Fe, Co, Ni, and Cu. They have been able to produce several new alloys in the BMG state and also increase the critical (or maximum) diameter of the BMG alloy rods during the last nearly 25 years. The results of these investigations will be referred to in the subsequent chapters. There have been significant advances in the synthesis of BMG alloys by other researchers also. The maximum diameters obtained in the different alloy systems and the years in which they were discovered are presented in Figure 2.7.

2.8 High-Entropy and Pseudo-High-Entropy Bulk Metallic Glasses

In recent years, there has been intense activity on another kind of novel material, commonly referred to as *high-entropy alloys* (HEAs) [77,78]. A few good reviews and a couple of monographs have also appeared on this topic [79–85]. Tomilin and Kaloshkin [86] believe that the efficiency of this approach to the development of "a new era" of HEAs is highly exaggerated, since the theory of regular solutions, on which the concept of HEAs is based, is not appropriate. The idea of random distribution of atoms in HEAs is appropriate only for ideal solutions.

The HEAs are characterized by five or more elements in equal or nearly equal atomic percentages, each with a concentration of 5–35 at.%. Thus, these HEAs (also referred to as *multicomponent alloys*) are complex, and thus, they should also be able to form metallic glasses, if not the BMGs. Arguing that the crystallization process of the multicomponent supercooled liquid tends to be more sluggish than that of the simple alloy systems, Ma et al. [87] synthesized the first BMG rod with equal amounts of the constituent elements and a 1.5 mm diameter in the $Ti_{20}Zr_{20}Hf_{20}Ni_{20}Cu_{20}$ system by copper mold casting. This development went practically unnoticed, and independent of this, a huge amount of activity was

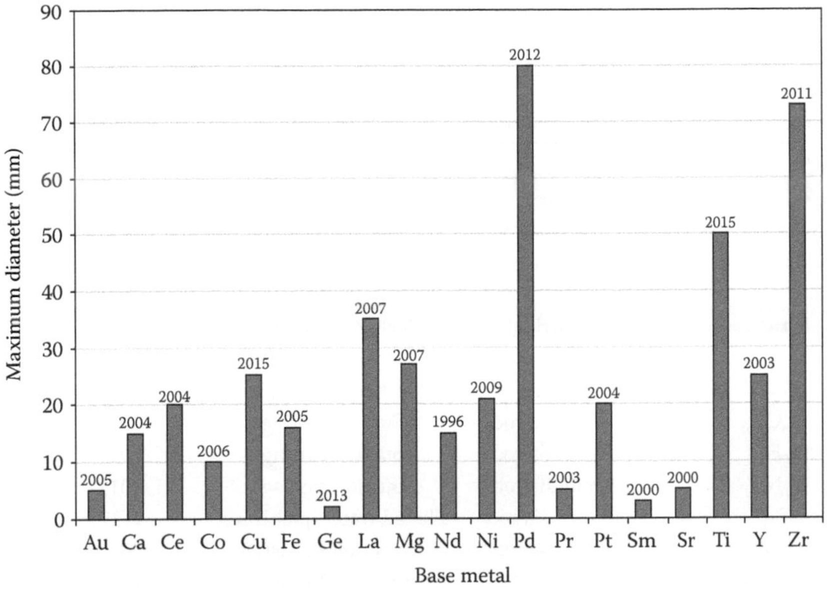

FIGURE 2.7
Maximum diameters of the BMG rods achieved in different alloy systems and the years in which they were discovered.

undertaken all over the world to produce multicomponent *crystalline* alloys with a simple single-phase structure, for example, body-centered cubic (bcc), fcc, or hexagonal close packed (hcp). These alloys possess an excellent combination of properties, such as very fine grain size, high thermal stability of the fine-grained structure, low atomic diffusivity of the constituent elements, high tensile strength, large elongation, high elevated temperature hardness, and high corrosion resistance. The reviews [79–83] and monographs [84,85] referenced at the beginning of this section summarize the present results.

A number of papers on high-entropy (multicomponent) BMGs have appeared since about 2011, but the list is still relatively short. Table 2.1 summarizes the results reported to date, along with the maximum diameter of the BMG rods and the technique used to produce them. One can notice that the high-entropy-type BMGs can be classified into the following two groups:

1. Ordinary HEAs with equal or nearly equal atomic compositions of the constituent elements.
2. Pseudo-HEAs with a total of more than five elements, but 40–60 at.% for the main constituent elements, and 5–15 at.% for each solute element. The advantage of this type of pseudo-high-entropy BMG is the achievement of similar useful engineering properties as those of ordinary BMGs.

It may also be noted from Table 2.1 that the ideal high-entropy BMGs are derived from the basic ternary BMG alloy systems by partial replacement of the major component elements with elements from a similar group in the periodic table. However, the maximum diameter of the BMGs produced from HEAs is relatively small, the maximum being only about 30 mm in a $Ti_{20}Zr_{20}Hf_{20}Be_{20}Cu_{7.5}Ni_{12.5}$ alloy produced by copper mold casting [98]. On the other hand, the largest diameter obtained in a regular BMG is 80 mm in a $Pd_{42.5}Cu_{30}Ni_{7.5}P_{20}$ alloy [68]. This

TABLE 2.1
Typical High-Entropy BMG Alloys

System/Composition	Shape and Size	Method	Year of Discovery	References
$AlCoCrFeNiZr_{0.6}$	Ribbon	Melt spinning	2013	[88]
$CoCrCuFeNiZr_{0.6}$	Ribbon	Melt spinning	2013	[88]
$Er_{20}Tb_{20}Dy_{20}Ni_{20}Al_{20}$	2 mm dia rod	Copper mold casting	2011	[89]
$Fe_{25}Co_{25}Ni_{25}B_{15}Si_{10}$	1 mm dia rod	Copper mold casting	2015	[90]
$Fe_{25}Co_{25}Ni_{25}B_{17.5}Si_{7.5}$	1.5 mm dia rod	Copper mold casting	2015	[90]
$Fe_{20}Si_{20}B_{20}Al_{20}Nb_{20}$	Powder	Mechanical alloying	2014	[91]
$Fe_{20}Si_{20}B_{20}Al_{20}Ni_{20}$	Powder	Mechanical alloying	2014	[91]
$Fe_{20}Cr_{20}Ni_{20}Zr_{20}Si_{20}$	Powder	Mechanical alloying	2015	[92]
$Gd_{20}Tb_{20}Dy_{20}Al_{20}Co_{20}$	1 mm dia rod	Suction casting	2015	[93]
$Gd_{20}Tb_{20}Dy_{20}Al_{20}Fe_{20}$	1 mm dia rod	Suction casting	2015	[93]
$Gd_{20}Tb_{20}Dy_{20}Al_{20}Ni_{20}$	1 mm dia rod	Suction casting	2015	[93]
$Pd_{20}Pt_{20}Cu_{20}Ni_{20}P_{20}$	10 mm dia rod	Fluxed water quenching	2011	[94]
$Sr_{20}Ca_{20}Yb_{20}Mg_{20}Zn_{20}$	5 mm dia rod	Copper mold casting	2011	[89]
$Sr_{20}Ca_{20}Yb_{20}Mg_{20}Zn_{10}Cu_{10}$	2 mm plate	Copper mold casting	2011	[89]
$Sr_{20}Ca_{20}Yb_{20}Mg_{9}Zn_{20}Li_{11}$	2 mm plate	Copper mold casting	2011	[89]
$Ti_{20}Zr_{20}Hf_{20}Cu_{20}Co_{20}$	Ribbon	Melt spinning	2002	[87]
$Ti_{20}Zr_{20}Hf_{20}Cu_{20}Fe_{20}$	Ribbon	Melt spinning	2002	[87]
$Ti_{20}Zr_{20}Hf_{20}Cu_{20}Ni_{20}$	1.5 mm dia rod	Copper mold casting	2002	[87]
$Ti_{20}Zr_{20}Cu_{20}Ni_{20}Be_{20}$	3 mm dia rod	Copper mold casting	2013	[95]
$Ti_{20}Zr_{20}Cu_{20}Ni_{20}Be_{20}$	3 mm dia rod	Copper mold casting	2015	[96]
$Ti_{20}Zr_{20}Hf_{20}Be_{20}Cu_{20}$	12 mm dia rod	Copper mold casting	2015	[97]
$Ti_{20}Zr_{20}Hf_{20}Be_{20}Cu_{17.5}Ni_{2.5}$	12 mm dia rod	Copper mold casting	2015	[98]
$Ti_{20}Zr_{20}Hf_{20}Be_{20}Cu_{15}Ni_{5}$	15 mm dia rod	Copper mold casting	2015	[98]
$Ti_{20}Zr_{20}Hf_{20}Be_{20}Cu_{12.5}Ni_{7.5}$	20 mm dia rod	Copper mold casting	2015	[98]
$Ti_{20}Zr_{20}Hf_{20}Be_{20}Cu_{10}Ni_{10}$	25 mm dia rod	Copper mold casting	2015	[98]
$Ti_{20}Zr_{20}Hf_{20}Be_{20}Cu_{7.5}Ni_{12.5}$	30 mm dia rod	Copper mold casting	2015	[98]
$Ti_{20}Zr_{20}Hf_{20}Be_{20}Cu_{5}Ni_{15}$	20 mm dia rod	Copper mold casting	2015	[98]
$Ti_{20}Zr_{20}Hf_{20}Be_{20}Cu_{2.5}Ni_{17.5}$	15 mm dia rod	Copper mold casting	2015	[98]
$Ti_{20}Zr_{20}Hf_{20}Be_{20}Ni_{20}$	15 mm dia rod	Copper mold casting	2015	[98]
$Ti_{16.7}Zr_{16.7}Hf_{16.7}Cu_{16.7}Ni_{16.7}Be_{16.7}$	15 mm dia rod	Injection mold casting	2014	[99]
$Zn_{20}Ca_{20}Sr_{20}Yb_{20}(Li_{0.55}Mg_{0.45})_{20}$	>3 mm dia rod	Copper mold casting	2011	[100]

observation suggests that high mixing entropy is neither a sufficient nor a necessary condition to achieve a high GFA in alloys. But, this is inconsistent with the previous interpretation that the increase in entropy can stabilize the supercooled liquid and enhance the GFA. This may be because the minor addition of soluble elements belonging to the same periodic group is effective for the enhancement of GFA, while the addition of elements with *equivalent* atomic amounts from a similar group is less effective. The reason for the discrepancy is presumably that the ability to form highly densely packed medium-range ordered atomic configurations is almost independent of the replacement amount of the host element by the elements in the same alloy group. At present, little is known about the details of the structure of the regular high-entropy-type BMGs and the understanding of the atomic configuration in high-entropy crystalline alloys will probably shed new light. This will also help in the development of new BMGs with higher GFA by multiplication of the alloy components.

The high-entropy BMGs exhibit some interesting characteristics, such as high thermal stability, good corrosion resistance, high magnetocaloric characteristics, low growth rate of precipitates, easy formation of nanocrystalline structure, high elevated temperature strength, and so on. The increase in the freedom in alloy design for high-entropy BMGs is expected to reveal novel and more useful characteristics that cannot be obtained for regular BMGs.

It is important to highlight some salient differences between the pseudo-high-entropy-type BMGs in the as-cast state and the regular BMGs belonging to similar metal-based alloy groups. The differences that the pseudo-high-entropy BMGs show are (1) a slight increase in T_g, (2) a significant decrease in T_x, (3) a significant reduction of the supercooled liquid region ($T_x - T_g$), (4) a very similar Vickers hardness and yield strength, (5) a lack of obvious change in elastic elongation and plastic strain, (6) a slight improvement of corrosion resistance in the case of the dissolution of noble and/or high passivated metallic elements, and (7) a lack of appreciable change in the bending ductility for melt-spun ribbons. These features are significantly different from those for the partially crystallized BMGs, and the difference may be due to the formation of sub-nanoscale fine clustered structures that cannot be formed in regular BMGs. A detailed study on the differences in the features seems to be an important subject for the establishment of the scientific and engineering value of high-entropy and pseudo-high-entropy-type BMGs to coexist with regular BMGs.

2.9 Potential Resources in the Literature on Metallic Glasses

A number of excellent reviews, emphasizing the different aspects of metallic glasses, and also books on metallic glasses, have been published periodically. A listing of such reviews and books, which is by no means exhaustive (and deals mostly with melt-spun glassy alloys), is provided in Refs. [101–120]. In addition to these, triennial international conferences have been held on rapidly quenched metals (RQ series). The proceedings of these conferences have been published [121–125]. However, starting with the sixth international conference (RQ6) held in Montréal, Canada, the proceedings of the conferences have been published in the *International Journal of Materials Science and Engineering A*, published by Elsevier [126–135]. These conferences have been redesignated as *Rapidly Quenched and Metastable Materials* since RQ-8 in Sendai, Japan. Fifteen RQ international conferences have been held so far, and RQ-16 will be held in Leoben, Austria, from August 27 to September 1, 2017. Additionally, the *Proceedings of the Annual International Symposia on Metastable, Mechanically Alloyed, and Nanocrystalline Materials,* now known as the *International Symposium on Metastable and Nano-Materials* (*ISMANAM*), also contain many papers on the different aspects of metallic glasses (including those on BMGs). These conference proceedings have been published by Trans Tech, Switzerland, first in their *Materials Science Forum* series up to *ISMANAM-97* [136–139], then simultaneously in *Materials Science Forum* and the *Journal of Metastable and Nanocrystalline Materials,* up to *ISMANAM-2001* [140–143], and later only in the *Journal of Metastable and Nanocrystalline Materials* [144–146]. Subsequently, proceedings of the *ISMANAM* conferences were published in the *Journal of Alloys and Compounds* by Elsevier [147–157]. However, proceedings of the *ISMANAM-2006* conference held in Warsaw, Poland, were published in *Reviews on Advanced Materials Science* by the Russian Academy of Sciences, St. Petersburg, Russia [148]. Twenty-three such conferences have been held so far, and the 24th ISMANAM conference will be held in San Sebastian,

Spain, during June 18–23, 2017. One can also expect to see the published proceedings of these conferences soon.

Apart from these, several research papers have been published in conference proceedings and archival journals in the fields of solid-state physics and materials science and engineering.

Activity in the area of BMGs started in the late 1980s, and gathered great momentum during the mid-1990s, and it continues to be an active field of research in materials science and engineering and solid-state physics. A very large number of research papers are published annually, with one estimate putting the figure at more than 1000. The majority of the research papers in this area are published in international journals such as the *Journal of Non-Crystalline Solids, Intermetallics, Materials Science and Engineering A, Applied Physics Letters, Journal of Applied Physics, Journal of Materials Research, Acta Materialia,* and *Scripta Materialia*. A significant number of papers are also published in the monthly journal *Materials Transactions*, published by the Japan Institute of Metals in Sendai. Dedicated international conferences focusing exclusively on BMGs are also held, and their conference proceedings have been published. Eleven of them have been held so far (the most recent one was in St. Louis, MO, USA; June 5–9, 2016) and the conference proceedings of some have been published [158–165]. Apart from these, Professor Peter Liaw from the United States has also been organizing annual symposia on BMGs during the Minerals, Metals, and Materials Society (TMS) annual meetings, and the proceedings of these symposia are published in the journal *Metallurgical and Materials Transactions A*, published by Springer [166–174], except for the fifth conference, whose proceedings were published in the *Journal of Advanced Engineering Materials*, published by VCH [167]. Several reviews [175–209] and other conference proceedings [210–214] have also been published on BMGs in recent years. An edited monograph covering some aspects of BMGs has recently been published [215], and a book on BMGs was also published [216], the second edition of which is the current book. Two excellent monographs, mostly dealing with work completed before 1998 at the Tohoku University in Sendai, Japan, and authored by Akihisa Inoue, were published by Trans Tech Publications in Switzerland [217,218]. A few other monographs have also been published in recent years [219–223].

References

1. Klement, Jr, W., R.H. Willens, and P. Duwez (1960). Non-crystalline structure in solidified gold–silicon alloys. *Nature* 187: 869–870.
2. Duwez, P. (1981). Metallic glasses—historical background. In *Glassy Metals I*, eds. H.-J. Güntherodt and H. Beck, pp. 19–23. Berlin, Germany: Springer-Verlag.
3. Jones, H. and C. Suryanarayana (1973). Rapid quenching from the melt: An annotated bibliography 1958–72. *J. Mater. Sci.* 8: 705–753.
4. Suryanarayana, C. (1980). *Rapidly Quenched Metals—A Bibliography 1973–79*. New York: IFI/Plenum.
5. Allen, S.M. and E.L. Thomas (1999). *The Structure of Materials*. New York: John Wiley & Sons.
6. Kelly, A., G.W. Groves, and P. Kidd (2000). *Crystallography and Crystal Defects*. New York: Wiley.
7. De Graef, M. and M.E. McHenry (2007). *Structure of Materials: An Introduction to Crystallography, Diffraction and Symmetry*. Cambridge, UK: Cambridge University Press.
8. Shechtman, D., I. Blech, D. Gratias, and J.W. Cahn (1984). Metallic phase with long-range orientational order and no translational symmetry. *Phys. Rev. Lett.* 53: 1951–1953.

9. Suryanarayana, C. and H. Jones (1988). Formation and characteristics of quasicrystalline phases: A review. *Int. J. Rapid Solidif.* 3: 253–293.
10. Kelton, K.F. (1993). Quasicrystals: Structure and stability. *Int. Mater. Rev.* 38: 105–137.
11. Sastry, G.V.S., C. Suryanarayana, M. Van Sande, and G. Van Tendeloo (1978). A new ordered phase in the Al–Pd system. *Mater. Res. Bull.* 13: 1065–1070.
12. Chen, H.S. and D. Turnbull (1967). Thermal evidence of a glass transition in gold–silicon–germanium alloy. *Appl. Phys. Lett.* 10: 284–286.
13. Chen, H.S. and D. Turnbull (1968). Evidence of a glass–liquid transition in a gold–germanium–silicon alloy. *J. Chem. Phys.* 48: 2560–2571.
14. Inoue, A., T. Zhang, W. Zhang, and A. Takeuchi (1996). Bulk Nd–Fe–Al amorphous alloys with hard magnetic properties. *Mater. Trans., JIM* 37: 99–108.
15. Angell, C.A. (1995). Formation of glasses from liquids and biopolymers. *Science* 267: 1924–1935.
16. Kauzmann, W. (1948). The nature of the glassy state and the behavior of liquids at low temperatures. *Chem. Rev.* 43: 219–256.
17. Lu, I.-R., G. Wilde, G.P. Görler, and R. Willnecker (1999). Thermodynamic properties of Pd-based glass-forming alloys. *J. Non-Cryst. Solids* 250–252: 577–581.
18. Johnson, W.L. (1986). Thermodynamic and kinetic aspects of the crystal to glass transformation in metallic materials. *Prog. Mater. Sci.* 30: 81–134.
19. Leuzzi, L. and T.M. Nieuwenhuizen (2008). *Thermodynamics of the Glassy State.* Boca Raton, FL: CRC Press.
20. Turnbull, D. (1956). Phase changes. *Solid State Phys.* 3: 225–306.
21. Gaskell, D.R. (2008). *Introduction to the Thermodynamics of Materials,* 5th edn. New York: Taylor & Francis Group, LLC.
22. DeHoff, R. (2006). *Thermodynamics in Materials Science,* 2nd edn. Boca Raton, FL: CRC Press.
23. Cohen, M.H. and D. Turnbull (1959). Molecular transport in liquids and glasses. *J. Chem. Phys.* 31: 1164–1169.
24. Adam, G. and J.H. Gibbs (1965). On the temperature dependence of cooperative relaxation properties in glass-forming liquids. *J. Chem. Phys.* 43: 139–146.
25. Turnbull, D. (1969). Under what conditions can a glass be formed? *Contemp. Phys.* 10: 473–488.
26. Uhlmann, D.R. (1972). A kinetic treatment of glass formation. *J. Non-Cryst. Solids* 7: 337–348.
27. Davies, H.A. (1983). Metallic glass formation. In *Amorphous Metallic Alloys,* ed. F.E. Luborsky, pp. 8–25. London, UK: Butterworths.
28. Cahn, R.W. (1991). Metallic glasses. In *Glasses and Amorphous Materials,* ed. J. Zarzycki, pp. 493–548. Vol. 9 of *Materials Science and Technology: A Comprehensive Treatment.* Weinheim, Germany: VCH.
29. Kramer, J. (1934). Nonconducting modifications of metals. *Ann. Phys. (Berlin, Germany)* 19: 37.
30. Kramer, J. (1936). The amorphous state of metals. *Z. Phys.* 106: 675–691.
31. Buckel, W. and R. Hilsch (1952). Superconductivity and resistivity of tin with lattice defects. *Z. Phys.* 132: 420–442.
32. Buckel, W. (1954). Electron diffraction patterns of thin metal layers at low temperatures. *Z. Phys.* 138: 136–150.
33. Colligon, J.S. (1999). Physical vapor deposition. In *Non-Equilibrium Processing of Materials,* ed. C. Suryanarayana, pp. 223–253. Oxford, UK: Pergamon.
34. Dahlgren, S.D. (1978). Vapor quenching techniques. In *Proceedings of Third International Conference on Rapidly Quenched Metals (RQ III),* ed. B. Cantor, Vol. 2, pp. 36–47. London, UK: Metals Society.
35. Bickerdike, R.L., D. Clark, J.N. Eastabrook, G. Hughes, M.N. Mair, P.G. Partridge, and H.C. Ranson (1984–1985). Microstructure and tensile properties of vapour-deposited aluminium alloys. Part 1. Layered microstructures. *Int. J. Rapid Solidif.* 1: 305–325.
36. Bickerdike, R.L., D. Clark, J.N. Eastabrook, G. Hughes, M.N. Mair, P.G. Partridge, and H.C. Ranson (1986). Microstructure and tensile properties of vapour-deposited aluminium alloys. Part 2. Co-deposited alloys. *Int. J. Rapid Solidif.* 2: 1–19.

37. Liu, B.X. (1999). Ion mixing. In *Non-Equilibrium Processing of Materials*, ed. C. Suryanarayana, pp. 197–222. Oxford, UK: Pergamon.
38. Liu, B.X. and O. Jin (1997). Formation and theoretical modeling of non-equilibrium alloy phases by ion mixing. *Phys. Stat. Sol. (a)* 161: 3–33.
39. Liu, B.X., W.S. Lai, and Z.J. Zhang (2001). Solid-state crystal-to-amorphous transition in metal-metal multilayers and its thermodynamic and atomistic modeling. *Adv. Phys.* 50: 367–429.
40. Li, J.H., Y. Dai, Y.Y. Cui, and B.X. Liu (2011). Atomistic theory for predicting the binary metallic glass formation. *Mater. Sci. Eng.* R72: 1–28.
41. Motorin, V.I. (1983). Vitrification kinetics of pure metals. *Phys. Stat. Sol. (a)* 80: 447–455.
42. Zhong, L., J.W. Wang, H.W. Sheng, Z. Zhang, and S.X. Mao (2014). Formation of monatomic metallic glasses through ultrafast liquid quenching. *Nature* 512: 177–180.
43. Brenner, S.S., D.E. Couch, and E.K. Williams (1950). Electrodeposition of alloys of phosphorus with nickel or cobalt. *J. Res. Natl. Bur. Stand.* 44: 109–122.
44. Brenner, S.S. (1963). *Electrodeposition of Alloys: Principles and Practice*, 2nd edn. New York: Academic Press.
45. Dietz, G. (1977). Amorphous ferromagnetic materials deposited from vapor or liquids. *J. Magn. Magn. Mater.* 6: 47–51.
46. Suryanarayana, C. (2001). Mechanical alloying and milling. *Prog. Mater. Sci.* 46: 1–184.
47. Suryanarayana, C. (2004). *Mechanical Alloying and Milling*. New York: Marcel Dekker.
48. Ermakov, A.E., E.E. Yurchikov, and V.A. Barinov (1981). The magnetic properties of amorphous Y-Co powders obtained by mechanical comminution. *Phys. Met. Metallogr.* 52(6): 50–58.
49. Koch, C.C., O.B. Cavin, C.G. McKamey, and J.O. Scarbrough (1983). Preparation of "amorphous" $Ni_{60}Nb_{40}$ by mechanical alloying. *Appl. Phys. Lett.* 43: 1017–1019.
50. Yeh, X.L., K. Samwer, and W.L. Johnson (1983). Formation of an amorphous metallic hydride by reaction of hydrogen with crystalline intermetallic compounds—A new method of synthesizing metallic glasses. *Appl. Phys. Lett.* 42: 242–243.
51. Aoki, K. (2001). Amorphous phase formation by hydrogen absorption. *Mater. Sci. Eng. A* 304–306: 45–53.
52. Schwarz, R.B. and W.L. Johnson (1983). Formation of an amorphous alloy by solid-state reaction of the pure polycrystalline metals. *Phys. Rev. Lett.* 51: 415–418.
53. Atzmon, M., J.D. Verhoven, E.D. Gibson, and W.L. Johnson (1984). Formation and growth of amorphous phases by solid-state reaction in elemental composites prepared by cold working. *Appl. Phys. Lett.* 45: 1052–1053.
54. Dinda, G.P., H. Rösner, and G. Wilde (2007). Cold-rolling induced amorphization in Cu–Zr, Cu–Ti–Zr and Cu–Ti–Zr–Ni multilayers. *J. Non-Cryst. Solids* 353: 3777–3781.
55. Battezzati, L. (1990). Thermodynamics of metastable phase formation. *Phil. Mag.* B 61: 511–524.
56. Suslick, K.S., S.-B. Choe, A.A. Cichowlas, and M.W. Grinstaff (1991). Sonochemical synthesis of amorphous iron. *Nature* 353: 414–416.
57. Fecht, H.-J. and W.L. Johnson (1990). Destabilization and vitrification of crystalline matter. *J. Non-Cryst. Solids* 117/118: 704–707.
58. Segal, V.M. (1995). Materials processing by simple shear. *Mater. Sci. Eng. A* 197: 157–164.
59. Valiev, R.Z., R.K. Islamgaliev, and I.V. Alexandrov (2000). Bulk nanostructured materials from severe plastic deformation. *Prog. Mater. Sci.* 45: 103–189.
60. Valiev, R.Z. (2007). The new trends in fabrication of bulk nanostructured materials by SPD processing. *J. Mater. Sci.* 42: 1483–1490.
61. Sun, Y.F., Y. Todaka, M. Umemoto, and N. Tsuji (2008). Solid-state amorphization of Cu+Zr multi-stacks by ARB and HPT techniques. *J. Mater. Sci.* 43: 7457–7464.
62. Sun, Y.F., T. Nakamura, Y. Todaka, M. Umemoto, and N. Tsuji (2009). Fabrication of CuZr(Al) metallic glasses by high pressure torsion. *Intermetallics* 17: 256–261.
63. Saito, Y., N. Tsuji, H. Utsunomiya, T. Sakai, and R.G. Hong (1998). Ultra-fine grained bulk aluminum produced by accumulative roll-bonding (ARB) process. *Scr. Mater.* 39: 1221–1227.
64. Tsuji, N. (2011). Bulk nanostructured metals and alloys produced by accumulative roll bonding. In *Nanostructured Metals and Alloys*, ed. S.H. Whang, pp. 40–58. Cambridge: Woodhead Publishing.

65. Ohsaki, O, S. Kato, N. Tsuji, T. Ohkubo, and K. Hono (2007). Bulk mechanical alloying of Cu–Ag and Cu/Zr two-phase microstructures by accumulative roll bonding process. *Acta Mater.* 55: 2885–2895.
66. Sun, Y.F., N. Tsuji, S. Kato, S. Ohsaki, and K. Hono (2007). Fabrication of bulk metallic glass sheet in Cu–47 at% Zr alloys by ARB and heat treatment. *Mater. Trans.* 48: 1605–1509.
67. Nishiyama, N. and A. Inoue (2002). Glass-forming ability of $Pd_{42.5}Cu_{30}Ni_{7.5}P_{20}$ alloy with a low critical cooling rate of 0.067 K/s. *Appl. Phys. Lett.* 80: 568–570.
68. Nishiyama, N., K. Takenaka, H. Miura, N. Saidoh, Y. Zeng, and A. Inoue (2012). The world's biggest glassy alloy ever made. *Intermetallics* 30: 19–24.
69. Chen, H.S. (1974). Thermodynamic considerations on the formation and stability of metallic glasses. *Acta Metall.* 22: 1505–1511.
70. Chen, H.S. (1978). The influence of structural relaxation on the density and Young's modulus of metallic glasses. *J. Appl. Phys* 49: 3289–3291.
71. Drehman, A.J., A.L. Greer, and D. Turnbull (1982). Bulk formation of a metallic glass: $Pd_{40}Ni_{40}P_{20}$. *Appl. Phys. Lett.* 41: 716–717.
72. Kui, H.W., A.L. Greer, and D. Turnbull (1984). Formation of bulk metallic glass by fluxing. *Appl. Phys. Lett.* 45: 615–616.
73. Inoue, A., T. Zhang, and T. Masumoto (1989). Al–La–Ni amorphous alloys with a wide supercooled liquid region. *Mater. Trans., JIM* 30: 965–972.
74. Inoue, A., M. Kohinata, A.-P. Tsai, and T. Masumoto (1989). Mg–Ni–La amorphous alloys with a wide supercooled liquid region. *Mater. Trans., JIM* 30: 378–381.
75. Inoue, A., K. Kita, T. Zhang, and T. Masumoto (1989). An amorphous $La_{55}Al_{25}Ni_{20}$ alloy prepared by water quenching. *Mater. Trans., JIM* 30: 722–725.
76. Peker, A. and W.L. Johnson (1993). A highly processable metallic glass: $Zr_{41.2}Ti_{13.8}Cu_{12.5}Ni_{10.0}Be_{22.5}$. *Appl. Phys. Lett.* 63: 2342–2344.
77. Cantor, B., I.T.H. Chang, P. Knight, and A.J.B. Vincent (2004). Microstructural development in equiatomic multicomponent alloys. *Mater. Sci. Eng. A* 375–377: 213–218.
78. Yeh, J.-W., S.K. Chen, S.J. Lin, J.Y. Gan, T.S. Chin, and T.T. Shun (2004). Nanostructured high entropy alloys with multiple principal elements: Novel alloy design concepts and outcomes. *Adv. Eng. Mater.* 6: 299–303.
79. Yeh, J.-W. (2013). Alloy design strategies on high-entropy alloys. *JOM* 65: 1759–1771.
80. Cantor, B. (2014). Multicomponent and high entropy alloys. *Entropy* 16: 4749–4768.
81. Miracle, D.B., J.D. Miller, O.N. Senkov, C. Woodward, M.D. Uchic, and J. Tiley (2014). Exploration and development of high entropy alloys for structural applications. *Entropy* 16: 494–525.
82. Wang, W.H. (2014). High-entropy metallic glasses. *JOM* 66: 2067–2077.
83. Zhang, Y., T.T. Zuo, Z. Tang, M.C. Gao, K.A. Dahmen, P.K. Liaw, and Z.P. Lu (2014). Microstructures and properties high-entropy alloys. *Prog. Mater. Sci.* 61: 1–93.
84. Murty, B.S., J.-W. Yeh, and S. Ranganathan (2014). *High Entropy Alloys*. London, UK: Butterworth-Heinemann.
85. Gao, M.C., J.-W. Yeh, P.K. Liaw, and Y. Zhang, ed. (2016). *High Entropy Alloys: Fundamentals and Applications*. Cham, Switzerland:. Springer.
86. Tomilin, I.A. and S.D. Kaloshkin (2015). 'High entropy alloys'—'semi-impossible' regular solid solutions? *Mater. Sci. Technol.* 31: 1231–1234.
87. Ma, L., L. Wang, T. Zhang, and A. Inoue (2002). Bulk glass formation of Ti-Zr-Hf-Cu-M (M = Fe,Co,Ni) alloys. *Mater. Trans.* 43: 277–280.
88. Guo, S., Q. Hu, C. Ng, and C.T. Liu (2013). More than entropy in high-entropy alloys: Forming solid solutions or amorphous phase. *Intermetallics* 41: 96–103.
89. Gao, X.G., K. Zhao, H.B. Ke, D.W. Ding, W.H. Wang, and H.Y. Bai (2011). High mixing entropy bulk metallic glasses. *J. Non-Cryst. Solids* 357: 3557–3560.
90. Qi, T.L., Y.H. Li, A. Takeuchi, G.Q. XIe, H. Miao, and W. Zhang (2015). Soft magnetic $Fe_{25}Co_{25}Ni_{25}(B,Si)_{25}$ high entropy bulk metallic glasses. *Intermetallics* 66: 8–12.
91. Wang, J., Z. Zheng, J. Xu, and Y. Wang (2014). Microstructure and magnetic properties of mechanically alloyed FeSiBAlNi(Nb) high entropy alloys. *J. Mag. Mag. Mater.* 355: 58–64.

92. Vaidya, M., S. Armugam, S. Kashyap, and B.S. Murty (2015). Amorphization in equiatomic high entropy alloys. *J. Non-Cryst. Solids* 413: 8–14.
93. Huo, J.T., L. Huo, H. Men, X.M. Wang, A. Inoue, J.Q. Wang, C.T. Chang, and R.-W. Li (2015). The magnetocaloric effect of Gd-Tb-Dy-Al-M (M – Fe, Co, and Ni) high-entropy bulk metallic glasses. *Intermetallics* 58: 31–35.
94. Takeuchi, A., N. Chen, T. Wada, Y. Yokoyama, H. Kato, A. Inoue, and J.-W. Yeh (2011). $Pd_{20}Pt_{20}Cu_{20}Ni_{20}P_{20}$ high-entropy alloy as a bulk metallic glass in the centimeter. *Intermetallics* 19: 1546–1554.
95. Ding, H.Y. and K.F. Yao (2013). High entropy $Ti_{20}Zr_{20}Cu_{20}Ni_{20}Be_{20}$ bulk metallic glass. *J. Non-Cryst. Solids* 364: 9–12.
96. Gong, P., S.F. Zhao, H.Y. Ding, K.F. Yao, and X. Wang (2015). Nonisothermal crystallization kinetics, fragility and thermodynamics of $Ti_{20}Zr_{20}Cu_{20}Ni_{20}Be_{20}$ high entropy bulk metallic glass. *J. Mater. Res.* 30: 2772–2782.
97. Zhao, S.F., G.N. Yang, H.Y. Ding, and K.F. Yao (2015). A quinary Ti-Zr-Hf-Be-Cu high entropy bulk metallic glass with a critical size of 12 mm. *Intermetallics* 61: 47–50.
98. Zhao, S.F., Y. Shao, X. Liu, N. Chen, H.Y. Ding, and K.F. Yao (2015). Pseudo-quinary $Ti_{20}Zr_{20}Hf_{20}Be_{20}(Cu_{20-x}Ni_x)$ high entropy bulk metallic glasses with large glass forming ability. *Mater. Des.* 87: 625–631.
99. Ding, H.Y., Y. Shao, P. Gong, J.F. Li, and K.F. Yao (2014). A senary TiZrHfCuNiBe high entropy bulk metallic glass with large glass forming ability. *Mater. Lett.* 125: 151–153.
100. Zhao, K., X.X. Xia, H.Y. Bai, D.Q. Zhao, and W.H. Wang (2011). Room temperature homogeneous flow in a bulk metallic glass with low glass transition temperature. *Appl. Phys. Lett.* 98: 141913 (3 pages).
101. Duwez, P. (1967). Structure and properties of alloys rapidly quenched from the liquid state. *Trans. ASM Qly.* 60: 607–633.
102. Anantharaman, T.R. and C. Suryanarayana (1971). Review: A decade of quenching from the melt. *J. Mater. Sci.* 6: 1111–1135.
103. Jones, H. (1973). Splat cooling and metastable phases. *Rep. Prog. Phys.* 36: 1425–1497.
104. Cargill, III, G.S. (1975). Structure of metallic alloy glasses. *Solid State Phys.* 30: 227–320.
105. Duwez, P. (1976). Structure and properties of glassy metals. *Ann. Rev. Mater. Sci.* 6: 83–117.
106. Chen, H.S. (1980). Glassy metals. *Rep. Prog. Phys.* 43: 353–432.
107. Suryanarayana, C. (1984). Metallic glasses. *Bull. Mater. Sci.* 6: 579–594.
108. Cahn, R.W. and A.L. Greer (1996). Metastable states of alloys. In *Physical Metallurgy*, 4th edn., eds. R.W. Cahn and P. Haasen, Vol. 2, pp. 1723–1830. Amsterdam, the Netherlands: Elsevier Science BV.
109. Gilman, J.J. and H.J. Leamy, eds. (1978). *Metallic Glasses*. Metals Park, OH: ASM.
110. Güntherodt, H.-J. and H. Beck, eds. (1981). *Glassy Metals I*. Berlin, Germany: Springer-Verlag.
111. Jones, H. (1982). *Rapid Solidification of Metals and Alloys*. London, UK: Institution of Metallurgists.
112. Beck, H. and H.-J. Güntherodt, eds. (1983). *Glassy Metals II*. Berlin, Germany: Springer-Verlag.
113. Luborsky, F.E., ed. (1983). *Amorphous Metallic Alloys*. London, UK: Butterworths.
114. Anantharaman, T.R., ed. (1984). *Metallic Glasses: Production, Properties, and Applications*. Zürich, Switzerland: Trans Tech Publications.
115. Anantharaman, T.R. and C. Suryanarayana (1987). *Rapidly Solidified Metals: A Technological Overview*. Zürich, Switzerland: Trans Tech Publications.
116. Liebermann, H.H., ed. (1993). *Rapidly Solidified Alloys: Processes, Structures, Properties, Applications*. New York: Marcel Dekker.
117. Srivatsan, T.S. and T.S. Sudarshan, eds. (1993). *Rapid Solidification Technology: An Engineering Guide*. Lancaster, PA: Technomic Publishing Company.
118. Beck, H. and H.-J. Güntherodt, eds. (1994). *Glassy Metals III*. Berlin, Germany: Springer-Verlag.
119. Otooni, M.A., ed. (1995). *Science and Technology of Rapid Solidification and Processing*. Dordrecht, the Netherlands: Kluwer.
120. Otooni, M.A., ed. (1998). *Elements of Rapid Solidification: Fundamentals and Applications*. Berlin, Germany: Springer.

121. *Metastable Metallic Alloys* (1970). *Proceedings of the First International Conference on Rapidly Quenched Metals*, Brela, Yugoslavia, September 28–30, 1970; *Fizika*, 2 (Suppl. 2), 1970.
122. Grant, N.J. and B.C. Giessen, eds. (1976). *Proceedings of the Second International Conference on Rapidly Quenched Metals (RQ II)*, Boston, MA, November 17–19, 1975; *Rapidly Quenched Metals*, Part I, pp. 1–543. Cambridge, MA: MIT Press; Part II in *Mater. Sci. Eng.* 23: 81–324.
123. Cantor, B., ed. (1978). *Proceedings of the Third International Conference on Rapidly Quenched Metals (RQ III)*, Brighton, UK, July 3–7, 1978; *Rapidly Quenched Metals*, in two volumes, Vol. 1, pp. 1–446, Vol. 2, pp. 1–481. London, UK: Metals Society.
124. Masumoto, T. and K. Suzuki, eds. (1982). *Proceedings of the Fourth International Conference on Rapidly Quenched Metals (RQ IV)*, Sendai, Japan, August 24–28, 1981; *Rapidly Quenched Metals IV*, in two volumes, Vol. I, pp. 1–784, Vol. II, pp. 785–1656. Sendai, Japan: Japan Institute of Metals.
125. Steeb, S. and H. Warlimont, eds. (1985). *Proceedings of the Fifth International Conference on Rapidly Quenched Metals (RQ V)*, Wurzburg, Germany, September 3–7, 1984, in two volumes, pp. 1–1112. Amsterdam, the Netherlands: North-Holland Publishing/Elsevier.
126. Cochrane, R.W. and J.O. Ström-Olsen, eds. (1988). *Proceedings of the Sixth International Conference on Rapidly Quenched Metals (RQ 6)*, Montreal, Canada, August 3–7, 1987; *Mater. Sci. Eng.* 97: 1–550; 98: 1–560; 99: 1–575.
127. Frederiksson, H. and S.J. Savage, eds. (1991). *Proceedings of the Seventh International Conference on Rapidly Quenched Metals (RQ 7)*, Stockholm, Sweden, August 13–17, 1990; *Mater. Sci. Eng. A.* A 133: 1–858; A 134: 859–1445.
128. Masumoto, T. and K. Hashimoto, eds. (1994). *Proceedings of the Eighth International Conference on Rapidly Quenched and Metastable Materials (RQ 8)*, Sendai, Japan, August 22–27, 1993; *Mater. Sci. Eng. A.* Part I, A 179–180: 1–716; Part II, A 181–182: 717–1468.
129. Duhaj, P., P. Mrafko, and P. Švec, eds. (1997). *Proceedings of the Ninth International Conference on Rapidly Quenched and Metastable Materials (RQ 9)*, Bratislava, Slovakia, August 25–30, 1996; *Mater. Sci. Eng. A.* A 226–228: 1–1101.
130. Chattopadhyay, K. and S. Ranganathan, eds. (2001). *Proceedings of the 10th International Conference on Rapidly Quenched and Metastable Materials (RQ 10)*, Bangalore, India, August 23–27, 1999; *Mater. Sci. Eng. A.* A 304–306: 1–1109.
131. O'Reilly, K., P. Warren, P. Schumacher, and B. Cantor, eds. (2004). *Proceedings of the 11th International Conference on Rapidly Quenched and Metastable Materials (RQ 11)*, Oxford, UK, August 25–30, 2002; *Mater. Sci. Eng. A.* A 375–377: 1–1320.
132. Kim, D.H., W.T. Kim, K. Hono, and G. Kostorz, eds. (2007). *Proceedings of the 12th International Conference on Rapidly Quenched and Metastable Materials (RQ 12)*, Jeju, South Korea, August 21–26, 2005; *Mater. Sci. Eng. A.* A 449–451: 1–1184.
133. Schultz, L., J. Eckert, L. Battezzati, and M. Stoica, eds. (2009). *Proceedings of the 13th International Conference on Rapidly Quenched and Metastable Materials (RQ 13)*, Dresden, Germany, August 24–29, 2008; *J. Phys. Conf. Ser.* 144: 011001–012120.
134. Jorge, Jr., A.M., C. Bolfarini, C.S. Kiminami, J. Eckert, and W.J. Botta, eds. (2012). *Proceedings of the 14th International Conference on Rapidly Quenched and Metastable Materials (RQ 14)*, Salvador, Brazil, August 24–28, 2011; select papers in *Int. J. Mater. Res.* 103 (9): 1082–1165 and others in *Materials Research* (Ibero-American Journal of Materials) 15 (5): 705–820.
135. Zhang, D., ed. (2015). *Proceedings of the 15th International Conference on Rapidly Quenched and Metastable Materials (RQ 15)*, Shanghai, China, August 24–28, 2014; Select papers in *Materials Research* (Ibero-American Journal of Materials) 18 (1): 1–163.
136. Yavari, A.R., ed. (1995). *Proceedings of the International Symposium on Metastable, Mechanically Alloyed, and Nanocrystalline Materials (ISMANAM-94)*, Grenoble, France, June 27–July 1, 1994. Zürich, Switzerland: Trans Tech Publications; *Mater. Sci. Forum* 179–181: 1–879.
137. Schulz, R., ed. (1996). *Proceedings of the Second International Symposium on Metastable, Mechanically Alloyed, and Nanocrystalline Materials (ISMANAM-95)*, Quebec, Canada, July 24–28, 1995. Zürich, Switzerland: Trans Tech Publications; *Mater. Sci. Forum* 225–227: 1–958.

138. Fiorani, D. and M. Magini, eds. (1997). *Proceedings of the Third International Symposium on Metastable, Mechanically Alloyed, and Nanocrystalline Materials (ISMANAM-96)*, Rome, Italy, May 20–24, 1996. Zürich, Switzerland: Trans Tech Publications; *Mater. Sci. Forum* 235–238: 1–1013.
139. Baró, M.D. and S. Suriñach, eds. (1998). *Proceedings of the Fourth International Symposium on Metastable, Mechanically Alloyed, and Nanocrystalline Materials (ISMANAM-97)*, Sitges (Barcelona), Spain, August 31–September 5, 1997. Zürich, Switzerland: Trans Tech Publications; *Mater. Sci. Forum* 269–272: 1–1080.
140. Calka, A. and D. Wexler, eds. (1999). *Proceedings of the Fifth International Symposium on Metastable, Mechanically Alloyed, and Nanocrystalline Materials (ISMANAM-98)*, Wollongong (Sydney), Australia, December 7–12, 1998. Zürich, Switzerland: Trans Tech Publications; *Mater. Sci. Forum* 312–314: 1–667, 1999 (and also in *J. Metastable Nanocryst. Mater.* 2–6, 1999).
141. Eckert, J., H. Schlörb, and L. Schultz, eds. (2000). *Proceedings of the Sixth International Symposium on Metastable, Mechanically Alloyed, and Nanocrystalline Materials (ISMANAM-99)*, Dresden, Germany, August 30–September 3, 1999. Zürich, Switzerland: Trans Tech Publications; *Mater. Sci. Forum* 343–346: 1–990 (and also in *J. Metastable Nanocryst. Mater.* 8, 2000).
142. Schumacher, P., P. Warren, and B. Cantor, eds. (2001). *Proceedings of the Seventh International Symposium on Metastable, Mechanically Alloyed, and Nanocrystalline Materials (ISMANAM-2000)*, Oxford, UK, July 9–14, 2000. Zürich, Switzerland: Trans Tech Publications; *Mater. Sci. Forum* 360–362: 1–662 (and also in *J. Metastable Nanocryst. Mater.* 10, 2001).
143. Ma, E., M. Atzmon, and C.C. Koch, eds. (2002). *Proceedings of the Eighth International Symposium on Metastable, Mechanically Alloyed, and Nanocrystalline Materials (ISMANAM-2001)*, Ann Arbor, MI, July 24–29, 2001. Zürich, Switzerland: Trans Tech Publications; *Mater. Sci. Forum* 386–388: 1–659 (and also in *J. Metastable Nanocryst. Mater.* 13, 2002).
144. Ahn, J.-H. and Y.D. Hahn, eds. (2003). *Proceedings of the Ninth International Symposium on Metastable, Mechanically Alloyed, and Nanocrystalline Materials (ISMANAM-2002)*, Seoul, South Korea, September 8–12, 2002; *J. Metastable Nanocryst. Mater.* Trans Tech Publications 15–16: 1–779.
145. Kiminami, C.S., C. Bolfarini, and W.J. Botta Filho, eds. (2004). *Proceedings of the 10th International Symposium on Metastable, Mechanically Alloyed, and Nanocrystalline Materials (ISMANAM-2003)*, Foz do Iguaçu, Brazil, August 24–28, 2003; *J. Metastable Nanocryst. Mater.* Trans Tech Publications 20–21: 1–803.
146. Inoue, A., ed. (2005). *Proceedings of the 11th International Symposium on Metastable, Mechanically Alloyed, and Nanocrystalline Materials (ISMANAM-2004)*, Sendai, Japan, August 22–26, 2004; *J. Metastable Nanocryst. Mater.* Trans Tech Publications 24–25: 1–742.
147. Yavari, A.R., E. Gaffet, G. LeCaer, K. Hajlaoui, and M. Stoica, eds. (2007). *Proceedings of the 12th International Symposium on Metastable and Nano-Materials (ISMANAM-2005)*, Paris, France, July 3–7, 2005; *J. Alloys Compd.* 434–435: 1–876.
148. Kulik, T., D. Oleszak, and J. Ferenc, eds. (2008). *Proceedings of the 13th International Symposium on Metastable and Nano-Materials (ISMANAM-2006)*, Warsaw, Poland, August 27–31, 2006; *Rev. Adv. Mater. Sci.* 18: 1–772.
149. Evangelaksi, G.A., P. Patsalas, and Ph. Komninoued (2009). *Proceedings of the 14th International Symposium on Metastable and Nano-Materials (ISMANAM-2007)*, Corfu Island, Greece, August 26–30, 2007; *J. Alloys Compounds* 483 (1–2): 1–700.
150. Audebert, F., L. Damonte, and M. Galano, eds. (2010). *Proceedings of the 15th International Symposium on Metastable, Amorphous and Nanostructured Materials (ISMANAM-2008)*, Buenos Aires, Argentina, July 6–10, 2008; *J. Alloys Compounds* 495 (2): 293–680.
151. Zhang, T., W.H. Wang, and A.R. Yavari, eds. (2010). *Proceedings of the 16th International Symposium on Metastable, Amorphous and Nanostructured Materials (ISMANAM-2009)*, Beijing, China, July 5–9, 2009; *J. Alloys Compounds* 504 (Supplement 1): S1–S542.
152. Loffler, J.F. and V. Wessels, eds. (2011). *Proceedings of the 17th International Symposium on Metastable, Amorphous and Nanostructured Materials (ISMANAM-2010)*, Zurich, Switzerland, July 4–9, 2010; *J. Alloys Compounds* 509 (Supplement 1): S1–S514.

153. Gorria, P. and J.A. Blanco, eds. (2012). Proceedings of the 18th International Symposium on Metastable, Amorphous and Nanostructured Materials (ISMANAM-2011), Gijo, Spain, June 26–July 1, 2011; *J. Alloys Compounds* 536 (Supplement 1): S1–S586.
154. Kaloshkni, S.D., V. Tcherdyntsevm, and V. Khovaylo, eds. (2014). Proceedings of the 19th International Symposium on Metastable, Amorphous and Nanostructured Materials (ISMANAM-2012), Moscow, Russia, June 18–22, 2012; *J. Alloys Compounds* 586 (Supplement 1): S1–S558.
155. Battezzati, L. and P. Tiberto, eds. (2014). *Proceedings of the 20th International Symposium on Metastable, Amorphous and Nanostructured Materials (ISMANAM-2013),* Torino, Italy, June 30–July 5, 2013; *J. Alloys Compound* 615 (Supplement 1): S1–S724.
156. Martínez-Pérez, C.A., J.R. Vargas Garcia, R.M. Sanchez, and P.E. Garcia Casillas, eds. (2015). *Proceedings of the 21st International Symposium on Metastable, Amorphous and Nanostructured Materials (ISMANAM-2014),* Cancun, Mexico, June 29–July 4, 2014; *J. Alloys Compounds* 643 (Supplement 1): S1–S302.
157. Yavari, A.R. and K. Georgarakis, ed. (2015). Proceedings of the 22nd International Symposium on Metastable, Amorphous and Nanostructured Materials (ISMANAM-2015), Paris, France, July 13–17, 2015.
158. Liu, C.T., G.L. Chen, A. Inoue, T.G. Nieh, S.K. Wu, and R.O. Ritchie, eds. (2000). *Proceedings of the First International Conference on Bulk Metallic Glasses (BMG-I),* Singapore, September 2000.
159. Nieh, T.G., ed. (2002). *Proceedings of the Second International Conference on Bulk Metallic Glasses (BMG-II),* Keelung, Taiwan, March 24–28, 2002; *Intermetallics* 10 (11–12): 1035–1296, November–December 2002.
160. Nieh, T.G., C.T. Liu, W.H. Wang, and M.X. Pan, eds. (2004). *Proceedings of the Third International Conference on Bulk Metallic Glasses (BMG-III),* Beijing, China, October 12–16, 2003; *Intermetallics* 12: 1031–1283.
161. Nieh, T.G., R.A. Buchanan, and C.T. Liu, eds. (2006). *Proceedings of the Fourth International Conference on Bulk Metallic Glasses (BMG-IV),* Gatlinburg, TN, May 1–5, 2005; *Intermetallics* 14: 855–1112.
162. Inoue, A., Y. Hirotsu, T.G. Nieh, and K. Hono, eds. (2007). Proceedings of the Fifth International Conference on Bulk Metallic Glasses (BMG-V), Awaji Island, Japan, October 1–5, 2006; *Mater. Trans.* 48: 1579–1975.
163. Nieh, T.G., T. Zhang, W.H. Wang, C.L. Ma, Z.P. Lu, and J. Sun, eds. (2009). *Proceedings of the Sixth International Conference on Bulk Metallic Glasses (BMG-VI),* Xi'an, China, May 11–15, 2008; *Intermetallics* 17 (4): 183–284.
164. Kim, D.H., W.T. Kim, T.G. Nieh, and C.T. Liu, eds. (2010). *Proceedings of the Seventh International Conference on Bulk Metallic Glasses (BMG-VII),* Busan, Korea, November 1–5, 2009; *Intermetallics* 18 (10): 1795–2023.
165. Lu, J., C.T. Liu, T.G. Nieh, and Z.P. Lu, eds. (2012). *Proceedings of the Eighth International Conference on Bulk Metallic Glasses (BMG-VIII),* Busan, Korea, November 1–5, 2009; *Intermetallics* 30: 1–158.
166. Liaw, P.K., W.H. Jiang, G.J. Fan, H. Choo, and Y.F. Gao, eds. (2008). Bulk Metallic Glasses IV. Proceedings of a Symposium held during the TMS Annual Meeting in Orlando, FL, February 25–March 1, 2007; *Metall. Mater. Trans. A* 39 (8): 1761–1962.
167. Liaw, P.K. and W.H. Jiang, eds. (2008). Bulk Metallic Glasses V. Proceedings of a Symposium held during the TMS Annual Meeting in New Orleans, LA, March 9–13, 2008; *Adv. Eng. Mater.* 10 (11): 995–1067.
168. Liaw, P.K., H. Choo, Y.F. Gao, and G.Y. Wang, eds. (2010). Bulk Metallic Glasses VI. Proceedings of a Symposium held during the TMS Annual Meeting in San Francisco, CA, February 15–19, 2009; *Metall. Mater. Trans. A* 41 (7): 1627–1804.
169. Liaw, P.K., G.Y. Wang, H. Choo, and Y.F. Gao, eds. (2011). Bulk Metallic Glasses VII. Proceedings of a Symposium held during the TMS Annual Meeting in Seattle, WA, February 14–18, 2010; *Metall. Mater. Trans. A* 42 (6): 1449–1533.
170. Liaw, P.K., G.Y. Wang, H. Choo, and Y.F. Gao, eds. (2012). Tin San Diego, CA, February 27–March 3, 2011; *Metall. Mater. Trans. A* 43 (8): 2591–2741.

171. Liaw, P.K., G.Y. Wang, H. Choo, and Y.F. Gao, eds. (2013). Bulk Metallic Glasses IX. Proceedings of a Symposium held during the TMS Annual Meeting in Orlando, FL, March 11–15, 2012; *Metall. Mater. Trans. A* 44 (5): 1979–2030.
172. Liaw, P.K., G.Y. Wang, H. Choo, and Y.F. Gao, eds. (2014). Bulk Metallic Glasses X. Proceedings of a Symposium held during the TMS Annual Meeting in San Antonio, TX, March 3–7, 2013; *Metall. Mater. Trans. A* 45 (5): 2351–2404.
173. Liaw, P.K., G.Y. Wang, H. Choo, and Y.F. Gao, eds. (2015). Bulk Metallic Glasses XI. Proceedings of a Symposium held during the TMS Annual Meeting in San Diego, CA, February 16–20, 2014; *Metall. Mater. Trans. A* 46 (6): 2380–2448.
174. Liaw, P.K., G.Y. Wang, H. Choo, and Y.F. Gao, eds. (2016). Bulk Metallic Glasses XII. Proceedings of a Symposium held during the TMS Annual Meeting in Orlando, FL, March 15–19, 2015; *Metall. Mater. Trans. A.*
175. Johnson, W.L. (1998). Bulk glass-forming metallic alloys: Science and technology. *MRS Bull.* 24(10): 42–56.
176. Inoue, A. (1999). Bulk amorphous alloys. In *Non-Equilibrium Processing of Materials,* ed. C. Suryanarayana, pp. 375–415. Oxford, UK: Pergamon.
177. Inoue, A. (2000). Stabilization of metallic supercooled liquid and bulk amorphous alloys. *Acta Mater.* 48: 279–306.
178. Schneider, S. (2001). Bulk metallic glasses. *J. Phys.: Condens. Matter* 13: 7723–7736.
179. Inoue, A. (2001). Bulk amorphous alloys. In *Amorphous and Nanocrystalline Materials; Preparation, Properties, and Applications,* eds. A. Inoue and K. Hashimoto, pp. 1–51. Heidelberg, Germany: Springer-Verlag.
180. Basu, J. and S. Ranganathan (2003). Bulk metallic glasses: A new class of engineering materials. *Sadhana* 28: 783–998.
181. Löffler, J.F. (2003). Bulk metallic glasses. *Intermetallics* 11: 529–540.
182. Telford, M. (2004). The case for bulk metallic glasses. *Mater. Today* 7(3): 36–43.
183. Wang, W.H., C. Dong, and C.H. Shek (2004). Bulk metallic glasses. *Mater. Sci. Eng. Rep. R* 44: 45–89.
184. Inoue, A. (2005). Bulk glassy and nonequilibrium crystalline alloys by stabilization of supercooled liquid: Fabrication, functional properties and applications. *Proc. Japan Acad.* B 81: 156–188.
185. Wang, W.H. (2007). Roles of minor additions in formation and properties of bulk metallic glasses. *Prog. Mater. Sci.* 52: 540–596.
186. Schuh, C.A., T. Huffnagel, and U. Ramamurty (2007). Mechanical behavior of amorphous alloys. *Acta Mater.* 55: 4067–4109.
187. Chen, M.W. (2008). Mechanical behavior of metallic glasses: Microscopic understanding of strength and ductility. *Ann. Rev. Mater. Res.* 38: 445–469.
188. Wang, G.Y., P.K. Liaw, and M.L. Morrison (2009). Progress in studying the fatigue behavior of Zr-based bulk metallic glasses and their composites. *Intermetallics* 17: 579–590.
189. Miracle, D.B., D.V. Louzguine-Luzgin, L.V. Louzguina-Luzgina, and A. Inoue (2010). An assessment of binary metallic glasses: Correlations between structure, glass forming ability and stability. *Int. Mater. Rev.* 55: 218–256.
190. Trexler, M.M. and N.N. Thadhani (2010). Mechanical properties of bulk metallic glasses. *Prog. Mater. Sci.* 55: 759–839.
191. Cheng, Y.Q. and E. Ma (2011). Atomic-level structure and structure-property relationship in metallic glasses. *Prog. Mater. Sci.* 56: 379–473.
192. Egami, T. (2011). Atomic level stresses. *Prog. Mater. Sci.* 56: 637–653.
193. Takeuchi, S. and K. Edagawa (2011). Atomistic simulation and modeling of localized shear deformation in metallic glasses. *Prog. Mater. Sci.* 56: 785–816.
194. Axinte, E. (2012). Metallic glasses from "alchemy" to pure science: Present and future of design, processing, and applications of glassy metals. *Mater. Des.* 35: 518–556.
195. Wang, G.Y., Q.M. Feng, B. Yang, W.H. Jiang, P.K. Liaw, and C.T. Liu (2012). Thermographic studies of temperature evolutions in bulk metallic glasses: An overview. *Intermetallics* 30: 1–11.

196. Li, J.H., Y. Dai, and X.D. Dai (2012). Long-range n-body potential and applied to atomistic modeling the formation of ternary metallic glasses. *Intermetallics* 31: 292–320.
197. Suryanarayana, C. (2012). Mechanical behavior of emerging materials. *Mater. Today* 15: 486–498.
198. Wang, W.H. (2012). The elastic properties, elastic models and elastic perspectives of metallic glasses. *Prog. Mater. Sci.* 57: 487–656.
199. Greer, A.L., Y.Q. Cheng, and E. Ma (2013). Shear bands in metallic glasses. *Mater. Sci. Eng. R* 74: 71–132.
200. Kim, D.H., W.T. Kim, E.S. Park, N. Mattern, and J. Eckert (2013). Phase separation in metallic glasses. *Prog. Mater. Sci.* 58: 1103–1172.
201. Schroers, J. (2013). Bulk metallic glasses. *Phys. Today* 66 (2): 32–37.
202. Suryanarayana, C. and A. Inoue (2013). Iron-based bulk metallic glasses. *Int. Mater. Rev.* 58: 131–166.
203. Dambatta, M.S., S. Izman, B. Yahaya, J.Y. Lim, and D. Kurniawan (2015). Mg-based bulk metallic glasses for biodegradable implant materials: A review on glass forming ability, mechanical properties, and biocompatibility. *J. Non-Cryst. Solids* 426: 110–115.
204. Inoue, A., F.L. Kong, S.L. Zhu, E. Shalaan, and F.M. Al-Marzouki (2015). Production methods and properties of engineering glassy alloys and composites. *Intermetallics* 58: 20–30.
205. Sun, B.A. and W.H. Wang (2015). The fracture of bulk metallic glasses. *Prog. Mater. Sci.* 74: 211–307.
206. Wang, W.H., Y. Yang, T.G. Nieh, and C.T. Liu (2015). On the source of plastic flow in metallic glasses: Concepts and models. *Intermetallics* 67: 81–86.
207. Wen, L.X., Z. Wang, Y. Han, and Y.C. Xu (2015). Structure and magnetic properties of Si-rich FeAlSiBNbCu alloys. *J. Non-Cryst. Solids* 411: 115–118.
208. Huffnagel, T.C., C.A. Schuh, and M.L. Falk (2016). Deformation of metallic glasses: Recent developments in theory, simulations, and experiments. *Acta Mater.* 109: 375–393.
209. Qiao, J.W., H.L. Jia, and P.K. Liaw (2016). Metallic glass matrix composites. *Mater. Sci. Eng.* R100: 1–69.
210. Farmer, J.C., P.E.A. Turchi, and J.H. Perepezko, eds. (2009). Iron-based amorphous metals—An important family of high-performance corrosion-resistant materials. *Proceedings of a Symposium held during MS&T07 in Detroit, MI*, September 16–20, 2007; *Metall. Mater. Trans. A* 40 (6): 1288–1354.
211. Johnson, W.L., A. Inoue, and C.T. Liu, eds. (1999). Bulk metallic glasses. *Materials Research Society Symposium Proceedings*, Boston, MA, December 1–3, 1998, Vol. 554, pp. 1–446. Warrendale, PA: Materials Research Society.
212. Inoue, A., A.R. Yavari, W.L. Johnson, and R.H. Dauskardt, eds. (2001). Supercooled liquid, bulk glassy, and nanocrystalline states of alloys. *Materials Research Society Symposium Proceedings*, Boston, MA, November 27–30, 2000, Vol. 644. Warrendale, PA: Materials Research Society.
213. Egami, T., A.L. Greer, A. Inoue, and S. Ranganathan, eds. (2003). Supercooled liquids, glass transition, and bulk metallic glasses. *Materials Research Society Symposium Proceedings*, Boston, MA, December 2–6, 2002, Vol. 754, pp. 1–483. Warrendale, PA: Materials Research Society.
214. Busch, R., T.C. Hufnagel, J. Eckert, A. Inoue, W.L. Johnson, and A.R. Yavari, eds. (2004). Amorphous and nanocrystalline metals. *Materials Research Society Symposium Proceedings*, Boston, MA, December 1–4, 2003, Vol. 806, pp. 1–416. Warrendale, PA: Materials Research Society.
215. Miller, M. and P.K. Liaw, eds. (2008). *Bulk Metallic Glasses*. New York: Springer.
216. Suryanarayana, C. and A. Inoue (2011). *Bulk Metallic Glasses*. Boca Raton, FL: CRC Press.
217. Inoue, A. (1998). *Bulk Amorphous Alloys: Preparation and Fundamental Characteristics*, Vol. 4 of *Materials Science Foundations*, p. 116. Uetikon-Zürich, Switzerland: Trans Tech Publications.
218. Inoue, A. (1999). *Bulk Amorphous Alloys: Practical Characteristics and Applications*, Vol. 6 of *Materials Science Foundations*, p. 148. Uetikon-Zürich, Switzerland: Trans Tech Publications.
219. Wang, Y.M. (2010). *Bulk Metallic Glass-Forming Alloys: Phase Transformation and Composition Design*. Saarbrücken: VDM Verlag Dr. Müller.
220. Chishty, M.A.G., S. Safdar, and S.A. Sheikh (2011). *A Novel Iron Based Bulk Metallic Glass: Development & Characterization*. LAP Lambert Academic Publishing.

221. George, T.F., R.R. Letfullin, and G.P. Zhang, eds. (2013). *Bulk Metallic Glasses*. New York: Nova Science Publishers.
222. Hoffmann, D.C. (2013). *Bulk Metallic Glasses and Composites for Optical and Compliant Mechanisms*. NASA.
223. Hirata, A., K. Matsue, and M.W. Chen (2016). *Structural Analysis of Metallic Glasses with Computational Homology*. Tokyo: Springer.

3

Glass-Forming Ability of Alloys

3.1 Introduction

In Chapter 2, we reviewed the principles of glass formation and realized that a metallic melt needs to be significantly undercooled to a temperature below T_g, the glass transition temperature, for it to transform into the glassy state. It was also briefly mentioned that not all metallic alloys can be transformed into the glassy state. Despite this, a very large number of metallic glasses have been synthesized in binary, ternary, quaternary, and higher-order alloy systems. Two dated compilations can be referred to for the alloy systems, compositions, and other details of the formation of metallic glasses in the form of thin ribbons [1,2]. There were also a few reports of the formation of glassy phases in pure metals rapidly solidified from their molten state (see, e.g., Ref. [3]). But later, careful analysis revealed that the glassy phase in these pure metals was mainly stabilized by the presence of impurity atoms and that a minimum concentration of solute atoms was necessary for glass formation. The pure metals Bi and Sn were, however, produced earlier in the amorphous state in thin film form by evaporation onto a substrate maintained at a low temperature of 4 K [4–6]. Even though some basic empirical rules and some of the thermodynamic conditions that need to be satisfied to form a glass were known, there was no rigorous scientific basis, at least in the beginning, for choosing the alloy compositions that could be formed as glasses. Many glasses were produced more or less by trial and error. In fact, the very first synthesis of a metallic glass in the Au–25 at.% Si alloy by rapid solidification processing (RSP) in 1960 by Pol Duwez and his students [7] was by accident (in more than one way!).

To produce metallic glasses in a reasonable and reliable way, and also to produce them in large quantities and in a reproducible way, it is essential that we understand the basic reasons for glass formation from liquids. There was reasonable success in predicting the compositions and alloy systems in which glasses could be synthesized during the 1970s and 1980s. But, with the discovery of bulk metallic glasses (BMGs), the level of activity in this area has increased in recent times. Both the alloy systems and their compositions that are likely to be transformed into the glassy condition have been predicted. The ability of a metallic alloy to transform into the glassy state is defined in this chapter as the *glass-forming ability* (GFA). (This term should not be confused with the actual forming operations on a glass, such as making different shapes and objects with a silicate glass, or forming different components with any type of glassy material.) This chapter is devoted to a description of the different criteria that have been developed to explain glass formation in metallic systems in general, and more specifically with reference to BMGs.

We will first cover the criteria that were developed to predict glass formation by RSP methods, that is, at high solidification rates. It should also be appreciated that some of the major and important criteria that were developed to explain the glass formation behavior in melt-spun ribbons (by RSP methods) are equally applicable to the case of BMGs. However, because of the nature of the differences between the melt-spun ribbons and BMGs (most importantly in section thickness), we will move on to cover additional and specific criteria that are applicable to BMGs. Toward the end of the chapter, we will also briefly cover the criteria for the formation of glasses by other (non-solidification) methods, such as mechanical alloying (MA) and/or milling.

3.2 Critical Cooling Rate

Glass formation can be achieved only if the formation of detectable crystal nuclei can be completely suppressed, and it has been generally accepted that the minimum volume fraction of the crystals that can be detected is about 10^{-6}. Such a situation can be realized only when the liquid alloy is undercooled significantly to a temperature below T_g. One simple way of obtaining this large amount of undercooling is to rapidly solidify the liquid alloy. Chapter 2 showed that glass formation is possible only when the liquid is rapidly solidified above a critical cooling rate, R_c, that is dependent on the alloy system and its composition.

If an alloy melt is solidified from a temperature above the liquidus temperature, T_ℓ, to below T_g, then the volume fraction of the solid crystalline phase, X, formed under non-isothermal crystallization conditions can be given by the equation [8,9]:

$$X(T) = \frac{4\pi}{3R^4} \int_{T_\ell}^{T_g} I(T') \left[\int_{T''}^{T_g} U(T'')dT'' \right]^3 dT' \tag{3.1}$$

where:

I and U	are the steady-state nucleation frequency and crystal growth rate, respectively
R	is the cooling rate
T'	represents the constant cooling rate imposed on the melt (In reality, the cooling rate is not constant, but it is assumed to be constant for calculation purposes.)

If the volume fraction of the crystalline phase formed is selected as a very small value, say $X = 10^{-6}$, a condition normally chosen for glass formation, then the critical cooling rate, R_c, can be derived from Equation 3.1 as

$$R_c^4 = \frac{4\pi}{3 \times 10^{-6}} \int_{T_\ell}^{T_g} I(T') \left[\int_{T''}^{T_g} U(T'')dT'' \right]^3 dT' \tag{3.2}$$

Since the equations for I and U contain terms such as the viscosity of the supercooled liquid, η, the entropy of fusion, ΔS_f, and so on, the critical cooling rate, R_c, decreases with increasing η and ΔS_f, and decreasing liquidus temperature, T_ℓ. The best way to experimentally determine R_c is by constructing time–temperature–transformation (*T–T–T*) diagrams.

3.2.1 *T–T–T* Diagrams

The concept of a critical cooling rate can be easily understood with reference to Figure 3.1, which shows a schematic time–temperature–transformation (*T–T–T*) diagram for a hypothetical alloy.* The temperature is represented on the *y*-axis and the time on the *x*-axis. Since the time required for the transformation to be completed is usually very long, the time is plotted on a logarithmic scale.

The transformation curve, which has a C-shape, represents, at any given temperature, the time required to start the formation of the stable (crystalline) solid phase. If the alloy is cooled from the liquid state under equilibrium conditions, that is, extremely slowly, solidification will require a very long time and the product of solidification will always be a crystalline solid. Even if the liquid alloy is cooled a little more rapidly (represented by curve "1" in Figure 3.1), solidification occurs at temperature, T_1, and time, t_1, and the product of the transformation is still a crystalline solid. A similar situation would occur even if the alloy were to be solidified at a higher solidification rate. But, if the liquid alloy is solidified at a rate faster than the cooling rate represented by curve "2," which represents a tangent to the C-curve at its nose (i.e., the temperature at which the formation of a crystalline phase takes place in the shortest time), then crystal formation will not take place. Instead, the liquid will be retained in the supercooled (or undercooled) condition. If the temperature of this supercooled liquid is further decreased (without allowing crystallization to take place), the viscosity of the liquid will continue to increase, and reach a value of 10^{12} Pa s, which is typical of the solid state. In other words, the supercooled liquid is now frozen-in and a glassy phase forms, at temperatures below T_g.

The cooling rate represented by curve "2" is referred to as the *critical cooling rate*, and is commonly designated by the symbol R_c. The significance of this value is that if the liquid alloy is cooled above this rate, then it is possible for a glass to form completely, provided that the supercooled liquid is cooled to a temperature below T_g, without any changes occurring in the liquid. But, a homogeneous glassy phase will not form if the alloy melt is cooled at a rate slower than the critical cooling rate. Thus, the simplest and most logical criterion one could think of to predict glass formation is that *the liquid alloy should be cooled at a rate faster than* R_c *and to a temperature below* T_g.

The form of the *T–T–T* curve and the temperature and time at the nose are determined by the competition between the increasing driving force for nucleation (due to increased supercooling) and the decreased atomic mobility (due to the lowering of temperature). It is possible to theoretically calculate the critical cooling rates for different alloy systems using the theory of isothermal crystallization kinetics (see Chapter 2). If the temperature and

* *T–T–T* diagrams are determined by allowing the liquid-to-solid (or solid-to-solid) transformation to occur isothermally (at a constant temperature). At each temperature, the times required to start and finish the transformation are noted. The locus of all these "start" and "finish" points at different temperatures results in two curves: one representing the start and the other the finish of the transformation. These curves will have a "C" shape for thermodynamic and kinetic reasons. (For our discussions here, however, we will simplify the figures [see, e.g., Figure 3.1] by showing only one curve to represent the process of solidification, i.e., conversion of the liquid to the solid phase. Note that this is only an approximation.) The interested reader should refer to standard books (see, e.g., Refs. [10,11]) for details on how to construct and interpret *T–T–T* diagrams. In practice, however, solidification takes place upon continuously reducing the temperature of the melt from the liquid state down to room temperature. Diagrams representing such continuous liquid-to-solid transformations are referred to as *continuous cooling transformation* (*C–C–T*) diagrams. There is a slight difference between the "isothermal" and "continuous" cooling transformation diagrams as regards both the temperatures and times at which the transformations take place (see Figure 2.3 as an example). But, for our discussions here, we will not differentiate between these two cases.

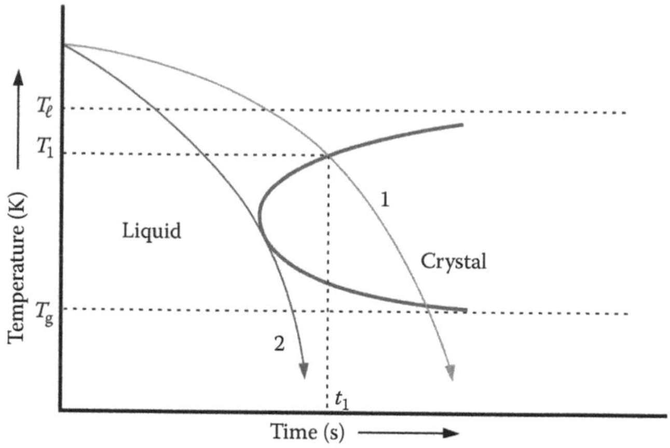

FIGURE 3.1
Schematic time–temperature–transformation (*T–T–T*) diagram for a hypothetical alloy system. When the liquid alloy is cooled from above the liquidus temperature, T_ℓ, at a rate indicated by curve "1," solidification starts at a temperature T_1 and time t_1. The resultant product is a crystalline solid. However, if the same liquid alloy is cooled, again from T_ℓ, at a rate faster than the rate indicated by curve "2," the liquid will continue to be in the undercooled state, and when cooled below the glass transition temperature, T_g, the liquid is "frozen-in" and a glassy phase is formed. The cooling rate represented by curve "2" is referred to as the *critical cooling rate*, R_c.

time at the nose of the *C*-curve are T_n and t_n, respectively, and the liquidus temperature is T_ℓ, then

$$R_c \cong \frac{T_\ell - T_n}{t_n} \tag{3.3}$$

However, this expression overestimates R_c, since it assumes that the crystallization rate corresponds to the nose of the *T–T–T* curve throughout the whole temperature interval of T_ℓ to T_n, and therefore results in a value somewhat higher than the experimentally determined value [12]. Another major difficulty in this approach is in estimating the viscosity of the supercooled melt, but it is possible to get a rough estimate of its value from the available empirical relationships. From such calculations, Davies [13] showed that the values of R_c for the pure metal nickel, and $Au_{78}Ge_{14}Si_8$, $Pd_{82}Si_{18}$, and $Pd_{78}Cu_6Si_{16}$ alloys are ~10^{10}, ~10^6, ~3×10^3, and ~35 K s^{-1}, respectively. First-principle calculations suggested that the minimum cooling rates for the formation of glassy phases in Ag, Cu, Ni, and Pb were about 10^{12}–10^{13} K s^{-1} [14]. Zhong et al. [15] have most recently generated very high quenching rates of 10^{14} K s^{-1} by locally heating two protruded nano-tips through joule heating via electric pulse, which melted the sample in the middle. On subsequent instantaneous cessation of the electric pulse, heat dissipated rapidly through the solidifying piece and the conductive heat reservoir. Using this method, the authors produced metallic glassy samples, 15 nm wide and 20 nm long, in pure metals Ta, V, W, and Mo, confirmed by *in situ* transmission electron microscopy (TEM) methods.

Equation 3.3 and the calculations for R_c are valid only for isothermal processes. But, in reality, glass formation occurs under continuous cooling conditions. Therefore, these equations were modified for continuous cooling transformations [12,13,16]. By performing calculations for continuous cooling transformation behavior, Barandiarán and Colmenero [16] derived an equation for the critical cooling rate for glass formation, R_c, as

$$\ln R = A - \frac{B}{(T_\ell - T_{xc})^2} \tag{3.4}$$

where:

R	is the cooling rate
A and B	are constants
T_ℓ	is the liquidus temperature
T_{xc}	is the onset temperature of solidification upon cooling at a rate R

The undercooling ΔT_c $(=T_\ell - T_{xc})$, at which the derivative of transformed fraction, X, with respect to temperature, T, is a maximum, varies with the cooling rate. When ΔT_c increases to infinity, no crystallization occurs and $A = \ln R_c$. The critical cooling rate, R_c, will be obtained by an extrapolation of the fitting of experimental values with Equation 3.4. Such a method has also been used by others to estimate R_c by measuring the values of T_ℓ and T_{xc} at a given cooling rate [17].

Lin and Johnson [18] derived another equation to estimate the critical cooling rate, R_c (K s^{-1}), required to form a glassy structure from the liquid as

$$R_c = \frac{10}{d^2} \tag{3.5}$$

where d is the sample dimension (e.g., diameter or thickness) in centimeters.

Thus, apart from experimental observations, R_c has also been estimated using different equations based on the thermal behavior of the liquid solidifying into the glass.

As is to be expected, the value of R_c is different for different alloy systems and is also different for different compositions in the same alloy system. Generally speaking, glass formation in pure metals requires an extremely high R_c (typically $>10^{12}$ K s^{-1}), and since it is not easy to achieve such high solidification rates, it has not been possible to produce a pure metal in the glassy state by rapidly quenching it from the molten state. But, for binary metallic liquids, the value of R_c is typically in the region of 10^4–10^6 K s^{-1}. Furthermore, as the number of components in the alloy system increases, the value of R_c will further decrease; consequently, for multicomponent alloys, R_c is typically about 10^2 K s^{-1} or less. Since BMGs are typically multicomponent alloy systems, they can be cast into the glassy state at relatively low R_c values. Very low critical cooling rates of 1.3×10^{-2} K s^{-1} for a $Pd_{37.5}Cu_{32.5}Ni_{10}P_{20}$ alloy [19] and 0.067 K s^{-1} for a $Pd_{30}Pt_{17.5}Cu_{32.5}P_{20}$ alloy [20] have been reported as sufficient to form BMGs. Oxide glasses, on the other hand, can be produced at extremely low cooling rates of the order of 10^{-5} to 10^{-4} K s^{-1}.

3.2.2 Measurement of R_c

The measurement of R_c is an involved and time-consuming process. One has to take a liquid alloy of a given composition, allow it to solidify at different cooling rates, and determine the nature and amount of phases formed after solidification. One has to then consider the solidification rate just above which the material is fully glassy. Whether the solidified material is fully glassy or not can be determined by techniques such as x-ray diffraction (XRD) and TEM. The cooling rate just above which the metallic alloy melt has transformed into a glass is then designated the critical cooling rate, R_c. Such experiments have to be repeated for different compositions in the same alloy system to determine the variation of R_c with composition. Inoue and Nishiyama [21,22] used a differential thermal analyzer (DTA) to estimate the critical cooling rates for the formation of glassy phases in Pd–Ni–Cu–P alloys. They allowed the molten alloys to solidify inside a wedge-shaped copper mold. By measuring the temperature as a function of time at different positions in the mold, they were able to calculate the solidification rates. Further, by observing whether recalescence (the phenomenon in which heat is released, that is, an increase in temperature is observed, when nucleation of a crystalline phase occurs in an undercooled melt; in other words, recalescence occurs only when the solidified alloy contains a crystalline phase and not when it is fully glassy) occurred or not, they were able to decide whether the solid formed was fully glassy or not.

Apart from this thermal analysis method to determine R_c, one could also use microscopic techniques to confirm the glassy nature of the material. One can prepare cross sections of the solid alloys produced at different solidification rates, observe the microstructures, and determine if any crystalline phase is present. Polarized light microscopy may be especially useful for such a purpose. The accuracy with which the presence of a crystalline phase can be detected will depend on the resolution of the microscope used and also the volume fraction of the crystalline phase(s) present. Lower-resolution instruments, such as optical/scanning electron microscopes, are unlikely to detect the presence of a small fraction of the crystalline phase, especially if it is present as fine particles.

Table 3.1 lists the R_c values, either estimated or determined experimentally, for some of the alloy compositions that have been solidified into the glassy state (either as ribbons or BMGs). The values seem to be overestimated or underestimated in some cases. For example, the R_c value appears to be overestimated for the $Au_{77.8}Ge_{13.8}Si_{8.4}$ and $Fe_{40}Ni_{40}P_{14}B_6$ glasses, while it seems to be underestimated for $Hf_{70}Pd_{20}Ni_{10}$. Further, some interesting observations may be made on the values listed in Table 3.1. First, R_c can be significantly reduced by fluxing treatment, in which a flux, such as B_2O_3, is added to remove all the impurities that act as heterogeneous nucleation sites. R_c for glass formation by this treatment may be brought down by at least 1–2 orders of magnitude. For example, R_c for glass formation in the $Pd_{40}Cu_{30}Ni_{10}P_{20}$ alloy is 1.58 K s^{-1} without fluxing [21], whereas it is only 0.1 K s^{-1} with fluxing [19]. Second, the R_c value may also be reduced by increasing the number of components in the alloy system. Thus, while a binary $Pd_{82}Si_{18}$ alloy requires 1.8×10^3 K s^{-1} for glass formation [39], a ternary $Pd_{78}Cu_6Si_{16}$ alloy requires only 550 K s^{-1} for glass formation [38]. Similarly, a binary $Cu_{50}Zr_{50}$ alloy needs to be cooled at rates higher than 250 K s^{-1}, whereas the ternary $Cu_{48}Zr_{48}Al_4$ alloy needs $\leq$40 K s^{-1} to be solidified into the glassy state [26]. Third, the nature of the alloying addition also determines the R_c value. Thus, an $Mg_{65}Cu_{25}Y_{10}$ alloy requires 100 K s^{-1} for glass formation [32]. But, if Y is replaced by Gd, the $Mg_{65}Cu_{25}Gd_{10}$ alloy requires only 1 K s^{-1} for glass formation [17]. By increasing the number of components in the alloy system, it can be further brought down to 0.7 K s^{-1} in $Mg_{65}Cu_{15}Ag_5Pd_5Gd_{10}$ [33]. Thus, these effects seem to be additive and not exclusive of each other.

3.2.3 Effect of Alloying Elements

It is well known from traditional steel metallurgy that the addition of alloying elements has a significant effect on the position (specifically the time axis) of the C-curves [11,45]. The majority of the alloying elements in steel shift the C-curve to the right, that is, the transformations are delayed and occur at longer times; a few of the alloying elements shift the C-curves to the left, that is, to shorter times. When the C-curve for the liquid–solid transformation is shifted to the right, it suggests that the liquid can be retained in the supercooled condition for a longer period of time at any temperature,

TABLE 3.1

Some Representative Critical Cooling Rates (R_c) for Formation of Glassy Phases in Different Alloy Systems

Alloy Composition	R_c (K s^{-1})	References
$Au_{77.8}Ge_{13.8}Si_{8.4}$	3×10^6	[23]
$Ca_{60}Mg_{25}Ni_{15}$	24	[24]
$Ca_{65}Mg_{15}Zn_{20}$	<20	[25]
$Cu_{50}Zr_{50}$	250	[26]
$Cu_{48}Zr_{48}Al_4$	<40	[26]
$Cu_{42}Zr_{42}Ag_8Al_8$	4.4	[27]
$Fe_{43}Cr_{16}Mo_{16}C_{10}B_5P_{10}$	100	[28]
$Fe_{40}Ni_{40}P_{14}B_6$ (Metglas 2826)	4.4×10^7	[29]
$Hf_{70}Pd_{20}Ni_{10}$	124	[30]
$La_{55}Al_{25}Cu_{20}$	58	[31]
$La_{55}Al_{25}Ni_{20}$	69	[31]
$Mg_{65}Cu_{25}Gd_{10}$	1	[17]
$Mg_{65}Cu_{25}Y_{10}$	100	[32]
$Mg_{65}Cu_{15}Ag_5Pd_5Gd_{10}$	0.7	[33]
$Mg_{65}Cu_{7.5}Ni_{7.5}Ag_5Zn_5Gd_5Y_5$	20	[34]
$Nd_{60}Co_{30}Al_{10}$	4	[35]
$Ni_{62}Nb_{38}$	57	[36]
$Ni_{65}Pd_{15}P_{20}$	10^5	[37]
$Pd_{78}Cu_6Si_{16}$	550	[38]
$Pd_{40}Ni_{40}P_{20}$	128	[21]
$Pd_{42.5}Cu_{30}Ni_{7.5}P_{20}$	0.067	[19]
$Pd_{40}Cu_{25}Ni_{15}P_{20}$	0.150	[19]
$Pd_{40}Cu_{30}Ni_{10}P_{20}$ (without fluxing)	1.58	[21]
$Pd_{40}Cu_{30}Ni_{10}P_{20}$ (with flux treatment)	0.1	[19]
$Pd_{30}Pt_{17.5}Cu_{32.5}P_{20}$	0.067	[20]
$Pd_{82}Si_{18}$	1.8×10^3	[39]
$Pd_{82}Si_{18}$	800	[40]
$Pd_{81}Si_{19}$ (with flux treatment)	6	[41]
$Zr_{65}Al_{7.5}Ni_{10}Cu_{17.5}$	1.5	[42]
$Zr_{41.2}Ti_{13.8}Cu_{12.5}Ni_{10.0}Be_{22.5}$	1.4	[43]
$Zr_{46.25}Ti_{8.25}Cu_{7.5}Ni_{10.0}Be_{27.5}$	28	[43]
$Zr_{57}Cu_{15.4}Ni_{12.6}Al_{10}Nb_5$	10	[44]

and therefore the value of R_c for glass formation is lower. As a result, it becomes easier to form the glass. The farther the C-curve is shifted to the right, the lower the value of R_c. This is the concept used in synthesizing BMGs, which contain a large number of alloying elements [46]. Since the C-curve for solidification is shifted far to the right due to the presence of a large number of alloying elements, the glassy phases are formed at rates as low as 1–100 K s^{-1}, and sometimes even lower. The effect of alloying elements on the required critical cooling rate for glass formation is shown schematically in Figure 3.2.

An important point to remember about this criterion is that irrespective of whether we are considering melt-spun ribbons with a thickness of a few micrometers or BMG rods with a diameter of a few millimeters to a few centimeters, the criterion of R_c is valid. That

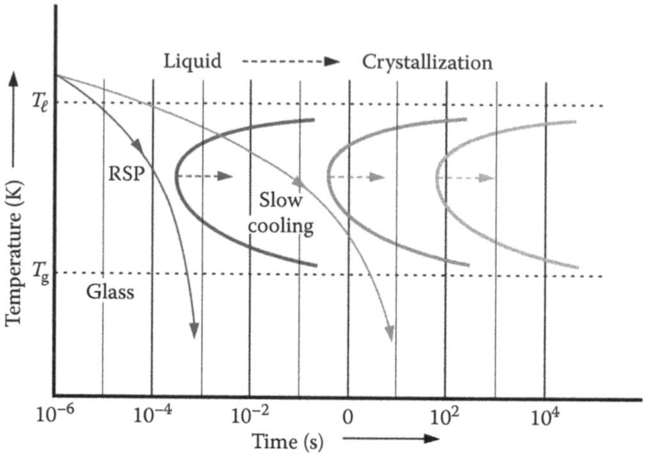

FIGURE 3.2
Position of the *T–T–T* curves with the addition of a large number of alloying elements. The C-curve shifts to the right with increasing numbers of alloying elements and consequently, the glassy phase can be synthesized at slow solidification rates. The left-most C-curve represents a typical situation of an alloy system where a glassy phase is obtained by RSP from the liquid state. The middle C-curve represents an alloy composition where a glassy phase can be obtained by slow cooling. The right-most C-curve represents a situation when an alloy can be very easily produced in a glassy state.

is, the critical cooling rate must always be exceeded to achieve glass formation. Once the critical cooling rate is exceeded, glass formation will occur, provided we have reached a temperature below T_g. Glass formation will, of course, occur even if a higher solidification rate is employed.

It is easy to appreciate that the solidification rate achieved in the material is a function of the section thickness. That is, the cooling rate is highest on the surface of the sample (which is in contact with the heat sink), and it decreases as one goes deeper into the interior of the sample. Thus, for a given section thickness or diameter, the cooling rate will decrease from the surface to the interior. Consequently, a situation may arise in which the solidification rate is higher than R_c on the surface, but is lower than R_c in the interior. In such a case, the glassy phase will form on the surface only up to a limited depth, and not throughout the interior of the sample. However, if the cooling rate is also higher than R_c in the center of the rod sample, that is, throughout the section thickness, then the whole sample will be glassy. (Sometimes it is noted that the solidified rods are glassy in the interior and a thin crystalline layer is obtained on the surface. This is because of surface nucleation of a crystalline phase, mainly due to the process of heterogeneous nucleation, and has nothing to do with the R_c value.)

In recent years, several investigations have confirmed that choosing a proper alloying element in the appropriate amount has an important role in decreasing the critical cooling rate required for glass formation, that is, in enhancing the GFA. For example, it was mentioned earlier that the GFA of Cu–Zr alloys was enhanced by the addition of Al [26]. Similarly, the addition of Al increased the GFA of U–Co alloys [47]. Further, the addition of 1 at.% Si to a Cu–Zr–Al alloy suppressed the precipitation of crystalline phases and improved the GFA [48]. Addition of Ni was also shown to increase the GFA of Zr–Cu–Al alloys [49]. Further, by combining experimental measurements with theoretical calculations, Guo and Yang [50] reported that minor additions of Ag to a Zr–Cu–Al alloy led to the enhancement of GFA. It was explained that minor additions of Ag lead to (i) a high fraction of Ag-centered icosahedral-like local structures, (ii) atomic packing efficiency, and (iii) structural regularity in Zr and Cu centered major clusters, and that all these effects can stabilize the amorphous structure and increase the GFA. Yu et al. [51] explained the good GFA of Zr–Cu–Al alloys using an atomistic approach derived from first-principle calculations based on the density functional theory and molecular dynamic simulations. Based on similar principles, Yu et al. [52] also explained the difference in the good GFA behavior of Zr–Cu and the poor GFA behavior of Zr–Ni binary glasses in terms of both the atomic configurations and the relative stability of their intermetallic phases.

On the other hand, Ma et al. [53] have shown that substitution of Al by Ti in a Cu–Zr–Al alloy decreased the GFA. They explained this observation by suggesting that Ti addition increased the distance from the eutectic point, and this reduced the ΔT_x and T_{rg} values. The plasticity, however, increased significantly.

Improved GFA has resulted from the addition of Cu to Ti–Zr–Be BMG alloys. Whereas the maximum diameter of a glassy rod that could be obtained in a $Ti_{41}Zr_{25}Be_{34}$ glass was 5 mm [54], the maximum diameter could be increased to 15 mm by adding 6 at.% Cu to replace Be [55]. Cu addition has also improved the GFA of Ti–Zr–Be–Fe alloys. For example, the maximum diameter of the glassy rod could be increased to >20 mm by adding 7–13 at.% Cu, from <10 mm when no Cu was present in the alloy [56]. The critical diameter of these BMG rods could be further increased to over 50 mm by fine-tuning the composition in the Ti–Zr–Cu–Fe–Be [57] and Ti–Zr–Cu–Ni–Be [58] systems.

Similar results were also reported in Zr-based [59–61], Fe-based [62–67], Ni-based [68], Mg-based [69,70], and other alloy systems as well [71–73].

The addition of Cu is usually made to Fe-based glassy alloys to produce improved soft magnetic properties in the FINEMET alloys upon subsequent crystallization [74]. It has been shown that the addition of Cu decreases the GFA of Fe-based alloys slightly [75]; however, with the advantage that it helps in separating the precipitation of α-Fe(Si), Fe_2B, and $Fe_3(B,P)$ in $(Fe_{0.76}Si_{0.09}B_{0.1}P_{0.05})_{99}Nb_1$ alloy, resulting in the enhancement of soft magnetic properties.

The purity of the base element Ce in Ce–Ga–Cu BMGs has been shown to be very important in determining the GFA [76]. It was shown that when high-purity (99.87 wt.% pure) Ce was used, the maximum diameter of the fully glass rod was only 1 mm. However, when the purity of Ce was reduced to 98.51 wt.%, the critical diameter could be increased to 10 mm, and when the purity was further reduced to 97.52 wt.%, the critical diameter was ≥20 mm. Similarly, Wang et al. [77] reported that the GFA of binary Zr–Cu alloys was improved when the alloys had a high oxygen content. Thus, oxygen can be considered an alloying element, rather than an impurity. It was also noted that when the oxygen content was high, the best glass-forming composition of the alloy had shifted from that under low oxygen-content conditions. Thus, by choosing the appropriate composition, high GFA can be obtained even in low-purity materials. These investigations suggest that it should be possible to produce BMGs using commercially pure elements and at low production cost.

In addition to the composition and purity of the elements in the alloy, the atmosphere under which the alloy is cast also seems to be important. While investigating the GFA of $Zr_{66}Cu_{22}Al_{12}$, Granata et al. [78] showed that the critical casting diameter for a fully glassy material was 2.8 mm when cast under Ar or He. This diameter increased to 4.9 mm when cast under a 95% Ar/5% H_2 mixture. But, the maximum diameter of 6 mm was achieved only when cast under the 95% Ar/5% H_2 mixture at the higher casting pressure of 850 mbar. Under these conditions, the hydrogen concentration was found to be 277 ± 5 wt. ppm. Thus, the authors concluded that hydrogen behaved as a microalloying element. The increased GFA was explained on the basis that the increased hydrogen content enhanced the liquid stability due to a decrease in the liquidus temperature, and consequently, the composition shifted closer to the eutectic. The smaller size of the hydrogen atom may also have led to a higher packing efficiency.

As previously discussed, R_c is a very effective indicator of the GFA of the alloy melt. But, its measurement is quite involved and further, it can only be determined when the alloy composition in which glass formation can occur has been synthesized. Additionally, it is only possible to determine if a glass is formed or not after doing several experiments. Therefore, investigators have been searching for other criteria to see if it is possible to *predict* the ability of glass formation (i.e., GFA) in metallic alloy systems. But, before discussing these predictive criteria for the formation of BMGs, let us first look at some of the early criteria that were used to explain glass formation in metallic glassy ribbons produced by RSP methods. We will also make appropriate comments about their applicability, or otherwise, with respect to BMGs.

3.3 Reduced Glass Transition Temperature

When a liquid alloy is cooled from the molten state down to a temperature below T_g, the viscosity of the melt increases to a high value and a glass is formed. Since the viscosity at T_g is chosen to be fixed at 10^{12} Pa s, Turnbull [79] suggested, purely on the kinetics of crystal nucleation and the viscosity of melts, that the ratio of the glass transition temperature, T_g, to the liquidus temperature of the alloy, T_ℓ, should be a good indicator of the GFA of the alloy. The higher this value, the higher the viscosity and, therefore, the alloy melt could be easily solidified into the glassy state at a low critical cooling rate. In other words, an alloy composition with as high a value of T_g and as low a value of T_ℓ as possible will promote easy glass formation. This ratio has been designated the reduced glass transition temperature, T_{rg}. That is,

$$T_{rg} = \frac{T_g}{T_\ell} \tag{3.6}$$

Based on the nucleation theory, Turnbull suggested that at $T_{rg} \geq 2/3$, homogeneous nucleation of the crystalline phase is completely suppressed. Most typically, a minimum value of $T_{rg} \cong 0.4$ has been found to be necessary for an alloy to become a glass, but the higher the T_{rg} value, the easier it is for the glass to form. Table 3.2 lists the T_{rg} values for some of the BMGs that have been reported. Data for a few of the melt-spun ribbons have also been included, since no differences were observed between the T_g, T_x, and T_ℓ values of melt-spun ribbons and BMGs. That is, the T_{rg} values are independent of the size of the glassy sample. This brings up an interesting point. The T_{rg} criterion predicts that an alloy can be transformed into the glassy state if its T_{rg} is reasonably high. But, the actual process of obtaining the glass is possible only when the critical cooling rate for glass formation is exceeded during solidification. If the molten alloy is solidified at a slower rate than the critical cooling rate for glass formation, the alloy will not form glass even if its T_{rg} is very high.

From Table 3.2, it may be noted that the value of T_{rg} is reasonably high for all glass-forming alloys, and is in fact about 0.6 for most of the compositions listed. Generally speaking, the T_{rg} value is higher for alloy systems that contain a large number of alloying elements.

TABLE 3.2

Reduced Glass Transition Temperatures (T_{rg}) for Different Glass-Forming Alloys

Alloy Composition	T_{rg}	References
$Au_{49}Ag_{5.5}Pd_{2.3}Cu_{26.9}Si_{16.3}$	0.622	[80]
$Ca_{65}Al_{35}$	0.69	[81]
$Ca_{57}Mg_{19}Cu_{24}$	0.64	[82]
$Ca_{60}M_{g2}5Ni_{15}$	0.631	[83]
$Ce_{60}Al_{15}Ni_{15}Cu_{10}$	0.61	[84]
$Cu_{65}Hf_{35}$	0.62	[85]
$Cu_{49}Hf_{42}Al_{9}$	0.62	[86]
$Cu_{64}Zr_{36}$	0.64	[87]
$(Cu_{47}Zr_{45}Al_{8})_{96}Dy_{4}$	0.634	[88]
$(Fe_{0.7}Mn_{0.3})_{65}Zr_{4}Nb_{4}Mo_{3}B_{24}$	0.63	[89]
$Fe_{66.24}Dy_{6.72}B_{23.04}Nb_{4}$	0.636	[90]
$Fe_{66.24}Mo_{3.68}B_{22.08}Gd_{8}$	0.632	[91]
$La_{55}Al_{25}Ni_{20}$	0.71	[31]
$La_{62}Al_{15.5}(Cu,Ni)_{22.3}$	0.58	[92]
$La_{50.2}Al_{20.5}(Cu,Ni)_{29.3}$	0.47	[92]
$Ni_{62}Nb_{38}$	0.60	[36]
$Ni_{61}Nb_{33}Zr_{6}$	0.49	[93]
$Ni_{52}Pd_{26}P_{20}B_{2}$	0.62	[94]
$Pd_{40}Ni_{40}P_{20}$	0.67	[95]
$Pd_{40}Cu_{30}Ni_{10}P_{20}$	0.67	[19,96]
$Ti_{20}Zr_{20}Hf_{20}Be_{20}Cu_{7.5}Ni_{12.5}$	0.608	[97]
$Zr_{41.2}Ti_{13.8}Cu_{12.5}Ni_{10}Be_{22.5}$	0.67	[98]
$Zr_{46}Cu_{30.14}Ag_{8.36}Al_{8}Be_{7.5}$	0.64	[99]

That is why it has been easier to produce a multicomponent alloy in the glassy state. For example, the T_{rg} value for an $La_{55}Al_{25}Ni_{20}$ alloy is 0.71 [31], while it is 0.69 for $Ca_{65}Al_{35}$ [81], and 0.67 for $Pd_{40}Ni_{40}P_{20}$ [95] and $Pd_{40}Ni_{10}Cu_{30}P_{20}$ [96] glassy alloys. Lu et al. [100] noted that even for BMGs, a strong correlation existed between T_{rg} and R_c, as also T_{rg} and the maximum section thickness or diameter of the glassy rod, t_{max}. However, such correlations were found only when they estimated T_{rg} as T_g/T_ℓ (Equation 3.6) and not when it was calculated from the relationship $T_{rg} = T_g/T_m$, where T_m is the solidus temperature of the alloy (i.e., the temperature corresponding to the onset of melting). This observation clearly suggests the importance of noting the correct temperature for the completion of melting. That is, in the DSC curves, one should note the liquidus temperature as the point at the end of the liquid formation, and not at the beginning of melting.

The T_{rg} criterion has not been found to be valid in all the glasses that have been studied; quite a few exceptions and discrepancies have been reported. For example, binary Cu–Zr BMGs have been produced in the composition range of Cu–34 to 40 at.% Zr alloys. It was noted that the best glass-forming composition ($Cu_{64}Zr_{36}$) did not correspond to the highest T_{rg} value [87]. A similar discrepancy was also reported by Tan et al. [92], who investigated glass formation in La–Al–(Cu,Ni) pseudo-ternary alloys and observed that the best GFA was exhibited by an off-eutectic composition and that the GFA did not correlate well with

T_{rg}. In fact, glass formation has been observed at off-eutectic compositions in many alloy systems. Exceptions to the T_{rg} criterion have been reported in other cases also.

The concept of T_{rg} was introduced for kinetic reasons with the need to avoid crystallization [79]. It is known (and we will discuss this in some detail in Section 3.4) that T_g is a very weak function of the solute content, that is, T_g varies much more slowly with solute concentration than does the liquidus temperature, T_ℓ. Consequently, the value of T_{rg} increases with increasing solute content, up to the eutectic composition, and therefore it becomes easier to avoid crystallization of the melt at the eutectic composition [13]. This reasoning seems to work well in simple binary alloy systems. But, in the case of multicomponent alloy systems, such as the BMG compositions, the values of T_g and T_ℓ vary significantly. Since the variation of viscosity with temperature is different for different alloy systems (and depends on whether a glass is strong or fragile), T_g alone may not provide information about the variation of viscosity with temperature and, therefore, the T_{rg} criterion may not be valid in some systems.

Easy glass formation at compositions corresponding to high T_{rg} can be readily accomplished in alloy systems that feature deep eutectic reactions in their phase diagrams, and this is further explained in Section 3.4.

3.4 Deep Eutectics

It has been experimentally shown that the value of T_g changes slowly with solute content (see, e.g., Ref. [101] for results on the Pd–Si system). On the other hand, the liquidus temperature (T_ℓ) of an alloy usually decreases with increasing solute content in most alloy systems. But, there are some phase diagrams in which the liquidus curves drop very steeply with solute content. An alloy system in which the eutectic temperature is significantly lower than the melting points of the individual components is referred to as a *deep eutectic*. In such cases, the T_{rg} ($=T_g/T_\ell$) value around the "deep" eutectic composition is a strong function of the alloy composition and reaches its highest value at the eutectic composition. Therefore, it should be possible to quench this alloy composition easily into the glassy state. On the other hand, if the liquidus temperature of an alloy also decreases slowly with the solute content, then the T_{rg} value at the eutectic composition may not be high. In this situation, it will be difficult to produce such an alloy in the glassy condition (Figure 3.3). Thus, the Turnbull criterion of high T_{rg} and deep eutectics converge at the deep eutectic compositions.

A number of binary phase diagrams, for example, those of Au–Si, Pt–Si, and Fe–B, show variation of T_ℓ with composition similar to that in the Pd–Si system. Therefore, an alloy with the eutectic composition, especially if it is a deep eutectic, will form the glassy phase most easily in the alloy system [102]. This empirical criterion has been most helpful in identifying glass-forming compositions in simple binary and ternary alloy systems in the early years of RSP research. Even though this criterion can also be used to identify alloy compositions to form BMGs, it should be realized that phase diagrams for many of the multicomponent alloy systems are not available. One way of overcoming this lack of information is to look for the occurrence of eutectic reactions in the constituent binary and ternary phase diagrams. If eutectic reactions occur in these systems, it is likely (but not necessary) that eutectic reactions will occur in the multicomponent alloy systems as well. It is, however, possible to detect whether melting in a given alloy composition occurs at

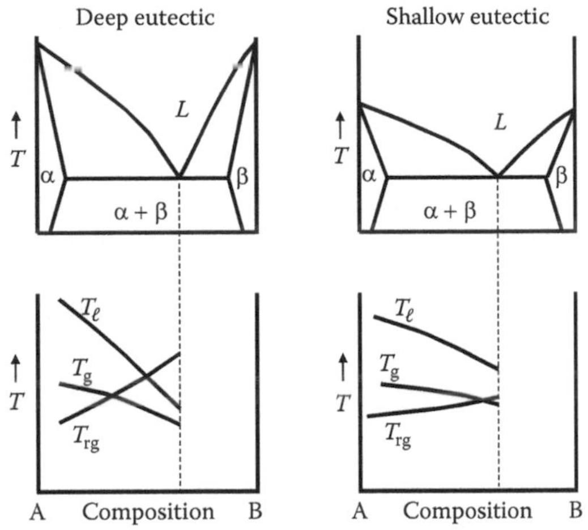

FIGURE 3.3
Schematic variation of the glass transition temperature, T_g, liquidus temperature, T_l, and the reduced glass transition temperature, T_{rg}, in two different types of eutectic systems—deep eutectic and shallow eutectic.

a well-defined temperature or over a range of temperatures during differential scanning calorimetry (DSC) or DTA. If melting occurs at a well-defined temperature, then that composition corresponds to the eutectic composition. On the other hand, if melting occurs over a range of temperatures, it represents an off-eutectic composition.

Both thermodynamic and kinetic arguments have been put forward to explain why it is easiest to obtain metallic glasses at the eutectic composition. Thermodynamically, the (disordered) liquid state is preferred energetically over the solid (ordered) crystalline phases at the deep eutectic composition, either through the destabilization of the crystalline phases or through the stabilization of the liquid state. Further, the driving force for the nucleation and growth of the crystalline phases below the eutectic temperature (which is the energy difference between the liquid and crystalline phases) is relatively small at the eutectic composition, compared with the off-eutectic compositions. Therefore, it becomes easier to quench the eutectic liquid into the glassy phase before a detectable fraction of the crystalline phase can be observed. Since a number of crystalline ordered phases, especially in multicomponent alloys, compete with each other for nucleation and growth, the crystallization of the liquid requires extensive rearrangement of the different types of atoms to form the new phases. Thus, kinetically also, glass formation is preferred at the eutectic composition.

Ribbons of about 20–50 μm in thickness were produced by melt-spinning techniques in a number of binary alloy systems near eutectic compositions and they were confirmed to be glassy. Some of the most investigated eutectic compositions are found in the Fe–B, Pd–Si, Cu–Zr, Ni–Nb, and Ni–Ta alloy systems [1,2]. There have been some reports in recent years of "bulk" (?) metallic glass synthesis in binary alloy systems such as Ca–Al [103], Cu–Hf [85], Cu–Zr [87], Ni–Nb [36], and Pd–Si [41]. But, the maximum diameters of these binary alloy glassy rods were only 1 or 2 mm, and even then, some of these binary alloy glasses contained nanocrystalline particles embedded in the glassy matrix [104]. This point will be further discussed in detail in later chapters. But, the important point here is that the "bulk" glassy phase is produced not at the eutectic composition, but at off-eutectic compositions. Further, the highest GFA, that is, the composition at which glass formation was easiest or the maximum diameter of the BMG rod could be obtained, was located away from the eutectic composition. The best glass-forming compositions have been reported to be at 35 at.% Hf in Cu–Hf, 36 at.% Zr in Cu–Zr, 38 at.% Nb in Ni–Nb, and 19 at.% Si in Pd–Si alloy systems. But, the eutectic compositions in these alloy systems are at 33.0 and 38.6 at.% Hf in Cu–Hf, 38.2 at.% Zr in Cu–Zr, 40.5 at.% Nb in Ni–Nb, and 17.2 at.% Si in Pd–Si systems [105]. Further, Lee et al. [106] showed that a 3 mm glassy rod could be produced in the quaternary eutectic alloy ($Zr_{49}Ti_{17}Ni_{20}Cu_{14}$). But, the best glass-forming composition was found to be at an off-eutectic composition ($Zr_{49}Ti_{14}Ni_{20}Cu_{17}$), at which composition a 4 mm glassy rod could be produced. Similarly, it was reported that while a nearly fully glassy rod with 12 mm diameter could be obtained at an off-eutectic composition near $La_{62}Al_{15.7}(Cu,Ni)_{22.3}$, only a 1.5 mm diameter rod could be obtained in a fully glassy condition for the eutectic alloy of $La_{66}Al_{14}(Cu,Ni)_{20}$ [92].

Zhang et al. [107] analyzed the GFA of $La_xAl_{14}(Cu,Ni)_{86-x}$ (x = 57–70) alloys by measuring their thermal properties. They noted that neither the width of the supercooled liquid region, ΔT_x, nor the activation energy for crystallization measured from the first peak correlated with the best glass-forming composition. They also noted that the GFA did not correlate satisfactorily with T_{rg}, γ (Section 3.10.1), ΔT^*, or σ (Section 3.12.5) parameters (to be discussed later). Thus, they concluded that the GFA parameters based on T_ℓ, the liquidus temperature of the alloy, are not effective in predicting the GFA of alloys. Instead, they suggested that the GFA of the best glass-forming composition could be related to T_s, the solidification start temperature.

Wang et al. [26] also analyzed the glass formation behavior in the $Cu_{50}Zr_{50}$ alloys and noted that fully glassy rods up to 2 mm in diameter could be cast from the liquid state. But, in larger-diameter rods, they observed the formation of the crystalline $Cu_{51}Zr_{14}$ phase and interpreted the GFA of $Cu_{50}Zr_{50}$ alloys in terms of a deep metastable eutectic between $Cu_{51}Zr_{14}$ and β-Zr. This suggests that the best GFA composition should be displaced from the eutectic compositions shown on the equilibrium diagram. By destabilizing the $Cu_{51}Zr_{14}$ phase, through the addition of 4 at.% Al, the authors were able to obtain fully glassy rods of 5 mm diameter in the $Cu_{48}Zr_{48}Al_4$ alloy.

It can be argued that glass formation in an alloy system is possible not only when the nucleation of the crystalline phase(s) is completely suppressed (or avoided altogether), but also when the growth rate of the crystalline phase(s) is suppressed. Consequently, a glass will form if its T_g isotherm is higher than the growth temperature of any of the competing crystalline phase(s). Based on these concepts, Ma et al. [108] have proposed a model to calculate the composition range (the minimum and maximum solute contents) in which a glassy phase could form. They suggested that the composition at which the GFA is highest will include the eutectic composition if it is a regular eutectic system, but will be outside the eutectic composition if it is an irregular eutectic. It has also been suggested that in cases when the best glass-forming composition is at off-eutectic values, the composition having the highest GFA would be on the side of the eutectic where the phase diagram exhibits a steeper liquidus line [92]. But, exceptions to this suggestion have been noted in Cu–Hf, where the best GFA is at 35 at.% Hf, and this lies on the shallower side of the liquidus slope [85]. It is also of importance to note that the eutectic reactions in the systems mentioned herein are not really of the "deep" eutectic type.

The off-eutectic composition at which the highest GFA occurs in alloy systems was suggested to be related to the kinetic considerations, and more specifically, the variation of T_g with composition, $T_g(C)$ [109]. It was proposed that the optimum composition at which GFA is highest, C^*, is shifted from the eutectic composition, C_{eu}, in the direction of increasing $T_g(C)$. That is,

$$\mathrm{Sign}\left(C^{*}-C_{eu}\right)-\mathrm{Sign}\left[\frac{dT_{g}(C)}{dC}\right]_{C_{eu}} \tag{3.7}$$

3.5 Topological Models

The empirical criteria (high values of T_{rg} and deep eutectics) mentioned in the preceding section have been extensively employed in synthesizing micrometer-thick metallic glasses by melt-spinning and other related RSP techniques, even though some limitations have been noted. Additionally, some theoretical formulations have been proposed to explain the glass formation in alloy systems by the RSP methods. The majority of these have been concerned with the atomic sizes of the constituent elements and their topological arrangement. So, let us now look at these aspects of glass formation also.

Metallic glasses produced by RSP methods in the form of thin ribbons have been traditionally classified into two groups, *viz.*, metal–metalloid and metal–metal types. Structural models of the metal–metalloid-type metallic glasses have identified that the best composition to form a glass is one that contains about 80 at.% of the metal component and 20 at.% of the metalloid component. The actual glass composition ranges observed are 75–85 at.% of the metal and 15–25 at.% of the metalloid. As stated in Chapter 2, the 80 at.% composed of metal can be either a single transition metal, a combination of transition metals, or one or a combination of noble metals. Similarly, the 20 at.% metalloid content could be made up of just one component or a mixture of a number of components. In the case of metal–metal types, however, there is no such restriction on compositions. Metal–metal-type metallic glasses have been observed to form over a wide range of compositions, starting from as low as 9 at.% of solute. Some typical compositions in which metal–metal-type glasses have been obtained are $Cu_{25-72.5}Zr_{27.5-75}$, $Fe_{89-91}Zr_{9-11}$, $Mg_{68-75}Zn_{25-32}$, $Nb_{55}Ir_{45}$, and $Ni_{58-67}Zr_{33-42}$ [110].

The requirement of a 20 at.% concentration of a metalloid component to form a glass in a metal–metalloid-type system arises because the metalloid atom can occupy the interior of the tetrahedron of four metal atoms ("stuffing" the voids in the Bernal dense random packing structure) [111] and thereby stabilize the structure against ready crystallization. Consequently, the typical alloy composition at which the metallic glass forms is approximately A_4B (where A represents the metal and B the metalloid). But, Gaskell [112] criticized this idea by stating that the actual size of the voids in the metal tetrahedra in the Bernal structure is too small for any metalloid atom to be accommodated there. It was also shown that the concentration of "holes" of the required sizes to accommodate the metalloid atoms was too low to account for the observed glass compositions [113]. Turnbull [114], however, has shown that the effective size of the metalloid atom depends on the nature of the metal in which it is dissolved. Thus, the real reason for the requirement of 20 at.% of metalloid atoms to form glasses is still not very clear.

Irrespective of the actual size of the voids and whether the aforementioned model is valid or not, it is of interest to note that the metal–metalloid-type binary phase diagrams exhibit deep eutectics at around a composition of 15–25 at.% metalloid. Some typical examples are Fe–B (17 at.% B), Au–Si (18.6 at.% Si), and Pd–Si (17.2 at.% Si). Therefore, the concepts of deep eutectics and structural models also seem to converge in obtaining glasses in the (transition or noble) metal–metalloid types. But, Nielsen [115] conducted a detailed analysis of the distribution of eutectic compositions in binary Fe-based alloy phase diagrams and pointed out that the maximum number of eutectic reactions occurs in binary phase diagrams at a metalloid content of about 13–14 at.% and that these compositions are not known to form glasses. But, if one conducts an analysis using only systems that exhibit deep eutectic reactions, the situation may be different. However, at this stage, it is not clear whether deep eutectics will necessarily be good glass formers. Also, as already noted, there are a few cases where the GFA appears to be higher at off-eutectic compositions than at eutectic compositions. Therefore, it may be fair to state at this stage that compositions at and around deep eutectics favor glass formation.

3.5.1 Atomic Size Mismatch

Analysis of a large amount of data on binary metallic glasses has shown that the atomic size ratios of the majority to the minority atoms vary between 0.79 and 1.41, with many of the glasses in the range 0.85–1.15. Cahn [101] has termed this the *anti-Hume-Rothery criterion*, since Hume-Rothery had earlier pointed out that the radius mismatch should not exceed ±15% to form extensive solid solutions in alloy systems based on noble metals [116]. Thus, it appears that the sizes of constituent atoms have a significant role to play in glass formation. Accordingly, several attempts have been made to explain glass formation based on the atomic size mismatch between the constituent elements in the alloy.

3.5.2 Egami and Waseda Criterion

One of the possible ways by which a crystalline metallic material can become glassy is through the introduction of lattice strain [117]. The lattice strain introduced disturbs the crystal lattice and, once a critical strain is exceeded, the crystal becomes destabilized and glassy. In fact, Egami is at pains to state that "In general, alloying makes glass formation easier, not because alloying stabilizes a glass, but because it destabilizes a crystal" [118, p. 576]. Using the atomic scale elasticity theory, Egami and Waseda [119] calculated the atomic-level stresses in the solid solution (the solute atoms are assumed to occupy the substitutional lattice sites in the solid solution) and the glassy phase. They observed that in a glass, neither the local stress fluctuations nor the total strain energy vary much with solute concentration, when normalized with respect to the elastic moduli. But, in a solid solution, the strain energy was observed to increase continuously and linearly with solute content. Thus, beyond a critical solute concentration, the glassy alloy becomes energetically more favorable than the corresponding crystalline lattice. From the vast literature available on the formation of binary metallic glasses obtained by RSP methods, the authors noted that a minimum solute concentration was necessary in a binary alloy system to obtain the stable glassy phase by RSP methods. In addition, this minimum solute concentration, C_B^{min} was found to be inversely correlated with the atomic volume mismatch, $(V_A - V_B)/V_A$, where V_A is the atomic volume of the solvent and V_B is the atomic volume of the solute. The authors found that the minimum solute concentration required can be obtained from the relation:

$$\left|\frac{(V_A - V_B)}{V_A}\right| C_B^{min} = 0.1 \tag{3.8}$$

This relationship suggests that the minimum solute concentration decreases as the difference between the atomic sizes of the solute and solvent atoms increases. These concepts have been subsequently expanded to cover the cases of multicomponent alloys frequently encountered in BMGs, and we will discuss these points in Section 3.12.

Using the long-range empirical potential, Dai et al. [120] have developed a model to describe the interatomic interactions in the bcc metals. Later, this model was shown to be applicable to fcc, bcc, and hcp transition metals and their alloys [121]. Luo et al. [121] carried out Monte-Carlo simulations, identified the favored compositions for glass formation and reported that the crystalline lattice collapsed when the solute concentration exceeded a critical value.

3.6 Bulk Metallic Glasses

Since 1989, intense research has been carried out in synthesizing and characterizing BMGs with a section thickness or diameter of a few millimeters to a few centimeters. It may be argued that the criteria described for glass formation in the case of melt-spun ribbons could also be used to rationalize glass formation in the new multicomponent BMG alloy systems. In fact, they have been used with some success; however, this has not been completely satisfactory for several reasons. First, phase diagrams are not available for the multicomponent alloy systems. Since the complexity of phase diagrams increases with the number of components in the alloy system, it becomes very difficult to experimentally determine phase diagrams of multicomponent alloy systems, even though the CALculation of PHAse Diagrams (CALPHAD) approach has been undertaken in recent years [122]. Consequently, experimentally determined phase diagrams are available for systems with only up to three or four components. Phase diagrams for many of the higher-order systems have not been determined. Therefore, we do not know where the eutectic compositions lie, and much less about deep eutectics. Second, because the number of components is really large, determining the minimum solute content necessary to obtain a glassy phase will be a formidable problem, since the contribution of each component to the volumetric strain is going to be different depending on their atomic sizes. The situation becomes more complex when we realize that the magnitude and the sign of stress could be different for different alloying elements. Even though the concept of high T_{rg} could be used, one again needs to first produce the glassy alloy and determine the glass transformation temperatures (T_g and T_ℓ): only then can the T_{rg} value be determined.

Therefore, newer criteria have been proposed to explain glass formation in BMGs in view of the large number of components present. The first set of criteria was proposed by Inoue, who systematically investigated glass formation in several multicomponent alloy systems. The three empirical rules formulated by Inoue [123] were the mainstay for a long time, and using these criteria, it was possible to produce BMGs in a number of alloy systems. In fact, several hundreds of alloy compositions were synthesized in the glassy state using these empirical rules. The following paragraphs discuss the criteria that have been developed to explain the GFA of BMGs.

3.7 Inoue Criteria

Based on the extensive data generated on the synthesis of BMGs, Inoue [123–125] formulated three basic empirical rules for the formation of BMGs. These may be stated as follows:

1. The alloy must contain at least three components. The formation of glass becomes easier with an increasing number of components in the alloy system.
2. A significant atomic size difference should exist between the constituent elements in the alloy. It is suggested that the atomic size differences should be above about 12% between the main constituent elements.
3. There should be negative heat of mixing for the (major) constituent elements in the alloy system.

Considering these criteria, the first is based on the thermodynamic and kinetic aspects of glass formation, and the second on the topological aspects (structure and packing of atoms). The third criterion is essential for the mixing of atoms (alloying) to occur and for the formation of a homogeneous glassy phase. Let us now look at the rationale for formulating the criteria in some detail.

As we discussed earlier, the formation of a glassy phase requires that its free energy is reduced to a level lower than that of the competing crystalline phase(s). This free energy reduction could be achieved by either increasing the entropy of fusion, ΔS_f, or decreasing the enthalpy of fusion, ΔH_f, or both. Since the value of ΔS_f can be significantly increased by increasing the number of components in the alloy, it has been relatively easy to produce BMGs in multicomponent alloys. Since an increase in ΔS_f also leads to an increase in the degree of the dense random packing of atoms, this results in a decrease in ΔH_f, and also an increase in the solid–liquid interfacial energy, σ. Both these factors contribute to a decrease in the free energy of the system.

We can appreciate the reasons for the formation of BMGs in multicomponent alloys from a kinetic point of view also. Since the equation for the homogeneous nucleation rate for the formation of crystalline nuclei from a supercooled melt (Equation 2.4) contains η, α, and β, controlling these parameters can lead to a reduction in the nucleation rate. For example, a reduction in ΔH_f, and an increase in σ and/or ΔS_f, can be achieved by an increase in α and β values. This, in turn, will decrease the nucleation rate and consequently promote glass formation. An increase in the viscosity of the melt will also lead to a reduction in both the nucleation and growth rates.

The second Inoue criterion is related to the topological aspects of glass formation. It was mentioned that the constituent atoms in multicomponent BMGs should exhibit a significant difference in their atomic sizes, reaching more than about 12%. The atomic sizes of the atoms in the periodic table can be classified into three groups—large, medium, and small—as shown in Figure 3.4.

The combination of the significant differences in atomic sizes between the constituent elements and the negative heat of mixing is expected to result in efficient packing of clusters (see Section 3.12.2) and consequently increased density of random packing of atoms in the supercooled liquid state. This, in turn, leads to increased liquid–solid interfacial energy, σ, and decreased atomic diffusivity, both contributing to enhanced glass formation.

The formation of a higher degree of dense randomly packed structures in BMGs has been well documented in the literature, through XRD studies, in several Zr-, La-, and

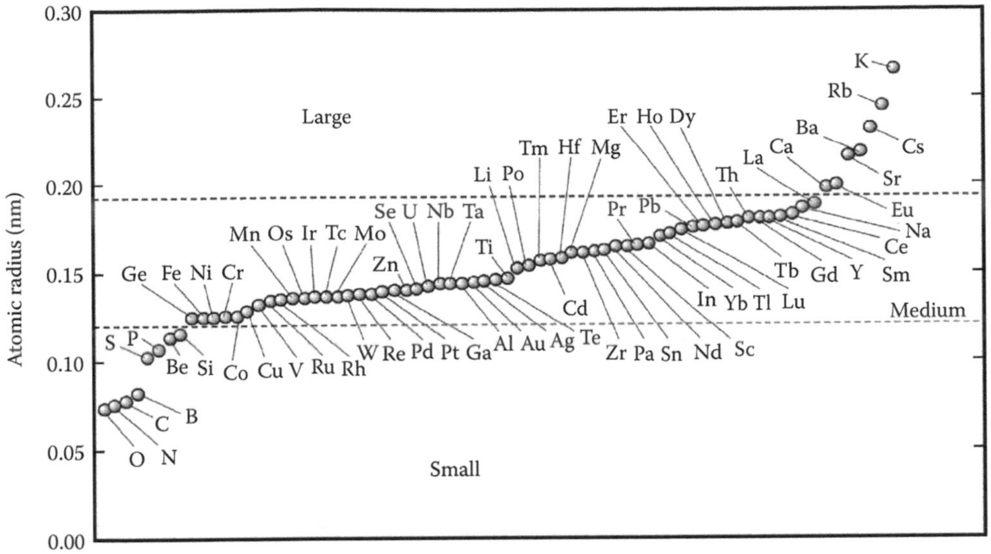

FIGURE 3.4
Atomic diameters of the elements that constitute bulk metallic glasses. These can be classified into three major groups of large, medium, and small sizes.

Mg-based BMG alloys. Table 3.3 summarizes the interatomic distances and the coordination numbers of the different atomic pairs for a glassy $Zr_{60}Al_{15}Ni_{25}$ alloy both in the as-quenched and crystallized conditions [126].

It may be noted from Table 3.3 that the nearest neighbor distances and the coordination number of atoms in the as-quenched and crystallized conditions are not very different for Zr–Ni and Zr–Zr atom pairs, testifying to the presence of dense random packing of atoms in the glassy state also. But, there is a significant change in the coordination number of Zr–Al atomic pairs on crystallization, suggesting that the local atomic configurations in the glassy state are quite different from those in the corresponding crystalline phase. This suggests that there is a necessity for long-range diffusion of Al atoms around Zr atoms during crystallization, which is difficult to achieve due to the presence of dense randomly packed clusters.

The presence of dense randomly packed atomic configurations in the glassy state of BMGs can also be inferred from the small changes in the relative densities of the fully glassy and the corresponding fully crystalline alloys (see Table 6.1). It is noted that the densities of the glassy alloys are lower than those in the crystallized state. The difference between the fully glassy and fully crystalline alloys is typically about 0.5%, but is occasionally as high as 1% (see, e.g., Ref. [127]). Further, the density difference between the structurally relaxed and fully glassy states is about 0.11%–0.15%. Thus, the small density differences between the glassy and crystallized conditions suggest that the glassy alloys contain dense randomly packed clusters in them.

3.7.1 The ΔT_x Criterion

The large width of the supercooled liquid region, $\Delta T_x = T_x - T_g$, in BMGs suggests that the glassy phase produced is very stable and that it resists crystallization. Using this as the basis, Inoue [128–130] proposed that the GFA of alloys is directly related to ΔT_x. Accordingly, it was noted that the critical cooling rate for glass formation decreases

TABLE 3.3

Nearest Neighbor Distances (r) and Coordination Numbers (N) of the Different Atomic Pairs in a Glassy $Zr_{60}Al_{15}Ni_{25}$ Alloy in Both the As-Quenched and Crystallized States

Condition		r_1 (nm)	$N_{Zr\text{-}Ni}$	r_2 (nm)	$N_{Zr\text{-}Zr}$	$N_{Zr\text{-}Al}$
As-quenched	(a)	0.267 ± 0.002	2.3 ± 0.2	0.317 ± 0.002	10.3 ± 0.7	−0.1 ± 0.9
	(b)	0.267 ± 0.002	2.1 ± 0.2	—	—	—
	(c)	0.269 ± 0.002	2.3 ± 0.2	—	—	—
Crystallized	(a)	0.268 ± 0.002	3.0 ± 0.2	0.322 ± 0.002	8.2 ± 0.7	0.8 ± 0.9
	(b)	0.267 ± 0.002	3.0 ± 0.2	—	—	—
	(c)	0.273 ± 0.002	2.3 ± 0.2	—	—	—

Source: Matsubara, E. et al., *Mater. Trans., JIM*, 33, 873, 1992 [126]. With permission.

Note: Data from (a) ordinary radial distribution function (RDF), (b) conventional RDFs for Zr, and (c) conventional RDFs for Ni. "—" means that no values were given in the original publication.

with an increase in the ΔT_x value (Figure 3.5). This early observation has been confirmed in a number of instances. However, with the increasing number of BMGs synthesized in the last few years in a variety of alloy systems and by different investigators, some exceptions to this criterion have been noted. For example, by plotting the R_c values determined for different Pd-based alloys obtained using calorimetric measurements, it was reported [19] that the $Pd_{40}Cu_{30}Ni_{10}P_{20}$ alloy with the maximum value of ΔT_x (87 K) did not correspond to the alloy composition that exhibited the lowest R_c value of 0.067 K s^{-1}. A similar result was also reported by Zhang et al. [107] who did not find a good correlation between GFA and ΔT_x in $La_xAl_{14}(Cu,Ni)_{86-x}$ alloys with $x = 57$–70 at.% Al. A similar situation was also obtained in Ti–Zr–Ni–Be–Cu BMGs synthesized by water quenching [58]. Therefore, even though it was concluded that the GFA of alloys cannot be evaluated directly in terms of ΔT_x, it can be safely stated that, qualitatively, most BMGs have a large value of ΔT_x. However, it was suggested that ΔT_x could be used only

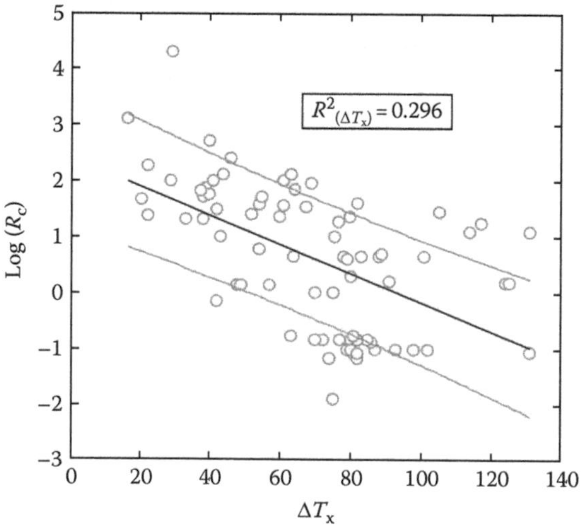

FIGURE 3.5
Variation of the critical cooling rate, R_c, with the width of the supercooled liquid region, ΔT_x, for a number of multicomponent bulk metallic glasses. Data for some of the binary and ternary metallic glasses reported earlier are also included for comparison.

to evaluate the thermal stability of the supercooled liquid. This aspect will be further discussed later.

3.7.2 The t_{max} Criterion

As emphasized in Chapter 2, a glass can be formed in an alloy system only when the critical cooling rate for glass formation is exceeded. Since the cooling rate at any point in the specimen is a function of the distance from the surface of the sample (from where the heat is extracted), the maximum diameter, t_{max}, of the rod that is solidifying into a fully glassy state is also another important parameter to evaluate the GFA. This is just like determining the hardenability of steels [11]. If rods of different diameters are solidified from the same melt temperature, then it is easy to understand that the solidification rate at the center of the rod will be higher the smaller the diameter. Therefore, if an alloy has a high GFA, that is, it is possible to obtain the glassy phase at a relatively low cooling rate, then the maximum diameter of the rod that can be made glassy should also be larger. This maximum diameter of the rod, or the section thickness, t_{max}, has been shown to increase with a decrease in the critical cooling rate for glass formation, R_c (Figure 3.6). Additionally, it is important to keep in mind that the solidification rate obtained in a sample is also dependent on the technique used to produce the glass. Keeping all the other parameters constant, the high-pressure die casting method will generate higher solidification rates than the copper mold casting method. On the other hand, the copper mold casting technique will generate higher solidification rates than the water-quenching method. Thus, while determining the maximum section thickness that could be cast into the fully glassy state, it is important to specify the technique used to synthesize the BMG sample (or the cooling rate achieved) to synthesize the BMG sample.

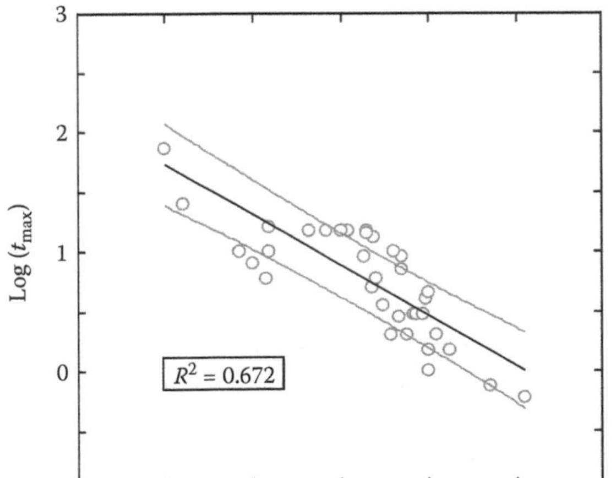

FIGURE 3.6
Relationship between t_{max}, the maximum diameter or section thickness of the glassy alloy and the critical cooling rate, R_c, required to form the glassy phase. The value of t_{max} increases with a decrease in the R_c value.

3.8 Exceptions to the Inoue Criteria

Even though the three criteria proposed by Inoue have been successfully applied to identify alloy compositions that could be produced in a bulk glassy condition, some apparent exceptions have been noted in a few alloy systems.

3.8.1 Less Than Three Components in an Alloy System

One of the apparent exceptions to this empirical rule appears to be that BMGs have been produced in binary alloy systems, such as Ca–Al [103], Cu–Hf [85], Cu–Zr [87], Ni–Nb [36], and Pd–Si [41]. Two important points need to be noted in these binary BMG alloys. One is that the maximum diameter of the glassy rods obtained in these binary alloys is relatively small, that is, a maximum of only about 2 mm. It is true of course, that any alloy greater than 1 mm in diameter is normally considered "bulk." But, the maximum diameter of the BMGs in these binary alloys is really at the low end. The second and more important point is that the "glassy" rods of the binary BMG alloys often seem to contain some nanocrystalline phases. For example, Inoue et al. [104] noted that 1 mm diameter cast $Cu_{50}Zr_{50}$ rods were found to be glassy, as indicated by the presence of a broad diffuse peak in the XRD patterns. However, high-resolution TEM studies showed that spherical nanocrystals of the Cu_5Zr phase with a size of about 5 nm were present dispersed in the glassy matrix. Thus, even small-diameter rods of binary bulk (?) "glassy" alloys are not fully glassy. (The presence of such crystals, however, has a beneficial effect in increasing the compressive ductility to about 50%. This aspect will be discussed further in Chapter 8.) A similar result was also reported by Chen et al. [131], who synthesized 1 mm diameter $Ni_{60.25}Nb_{39.75}$ glassy alloys by injecting the molten alloy into a copper mold. The XRD pattern showed that the alloy was "amorphous." But, high-resolution TEM studies again confirmed that small crystals, 10–15 nm in diameter, were present dispersed in the glassy matrix.

Even though the solidification rate achieved during cooling determines whether the rod will be fully glassy or not, the observation of crystalline phases present in some of these alloys, of even small sizes, suggests that the maximum achievable diameter of a "fully glass" rod is not really large. It may turn out that this value is only slightly larger than what is obtained by RSP methods, and therefore it is doubtful if we could designate these binary alloy glasses as BMGs. In fact, observations of the presence of nanocrystals dispersed in a glassy matrix were also made in ternary 3 mm diameter $Cu_{60}Zr_{30}Ti_{10}$ and 4 mm diameter $Cu_{60}Zr_{30}Hf_{10}$ alloys [132]. In this context, it will be useful to remember that the addition of alloying elements to these binary alloys improves the GFA, and fully glassy rods with larger diameters could be obtained. For example, as mentioned earlier, Wang et al. [26] were able to produce only 2 mm diameter rods in the binary $Cu_{50}Zr_{50}$ alloy composition; however, by adding just 4 at.% Al to this alloy, the maximum diameter of the fully glassy rod could be increased to at least 5 mm. Such observations are noted in other alloy systems as well. Therefore, even though glassy (BMG) alloys of 1 or 2 mm diameter are produced in binary alloy compositions, their GFA improves dramatically with the addition of a third component. This observation again proves that a minimum of three components is required to produce a BMG alloy with a reasonably large diameter.

As discussed earlier, one needs at least three components to synthesize BMGs. It has also been mentioned that binary BMGs frequently contain some crystalline inclusions and therefore they may not be truly glassy. Therefore, it was most surprising to see that Zhang and Zhao [133] reported that pure metal Zr could be produced in the "bulk" glassy condition. The authors had subjected polycrystalline zirconium specimens of 1.0 mm diameter and 0.5 mm thickness to high pressures of up to 17 GPa and 1000°C. The authors noted that the α-phase had transformed to the ω-phase by room temperature compression. When the sample containing the ω-phase (hexagonal structure, but not close-packed) was heated to 650°C at 5.3 GPa, amorphous zirconium had formed. Upon cooling this material, the crystallization of the amorphous zirconium into the ω-phase was observed at 4.8 GPa and 450°C. At pressures higher than 9 GPa, zirconium was only partially transformed to the amorphous phase. These conclusions were reached by observing the disappearance of Bragg peaks in the energy-dispersive XRD patterns and the time of flight neutron scattering experiments.

However, careful observations using an angular dispersive method and imaging-plate detector, together with x-ray transparent anvils have revealed that their conclusions were erroneous. They were convinced that the disappearance of the diffraction peaks in zirconium noted in their earlier investigation should be interpreted as rapid crystal growth at temperatures above that of the ω–β phase transformation. Therefore, the authors later retracted their paper [134].

Wang et al. [135] conducted high-pressure experiments on pure metal titanium and reported that amorphous titanium formed at pressures and temperatures close to the α–β–ω triple point in the pressure–temperature phase diagram for pure titanium.

Hattori et al. [136] conducted very careful high-pressure experiments on elemental Zr and Ti using a newly developed *in situ* angle-dispersive XRD using a two-dimensional detector and x-ray transparent anvils. These authors noted that despite the disappearance of all the Bragg peaks in the one-dimensional energy-dispersive data, two-dimensional angle-dispersive data showed several intense Bragg spots even at the conditions where amorphization was reported in these two metals. This investigation clearly confirms that it is very difficult to amorphize pure metals.

This investigation brings out an important point. The experimental data must be carefully analyzed to correctly interpret the condition of the material. It appears desirable to use more than one technique before viable conclusions can be drawn. In general, it is important to carefully characterize the glassy/amorphous structures, not just by XRD methods, but also by high-resolution techniques such as TEM. Further, while the XRD technique is easy and convenient to use, the resolution is limited, and only the average structure can be determined. But, the presence of a small volume fraction of a second crystalline phase, especially if it is of nanometer dimensions, may be completely missed by XRD methods. On the other hand, TEM studies are time-consuming, but they help in unambiguously identifying the presence of tiny crystalline particles of even a few nanometers in size.

3.8.2 Negative Heat of Mixing

In addition to these observations, where the presence of nanocrystals was noted in binary BMGs, phase separation was reported in ternary BMGs [137]. This suggests that negative heat of mixing between the constituent elements may not always be necessary to form BMGs. It should, of course, be pointed out that these authors had clearly mentioned that the GFA was lower when phase separation occurred in the glassy state. Since such observations have been made only in a few limited cases and phase separation was also observed in metallic glasses obtained in melt-spun ribbons, for example, in $Cu_{50}Zr_{50}$ alloys [138] and in $Zr_{36}Ti_{24}Be_{40}$ alloys [139], this need not be of major concern.

The phenomenon of phase separation will be dealt with in detail in Chapter 5, when we discuss the crystallization behavior of glassy alloys. But, let us briefly mention a couple of salient features to counter the argument that the negative heat of mixing between the constituent elements is not a requirement for the formation of BMGs. Phase separation is generally expected to occur in alloy systems containing elements that exhibit a positive heat of mixing. This is indicated by the presence of a miscibility gap in the corresponding phase diagram. Therefore, if phase separation has occurred, one immediately concludes that the constituent elements have a positive heat of mixing.

It has been suggested that it is theoretically possible to observe phase separation in alloy systems containing three or more elements, even though the heat of mixing is negative between any two elements in the alloy system. According to Meijering [140,141], a ternary alloy phase, consisting of components A, B, and C, can decompose into two phases with different compositions even when the enthalpy of mixing between any two components is negative. This is possible when the enthalpy of mixing, ΔH, for one of the three possible binary alloy systems is significantly more negative than the others. For example, it is possible that in a ternary alloy system A–B–C, $\Delta H_{A\text{–}B}$ is much more negative than $\Delta H_{B\text{–}C} \approx \Delta H_{A\text{–}C}$. This argument suggests that a miscibility gap could be present in a ternary (or higher-order) BMG alloy system even when all the constituent elements have a negative enthalpy of mixing. In other words, phase separation is possible even in an alloy with a reasonably good GFA.

3.8.3 Other Exceptions

Waniuk et al. [43] reported that in Zr–Ti–Cu–Ni–Be alloys (the Vitreloy series), the GFA correlated well with T_{rg}, but not with ΔT_x. In fact, alloy compositions with the highest ΔT_x were found to be poor glass formers. A similar situation was also reported in the ternary Cu–Zr–Ti and Cu–Hf–Ti BMGs [142], where the highest GFA could be related to T_{rg}, but not to the ΔT_x value.

On the other hand, there are also reports that suggested that the GFA of alloys was more closely related to ΔT_x, but not to T_{rg}. Shen et al. [143] reported that the GFA in their water-quenched $Pd_{40}Ni_{40-x}Fe_xP_{20}$ $(0 \leq x \leq 20)$ alloy glasses did not correlate with T_{rg} values at all. Instead, they concluded that the ΔT_x was a simple and reliable gauge for quantifying the GFA in this alloy and also in the $(Fe_{66}Cr_4Mo_4Ga_4P_{12}C_5)_{100-x/95}B_x$ system [144].

There have also been reports where the GFA of alloys did not correlate with either T_{rg} or ΔT_x. Xu et al. [87] reported the formation of metallic glasses in the binary Cu–Zr system by the copper mold casting technique and observed a strong dependence of the critical casting thickness with the solute content. They noted that the maximum casting thickness of 2 mm was achieved at 36 at.% Zr. This alloy composition did not correspond to either the largest ΔT_x value or the highest T_{rg} value. A similar result was also reported in $La_xAl_{14}(Cu,Ni)_{86-x}$ (with $x = 57$–70 at.%) alloys by Zhang et al. [107]. Tang et al [58] produced BMG rods of >50 mm in diameter in Ti–Zr–Ni–Be–Cu alloys and showed that ΔT_x and T_{rg} were not effective in predicting the GFA correctly. For example, compositions that produced 50 mm diameter rods had a smaller value of T_{rg} and ΔT_x than alloys that could be produced only in smaller-diameter sizes. BMGs with a 7 mm diameter were produced in ternary Mg–Cu–Y alloys by the high-pressure die casting method and it was noted that the critical diameter of the rod increased with increasing values of ΔT_x [145]. But, when similar alloys were melt-spun to produce thin glassy ribbons, it was noted that the GFA could not be directly correlated with any one parameter (ΔT_x or T_{rg}) for all the alloys. Different parameters were satisfying the GFA depending on the alloy composition and the wheel speed used to produce the melt-spun ribbons [146]. Similarly, Liu et al. [147] noted that in the Fe–Ni–Mo–P–C–B BMG alloys synthesized by them, the GFA could not be related to either ΔT_x, T_g, or γ (to be discussed in Section 3.10.1).

From this discussion, it is clear that the description of the GFA of alloys using the ΔT_x parameter as a criterion has not been found universally applicable in all situations and for all alloy systems. Some exceptions have been certainly noted. It should, however, be emphasized in this context that this was one of the most successful parameters in the early years of research on BMGs.

3.9 New Criteria

The observations of the preceding section summarize some of the exceptions that have been noted in relating the GFA of alloys to the criteria used earlier. Some of them are only apparent exceptions and can be resolved easily to show that they satisfy the criteria proposed earlier, while the others are real. Therefore, it was felt that some new attempts needed to be made to develop better and more precise criteria to predict the GFA of alloy systems. An early attempt in this direction was by Li et al. [148], who analyzed the thermal data of a number of BMG alloys based on Zr, Mg, La, Nd, Ti, Cu, and Fe in terms of five established GFA parameters, *viz.*, T_{rg}, ΔT_x, ΔT^* [149], K_{gl} [150], and S [151]. Among other findings, they noted that ΔT_x represented a good measure of the GFA, while T_{rg} did not. Later, Lu and Liu [152] published a paper proposing a new indicator, which they called the γ parameter, to explain the GFA of alloy systems. Since this publication, a plethora of papers have been published by different researchers in the open literature that propose a number of different criteria for determining the GFA of alloys. Many of these criteria are somewhat related to each other, with some minor differences in the details of their formulations.

Fan et al. [153] identified a four-step approach consisting of (1) deep eutectics, (2) mutual solubility, (3) atomic size and thermodynamic effects, and (4) microalloying to fabricate Ni-based BMGs. It was suggested that this approach could progressively increase the GFA of liquids by (a) increasing the atomic packing density of the liquids, (b) stabilizing the liquid structure, and (c) destabilizing the crystalline structure of the competing phases.

All the new criteria that have been proposed in recent years to explain the high GFA of BMGs can be broadly grouped into the following categories:

1. *Transformation temperatures of glasses.* In this group, the GFA is explained on the basis of the characteristic transformation temperatures of the glasses, such as T_g, T_x, and T_ℓ, and the different combinations of these three parameters. The values of all these parameters can only be obtained after the glass has been synthesized and the transformation temperatures are measured using a DSC/DTA apparatus during reheating of the glassy alloy. The parameters suggested to explain the high GFA of BMGs in this group include the parameters α, β, *new* β, γ, γ_m, δ, T_{rx}, $K_{g\ell}$, ω, ω_m, and ϕ.
2. *Thermodynamic modeling.* Thermodynamic parameters, such as heat of mixing, are used in this group to predict glass formation and evaluate GFA in a given alloy system.
3. *Structural and topological parameters.* In this group, consideration is given to the atomic sizes of the constituent elements, their electronegativity, electron-to-atom ratio, and heat of mixing. The majority of the work in this area has been due to Egami [117,154] and Miracle [155–159].

4. *Physical properties of alloys.* This group considers the physical properties of materials, such as the viscosity of the melt, heat capacity, activation energies for glass formation and crystallization, bulk modulus, and density.
5. *Computational approaches.* These methods help in predicting the GFA of alloys from basic thermodynamic data [160,161], without the necessity to actually conduct any experiments to synthesize the glass and determine the GFA.

The first two groups attracted the maximum attention of researchers and a large number of papers have been published in these areas. Let us now look at these different criteria in some detail.

3.10 Transformation Temperatures of Glasses

The majority of the early researchers in the field of metallic glasses considered only the ability of the melt to transform into the glassy state on rapid solidification and termed the efficiency of this process the *glass-forming ability* (GFA). The GFA was usually evaluated in terms of the critical cooling rate, R_c, needed to form the glassy ribbons. Subsequently, when BMGs were synthesized, researchers also considered the maximum section thickness, t_{max}, of the fully glassy rods as a measure of GFA. It is easy to appreciate that GFA is higher, the lower the critical cooling rate or the higher the section thickness. However, Lu and Liu [152] argued that glass formation involves two different aspects. One is the *stability of the liquid phase* and the other is the *resistance to crystallization* of the glass that has formed. Although the actual term *glass-forming ability* should only refer to the formation of the glass, as was done by the earlier researchers, it is also important to consider the thermal stability of the glass. For example, it was reported that the first (Au–25 at.% Si) metallic glass, produced by Klement Jr. et al. [7] using RSP, had completely crystallized at room temperature in 24 h. Based on these two aspects, these authors developed a new parameter, γ, as an indicator to evaluate the GFA of alloys. Subsequent to this publication, a very large number of other parameters have also been proposed as measures of GFA. These later parameters also involve different combinations of the glass transformation temperatures. We will now discuss the different parameters and also comment on their suitability in different situations.

3.10.1 The γ Parameter

Figure 3.7 presents the schematic *T–T–T* diagrams for two different alloys—one that can be easily produced in the glassy state (Alloy #1), and another that requires a higher solidification rate to produce the glassy phase (Alloy #2) [162]. As mentioned earlier, glass formation requires that the alloy melt is cooled (from above the liquidus temperature, T_ℓ) fast enough (above the critical cooling rate) through the glass transition temperature, T_g, without intersecting the *T–T–T* curve. That is, the cooling curve should not touch the nose of the *T–T–T* curve; it needs to be avoided. Thus, the GFA of the melt is directly related to the location of the *T–T–T* curve on the temperature and time coordinates. The temperature corresponding to the nose of the *T–T–T* curve, T_n, is located on the temperature axis at $\alpha(T_g + T_\ell)$, where the value of α varies between 0.45 and 0.55, and it is generally close to 0.5, that is, $T_n \cong (1/2)(T_g + T_\ell)$.

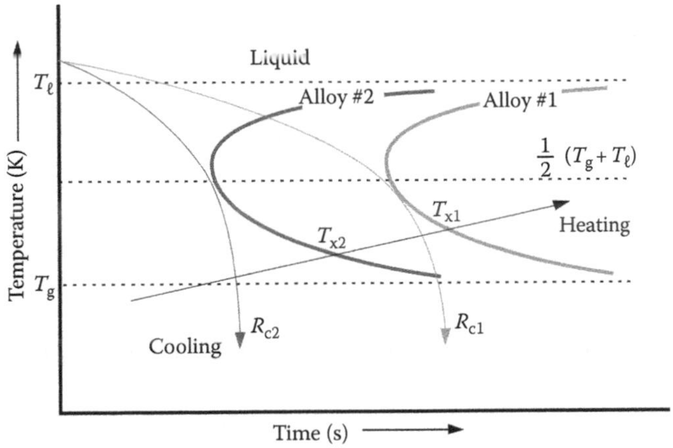

FIGURE 3.7
Schematic *T–T–T* curves for two different alloys. While Alloy #1 can be easily produced in a glassy state, it is more difficult to produce Alloy #2 in the glassy state. That is, the critical cooling rate, R_c, is higher for Alloy #2 than for Alloy #1, $R_{c2} > R_{c1}$. Similarly, on continuous heating, the crystallization temperature, T_x, for Alloy #1 becomes higher than for Alloy #2, that is, $T_{x1} > T_{x2}$. (Reprinted from Lu, Z.P. and Liu, C.T., *Phys. Rev. Lett.*, 91, 115505-1, 2003. With permission.)

When the glass is reheated from below T_g to higher temperatures, the sample will start crystallizing at a temperature T_x. For the two different alloys considered, it may be noted that T_x is higher for Alloy #1 than for Alloy #2. Thus, the stability of the glass is determined by the time it requires to start crystallizing, that is, the position of the *T–T–T* curve on the time coordinate, t_n.

From a physical metallurgy point of view, the stability of the liquid phase can be considered in terms of the thermodynamic (or equilibrium) stability of the liquid above the liquidus temperature, T_ℓ, and the stability of the liquid during cooling, that is, in the metastable state. Thus, one should consider the interplay between both T_ℓ and T_g. For two glass-forming melts that have the same T_g, but different T_ℓ, the relative liquid phase stability is determined by their T_ℓ values. Therefore, one can consider that the melt is thermodynamically more stable if the liquidus temperature, T_ℓ, of the alloy is lower. On the other hand, when the two melts have the same T_ℓ, but different T_g, then their liquid phase stability is determined by their metastable states. Consequently, the melt is more stable the lower the T_g value. However, if the two melts have different T_ℓ and T_g values, which is likely to be the more common case, then the overall stability of the melt can be measured by (1/2)($T_g + T_\ell$), which represents the average stability of the melt both in the stable and metastable states. Thus, the lower this value, the higher the stability of the glass-forming melt, and consequently the higher the GFA.

The resistance to crystallization, on the other hand, is determined by the value of the crystallization temperature, T_x. It is easy to appreciate that the stability of the glass is higher, the higher the T_x. That is, T_x alone will be useful, only if all liquids have the same phase stability. Stated differently, for liquids with the same stability, that is, the same value of (1/2)(T_g + T_ℓ), the stability of the glass is represented by T_x. But, if the stability of the liquids is different, T_x alone will not be able to represent the stability of the glass. Then, to evaluate the relative GFA of different liquids, T_x should be normalized with respect to the average position of the *T–T–T* curve on the temperature axis.

Combining these two stabilities into one parameter, Lu and Liu [152] proposed the use of the γ parameter to evaluate the GFA of different compositions as

$$\gamma \propto T_x\left[\frac{1}{2(T_g+T_\ell)}\right] \propto \frac{T_x}{T_g+T_\ell} \tag{3.9}$$

These authors have subsequently extended this concept to oxide glasses and cryoprotective solutions and demonstrated that the same γ parameter could be used to explain the GFA of these glasses also [162]. It was thus claimed that this γ parameter could be used to explain the GFA of any type of glass. The concept in deriving the γ parameter to explain the GFA of alloys is summarized in Figure 3.8 [163].

Figure 3.9 shows the correlations between the γ parameter and the critical cooling rate (R_c), in Kelvin per second (K s^{-1}), and the maximum section thickness (t_{max}), in mm. A linear relationship was obtained between γ and $\log_{10}R_c$ or $\log_{10}t_{max}$. The relationships between $\log_{10} R_c$ and γ were expressed as

$$\log_{10}R_c = (21.71 \pm 1.97) - (50.90 \pm 0.71)\gamma \tag{3.10}$$

or

$$R_c = R_0\exp\left[\left(-\frac{\ln R_0}{\gamma_0}\right)\gamma\right] \tag{3.11}$$

where R_0 and γ_0 are constants with the values of 5.1×10^{21} K s^{-1} and 0.427, respectively. These values will be different for different types of glasses. While these values are for metallic glasses, they will be different for oxide glasses or for cryoprotective glasses.

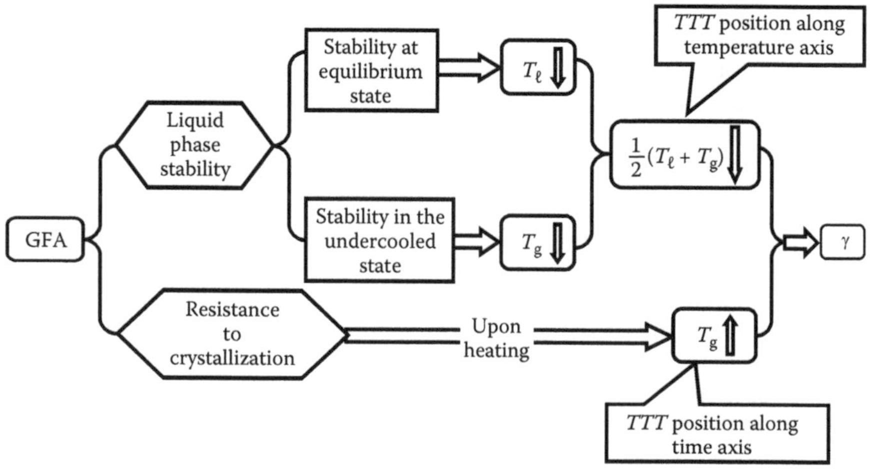

FIGURE 3.8
Schematic to illustrate the different factors involved in deriving the γ parameter to explain the GFA of alloys. (Reprinted from Lu, Z.P. and Liu, C.T., *Intermetallics*, 12, 1035, 2004. With permission.)

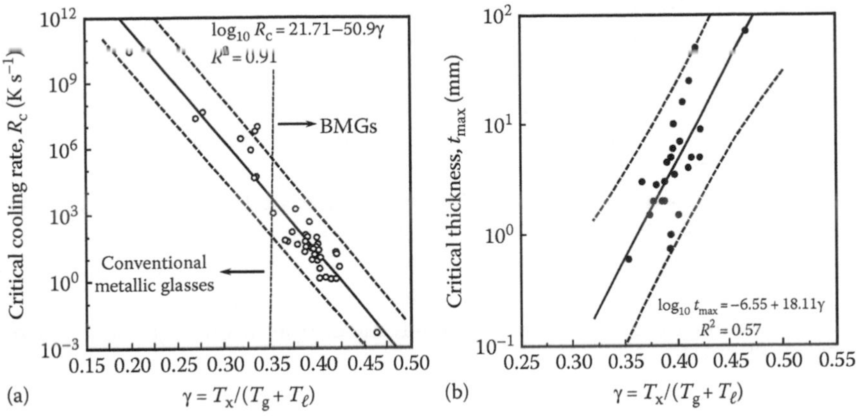

FIGURE 3.9
(a) Correlation between the critical cooling rate (R_c) and the γ parameter for BMGs. (b) Correlation between the maximum section thickness (t_{max}) and the γ parameter for BMGs. (Reprinted from Lu, Z.P. and Liu, C.T., *Acta Mater.*, 50, 3501, 2002. With permission.)

In Equation 3.11, R_0 represents the critical cooling rate for a material with $\gamma=0$, and γ_0 is the corresponding γ value for a material having a critical cooling rate of 1 K s^{-1}. Using these values, the expression reduces to

$$R_c = 5.1\times10^{21}\exp(-117.2\gamma) \tag{3.12}$$

(Equation 3.12 was represented with slightly different values in Ref. [163].) Similarly, the linear relationship between γ and $\log_{10}t_{max}$ was expressed as

$$\log_{10}t_{max} = (-6.55\pm1.07)+(18.11\pm0.70)\gamma \tag{3.13}$$

or

$$t_{max} = t_0\exp\left[\left(-\frac{\ln t_0}{\gamma_1}\right)\gamma\right] \tag{3.14}$$

where t_0 (in mm) and γ_1 are constants with the values of $t_0=2.80\times10^{-7}$ mm and $\gamma_1=0.362$. The t_0 parameter represents the attainable maximum diameter for a material with $\gamma=0$, and γ_1 represents the γ value for a material that can form glass of at least 1 mm in section thickness or diameter. Substituting these values into Equation 3.14, we get

$$t_{max} = 2.80\times10^{-7}\exp(41.7\gamma) \tag{3.15}$$

The statistical correlation parameters, R^2 values, for these relationships were computed using regression programs and were found to be 0.91 for R_c and 0.57 for t_{max}, respectively. The R^2 value can range from 0 to 1 and is an indicator of the reliability of the correlation. The higher the R^2 value, the more reliable the regression is. Even though the correlation between γ and R_c is apparently very good, it should be kept in mind that not many researchers have actually determined the R_c values for the metallic glasses they have produced. In the majority of the cases, these values have been estimated from some empirical relationships, as described in Section 3.2.

The wide scatter for the t_{max} correlation (and consequently lower R^2 value) was attributed to the different section thicknesses obtained by the different techniques used to synthesize BMGs. For example, as will be discussed in detail in Chapter 4, BMGs can be synthesized by many different techniques, such as water quenching, suction casting, and copper mold casting. The use of these different techniques can account for the differences in the section thicknesses of BMGs of the same composition. For example, the t_{max} value for $Mg_{65}Cu_{25}Gd_{10}$ was reported to be 8 mm if a copper mold wedge casting method was used to produce the BMG [17]. But, its value decreased to 6 mm when the copper mold injection casting method was used [164]. These results are from two different groups of investigators. For an appropriate and fair comparison, it would be ideal if the same group of researchers produced these alloys by different methods and evaluated the GFA. Further, the section thicknesses of the BMGs are almost always reported in integer numbers, and this need not be true in a real situation. Both these factors can account for the larger deviation in the t_{max} correlation and, therefore, the low value of R^2.

A point of interest is that the γ parameter for an alloy composition can be determined from one single DSC curve, from which one can get the values of T_g, T_x, and T_ℓ. However, these values also depend on the heating rate employed during the DSC run. These characteristic transformation temperatures are higher when the heating rate employed during measurement is higher. Therefore, it becomes important to use the data obtained at the same heating rate to draw comparisons. Fortunately, however, γ is not too sensitive to heating rates.

As is well known, the GFA of alloys is dependent on the alloy composition, the technique used to synthesize the glass, the cleanliness of the melt, the presence of oxygen (especially in Zr-based and other reactive-metal-based BMGs), and other process variables. Subsequent to the proposal of the γ parameter to explain the GFA of alloys, several other parameters have also been proposed. Lu et al. analyzed the available results in terms of these different parameters reported by different authors [163,165] and came to the conclusion that all the available results on GFA could be most satisfactorily explained on the basis of the γ parameter.

While analyzing glass formation and GFA in the $La_{100-x}[Al_{0.412}(Cu,Ni)_{0.588}]_x$ (with $x = 30$–44.6 at.%) system, Lu et al. [166] noted that, while γ was capable of representing the GFA of most alloys in the eutectic systems, it could be less accurate in assessing the GFA of compositions whose decisive competing crystalline phase during cooling was different from that upon the crystallization of the metallic glass. In the latter cases, the measured T_x value upon devitrification was not sufficient to indicate the crystallization resistance of the real competing phase during cooling.

Fan et al. [167,168] investigated the effect of quenching temperature on the GFA of Zr-based Zr–Cu–Au (or Pt)–Al glassy alloy ribbons on thermophysical properties and noted that T_g, T_x, and ΔT_x increased with increasing melt quenching temperature. They also reported that the GFA increased with increasing melt quenching temperature, as indicated by the increased γ value. Further, the γ criterion was found to better reflect the GFA than the T_{rg} value.

Since the announcement of the γ parameter to evaluate the GFA of alloys by Lu and Liu [152,162], a number of other investigators have come up with several other different criteria. The efficiency of these criteria to explain the GFA was statistically analyzed by all the proposers—all claimed that their criterion was better than others, and almost all claimed that their criterion was *"the best."* Let us now look at the other parameters that were proposed to explain the enhanced GFA of BMGs.

3.10.2 The γ_m Parameter

In an attempt to improve the correlation between γ and the transformation temperatures of BMGs, Du et al. [169] also considered the width of the supercooled liquid region, $\Delta T_x = T_x - T_g$, in addition to T_ℓ and T_x in their analysis and came up with a modified parameter, γ_m, to explain the high GFA of BMG alloys. The parameter, γ_m, is expressed as

$$\gamma_m = \frac{2T_x - T_g}{T_\ell} \tag{3.16}$$

By plotting γ_m as a function of the critical cooling rate, R_c, and conducting regression analysis, these authors derived a linear relationship between $\log_{10} R_c$ and γ_m, which was stated as

$$\log_{10} R_c = 14.99 - 19.441\gamma_m \tag{3.17}$$

The statistical correlation factor, R^2, was calculated by these authors as 0.931. In comparison to the other criteria such as T_{rg}, ΔT_x, and γ, the authors claimed that the γ_m parameter gave the highest R^2 value, and consequently, they claimed that the *γ_m parameter exhibited the best correlation with GFA of all the parameters suggested so far.*

3.10.3 The α and β Parameters

Arguing along lines similar to those used to derive the γ and γ_m parameters, Mondal and Murty [170] considered T_ℓ to be a measure of the stability of the liquid phase, and T_x a measure of the thermal stability of the glass that had formed. In other words, a high T_x relates to the higher stability of the glass and a low T_ℓ corresponds to the higher stability of the supercooled liquid. By combining these two aspects of glass formation, they defined a new parameter, α, as

$$\alpha = \frac{T_x}{T_\ell} \tag{3.18}$$

Since ΔT_x is defined as $T_x - T_g$, Equation 3.18 may be rewritten as

$$\alpha = \frac{T_x - T_g + T_g}{T_\ell} = \frac{\Delta T_x}{T_\ell} + \frac{T_g}{T_\ell} \tag{3.19}$$

which includes both the stability of the glass ($\Delta T_x/T_\ell$) and the GFA, T_g/T_ℓ ($=T_{rg}$), as proposed by Turnbull [79].

Incidentally, Wakasugi et al. [171] suggested earlier that the T_x/T_ℓ ratio increases with increasing viscosity of the supercooled liquid, fusion entropy, activation energy for viscous flow, heating rate employed during heating of the glassy alloy, and with decreasing T_ℓ value. Since all these parameters affect the critical cooling rate, R_c, for the formation of the glassy phase, it was suggested that T_x/T_ℓ is a reasonable indicator of the GFA based on the crystallization behavior of an undercooled liquid. That is, a system with a larger T_x/T_ℓ

value could have a lower R_c, and thereby a higher GFA. In fact, a linear relationship was found between T_x/T_ℓ and $\log R_c$ [171].

Mondal and Murty [170] also proposed another parameter β, by considering the two important aspects of glass formation, *viz.*, the ability to form a glass during cooling from the liquid state ($T_{rg}=T_g/T_\ell$), and the stability of the glass (T_x/T_g) as

$$\beta = \frac{T_x}{T_g} + \frac{T_g}{T_\ell} \tag{3.20}$$

Should a glass not exhibit T_g during heating to higher temperatures (as is often the case in melt-spun glassy ribbons), T_x may be taken as T_g. Therefore, when $T_g=T_x$, Equation 3.20 can be rewritten as

$$\beta = 1 + \frac{T_x}{T_\ell} = 1 + \alpha \tag{3.21}$$

Thus, both the parameters, α and β, contain the two important aspects of the ease of glass formation as well as the thermal stability of the glass.

Plotting the α parameter against the logarithm of the critical cooling rate, R_c, and the maximum section thickness, t_{max}, the authors noted a linear relationship in both cases that can be expressed as

$$R_c = A\exp(B\alpha) \tag{3.22}$$

and

$$t_{max} = C\exp(D\alpha) \tag{3.23}$$

with the values of $A=1.83\times10^{16}$ K s^{-1}, $B=-54.14$, $C = 1.39 \times 10^{-4}$ mm, and $D=16.18$. Similar equations were also derived for the variation of R_c and t_{max} with the β parameter, but with different values for the constants. These authors calculated the R^2 values for both the α and β parameters and noted that the β parameter gave an R^2 value of 0.93, which was *the highest among all the GFA criteria* (including α, γ, ΔT_x, and T_{rg}). In decreasing order of importance, the effective parameters were γ, α, T_{rg}, and ΔT_x. As with the γ parameter, the correlation with the maximum section thickness, t_{max}, for the α and β parameters was also poor.

Mondal and Murty [170] also noted that the heating rate employed during reheating of the glass to obtain the transformation temperatures did not significantly affect the α and β parameters previously proposed. The biggest advantage of these two parameters appears to be that these criteria could be used to predict bulk glass-forming compositions from the results obtained on melt-spun ribbons. This is because one can measure the T_x and T_ℓ on melt-spun ribbons, evaluate the α and β parameters, and use this information to identify compositions with the highest GFA. Further, it has frequently been reported and repeatedly confirmed that the transformation temperatures are not a function of the section thickness. That is, these temperatures are the same whether measured on melt-spun ribbons or BMG rods. Alloys with the compositions determined to have the highest value of α or β using the melt-spun ribbons can then be made into BMGs.

3.10.4 The δ Parameter

Realizing that glass formation requires that the rates of nucleation and growth of the crystalline phase are suppressed (or extremely low), to completely avoid the formation of a crystalline phase, and that these two processes are dependent on the degree of undercooling the melt experiences, Chen et al. [172,173] argued that the GFA of alloys should be inversely proportional to the rates of nucleation and growth. Accordingly, they noted that GFA $\propto [T_g/(T_\ell - T_g)]$. Further, examining the dependency of GFA on the viscosity of the melt, they also noted that GFA $\propto T_x/T_g$. Combining these two, they proposed that the GFA of an alloy should be determined by a parameter, δ, defined as

$$\delta = \frac{T_x}{T_\ell - T_g} \tag{3.24}$$

Chen et al. have also conducted a statistical analysis of the T_{rg}, γ, and δ parameters against the maximum section thickness (diameter), t_{max}, of the glassy alloys produced (not the logarithmic value of t_{max} as other researchers had done) and noted that the R^2 *value was the highest (for δ) among all the parameters,* and therefore strongly represented the GFA of the alloys. The R^2 value for the δ parameter was calculated as 0.46, as against 0.3 for both the T_{rg} and γ parameters.

3.10.5 The T_{rx} Parameter

Since the γ parameter was not able to predict the GFA accurately in at least some alloy systems [24,174,175], Kim et al. [176] came up with the concept of reduced crystallization temperature, T_{rx}, defined as

$$T_{rx} = \frac{T_x}{T_s} \tag{3.25}$$

where T_s represents the onset temperature of solidification. While the value of T_x can be obtained during reheating of the glassy alloy in either a DSC or DTA apparatus, the value of T_s can be obtained as the solidification start temperature during cooling of the liquid at a cooling rate *R*, lower than the critical cooling rate, R_c. The value of T_x can be obtained as the point of intersection of the *T–T–T* curve with the heating rate curve during continuous heating. Thus, the T_{rx} value is dependent upon both the heating and cooling rates employed.

The way T_{rx} is measured is as follows. First, at a given cooling rate, the T_s value is measured, as the temperature at which solidification starts, as the point of intersection of the cooling curve with the *T–T–T* curve. This will also correspond to a time t_{so}. The corresponding isochronal crystallization onset temperature, T_{xo}, value at t_{so} is then directly obtained from the heating-rate dependence of T_x. Then the reduced crystallization temperature, T_{rx}, is obtained as $T_{rx} = T_{xo}/T_s$.

As stated, both T_s and T_x depend on the cooling rate and heating rate employed, respectively. At extremely slow cooling rates, T_s will merge with T_ℓ, the liquidus temperature, and it decreases with an increasing solidification rate. The value of T_x is extremely low at low heating rates and increases with increasing heating rate. With a decrease in heating rate, T_x will move more closely to T_g, since T_x is more strongly dependent on the heating rate than

T_g. Thus, the highest value that the T_{rx} parameter can have is 1, at the highest cooling and heating rates, at the nose of the *T–T–T* curve. Under extremely slow cooling and heating conditions, $T_{rx} = T_g/T_\ell$, which is nothing but the reduced glass transition (T_{rg}) temperature. Hence, the T_{rx} parameter will always lie in the range $T_{rg} \leq T_{rx} \leq 1$. Also, it is important to note that the higher the T_{rx} parameter, the higher the GFA of the alloy.

These authors had analyzed the results of T_{rx} on nine different alloy systems with a wide range of GFA, as expressed by the critical cooling rate varying over three to four orders of magnitude and the maximum section thickness varying by two orders of magnitude, and came to the conclusion that T_{rx} exhibits the best correlation with R_c and t_{max}, that is, the GFA of the alloys. In fact, this parameter was found to represent the GFA better than the ΔT_x, T_{rg}, and γ parameters.

3.10.6 The K_{gl} Parameter

By analyzing the DTA plots of glasses based on As, Te, and Cd, a simple method was developed by Hrubý [150] to evaluate the GFA of alloys. Arguing that the temperature interval $T_x - T_g$ is directly proportional and $T_m - T_x$ is indirectly proportional to the glass-forming tendency of the melt, they proposed a numerical measure of the glass-forming tendency, K_{gl}, defined as

$$K_{gl} = \frac{T_x - T_g}{T_m - T_x} \tag{3.26}$$

where T_m is the temperature at which melting starts (solidus temperature). The author suggested that it is difficult for the glass to form when $K_{gl} = 0.1$; a solidification rate of about 10^2 K s^{-1} is required. Higher values of K_{gl} translate to lower cooling rates. It was mentioned that free cooling of the melt in air is sufficient to produce the glass if $K_{gl} = 0.5$. Note, however, that the numerator in this equation, $T_x - T_g$ ($=\Delta T_x$), is the width of the supercooled liquid region, and thus, the GFA is higher, the higher the ΔT_x value.

3.10.7 The ϕ Parameter

Fan et al. [177] noted that experimentally determined R_c values did not correlate well with the T_{rg} parameter proposed by Turnbull [79] and concluded that the T_{rg} parameter may not be reliable in predicting the GFA of alloys. To solve this anomaly, they employed a model glass-forming system, with a constant melting temperature of 1000 K, and viscosity, η, obeying the Vogel–Fulcher–Tammann (VFT) relationship:

$$\eta(T) = \eta_0 \exp\left(\frac{D^* T_0}{T - T_0}\right) \tag{3.27}$$

where:

- η_0 is a constant (~10^{-5} Pa s)
- D^* is the fragility index, which varies from 5 to 100
- T_0 is the temperature at which the liquid ceases to flow [178]

In the model glass system, the authors chose different fragility index values to represent the strength of the glass (the glass is considered fragile when the D^* value is low and strong when the D^* value is high) and different T_{rg} values. By calculating the viscosity of the liquids using Equation 3.27 and plotting it as a function of temperature for liquids with different fragility indices and T_{rg} values, they noted that a fragile liquid did not necessarily exhibit a low T_{rg} value. Further, over widely ranging values of undercooling, the viscosity was found to be lower for an alloy with a high T_{rg} parameter, and consequently, the R_c value was higher in comparison to a liquid that had a lower T_{rg} parameter (and also a lower R_c value). That is, the T_{rg} had severely overestimated the GFA of the liquid. In some instances, the reverse situation was observed, that is, T_{rg} underestimated the GFA of the liquid. Since the variation of viscosity with temperature is determined by the fragility of the liquid and the liquid stability (indicated by ΔT_x, the width of the supercooled liquid region), Fan et al. [177] introduced a variable (ΔT_x normalized by T_g) to correct this overestimation (or underestimation) of T_{rg} and to properly predict the GFA. Hence, they proposed a parameter, ϕ, expressed as

$$\phi = T_{rg}\left(\frac{\Delta T_x}{T_g}\right)^{0.143} \tag{3.28}$$

to correctly predict the GFA of different alloys. The validity of the ϕ parameter was established by plotting the available R_c values for network (oxide, fluoride, and semiconductor), selected metallic, and molecular glasses against the ϕ parameter and noting a linear relationship, described by the equation:

$$R_c = 10^{8.638\pm0.475-(18.0\pm1.055)\phi} \tag{3.29}$$

3.10.8 The *New* β Parameter

Yuan et al. [179] conducted a detailed analysis of the several hundreds of BMGs synthesized in different alloy systems and noted that the existing criteria (*viz*., α, β, γ, δ, γ_m, ϕ, etc.) for predicting GFA were not satisfactory. They came up with a different criterion that they called the β parameter (note that this β parameter is different from the β parameter proposed by Mondal and Murty [170]) and therefore we will designate this the *new* β parameter, defined as

$$New\,\beta = \frac{T_x \times T_g}{(T_\ell - T_x)^2} \tag{3.30}$$

By performing statistical analysis of the available t_{max} data against T_{rg}, γ, δ, ϕ, and the *new* β parameters, Yuan et al. [179] noted that *the new* β *parameter had the strongest ability to represent the GFA, since the correlation coefficient was the highest among all the parameters studied.* It should, however, be noted in this context that these authors had plotted the values of t_{max} (and not their logarithmic values) to calculate the correlation coefficients. The linear relationship between the *new* β and t_{max} was expressed by the equation:

$$t_{max} = (4.555 \pm 0.270)\,new\,\beta - (7.644 \pm 0.822) \tag{3.31}$$

An important point highlighted by the authors was that the *new* β parameter had a very wide range, while the other parameters had a very narrow range for the widely different critical diameters of BMG specimens. For example, the critical diameters for $Pd_{40}Cu_{30}Ni_{10}P_{20}$ and $Pd_{40}Ni_{40}P_{20}$ BMGs were reported to be 72 [96] and 25 mm, [180], respectively. While the *new* β parameters for these two alloys were calculated to be 11.65 and 3.866, respectively, the other parameters calculated were very close to each other. For example, the γ parameters were 0.464 and 0.424, respectively. Thus, the authors concluded that the *new* β parameter was more sensitive to GFA than the other proposed parameters.

Table 3.4 lists the correlation parameters obtained by Yuan et al. [179] for different alloy systems and for different GFA parameters. It may be noted that, for all the parameters, the highest correlation was obtained for the Pd-based BMGs, while the lowest correlation was found for the Zr-based BMGs. Another interesting point to be noted is that, irrespective of the different criteria, the order of the different correlation parameters was almost the same. Thus, except for quantitative differences, it is difficult to decide which of these parameters best represents the GFA of alloys.

3.10.9 The ω Parameter

Long et al. [181,182] proposed yet another criterion to explain the GFA of alloys. They argued that a glass-forming liquid should have a higher liquid phase stability, and hence higher GFA, if the T_n temperature, defined as the average of T_g and T_ℓ, that is, $T_n = (T_g + T_\ell)/2$, is lower. Note that this T_n corresponds to the temperature at the nose of the C-curve as mentioned earlier. Since different liquids will have different T_g and T_ℓ values, T_n was normalized with respect to T_g and, therefore, it was suggested that GFA should be proportional to T_g/T_n. Combining this with the suggestion of Inoue that the

TABLE 3.4

Values of Linear Correlation Coefficients Corresponding to Plots of t_{max} against T_{rg}, γ, δ, ϕ, and the *New* β Parameters for Different Alloy Systems

	Correlation of t_{max} with						
Base Metal	T_{rg}	γ	δ	ϕ	*New* β	**Number of Samples**	$R_{min}^{95\%}$
Cu	0.508	0.773	0.682	0.719	0.789	51	<0.372
Ca	0.575	0.748	0.673	0.738	0.752	52	0.354
Mg	0.541	0.589	0.630	0.596	0.640	40	<0.418
La	0.677	0.660	0.732	0.673	0.759	27	0.487
Ti	0.773	0.756	0.799	0.712	0.809	20	0.561
Pd	0.941	0.965	0.971	0.957	0.981	9	0.798
Zr	0.765×10^{-4}	−0.144	−0.064	−0.135	−0.142	23	0.413
Fe	−0.253	−0.3323	−0.292	−0.232	−0.315	25	0.413
Ni	−0.025	0.330	0.064	0.376	0.172	16	0.497

Source: Yuan, Z.-Z. et al., *J. Alloys Compd.*, 459, 251, 2008. With permission.

Note: $R_{min}^{95\%}$ is the minimum value of the linear correlation coefficient corresponding to the 95% confidence level.

GFA should be proportional to ΔT_x, the width of the supercooled liquid region, a new dimensionless parameter, ω, was developed by these authors, defined as

$$\omega = \frac{T_g}{T_x} - \frac{2T_g}{T_g + T_\ell} \tag{3.32}$$

It was claimed that this new criterion showed the strongest correlation, of all the parameters, with R_c and t_{max} in a number of BMG compositions.

3.10.10 The ω_m Parameter

Błyskun et al. [183] analyzed the critical diameters of four Zr-based glasses in terms of a number of the existing GFA parameters (ΔT_x, T_{rg}, K_{gl}, α, β, γ, γ_m, δ, and others) and noted the correlation coefficients were poor in some cases and better in others. Of all the parameters analyzed, γ_m was shown to have the highest correlation with the critical diameter. By tweaking the parameters, the authors were able to come up with a different criterion, designated ω; however, we will designate this as modified ω, or ω_m, since Long et al. [181,182] already described a ω parameter. ω_m is defined as

$$\omega_m = \frac{2T_x - T_g}{T_x + T_\ell} \tag{3.33}$$

This parameter gave the best correlation coefficient of 0.882. It was also shown that the solidus temperature was less highly correlated with the GFA and so it was suggested that this should not be used as a parameter to evaluate GFA.

3.10.11 The θ Parameter

Zhang and Chou [184] analyzed the transformation temperatures for 53 different alloys and found that the current criteria did not satisfactorily explain the GFA of alloys. Consequently, they proposed a new criterion, called the θ criterion, defined as

$$\theta = \frac{(T_x + T_g)}{T_\ell} \times \left[\frac{(T_x - T_g)}{T_\ell}\right]^{\alpha} \tag{3.34}$$

where α is a positive parameter that can be obtained by conducting a regression analysis on the experimental data. Through linear regression analysis of $\ln R_c$ and θ, the value of α was determined as 0.0728. The authors validated the effectiveness of this parameter by evaluating the GFA in terms of R_c and found that there exists a linear relationship between $\ln R_c$ and the parameter θ, with a coefficient of determination $R^2 = 0.942$. The authors also analyzed the experimental data in terms of the other parameters ϕ, γ, T_{rg}, new β, and δ, and concluded that "the θ criterion has the best ability for evaluating the GFA of alloys."

3.10.12 The ζ Parameter

Arguing again that the GFA is related to both the liquid phase stability and the resistance to crystallization of the glassy phase, Du and Huang [185] proposed a new parameter to evaluate the GFA. This parameter, referred to as the ζ *parameter*, is defined as

$$\xi = \frac{T_g}{T_\ell} + \frac{\Delta T_x}{T_x} \tag{3.35}$$

It was claimed that this ζ parameter could be used to predict the GFA of metallic glasses, oxide glasses, and cryoprotective glasses. In comparison to other criteria, such as T_{rg}, ΔT_x, α, γ, and δ, the authors claimed that the ζ parameter had the best R^2 values for all types of glasses and that it was simple and user-friendly.

3.10.13 An Overview of All the Criteria Based on Transformation Temperatures

As already discussed, a very large number of parameters have been proposed to explain the formation and thermal stability of BMGs produced in a large number of alloy systems. These parameters are summarized in Table 3.5. Almost all the proposers claimed that the parameter proposed by them was the best on the basis of the data (available to them!). It is a pity that the data listed and the data in the references quoted do not match in some cases, and sometimes the data quoted were wrong. Further, new data are continuously being generated in different laboratories around the world. If you are interested, Refs. [152,157,173,179,181,182,184], among others, list a large amount of data on the transformation temperatures of BMGs for further analysis. But, it should be cautioned here that not all the authors have actually collected the data from the original sources. Therefore, in some cases, one may find the data to be erroneous.

Therefore, to make a valid comparison of the efficiency of the different criteria that were proposed to explain the GFA of BMG alloys, all the data available till the end of 2007 were collected (from the original sources) and analyzed in terms of all the parameters proposed by that time, *viz.*, T_{rg}, ΔT_x, α, β, γ, γ_m, δ, ϕ, and *new* β. The statistical correlation functions are listed in Table 3.6 for the different alloy systems and all the data put together [186]. On the basis of this analysis, it was concluded that a single parameter cannot satisfactorily explain the GFA of BMGs in *all* the alloy systems. Different parameters were able to give a strong correlation for different alloy systems. Further, a very strong correlation was observed only in the case of Au-, Co-, Mg-, Nd-, Pd-, and Pr-based BMGs (with $R^2 > 0.8$), and a reasonably strong correlation in Ca-, Ce-, Cu-, La-, and Ti-based BMGs (with $0.4 < R^2 < 0.8$), and very poor correlation in Fe-, Ni-, Pt-, and Zr-based BMGs (with $R^2 \leq 0.1$).

A few specific points need to be highlighted at this stage. First, all the parameters discussed consider both the stability of the liquid phase and the resistance of the glassy phase to crystallization. Therefore, all of these parameters should be giving similar results, but they do not. The majority of the parameters have relevance and validity only in some cases. This may be attributed, at least partially, to the data that were chosen for analysis by the individual authors.

Second, from a kinetic point of view, for a highly glass-formable alloy, T_g should be as high as possible and T_ℓ should be as low as possible. This will provide a very high T_{rg} value. But, in some cases, for example, melt-spun ribbons, T_g may not be observed and, therefore, one could consider that $T_x = T_g$. Extending this argument further, we can state

TABLE 3.5
Summary of the Quantitative Criteria Proposed to Evaluate the Glass-Forming Ability of Liquid Alloys and Their Values under the Ideal Situation When $T_g = T_x = T_\ell$

Criterion/Parameter	Equation	References	Ideal Value
Reduced glass transition temperature	$T_{rg} = \frac{T_g}{T_\ell}$	[79]	1.0
ΔT_x parameter	$\Delta T_x = T_x - T_g$	[128]	0.0
α parameter	$\alpha = \frac{T_x}{T_\ell}$	[170]	1.0
β parameter	$\beta = 1 + \frac{T_x}{T_\ell} = 1 + \alpha$	[170]	2.0
New β parameter	$\beta = \frac{T_x \times T_g}{(T_\ell - T_x)^2}$	[179]	∞
γ parameter	$\gamma = \frac{T_x}{T_g + T_\ell}$	[152,162]	0.5
γ_m parameter	$\gamma_m = \frac{2T_x - T_g}{T_\ell}$	[169]	1.0
δ parameter	$\delta = \frac{T_x}{T_\ell - T_g}$	[173]	∞
K_{gl} parameter	$K_{gl} = \frac{T_x - T_g}{T_m - T_x}$	[150]	∞
ϕ parameter	$\phi = T_{rg}\left(\frac{\Delta T_x}{T_g}\right)^{0.143}$	[177]	0.0
T_{rx} parameter	$T_{rx} = \frac{T_x}{T_s}$	[176]	1.0
ω parameter	$\omega = \frac{T_g}{T_x} - \frac{2T_g}{T_g + T_\ell}$	[181,182]	0.0
ω_m parameter	$\omega_m = \frac{2T_x - T_g}{T_x + T_\ell}$	[183]	0.0
θ parameter	$\theta = \frac{(T_x + T_g)}{T_\ell} \times \left[\frac{(T_x - T_g)}{T_\ell}\right]^{\alpha}$	[184]	0.0
ζ parameter	$\xi = \frac{T_g}{T_\ell} + \frac{\Delta T_x}{T_x}$	[185]	1.0

(Continued)

TABLE 3.5 (CONTINUED)
Summary of the Quantitative Criteria Proposed to Evaluate the Glass-Forming Ability of Liquid Alloys and Their Values under the Ideal Situation When $T_g = T_x = T_\ell$

Criterion/Parameter	Equation	References	Ideal Value
γ^* parameter	$\gamma^* = \frac{\Delta H^{glass}}{\Delta H^{inter} - \Delta H^{glass}}$	[36]	
γ^{FP} parameter	$\gamma^{FP} = \gamma^*/p_{S_{ea}} = \left(\frac{\Delta H^{glass}}{\Delta H^{inter} - \Delta H^{glass}}\right)\Big/ p_{S_{ea}}$	[194]	
θ^* parameter	$\theta^* = \Delta S_{mix} \times S_\sigma / k_B$	[198]	
G_{CE} a	$G_{CE} = (\Delta H^{chem} - \Delta H^{elastic})$	[192]	
P_{HSS} parameter	$P_{HSS} = \Delta H^c (\Delta S_\sigma / k_B)(\Delta S_c / R)$	[197]	
Q parameter	$Q = [(T_x + T_\ell)/T_g] \times (\Delta E/\Delta H)$	[199]	

TABLE 3.6
Statistical Correlation Functions

	R^2 for Log t_{max} against Different Criteria								
Base	T_{rg}	ΔT_x	α	β	Γ	γ_m	δ	ϕ	*New* β
Au	0.820	0.575	**0.988**	**0.988**	0.979	0.958	0.957	0.976	0.963
Ca	0.313	0.127	0.451	0.451	0.476	**0.478**	0.377	0.455	0.365
Ce	0.315	0.299	0.318	0.318	0.322	0.318	0.303	**0.331**	0.233
Co	**0.990**	0.293	0.573	0.573	0.498	0.475	0.695	0.439	0.654
Cu	0.266	0.274	**0.458**	**0.458**	0.447	0.433	0.398	0.390	0.420
Fe	0.005	0.019	0.012	0.012	0.020	0.025	0.006	**0.031**	0.011
La	0.297	0.179	0.438	0.438	0.426	0.417	0.389	**0.458**	0.415
Mg	0.297	0.038	0.439	**0.802**	0.367	0.320	0.370	0.390	0.350
Nd	0.859	0.888	0.875	0.875	0.981	0.971	**0.992**	0.962	0.990
Ni	0.006	0.077	0.038	0.038	0.067	0.075	0.009	**0.102**	0.013
Pd	0.742	0.662	0.803	0.803	0.815	**0.818**	0.753	0.809	0.787
Pr	0.371	0.594	0.733	0.733	0.816	**0.855**	0.614	0.844	0.794
Pt	0.082	0.044	0.078	0.078	0.073	0.077	**0.098**	0.070	0.146
Ti	0.394	0.010	0.432	0.432	0.413	0.392	0.419	0.381	**0.472**
Zr	0.001	0.007	0.021	**0.021**	0.006	0.005	0.001	0.009	0.001
All	0.112	0.184	0.295	0.295	0.349	0.354	0.207	**0.366**	0.227

Source: Suryanarayana, C. et al., *J. Non-Cryst. Solids*, 355, 355, 2009. With permission.
Notes: R^2 for the maximum critical diameter, t_{max}, against the different criteria proposed to explain the GFA of bulk metallic glasses. The higher the R^2 value the better is the correlation [131]. The numbers in bold indicate the best correlation with that parameter for the alloy. For example, the best correlation was obtained with the T_{rg} parameter for the Co-based BMGs and with the β parameter for Mg-based BMGs.

that for the highest GFA, the T_g (and T_x) should be very high and T_ℓ very low. In other words, under ideal conditions, all three parameters should be equal to each other, that is, $T_g = T_x = T_\ell$. Ideal values for the different GFA parameters under this condition also are listed in Table 3.5. Thus, any alloy composition that exhibits these ideal values should exhibit the highest GFA. For example, an alloy with $T_{rg} = 1$ or with $\gamma = 0.5$ should be an excellent glass former. A similar argument can be made for the other parameters as well. Like in any other situation, it will be impossible to achieve this ideal value for any "real" alloy.

Third, given that $T_g = T_x = T_\ell$, it will be difficult to use the ΔT_x and ϕ parameters as indicators of GFA, since their ideal values are 0. The problem with the ΔT_x parameter is that it considers only the stability of the glassy phase against crystallization and not the ease of formation of the glass. Similarly, T_{rg} only considers the ease of glass formation and not the stability of the glass. A product of these two parameters also did not yield any satisfactory correlation. This is the reason why so many exceptions have been noted in describing the GFA of alloys by these criteria. Even though the ϕ parameter also considers the ΔT_x value, it also includes the T_{rg} parameter, which considers the ease of glass formation. That is, the factors responsible for both glass formation and glass stability are included, and this may be the reason why ϕ was able to show better correlation than either the T_{rg} or ΔT_x parameters alone.

Lastly, the ideal values for the δ, *new* β, and K_{gl} parameters are ∞. This means that the range of possible values is very large and, therefore, these parameters will be highly sensitive to small variations in composition or processing conditions to produce the glasses. On the other hand, for the other parameters, the allowed range is small and, therefore, even small variations in these parameters assume significance. But, it should also be noted that the δ and *new* β parameters do not show a strong correlation for many alloy systems.

Let us now look at the three criteria—α, β, and γ—that seem to show a good correlation with the majority of the published data on t_{max} and R_c. Table 3.7 compares the parameter values and the R_c required to produce 1, 10, and 100 mm thick bulk glassy alloy rods. All three criteria predict that an increase in the parametric value allows the production of larger-diameter rods. But, surprisingly, the largest diameter of rod that can be produced at the ideal value of these parameters is about 1480, 680, and 320 mm for α, β, and γ, respectively. It is not clear why there should be such a large difference in these values. Further, there is not much difference in the R_c value predicted by any of these parameters for different thicknesses. If one considers that 10 mm diameter for a glass is "bulk," then the R_c value predicted to achieve this section thickness by all three parameters is about 1 K s^{-1}. In this sense, all three parameters have similar predictive capabilities.

From this discussion, it becomes clear that almost all the parameters show a similar trend. That is, if one parameter is showing a good correlation, then the other parameters also show a reasonably good correlation and there is not too much to choose between these parameters. Therefore, practically any one of them could be used to obtain guidelines in choosing an appropriate alloy composition to obtain a BMG. Analyzing all the data available, plots of t_{max} and R_c against the three most accepted GFA criteria, *viz.*, T_{rg}, ΔT_x, and γ, are plotted in Figure 3.10. Note that the R^2 values in all cases are rather low.

3.11 Thermodynamic and Computational Approaches

All the criteria previously discussed require knowledge of the transformation temperatures of the glassy alloy, for example, T_g, T_x, T_ℓ, and so on. But, these temperatures can be obtained only *after* the alloy has been solidified into the glassy state and the glassy sample is reheated in DSC/DTA equipment to determine its transformation temperatures. Only then can one can say whether the glass that was produced has a high GFA or not. In this respect, these methods are similar to determining the R_c, t_{max}, and other values to predict GFA. This is a roundabout process, and it would be much more

TABLE 3.7

Values of α, β, and γ and the Critical Cooling Rate (R_c) Required to Produce 1, 10, and 100 mm Thick Bulk Glassy Alloy Rods

Maximum Section Thickness (mm)	α Parameter		β Parameter		γ Parameter	
	α	R_c (K s^{-1})	β	R_c (K s^{-1})	γ	R_c (K s^{-1})
1	0.549	2.3×10^3	1.579	2.5×10^3	0.362	1.9×10^3
10	0.691	1	1.727	2	0.417	3
100	0.834	4.5×10^{-4}	1.876	1.3×10^{-3}	0.472	4.8×10^{-3}

beneficial if one could predict the GFA of an alloy without actually doing an experiment to produce the glass. That is, we require a predictive approach, and thermodynamic and computational approaches have been developed in recent years to solve this problem. (Thermodynamic modeling is different from the computational approach in the sense that we will be using the measured or calculated heats of mixing or other thermodynamic parameters, whereas in the computational approach, we will be utilizing the CALPHAD method.)

3.11.1 Miedema Model

The stability of any given phase is determined by its free energy with reference to the competing phases. Thus, a glassy phase will be stable if its free energy is lower than that of the competing crystalline phases. Miedema [187–189] developed a method to calculate the enthalpies of the formation of solid solutions and amorphous phases in binary alloys based on the chemical, elastic, structural, and topological effects. This approach has subsequently been extended to the ternary alloy systems [190].

The enthalpy of formation of a ternary solid solution of transition metals A, B, and C is given by

$$\Delta H_{\mathrm{ABC}}^{\mathrm{sol}} = \Delta H_{\mathrm{ABC}}^{\mathrm{c}} + \Delta H_{\mathrm{ABC}}^{\mathrm{e}} + \Delta H_{\mathrm{ABC}}^{\mathrm{s}} \tag{3.36}$$

where the superscripts c, e, and s represent the chemical, elastic, and structural contributions to the enthalpy of formation. The chemical contribution is due to electron redistribution when the alloy is formed, the elastic contribution is due to atomic size mismatch, and the structural contribution is due to structure-dependent enthalpies. These will vary systematically with the (average) number of valence electrons per atom in solid solutions of transition metals, as long as the two metals form a common bond of *d*-type electron states. Since the elastic and structural contributions are absent in the glassy phase, only the chemical contribution needs to be considered. Thus, the enthalpy of formation of the glassy phase is

$$\Delta H_{\mathrm{ABC}}^{\mathrm{glass}} = \Delta H_{\mathrm{ABC}}^{\mathrm{c}} + x_{\mathrm{A}}\Delta H_{\mathrm{A}}^{\mathrm{g-s}} + x_{\mathrm{B}}\Delta H_{\mathrm{B}}^{\mathrm{g-s}} + x_{\mathrm{C}}\Delta H_{\mathrm{C}}^{\mathrm{g-s}} \tag{3.37}$$

where:

x_i is the atomic fraction of the *i*th component

$\Delta H_i^{\mathrm{g-s}}$ is the enthalpy difference between the glassy and crystalline states of the pure element *i*

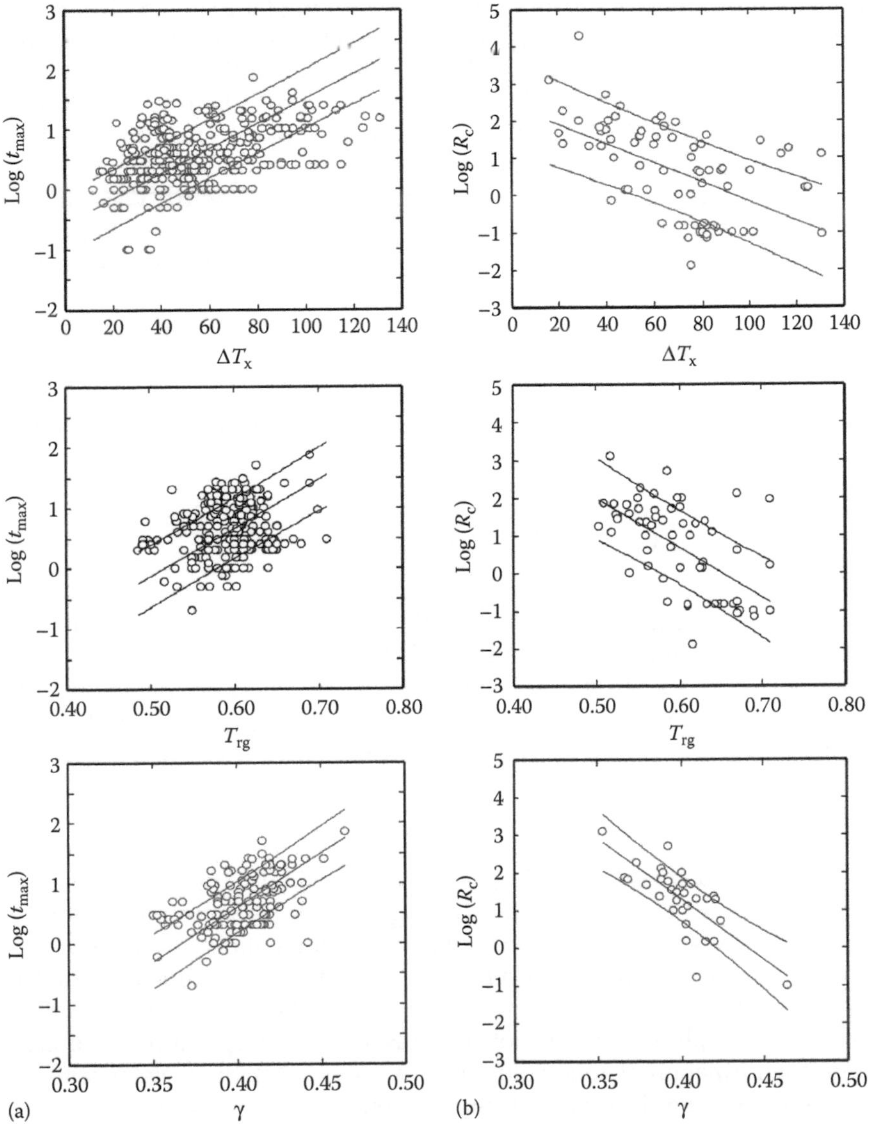

FIGURE 3.10
Plots of (a) t_{max} and (b) R_c against some of the commonly accepted GFA indicators (ΔT_x, T_{rg}, and γ). Note that with the large amount of data currently available, the correlation between the different parameters and t_{max} or R_c is not as good as it was with the limited amount of data reported earlier.

By calculating the enthalpy of formation of the solid solution and the glassy phases as a function of composition, the relative stabilities of the two phases could be determined at any given composition and/or temperature. In addition, using the common tangent construction, the composition range wherein the glassy phase is stable can be determined.

Strictly speaking, it is the free energy and not the enthalpy that should be considered to determine the stability of a phase. However, the contribution from entropy is much smaller than that from enthalpy in solid compounds, and therefore the entropy contribution to the free energy term is neglected, and only the enthalpy term is used.

Takeuchi and Inoue [191] used the enthalpies of mixing tabulated by Miedema [187,188] to determine the stability of the glassy phase and estimate the GFA. By performing calculations for 338 glass-forming ternary alloy systems, the authors showed that the calculated results were in agreement with the experimental results in Cu–Ni–X and Al–Ti–X ternary systems. But, the calculated glass-forming composition range in Zr-, La-, Fe-, and Mg-based alloy systems was overestimated due to the simplification inherent in their model. They noted that the melting temperature of the alloy and the viscosity at the melting temperature, elastic enthalpy arising in the solid solution, SRO present in the glassy phase, and the three Inoue empirical rules were found to be important in evaluating the GFA of alloys.

3.11.2 The G_{CE} Parameter

On the basis of a postulation by Takeuchi and Inoue [191], Sarwat et al. [192] developed a new parameter designated G_{CE} and defined it as

$$G_{CE} = \Delta H^{chemical}(\text{amorphous}) - \Delta H^{elastic} \quad (3.38)$$

The more negative the difference, the higher the GFA. By comparing the calculated value of G_{CE} with the experimentally determined critical diameters, T_{rg}, ΔT_x, and other parameters, the authors have shown that this parameter has a good predictability. The authors have also shown that a modified parameter, $G_{CEE} = (G_{CE})(\Delta S_{config})$, where ΔS_{config} is the configurational entropy, can satisfactorily explain the GFA in higher-order systems.

3.11.3 The γ^* Parameter

Xia et al. [36] also argued that metallic glass formation should involve two aspects. One is that there should exist a thermodynamic driving force for glass formation and the second is that the formed glass should resist crystallization. (Incidentally, these are very similar to those discussed by Lu and Liu [152].) While glass formation requires that the free energy of the hypothetical glassy phase is lower than that of the competing crystalline phases (mostly intermetallic phases because of the presence of a large number of components, even though the formation of solid solutions is a possibility), the resistance to crystallization is dependent on the difference in the free energies of the glassy and intermetallic phases. The first factor can be indicated by $-\Delta H^{glass}$, the enthalpy for glass formation, which is the driving force for glass formation. The larger this value, the easier it is for the glass to form. The resistance of the glass formed to crystallization is determined by the difference between the enthalpies of the glassy and intermetallic phases, that is, $\Delta H^{glass} - \Delta H^{inter}$. The smaller this value, the higher the stability of the glass. Based on this, the authors proposed the γ^* parameter as a measure of GFA of alloy systems, defined as

$$\gamma^* = \text{GFA} \propto \frac{-\Delta H^{glass}}{\Delta H^{glass} - \Delta H^{inter}} = \frac{\Delta H^{glass}}{\Delta H^{inter} - \Delta H^{glass}} \quad (3.39)$$

The higher the value of γ^*, the higher the GFA of the alloy. Strictly speaking, one should, however, use free energy instead of the enthalpy. But, as mentioned earlier, the entropy contribution in solids is small and therefore it is neglected.

From the available data on enthalpies, one can calculate ΔH as a function of solute content for all the competing phases. Using the common tangent approach, one could then determine the composition ranges in which different phases are stable. By calculating the γ^* parameter within the composition range in which the glassy phase is stable, the best glass-forming composition can be determined.

Xia et al. [36] calculated the γ^* parameter for different compositions in the Ni–Nb system, and a plot of γ^* versus composition (presented in Figure 3.11) shows that γ^* has the highest value corresponding to a composition of $Ni_{61.5}Nb_{38.5}$, which is an off-eutectic composition. By actually conducting experiments on $Ni_{100-x}Nb_x$ alloys, at short intervals of 0.5 at.% in the composition range $x = 37.5–40.5$ at.% Nb, the authors showed that the alloy $Ni_{62}Nb_{38}$ had the highest GFA of the investigated compositions. The small difference in the experimental and predicted compositions was attributed to the fact that the entropy term was neglected in the calculations.

Similarly, when Cu–Hf alloys with 34–40 at.% Hf were quenched from the liquid state, a completely glassy phase was observed in an alloy with 35 at.% Hf. Calculation of the γ^* parameter showed that alloys with 35–36 at.% Hf were the best glass formers, confirming the experimental observations [85]. Such an analysis was also successfully applied to the Cu–Zr alloy system [193].

Since ΔH values for each composition may not be readily available, the ΔH values for the required compositions were calculated using the values for the neighboring intermetallic compounds with the Miedema method [187,188]. For example, in the Cu–Hf system,

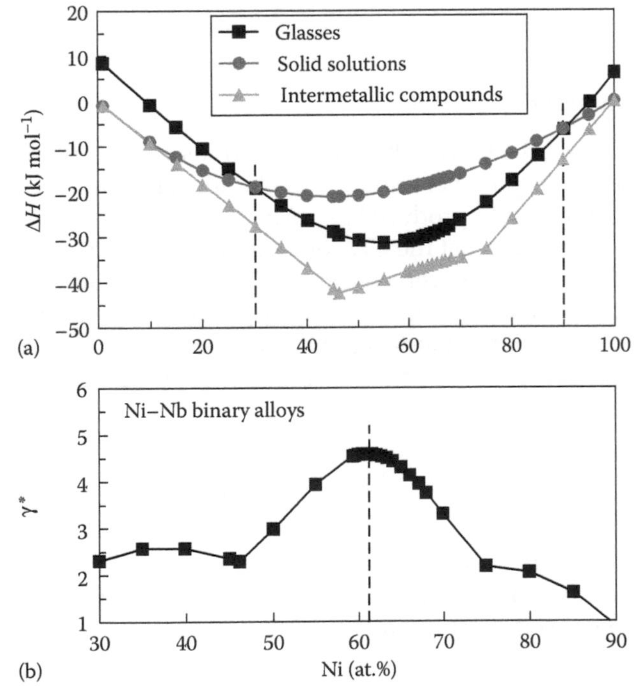

FIGURE 3.11
(a) Plot of calculated enthalpies of solid solutions, intermetallic compounds, and metallic glasses in the Ni–Nb system. At any given composition, the phase with the lowest enthalpy will be the most stable phase. (b) Plot of γ^* as a function of composition in the binary Ni–Nb system. Note that the largest value of γ^* corresponds to the best glass-forming composition. (Reprinted from Xia, L. et al., *J. Appl. Phys.*, 99, 026103-1, 2006. With permission.)

the ΔH values were calculated using the values for the intermetallic compounds Cu_8Hf_3 and $Cu_{10}Hf_7$, and taking the appropriate weightage for the individual compounds. The γ^* parameter was then calculated using these ΔH values and Equation 3.39. It is important to realize that formation of intermetallics can dramatically influence the calculations and, consequently, the GFA of the alloys.

3.11.4 The γ^{FP} Parameter

The γ^* parameter considers only the thermodynamic parameters required for glass formation. Arguing that both thermodynamic and kinetic parameters determine the GFA of alloys, Yu et al. [194] formulated a new expression from first-principles calculations to predict the GFA of binary Zr–Cu glassy alloys. This parameter, designated γ^{FP}, is defined as

$$\gamma^{FP} = \gamma^* / p_{S_{ea}} = \left(\frac{\Delta H^{glass}}{\Delta H^{inter} - \Delta H^{glass}} \right) \Big/ p_{S_{ea}} \tag{3.40}$$

where $p_{S_{ea}}$ represents the kinetic effect which can be estimated in terms of the ratio of the total excessive atomic spaces, S_{ea}, by the total atom volume, V_{tot}, in the supercooled liquid near T_g. This parameter was shown to follow the exact trend of critical diameter with composition in the binary Zr–Cu system. This has been subsequently extended to the ternary $(Zr_{50}Cu_{50})_{100-x}Al_x$ system and it was shown that the experimentally determined critical diameters as a function of the Al content followed the trend of γ^{FP} [195].

3.11.5 The P_{HSS} Parameter

Bhatt et al. [196] combined the thermodynamic and topological approaches and developed a parameter, wherein the product of enthalpy of chemical mixing, ΔH^{chem}, and entropy of mismatch, S_σ, were combined to produce $P_{HS} = \Delta H^{chem} \times (S_\sigma/k_B)$, where k_B is the Boltzmann constant. They were able to predict the GFA of a number of ternary Zr-, Cu-, and Fe-based glasses. These authors have, however, considered the number of components in the system only in a qualitative manner and, thus, its applicability to multicomponent systems will be limited. But, to make the model applicable to a variety of systems, Ramakrishna Rao et al. [197] incorporated the configurational entropy also and developed the P_{HSS} parameter, defined as

$$P_{HSS} = \Delta H^{chem} \times (\Delta S_\sigma / k_B) \times (\Delta S_C / R) \tag{3.41}$$

where R is the gas constant. It was reported that the P_{HSS} parameter was able to better predict the GFA in multicomponent Fe-based systems, whereas both P_{HS} and P_{HSS} were equally applicable to Fe-based ternary alloys.

3.11.6 The θ^* Parameter

Arguing that the mismatch entropy has an effect on the stability of the structure and that an increase in S_σ/k_B can lead to a more random structure resulting from a large difference in the atomic sizes, Hu et al. [198] proposed another criterion, θ^* defined as

$$\theta^* = \Delta S_{mix} \times (S_\sigma / k_B) \tag{3.42}$$

(Since a θ parameter was proposed earlier [184], we have redesignated the current parameter as θ*.) They also used the fragility parameter of superheated liquid, *M*, proposed by Bian et al. [199] to refine the parameter θ*, and proposed the parameter θ*/*M*, and observed a good correlation between θ*/*M* and GFA, and this was noted to be better than the simple θ* parameter to evaluate the GFA of Al-based alloys.

3.11.7 The *Q* Parameter

Based on the consideration of liquid phase stability, resistance to crystallization, and the glass transition enthalpy, Suo et al [200] developed a new criterion, the *Q* parameter, to evaluate the GFA of BMGs. *Q* is defined as

$$Q = [(T_x + T_\ell)/T_g] \times (\Delta E / \Delta H) \tag{3.43}$$

where:
- ΔE is the crystalline enthalpy
- ΔH is the melting enthalpy

In comparison to other criteria, such as γ, T_{rg}, and ΔT_x, the *Q* parameter was shown to exhibit better correlation with the maximum cross-section thickness.

Srinivasa Rao et al. [201] tried to establish correlations between T_{rg} and different thermodynamic parameters such as $T\Delta S$, ΔH^e, ΔH^c, and ΔG for a number of Zr-based BMG alloys, and noted that ΔG showed the best correlation. The authors had calculated the free energy change between the glassy and solid solution phases using the Miedema approach, combined with the atomic size mismatch, and suggested that the composition that represented the most negative ΔG value would be the best glass-forming composition.

Ge et al. [202] also used a similar thermodynamic approach to evaluate the GFA of alloys. By combining the CALPHAD approach with the kinetic approach, they evaluated the GFA through calculation of the critical cooling rates (R_c) for glass formation. The driving force for crystallization was calculated from the undercooled melt using the Turnbull [203] and Thompson and Spaepen [204] free energy equations. *T–T–T* curves for these alloys were also obtained using the Davies–Uhlmann kinetic equations. The critical cooling rates, R_c, were calculated for 9 compositions in the binary Cu–Zr system and 13 compositions in the ternary Cu–Zr–Ti system. The R_c values calculated using the Thompson–Spaepen equations were in the range 4.32×10^2–3.63×10^4 K s^{-1} for the Cu–Zr binary alloys and 0.64–23.98 K s^{-1} for the Cu–Zr–Ti ternary alloys, except for one composition, for which it was estimated to be 1.36×10^4 K s^{-1}. But, the values calculated by the Turnbull equation were in the range 9.78×10^3–8.23×10^5 K s^{-1} for the binary and 1.38×10^2 – 4.82×10^3 K s^{-1} for the ternary alloys, again except for one composition, for which it was much higher. Even though the trend of the cooling rates required for glass formation was the same in both approaches, it was noted that the R_c values determined by the Thompson–Spaepen equations were closer to the experimentally determined R_c values. Thus, it appears that combined thermodynamic and kinetic modeling may provide an effective method for the prediction of GFA in alloys.

Hu et al. [205] used the CALPHAD approach to calculate the stability of the liquid phase in the constituent binary and ternary phase diagrams of W-based alloys, and identified several plausible W-rich compositions to form glasses in the W–Fe–Si–C system.

3.11.8 Computational Approaches

Chang and colleagues [161,206] used the concept that deep eutectics, or for that matter, compositions around eutectic points, are good glass formers and employed a thermodynamic approach to identify alloy compositions that correspond to eutectic compositions in multicomponent alloy systems. Since a strong liquid favors BMG formation, and also because a strong liquid is more exothermic in the excess Gibbs free energy than its competing crystalline solid phases, which decreases the driving force for crystallization, the authors calculated the excess Gibbs free energy for the competing phases using a regular solution model. That is,

$$G^{\ell,xs} = x^{\ell}\left(1-x^{\ell}\right)\varepsilon^{\ell} \tag{3.44}$$

$$G^{\alpha,xs} = x^{\alpha}(1-x^{\alpha})\varepsilon^{\alpha} \tag{3.45}$$

where:

G	represents the Gibbs free energy
ℓ and α	refer to the liquid and solid phases, respectively
x	is the mole fraction
ε	is the regular solution parameter

By varying the values of ε^{ℓ} and ε^{α}, different topologies of phase diagrams can be obtained. For example, when ε^{ℓ} has a large negative value and ε^{α} has a large positive value, the phase diagram will exhibit a deep eutectic. Under these circumstances, the excess Gibbs energy of the liquid phase is more exothermic than that of the solid phase. This is why a eutectic liquid is energetically more stable and consequently a good glass former. Thus, if phase diagrams of multicomponent alloy systems are known as functions of temperature and composition, potential alloy compositions corresponding to low-lying-liquidus surfaces, which favor BMG formation, can be identified.

The thermodynamic properties of multicomponent alloy systems are obtained by building up thermodynamic information from the lower-order to higher-order alloy systems. The thermodynamic parameters of the phases in a binary alloy system are obtained based on thermodynamic and phase diagram data determined experimentally. Using the thermodynamic descriptions of the three binaries and any ternary experimental information, one can describe the thermodynamic behavior of the ternary system. An extrapolation method can then be used to obtain a thermodynamic description of the quaternary or higher-order systems, without the need for further experimental data. This method is acceptable as long as new phases do not form in the higher-order systems. A good correspondence was found between the thermodynamically predicted [161] and experimentally observed [18] composition ranges for glass formation in the Zr–Ti–Cu–Ni system.

Since it is difficult to visualize multicomponent phase diagrams, Ma et al. [206] proposed the Scheil solidification simulation method to identify the invariant eutectic reaction associated with BMG formation. In the Scheil method, the solidification path of an alloy is calculated by plotting the solidification temperature as a function of the volume fraction of the solid phase.

This approach has some merits. Since eutectics are determined only on the basis of thermodynamic equilibrium (and kinetics do not play any role), the eutectic compositions can be predicted using thermodynamic data. However, there is a snag with this approach. Since phase equilibrium is associated with the differences in the Gibbs free energy values, it is important to obtain very accurate values, often of the order of 100 J mol^{-1}. But, the thermodynamic quantities either measured experimentally or calculated from a first-principles approach using quantum mechanics are not accurate enough to calculate the correct phase diagrams in most cases.

Kim et al. [207] used another thermodynamic approach to identify glass-forming compositions in multicomponent alloy systems. Since glass formation involves suppression of the nucleation and growth of a crystalline phase from the supercooled liquid, these authors calculated the driving force for the formation of a crystalline phase from the liquid state. That alloy composition at which the driving force was minimum was selected as the best glass-forming composition. The thermodynamic calculations were performed using the CALPHAD method [122]. Choosing the Cu–Ti–Zr system as a typical example, the authors calculated the driving forces of the individual crystalline phases as a function of Ti content in the temperature range 600°C–800°C, where the alloys exist in the supercooled liquid state. Two local minima were located for the driving forces of the crystalline phases in the $Cu_{55}Zr_{45-x}Ti_x$ system—one at the Zr-rich region ($x = 7$–10) and the other at the Ti-rich region ($x = 28$–29)—and these should have corresponded to alloys with high GFA. Indeed, such compositions were shown to be good glass formers—$Cu_{60}Zr_{30}Ti_{10}$ by Inoue et al. [142] and $Cu_{55}Ti_{35}Zr_{10}$ by Lin and Johnson [18].

The authors point out that it is possible that the low-lying-liquidus surfaces method [161] also give a similar result. But, this is possible only when the degree of undercooling is small, that is, near eutectic compositions. But, in deeply undercooled liquid melts, the composition dependence of driving forces for crystalline phases is different from that of liquidus temperature. Therefore, the authors feel that this method of predicting the best glass former on the basis of driving force is a more general criterion.

3.12 Structural and Topological Parameters

Let us now look at some of the structural parameters and the topological criteria that have been used to explain the high GFA of bulk glass-forming alloys. Let us recall that, according to the Inoue criteria (Section 3.7), the atoms of the constituent elements in the BMG should differ by at least about 12%. Additionally, even if some apparent exceptions were noted for these criteria, the atomic size criterion is never disobeyed in the formation of metallic glasses. It is useful to note that all BMGs contain at least one very large atom and the other atoms are either medium or small in size. However, as pointed out earlier by Inoue [125], the majority of the BMGs contain at least three different sizes of atoms—large, medium, and small. Therefore, it is not surprising that all the models in this category consider the atomic size as the most important basis. Quite a few models have been proposed in this category and these are reviewed in the following sections.

3.12.1 Egami's Atomistic Approach

Egami and Waseda [119] had earlier proposed the atomic volume strain criterion to predict the glass formation range in binary metallic glasses produced by RSP methods. They suggested that a minimum solute concentration was necessary to achieve this critical strain to destabilize the crystal lattice (see Section 3.5.1 for further details). Yan et al. [208] and Ueno and Waseda [209] used this concept to calculate the λ value in the case of multicomponent alloy systems. Egami and Waseda [119] calculated the λ parameter for a binary alloy using Equation 3.8. Similarly, for a ternary alloy, Ueno and Waseda [209] suggested that the minimum solute concentration could be calculated using the equation:

$$\lambda_0' = \left|\frac{\Delta V_{AB}}{V_A}\right| \cdot C_B + \left|\frac{\Delta V_{AC}}{V_A}\right| \cdot C_C^{min} \tag{3.46}$$

where ΔV_{AC} is the difference in atomic volumes between A and C. For a multicomponent glass-forming alloy system, Yan et al. [208] suggested that λ_n can be calculated using the equation:

$$\lambda_n = \sum_{B=1}^{n-1} \left|\frac{\Delta V_{AB}}{V_A}\right| \cdot C_B = \sum_{B=1}^{n-1} \left|\left(\frac{R_B}{R_A}\right)^3 - 1\right| \cdot C_B \tag{3.47}$$

where:

R_A and R_B	are the radii of the solvent and solute atoms, respectively
C_B (in atomic percent)	is the solute concentration of element B
V_A	is the atomic volume of A
ΔV_{AB}	is the difference in atomic volumes between A and B

Based on observations in a number of glass-forming alloy systems, they noted that the diameter of the rod that could be cast into a fully glassy state increased with increasing values of λ_n, and at the maximum possible diameter, the λ_n value approached 0.18. Therefore, the authors suggested that the value of λ_n for the best glass-forming alloys is about 0.18. Incidentally, Ueno and Waseda [209] had also arrived at a similar value of 0.1 based on their lattice strain concept used for the binary alloys. However, the contribution of Yan et al. [208] was that they had used the mathematical description of regular polytopes and the dense random packing of hard spheres (DRPHS), a cluster model to describe the structure of metallic glasses, as the basis to reach this value. They suggested that the glass structure would have an optimum defect concentration when $\lambda_n \approx 0.18$.

Botta et al. [210] developed a new criterion to predict glass formation by combining the average electronegativity difference with the topological instability criterion. The topological instability was calculated from the atomic radii (in simple solid solutions) or molar volumes of the compounds according to the equation:

$$\lambda \cong \sum x_i \left|\frac{V_{mi}}{V_{mo}} - 1\right| \tag{3.48}$$

where:

x_i and V_{mi}	are the molar fraction and molar volume, respectively, of the solute element in the compound
V_{mo}	is the molar volume of the compound

Large values of λ indicate a greater topological instability of the phase under consideration. The values of λ are calculated as a function of the composition. After such calculations for all phases for a given alloy, the minimum value of λ, that is, λ_{min}, is found, and these λ_{min} values are plotted as a function of the composition for a given binary system. The peaks in the λ_{min} plot then correspond to compositions where the topological instability reaches a local maximum and, therefore, a better GFA. The authors also calculated the average electronegativity difference of an element and its surrounding neighbors, $\overline{\Delta e}$, using the equation:

$$\overline{\Delta e} = \sum x_i \sum S_j \left| e_i - e_j \right| \tag{3.49}$$

where:

e_i is the Pauling electronegativity of a central atom
e_j is the electronegativity of each neighbor
S_j is the surface concentration

S_j is given by the equation:

$$S_j = \frac{x_i (V_{mj})^{2/3}}{\sum x_i (V_{mj})^{2/3}} \tag{3.50}$$

By plotting this parameter ($\lambda_{min}\overline{\Delta e}$) as a function of the alloy composition, the authors identified good glass-forming compositions, which match very well with those experimentally observed in the Zr–Cu system. This was further extended to a number of other systems.

Even though λ_{min} appears to be a good parameter to predict the GFA among the competing phases in a given phase diagram, it was not found to be efficient for tracking the GFA generally. Accordingly, to better predict the GFA of alloys, de Oliveira [211] used an electronic parameter, Δh, defined as

$$\Delta h = (\Delta\varphi)^2 - k\left(\Delta n_{ws}^{1/3}\right)^2 \tag{3.51}$$

where $\Delta\varphi$ and $\Delta n_{ws}^{1/3}$ are the average difference of the work function and the average difference of the cube root of the electronic density among the constituent elements of the alloy. By combining λ_{min} and Δh into one equation, the author observed that the simplest combination of $\lambda_{min} + (\Delta h)^{1/2}$ was found to be the best and simplest combination. He noted that the critical cooling rate, R_c, can be related to this parameter by an equation:

$$\log R_c = 8.21 - 14.06\ (\lambda_{min} + \sqrt{\Delta h}) \tag{3.52}$$

When compared with the $\lambda_{min}\overline{\Delta e}$ criterion ($R^2 = 0.37$), the new criterion produced a much better linear correlation ($R^2 = 0.76$) when examined for 68 alloys in 30 different metallic systems. It was claimed that "the new proposed criterion is a fast and easy-to-use tool for guiding the search for new glass-forming alloys, saving time and reducing the number of experimental trials."

The concept of volume strain destabilizing the crystalline lattice was later extended by Egami [154] to explain the formation of BMGs. By using a local packing model of solvent atoms surrounding a solute atom and obtaining the average coordination number, Egami [212] calculated the strain introduced into the lattice. When this strain exceeded a critical value (about 6%), the crystal lattice became unstable and the glass had formed. Recently, Han et al. [213] investigated the GFA based on molecular dynamics simulations for binary alloys and showed that the size effect is accompanied by a change of phase diagram from solid solution to eutectic and change in the local stress/internal energy, which could contribute more to GFA than kinetic factors such as efficient packing.

When a solute atom of a different size is introduced into the lattice of a metal, the consequences are completely different depending on whether we are dealing with a solid or a liquid. A solute atom in a crystal lattice produces a long-range stress field around it since the topology of the atomic structure in a crystalline solid is fixed. But in a liquid, the environment relaxes instantaneously to accommodate the size difference by altering the local topology.

It is well known that when atoms of equal size are packed together, the coordination number, N, is 6 in two dimensions and 12 in three dimensions. But, when atoms of different sizes are packed together, the coordination numbers will be different. It was shown by Egami and Aur [214] that, in three dimensions, the average coordination number, N_C^A, of an A atom with a radius R_A, embedded in a glass, B, with radius R_B, is given by

$$N_C^A(R) = \frac{4\left[1-\frac{\sqrt{3}}{2}\right]}{\left[1-\frac{\sqrt{R(R+2)}}{R+1}\right]} \tag{3.53}$$

where $R = R_A/R_B$. This coordination number increases with increasing values of R, and at $R=1$, $N_C = 4\pi$. When the value of R changes continuously, the local coordination number, N_C, also changes, but it cannot change continuously, since it is an integer number. A stepwise or discontinuous change will occur when the corresponding value of N_C is changed by 0.5. The amount of change, ΔR_C, necessary to introduce a change of 0.5 in N_C was calculated by Egami [154] to be

$$\Delta R_C = \frac{1/2}{\partial N_C^A(R)/\partial R} \tag{3.54}$$

From this equation, the critical volume strain for a single-component system was calculated as

$$\varepsilon_V^{crit} = \frac{3}{2}\Delta R_C = \frac{3}{2}\times\frac{1/2}{\partial N_C^A(R)/\partial R} = \frac{6\sqrt{3}-9}{8\pi} = 0.0554 \tag{3.55}$$

Thus, the crystal lattice destabilizes when the critical volume strain exceeds 0.0554, and a glass is formed.

Difficulties arise when this concept has to be extended to a system containing more than one solute. For example, different solute elements will contribute differently to the critical strain value and therefore, the instability conditions could be different. But, this concept was subsequently extended by Egami [215] to a binary transition metal–metalloid-type metallic glass, and the results were generalized for multicomponent glasses. It was noted that glass transition in a multicomponent glass does not occur uniformly, but gradually over a range of temperature. This is because each component has a local glass transition temperature. A wide dispersion of such local glass transition temperatures results in a strong liquid behavior (in the scheme of Angell [178]) and increases the viscosity of the liquid. It was also suggested, for example, in the binary metal–metalloid-type glass, that a repulsive potential between the small metalloid atoms and attraction between the small metalloid and large transition metal atoms result in a much higher global transition temperature than the local glass transition temperature of the larger atoms. This creates atomic clusters around the larger atoms in the liquid and increases the viscosity of the liquid and therefore the GFA.

Since glass formation involves both kinetic aspects (indicated by the distributed local glass transition temperatures) and energetic aspects (addressed by the local atomic-level strains), Egami [215] identified that BMG formation is favored when

1. The atomic size ratio of the constituent elements increases
2. The number of elements involved is larger
3. The interaction between the small and large atoms is enhanced
4. Repulsive interactions are introduced between the small atoms

3.12.2 Miracle's Topological Criterion

Considering purely the geometry of the packing of spheres of different diameters, Miracle and coworkers [216–218] came up with atomic size distribution (ASD) plots. In these plots, the concentration of each element in the alloy is plotted against *R*, defined as the radius ratio of the solute to the solvent atom. For conventional glass-forming alloys (that is, glassy alloys that can be produced either only in the form of thin ribbons or those that have a low GFA), it was noted that the plot showed a concave downward shape. As an example, Figure 3.12a shows the ASD plot for some Fe-based metallic glassy alloys. Note that the shape of this curve is concave downward and has a single peak. Further, the glassy alloy has at least one solute element that is smaller and at least one element that is larger in size than the solvent atom. Such typical concave downward plots have been noted for all glassy alloys that require a high solidification rate for glass formation.

On the other hand, for alloys that could easily be solidified into the glassy state, for example, BMGs, the ASD plot shows a concave upward shape, as shown in Figure 3.12b. Note that, in this case, all the solute atoms have a size smaller than that of the solvent atom. Further, while the solvent atom has the highest concentration in the alloy, the smallest atom in the alloy has the next-highest concentration among all the solute atoms, and the atoms with intermediate sizes have intermediate concentrations. In other words, the concentrations of the solute elements decrease, or approach a minimum, at an intermediate size, and then increase with decreasing atomic size, producing a specific concave upward shape of concentration versus atomic size plot.

It was noted that the shape of the ASD plot changes from concave downward to concave upward when the critical cooling rate for glass formation decreases to about 10^2–10^3 K s^{-1} [216]. These ASDs seem to provide for more efficient atomic packing in the liquid in comparison to those that exhibit a concave downward shape. As mentioned earlier, a more

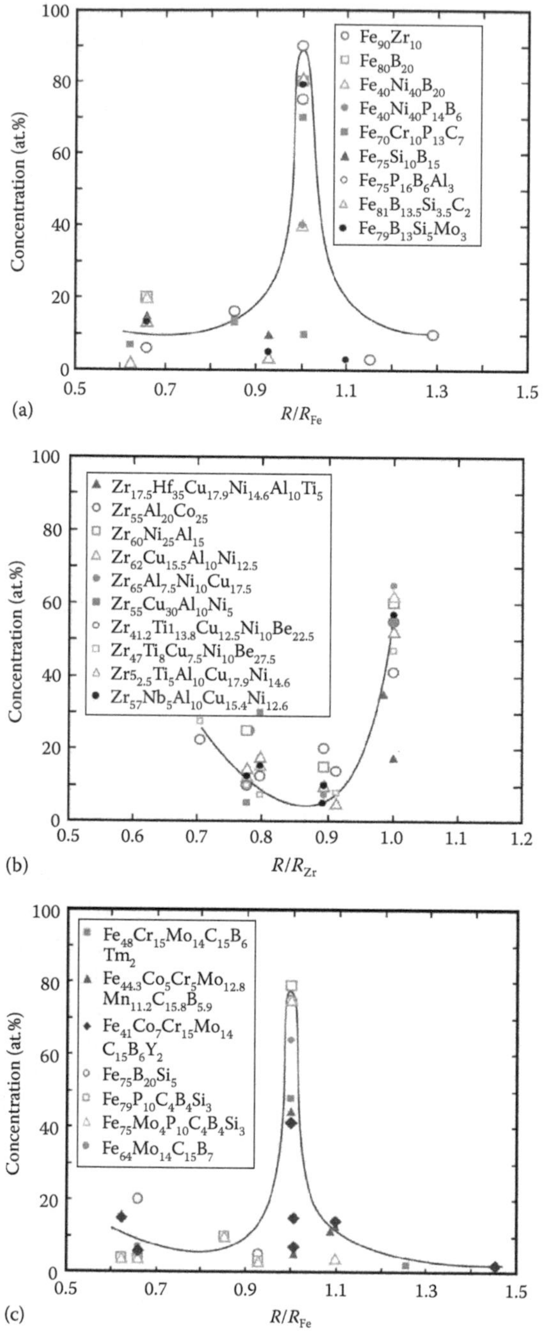

FIGURE 3.12
Atomic size distribution plots (ASDP), that is, concentration versus radius ratio plots for (a) Fe-based melt-spun glasses, (b) Zr-based BMGs, and (c) Fe-based BMGs. A solid trend line is drawn as a visual guide. It was suggested that melt-spun glassy alloys exhibit a concave downward shape and that BMGs, which are characteristic of alloys that can form glasses at relatively low critical cooling rates, display a concave upward trend. But, data for Fe-based BMGs, especially for very good glass formers, shown in (c) clearly reveal a concave downward shape. It appears that this concave downward or upward shape is associated with whether we are dealing with metal–metal or metal–metalloid glasses and not with the critical cooling rates at which these form.

compact structure has a higher viscosity and lower diffusivity and, consequently, the formation of crystalline phases is impeded, favoring glass formation.

Figure 3.12c shows the ASD plot for several Fe-based BMG alloys that have been shown to have very high GFA. Some of these have been produced with a critical diameter as large as 16 mm. But, the important point to note here is the concave downward shape, and not the expected concave upward shape of the ASD plot. From Figure 3.12a through c, it appears that the concave upward or downward shape relates to the metal–metal or metal–metalloid system and not the critical cooling rate at which the glass forms. Note that, although the Be atom is small in size, it is not very much smaller than Fe, Co, or Ni.

An interesting point that emerges from such ASD plots is that there are some critical values of *R*, labeled *R**, which seem to populate the solute elements. These have been identified to be concentrated around 0.62, 0.71, 0.80, and 0.89. (Because of the uncertainties involved in the determination of atomic radii, the *R** values can deviate by up to 3% [219]. To validate this point, Senkov and Scott [220] designed BMG alloy compositions and successfully produced glassy alloys based on a simple metal, Ca.)

As mentioned earlier, Egami and Waseda [119] also used the concept of the packing of atoms using geometry principles to explain glass formation. But, they had only considered the case when the solute atoms occupy substitutional positions. If the difference in the atomic sizes of the solute and solvent atoms continues to increase (e.g., when both metal and metalloid atoms are present), the substitutional occupancy of the solute atoms is no longer energetically favorable and, therefore, the solute atoms may occupy interstitial lattice sites. Therefore, Miracle and Senkov [217,218] proposed that lattice strain is produced when the solute atoms occupy both the substitutional (s) and interstitial (i) lattice sites. The lattice strain generated in this case will be different from that calculated assuming substitutional occupancy alone. It was suggested that in the case of BMGs, which contain solute atoms that are smaller than the solvent atom, interstitial atoms produce a positive lattice strain (expansion), while substitutional atoms create a negative strain (compression) [216,217]. They also considered that glass formation is achieved when the crystalline lattice experiences a critical volume strain and destabilizes. Using this as the basis, Miracle and Senkov [217,218] proposed a topological criterion for metallic glass formation.

Based on the fact that the energy states decide the position of the solute atom either in the substitutional or interstitial sites, Miracle and Senkov [218] calculated the fraction of solute atoms that occupy these sites using the equations:

$$X_s = \frac{n_s}{CN_A} = \left[1 + \exp\left(\frac{\Delta E}{kT}\right)\right]^{-1} \tag{3.56}$$

$$X_i = \frac{n_i}{CN_A} = \left[1 + \exp\left(-\frac{\Delta E}{kT}\right)\right]^{-1} \tag{3.57}$$

where:

X_s and X_i represent the fraction of solute atoms in the substitutional and interstitial positions, respectively

n_s and n_i represent the number of solute atoms in the substitutional (s) and interstitial (i) positions, respectively

C is the concentration of solute atoms

N_A	is Avogadro's number
k	is the Boltzmann constant
T	is the temperature
$\Delta E = E_s - E_i$	is the difference in energy states of the substitutional and interstitial sites

It may be noted that $X_s + X_i = 1$. It was also assumed that the chemical effects between the solute and solvent atoms do not affect the location of the solute atoms, and therefore ΔE is equal to the elastic energy difference between the substitutional and interstitial sites. Thus, knowing the elastic energies of these sites, the distribution of solute atoms between the substitutional and interstitial sites can be calculated in terms of the reduced radius, R (R_B/R_A = radius of the solute/radius of the solvent) as

$$X_s = \left[1 + \frac{\exp(\alpha/R^3)}{\beta}\right]^{-1} \tag{3.58}$$

$$X_i = \left[1 + \frac{\beta}{\exp(\alpha/R^3)}\right]^{-1} \tag{3.59}$$

The parameters α and β depend on the elastic constants of both the solute and solvent atoms, the atomic volume of the host atom, V_A, and temperature, and their values will be different for different alloying elements and at different temperatures. Figure 3.13 shows the variation of the fraction of solute atoms located in the interstitial sites as a function of the reduced radius, temperature, and elastic constants calculated according to Equation 3.59.

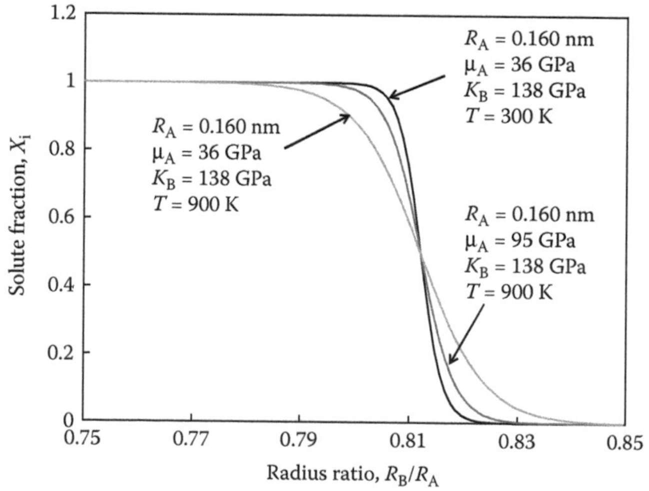

FIGURE 3.13
Fraction of solute atoms located in the interstitial sites as a function of reduced radius, temperature, and elastic constants calculated according to Equation 3.43. (Reprinted from Senkov, O.N. and Miracle, D.B., *J. Non-Cryst: Solids*, 317, 34, 2003. With permission.)

A few important points become clear from this plot. First, at room temperature, all the solute atoms occupy the interstitial sites if the reduced radius $R = R_B/R_A$ is less than 0.8. But, almost all of them are located in the substitutional sites if $R > 0.83$. In the intermediate range $0.8 < R < 0.83$, the solute atoms are located both in the interstitial and substitutional sites. At $R = 0.812$, the solute atoms are equally distributed between the substitutional and interstitial sites, that is, $X_s = X_i = 0.5$. The transition between the substitutional and interstitial sites becomes wider when R_A, μ_A, and/or K_B decrease and/or temperature increases. Here, μ and K represent the shear modulus and bulk modulus, respectively, and the subscripts A and B represent the solvent and solute, respectively. According to Equations 3.58 and 3.59, in a Zr-based BMG containing Be, Ni, Cu, Al, and Ti, the preferred occupancies are substitutional for Ti and Al and interstitial for Cu, Ni, and Be.

The local volume strain produced by the solute atoms can now be calculated in terms of the reduced radius as

$$\varepsilon_A^V = \frac{C\xi\gamma\left[X_s\left(R^3-1\right)+X_i\left(R^3-\eta^3\right)\right]}{\left[1+CX_s\left(R^3-1\right)\right]} \tag{3.60}$$

In this equation, ξ is the coefficient of compaction of the crystalline lattice (e.g., 0.74 for fcc or hcp and 0.68 for bcc-type packing), and

$$\gamma = \frac{1+4\mu_A/3K_A}{1+4\mu_A/3K_B} \tag{3.61}$$

with $\eta = \sqrt{2}-1 = 0.4142$ for an octahedral site in a close-packed structure. Due to these local volume strains, the distances between the nearest neighbor atoms in the crystalline lattice change, and this uniform contraction (or expansion) was considered by Egami [118] as a decrease (or an increase) in the apparent atomic radius of the atom. This, in turn, should lead to a change in the local coordination number once the apparent atomic radius reaches a critical value. For a single-component system, the relative expansion has been calculated as

$$\delta_c = \frac{1}{2}\frac{\partial R}{\partial N^{th}} = 0.036 \tag{3.62}$$

where N^{th} represents the theoretical (i.e., continuously changed with R) coordination number, which is an upper bound. In reality, the actual coordination number is smaller than N^{th}. For a packing of spheres in three dimensions, N^{th} can be calculated using the equation [212]:

$$N^{th} = \frac{4\pi\left(1-\sqrt{3}/2\right)}{\left(1-\sqrt{R(R+2)}/(R+1)\right)} \tag{3.63}$$

Since this equation underestimates the actual values of the coordination numbers, Miracle et al. [219] modified this equation to

$$N^{\text{th}} = \frac{4\pi}{\pi(2-n)+(2n)\arccos\left[\sin\left(\frac{\pi}{n}\right)\frac{\sqrt{R(R+2)}}{(R+1)}\right]} \tag{3.64}$$

where:
$n = 5$ at $R \geq 0.902$
$n = 4$ at $0.414 \leq R \leq 0.902$

This equation correctly predicts that an ideal icosahedron ($N = 12$) is formed at $R = 0.902$ and that an ideal octahedron ($N = 6$) is formed at $R = 0.414$.

The critical volume strain on this matrix when the crystalline lattice becomes destabilized is then estimated using the equation [118]:

$$\varepsilon_{cr} = \tfrac{3}{2}\delta_c = 0.054 \tag{3.65}$$

Since the critical strain is a function of the atomic size ratio and also the concentration of solute atoms, a minimum solute concentration is required to achieve the critical strain to destabilize the crystalline lattice, and it was calculated [217,218] as

$$C_{\min} = \frac{\chi}{\left[X_s(1-\chi)(R^3-1)+X_i(R^3-\eta^3)\right]} \tag{3.66}$$

where $\chi = \varepsilon_{cr}/\gamma\xi$.

A plot of the minimum solute concentration against the reduced radius ratio, R, is shown in Figure 3.14 for both the cases when the solute atom occupies only the substitutional sites (Egami and Waseda model) and when it may occupy either the substitutional and/or interstitial sites. An important difference is noted between these two cases. For the Egami and Waseda model, the minimum solute concentration decreases continually as the relative radius ratio increases. But, in the Miracle and Senkov model, with an increase in the difference between the atomic sizes of the solvent and solute atoms, the minimum solute concentration decreases when $R > 0.82$, where the solute atoms mostly occupy the substitutional sites. However, the minimum solute concentration increases when $R < 0.8$, when the solute atoms predominantly occupy the interstitial sites. In the range $0.8 < R < 0.82$, the minimum solute concentration increases rapidly to infinity at $R \cong 0.812$, where the positive volume change due to interstitial atoms is compensated by the negative volume change from the substitutional atoms. The significance of this large increase in solute concentration is presently not well understood, but it could be an artifact of the assumption made in the model, *viz.*, the strains of opposite sign get cancelled out; such an assumption is not supported by experimental data for binary metallic glasses.

The minimum solute concentration required to destabilize the crystalline lattice is normally small. But, in the case of BMGs, it is noted that the total solute content is usually above about 20–40 at.%. This was explained on the basis that the optimum concentration of

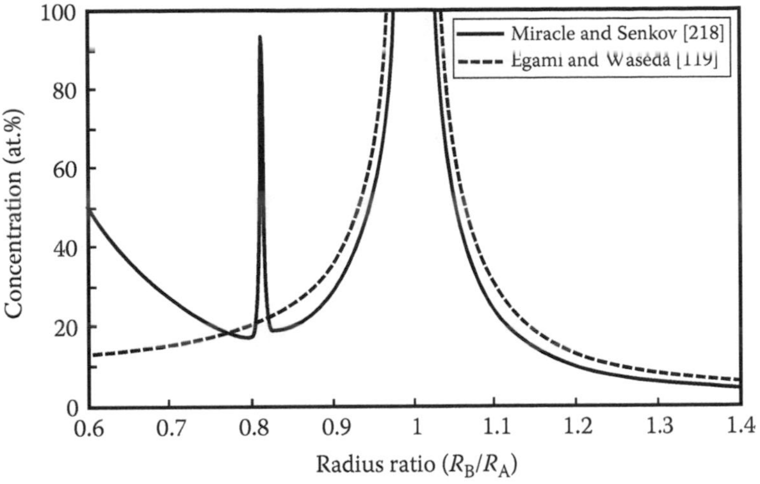

FIGURE 3.14
Minimum (critical) solute concentration versus relative radius for the two cases when the solute atom occupies only the substitutional sites (Egami and Waseda model) and when it may occupy substitutional and/or interstitial sites (Miracle and Senkov model). (Reprinted from Miracle, D.B. and Senkov, O.N., *Mater. Sci. Eng. A*, 347, 50, 2003. With permission.)

an individual alloying element in a BMG is less than the minimum solute concentration of an element with the same atomic size in a binary alloy. Therefore, the destabilizing effect of each solute element was suggested to be additive in these models.

The topological model proposed by Miracle and Senkov suffers from two limitations. First, the volume change of a specimen is calculated by linear superposition of the volume changes caused by the addition of each of the solute atoms in a single-component matrix and this restricts the application of the model to systems containing no more than 30–40 at.% of the solute atoms. Second, the chemical effects on glass formation have not been taken into consideration. Therefore, the topological criterion may be a necessary, but not a sufficient, condition for glass formation.

Since the density difference between BMGs and the corresponding crystalline solids is <0.5%, it is only proper to expect that the BMGs will contain units that are densely packed. This suggestion would also be in tune with the high viscosity and easy glass formability of the melts. Therefore, Miracle proposed a structural model for metallic glasses based on the efficient packing of clusters [155,156]. He considered solute-centered clusters as representative structural elements. Considering these idealized clusters as spheres, a scheme was suggested to efficiently pack these clusters to fill space.

A schematic of the proposed cluster is shown in Figure 3.15. Basically, a cluster consists of the primary solute, α, which is surrounded by the solvent atoms, Ω. Additionally, two topologically distinct solutes exist. These are the secondary β solute atoms that occupy the cluster-octahedral interstices and the tertiary γ solute atoms that occupy the cluster-tetrahedral interstices. (The positions of the cluster-tetrahedral interstices are not shown in Figure 3.15.) Therefore, this structural model contains only four topologically distinct atomic constituents—Ω, α, β, and γ. The solutes have specific and predictable sizes relative to the solvent atom. The efficient packing of these clusters is achieved by choosing the solute atoms that have a radius ratio close to the R^* values mentioned earlier. Solute atoms that have atomic radii within ±2% are considered topologically equivalent. Accordingly,

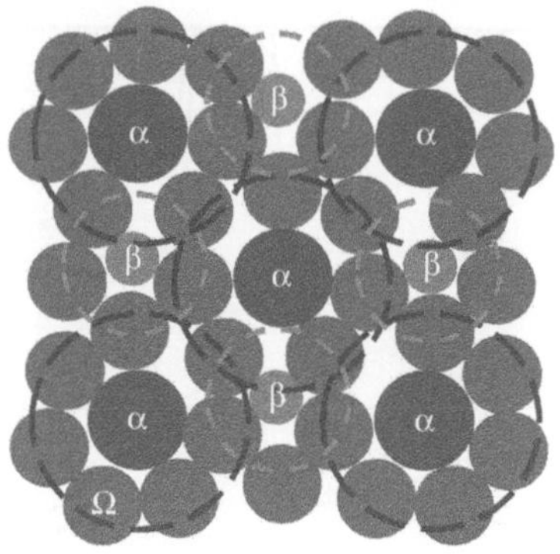

FIGURE 3.15
Schematic two-dimensional representation of a dense cluster-packing structure of clusters in a (1 0 0) plane. The figure illustrates the features of interpenetrating clusters and efficient atomic packing around each solute atom. The solvent spheres form relaxed icosahedra around each α solute. There is no orientational order among the icosahedral clusters. (Reprinted from Miracle, D.B., *Nat. Mater.*, 3, 697, 2004. With permission.)

it is shown that irrespective of the number of chemically distinct solute species, the structural model of Miracle for BMGs contains no more than three topologically distinct solutes.

This model has been termed the *efficient cluster packing (ECP) model*, in which adjacent clusters share solvent atoms in common faces, edges, or vertices so that neighboring clusters overlap in the first coordination shell. That is, the ECP structure is comprised of interpenetrating arrays of efficiently packed solute-centered clusters. By considering the atomic order of the solute atoms and placing the solvent atoms randomly around these solute atoms, the structure of the glass could be explained. Because of the ordering of the solute atoms in the cluster, a cluster unit cell is also defined. However, because of the randomness present in the glass, the lengths of the [100], [110], and [111] directions of the unit cell would be different, and the cluster unit cell length is taken as the largest value for face-sharing adjacent unlike clusters.

The stoichiometry or concentration of solutes and the relative sizes of the constituent atoms provides useful information about the structure of the glass. For example, let us consider a system that contains α solute atoms that have $R^*_\alpha = 0.902$ so that the local coordination number is $N_\alpha = 12$. Solutes that occupy β and γ sites are so chosen as to have $R^*_\beta = 0.799$ and $R^*_\gamma = 0.710$. The atoms of this size ensure that the cluster has the most efficient packing of spheres in the first coordination shell of the solute-centered cluster, and then $N_\beta = 10$ and $N_\gamma = 9$ [219]. According to Miracle [155], this model system is designated <12-10-9>, representing the coordination numbers of the α, β, and γ solutes, respectively. An fcc arrangement of α clusters contains one octahedral (β) site and two tetrahedral (γ) sites for each α site. Since the solvent (Ω) atoms are shared between two neighboring α clusters, only six of them belong to one cluster. Thus, each cluster contains six solvent (Ω) atoms, one α atom, one β atom, and two γ atoms. Therefore, the composition of this glass will be $\Omega_6\alpha_1\beta_1\gamma_2$. If some of the sites are not occupied, or if the cluster contains antisite defects, the composition of the glass will be appropriately modified. Further, Miracle [155] suggested that if the cluster does not exhibit fcc symmetry, that is, if $N_\alpha \neq 12$, then the number of Ω atoms per α solute is generalized as

$$N_\Omega = \frac{N_\alpha}{1+(12/N_\alpha)} \tag{3.67}$$

where 12 represents the number of nearest neighbor α clusters in an fcc lattice. A major success of the Miracle model is that it accurately predicts the chemical compositions of a number of metallic glasses in a wide range of simple and complex alloys.

In 1935, Stockdale [221] suggested that the compositions of eutectic points in binary alloy systems should correspond with simple integer ratios of the two constituent atoms. Hume-Rothery and Anderson [222] plotted the frequency of the occurrence of binary eutectic compositions versus alloy compositions and observed clear maxima around compositions that corresponded to A:B atomic ratios of 8:1, 5:1, 3:1, 2:1, and 3:2. Yavari [223] suggested that it is possible to solve the Stockdale/Hume-Rothery eutectic puzzle using the structural model of Miracle [155]. Let us consider the $\Omega_6\alpha_1\beta_1\gamma_2$ composition corresponding to the <12-10-9> cluster packing. If we replace the β and γ atoms (in the octahedral and tetrahedral sites) with additional α atoms, one obtains the composition $\Omega_3\alpha_2$. On the other hand, if the γ sites are filled by α atoms leaving the β sites empty, the composition will correspond to $\Omega_2\alpha$. Other stoichiometries can also be obtained by considering suitable clusters and replacing the sites with atoms of a different kind.

Sheng et al. [224] have recently verified the presence of solute-centered structural entities and the fact that efficient packing of atoms reduces the energy of the system based on systematic experimental and computational analyses.

One of the major deficiencies of the ECP model is that it is not truly predictive, since it does not directly account for chemical and internal strain contributions to the stability of the glass. To account for this, Wang et al. [225] modified this model and called it the modified ECP (MECP) model. The characteristic features of MECP are as follows:

First, it considers ternary alloy systems in the sense that there are three topologically different species, that is, they contain one dominant solvent (Ω) and two solute elements (α and β). Second, unlike in the Miracle model, where the larger size of the solute atoms was considered, the MECP model considers the solute atom with the larger negative heat of mixing with the solvent Ω, as the primary cluster-forming solute (α). The other solute atom is taken as the secondary solute atom (β). Third, the β solutes share Ω with α, forming β-centered clusters alternating with α-centered clusters. This can be viewed as β entering the cluster-octahedral interstitial sites. Lastly, solute atoms with atomic radii within ±2% of one another are considered equivalent only when their heat of mixing is zero or nearly zero.

According to the MECP model, a critical concentration of β is achieved by two opposing effects. When the β concentration is too low, some of the cluster-interstitial sites are left unfilled. Consequently, efficient packing is not fully realized. Therefore, a higher concentration of β is favored. However, since the β atom is, in general, not the right size to fit the interstitial site perfectly, increasing the β content leads to increased strain and, consequently, energy cost associated with each β solute atom is added. Too large a concentration of β would make the cluster highly strained, and the ECP scheme becomes topologically unsustainable. Therefore, the optimum concentration of β is one where ECP is maximized.

By comparing the predicted alloy compositions with the experimentally observed BMG compositions, the authors claimed an impressive match, since the differences were, in general, less than a few atomic percent (at.%). For example, it was shown that for the Be-containing BMGs, the predicted composition was close to that of the Zr-based BMG (Vitreloy 1), while the original ECP model gave a discrepancy of about 20 at.% from the experimental composition.

For systems with a very high concentration of solute atoms, the solute-centered cluster packing models (ECP and MECP) will not be relevant.

3.12.3 Average Bond Constraint Model

It was mentioned in Section 3.4 that alloy compositions near deep eutectics are favorable for glass formation, but that the actual compositions at which the highest GFA was achieved were away from the eutectic compositions. In order to produce glasses with the largest section thickness, it would be desirable to know the direction and magnitude of the shift from the eutectic composition. Since the optimum composition was found to shift in the direction of increasing $T_g(C)$, according to Equation 3.7, the average bond constraint (ABC) model for estimating $T_g(C)$ near eutectic compositions was proposed [109]. This model is built upon analyses that have been validated for oxide glasses.

In this model, the authors have shown that T_g is proportional to the Kauzmann temperature, T_K (the temperature at which the liquid network becomes completely rigid), and calculated the compositional dependence of $T_K(C)$. In this process, the authors also introduced a degree of covalent bonding into the calculations. This was defined by the parameter μ_c, and $\mu_c=0$ and 1 for pure metallic bonding and mixed covalent/metallic bonding, respectively (this symbol μ_c is different from the symbol μ, normally used to represent the shear modulus of a material). Since the coordination number, Z, is also important, the variation of T_K as a function of composition and coordination number was evaluated and the minimum solute concentration, C_{min}, required for glass formation, was estimated as that solute content at which $T_K=0$. This minimum solute concentration was calculated using the equation:

$$C_{min} = \frac{3}{\left[(1+2\mu_c)Z(C)-3\mu_c\right]} \tag{3.68}$$

A plot of C_{min} calculated in this way was plotted against the coordination number Z (in the range 8–20, where glass formation was observed), as shown in Figure 3.16. It may be generally noted that the minimum solute concentration for glass formation is higher for alloys with metallic bonding than for alloys with some amount of covalent bonding. A few important points may be noted from Figure 3.16.

First, glass formation is predicted in the solute concentration range 15–40 at.% in alloys with purely metallic bonding, and in the range 5–15 at.% in alloys where covalent bonding dominates. Second, there is a conspicuous absence of glass formation for Z = 12–14, which corresponds to the range $0.88 \le R \le 1.12$, supporting the size criterion proposed by Inoue.

By comparing the experimentally determined minimum solute concentrations for glass formation with the values predicted by this model, it becomes clear that the $\mu_c=1$ lower bound provides an accurate representation of the experimentally observed lowest solute concentrations.

The present model suggests that in comparison with the model of Egami and Waseda [119], the numerical value of 0.1 in Equation 3.8 needs to be reduced to 0.07 to provide an accurate lower bound for the observed glass formation data. It is also to be noted that both the ABC and ECP models [155,156] show similar dependence of the glass composition on Z.

3.12.4 Mandal Model

Azad et al. [226] tried to approach the formation of BMGs from geometrical and electronic points of view. Based on the compositions of 200 metallic glasses in 77 different alloy systems, they calculated three parameters—the electron-to-atom ratio (e/a), the

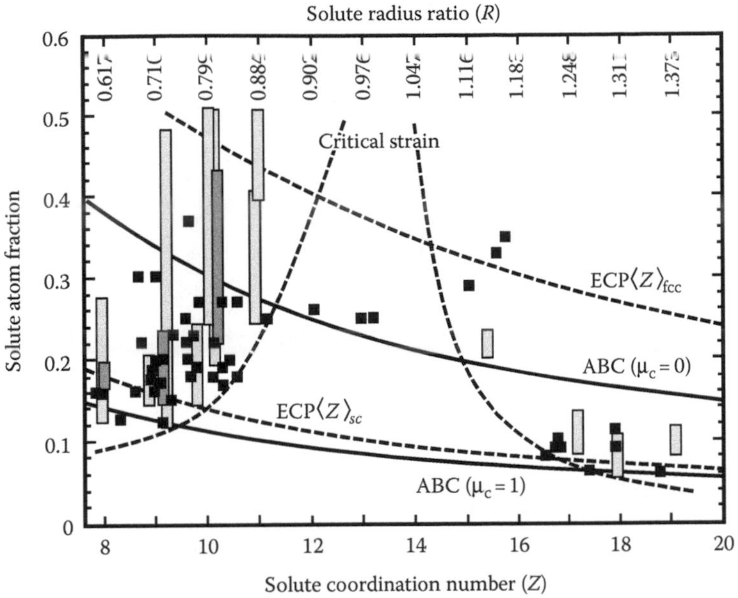

FIGURE 3.16
Solute atom fraction versus the coordination number of solvent atoms around solute atoms calculated according to the ABC model. The squares represent the measured minimum solute concentration and the bars represent solute concentration ranges for selected glasses. The solid curves represent bounds from the ABC model, and dashed lines represent bounds from the ECP model. The minimum solute concentration from the modified critical strain model is indicated by the pair of dotted curves. Radius ratios, R, corresponding to integral values of Z are shown at the top on a nonlinear scale. (Reprinted from Gupta, P.K. and Miracle, D.B., *Acta Mater.*, 55, 4507, 2007. With permission.)

radius ratio of the largest to the smallest size atoms (R_R), and the ratio of volume per atom of the alloy to the volume of the largest atom (V_R). R_R is constant for a given alloy system and does not depend on the relative amounts of the constituent atoms. On the other hand, both e/a and V_R are functions of the alloy composition in a given alloy system. By calculating these three different parameters for the available metallic glasses, they showed that all the BMG compositions satisfied at least one of the three following criteria:

1. The e/a ratio lies between 1 and 3.
2. The R_R value is in the range 1.25–1.65.
3. The V_R value is in the range 0.7–0.85.

Figure 3.17 plots these three values in the form of a Venn diagram, from which a few important points become clear. First, it is noted that 75 of the 77 systems satisfied at least one of the three criteria. Second, 67 of the 77 systems satisfied the e/a criterion, and 66 of the systems satisfied the R_R criterion. Third, 43 systems satisfied all three criteria, and they lie in the intersection region of the three circles, and 65 systems satisfied the e/a and R_R values. Lastly, none of the 43 systems lying in the intersection of the three circles contained any metalloid atoms, such as boron, carbon, silicon, and germanium. But, the significance of the alloys lying in the intersecting region of the three circles is not

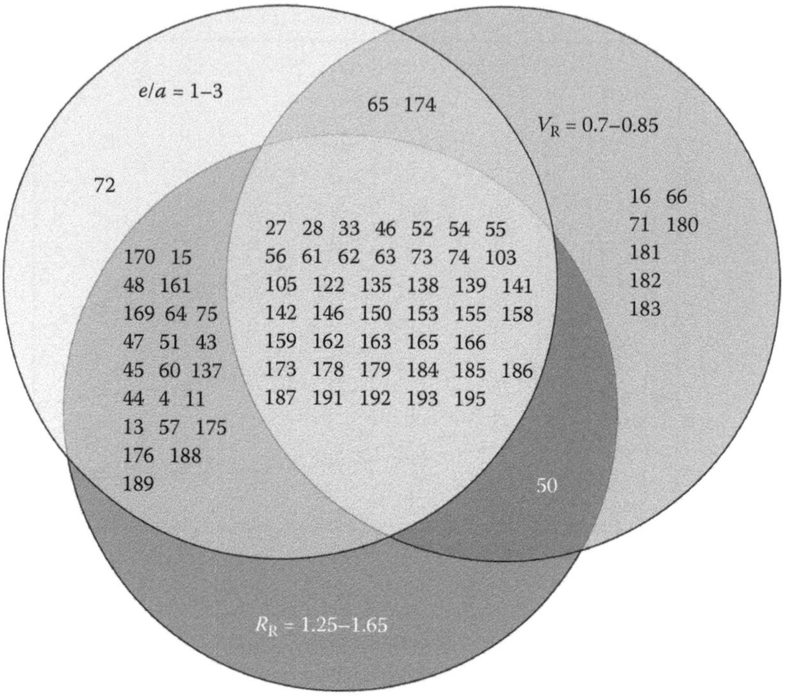

FIGURE 3.17
Venn diagram depicting the three important parameters, *e/a*, R_R, and V_R for 77 different BMG systems. (Reprinted from Azad, S. et al., *Mater. Sci. Eng. A*, 458, 348, 2007. With permission.)

clear. For example, the best glass former that could be cast into a 72 mm diameter rod ($Pd_{40}Cu_{30}Ni_{10}P_{20}$, Alloy #67 in the listing of the authors) is outside the *e/a* range required, with $e/a = 3.0–3.45$, and does not lie in the intersecting region of the three circles. Even the moderately good glass former, *viz.*, $Pd_{40}Ni_{40}P_{20}$, Alloy #72 in their listing, satisfies only the *e/a* criterion. Thus, while this diagram has its usefulness, it appears that its applicability may be limited to only BMG compositions that do not contain any metalloid atoms.

Based on the diffraction evidence from BMGs, these authors have suggested that BMGs contain essentially five different types of atomic clusters, which have close relationship to intermetallic phases with the space groups *I4/mcm* (tI12), *P6/m* (hP68), *Fddd* (oF48), *Fd3m* (cF24), and $Al_{11}Sm_3$, whose space group is unknown. (The symbols in parentheses represent the Pearson symbols for these structures.)

3.12.5 The σ Parameter

As discussed in the preceding sections, the atomic sizes of the constituent elements play an important role in the formation of both conventional metallic glasses and BMGs. The atomic sizes and their mismatch were shown to be important in both the Egami and Waseda [119] and Miracle and Senkov [217,218] models earlier. Continuing this approach, Park et al. [227] considered a combination of the melting point depression of the liquid due to the presence of different alloying elements and the atomic size mismatch among the different alloying elements to evaluate the GFA of alloys.

The melting point depression, ΔT^*, as suggested by Donald and Davies [149], was calculated using the equation:

$$\Delta T^* = \frac{T_m^{mix} - T_\ell}{T_m^{mix}} \tag{3.69}$$

where:

T_ℓ is the liquidus temperature

$T_m^{mix} = \sum_i^n x_i . T_m^i$ with x_i and T_m^i representing the mole fraction and melting temperature, respectively, of the ith constituent in the n-component alloy

The atomic size mismatch parameter for a binary alloy was calculated by the method of Egami and Waseda [119] in terms of the atomic volumes of the constituent elements as

$$\text{Atomic size mismatch} = \frac{V_B - V_A}{V_A} \tag{3.70}$$

where the subscripts A and B represent the solvent and solute, respectively. This concept of calculating the atomic strain was extended to ternary alloys [227] using the P parameter, defined as follows:

$$P = x_B \left| \frac{(V_B - V_A)}{V_A} \right| + x_C \left| \frac{(V_C - V_A)}{V_A} \right| \tag{3.71}$$

This atomic size mismatch was normalized with respect to the total amount of the solute as

$$P' = \frac{x_B}{x_B + x_C} \left| \frac{(V_B - V_A)}{V_A} \right| + \frac{x_C}{x_B + x_C} \left| \frac{(V_C - V_A)}{V_A} \right| \tag{3.72}$$

that is,

$$P' = \frac{P}{x_B + x_C} \tag{3.73}$$

The P' parameter represents the effective atomic size mismatch of each solute atom, and is dependent on the alloy system and composition. The GFA of the alloy was then considered to be proportional to the product of these two parameters, and so the authors proposed the σ parameter as

$$\sigma = \Delta T^* \times P' \tag{3.74}$$

It was suggested that the larger the σ value, the easier it is for the melt to become glassy. These authors also plotted the largest diameter of the glassy rod (a measure of the GFA), t_{max}, against the different proposed parameters—ΔT_x, T_{rg}, γ, σ, and ΔT^*—and observed that the correlation function, R^2, value for the σ parameter was 0.903, the highest among the different parameters. The R^2 value was only 0.478 for the T_{rg} parameter and 0.501 for the γ parameter. The scatter was too much to obtain any correlation for the other parameters.

The authors [227] expressed the relationship between the σ parameter and log t_{max} by the equation:

$$\log_{10} t_{max} = (-0.70 \pm 0.19) + (8.69 \pm 1.14)\sigma \quad (3.75)$$

Lu et al. [165] were quite critical of the suitability of the σ parameter in explaining the GFA of alloys, and commented (1) that it was useful to predict the GFA of only binary or ternary alloys and (2) that the effectiveness of the σ parameter was assessed based on very limited data. By plotting the data and calculating the correlation factors between the γ parameter and σ parameter with t_{max}, Lu et al. [165] claimed that the R^2 value for $\gamma - t_{max}$ was around 0.60, while it was only 0.21 for $\sigma - t_{max}$. It is difficult to resolve this conflict of two entirely different values of R^2 for the same set of data, unless independent and impartial calculations of the R^2 values are made again.

In defense of Park et al. [227], it should, however, be mentioned that just as they extended the strain calculation from a binary alloy to a ternary alloy, it should also be possible to extend the atomic size mismatch calculation to higher-order systems. But, of course, a major assumption in this case is that the individual solute elements do not interact and that the strain contributed by each of the solute elements is additive in nature.

A serious shortcoming of the σ parameter that was pointed out by Lu et al. [165] is that it does not take into consideration the effect of melt treatment on the GFA of the alloy systems. For example, in $Pd_{40}Ni_{40}P_{20}$, the t_{max} value can be increased from 3 mm without the flux treatment to 10 mm with a flux melting treatment. While the T_g and T_ℓ values remain the same with the flux treatment, the T_x value increases by 29 K [21]. Consequently, the γ value increases from 0.415 to 0.432, but there is no effect of this increase in T_x on the σ parameter. Similarly, the addition of small amounts (at the ppm level) of appropriate alloying elements to Zr-based alloys is known to increase the T_x, but not T_g and T_ℓ [228,229]. Here again, the γ value is higher, and reflects the increased GFA; but, the σ parameter will be ineffective in predicting this increased GFA.

3.12.6 Fang's Criterion of Bond Parameters

It was mentioned earlier that the magnitude of the supercooled liquid region, $\Delta T_x (= T_x - T_g)$ is related to the GFA of alloys [125]; the GFA increases with increasing values of ΔT_x. Exceptions to this criterion have been noted in some cases, as described in Section 3.7.1. Assuming that bond parameters such as differences in electronegativity and atomic sizes of constituent elements should also affect the GFA of alloys, Fang et al. [230] calculated the electronegativity differences (Δx) and atomic size parameters (δ) using the following equations:

$$\Delta x = \sqrt{\sum_{i=1}^{n} C_i \cdot (x_i - \overline{x})^2} \quad (3.76)$$

where:

- n is the number of components in the alloy
- x_i is the Pauling electronegativity of element i, obtained from the literature
- C_i is the atomic percentage of element i in the alloy
- $\overline{x}$ is the arithmetical mean of the electronegativity of the compound, calculated from the relationship:

$$\bar{x} = \sum_{i=1}^{n} C_i x_i \tag{3.77}$$

The atomic size parameter, δ, is calculated from the relationship:

$$\delta = \sqrt{\sum_{i=1}^{n} C_i \cdot \left(1 - \frac{R_i}{\bar{R}}\right)^2} \tag{3.78}$$

where:
R_i is the covalent atomic radius of element i
$\bar{R}$ is the arithmetical mean of the covalent atomic radius of the alloy, calculated as

$$\bar{R} = \sum_{i=1}^{n} C_i \cdot R_i \tag{3.79}$$

By plotting the ΔT_x values for several Mg-based BMGs against the electronegativity difference (Δx) and also the atomic size parameter (δ), the authors noted that ΔT_x (and consequently the GFA) increased with increasing Δx and δ values.

Assuming that ΔT_x is proportional to the two bond parameters, Δx and δ, Fang et al. [230] derived an equation for ΔT_x as

$$\Delta T_x = a_0 + a_1 \cdot \Delta x + a_2 \cdot \delta \tag{3.80}$$

where a_0 is a constant that represents the ΔT_x value of a pure metal ($\Delta x = 0$ and $\delta = 0$). For Mg-based BMGs, the authors found that

$$\Delta T_x = -13.52254 + 113.35386 \times \Delta x + 264.37921 \times \delta \tag{3.81}$$

The negative value of a_0 suggests that it is impossible to produce the pure metal Mg in a glassy condition. Since the ΔT_x value of BMGs is always finite, the authors also noted that it is possible to produce BMGs in Mg-based alloys, when

$$8.38 \times \Delta x + 19.55 \times \delta > 1 \tag{3.82}$$

These authors have subsequently extended this concept to Fe-based BMGs [231] by also including the valence electron difference (Δn) defined as

$$\Delta n^{1/3} = \sum_{i=1}^{n} C_i \left[n_i^{1/3} - (\bar{n})^{1/3} \right] \tag{3.83}$$

For the transition metals, the element electron concentration, n_i, was used as the number of (s + d) electrons, and for elements with a p electronic structure, the number of (s + p)

electrons was taken. The number of mean valence electrons among different atoms is represented by n and is given by

$$\bar{n} = \sum_{i=1}^{n} C_i n_i \tag{3.84}$$

Using these parameters, the correlation between ΔT_x and the three bond parameters for Fe-based BMGs was found to be [231]

$$\Delta T_x = 29.6 + 244.6 \times \Delta x^2 + 1001.5 \times \delta^2 + 154.0 \times \left(\Delta n^{1/3}\right)^2 \tag{3.85a}$$

Similar equations were derived for Mg- and Pd-based BMGs as

$$\Delta T_x = 12.08 + 130.2 \times \Delta x^2 + 711.8 \times \delta^2 + 0.44 \times \left(\Delta n^{1/3}\right)^2 \tag{3.85b}$$

for Mg-based BMGs, and

$$\Delta T_x = -29.93 + 2504.0 \times \Delta x^2 + 9067.9 \times \delta^2 + 128.4 \times \left(\Delta n^{1/3}\right)^2 \tag{3.85c}$$

for Pd-based BMGs.

(Note that the value of a_0 was mentioned as 12.08 for Mg-based BMGs in Equation 3.85b, and this value was shown in Equation 3.81 to be −13.52. This was probably an error, and the value in Equation 3.85b should also be −12.08.) The negative sign of this constant was interpreted to mean that it was impossible to produce the pure metal Mg in a glassy condition. The a_0 value for the Fe-based BMGs was estimated as 29.6. This does not mean, however, that it is possible to obtain pure Fe in the glassy state. The physical significance of a_0 and its sign is presently not clear.

The differences in the valence electron concentration $\Delta n^{1/3}$ for Fe-based BMGs were noted to be in the range 0.1547–0.2525. Substituting this range in Equation 3.85a for Fe-based BMGs, the authors showed that for BMGs to be obtained with a ΔT_x of at least 50 K, one needs to satisfy the criterion:

$$\frac{\Delta x^2}{(0.21)^2 - (0.26)^2} + \frac{\delta^2}{(0.1)^2 - (0.13)^2} \geq 1 \tag{3.86a}$$

Similar equations were also derived for Mg- and Pd-based BMGs, which are

$$\frac{\Delta x^2}{(0.54)^2} + \frac{\delta^2}{(0.23)^2} \geq 1 \quad \text{for Mg-based BMGs} \tag{3.86b}$$

and

$$\frac{\Delta x^2}{(0.17)^2} + \frac{\delta^2}{(0.09)^2} \geq 1 \quad \text{for Pd-based BMGs} \tag{3.86c}$$

Figure 3.18 shows the δ–Δx parameter plots for Fe-, Mg-, and Pd-based BMGs. The regions outside the quarter-arc ellipse show the regions where BMGs with a ΔT_x value of at least 50 K can be produced. In accordance with the available experimental results, it appears from Figure 3.18 that it is easiest to produce BMGs in the Pd-based alloy systems.

Lu et al. [165], however, noted that there were no clear trends between these two parameters and ΔT_x for Zr-based BMGs. It will be interesting to check if good correlations can be found for other BMGs. Further, it has been shown clearly that there is no direct and clear correlation between ΔT_x and the GFA of alloys. Therefore, it is not clear whether a new criterion based on a parameter that has been shown to be ineffective in predicting (or at least correlating) the GFA of alloys will prove useful.

As mentioned earlier, Fang et al. [230,231] established a linear correlation between GFA (as indicated by ΔT_x) and the atomic size parameter, δ, electronegativity difference, Δx, and also the valence electron difference, Δn. But, they did not consider the relationship between these parameters and t_{max}, the most popular criterion used by many other researchers. Therefore, Liu et al. [232] analyzed the available results in BMGs based on Ca, Mg, Zr, Ti, Cu, Au, Pt, Pd, Fe, Ni, and RE (rare earth) elements in terms of differences between atomic sizes, electronegativity, and valence electron concentration. These authors noted a poor correlation of these parameters with both ΔT_x and t_{max} in the BMGs chosen.

Liu et al. [232] argued that too small a difference in atomic sizes or electronegativities is likely to produce solid solutions and too large a difference in these parameters is likely to result in the formation of compounds. Hence, it was suggested that these parameters should be optimized. Therefore, to obtain a better correlation, they combined two

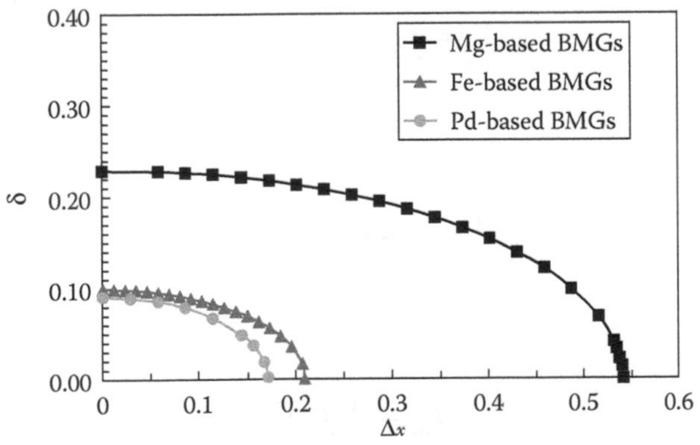

FIGURE 3.18
δ–Δx parameter plots for Fe-, Mg-, and Pd-based BMGs. The region outside the quarter-arc ellipse shows the BMG formation region with a ΔT_x value of at least 50 K. (Reprinted from Fang, S.S. et al., *Intermetallics*, 12, 1069, 2004. With permission.)

electronegativity difference parameters (L and L') and three atomic size ratio parameters (W, W', and λ_n), as follows:

$$L = \sum_{i=1}^{n} C_i \left| x_i - \bar{x} \right| \tag{3.87a}$$

$$L' = \sum_{i=1}^{n} C_i \left| 1 - \frac{x_i}{\bar{x}} \right| \tag{3.87b}$$

$$W = \sum_{i=1}^{n} C_i \left| R_i - \bar{R} \right| \tag{3.87c}$$

$$W' = \sum_{i=1}^{n} C_i \left| 1 - \frac{R_i}{\bar{R}} \right| \tag{3.87d}$$

$$\lambda_n = \sum_{i=2}^{n} C_i \left| 1 - \left(\frac{R_i}{R_1} \right)^3 \right| \tag{3.87e}$$

where x and R represent the electronegativity and atomic radius, respectively. The authors suggested that the GFA of the alloy will be limited if either the atomic size and/or the electronegativity are unfavorable. But, they felt that even if these are favorable, additional factors may need to be considered. Accordingly, they considered the valence electron difference, Y, and the reduced melting temperature, T_{rm}, defined as

$$Y = \sum_{i=1}^{n} C_i \left| n_i^{1/3} - \bar{n}^{1/3} \right| \tag{3.88}$$

$$T_{\mathrm{rm}} = \sum_{i=1}^{n} C_i \left| 1 - \frac{T_{\mathrm{m}i}}{\bar{T}_{\mathrm{m}}} \right| \tag{3.89a}$$

with

$$\bar{T}_{\mathrm{m}} = \sum_{i=1}^{n} C_i T_{\mathrm{m}i} \tag{3.89b}$$

where:

- C is the mole fraction
- T_{m} is the melting temperature
- $\bar{T}_{\mathrm{m}}$ is the melting temperature calculated from the simple rule of mixtures

By combining Equations 3.87a through 3.89a, the authors obtained an equation of the type:

$$\ln f(u) = A_0 + A_1 L + A_2 L' + T_{rm}\left(A_3 W + A_4 W' + A_5 \lambda_n\right) + A_6 Y \tag{3.90}$$

where:

$f(u)$ represents the as-calculated value (for $u = t_{max}$, R_c, ΔT_x, T_g, T_x, or T_ℓ)

A_i $(i = 1–6)$ represents the coefficients for L, L', W, W', λ_n, and Y, respectively, which can be calculated from the data available in the literature

These coefficients are different for different GFA indicators such as t_{max}, R_c, ΔT_x, and so on. Analyzing the BMG data, these authors came up with empirical formulae for different parameters. For example, the empirical relationships for the t_{max} and R_c are

$$\ln t_{max} = -1.50 + 27.84L - 36.43L' + T_m(-1995.35W + 475.55W' - 25.15\lambda_n) + 11.10Y \tag{3.91a}$$

$$\ln R_c = 10.81 - 136.15L + 177.69L' + T_{rm}(-3337.61W + 288.21W' - 48.64\lambda_n) - 17.38Y \tag{3.91b}$$

Using this approach, the authors reported that the correlation coefficient between the calculated and experimental t_{max} values was as high as 0.85, suggesting a strong correlation.

3.13 Physical Properties

While discussing the kinetics of glass formation, it was emphasized that the viscosity of the liquid plays an important role in determining the GFA of alloys. While it is relatively easy to determine the viscosity of the BMGs near the glass transition temperature, T_g, it is much more difficult to obtain such data near the melting point, where the critical cooling rate required to form the glass is likely to be determined. Cohen and Grest [233] showed that a correlation exists between the free volume and viscosity. Therefore, the measurement of either the free volume or the viscosity of the alloy in both the crystalline and glassy states should enable one to establish the criteria for glass formation.

3.13.1 Minimum Molar Volume

Varley [234] calculated the misfit energy in a liquid and noted that the atomic volumes and isothermal compressibilities of the constituent elements are important. Using this approach, Ramachandrarao [235], estimated the critical composition in a binary alloy system at which the decrease in mean atomic volume is highest (at which it is also expected to have the highest viscosity) as

$$X_B = \frac{1}{1+\sqrt{\frac{V_A \chi_A}{V_B \chi_B}}} \tag{3.92}$$

where:

X_B	represents the critical solute concentration
V_A and V_B	are the atomic volumes of elements *A* (solvent) and *B* (solute), respectively
χ_A and χ_B	are the isothermal compressibilities of pure A and pure B, respectively, in their liquid states

Based on an analysis of over 35 binary metal–metal systems, which were shown to be glass formers by RSP and vapor deposition methods, Ramachandrarao showed that the composition at which the binary liquid alloy volume was minimum was always within the experimentally determined glass formation range. In other words, if one could calculate the solute content at which the molar volume was the minimum, then the glass-forming composition range would be centered on that value. The author also showed that the glass-forming composition range was asymmetrical with respect to the minimum value calculated (X_B). This seems to be particularly true when the solvent atom has an unusually large atomic volume, typical examples shown being Ca and Te. This asymmetry in the glass-forming composition range was explained on the basis of the atomic volumes and compressibilities of the elements. Elements having comparable $V \times \chi$ or χ values were shown to form glasses near equiatomic composition, whereas alloys consisting of elements with widely different values of these parameters formed glasses at the solvent-rich compositions.

A similar idea was also developed by Yavari [236] who calculated the change in volume, ΔV_s, from the known densities of glassy, liquid, and crystalline states. He showed that glass formation by rapid solidification was easy if the volume change was either zero or negative and that it was difficult if the volume change was between 2% and 6% of the volume per atom in the crystalline state. Cahn [101] has pointed out that both approaches are equivalent when the alloy having the minimum molar volume in the liquid state also has the lowest value of the volume change, ΔV_s.

Ma et al. [237] reported that the excess molar volumes of metalloid-free BMGs were essentially zero, suggesting that the original volumes of "mixed" constituent elemental metals are conserved after glass formation. This was attributed to the lack of solute–solute bonds (that is, solute-centered clusters in the glass) and the ideal mixing nature of the solvent–solute bonding upon amorphization.

3.13.2 Density

As mentioned earlier, GFA is expected to be correlated with the degree of dense packing. It is also known that metallic glasses have a high degree of densely packed atomic structures [130,238,239] and several models of the glasses are based on the concept of efficient dense random packing of atoms [119,155,156,224,240]. However, experimental evidence to establish a clear relationship between GFA and density of glasses has not been available. Xu et al. [241] investigated the glass formation in ternary $Ce_{70}Ga_xCu_{30-x}$ alloys with $x = 4–15$ and noted that the best glass formers were in the Ga composition range of $x = 6–10$. By measuring the density of the glasses using the Archimedes method, the authors found that best glass-forming compositions have the highest density values. In this composition range, the atomic packing fraction also was the highest and the molar volume the lowest. It was further stated that efficient packing of the atoms was determined not only by the atomic radii of the constituent atoms but also by their chemical interaction effects, which will significantly change the atomic distance/volume or coordination number, and consequently, the density. In fact, Yang et al. [242] noted that bond shortening occurs in the atomic pairs in $Zr_{48}Cu_{45}Al_7$ metallic glass, due to the strong interaction between the Al atoms and their neighbors, which leads to the atomic- and cluster-level dense packing of atoms, resulting in increased GFA of alloys. However, it is useful to remember that the use of effective atomic radii (due to the interaction between different elements) will better predict the GFA of alloys, as shown in Al–Ni–RE (RE = La, Y, Ce, Gd, and Dy) alloys [243].

3.13.3 Viscosity

By measuring the specific volume and viscosity of four different BMGs of widely differing GFAs (the critical cooling rates were very different and varied from 2 to 250 K s^{-1}) at different temperatures, Mukherjee et al. [44] observed that the viscosity of the best glass-forming alloy was an order of magnitude larger than that of the worst glass former, at their respective liquidus temperatures. Further, the best glass-forming alloy showed the smallest volume change upon crystallization and the worst glass former showed the largest change. Therefore, it could be concluded that a high viscosity and correspondingly small free volume in the liquid at the melting temperature contribute significantly to improve the GFA of alloys.

The variation of viscosity, η, with temperature, T can be expressed by the VFT equation:

$$\eta(T) = \eta_0 \exp\left(\frac{D^* T_0}{T - T_0}\right) \tag{3.27}$$

where:

T_0 is referred to as the *VFT temperature*
D^* is called the *fragility parameter*
η_0 is the high-temperature limit of viscosity

The fragility parameter provides an index of the stability of the glass; the higher the value of D^*, the higher is the stability of the glass. Melts having a large D^* value are referred to as *strong liquids*. Mukherjee et al. [44] reported that the fragility parameter is also a direct measure of the GFA. This was found to be true in BMGs with all metallic elements [244,245], but not in BMGs containing metalloid elements [246]. For example, the BMG with the composition $Pd_{43}Ni_{10}Cu_{27}P_{20}$, despite having the best GFA reported so far with $R_c < 0.1$ K s^{-1}, is relatively fragile compared with other BMGs that contain all metallic elements. This difference in behavior between these two types of glasses was attributed to the difference in their liquid structures.

3.13.4 The *New* δ Parameter

Arguing that the specific volume of a metallic liquid cannot be smaller than that of the corresponding crystal, if a metal contracts on melting, Louzguine-Luzgin and Inoue [247] suggested that the intersection point of specific volume versus temperature plots for the supercooled liquid and the solid could be treated as an ideal glass transition temperature. Therefore, the authors proposed a new criterion for the ease of glass formation as

$$\text{GFA} = \alpha_\ell \times V_\ell \times \frac{T_m - 298}{\Delta V_{\ell-s}} \tag{3.93}$$

where:

- α_ℓ is the coefficient of thermal expansion of the liquid
- V_ℓ is the slope of the variation of liquid volume with temperature
- T_m is the melting temperature
- $\Delta V_{\ell-s}$ is the volume change from the liquid–solid transformation

However, since the variation of density with temperature is almost linear (and the specific volume shows some deviation from linearity), the authors rewrote their criterion as

$$new\,\delta = \alpha_\ell \times \rho_\ell \times \frac{(T_m - 298)}{\Delta\rho_{s-\ell}} \tag{3.94}$$

where:

- ρ_ℓ represents the slope of the resistivity of the liquid versus temperature plot, that is, $d\rho\ell/dT$
- $\Delta\rho_{s-\ell}$ represents the change in density during the solid–liquid transition

(Note that this parameter is redesignated *new* δ, since Chen et al. [172,173] had earlier proposed a δ criterion to explain the GFA of alloys.)

A large value of *new* δ is an indication of a high T_g and consequently increased GFA. As the authors point out, the *new* δ criterion indicates how fast the liquid can reach the glass transition temperature, but it does not provide information on the stability of the resulting glass against crystallization.

This criterion was based on data for pure metals. But, a similar trend is expected to be followed by alloys. Therefore, for an alloy to have a large GFA, it is necessary that it should have a large *new* δ value.

3.14 Miscellaneous Criteria

3.14.1 Enthalpy of Vacancy Formation

Buschow and Beekmans [248] studied the crystallization behavior of a number of binary metal–metal glasses produced by RSP methods using DSC and electrical resistivity methods. They noted that there was a linear relationship between the experimentally determined crystallization temperature and the energy for the formation of a vacancy, ΔH_v. Therefore, it was suggested that the crystallization temperatures of glasses could be estimated even in those alloys that have not been investigated.

Recently, it has also been shown that the T_x of BMG alloys based on Ca, Mg, Zr, and Cu varies linearly with ΔH_v [249]. Further, the Gibbs free energy will be an indicator of the melting point, T_ℓ, of the alloys. Combining these two effects, the authors [249] proposed that the GFA of BMG alloys can be estimated from the values of $\Delta H_v \Delta H^{mix} S_{config}$, where ΔH^{mix} is the enthalpy of mixing and S_{config} is the configurational entropy. It was suggested that the best glass-forming composition, that is, the composition that shows the highest GFA, has the lowest value of $\Delta H_v \Delta H^{mix} S_{config}$. This has been confirmed by plotting the values of $\Delta H_v \Delta H^{mix} S_{config}$ in the ternary phase diagram and noting that the compositions having the lowest $\Delta H_v \Delta H^{mix} S_{config}$ values coincide with the best glass-forming composition range. It was, however, noted that the best GFA compositions could be predicted by computing the $\Delta H_v \Delta H^{mix} S_{config}$ values in the 45–65 at.% range of the main constituent element.

Some attempts have also been made to theoretically predict the GFA of BMG alloys through a combination of the electronic theory of alloys in the pseudo-potential approximation and the statistical thermodynamic theory of liquid alloys. Suer et al. [250] calculated the magnitude of ordering energies, based on the electronic theory of alloys, and used this to calculate the key thermodynamic parameters such as enthalpy, entropy, and Gibbs free energy of mixing, viscosity, and critical cooling rate for glass formation in binary Ti–Cu and ternary Ti–Cu–X alloys. The alloying elements (X) were grouped into two categories based on their ability to improve the GFA. Relative to the binary alloy, the first group of elements caused an increase in the negative heat of mixing, ΔH^{mix}, and the second group caused a decrease in the critical cooling rate, R_c. While a decrease in the value of R_c in improving the GFA is obvious, it was suggested that a large negative heat of mixing would lead to a rapid increase in viscosity on undercooling due to a decrease in the atomic diffusivity, caused by the addition of alloying elements. Further, even though both aspects mentioned seem to be important, the increase in the negative heat of mixing appears to be less significant in determining the GFA of alloys.

3.14.2 Electron Concentration

Similar to the Hume-Rothery alloys, Nagel and Tauc [251, 252] proposed a mechanism by which they were able to find conditions at which the metallic glasses would be stable and the reason for the observation of metal–metalloid glasses forming at the 80:20 composition ratio. They showed that a metallic glass will be stable when its electronic energy lies in a local metastable minimum with respect to composition change. They showed that if the structure factor corresponding to the first strong peak of the diffuse scattering curve, K_p, satisfies the relationship $K_p = 2\,k_F$, where k_F is the wave vector at the Fermi energy, then the electronic energy does indeed occupy a local minimum in the density of states curve, that is, the Fermi sphere with the diameter $2k_F$ will touch the Brillouin zone defined by K_P. The number of valence electrons per atom (e/a) at which this happens was estimated to be 1.7. Some of the other theoretical models have been summarized by Hafner [253].

This e/a value has been estimated differently by different investigators. For example, Häussler [254] suggested a value of 1.8, and these glasses were referred to as *ideal glasses* by him. On the other hand, values of 1.5 [255] and 3.5 [256] were also suggested. The assignment of e/a values is complicated by sp-d hybridization effects, especially in transition metals. To avoid this confusion, Jiao et al. [257] analyzed the GFA of Ca–Mg–Cu alloys in terms of the critical diameter for glass formation and noted that the best GFA was observed when the e/a ratio was 1.75, which is close to the "ideal" value mentioned earlier. Han et al. [258] combined the "cluster-plus-glue-atom model" [259] with Haussler's global resonance model to develop the cluster-resonance model and derived equations for the calculation of e/a values for ideal metallic glasses.

3.14.3 Local Ordering

Bian et al. [260] also noted that there existed a close relationship between the degree of local ordering and the GFA of Cu–Hf alloys. By determining the local atomic ordering in a range of Cu-Hf alloys, they concluded that the $Cu_{70}Hf_{30}$ alloy had the greatest degree of local atomic ordering among all the samples studied. By measuring the glass transformation temperatures using a DSC, it was shown that Hf additions decrease the GFA of the alloys and that $Cu_{70}Hf_{30}$ has the highest GFA. It was suggested that a strong tendency to show chemical short-range order (SRO) (or icosahedral SRO) improves the GFA of the corresponding alloy. This is because the local ordered structures act as random fields against crystallization when the primary crystallization mechanism for the undercooled melts results in the formation of intermetallic crystals.

3.14.4 Fragility Parameter

In Angell's classification [178], liquids are classified as "strong" or "fragile." depending on the variation of viscosity with temperature. It has been perceived that good glass-forming alloys are strong liquids with a high viscosity, while the fragile liquids are only marginal glass formers [261,262]. It has also been suggested that GFA can be a complex function of fragility and the reduced glass transition temperature, T_{rg} [263] or fragility and the onset driving force for crystallization [264]. The fragility provides an indicator of the sensitivity of the structure of a liquid to temperature changes, and a convenient method to measure the fragility of glass-forming liquids is through the fragility index, *m*, defined as [265]

$$m = \frac{d\left[\ln \tau(T)\right]}{d\left(T_g / T\right)} \Big|_{T_g} \tag{3.95}$$

where $\tau(T)$ is the temperature-dependent relaxation time.

Liquids with $m \approx m_{min}$ show the Arrhenius behavior (strong liquids), and a larger departure of m from m_{min} indicates higher fragility. Senkov [266] analyzed the relaxation times of the glass-forming liquids at near-liquidus temperatures and identified a correlation between the critical cooling rate for glass formation, R_c, fragility index of the glass-forming liquid, m, and the reduced glass transition temperature, T_{rg}. A GFA parameter, F_1, defined as

$$F_1 = 2\left[\frac{m}{m_{min}}\left(\frac{1}{T_{rg}} - 1\right) + 2\right]^{-1} \tag{3.96a}$$

or

$$F_1 = \frac{2T_{rg}D}{D\left(1+T_{rg}\right) + m_{min}\left(1-T_{rg}\right)\ln 10} \tag{3.96b}$$

was proposed to explain the GFA of alloys. The value of F_1 varies with T_{rg}, R_c, and m. For example, it increases with a decrease in R_c, with an increase in T_{rg}, and a decrease in m. Its value varies from ~0 for fragile liquids to $2T_{rg}/(1+T_{rg})$ in the case of extremely strong liquids. By plotting F_1 as a function of R_c, the author identified an exponential relationship between these two, expressed as

$$R_c = R_{co}\exp(-AF_1) \tag{3.97}$$

where

$$R_{co} \approx 2.7 \times 10^{11} K^{s-1}$$

$$A = 48.7$$

For an extremely fragile liquid ($F_1 = 0$), $R_c \approx 2.7 \times 10^{11}$ K s^{-1}, and this value is in good agreement with the critical cooling rates required to produce pure metals in the glassy state. On the other hand, for extremely strong liquids, when $T_{rg} \approx 2/3$ and $F_1 \approx 0.8$, the calculated $R_c \approx 2.3 \times 10^{-6}$ K s^{-1} is comparable to the cooling rate required to produce glassy SiO_2. The validity of this parameter was verified in a number of BMG alloys. Using a similar approach, Zheng et al. [267] analyzed the relaxation times, based on the VFT equation, in a series of Mg–Cu(Ag)–Gd alloys and reported the formation of BMGs with critical diameters (t_{max}) in the range 20–27 mm.

Recently, Kozmidis-Petrovic [268] considered factors that could affect GFA, in addition to those considered by Senkov [266], and came up with a new parameter, F_2, defined as

$$F_2 = 16F_1 - \log(0.5)(T_\ell - T_g) - 14 \tag{3.98}$$

Even though F_2 has similar correlations with R_C as F_1, F_2 has the advantage that it is more sensitive to variations in R_c. Two other parameters, F_K and F_{KA}, were also presented as

$$F_K = (T_g - T_o)/(T_\ell - T_o) \tag{3.99}$$

and

$$F_{KA} = (16/17) \times \left[(T_g - T_o)/(T_\ell - T_o)\right] + (T_{rg}/17) \tag{3.100}$$

The parameter F_K was obtained by assuming that the GFA is proportional to liquid viscosity, but there is no correlation between F_K and T_{rg}. F_{KA}, on the other hand, takes into account the influence of T_{rg} on GFA. It was shown that glasses with the same T_{rg} will have higher GFA if they have lower fragility.

3.14.5 Electrical Transport Properties

Based on Anderson's theory and the Mott-CFO model for electrical transport properties of disordered materials, Wang et al. [269] developed an empirical criterion to predict the GFA of glassy alloys. The relative electrical resistivity difference between the glassy and fully crystallized states of the alloy, that is,

$$\Delta\rho = \left(\rho_{\text{glass}}^{\text{RT}} - \rho_{\text{cryst}}^{\text{RT}}\right) / \rho_{\text{cryst}}^{\text{RT}} \tag{3.101}$$

where ρ represents the electrical resistivity and RT the room temperature was shown to be a good indicator of the GFA of the glass. A higher Δρ value of an alloy correlates with a better GFA, unambiguously confirmed in different alloy systems.

3.14.6 Miscellaneous

Liu et al. [270] investigated the effect of melt superheat on the GFA of $Gd_{55}Al_{25}Cu_{10}Co_{10}$ alloys. They increased the melt temperature from 919°C to 1279°C and noted that the transformation temperatures, ΔT_x, and γ values peaked when the melt was solidified from 1105°C. From these observations, they concluded that the GFA increased with increasing superheat in the liquid state and reached its highest value at 1105°C. Using HRTEM studies, they noted that the average size of the local ordered regions was smallest at 1105°C, suggesting that there is an inverse relationship of GFA with the size of the local ordered regions. A similar result was also reported by Cui et al. [271].

Louzguine-Luzgin et al. [272] considered the different factors that affect the GFA of alloys and classified them into intrinsic and extrinsic factors. The intrinsic factors (belonging to the glass itself) assume that homogeneous nucleation competes with glass formation and include a number of fundamental and derived thermal parameters (T_g, T_x, T_ℓ, T_{rg}); physical properties such as heat capacity, thermal diffusivity, and coefficient of thermal expansion, compositional proximity to a deep eutectic composition, and a topological contribution from efficient atomic packing in the structure. The extrinsic factors (depending on external conditions) are usually operative when heterogeneous nucleation intervenes during solidification. These include inclusions or dissolved impurities in the melt, poor mold surface finish or cleanliness, turbulence during solidification, and the degree of liquid metal superheat. Based on their analysis, they concluded that even though an alloy has a high intrinsic GFA, extrinsic factors could severely limit the GFA. Thus, it becomes very important to monitor both the intrinsic and extrinsic factors in evaluating the GFA of alloys.

3.15 Criteria for Glass Formation by Non-Solidification Methods

As mentioned earlier, glasses (or, to be more accurate, amorphous or noncrystalline alloys) can be produced by non-solidification methods also. These include, among others, MA, irradiation, electrodeposition, interdiffusion of multilayers of thin films, application of pressure, and hydrogen absorption (see, e.g., Refs. [273,274]). Of all these methods, MA has been the most popular to synthesize amorphous alloys in binary, ternary, and higher-order systems. References [275,276] may be consulted by the interested reader for a list of the compositions that have been produced in the amorphous condition, including those of BMG compositions.

In general, there have not been many systematic investigations to identify the factors that contribute to amorphous phase formation by MA. This is because MA is a complex process and involves a number of process parameters that include alloy composition, powder particle size, type of mill, type of milling medium, ball-to-powder weight ratio, milling time, milling temperature, milling atmosphere, and the nature and amount of process control agent added, among others. Therefore, there have been only a few parametric studies conducted to relate the GFA to the ball-to-powder weight ratio, power absorption, impact energy, and total energy. Some modeling efforts have also been carried out, but with only a limited success [277]. Further, a number of studies have also attempted to reproduce the formation of amorphous phases by MA in alloy compositions that have already been proven to form glasses by solidification methods.

The mechanism by which amorphous phase formation occurs in MA has not been unambiguously determined. The earliest mechanism that was proposed was of a substantial local rise in temperature causing the powder particles to melt locally. The large volume of cold powder surrounding the molten particles would cause rapid solidification of the melt and result in the formation of an amorphous phase. If this were to be the case, the composition range in which an amorphous phase is produced should be the same in both RSP and MA processes. But, experimentally it has been shown in a number of cases that the composition range in which amorphous phases form by MA and RSP methods are not necessarily the same; in fact, they are quite different. Further, the composition range in which an amorphous phase formed was much wider by MA than by RSP methods. Therefore, this mechanism has been discarded. Further, it was also noted that the rise in temperature was not enough to cause melting of the constituent elements in the majority of the cases. The estimated global rise in temperature was only about 100–200 K.

Phenomenologically, it has been shown that an amorphous phase is formed by MA methods when the milled material reaches very small grain sizes of a few nanometers. This minimum value, below which it becomes amorphous, has been mentioned to be in the range 5–20 nm [275,276,278,279].

Another mechanism that is currently accepted by most investigators is that, due to heavy deformation experienced by the powder particles, a large number of different crystal defects and lattice strain are introduced into the system. These defects raise the free energy of the crystalline system to a level above that of the hypothetical amorphous phase. Consequently, the crystalline lattice becomes destabilized and the amorphous phase becomes stabilized. This situation can be expressed in terms of a simple equation:

$$\Delta G_c + \Delta G_d > \Delta G_a \tag{3.102}$$

where:

ΔG_c is the free energy of the crystalline lattice
ΔG_d is the increase in free energy due to incorporation of crystal defects
ΔG_a is the free energy of the hypothetical amorphous phase

The increase in free energy is mainly achieved through generation and increase in the concentration of vacancies, dislocations, grain boundaries, and lattice strain introduced by the incorporation of atoms of different sizes into the host lattice. Since the defects can be annealed out at higher temperatures, amorphous phase formation is favored at relatively lower temperatures.

A number of modeling studies have also been undertaken in recent times to understand the kinetics of amorphization achieved by MA methods [277,280].

Some important differences have also been noted in the amorphous phase formation by RSP and MA methods. These differences include the composition ranges in a given alloy system (as previously mentioned) and the type of alloy system. For example, alloy systems with deep eutectics have been noted to be good glass formers by RSP methods. But, such is not the case by MA methods. This has been attributed to the fact that MA is carried out at or near room temperature and, therefore, the nature of the phase diagrams is not expected to be an important consideration. We will, however, see later that this is not necessarily true. Further, the composition range over which an amorphous phase is formed is wider in alloys processed by MA in comparison to those processed by RSP methods.

Coming to the specific case of BMGs, it has been noted that while some compositions can be produced in the amorphous state both by MA and solidification methods, there have been reports of compositions that were made glassy by solidification methods but that could not be made amorphous by MA methods. For example, $Fe_{72}Al_5Ga_2C_6B_4P_{11}$ is a well-known soft magnetic alloy [281]. Attempts were made to produce this alloy composition in the amorphous state. While Schlorke et al. [282] reported the formation of an amorphous phase by milling in a Retsch PM4000 planetary ball mill at 250 rpm speed and a ball-to-powder weight ratio of 15:1, Chueva et al. [283] reported that it was difficult to produce this alloy in the amorphous state from blended elemental powders due to the widely different mechanical properties of the constituent elements. Instead, they were able to produce an amorphous phase in this composition either by the mechanical milling (MM) of a prealloyed ingot of this composition or by the MA of a mixture of two prealloyed compounds of $Fe_{66}Al_{10}Ga_4P_{20}$ and $Fe_{78}C_{12}B_8Si_2$ compositions. Reasonably significant differences were noticed in the structural and thermal behavior of these MA/MM alloys and the rapidly solidified ribbon samples. The amorphous phase formed by rapid solidification was found to be more homogeneous, as indicated by the narrow width of the halo present in the XRD patterns. The T_g and T_x temperatures of the glasses produced by solidification and non-solidification methods were also found to be different.

Since thermodynamics plays an important role in determining the phase stabilities of alloy systems, the enthalpies of the competing phases (solid solutions and amorphous phases) have been calculated using the Miedema method [187,188]. By plotting the enthalpies as a function of composition and by determining the composition range in which the amorphous phase has a lower enthalpy than the (crystalline) solid solution phase, it was possible to determine the composition range in which an amorphous phase could be produced. Even though exact correspondence has not been obtained between the theoretical predictions and experimental observations, reasonable success has been achieved in identifying alloy compositions that could be amorphized. An expression, developed to calculate the free energy (ΔG) of an undercooled liquid [284], that combines the thermodynamic factors (chemical enthalpy and mismatch entropy) and topological factors (elastic enthalpy, which takes into consideration the atomic sizes of the constituent elements), was found to relate well with the T_{rg} parameter of a number of glasses [201]. It was, however, noted that the topological parameters are perhaps more important in determining the GFA of alloys even by MA methods.

Lattice strain is generated by the introduction of solute atoms of different sizes into the solvent lattice, and when this lattice strain reaches a critical value, it has been shown that the crystalline lattice becomes destabilized. This model, by Egami, [119,154,212] can be successfully applied to explain the formation of amorphous phases by solid-state processing methods also. It was recently shown [285] that the minimum lattice strain concept could be used to explain the formation of amorphous phases in Fe–Ni–Zr–Nb–B alloys by MA.

Another empirical relationship that was noted in evaluating the GFA of alloys was the type of phase diagram under equilibrium conditions. It was reported [286] that amorphization in an alloy system was not possible when the equilibrium phase diagram showed the presence of a solid solution phase over a wide composition range. On the other hand, amorphous phase formation was possible when the phase diagram contained a large number of intermetallics. In fact, it was reported that it was easier to produce the amorphous phase when the number of intermetallics was large. This was explained on the basis that the disordering of intermetallics contributed a significant amount of energy increase to the system and this raised the free energy of the crystalline phases (Equation 3.102) to a level higher than that of the amorphous phase. This increased free energy of the crystalline phase destabilized it, and consequently, the amorphous phase became stabilized.

3.16 Concluding Remarks

We have provided brief descriptions of the different models/criteria available to predict the GFA of liquid alloys. We first described the criteria that are well accepted and successful in describing the GFA of melt-spun metallic glasses, followed by the new ones developed to explain the GFA of BMGs. These models were grouped under three major categories: transformation temperatures, structural and topological models, and those based on physical properties. Thermodynamic and computational approaches have also been adopted. Looking at the analysis of the vast amount of data generated during the last 25 years or so, it has become clear that no one single parameter or criterion is able to satisfactorily explain the GFA of alloys. There were many variables that played an important role in coming to such a conclusion. The nature of the alloy system is important. Depending on the constituent atoms, either geometrical factors or transformation temperatures are more important. Even for the same alloy system, the GFA is different depending on the technique used to synthesize the BMG. Obviously, the technique capable of providing a higher solidification rate is able to increase the section thickness (or diameter) of the sample. In view of these different situations, a very large number of empirical relationships have been proposed. Of all the possible factors, the atomic sizes of the constituent elements and the chemical interaction among the solute elements seem to be the most important. Further, the three empirical rules proposed by Inoue have stood the test of time and are still applicable to predict glass formation, even though some apparent exceptions have been noted.

As discussed, almost every one of the parameters proposed to explain the GFA of alloys was claimed to be the best. But, unfortunately, the analysis in each individual proposal dealt with either a limited number of data, or focused only on some specific alloy system(s). Further, different researchers had obtained correlations for different parameters. Thus, it is difficult to strictly compare the different models and parameters and decide which of the large number of parameters that have been proposed are really worth pursuing further to predict alloy compositions to produce new BMGs. Ideally, all the available data would be analyzed using different parameters, and correlations established with the same parameters. This has partly been done. But, with new data generated continuously, such an exercise needs to be taken up periodically. Only then will it be possible to decide which of the parameters is best to predict the GFA of alloys.

In recent years, the concept of alloy design to predict the best glass-forming composition has been used; for example, selecting an alloy, solidified from the liquid state, in which it is noted that the product contains a mixture of the glassy and crystalline phases. The composition of the glassy phase in the composite can be analyzed, and that composition is then expected to be fully glassy on solidification from the liquid state. In fact, using such an approach, it was possible to increase the critical diameter, t_{max}, of a Ti-based glass to over 50 mm [57].

References

1. Jones, H. and C. Suryanarayana (1973). Rapid quenching from the melt: An annotated bibliography 1958–72. *J. Mater. Sci.* 8: 705–753.
2. Suryanarayana, C. (1980). *Rapidly Quenched Metals: A Bibliography 1973–79*. New York: IFI/Plenum.
3. Davies, H.A. and J.B. Hull (1976). The formation, structure and crystallization of non-crystalline nickel produced by splat-quenching. *J. Mater. Sci.* 11: 215–223.
4. Kramer, J. (1937). The amorphous state of metals. *Z. Phys.* 106: 675–691.
5. Buckel, W. and R. Hilsch (1952). Superconductivity and resistivity of tin with lattice defects. *Z. Phys.* 131: 420–442.
6. Buckel, W. and R. Hilsch (1956). Superconductivity and electrical resistivity of novel tin-bismuth alloys. *Z. Phys.* 146: 27–38.
7. Klement, Jr., W., R.H. Willens, and P. Duwez (1960). Non-crystalline structure in solidified gold–silicon alloys. *Nature* 187: 869–870.
8. Weinberg, M.C. (1996). Glass-formation and crystallization kinetics. *Thermochim. Acta* 280–281: 63–71.
9. Weinberg, M.C., D.R. Uhlmann, and E.D. Zanotto (1989). "Nose Method" of calculating critical cooling rates for glass formation. *J. Am. Ceram. Soc.* 72: 2054–2058.
10. Christian, J.W. (2002). *The Theory of Transformations in Metals and Alloys*. Oxford, UK: Pergamon.
11. Suryanarayana, C. (2011). *Experimental Techniques in Materials and Mechanics*. Boca Raton, FL: CRC Press.
12. Onorato, P.I.K. and D.R. Uhlmann (1976). Nucleating heterogeneities and glass formation. *J. Non-Cryst. Solids* 22: 367–378.
13. Davies, H.A. (1983). Metallic glass formation. In *Amorphous Metallic Alloys*, ed. F.E. Luborsky, pp. 8–25. London, UK: Butterworths.
14. Motorin, V.I. (1983). Vitrification kinetics of pure metals. *Phys. Stat. Sol.* 80: 447–455.
15. Zhong, L., J.W. Wang, H.W. Sheng, Z. Zhang, and S.X. Mao (2014). Formation of monatomic metallic glasses through ultrafast liquid quenching. *Nature* 512: 177–180.
16. Barandiarán, J.M. and J. Colmenero (1981). Continuous cooling approximation for the formation of a glass. *J. Non-Cryst. Solids* 46: 277–287.
17. Men, H. and D.H. Kim (2003). Fabrication of ternary Mg–Cu–Gd bulk metallic glass with high glass-forming ability under air atmosphere. *J. Mater. Res.* 18: 1502–1504.
18. Lin, X.H. and W.L. Johnson (1995). Formation of Ti–Zr–Cu–Ni bulk metallic glasses. *J. Appl. Phys.* 78: 6514–6519.
19. Nishiyama, N. and A. Inoue (2002). Direct comparison between critical cooling rate and some quantitative parameters for evaluation of glass-forming ability in Pd–Cu–Ni–P alloys. *Mater. Trans.* 43: 1913–1917.
20. Nishiyama, N., K. Takenaka, and A. Inoue (2006). $Pd_{30}Pt_{17.5}Cu_{32.5}P_2$ alloy with low critical cooling rate of 0.067 K/s. *Appl. Phys. Lett.* 88: 121908-1–121908-3.
21. Nishiyama, N. and A. Inoue (1996). Glass-forming ability of bulk $Pd_{40}Ni_{10}Cu_{30}P_{20}$ alloy. *Mater. Trans., JIM* 37: 1531–1539.

22. Inoue, A. and N. Nishiyama (1997). Extremely low critical cooling rates of new Pd–Cu–P base amorphous alloys. *Mater. Sci. Eng. A* 226–228: 401–405.
23. Davies, H.A. (1975). The kinetics of formation of a Au–Ge–Si metallic glass. *J. Non-Cryst. Solids* 17: 266–272.
24. Park, E.S. and D.H. Kim (2005). Effect of atomic configuration and liquid stability on the glass-forming ability of Ca-based metallic glasses. *Appl. Phys. Lett.* 86: 201912-1–201912-3.
25. Park, E.S. and D.H. Kim (2004). Formation of Ca–Mg–Zn bulk glassy alloy by casting into cone-shaped copper mold. *J. Mater. Res.* 19: 685–688.
26. Wang, W.H., J. Lewandowski, and A.L. Greer (2005). Understanding the glass-forming ability of $Cu_{50}Zr_{50}$ alloys in terms of a metastable eutectic. *J. Mater. Res.* 20: 2307–2313.
27. Zhang, Q.S., W. Zhang, and A. Inoue (2006). New Cu–Zr-based bulk metallic glasses with large diameters of up to 1.5 cm. *Scr. Mater.* 55: 711–713.
28. Pang, S.J., T. Zhang, K. Asami, and A. Inoue (2002). Synthesis of Fe–Cr–Mo–C–B–P bulk metallic glasses with high corrosion resistance. *Acta Mater.* 50: 489–497.
29. Anderson III P.M. and A.E. Lord Jr. (1980). Viscosity of METGLAS 2826 near the glass transition using rapid heating. *J. Non-Cryst. Solids* 37: 219–229.
30. Louzguine-Luzgin, D.V., L.V. Louzguina-Luzgina, A.R. Yavari, K. Ota, G. Vaughan, and A. Inoue (2006). Devitrification of Hf–Pd–Ni glassy alloy on heating. *Thin Solid Films* 509: 75–80.
31. Inoue, A., T. Nakamura, T. Sugita, T. Zhang, and T. Masumoto (1993). Bulky La–Al–TM (TM = transition metal) amorphous alloys with high tensile strength produced by a high-pressure die casting method. *Mater. Trans., JIM* 34: 351–358.
32. Men, H., Z.Q. Hu, and J. Xu (2002). Bulk metallic glass formation in the Mg–Cu–Zn–Y system. *Scr. Mater.* 46: 699–703.
33. Men, H., W.T. Kim, and D.H. Kim (2003). Fabrication and mechanical properties of $Mg_{65}Cu_{15}Ag_5Pd_5Gd_{10}$ bulk metallic glass. *Mater. Trans.* 44: 2141–2144.
34. Park, E.S. and D.H. Kim (2005). Formation of Mg–Cu–Ni–Ag–Zn–Y–Gd bulk glassy alloy by casting into cone-shaped copper mold in air atmosphere. *J. Mater. Res.* 20: 1465–1469.
35. Inoue, A. and T. Zhang (1997). Thermal stability and glass-forming ability of amorphous Nd–Al–TM (TM = Fe, Co, Ni or Cu) alloys. *Mater. Sci. Eng. A* 226–228: 393–396.
36. Xia, L., W.H. Li, S.S. Fang, B.C. Wei, and Y.D. Dong (2006). Binary Ni–Nb bulk metallic glasses. *J. Appl. Phys.* 99: 026103-1–026103-3
37. Nishi, Y., K. Suzuki, and T. Masumoto (1982). Glass-forming ability of transition metal-metalloid type alloys. In *Proceedings of Fourth International Conference on Rapidly Quenched Metals (RQIV)*, eds. T. Masumoto and K. Suzuki, Vol. I, pp. 217–220. Sendai, Japan: The Japan Institute of Metals.
38. Naka, M., Y. Nishi, and T. Masumoto (1978). Critical cooling rate for glass formation in Pd–Cu–Si alloys. In *Proceedings of Third International Conference on Rapidly Quenched Metals (RQIII)*, ed. B. Cantor, Vol. I, pp. 231–238. London, UK : Metals Society.
39. Davies, H.A. (1978). Rapid quenching techniques and formation of metallic glasses. In *Proceedings of Third International Conference on Rapidly Quenched Metals (RQIII)*, ed. B. Cantor, Vol. I, pp. 1–21. London, UK: Metals Society.
40. Drehman, A.J. and D. Turnbull (1981). Solidification behavior of undercooled $Pd_{83}Si_{17}$ and $Pd_{82}Si_{18}$ liquid droplets. *Scr. Metall.* 15: 543–548.
41. Yao, K.F. and F. Ruan (2005). Pd–Si binary bulk metallic glass prepared at low cooling rate. *Chin. Phys. Lett.* 22: 1481–1483.
42. Inoue, A., T. Zhang, N. Nishiyama, K. Ohba, and T. Masumoto (1993). Preparation of 16 mm diameter rod of amorphous $Zr_{65}Al_{7.5}Ni_{10}Cu_{17.5}$ alloy. *Mater. Trans., JIM* 34: 1234–1237.
43. Waniuk, T.A., J. Schroers, and W.L. Johnson (2001). Critical cooling rate and thermal stability of Zr–Ti–Cu–Ni–Be alloys. *Appl. Phys. Lett.* 78: 1213–1215.
44. Mukherjee, S., J. Schroers, Z. Zhou, W.L. Johnson, and W.-K. Rhim (2004). Viscosity and specific volume of bulk metallic glass-forming alloys and their correlation with glass forming ability. *Acta Mater.* 52: 3689–3695.

45. Bhadeshia, H.K.D.H. and R.W.K. Honeycombe (2006). *Steels: Microstructure and Properties*, 3rd edn. London, UK : Butterworths-Heinemann.
46. Johnson, W.L. (1999). Bulk glass-forming metallic alloys: Science and technology. *MRS Bull.* 24(10): 42–56.
47. Huang, H.G., H.B. Ke, Y.M. Wang, Z. Pu, P. Zhang, P.G. Zhang, and T.W. Liu. (2016). Stable U-based metallic glasses. *J. Alloys Compd.* 684: 75–83.
48. Malekan, M., S.G. Shabestari, W. Zhang, S.H. Seyedein, R. Gholamipour, A. Makino, and A. Inoue (2010). Effect of Si addition on glass-forming ability and mechanical properties of Cu–Zr–Al bulk metallic glass. *Mater. Sci. Eng. A* 527: 7192–7196.
49. Li, Y.H., W. Zhang, C. Dong, J.B. Qiang, M. Fukuhara, A. Makino, and A. Inoue (2011). Effects of Ni addition on the glass-forming ability, mechanical properties and corrosion resistance of Zr–Cu–Al bulk metallic glasses. *Mater. Sci. Eng. A* 528: 8551–8556.
50. Guo, G.Q. and L. Yang (2015). Structural mechanisms of the microalloying-induced high glass-forming abilities in metallic glasses. *Intermetallics* 65: 66–74.
51. Yu, C.Y., X.J. Liu, G.P. Zheng, X.R. Niu, and C.T. Liu (2015). Atomistic approach to predict the glass-forming ability in Zr–Cu–Al ternary metallic glasses. *J. Alloys Compd.* 627: 48–53.
52. Yu, C.Y., X.J. Liu, and C.T. Liu (2014). First-principles prediction of the glass-forming ability in Zr-Ni binary metallic glasses. *Intermetallics* 53: 177–182.
53. Ma, G.Z., B.A. Sun, S. Pauly, K.K. Song, U. Kühn, D. Chen, and J. Eckert (2013). Effect of Ti substitution on glass-forming ability and mechanical properties of a brittle Cu–Zr–Al bulk metallic glass. *Mater. Sci. Eng. A* 563: 112–116.
54. Gong, P., K.F. Gong, X. Wang, and Y. Shao (2013). A new centimeter-sized Ti-based quaternary bulk metallic glass with good mechanical properties. *Adv. Eng. Mater.* 15: 691–696.
55. Zhao, S.F., N. Chen, P. Gong, and K.F. Yao (2015). New centimeter-sized quaternary Ti–Zr–Be–Cu bulk metallic glasses with large glass forming ability. *J. Alloys Compd.* 647: 533–538.
56. Gong, P., X. Wang, Y. Shao, N. Chen, X. Liu, and K.F. Yao (2013). A Ti–Zr–Be–Fe–Cu bulk metallic glass with superior glass-forming ability and high specific strength. *Intermetallics* 43: 177–181.
57. Zhang, L., M.Q. Tang, Z.W. Zhu, H.M. Fu, H.W. Zhang, A.M. Wang, H. Li, H.F. Zhang, and Z.Q. Hu (2015). Compressive plastic metallic glasses with exceptional glass forming ability in the Ti–Zr–Cu–Fe–Be system. *J. Alloys Compd.* 638: 349–355.
58. Tang, M.Q., H.F. Zhang, Z.W. Zhu, H.M. Fu, A.M. Wang, H. Li, and Z.Q. Hu (2010). TiZr-base bulk metallic glass with over 50 mm in diameter. *J. Mater. Sci. Technol.* 26: 481–486.
59. Shi, B., Y.L. Xu, W.L. Ma, C. Li, C.V. Estefania, and J.G. Li (2015). Effect of Ti addition on the glass forming ability, crystallization, and plasticity of $(Zr_{64.13}Cu_{15.75}Ni_{10.12}Al_{10})_{100-x}Ti_x$ bulk metallic glasses. *Mater. Sci. Eng. A* 639: 345–349.
60. Qiao, D.C. and A. Peker (2012). Enhanced glass-forming ability in Zr-based bulk metallic glasses with Hf addition. *Intermetallics* 24: 115–119.
61. Tan, J., F.S. Pan, Y. Zhang, Z. Wang, M. Stoica, B.A. Sun, U. Kühn, and J. Eckert (2012). Effect of Fe addition on glass forming ability and mechanical properties in Zr–Co–Al– (Fe) bulk metallic glasses. *Mater. Sci. Eng. A* 539: 124–127.
62. Xu, K., H. Ling, Q. Li, J.F. Li, K. Yao, and S.F. Guo (2014). Effects of Co substitution for Fe on the glass forming ability and properties of $Fe_{80}P_{13}C_7$ bulk metallic glasses. *Intermetallics* 51: 53–58.
63. Yang, W.M., H.S. Liu, X.D. Fan, L. Xue, C.C. Dun, and B.L. Shen (2015). Enhanced glass forming ability of Fe-based amorphous alloys with minor Cu addition. *J. Non-Cryst. Solids* 419: 65–68.
64. Li, J.W., J.Y. Law, J.T. Hao, A. He, Q. Man, C.T. Chang, H. Men, J.Q. Wang, X.M. Wang, and R.W. Li (2015). Magnetocaloric effect of Fe–RE–B–Nb (RE = Tb, Ho or Tm) bulk metallic glasses with high glass-forming ability. *J. Alloys Compd.* 644: 346–349.
65. Seifoddini, A., M. Stoica, M. Nili-Ahmadabadi, S. Heshmati-Manesh, U. Kühn, and J. Eckert (2013). New $(Fe_{0.9}Ni_{0.1})_{77}Mo_5P_9C_{7.5}B_{1.5}$ glassy alloys with enhanced glass-forming ability and large compressive strain. *Mater. Sci. Eng. A* 560: 575–582.

66. Li, J.W., W.M. Yang, D. Estevez, G.X. Chen, W.G. Zhao, Q. Man, Y.Y. Zhao, Z.D. Zhang, and B.L. Shen (2014). Thermal stability, magnetic and mechanical properties of Fe–Dy–B–Nb–bulk metallic glasses with high glass-forming ability. *Intermetallics* 46: 85–90.
67. Li, Z.Z., A. Wang, C.T. Chang, Y.G. Wang, B.S. Dong, and A.X. Zhou (2014). FeSiBPNbCu alloys with high glass-forming ability and good soft magnetic properties. *Intermetallics* 54: 225–231.
68. Deo, L.P. and M.F. de Oliveira (2015). Y and Er minor addition effect on glass forming ability of a Ni–Nb–Zr alloy. *J. Alloys Compd.* 644: 729–733.
69. Liu, K.M., H.T. Zhou, B. Yang, D.P. Lu, and A. Atrens (2010). Influence of Si on glass forming ability and properties of the bulk amorphous alloy $Mg_{60}Cu_{30}Y_{10}$. *Mater. Sci. Eng. A* 527: 7475–7479.
70. Kyeong, J.S., D.H. Kim, J.I. Lee, and E.S. Park (2012). Effects of alloying elements with positive enthalpy of mixing in $Mg_{65}Cu_{25}Gd_{10}$ bulk-forming metallic glass. *Intermetallics* 31: 9–15.
71. Han, J.J., W.Y. Wang, C.P. Wang, X.D. Hui, X.J. Liu, and Z.K. Liu (2014). Origin of enhanced glass-forming ability of Ce-containing Al–Fe alloy: *Ab initio* molecular dynamics study. *Intermetallics* 46: 29–39.
72. Reyes-Retana, J.A. and G.G. Naumis (2015). *Ab initio* study of Si doping effects in Pd–Ni–P bulk metallic glass. *J. Non-Cryst. Solids* 409: 49–53.
73. Wu, C., P. Yu, and L. Xia (2015). Glass forming ability and magnetic properties of a $Gd_{55}Ni_{25}Al_{18}Zn_2$ bulk metallic glass. *J. Non-Cryst. Solids* 422: 23–25.
74. Yoshizawa, Y., S. Oguma, and K. Yamauchi (1988). New Fe-based soft magnetic alloys composed of ultrafine grain structure. *J. Appl. Phys.* 64: 6044–6046.
75. Li, Z.Z., A. Wang, C.T. Chang, Y.G. Wang, B.S. Dong, and S.X. Zhou (2014). FeSiBPNbCu alloys with high glass-forming ability and good soft magnetic properties. *Intermetallics* 54: 225–231.
76. Zhou, Y., Y. Zhao, B.Y. Qu, L. Wang, R.L. Zhou, Y.C. Wu, and B. Zhang (2015). Remarkable effect of Ce base element purity upon glass forming ability in Ce–Ga–Cu bulk metallic glasses. *Intermetallics* 56: 56–62.
77. Wang, Y.X., H. Yang, G. Lim, and Y. Li (2010). Glass formation enhanced by oxygen in binary Zr-Cu system. *Scr. Mater.* 62: 682–685.
78. Granata, D., E. Fischer, and J.F. Löffler (2015). Hydrogen microalloying as a visible strategy for enhancing the glass-forming ability of Zr-based bulk metallic glasses. *Scr. Mater.* 103: 53–56.
79. Turnbull, D. (1969). Under what conditions can a glass be formed? *Contemp. Phys.* 10: 473–488.
80. Schroers, J., B. Lohwongwatana, W.L. Johnson, and A. Peker (2005). Gold based bulk metallic glass., *Appl. Phys. Lett.* 87: 061912-1–061912-3.
81. Giessen, B.C., J. Hong, L. Kadacoff, D.E. Polk, R. Ray, and R. St. Amand (1978). Compositional dependence of the thermal stability and related properties of metallic glasses I: T_g for $Ca_{0.65}M_{0.35}$ and $Zr_{0.475}Cu_{0.475}M_{0.05}$ glasses. In *Proceedings of Third International Conference on Rapidly Quenched Metals (RQIII)*, ed. B. Cantor, Vol. I, pp. 249–260. London, UK : Metals Society.
82. Amiya, K. and A. Inoue (2002). Formation, thermal stability and mechanical properties of Ca-based bulk glassy alloys. *Mater. Trans.* 43: 81–84.
83. Park, E.S. and D.H. Kim (2005). Effect of atomic configuration and liquid stability on the glass-forming ability of Ca-based metallic glasses. *Appl. Phys. Lett.* 86: 201912-1–201912-3.
84. Zhang, B., M.X. Pan, D.Q. Zhao, and W.H. Wang (2004). "Soft" bulk metallic glasses based on cerium. *Appl. Phys. Lett.* 85: 61–63.
85. Xia, L., D. Ding, S.T. Shan, and Y.D. Dong (2006). The glass forming ability of Cu-rich Cu–Hf binary alloys. *J. Phys. Condens. Matter* 18: 3543–3548.
86. Jia, P., H. Guo, Y. Li, J. Xu, and E. Ma (2006). A new Cu–Hf–Al ternary bulk metallic glass with high glass forming ability and ductility. *Scr. Mater.* 54: 2165–2168.
87. Xu, D., B. Lohwongwatana, G. Duan, W.L. Johnson, and C. Garland (2004). Bulk metallic glass formation in binary Cu-rich alloy series—$Cu_{100-x}Zr_x$ ($x = 34$, 36, 38.2, 40 at.%) and mechanical properties of bulk $Cu_{64}Zr_{36}$ glass. *Acta Mater.* 52: 2621–2624.
88. Deng, L., B.W. Zhou, H.S. Yang, X. Jiang, B. Jiang, and X.G. Zhang (2015). Roles of minor rare-earth elements addition in formation and properties of Cu–Zr–Al bulk metallic glasses. *J. Alloys Compd.* 632: 429–434.

89. Ponnambalam, V., S.J. Poon, G.J. Shiflet, V.M Keppens, R. Taylor, and G. Petculescu (2003). Synthesis of iron-based bulk metallic glasses as nonferromagnetic amorphous steel alloys. *Appl. Phys. Lett.* 83: 1131–1133.
90. Li, J.W., W.M. Yang, D. Estevez, G.X. Chen, W.G. Zhao, Q. Man, Y.Y. Zhao, Z.D. Zhang, and B.L. Shen (2014). Thermal stability, magnetic and mechanical properties of Fe–Dy–B–Nb bulk metallic glasses with high glass forming ability. *Intermetallics* 46: 85–90.
91. Sarlar, K. and I. Kucuk (2016). Phase separation and glass-forming ability of $(Fe_{0.72}Mo_{0.04}B_{0.24})_{100-x}Gd_x$ (x = 4, 8) bulk metallic glasses. *J. Non-Cryst. Solids* 447: 198–201.
92. Tan, H., Y. Zhang, D. Ma, Y.P. Feng, and Y. Li (2003). Optimum glass formation at off-eutectic composition and its relation to skewed eutectic coupled zone in the La based La–Al–(Cu,Ni) pseudo ternary system. *Acta Mater.* 51: 4551–4561.
93. Chen, L.Y., H.T. Hu, G.Q. Zhang, and J.Z. Jiang (2007). Catching the Ni-based ternary metallic glasses with critical diameter up to 3 mm in Ni–Nb–Zr system. *J. Alloys Compd.* 443: 109–113.
94. Bazlov, A.I., A. Yu. Churyumov, S.V. Ketov, and D.V. Louzguine-Luzgin (2015). Glass-formation and deformation behavior of Ni–Pd–P–B alloy. *J. Alloys Compd.* 619: 509–512.
95. Chen, H.S. (1976). Glass temperature, formation and stability of Fe, Co, Ni, Pd and Pt based glasses. *Mater. Sci. Eng.* 23: 151–154.
96. Inoue, A., N. Nishiyama, and H.M. Kimura (1997). Preparation and thermal stability of bulk amorphous $Pd_{40}Cu_{30}Ni_{10}P_{20}$ alloy cylinder of 72 mm in diameter. *Mater. Trans., JIM* 38: 179–183.
97. Zhao, S.F., Y. Shao, X. Liu, N. Chen, H.Y. Ding, and K.F. Yao (2015). Psuedo-quinary $Ti_{20}Zr_{20}Hf_{20}Be_{20}(Cu_{20-x}Ni_x)$ high entropy bulk metallic glasses with large glass forming ability. *Mater. & Design* 87: 625–631.
98. Peker, A. and W.L. Johnson (1993). A highly processable metallic glass: $Zr_{41.2}Ti_{13.8}Cu_{12.5}Ni_{10.0}Be_{22.5}$. *Appl. Phys. Lett.* 63: 2342–2344.
99. Cao, Q.P., J.B. Jin, Y. Ma, X.Z. Cao, B.Y. Wang, S.X. Qu, X.D. Wang, D.X. Zhang, and J.Z. Jiang (2015). Enhanced plasticity in Zr–Cu–Ag–Al–Be bulk metallic glasses. *J. Non-Cryst. Solids* 412: 35–44.
100. Lu, Z.P., Y. Li, and S.C. Ng (2000). Reduced glass transition temperature and glass forming ability of bulk glass forming alloys. *J. Non-Cryst. Solids* 270: 103–114.
101. Cahn, R.W. (1991). Metallic glasses. In *Glasses and Amorphous Materials*, ed. J. Zarzycki, pp. 493–548. Vol. 9 of *Materials Science and Technology: A Comprehensive Treatment*. Weinheim, Germany: VCH.
102. Turnbull, D. (1974). Amorphous solid formation and interstitial solution behavior in metallic alloy systems. *J. Phys. Colloq.* 35: C4-1–C4-10.
103. Guo, F.Q., S.J. Poon, and G.J. Shiflet (2004). CaAl-based bulk metallic glasses with high thermal stability. *Appl. Phys. Lett.* 84: 37–39.
104. Inoue, A., W. Zhang, T. Tsurui, A.R. Yavari, and A.L. Greer (2005). Unusual room-temperature compressive plasticity in nanocrystal-toughened bulk copper–zirconium glass. *Philos. Mag. Lett.* 85: 221–229.
105. Massalski, T.B., ed. (1996). *Binary Alloy Phase Diagrams*. Materials Park, OH : ASM International.
106. Lee, D.M., J.H. Sun, D.H. Kang, S.Y. Shin, G. Welsch, and C.H. Lee (2012). A deep eutectic point in quaternary Zr–Ti–Ni–Cu system and bulk metallic glass formation near the eutectic point. *Intermetallics* 21: 67–74.
107. Zhang, Y., Y. Li, H. Tan, G.L. Chen, and H.A. Davies (2006). Glass forming ability criteria for La–Al– (Cu,Ni) alloys. *J. Non-Cryst. Solids* 352: 5482–5486.
108. Ma, D., H. Tan, D. Wang, Y. Li, and E. Ma (2005). Strategy for pinpointing the best glass-forming alloys. *Appl. Phys. Lett.* 86: 191906-1–191906-3.
109. Gupta, P.K. and D.B. Miracle (2007). A topological basis for bulk glass formation. *Acta Mater.* 55: 4507–4515.
110. Suryanarayana, C. (1980). Liquid-quenched metal-metal glasses. *Sci. Rep. Res. Inst. Tohoku Univ. A* 28: 143–154.
111. Polk, D.E. (1972). The structure of glassy metallic alloys. *Acta Metall.* 20: 485–491.

112. Gaskell, P.H. (1991). Models for the structure of amorphous solids. In *Glasses and Amorphous Materials*, ed. J. Zarzycki, pp. 175–278. Vol. 9 of *Materials Science and Technology: A Comprehensive Treatment*. Weinheim, Germany: VCH.
113. Frost, H.J. (1982). Cavities in dense random packings. *Acta Metall.* 30: 889–904.
114. Turnbull, D. (1977). On the gram-atomic volumes of metal–metalloid glass forming alloys. *Scr. Metall.* 11: 1131–1136.
115. Nielsen, H.J.V. (1979). The eutectic compositions as a basis for the formation of metallic glasses in the binary alloys of iron-group transition metals with metalloids. *Z. Metall.* 70: 180–184.
116. Hume-Rothery, W., G.W. Mabbott, and K.M. Channel-Evans (1934). The freezing points, melting points, and solid solubility limits of the alloys of silver and copper with the elements of the B sub-groups. *Philos. Trans. R. Soc. (Lond.) A* 233: 1–97.
117. Egami, T. (2011). Atomic level stresses. *Prog. Mater. Sci.* 56: 637–653.
118. Egami, T. (1996). The atomic structure of aluminum based metallic glasses and universal criterion for glass formation. *J. Non-Cryst. Solids* 205–207: 575–582.
119. Egami, T. and Y. Waseda (1984). Atomic size effect on the formability of metallic glasses. *J. Non-Cryst. Solids* 64: 113–134.
120. Dai, X.D., J.H. Li, and Y. Kong (2007). Long-range empirical potential for the bcc structured transition metals. *Phys. Rev.* B75: 052102-1–052102-4.
121. Luo, S.Y., J.H. Li, J.B. Liu, and B.X. Liu (2014). Atomic modeling to design favored compositions for the ternary Ni–Nb–Zr metallic glass formation. *Acta Mater.* 76: 482- 492.
122. Kaufman, L. and H. Bernstein (1970). *Computer Calculation of Phase Diagrams*. New York, NY: Academic Press.
123. Inoue, A. (1996). Recent progress of Zr-based bulk amorphous alloys. *Sci. Rep. Res. Inst. Tohoku Univ. A* 42: 1–11.
124. Inoue, A. (1997). Stabilization of supercooled liquid and opening-up of bulk glassy alloys. *Proc. Jpn. Acad. B* 73: 19–24.
125. Inoue, A. (1998). *Bulk Amorphous Alloys: Preparation and Fundamental Characteristics*. Vol. 4 of *Materials Science Foundations*. Uetikon-Zürich, Switzerland: Trans Tech Publications.
126. Matsubara, E., T. Tamura, Y. Waseda, A. Inoue, T. Zhang, and T. Masumoto (1992). Structural study of $Zr_{60}Al_{15}Ni_{25}$ amorphous alloys with a wide supercooled liquid region by the anomalous x-ray scattering (AXS) method. *Mater. Trans., JIM* 33: 873–878.
127. Inoue, A., T. Negishi, H.M. Kimura, T. Zhang, and A.R. Yavari (1998). High packing density of Zr- and Pd-based bulk amorphous alloys. *Mater. Trans., JIM* 39: 318–321.
128. Inoue, A. (1995). High strength bulk amorphous alloys with low critical cooling rates. *Mater. Trans., JIM* 36: 866–875.
129. Inoue, A. (1997). Bulk amorphous alloys with soft and hard magnetic properties. *Mater. Sci. Eng. A* 226–228: 357–363.
130. Inoue, A. (2000). Stabilization of metallic supercooled liquid and bulk amorphous alloys. *Acta Mater.* 48: 279–306.
131. Chen, L.Y., Z.D. Fu, W. Zeng, G.Q. Zhang, Y.W. Zeng, G.L. Xu, S.L. Zhang, and J.Z. Jiang (2007). Ultrahigh strength binary Ni–Nb bulk glassy alloy composite with good ductility. *J. Alloys Compd.* 443: 105–108.
132. Inoue, A., W. Zhang, and J. Saida (2004). Synthesis and fundamental properties of Cu-based bulk glassy alloys in binary and multi-component systems. *Mater. Trans.* 45: 1153–1162.
133. Zhang, J. and Y. Zhao (2004). Formation of zirconium metallic glass. *Nature (London)* 430: 332–335.
134. Zhang, J. and Y. Zhao (2005). Formation of zirconium metallic glass. *Nature (London)* 437: 1057.
135. Wang, Y., Y.Z. Fang, T. Kikegawa, C. Lathe, K. Saksl, H. Franz, J.R. Schneider et al. (2005). Amorphouslike diffraction pattern in solid metallic titanium. *Phys. Rev. Lett.* 95: 155501-1–155501-4.
136. Hattori, T., H. Saitoh, H. Kaneko, Y. Okajima, K. Aoki, and W. Utsumi (2006). Does bulk metallic glass of elemental Zr and Ti exist? *Phys. Rev. Lett.* 96: 255504-1–255504-4.

137. Abe, T., M. Shimono, K. Hashimoto, K. Hono, and H. Onodera (2006). Phase separation and glass-forming abilities of ternary alloys. *Scr. Mater.* 55: 421–424.
138. Schulz, R., K. Samwer, and W.L. Johnson (1984). Kinetics of phase separation in $Cu_{50}Zr_{50}$ metallic glasses. *J. Non-Cryst. Solids* 61–62: 997–1002.
139. Tanner, L. and R. Ray (1980). Phase separation in Zr–Ti–Be metallic glasses. *Scr. Metall.* 14: 657–662.
140. Meijering, J.L. (1950). Segregation in regular ternary solutions. 1. *Philips Res. Rep.* 5: 333–356.
141. Meijering, J.L. (1951). Segregation in regular ternary solutions. 2. *Philips Res. Rep.* 6: 183–210.
142. Inoue, A., W. Zhang, T. Zhang, and K. Kurosaka (2001). High-strength Cu-based bulk glassy alloys in Cu–Zr–Ti and Cu–Hf–Ti ternary systems. *Acta Mater.* 49: 2645–2652.
143. Shen, T.D., Y. He, and R.B. Schwarz (1999). Bulk amorphous Pd–Ni–Fe–P alloys: Preparation and characterization. *J. Mater. Res.* 14: 2107–2115.
144. Shen, T.D. and R.B. Schwarz (1999). Bulk ferromagnetic glasses prepared by flux melting and water quenching. *Appl. Phys. Lett.* 75: 49–51.
145. Inoue, A., T. Nakamura, N. Nishiyama, and T. Masumoto (1992). Mg–Cu–Y bulk amorphous alloys with high tensile strength produced by a high-pressure die casting method. *Mater. Trans., JIM* 33: 937–945.
146. Murty, B.S. and K. Hono (2000). Formation of nanocrystalline particles in glassy matrix in melt-spun Mg–Cu–Y based alloys. *Mater. Trans., JIM* 41: 1538–1544.
147. Liu, F.J., Q.W. Yang, S.J. Pang, C.L. Ma, and T. Zhang (2008). Ductile Fe-based BMGs with high glass forming ability and high strength. *Mater. Trans.* 49: 231–234.
148. Li, Y., S.C. Ng, C.K. Ong, H.H. Hng, and T.T. Goh (1997). Glass forming ability of bulk glass forming alloys. *Scr. Mater.* 36: 783–787.
149. Donald, I.W. and H.A. Davies (1978). Prediction of glass-forming ability for metallic systems. *J. Non-Cryst. Solids* 30: 77–85.
150. Hrubý, A. (1972). Evaluation of glass-forming tendency by means of DTA. *Czech. J. Phys. B* 22: 1187–1193.
151. Saad, M. and M. Poulain (1987). Glass forming ability criterion. *Mater. Sci. Forum* 19–20: 11–18.
152. Lu, Z.P. and C.T. Liu (2002). A new glass-forming ability criterion for bulk metallic glasses. *Acta Mater.* 50: 3501–3512.
153. Fan, G.J., J-C. Zhao, and P.K. Liaw (2010). A four-step approach to the multicomponent bulk-metallic glass formation. *J. Alloys & Compd.* 497: 24–27.
154. Egami, T. (2003). Atomistic mechanism of bulk metallic glass formation. *J. Non-Cryst. Solids* 317: 30–33.
155. Miracle, D.B. (2004). A structural model for metallic glasses. *Nat. Mater.* 3: 697–702.
156. Miracle, D.B. (2006). The efficient cluster packing model—an atomic structural model for metallic glasses. *Acta Mater.* 54: 4317–4336.
157. Miracle, D.B., D.V. Louzguine-Luzgin, L.V. Louzguina-Luzgina, and A. Inoue (2010). An assessment of binary metallic glasses: Correlations between structure, glass forming ability and stability. *Internat. Mater. Rev.* 55 : 218–256.
158. Miracle, D.B. (2012). A physical model for metallic glass structures: An introduction and update. *JOM* 64 : 846–855.
159. Laws, K.J., D.B. Miracle, and M. Ferry (2015). A predictive structural model for bulk metallic glasses. *Nature Commun.* 6: Article 9123 10.
160. Shao, G., B. Lu, Y.Q. Liu, and P. Tsakiropoulos (2005). Glass forming ability of multi-component metallic systems. *Intermetallics* 13: 409–414.
161. Yan, X.-Y., Y.A. Chang, Y. Yang, F.-Y. Xie, S.-L. Chen, F. Zhang, S. Daniel, and M.-H. He (2001). A thermodynamic approach for predicting the tendency of multicomponent metallic alloys for glass formation. *Intermetallics* 9: 535–538.
162. Lu, Z.P. and C.T. Liu (2003). Glass formation criterion for various glass-forming systems. *Phys. Rev. Lett.* 91: 115505–115505.
163. Lu, Z.P. and C.T. Liu (2004). A new approach to understanding and measuring glass formation in bulk amorphous materials. *Intermetallics* 12: 1035–1043.

164. Xi, X.K., D.Q. Zhao, M.X. Pan, and W.H. Wang (2005). On the criteria of bulk metallic glass formation in MgCu-based alloys. *Intermetallics* 13: 638–641.
165. Lu, Z.P., H. Bei, and C.T. Liu (2007). Recent progress in quantifying glass-forming ability of bulk metallic glasses. *Intermetallics* 15: 618–624.
166. Lu, Z.P., C.T. Liu, Y. Wu, H. Tan, and G.L. Chen (2008). Composition effects on glass-forming ability and its indicator γ. *Intermetallics* 16: 410–417.
167. Fan, C. , C.T. Liu, G. Chen, G. Chen, D. Chen, X. Yang, P.K. Liaw, and H.G. Yan (2013). Influence of the molten quenching temperature on the thermal physical behavior of quenched Zr-based metallic glasses. *Intermetallics* 38: 19–22.
168. Fan, C. , C.T. Liu, G. Chen, G. Chen, P.K. Liaw, and H.G. Yan (2013). Effect of molten quenching temperature on glass-forming ability of nanoquasi-crystal-forming Zr-based metallic glasses. *Scr. Mater.* 68: 534–537.
169. Du, X.H., J.C. Huang, C.T. Liu, and Z.P. Lu (2007). New criterion of glass forming ability for bulk metallic glasses. *J. Appl. Phys.* 101: 086108-1–086108-3.
170. Mondal, K. and B.S. Murty (2005). On the parameters to assess the glass forming ability of liquids. *J. Non-Cryst. Solids* 351: 1366–1371.
171. Wakasugi, T., R. Ota, and J. Fukunaga (1992). Glass-forming ability and crystallization tendency evaluated by the DTA method in the Na_2O–B_2O_3–Al_2O_3 system. *J. Am. Ceram. Soc.* 75: 3129–3132.
172. Chen, Q.J., J. Shen, H.B. Fan, J.F. Sun, Y.J. Huang, and D.G. McCartney (2005). Glass-forming ability of an iron-based alloy enhanced by Co addition and evaluated by a new criterion. *Chin. Phys. Lett.* 22: 1736–1738.
173. Chen, Q.J., J. Shen, D. Zhang, H.B. Fan, J.F. Sun, and D.G. McCartney (2006). A new criterion for evaluating the glass-forming ability of bulk metallic glasses. *Mater. Sci. Eng. A* 433: 155–160.
174. Lee, J.Y., D.H. Bae, J.K. Lee, and D.H. Kim (2004). Bulk glass formation in the Ni–Zr–Ti–Nb–Si–Sn alloy system. *J. Mater. Res.* 19: 2221–2225.
175. Kim, J.H., J.S. Park, H.T. Jeong, W.T. Kim, and D.H. Kim (2004). A new criterion for evaluating the glass-forming ability of bulk metallic glasses. *Mater. Sci. Eng. A* 386: 186–193.
176. Kim, J.H., J.S. Park, H.K. Lim, W.T. Kim, and D.H. Kim (2005). Heating and cooling rate dependence of the parameters representing the glass forming ability in bulk metallic glasses. *J. Non-Cryst. Solids* 351: 1433–1440.
177. Fan, G.J., H. Choo, and P.K. Liaw (2007). A new criterion for the glass-forming ability of liquids. *J. Non-Cryst. Solids* 353: 102–107.
178. Angell, C.A. (1995). Formation of glasses from liquids and biopolymers. *Science* 267: 1924–1935.
179. Yuan, Z.-Z., S.-L. Bao, Y. Lu, D.-P. Zhang, and L. Yao (2008). A new criterion for evaluating the glass-forming ability of bulk glass forming alloys. *J. Alloys Compd.* 459: 251–260.
180. He, Y., R.B. Schwarz, and J.I. Archuleta (1996). Bulk glass formation in the Pd–Ni–P system. *Appl. Phys. Lett.* 69: 1861–1863.
181. Long, Z.L., H.Q. Wei, Y.H. Ding, P. Zhang, G.Q. Xie, and A. Inoue (2009). A new criterion for predicting the glass-forming ability of bulk metallic glasses. *J. Alloys Compd.* 475: 207–219.
182. Long, Z.L., G.Q. Xie, H.Q. Wei, X.P. Su, J. Peng, P. Zhang, and A. Inoue (2009). On the new criterion to assess the glass-forming ability of metallic alloys. *Mater. Sci. Eng. A* 509: 23–30.
183. Błyskun, P., P. Maj, M. Kowalczyk, J. Latuch, and T. Kulik (2015). Relation of various GFA indicators to the critical diameter of Zr-based BMGs. *J. Alloys Compd.* 625: 13–17.
184. Zhang, G.-H. and K.-C. Chou (2009). A criterion for evaluating glass-forming ability of alloys. *J. Appl. Phys.* 106: 094902-1–094902-4.
185. Du, X.H. and J. C. Huang (2008). New criterion in predicting glass forming ability of various glass-forming systems. *Chinese Phys.* B17: 249–254.
186. Suryanarayana, C., I. Seki, and A. Inoue (2009). A critical analysis of the glass-forming ability of alloys. *J. Non-Cryst. Solids* 355: 355–360.
187. de Boer, F.R., R. Boom, W.C.M. Mattens, A.R. Miedema, and A.K. Niessen (1988). *Cohesion in Metals: Transition Metal Alloys*. Amsterdam, The Netherlands: North-Holland.

188. Niessen, A.K., F.R. de Boer, R. Boom, P.F. de Chatel, W.C.M. Mattens, and A.R. Miedema (1983). Model predictions for the enthalpy of formation of transition metal alloys II. *CALPHAD* 7: 51–70.
189. Bakker, H. (1998). *Enthalpies in Alloys: Miedema's Semi-Empirical Model.* Uetikon-Zurich, Switzerland: Trans Tech.
190. Gallego, L.J., J.A. Somoza, and J.A. Alonso (1990). Glass formation in ternary transition metal alloys. *J. Phys. Condens. Matter* 2: 6245–6250.
191. Takeuchi, A. and A. Inoue (2001). Calculations of amorphous-forming composition range for ternary alloy systems and analyses of stabilization of amorphous phase and amorphous-forming ability. *Mater. Trans.* 42: 1435–1444.
192. Sarwat, S.G., M. Ramya, P.S. Ali, B. Raj, and K.R. Ravi (2015). A new thermodynamic parameter G_{CE} for identification of glass forming compositions. *J. Alloys Compd.* 627: 337–343.
193. Xia, L., S.S. Fang, Q. Wang, Y.D. Dong, and C.T. Liu (2006). Thermodynamic modeling of glass formation in metallic glasses. *Appl. Phys. Lett.* 88: 171905-1–171905-3.
194. Yu, C.Y., X.J. Liu, J. Lu, G.P. Zheng, and C.T. Liu (2013). First-principles prediction and experimental verification of glass-forming ability in Zr-Cu binary metallic glasses. *Sci. Rep.* 3: Article 2124 5.
195. Yu, C.Y., X.J. Liu, G.P. Zheng, X.R. Niu, and C.T. Liu (2015). Atomistic approach to predict the glass-forming ability in Zr–Cu–Al ternary metallic glasses. *J. Alloys Compd.* 627: 48–53.
196. Bhatt, J., J. Wu, J. H. Xia, Q. Wang, C. Dong, and B.S. Murty (2007). Optimization of bulk metallic glass forming compositions in Zr–Cu–Al system by thermodynamic modeling. *Intermetallics* 15: 716–721.
197. Ramakrishna Rao, B., M. Srinivas, A.K. Shah, A.S. Gandhi, and B.S. Murty (2013). A new thermodynamic parameter to predict glass forming ability in iron based multi-component systems containing zirconium. *Intermetallics* 35: 73–81.
198. Hu, X.F., J. Guo, G.J. Fan, and T.T. Feng (2013). Evaluation of glass-forming ability for Al-based amorphous alloys based on superheated liquid fragility and thermodynamics. *J. Alloys Compd.* 574: 18–21.
199. Bian, X.F., B.A. Sun, L.N. Hu, and Y.B. Jia (2013). Fragility of superheated melts and glass-forming ability in Al-based alloys. *Phys. Lett.* A335: 61–67.
200. Suo, Z.Y., K.Q. Qiu, Q.F. Li, J.H. You, Y.L. Ren, and Z.Q. Hu (2010). A new parameter to evaluate the glass-forming ability of bulk metallic glasses. *Mater. Sci. Eng.* A528: 429–443.
201. Srinivasa Rao, B., J. Bhatt, and B.S. Murty (2007). Identification of compositions with highest glass forming ability in multicomponent systems by thermodynamic and topological approaches. *Mater. Sci. Eng. A* 449–451: 211–214.
202. Ge, L., X. Hui, E.R. Wang, G.L. Chen, R. Arroyave, and Z.K. Liu (2008). Prediction of the glass forming ability in Cu–Zr binary and Cu–Zr–Ti ternary alloys. *Intermetallics* 16: 27–33.
203. Turnbull, D. (1950). Formation of crystal nuclei in liquid metals. *J. Appl. Phys.* 21: 1022–1028.
204. Thompson, C.V. and F. Spaepen (1979). On the approximation of the free energy change on crystallization. *Acta Metall.* 27: 1855–1859.
205. Hu, Y.J., A.C. Lieser, A. Saengdeejing, Z.K. Liu, and L.J. Kecskes (2014). Glass formability of W-based alloys through thermodynamic modeling: W–Fe–Hf–Pd–Ta and W–Fe–Si–C. *Intermetallics* 48: 79–85.
206. Ma, D., H. Cao, and Y.A. Chang (2007). Identifying bulk metallic glass-formers from multicomponent eutectics. *Intermetallics* 15: 1122–1126.
207. Kim, D., B.-J. Lee, and N.J. Kim (2004). Thermodynamic approach for predicting the glass forming ability of amorphous alloys. *Intermetallics* 12: 1103–1107.
208. Yan, Z.J., J.F. Li, S.R. He, and Y.H. Zhou (2003). Evaluation of the optimum solute concentration for good glass forming ability in multicomponent metallic glasses. *Mater. Res. Bull.* 38: 681–689.
209. Ueno, S. and Y. Waseda (1987). Evaluation of the optimum solute concentration for good glass formability in multi-component alloys. *J. Mater. Eng.* 9: 199–204.

210. Botta, W.J., F.S. Pereira, C. Bolfarini, C.S. Kiminami, and M.F. de Oliveira. (2008). Topological instability and electronegativity effects on the glass forming ability of metallic alloys. *Phil. Mag. Lett.* 88: 785–791.
211. de Oliveira, M.F. (2012). A simple criterion to predict the glass forming ability of metallic alloys. *J. Appl. Phys.* 111: Article 023509 5.
212. Egami, T. (1997). Universal criterion for metallic glass formation. *Mater. Sci. Eng. A* 226–228: 261–267.
213. Han, H.-S., N. Park, J.-Y., Suh, H.-S. Nam, H.-K. Seok, W.T. Kim, Y.-C. Kim, and P.-R. Cha (2016). Reassessing the atomic size effect on glass forming ability: Effect of atomic size difference on thermodynamics and kinetics. *Intermetallics* 69: 123–127.
214. Egami, T. and S. Aur (1987). Local atomic structure of amorphous and crystalline alloys: Computer simulation. *J. Non-Cryst. Solids* 89: 60–74.
215. Egami, T. (2002). Nano-glass mechanism of bulk metallic glass formation. *Mater. Trans.* 43: 510–517.
216. Senkov, O.N. and D.B. Miracle (2001). Effect of the atomic size distribution on glass forming ability of amorphous metallic alloys. *Mater. Res. Bull.* 36: 2183–2198.
217. Senkov, O.N. and D.B. Miracle (2003). A topological model for metallic glass formation. *J. Non-Cryst. Solids* 317: 34–39.
218. Miracle, D.B. and O.N. Senkov (2003). Topological criterion for metallic glass formation. *Mater. Sci. Eng. A* 347: 50–58.
219. Miracle, D.B., W.S. Sanders, and O.N. Senkov (2003). The influence of efficient atomic packing on the constitution of metallic glasses. *Philos. Mag.* 83: 2409–2428.
220. Senkov, O.N. and J.M. Scott (2004). Specific criteria for selection of alloy compositions for bulk metallic glasses. *Scr. Mater.* 50: 449–452.
221. Stockdale, D. (1935). Numerical relationships in binary metallic systems. *Proc. R. Soc. A* 152: 81–104.
222. Hume-Rothery, W. and E. Anderson (1960). Eutectic compositions and liquid immiscibility in certain binary alloys. *Philos. Mag.* 5: 383–405.
223. Yavari, A.R. (2005). Solving the puzzle of eutectic compositions with "Miracle glasses." *Nat. Mater.* 4: 2–3.
224. Sheng, H.W., W.K. Luo, F.M. Alamgir, J.M. Bai, and E. Ma (2006). Atomic packing and short-to-medium-range order in metallic glasses. *Nature (London)* 439: 419–425.
225. Wang, A.P., J.Q. Wang, and E. Ma (2007). Modified efficient cluster packing model for calculating alloy compositions with high glass forming ability. *Appl. Phys. Lett.* 90: 121912-1–121912-3.
226. Azad, S., A. Mandal, and R.K. Mandal (2007). On the parameters of glass formation in metallic systems. *Mater. Sci. Eng. A* 458: 348–354.
227. Park, E.S., D.H. Kim, and W.T. Kim (2005). Parameter for glass forming ability of ternary alloy systems. *Appl. Phys. Lett.* 86: 061907-1–061907-3
228. Lu, Z.P. and C.T. Liu (2004). Role of minor alloying additions in formation of bulk metallic glasses: A review. *J. Mater. Sci.* 39: 3965–3974.
229. Wang, W.H. (2007). Roles of minor additions in formation and properties of bulk metallic glasses. *Prog. Mater. Sci.* 52: 540–596.
230. Fang, S.S., X.S. Xiao, L. Xia, W.H. Li, and Y.D. Dong (2003). Relationship between the widths of supercooled liquid regions and bond parameters of Mg-based bulk metallic glasses. *J. Non-Cryst. Solids* 321: 120–125.
231. Fang, S.S., X.S. Xiao, L. Xia, Q. Wang, W.H. Li, and Y.D. Dong (2004). Effects of bond parameters on the widths of supercooled liquid regions of ferrous BMGs. *Intermetallics* 12: 1069–1072.
232. Liu, W.Y., H.F. Zhang, A.M. Wang, H. Li, and Z.Q. Hu (2007). New criteria of glass forming ability, thermal stability and characteristic temperatures for various bulk metallic glass systems. *Mater. Sci. Eng. A* 459: 196–203.
233. Cohen, M.H. and G.S. Grest (1979). Liquid-glass transition, a free-volume approach. *Phys. Rev. B* 20: 1077–1098.
234. Varley, J.H.O. (1959). *The Physical Chemistry of Metallic Solutions and Intermetallic Compounds*, p. 2H. London, UK : HMSO.

235. Ramachandrarao, P. (1980). On glass formation in metal-metal systems. *Z. Metall.* 71: 172–177.
236. Yavari, A.R. (1983). Small volume change on melting as a new criterion for easy formation of metallic glasses. *Phys. Lett. A* 95: 165–168.
237. Ma, D., A.D. Stoica, and X.-L. Wang (2007). Volume conservation in bulk metallic glasses. *Appl. Phys. Lett.* 91: 021905-1–021905-3.
238. Greer, A.L. (1995). Metallic glasses. *Science* 267: 1947–1953.
239. Cheng, Y.Q. (2011). Atomic-level structure and structure-property relationship in metallic glasses. *Prog. Mater. Sci.* 56: 379–473.
240. Bernal, J.D. (1960). Geometry of the structure of monatomic liquids. *Nature* 185: 68–70.
241. Xu, B.C., R.J. Xue, and B. Zhang (2013). Superior glass-forming ability and its correlation with density in Ce–Ga–Cu ternary bulk metallic glasses. *Intermetallics* 32: 1–5.
242. Yang, L., T. Ge, C.Q. Guo, C.L. Huang, X.F. Meng, S.H. Wei, D. Chen, and L.Y. Chen (2013). Atomic and cluster level dense packing contributes to the high glass-forming ability in metallic glasses. *Intermetallics* 34: 106–111.
243. Zhang, Z., X.Z. Xiong, W. Zhou, X. Lin, A. Inoue, and J.F. Li (2013). Glass forming ability and crystallization behavior of Al–Ni–RE metallic glasses. *Intermetallics* 42: 23–31.
244. Busch, R., E. Bakke, and W.L. Johnson (1998). Viscosity of the supercooled liquid and relaxation at the glass transition of the $Zr_{46.75}Ti_{8.25}Cu_{7.5}Ni_{10}Be_{27.5}$ bulk metallic glass forming alloy. *Acta Mater.* 46: 4725–4732.
245. Waniuk, T.A., R. Busch, A. Masuhr, and W.L. Johnson (1998). Equilibrium viscosity of the $Zr_{41.2}Ti_{13.8}Cu_{12.5}Ni_{10}Be_{22.5}$ bulk metallic glass-forming liquid and viscous flow during relaxation, phase separation, and primary crystallization. *Acta Mater.* 46: 5229–5236.
246. Fan, G.J., H.-J. Fecht, and E.J. Lavernia (2004). Viscous flow of the $Pd_{43}Ni_{10}Cu_{27}P_{20}$ bulk metallic glass-forming liquid. *Appl. Phys. Lett.* 84: 487–489.
247. Louzguine-Luzgin, D.V. and A. Inoue (2007). An extended criterion for estimation of glass-forming ability of metals. *J. Mater. Res.* 22: 1378–1383.
248. Buschow, K.H.J. and N.M. Beekmans (1980). Thermal stability of amorphous alloys. *Solid State Commun.* 35: 233–236.
249. Ji, X.L. and Y. Pan (2008). Predicting alloy compositions of bulk metallic glasses with high glass-forming ability. *Mater. Sci. Eng. A* 485: 154–159.
250. Suer, S., A.O. Mekhrabov, and M.V. Akdeniz (2009). Theoretical prediction of bulk glass forming ability (BGFA) of Ti–Cu based multicomponent alloys. *J. Non-Cryst. Solids* 355: 373–378.
251. Nagel, S.R. and J. Tauc (1975). Nearly-free-electron approach to the theory of metallic glass alloys. *Phys. Rev. Lett.* 35: 380–383.
252. Nagel, S.R. and J. Tauc (1977). Correlations in binary liquid and glassy metals. *Solid State Commun.* 22: 129–132.
253. Hafner, J. (1981). Theory of the structure, stability and dynamics of simple-metal glasses. In *Glassy Metals I*, eds. Güntherodt, H.J. and H. Beck, pp. 93–140. Berlin, Germany: Springer Verlag.
254. Häussler, P. (1992). Interrelationships between atomic and electronic structures – liquid and amorphous metals as model systems. *Phys. Rep.* 222: 65–143.
255. Wang, Y.M., J.B. Qiang, C.H. Wong, C.H. Shek, and C. Dong (2003). Composition rule of bulk metallic glasses and quasicrystals using electron concentration criterion. *J. Mater. Res.* 18: 642–648.
256. Jiang, Q., B.Q. Chi, and J.C. Li (2003). A valence electron concentration criterion for glass-formation ability of metallic liquids. *Appl. Phys. Lett.* 82: 2984–2986.
257. Jiao, W., D.Q. Zhao, D.W. Ding, H. Bai, and W.H. Wang (2012). Effect of free electron concentration on glass-forming ability of Ca–Mg–Cu system. *J. Non-Cryst. Solids* 358: 711–714.
258. Han, G., J.B. Qiang, F.W. Li, L. Yuan, S. Quan, Q. Wang, Y.M. Wang, C. Dong, and P. Häussler (2011). The *e/a* values of ideal metallic glasses in relation to cluster formulae. *Acta Mater.* 59: 5917–5923.
259. Xia, J.H., J.B. Qiang, Y.M. Wang, Q. Wang , and C. Dong (2006). Ternary bulk metallic glasses formed by minor alloying of Cu_8Zr_5 icosahedron. *Appl. Phys. Lett.* 88: 101907-1–101907-3.
260. Bian, X.F., L. Hu, and C.D. Wang (2006). Degree of local structural ordering and its correlation with the glass forming ability of alloys. *J. Non-Cryst. Solids* 352: 4149–4154.

261. Perera, D.N. (1999). Compilation of the fragility parameters for several glass-forming metallic alloys. *J. Phys. Condens. Matter* 11: 3807–3812.
262. Lu, Z.P., Y. Li, and C.T. Liu (2003). Glass-forming tendency of bulk La–Al–Ni–Cu–(Co) metallic glass-forming liquids. *J. Appl. Phys.* 93: 286–290.
263. Tanaka, H. (2005). Relationship among glass-forming ability, fragility, and short-range bond ordering of liquids. *J. Non-Cryst. Solids* 351: 678–690.
264. Gorsse, S., G. Orveillon, O.N. Senkov, and D.B. Miracle (2006). Thermodynamic analysis of glass-forming ability in a Ca–Mg–Zn ternary alloy system. *Phys. Rev. B* 73: 224202-2-1–224202-9.
265. Böhmer, R., K.L. Ngai, C.A. Angell, and D.J. Plazek (1993). Nonexponential relaxations in strong and fragile glass formers. *J. Chem. Phys.* 99: 4201–4209.
266. Senkov, O.N. (2007). Correlation between fragility and glass-forming ability of metallic alloys. *Phys. Rev. B* 76: 104202-1–104202-6.
267. Zheng, Q., J. Xu, and E. Ma (2007). High glass-forming ability correlated with fragility of Mg–Cu(Ag)–Gd alloys. *J. Appl. Phys.* 102: 113519-1–113519-5.
268. Kozmidis-Petrovic, A.F. (2015). Dynamic fragility and reduced glass transition temperature as a pair of parameters for estimating glass forming ability. *J. Non-Cryst. Solids* 417–418: 1–9.
269. Wang, L.-F., Q.-D. Zhang, X. Cui, and F.-Q. Zu (2015). An empirical criterion for predicting the glass-forming ability of amorphous alloys based on electrical transport properties. *J. Non-Cryst. Solids* 419: 51–57.
270. Liu, J.T., J. Guo, X.F. Hu, S. Guo, W.J. Meng, and Q.F. Xiao (2013). Effect of melt superheated treatment on glass forming ability and thermal expansion of $Gd_{55}Al_{25}Cu_{10}Co_{10}$ alloys. *J. Alloys & Compds.* 581: 671–674.
271. Cui, X., Q.-D. Zhang, X.-Y. Li, and F.-Q. Zu (2016). Dependence of GFA and thermal stability of the $Cu_{50}Zr_{50}$ alloy on temperature-functioned different liquid states. *Intermetallics* 73: 79–85.
272. Louzguine-Luzgin, D.V., D.B. Miracle, and A. Inoue (2008). Intrinsic and extrinsic factors influencing the glass-forming ability of alloys. *Adv. Eng. Mater.* 10: 1008–1015.
273. Suryanarayana, C. (1999). *Non-Equilibrium Processing of Materials*. Oxford, UK: Pergamon.
274. Suryanarayana, C. (2001). Mechanical alloying and milling. *Prog. Mater. Sci.* 46: 1–184.
275. Suryanarayana, C. (2004). *Mechanical Alloying and Milling*. New York: Marcel-Dekker.
276. Johnson, W.L. (1986). Thermodynamic and kinetic aspects of the crystal to glass transformation in metallic materials. *Prog. Mater. Sci.* 30: 81–134.
277. Courtney, T.H. (1994). Modeling of mechanical milling and mechanical alloying. *Rev. Particulate Mater.* 2: 63–116.
278. Koch, C.C. (1995). Research on metastable structures using high energy ball milling at North Carolina State University. *Mater. Trans., JIM* 36: 85–95.
279. Bhatt, J. and B.S. Murty (2008). On the conditions for the synthesis of bulk metallic glasses by mechanical alloying. *J. Alloys Compd.* 459: 135–141.
280. Delogu, F. and G. Cocco (2007). Kinetics of amorphization processes by mechanical alloying: A modeling approach. *J. Alloys Compd.* 436: 233–240.
281. Inoue, A. and J.S. Gook (1995). Fe-based ferromagnetic glassy alloys with wide supercooled liquid region. *Mater. Trans., JIM* 36: 1180–1183.
282. Schlorke, N., J. Eckert, and L. Schultz (1999). Thermal and magnetic properties of bulk glass forming Fe–Al–P–C–B–(Ga) alloys. *J. Phys. D Appl. Phys.* 32: 855–861.
283. Chueva, T.R., N.P. Dyakonova, V.V. Molokanov, and T.A. Sviridova (2007). Bulk amorphous alloy $Fe_{72}Al_5Ga_2C_6B_4P_{10}Si_1$ produced by mechanical alloying. *J. Alloys Compd.* 434–435: 327–332.
284. Mondal, K., U.K. Chatterjee, and B.S. Murty (2003). Gibbs free energy for the crystallization of glass forming liquids. *Appl. Phys. Lett.* 83: 671–673.
285. Sharma, S. and C. Suryanarayana (2008). Effect of Nb on the glass-forming ability of mechanically alloyed Fe–Ni–Zr–B alloys. *Scr. Mater.* 58: 508–511.
286. Sharma, S., R. Vaidyanathan, and C. Suryanarayana (2007). Criterion for predicting the glass-forming ability of alloys. *Appl. Phys. Lett.* 90: 111915-1–111915-3.

4

Synthesis of Bulk Metallic Glasses

4.1 Introduction

Metallic glasses have been produced in the form of thin ribbons by rapidly solidifying metallic melts at cooling rates of about 10^6 K s^{-1}. The early excitement of being able to produce the normally crystalline metals in a glassy state, and their excellent mechanical, chemical, and magnetic properties, led to the development of a variety of techniques to obtain metallic glasses in different sizes and shapes (e.g., ribbons, wires, and powders). Some of these techniques have been described in earlier publications (see, e.g., Refs. [1–4]). But, the commercial requirements of large-size sheets for different applications resulted in the development of the planar flow casting method, wherein rapidly solidified sheets of about 30 cm in width could be produced. The quest for bulk glassy alloys for industrial applications eventually culminated in the discovery of bulk metallic glasses (BMGs). BMGs have been produced at relatively slow solidification rates of about 10^3 K s^{-1} or less. But, before describing the different techniques to synthesize BMGs, let us briefly look at the requirements for rapid solidification processing (RSP) and the important techniques used to produce rapidly solidified ribbons. This assumes relevance here because a lot of researchers in the field of BMGs still use the melt-spinning technique to produce ribbons and study their properties and crystallization behavior before proceeding to investigate the synthesis and characterization of BMGs.

4.2 Principles of Rapid Solidification Processing

Metallic glasses were first synthesized in an Au–25 at.% Si alloy by Pol Duwez and his colleagues in 1960 by rapidly solidifying their metallic melts [5]. They used the so-called gun technique [6] to solidify metallic melts at rates approaching 10^6 K s^{-1}. In this technique, a small quantity of the molten metal is ejected using a shock wave on to a conducting substrate. The molten metal spreads in the form of a thin layer, typically a few tens of micrometers (usually about 20–50 μm) in thickness, and the heat is extracted rapidly by the conducting copper substrate. Consequently, the molten alloy solidifies very rapidly. Thus, the basic requirements to achieve high solidification rates are as follows:

1. Thin layer formation (film or ribbon) by the molten metal.
2. Intimate thermal contact with a good heat-conducting substrate to rapidly extract the heat from the liquid metal.

It has generally been noted that the solidification rate achieved, R, is inversely proportional to the square of the thickness of the solidified molten layer. For a layer of thickness x that has solidified at a heat transfer coefficient of ∞, the solidification rate may be expressed as

$$R = \frac{A}{x^2} \tag{4.1}$$

where:

x	is the distance from the splat/substrate interface
Constant A	is a function of the material properties and initial temperatures, but is independent of x

The value of A is 8.1×10^{-3} m^2 K s^{-1} for ideal cooling (when the heat transfer coefficient is ∞) and it is less for nonideal cooling conditions. For example, assuming an average value of $A = 10^{-3}$ m^2 K s^{-1}, for rough estimates, the solidification rate achieved will be approximately 10^5 K s^{-1} for $x = 100$ µm and 10^9 K s^{-1} for $x = 1$ µm. The typical thickness of a rapidly solidified foil is about 50 µm, and therefore the foil would have solidified at a rate of approximately 10^6 K s^{-1}. These examples serve to illustrate that it is necessary to have as small a section thickness as possible to achieve high solidification rates.

Metals and alloys solidified at these high rates exhibited some very interesting properties. Huge departures from equilibrium were noted (see Chapter 1), resulting in the formation of a variety of metastable phases, including supersaturated solid solutions and novel intermetallic phases. In addition, in appropriate alloy systems and at suitable compositions, one could achieve formation of glassy phases and these were termed *metallic glasses*. Significant refinement of microstructural features and of segregation patterns was also noted. Developments in this area have been well documented in the literature, as described in Chapter 2. The basic principles of RSP have been covered in some earlier reviews (see, e.g., Ref. [7]).

4.3 General Techniques to Achieve High Rates of Solidification

As one can imagine, the cross section of the solidified metal layer obtained by the gun technique of rapid solidification is not uniform because of the free flow of the liquid metal under pressure. Hence, it was very difficult to determine the physical and mechanical properties of these novel materials. Consequently, the early years of research focused mostly on the structural aspects. To overcome this deficiency, a number of other techniques were developed to improve upon the uniformity of cross section of the solidified layers. The most significant milestone in this direction was the development of the chill block melt-spinning technique in the 1970s and several of its variants. With the introduction of these techniques, rapid progress was made in this field, and the technology of RSP has become an important branch of materials science and engineering. It will be difficult to describe all the different techniques that were developed to achieve high solidification rates in metallic melts, so we will describe only the melt-spinning technique, since this has been the most popular RSP technique used by a number of researchers all over the world. Its main advantages are that (1) it will be possible to produce ribbons of uniform cross section, (2) its process parameters have been optimized, and (3) melt spinners are commercially available. The reader is advised to refer to some of the standard books and reviews, and also to a very comprehensive (but not very recent) review and compilation of the information on different rapid solidification techniques available at that time [1].

It is important to note that metallic glasses (or amorphous alloys) have been synthesized by other methods also. Since metallic glasses are nonequilibrium phases, any method capable of processing materials at far from equilibrium conditions should, in principle, be able to produce metallic glasses. (The true definition of a glass is that it is a liquid that has been cooled into a state of rigidity without crystallizing. Accordingly, purists reserve this term only for those noncrystalline materials that have been obtained directly from the liquid state. Noncrystalline materials synthesized by other methods, such as starting from the solid or vapor states, are referred to as *amorphous materials*. But, the distinction between glasses and amorphous solids has not been strictly followed in the literature and there has been frequent interchange between these two terms. But, in this book we will try to use the term *glass* only for those noncrystalline solid materials that have been obtained on continuous cooling from the liquid state.)

Amorphous alloys have also been synthesized by a number of other nonequilibrium processing techniques [8]. These include mechanical alloying (MA) [9,10], irradiation methods, ion implantation and ion mixing, laser processing, plasma processing, physical and chemical vapor deposition processes, electrolytic deposition, thermal spray processes [11], application of high pressures [12], interdiffusion and reaction methods, and others. In all these methods, the common theme has been to increase the free energy of the system (by either raising the temperature, or pressure, or the input of mechanical energy, or by other means) and subsequently quenching the material to either retain the metastable phase or to use it as an intermediate step to achieve the desired microstructure and/or properties. Such techniques have been termed *energize and quench* methods by Turnbull [13]. Since these techniques are outside the scope of the present book and were briefly described in Chapter 2, we will not discuss these methods further.

In Section 4.4, we will briefly describe the melt-spinning technique. Following that, we will describe in detail the different techniques that have been used to synthesize BMGs in different alloy systems. An important difference between the technique of melt spinning and the techniques used to obtain BMGs is that the solidification rates required in the latter case are much lower. Typically, the critical cooling rate required to produce a BMG is $\leq 10^2$ K s^{-1} (see Table 3.1 for some typical values). Therefore, the techniques developed to produce the desired cooling rates are relatively simple.

4.4 Melt Spinning

The technique of melt spinning has been the most commonly used method to produce long and continuous rapidly solidified ribbons, wires, and filaments. In fact, it is the development of this technique that was mostly responsible for the accelerated progress of the RSP technology since the 1970s. Details of the melt-spinning technique, along with the details of other techniques to achieve rapid solidification rates, are available in earlier reviews [1,4,14]. Therefore, only a brief description and the important process variables in the technique will be provided here. The interested reader could refer to these reviews and/or the original publications for further details.

The melt-spinning process derives its name from the fact that it involves the extrusion of molten metal to produce fine filaments/ribbons in a way akin to that used for the manufacture of synthetic textile fibers. In the melt-spinning process, the molten metal is ejected through an orifice and the melt stream is allowed to solidify either in flight or against a chill. The process is referred to as *free-flight melt spinning* (FFMS) if the melt is allowed to solidify in flight, or *chill-block melt spinning* (CBMS) if it is allowed to solidify against a chill (substrate). But, it is the CBMS process that is most commonly used by researchers.

The CBMS technique was originally patented by Strange and Pim [15] and subsequently improved upon by Pond [16]. In this process, a molten metal jet is directed onto a cold, rotating heat sink where the jet is reshaped and allowed to solidify. The jet, on impingement with the heat sink, forms a melt puddle of thickness approximately equal to and of length about double that of the jet. As solidification begins, the ribbon is expelled from the surface of the fast-rotating heat sink. Figure 4.1 shows a schematic illustration of the melt-spinning process. Melt spinners are available commercially from Marco Materials in the United States, and Nisshin Giken Corporation and Makabe Corporation in Japan.

In the melt-spinning method, a small quantity of the alloy is melted inside a crucible or by levitation methods, and then ejected by pressurization through a fine nozzle onto a fast-rotating copper wheel. Every one of these parameters can be carefully controlled to obtain the desired size, shape, and thickness of the ribbon.

The crucible material is chosen based on its chemical compatibility with the melt, its temperature handling capability, its resistance to thermal shock, its low thermal conductivity, and its low porosity. Dense alumina and quartz are the most common crucible materials used. The nozzles are typically circular in cross section with diameters ranging from about 50 to 1250 μm. The nozzles are made of a suitable refractory material, the exact nature of which depends on the type of metal to be processed. Some of the nozzle materials used are alumina, graphite, SiC, sapphire, and Pyrex glass. The ejection pressures are

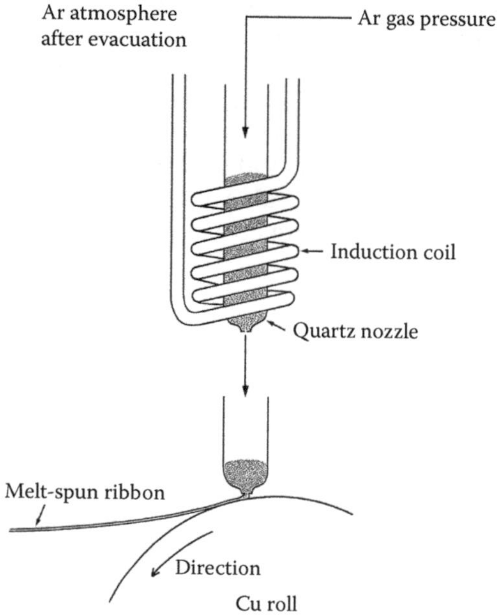

FIGURE 4.1
Schematic illustration of the melt-spinning process.

typically in the range of 5–70 kPa depending on the desired melt delivery rate. The use of higher ejection pressures results in an improvement of the wetting pattern and, hence, better thermal contact between the melt puddle and the substrate. The required ejection pressure increases as the nozzle diameter decreases.

Wheels for melt spinning have been made from a variety of materials, including copper, stainless steel, chromium, and molybdenum, although copper is the most popular one. The primary purpose of the wheel is to extract the heat from the ribbon as quickly as possible, while allowing the puddle to wet the wheel and form the ribbon. Cooling of the wheel is desirable for long runs. The outer surface of the wheel is generally polished to remove any surface roughness since the wheel side of the cast ribbon is almost an exact replica of the wheel surface. The wheel speed is an important parameter in determining the thickness of the ribbon; the faster the wheel rotates, the thinner is the ribbon. For example, an $Fe_{40}Ni_{40}B_{20}$ alloy cast on a 250 mm diameter copper wheel rotating at a substrate velocity of 26.6 m s^{-1} produced a ribbon of 37 μm thickness, while at a velocity of 46.5 m s^{-1}, the ribbon thickness was only 22 μm [17].

The melt-spinning operation is carried out in a vacuum, air, inert atmosphere, or reactive gas depending on the chemical and physical properties of the charge. Alloys susceptible to oxidation can be cast in vacuum or inert gas environments.

The solidification rates achieved in this process are typically about 10^5–10^6 K s^{-1}. Typical dimensions of the ribbons produced are about 2–5 mm in width and the thickness is in the range of 20–50 μm. Because of the high solidification rates, it will be possible to produce most alloys of appropriate composition in the glassy state by this technique. But the downside is that the section thickness is limited. The width of the ribbons can, however, be increased using the planar flow casting method, in which the nozzle has a rectangular cross section and occasionally more than one nozzle is also used to ensure overlap of the melt puddle to increase the width of the ribbon. But, the thickness cannot be increased much by any of the known melt-spinning methods.

We will now describe in detail the different methods that have been utilized by different researchers to produce BMGs. The advantages and disadvantages of the different techniques will also be highlighted. It is also appropriate to mention here that the majority of the researchers working on BMGs also use the melt-spinning technique to produce thin ribbons of the same composition as that of the BMG. These thin ribbons are used to measure the thermal properties (T_g, T_x, and T_ℓ) using the differential scanning calorimetry (DSC) and/or differential thermal analysis (DTA) methods. This is appropriate because the thermal properties of the glass do not depend on the dimensions of the glass specimen. It has been repeatedly shown that the transformation temperatures of the bulk samples are identical with those of the melt-spun ribbons.

4.5 Bulk Metallic Glasses

BMGs are those metallic glasses that have a section thickness of at least a few millimeters. As mentioned in Section 4.2, the solidification rate decreases with increasing section thickness. The present consensus appears to be that a glassy material is considered bulk only when the section thickness is at least a few millimeters. This number is left very vague, and depending on who defines it, it ranges from 1 mm up to as large as 10 mm. Therefore, if one wishes to obtain bulk glassy samples, it is necessary that the critical cooling rate for glass formation is very low. Achievement of slow solidification rates is not a problem;

but, the selection of an alloy system and the appropriate compositions that could produce glasses at these slow solidification rates needs to be determined wisely. As mentioned in Chapters 2 and 3, we now have some guidelines to help us in this direction.

Ignoring the limit of 10 mm section thickness of the glass for it to be considered as bulk, Chen [18] synthesized glassy alloys, 1–3 mm in diameter and several centimeters long in Pd- and Pt-based alloys by water quenching. Hence, these could be considered the first BMGs. These (bulk metallic) glasses could be produced in alloys with the general composition $(Pd_{1-x}M_x)_{0.835}Si_{0.165}$ where M = Ag, Au, Cu, Co, Fe, Ni, and Rh. Bulk glasses were produced with Ag up to 12 at.%, Au up to 12 at.%, Cu up to 16 at.%, Co up to 12 at.%, Fe up to 7.2 at.%, Ni up to 45 at.%, and Rh up to 4.8 at.%. BMGs were also produced in Pd–TM–P and Pt–TM–P alloys (where TM is a transition metal Fe, Co, or Ni). The $Pd_{40}Ni_{40}P_{20}$ and $Pt_{40}Ni_{40}P_{20}$ alloys subsequently became the basis for developing BMGs with larger and larger diameters, with 72 mm holding the record since 1997 [19]. More recently, by fine tuning the composition, the largest diameter of a BMG rod could be increased to 80 mm [20].

It was also highlighted in Chapter 2 that glass formation is possible only when the critical cooling rate for glass formation is exceeded and the liquid is supercooled to a temperature below T_g. Since heterogeneous nucleation plays an important role in nucleating crystalline phases on the surfaces, large super- (or under-) coolings cannot easily be achieved when heterogeneous nucleation of crystalline phases takes place well before reaching the T_g temperature. Under these circumstances, it will be difficult to undercool the melt to temperatures below T_g. If heterogeneous nucleation can be avoided, then the minimum cooling rate required for glass formation is determined by the homogeneous nucleation rate of the alloy. Further, it is known from the classical nucleation theory that alloys with a large value of the reduced glass transition temperature, T_{rg}, will have low homogeneous nucleation rates, and therefore glass formation should be easy at slow solidification rates. By subjecting the $Pd_{40}Ni_{40}P_{20}$ specimens to surface etching followed by a succession of heating and cooling cycles, Drehman et al. [21] were able to produce glasses that were 5–6 mm in diameter. And this was achieved by slowly cooling the melt at rates of about 1 K s^{-1}!

4.5.1 Flux Melting Technique

The work of Drehman et al. [21] clearly showed that removal of heterogeneous nucleation sites is very important in suppressing the nucleation of crystalline phases. Another way by which the impurities present in the alloy could be removed is to heat and cool the molten metal while it is immersed in a molten oxide flux. After gravity segregation, the impurities present in the alloy get dissolved in the molten oxide flux. If the flux (like the slag in foundries) containing the impurities from the alloy is maintained in the liquid state at the glass transition temperature of the alloy, T_g, that is, when the glass is formed, then glass formation can be achieved at slower solidification rates. These slow solidification rates translate to large section thicknesses. Using this logic, Kui et al. [22] produced a 10 mm-size glass in 1984. They used anhydrous boron oxide (B_2O_3) to remove the impurities from the $Pd_{40}Ni_{40}P_{20}$ alloy melt. At the glass transition temperature of $Pd_{40}Ni_{40}P_{20}$ (~600 K), B_2O_3 was still in the molten state. Thus, by heating this alloy to higher temperatures (~1273 K) and cooling it slowly to temperatures below the T_g of the alloy, these investigators were able to produce 10 mm-size glassy samples of $Pd_{40}Ni_{40}P_{20}$ using this technique. Incidentally, this was the first time that the term *bulk metallic glass* was used in the title of a publication. This method has now come to be known as the *fluxing* or *flux-melting technique* and continues to be popular to remove impurities from the melt and to avoid heterogeneous nucleation events. Most recently, researchers have used a mixture of B_2O_3 + CaO in the weight ratio 3.5:1 as a fluxing agent [23].

Even though these two investigations were reported by 1984, their importance was not immediately grasped by the materials science community, probably due to the excitement of the discovery of quasicrystals reported the same year and the scientific frenzy that immediately followed [24]. Systematic investigations by the group of Masumoto and Inoue at Tohoku University in Sendai, Japan (see, e.g., Ref. [25]) resulted in the development of alloys that could be produced in the glassy state at solidification rates of 10^3 K s^{-1} or lower. These included alloys in the Mg–Ln–TM, Ln–Al–TM, Zr–Al–TM (where Ln represents lanthanide), and Mg–Cu–Y [26] systems. Around the same time, Peker and Johnson [27] at Caltech reported on the formation of a highly processable metallic glassy alloy with the composition $Zr_{41.2}Ti_{13.8}Cu_{12.5}Ni_{10}Be_{22.5}$. This alloy, referred to as *Vitreloy 1* or *Vit 1*, became glassy at a solidification rate of about 1 K s^{-1} and large samples of up to 14 mm in diameter could be produced. Table 4.1 summarizes some of the early results on the synthesis of BMGs at slow solidification rates and large section thicknesses. These results have triggered the imagination of a large number of materials scientists, and today there is worldwide activity in this subarea of materials science and engineering.

The critical diameter of the glassy rods had continuously increased from a few millimeters to a few centimeters in the early years of research. This was possible initially by substitution of some of the metallic elements in the alloy based on intuition and past experience and results. But in recent years, the principles of alloy design based on the atomic sizes of the constituent elements, chemical interactions between the different elements, and phase diagram considerations have been successfully used to increase the glass-forming ability (GFA) of alloys and consequently the critical diameters of the fully glassy rods. The largest diameter of a fully glassy alloy rod is 80 mm in a $Pd_{42.5}Cu_{30}Ni_{7.5}P_{20}$ alloy established in 2012 [20].

The substitution of alloying elements has played a big role in these studies. Researchers have identified that elements having similar atomic diameters could be easily substituted for each other. Further, based on the chemical nature of alloying elements, they could be grouped into some categories [29]. For example, by replacing some Ni with Cu in the Pd–Ni–P alloy to a composition $Pd_{40}Ni_{10}Cu_{30}P_{20}$, Inoue et al. [28] were able to increase the diameter of the glassy rod to 40 mm. By subjecting this $Pd_{40}Ni_{10}Cu_{30}P_{20}$ alloy to a flux treatment (to remove the presence of any heterogeneous nucleation sites), they were able to further increase the diameter of the bulk metallic glassy alloy to 72 mm [19]. By fine tuning the composition, the critical diameter could be increased to 80 mm in a $Pd_{42.5}Cu_{30}Ni_{7.5}P_{20}$ alloy [20].

TABLE 4.1

Summary of Early Results on Discovery of BMGs

Alloy Composition	Critical Cooling Rate (K s^{-1})	Year of Discovery	Largest Section Thickness (mm)	References
$(Pd_{1-x}M_x)_{0.835}Si_{0.165}$ (M = Cu, Ag, Au, Fe, Co, Ni)	—	1974	1–3	[18]
$(Pd_{1-x}TM_x)_{1-xP}P_{xP}$ or $(Pt_{1-x}TM_x)_{1-xP}P_{xP}$ (TM = Fe, Co, or Ni)	—	1974	1–3	[18]
$Pd_{40}Ni_{40}P_{20}$	1	1982	5	[21]
$Pd_{40}Ni_{40}P_{20}$ (flux treated)	—	1984	10	[22]
$La_{55}Al_{25}Ni_{20}$	—	1989	1.2	[25]
$Mg_{65}Cu_{25}Y_{10}$	—	1992	7	[26]
$Zr_{41.2}Ti_{13.8}Cu_{12.5}Ni_{10}Be_{22.5}$	~1	1993	14	[27]
$Pd_{40}Ni_{10}Cu_{30}P_{20}$	1.57	1996	40	[28]
$Pd_{40}Ni_{10}Cu_{30}P_{20}$ (flux treated)	0.1	1997	72	[19]

4.5.2 Role of Contamination

As highlighted in the preceding section and in earlier chapters, the critical cooling rate, R_c, for the formation of BMGs is relatively low, the highest required being only about 10^2–10^3 K s^{-1}. Because of these relatively low critical cooling rates, the synthesis of BMGs does not require highly sophisticated equipment. The general casting equipment used routinely in the laboratory/industry can be easily adapted for this purpose. However, some alloy systems, for example, alloys based on zirconium and titanium, are very sensitive to the presence of impurities, and so care must be taken to make sure that such alloys do not come into contact with contaminants. For example, it has been shown that high oxygen content reduced the supercooled liquid region (ΔT_x) [30] and changed the crystallization behavior in the Zr-based BMGs [31]. Further, the nature of the crystalline phases formed on crystallization was different depending on the oxygen content in the glassy alloy; a quasicrystalline phase was observed in the early stages of crystallization by some investigators and not others.

de Oliveira et al. [32] investigated the crystallization behavior of a $Zr_{55}Al_{10}Ni_5Cu_{30}$ alloy by producing it in a melt-spun ribbon form and also as a 5 mm diameter rod by copper mold casting under a gettered argon atmosphere. They noted that while the melt-spun ribbon was completely glassy, the copper mold-cast alloy rod contained a mixture of a glassy phase and about 20 vol.% of a crystalline phase identified as Zr_4Cu_2O. A similar phase (along with three other phases) was observed in the ribbons that were fully crystallized. Based on these results, the authors concluded that oxygen, amounting to about 0.8 at.%, was responsible for the presence of the crystalline phase in the cast rod and also the decreased supercooled liquid region.

Lin et al. [30] showed that the level of oxygen in the alloy played a crucial role in the crystallization kinetics of $Zr_{52.5}Ti_5Cu_{17.9}Ni_{14.6}Al_{10}$ bulk glassy alloy that could be solidified into the glassy state at a critical cooling rate of 10 K s^{-1}. They showed that the crystallization incubation time decreased by orders of magnitude as one went from 250 to 5250 ppm of oxygen, suggesting that a sample of higher purity (<250 ppm oxygen) would exhibit even

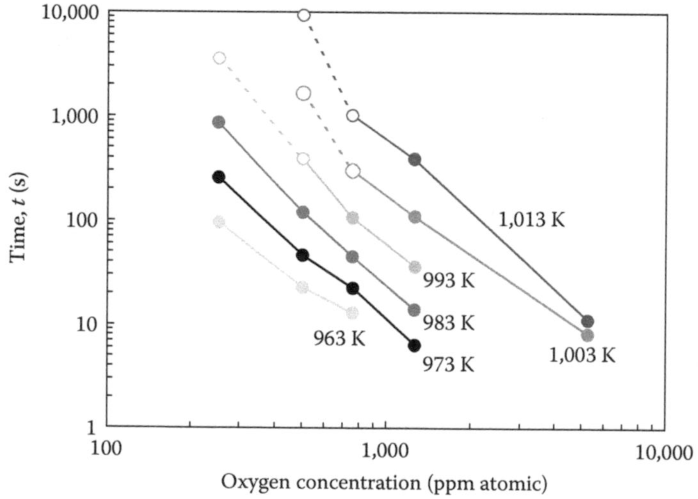

FIGURE 4.2
Crystallization incubation time at different temperatures as a function of oxygen content in a $Zr_{52.5}Ti_5Cu_{17.9}Ni_{14.6}Al_{10}$ bulk glassy alloy. (Reprinted from Lin, X.H. et al., *Mater. Trans., JIM*, 38, 473, 1997. With permission.)

greater undercooling and stability against crystallization. Figure 4.2 shows the crystallization incubation time at different temperatures as a function of oxygen content.

Gebert et al. [33] investigated the effect of oxygen content (0.28–0.60 at.%) on the thermal stability of the glassy phase in a bulk glass-forming $Zr_{65}Al_{7.5}Cu_{17.5}Ni_{10}$ alloy and also on the nature of crystalline phases formed on reheating the glassy alloy. They observed that the GFA of the alloy decreased with increasing oxygen content. This was interpreted from the observation that the volume fraction of the crystalline phase in the as-cast alloy increased with an increase in the oxygen content. Further, with increasing oxygen content, T_g of the glassy alloy increased and T_x decreased, causing a drastic reduction in the supercooled liquid region. Figure 4.3 shows the dependence of T_g, T_x, and ΔT_x on the oxygen content in the Zr-based glassy alloys.

This change in the transformation temperatures of the Zr-based glassy alloys due to oxygen contamination was also accompanied by a change in the crystallization sequence, which pointed to the reduced stability of the supercooled liquid [34]. Based on these observations, one has to be careful in directly correlating the extent of supercooled liquid region with the high GFA of alloys. This may be valid only when the oxygen concentration is low, especially in reactive alloy systems such as those based on zirconium. It is possible that other interstitial elements may also have a significant effect on the thermal stability and crystallization behavior of the BMG alloys. For example, Chen et al. [35] intentionally added different impurity elements such as Cr, Mn, C, Si, and oxygen and noted that the GFA and mechanical properties of the $Zr_{50}Ti_4Y_1Al_{10}Cu_{25}Ni_7Co_2Fe_1$ alloy were insensitive to these impurities. The large number of components present in the alloy appeared

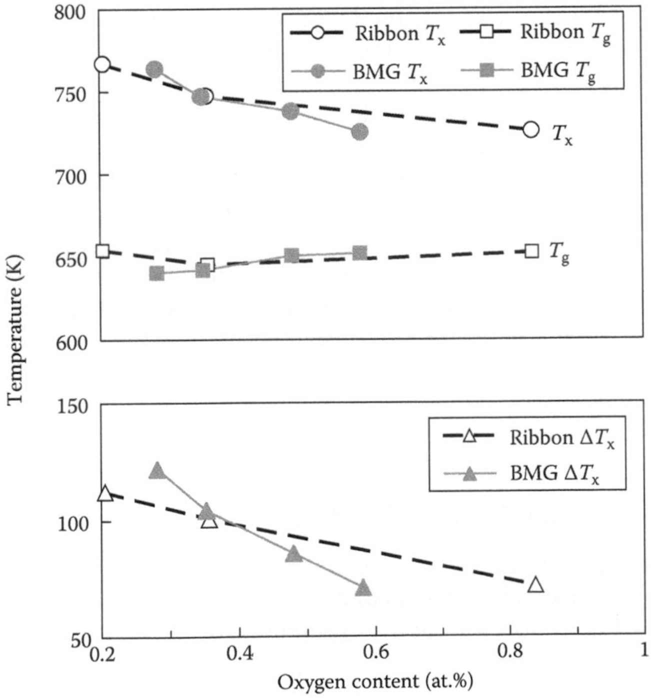

FIGURE 4.3
Variation of T_g, T_x, and ΔT_x as a function of the oxygen content in the Zr glassy alloys. Results for both bulk samples and melt-spun ribbons are shown. (Reprinted from Gebert, A. et al., *Acta Mater.*, 46, 5475, 1998. With permission.)

to be responsible for hindering the heterogeneous crystallization induced by nanoscale nuclei resulting from impurities.

The effect of oxygen content on melt-spun ribbons appears to be different. Eckert et al. [31] observed that the addition of oxygen to $Zr_{65}Cu_{17.5}Ni_{10}Al_{7.5}$ glassy ribbons resulted in the formation of a metastable quasicrystalline phase as against the formation of a stable intermetallic compound in an oxygen-free alloy. Murty et al. [36] also made a similar observation in a glassy $Zr_{65}Cu_{27.5}Al_{7.5}$ alloy and noted that the quasicrystalline phase was enriched with oxygen, suggesting that the quasicrystals were stabilized by oxygen. Sordelet et al. [37] investigated the effect of oxygen content on the structure of $Zr_{80}Pt_{20}$ glassy ribbons and noted that, depending on the oxygen content, different structures had formed ranging from crystalline to amorphous or quasicrystalline phases.

An entirely different effect was, however, noted by Scudino et al. [38], who investigated the influence of oxygen addition on the crystallization behavior of melt-spun $Zr_{57}Ti_8Nb_{2.5}Cu_{13.9}Ni_{11.1}Al_{7.5}$ glassy ribbons. They studied the nominally pure (0.03–0.05 at.% oxygen) and intentionally contaminated (with up to 0.24 wt.% oxygen) ribbons. DSC curves showed T_g and two exothermic peaks at T_{x1} and T_{x2}, the first one corresponding to the formation of a quasicrystalline phase. It was noted that, with increasing oxygen content, both the T_g and T_{x1} temperatures of the glassy ribbons increased but the T_{x2} value decreased. This observation suggests that while oxygen enhances the thermal stability of the glass/supercooled liquid against (quasi)crystallization, the stability of the quasicrystalline phase formed has decreased, as evidenced by the decreased temperature interval between T_{x1} and T_{x2}.

A completely different approach was taken up by Kündig et al. [39] to increase the GFA of Zr-based alloys. They added small amounts of C, Si, Ca, Sc, and La at levels of 0.1, 0.3, and 1.0 at.% and noted that the best GFA for the $Zr_{52.5}Cu_{17.9}Ni_{14.6}Al_{10}Ti_5$ alloy was achieved with 0.03–0.06 at.% Sc. This was confirmed by producing a maximum section thickness of 12 mm when 0.03 at.% Sc was added to the alloy. This effect was explained by the fact that Sc interacts with oxygen to form Sc_2O_3 and this was found to be less active as a nucleating agent than ZrO_2, which normally forms in the presence of oxygen and in the absence of other strong oxide-forming elements.

These examples clearly show that the purity (or cleanliness) of the melt is very important in achieving the desired effects. With a contaminated melt, the GFA of the alloy will be reduced, and at very high levels of impurity concentration, it may not be possible to achieve formation of a glassy phase at all in some cases. Even if it is achieved, the stability of the glassy phase and/or the nature and size of the crystallization products could be quite different from what one would have obtained in an oxygen-free alloy.

4.6 Bulk Metallic Glass Casting Methods

A number of different techniques have been developed to synthesize BMG alloys. These developments have occurred in different laboratories, and sometimes, for some specific applications or materials. Let us now look at the different techniques available to produce BMGs.

4.6.1 Water-Quenching Method

This is the simplest of the quenching methods used for centuries to harden steel (by transforming the soft austenite to the hard martensite phase). In hardening steel, the steel specimen is heated (in the solid state) to the austenitic phase field, held there for some predetermined time (austenitized), and then the hot specimen is transferred quickly to a quenching medium (usually water), maintained at a temperature below M_f (the martensite finish temperature). The quenching medium (water) extracts the heat rapidly from the hot steel specimen and as a result the austenite phase transforms to the martensitic condition in a diffusionless manner. The cooling rates achieved by the water-quenching method are inherently dependent on the heat transfer efficiency of the quenching medium, the size of the steel specimen, and its heat transfer properties. The cooling rates achieved by this method have been usually reported to be about 10–100 K s^{-1}.

Since some of the BMG alloy compositions require very low critical cooling rates for glass formation, it is possible to produce glasses in these alloys by this simple technique of water quenching.

The very first report of producing 1–3 mm diameter rods of $(Pd_{1-x}M_x)_{0.835}Si_{0.165}$, $(Pd_{1-x}TM_x)_{1-xP}P_{xP}$, and $(Pt_{1-x}Ni_x)_{1-xP}P_{xP}$ (with $0.20 \le x_P \le 0.25$ and where M = Cu, Ag, Au, Fe, Co, or Ni and TM = Fe, Co, or Ni) glassy alloys was by water quenching [18]. Further, the $Pd_{40}Ni_{40}P_{20}$ glassy alloys produced by Turnbull and his group in 5 and 10 mm size were also quenched by water [21,22]. Inoue et al. [25] produced 1.2 mm diameter $La_{55}Al_{25}Ni_{20}$ glassy rods by quenching the melt contained in a quartz capillary into water. Subsequently, this technique was used by many other researchers to produce BMGs in different alloy systems. Table 4.2 lists some of the alloys that have been produced in the glassy state by the water-quenching method. It should be mentioned, however, that this is a technique adopted by many researchers.

The technique of water quenching is quite simple. The alloys are made by conventional methods such as in an arc furnace or by induction melting. These prepared alloys are then placed inside a quartz tube, given the flux treatment (the addition of an oxide such as B_2O_3,

TABLE 4.2

Details of Bulk Metallic Glassy Rods Produced by the Water-Quenching Method

Alloy System	Rod Diameter (mm)	Critical Cooling Rate (K s^{-1})	Year	References
$(Pd_{1-x}M_x)_{0.835}Si_{0.165}$	1–3	<10^3	1974	[18]
$(Pd_{1-x}T_x)_{1-xP}P_{xP}$	1–3	<10^3	1974	[18]
$(Pt_{1-x}Ni_x)_{1-xP}P_{xP}$	1–3	<10^3	1974	[18]
$Pd_{40}Ni_{40}P_{20}$	5–6	~1	1982	[21]
$Pd_{40}Ni_{40}P_{20}$ (flux treated)	10	—	1984	[22]
$Zr_{65}Al_{7.5}Ni_{10}Cu_{17.5}$	<16	1.5	1993	[40]
$Zr_{41.2}Ti_{13.8}Cu_{12.5}Ni_{10}Be_{22.5}$	14	<10	1993	[27]
$Pd_{40}Cu_{30}Ni_{10}P_{20}$	40	1.57	1996	[28]
$Pd_{40}Cu_{30}Ni_{10}P_{20}$ (flux treated)	50–72	0.1	1997	[19]
$Pd_{40}Ni_{40}P_{20}$	7	100	1999	[41]
$Pd_{40}Ni_{32.5}Fe_{7.5}P_{20}$	7	100	1999	[41]
$Pd_{40}Ni_{20}Fe_{20}P_{20}$	7	100	1999	[41]
$Mg_{65}Y_{10}Cu_{15}Ag_5Pd_5$	12	—	2001	[42]
$Y_{56}Al_{24}Co_{20}$	1.5	—	2003	[43]
$Y_{36}Sc_{20}Al_{24}Co_{20}$	25	—	2003	[43]
$Pt_{60}Cu_{20}P_{20}$	<4	—	2004	[44]
$Pt_{60}Cu_{16}Co_2P_{22}$ (flux treated)	16	—	2004	[44]
$Pt_{57.5}Cu_{14.7}Ni_{5.3}P_{22.5}$ (flux treated)	16	—	2004	[44]
$Pt_{42.5}Cu_{27}Ni_{9.5}P_{21}$ (flux treated)	20	—	2004	[44]

which is known to improve the GFA of alloys by removing the impurities from the alloys), if necessary, and heated to a temperature above the liquidus temperature of the alloy to completely melt it. The quartz tube containing the molten alloy is then quenched into flowing or agitated water. The diameter of the quartz tube may be varied to obtain glassy rods of different diameters and also to determine the maximum diameter of the rod that could be quenched into the glassy state. The wall thickness of the quartz tube is typically about 1 mm. The length of the quartz tubes can be a few centimeters in length (up to 15). The cooling rate achieved is typically approximately 10^2 K s^{-1}, and so this technique can be used to produce glassy rods only in those systems that have a high GFA, that is, those alloys that possess low critical cooling rates for the formation of the glassy phase.

Even though quartz is the most common container material for the tubes used to melt the alloys, Amiya and Inoue [42] noted that when a quartz tube was used to melt their Mg alloys, they were not able to produce a glassy rod of even 4 mm diameter. Instead, when an iron tube was used to melt the Mg alloys, they were successful in producing glassy rods up to 12 mm in diameter. A careful analysis showed that when a quartz tube was used, Si dissolved in the Mg melt as an impurity and provided heterogeneous nucleation sites. Consequently, the GFA of the alloy was reduced. On the other hand, when an iron tube was used, there was no interaction between iron and the Mg melt and, therefore, large-diameter glassy rods could be produced. Thus, the compatibility between the melt and the crucible is also an important issue to be considered.

Figure 4.4 shows a photograph of the 80 mm diameter cylinder of $Pd_{42.5}Cu_{30}Ni_{7.5}P_{20}$ alloy produced by water quenching. The cylinder possesses good luster, typical of glassy samples, and its external appearance is smooth.

A distinct advantage of the water-quenching method is that, due to the slow solidification rates, the cast specimen contains much less residual stresses.

4.6.2 High-Pressure Die Casting

Die casting is a common method to produce different types of castings in the industry. Compared with the conventional sand-casting methods, die-casting methods offer higher

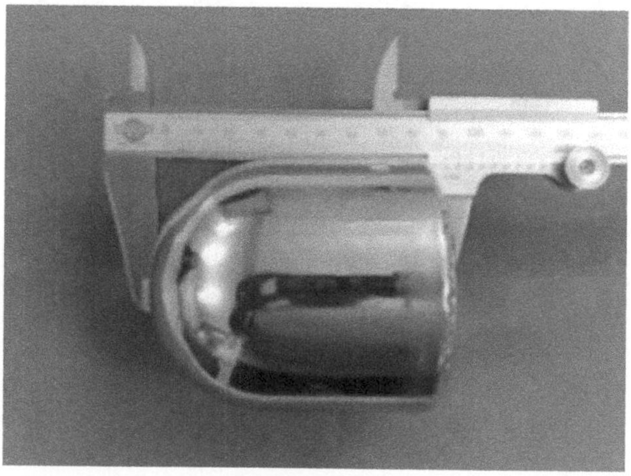

FIGURE 4.4
Photograph of the 80 mm diameter glassy cylinder of $Pd_{42.5}Cu_{30}Ni_{17.5}P_{20}$ produced by water quenching. Note the smooth appearance of the cylinder, which is typical of glassy alloys.

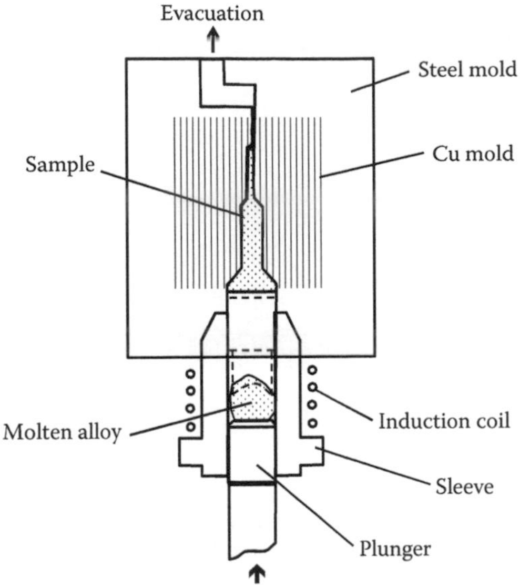

FIGURE 4.5
Schematic diagram of the high-pressure die casting equipment designed and used by Inoue et al. (Reprinted from Inoue, A. et al., *Mater. Trans., JIM*, 33, 937, 1992. With permission.)

solidification rates (because heat is extracted more rapidly by the metal mold) and more complex shapes can also be produced. Therefore, this method has been used by several researchers to synthesize BMGs in different alloy systems.

Figure 4.5 shows a schematic of the high-pressure die-casting equipment designed and used by Inoue et al. [26] to synthesize Mg-based BMGs. The main components of the equipment are a sleeve to melt the alloy, a plunger to push the molten alloy through hydraulic pressure into the copper mold, and a copper mold to solidify the melt. The whole unit is evacuated to prevent gas entrapment by the melt and consequent porosity in the casting.

The sleeve and plunger are made of a heat-resistant tool steel (SKD61). The metallic alloy is melted in the sleeve under an argon atmosphere with a high-frequency induction coil. After the alloy is melted, the plunger is moved by hydraulic pressure into the copper mold, and the molten liquid solidifies once it comes into contact with the highly conductive copper mold. Generally, a lubricant is not required. However, a lubricant can be coated inside the copper mold if it becomes difficult to remove the casting from the mold.

This equipment satisfies all the requirements of RSP. First, the casting (solidification) is complete in a few milliseconds, thus achieving high solidification rates, and also a high productivity. Second, the application of high pressure ensures good contact between the melt and the copper mold. This good contact results in a large heat transfer coefficient at the die/melt interface and consequently high cooling rates. In addition, casting defects such as shrinkage holes generated due to contraction of the liquid metal during solidification are also reduced. Last, using this technique, it is possible to produce complex shapes even in alloys with a high viscosity. Further, by changing the design of the mold, it should be possible to cast either rods or sheets. But, it should be emphasized that in contrast to conventional RSP methods, high-pressure die casting can synthesize alloys with much larger dimensions, and therefore this technique is eminently suited to producing BMGs.

Inoue et al. used this technique to produce BMGs in the Mg–Cu–Y [26] and La–Al–TM [45] alloy systems. They used a casting pressure of 63 MPa, a plunger velocity of 1.7 m s^{-1},

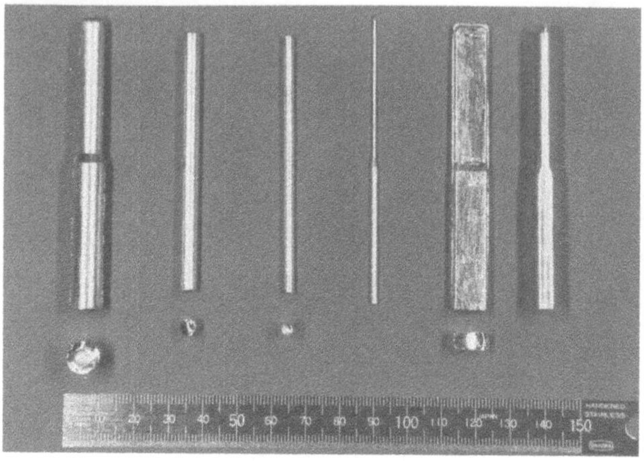

FIGURE 4.6
Photographs of the $Mg_{65}Cu_{25}Y_{10}$ rods and sheets (of different diameters) produced by the high-pressure die-casting technique. The length of the samples is 80 mm and the thickness or diameter varies from 0.5 to 9 mm. Note the bright and shiny appearance of both types of sample. (Reprinted from Inoue, A. et al., *Mater. Trans., JIM*, 33, 937, 1992. With permission.)

and a hold time of 5 s. The copper mold was cooled with flowing cold water. Rods with a length of 40 and 80 mm with diameters ranging from 1 to 9 mm could be produced with these conditions. Additionally, sheet samples with a size of 80 mm and thicknesses ranging from 0.5 to 9.0 mm could also be produced.

Figure 4.6 shows photomacrographs of the rods and sheets synthesized in the Mg–Cu–Y system using this technique. Note that the surface of the as-cast samples is bright and lustrous, as expected of a metallic glassy sample. Also, no surface defects are seen on these samples. By observing the cross sections of the as-cast samples, one can detect whether porosity, if any, is present in the center of the sample. If porosity, indicated by the presence of tiny dark spots, is present in the sample, it may become necessary to control the process parameters to ensure that the whole sample is fully dense and devoid of any porosity. It is important to obtain fully dense samples because the properties (especially the mechanical properties such as strength and modulus) are critically dependent on the amount of porosity present in the samples.

The cooling rates achieved in this technique are typically about 10^3 K s^{-1}. For example, based on the dendrite arm spacings measured in an Al-based Al–Si–Cu–Zn alloy (ADC-12), Inoue et al. [46] estimated the cooling rate as 2×10^4 K s^{-1} on the outer surface of a 3.5 mm cylinder and as 2.5×10^3 K s^{-1} in the central region. Since these cooling rates are relatively high, this technique could be used to produce BMGs in those alloy systems that require a reasonably high critical cooling rate to produce a glassy phase. Some of the alloy systems that were produced in the glassy state are Mg–Ni–Ln, Mg–Cu–Ln, La–Al–TM, and Zr–Al–TM. Of course, alloys that require relatively low cooling rates to produce a glassy phase can also be synthesized using this technique.

4.6.3 Copper Mold Casting

This appears to be perhaps the most common and popular method to produce BMGs in different alloy systems. It has been frequently used by the groups of Inoue (see, e.g., Ref. [47]) and Kim (see, e.g., Ref. [48]), among others.

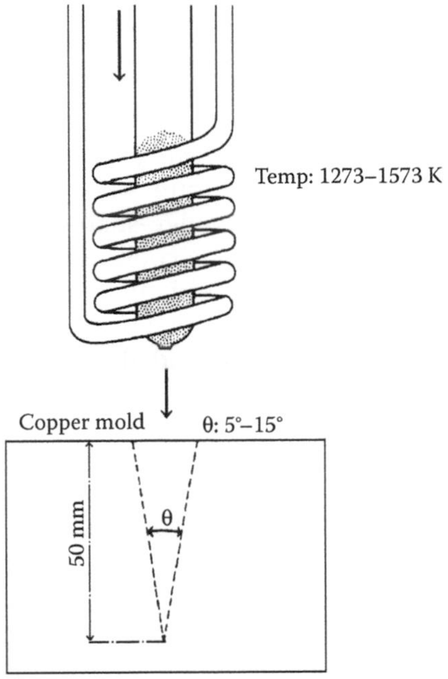

FIGURE 4.7
Schematic diagram of the equipment used to prepare bulk metallic glassy alloys by the copper mold wedge-casting technique. (Reprinted from Inoue, A. et al., *Mater. Trans., JIM*, 36, 1276, 1995. With permission.)

Figure 4.7 shows a schematic of the equipment normally used for copper mold casting of BMG alloys [49]. In simple terms, in this technique, the alloy is melted and poured into a copper mold, where it solidifies quickly because of the rapid heat extraction by the metal mold. The starting raw materials for alloy preparation could include either pure metals or master alloys of some of the elements and/or combinations of them. Melting is carried out in a crucible made up of quartz [50], boron nitride (BN) [51], graphite [48], BN-coated graphite [52], and others. Most frequently, the alloys are melted by induction-melting techniques, but sometimes they are prepared by arc melting the pure metals under a Ti-gettered argon atmosphere in a water-cooled copper crucible. The molten alloys are melted repeatedly several times to ensure compositional homogeneity. When melting of alloys with a high vapor pressure, for example, magnesium (or other materials), is carried out, about 5 wt.% excess magnesium is added to compensate for the loss of magnesium through evaporation during repeated melting operations.

The molten alloy is then poured into a copper mold. Generally, a low pressure (0.05 MPa = 50 kPa) is applied to eject the metal from the crucible toward the mold. The temperature of the molten metal can be maintained in such a way that it continues to be in the liquid state during the whole process of ejection until it fills the mold cavity. The casting can be done in air or vacuum, inert atmosphere, or under argon, if oxidation has to be avoided.

The mold used can take different forms. The most common and simple form of the mold has a cylindrical or rod-shaped cavity of a predetermined length. In this case, one has to use molds with cavities of different internal diameters to determine the maximum diameter of the sample that can be produced in a fully glassy state. The necessity of using molds of different internal diameters could be avoided by using a wedge- or cone-shaped mold. The advantage of using the wedge-shaped mold is that one can get specimens of different diameters in one experiment, and this helps in determining the maximum diameter of the rod that can be produced in the glassy state in one single experiment. The wedge-shaped mold used by the group of Inoue [49] had a depth of 50 mm and the included angle varied between 5° and 15°. In contrast, the cone-shaped mold used by Park and Kim [52] had a height of 45 mm. The mold opening at the top was 15 mm and at the bottom it was 6 mm.

Inoue et al. [49] placed platinum (Pt)—platinum–rhodium (Pt–Rh) thermocouples—at different positions along the length of the wedge-shaped mold to measure the cooling rates and eventually determine the continuous cooling transformation (CCT) diagrams. Samples could be collected from different positions in the mold for characterization by x-ray diffraction, optical microscopy, transmission electron microscopy, and DSC to determine the true glassy nature of the cast material and also to study its transformation behavior to the equilibrium crystalline state.

Figure 4.8a shows the phase relationship when the molten $Zr_{60}Al_{10}Ni_{10}Cu_{15}Pd_5$ alloy is ejected into the wedge-shaped copper mold cavity at a temperature of 1473 K. Figure 4.8a plots the constitution of the alloy as a function of the height from the bottom of the wedge-shaped cavity (d) and the vertical angle (θ). It may be noted that different phases are forming in different regions. A glassy phase is forming at all θ values up to a height of about 20 mm from the bottom of the wedge. Similarly, a glassy phase is found to form along the whole height of the wedge, up to a θ value of about 10°. But, when the angle is >10°, a fully glassy phase is not obtained along the whole length of the sample and a crystalline phase starts to form either at higher wedge angles or at larger heights from the bottom of

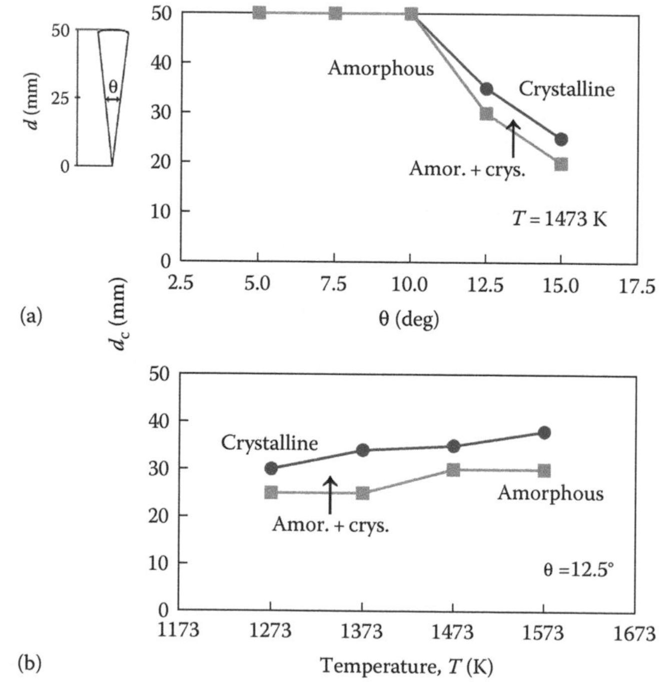

FIGURE 4.8

(a) Variation of the constitution of the alloy as a function of the height of the sample from the bottom of the wedge, d_c, and the vertical angle, θ. The figure shows the region of formation of the fully glassy phase when the $Zr_{60}Al_{10}Ni_{10}Cu_{15}Pd_5$ alloy was ejected into the copper mold cavity at a temperature of 1473 K. (b) Variation of d_c with ejection temperature of the molten metal for the $Zr_{60}Al_{10}Ni_{10}Cu_{15}Pd_5$ alloy cast into a wedge-shaped mold with a vertical angle θ = 12.5°. (Reprinted from Inoue, A. et al., *Mater. Trans., JIM*, 36, 1276, 1995. With permission.)

the wedge. Thus, one can easily plot the "phase contours" relating the height and angular relationships where glassy, crystalline, or a combination of these two phases could coexist. However, since we are only interested in the glassy phase, we can define the critical diameter (d_c) up to which a glassy phase is obtained as a function of the wedge angle. From Figure 4.8a, this value is 50 mm up to a θ value of 10°, and then reduces to 30 mm at θ = 12.5°, and further reduces to 20 mm at 15°. This phase diagram is for one alloy composition and one temperature from which the alloy melt was poured into the mold. It will be different for different combinations of these parameters.

It is easy to appreciate that this value of d_c should also depend on the temperature of the melt (since the viscosity of the melt is going to be different). Figure 4.8b shows the phase relationship between the critical diameter (d_c) as a function of the melt-pouring temperature for a fixed θ value of 12.5°. Here, again, we can see that a fully glassy phase is obtained up to a height of 25–30 mm, depending on the melt-pouring temperature, noting that the height is larger for higher pouring temperatures.

4.6.4 Cap-Cast Technique

The bulk metallic glassy alloy $Zr_{55}Cu_{30}Ni_5Al_{10}$ possesses excellent mechanical properties, such as high strength, high fracture toughness, and high fatigue strength, and has been suggested as a suitable material for applications for micro-geared motors, pressure sensors, golf clubs, and optical parts, among others. However, the maximum diameter of the fully glassy rod that is produced by conventional metallic mold casting has been limited to about 16 mm. In order to increase the diameter of the glassy rod further, Yokoyama et al. [53] have recently developed the cap-cast technique to increase the maximum diameter of the fully glassy rod that can be cast.

Figure 4.9 compares the arc-melting, tilt-casting, and cap-casting techniques. In the cap-casting technique, shown schematically in Figure 4.9c, the molten alloy is poured into a copper mold and is also allowed to solidify quickly from the top by bringing a metallic cap into contact with the molten metal. Simultaneously, a small pressure of about 1 kN is applied to push the cap down toward the molten metal. The biggest advantage of this technique is that high cooling rates are achieved not only on the sides and bottom of the casting since they are in contact with the metal mold, but even in the upper part of the cast specimen since this is in contact with the metal cap. Using this method, the authors were able to produce a fully glassy rod of 30 mm diameter in the $Zr_{55}Cu_{30}Ni_5Al_{10}$ alloy. High-resolution transmission electron microscopy of this alloy, at the center and 10 mm from

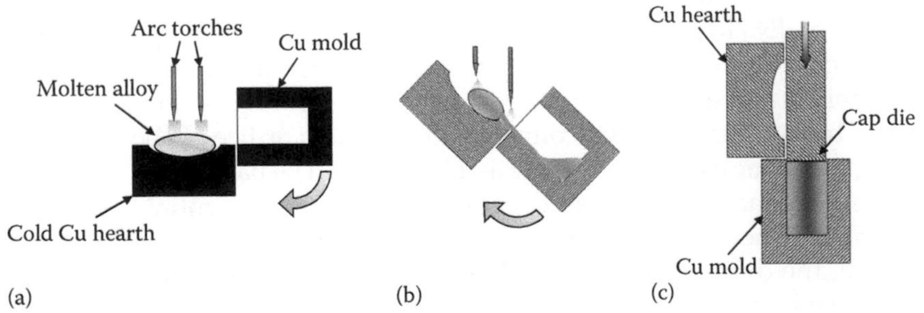

FIGURE 4.9
Schematic diagrams comparing the (a) arc-melting, (b) tilt-casting, and (c) cap-cast techniques used to produce bulk metallic glassy alloys.

FIGURE 4.10
High-resolution transmission electron micrographs of cap-cast $Zr_{55}Cu_{30}Ni_5Al_{10}$ glassy alloy of 30 mm diameter. The micrograph was recorded from the center of the sample at the site 10 mm from the bottom of the casting. No fringe marks are seen even on a nanometer scale, suggesting that the whole sample was fully glassy.

the bottom of the cast rod, did not show the presence of any crystalline inclusions, and not even a fringe pattern, indicating that the whole alloy is glassy (Figure 4.10).

4.6.5 Suction-Casting Method

This is another popular method of synthesizing BMGs. The principle involved in this method is to suck the molten alloy into a mold/die cavity by using a pressure differential between the melting chamber and the casting chamber. Some of the details of this technique may be found in Inoue and Zhang [54], Figueroa et al. [55], and Gu et al. [56], and a full description of the development was produced by Wall et al. [57].

The technique is referred to as the *drop-casting* technique if the molten alloy is just dropped into the mold instead of being sucked in through a pressure differential. Drop casting is typically used to process materials with diameters larger than 6 mm, while suction casting is used for casting of materials with diameters smaller than 6 mm. Pressure is applied in the suction-casting process of smaller diameter rods to essentially overcome the difficulty of the molten alloy to enter the small die cavity. The drop-casting method was used by Shen et al. [58] to produce rods of $Fe_{41}Co_7Cr_{15}Mo_{14}C_{15}B_6Y_2$ of up to 16 mm diameter.

The suction-casting system consists of two chambers—an upper chamber in which the alloy is melted and a lower chamber in which the casting is done in a copper mold (Figure 4.11). The two chambers are connected through an orifice of about 2 mm in diameter [56], though in some cases it is as large as 16 mm [54]. The base of the mold is connected to a vacuum source that, when released, causes a pressure differential that forces the melt into the chamber. When the piston separating the melting chamber and the casting chamber is removed, the differential pressure between these two chambers allows the molten metal to be sucked into the casting chamber, and then to be solidified in the copper mold.

In this method, the alloy is usually melted on a water-cooled copper hearth, which sits under a vacuum bell. The copper plate acts as the anode for the arc-current transfer by the cathode during arc melting under an Ar cover (at a pressure of 5×10^4 Pa).

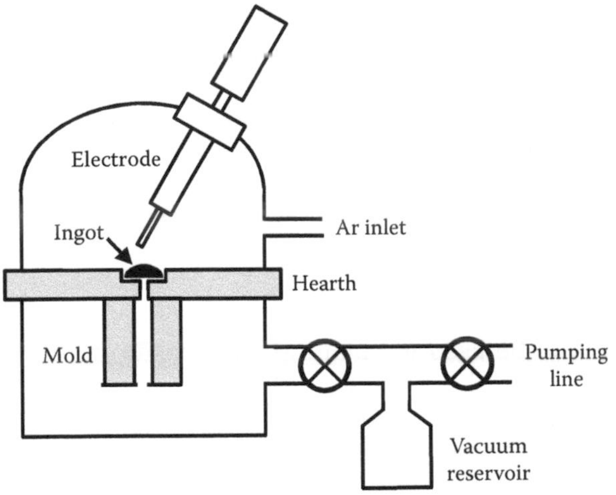

FIGURE 4.11
Schematic diagram of the arc melting/suction-casting apparatus. (Reprinted from Gu, X. et al., *J. Non-Cryst. Solids*, 311, 77, 2002. With permission.)

4.6.6 Squeeze-Casting Method

Sometimes, the cast BMG alloys have some porosity in them. Further, it would also be desirable to have net-shape forming capability during the casting process. Squeeze casting has great potential to achieve such features. The squeeze-casting process involves solidification of the molten metal under a high pressure within a closed die by utilizing a hydraulic press [59].

Zhang and Inoue [60] used this technique as early as 1998 to produce Zr–Ti–Al–Ni–Cu bulk glassy sheets with dimensions of 2.5 mm in thickness, 35 mm in width, and 80 mm in length. The cast sheet had a smooth surface and a good luster. No distinct contrast revealing the presence of a crystalline phase was seen on the surface of the sheet. Further, the sheet did not have any casting defects or cavities, and it was fully dense. The mechanical properties determined from different sections of the sheet showed nearly uniform mechanical properties, especially fracture toughness, which is highly sensitive to the presence of defects in the specimen. Further, these mechanical properties have been found to be nearly constant for every batch of the material. Since this technique has been found to be capable of producing a BMG with highly reliable properties, this has been used by Inoue et al. to produce 4–5 mm thick sheets for application as face material for golf clubs.

Kang et al. [61] used this technique to produce BMGs in the $Mg_{65}Cu_{15}Ag_{10}Y_{10}$ system. In this method, they remelted the alloy by an induction method in a graphite crucible under vacuum and bottom-filled a water-cooled copper mold that was 10 mm in diameter and 75 mm long utilizing a hydraulic press. After complete filling of the mold, the pressure, which reached 100 MPa, was maintained for 2 min until the liquid alloy completely solidified. The surfaces of the squeeze-cast BMGs showed a bright luster that closely mirrored the mold cavity surface. Neither holes nor cavities were seen on the surface.

The application of high pressure during solidification in the squeeze-casting process allows intimate contact between the liquid alloy and the mold wall. This results in rapid heat extraction from the liquid and, consequently, the solidification rates achieved are high. Additionally, since the melting temperature of alloys is increased by the application of pressure, according to the Clausius–Clapeyron equation, the melt also undercools to a larger degree. Further, since solidification takes place under high pressure, materials produced by the squeeze-casting method show additional advantages. These include (1) increased heat transfer coefficient between the molten metal and die surface, (2) undercooling to much below the equilibrium solidification temperature, and (3) complete elimination of shrinkage and/or gas porosity. Consequently, it should be possible to produce near-net-shape castings that closely mirror the inner mold surface and are dense and porosity-free as well.

4.6.7 Arc-Melting Method

This is a method that can be used to obtain glassy phases in alloy systems that require a low critical cooling rate for glass formation. The procedure used here is to arc melt the alloy on a copper hearth. Once the alloy is molten, the copper hearth acts like a heat sink and extracts the heat from the melt. A schematic of the technique is shown in Figure 4.9a. This is somewhat similar to the method of producing rapidly solidified alloys by laser processing, wherein the heat from the molten alloy is rapidly extracted by the large solid mass with which the molten pool is in contact.

This process is also akin to the conventional casting of alloys in a metal mold [62,63]. The solidification microstructure in a conventional cast alloy can be classified into three types. The first type that is in direct contact with the mold wall experiences a very high undercooling. Consequently, the nucleation rate is very high and therefore, the "chill" zone, consisting of extremely fine grains, is formed. Following this, one has a columnar structure due to the presence of an inverse temperature gradient in the liquid ahead of the solid/liquid interface. The last type is an equiaxed structure, again formed due to the occurrence of constitutional supercooling in alloys.

In the case of arc-cast alloys also, one can visualize these three different zones of solidification. The first two are similar to those described in the preceding paragraph, *viz.*, chill zone and columnar zone. But, instead of the central equiaxed zone in conventional cast alloys, we have a glassy phase region in the center of the ingot. It is to be noted here that a glassy phase forms in the inner region where the heat flux reinforced from the copper hearth disappears and the cooling rate is reduced. But, this low cooling rate is high enough for some of the alloys to become glassy. That is, those alloys for which the critical cooling rate for glass formation is lower than this cooling rate can be made glassy.

Optical micrographs of arc-melt cast samples show the presence of chill zone and columnar zone very clearly at the edges of the casting and the glassy phase in the center of the casting [64]. A point of interest in these micrographs is that the interface between the glassy and crystalline phases is very smooth. This is because the velocity with which the solid/liquid interface moves, v_i, is much higher than the ratio of the diffusivity of the constituent elements, D_i, to the diffusion distance required to rearrange the atoms to form the crystalline solid phase from the liquid, δ_i, that is, $v_i >> D_i/\delta_i$ [65].

A serious drawback of this method is that it is very difficult to completely avoid the formation of a small amount of the crystalline phase, at least on the surface. This is because of the ease of heterogeneous nucleation due to incomplete melting of the alloy at the bottom side that is in contact with the copper hearth.

4.6.8 Unidirectional Zone-Melting Method

The technique of zone melting is well known to produce highly pure single crystals for electronic applications [66]. In this method, a small region of a long and impure metal bar

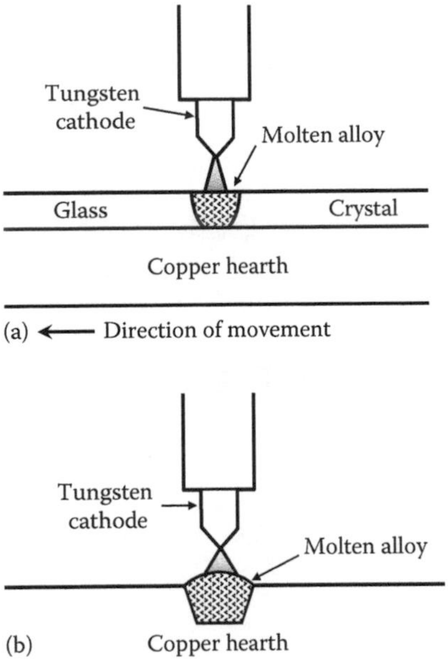

FIGURE 4.12
Schematic diagram of the zone-melting equipment used to produce bulk metallic glass ingot of a Zr-based alloy. (a) Front view and (b) side view. (Reprinted from Inoue, A. et al., *Mater. Trans., JIM*, 35, 923, 1994. With permission.)

is melted, and this molten zone traverses the length of the metal rod slowly. Because of solute partitioning effects, the solute concentration is (usually) higher in the liquid phase than in the solid phase. Accordingly, the impurities present in the metal get segregated to the liquid phase. As the molten zone travels slowly, more and more of the impurities get segregated into the neighboring liquid phase, leaving behind a purer solid phase. If this process is repeated a few times, one ends up with an extremely pure metal rod, with all the impurities segregated to one end of the rod. This method was originally employed to purify silicon and germanium crystals for application in transistors, but can be used for any material that needs to be purified. This technique has been successfully adopted to produce continuous BMG samples in a Zr-based alloy by Inoue et al. [67].

Figure 4.12 shows a schematic diagram of the zone-melting equipment used by Inoue et al. [68]. A rectangular parallelepiped $Zr_{60}Al_{10}Ni_{10}Cu_{15}Pd_5$ alloy ingot with 10 mm height, 12 mm width, and 170 mm length was prepared from prealloyed ingots and placed on a water-cooled copper hearth. The length of the usable rod was subsequently increased to 300 mm [68]. A small zone in this ingot was melted with the help of a tungsten cathode positioned at a distance of about 3 mm from the ingot surface. The cathode (and therefore the molten zone) moved at a velocity of 5.7 mm s^{-1}.

X-ray diffraction patterns from the "zone-melted" alloy ingot showed that it was glassy in nature. However, optical microscopy of the transverse cross section in the central portion of the ingot revealed that the sample contained a few crystals, about 200 μm in size, inside a predominantly glassy matrix (Figure 4.13). This was particularly true at the bottom of the ingot. This observation again highlights the fact that, because of the copper hearth, it will be impossible to completely suppress heterogeneous nucleation of the crystalline phase.

Yokoyama and Inoue [69] subsequently analyzed the solidification characteristics of the unidirectionally solidified Zr-based alloy. They measured the velocity of the liquid/solid

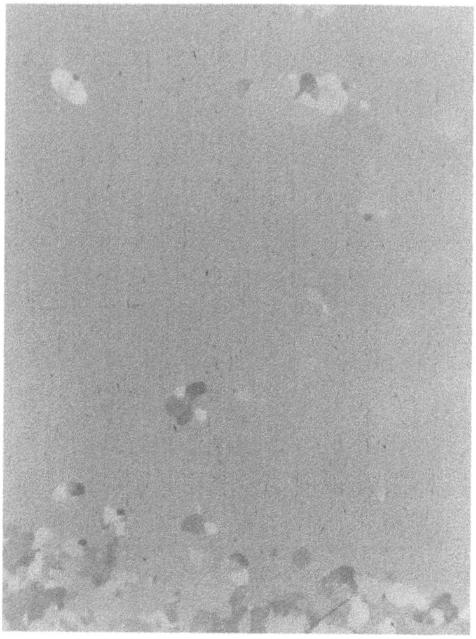

FIGURE 4.13
Optical micrograph of the unidirectionally zone-melted $Zr_{60}Al_{10}Ni_{10}Cu_{15}Pd_5$ alloy ingot showing the presence of some crystals in a glassy matrix. One can notice a large volume fraction of larger (about 200 μm in size) crystals in the region close to the bottom of the ingot that is about 2 mm away from the copper hearth.

interface (v) and the temperature gradient at the liquid/solid interface (G), and calculated the cooling rate as the product of $v \times G$. The calculated cooling rates were a maximum of 40 K s^{-1} at the solid/liquid interface, at which glass formation occurred. Therefore, they concluded that the critical cooling rate required for glass formation (R_c) was only 40 K s^{-1} for the $Zr_{60}Al_{10}Ni_{10}Cu_{15}Pd_5$ alloy, which was substantially lower than the value of 110 K s^{-1} reported earlier [49] on the basis of the CCT curves obtained from thermal analysis data. It should be noted that the cooling rate determined in the present study corresponds to the value at the liquid/solid interface, while in the CCT method it corresponds to the average cooling rate in the temperature range between the solidus and glass transition temperatures. In such a situation, the R_c value estimated at the liquid/solid interface should be higher than the value estimated from the CCT curve; but exactly the opposite trend was reported by these authors. This observation suggests that the number of cycles (N) that the melt goes through during arc melting (i.e., cleanliness of the melt) is also an important factor since it was reported earlier that R_c decreases from 100 K s^{-1} for $N = 3$ to 12 K s^{-1} for $N = 7$ in a $Zr_{65}Al_{10}Ni_{10}Cu_{15}$ alloy [70].

4.6.9 Electromagnetic Vibration Process

It has been well recognized and repeatedly emphasized in earlier sections that for an alloy melt to be solidified into the glassy state, it needs to be cooled at a rate higher than the critical cooling rate for glass formation. However, the role of other factors such as electric and magnetic fields has been largely ignored in the formation of metallic glasses. Tamura et al. [71] showed that the application of electromagnetic vibrations induced by the interaction of alternating electric current and stationary magnetic fields could act as powerful vibrating forces in the melt and affect the presence of clusters that could act as nucleation sites for the formation of crystalline phases. By eliminating such clusters in the liquid state, they demonstrated that the GFA of alloys could be substantially increased.

The $Mg_{65}Cu_{25}Y_{10}$ alloy that had been shown to be a good glass former was chosen for the investigation. An alloy sample with a diameter of 2 mm and a length of 12 mm was placed between two molybdenum electrodes in an alumina tube. The sample was melted, maintained at this temperature for 2 min, and then water was sprayed on the alumina tube to cool the melt. By varying the magnetic flux density and the time for which the electromagnetic vibrations were applied, the authors could determine the optimum processing conditions. They had demonstrated that while the alloy contained only the crystalline structures when no magnetic flux was applied (0 T), a fully glassy structure could be produced by increasing the magnetic flux density to 10 T. Similarly, a metallic glassy phase was produced by applying the electromagnetic vibrations for 10 s; shorter times produced either a glass + crystalline composite or a fully crystalline material. But, a different result was obtained by increasing the rest time after the melting, but before the onset of the water spray. The shorter the rest time, the better it was for glass formation. Thus, the authors could optimize the different process variables.

The element Mg is known to form clusters of strong local order in the liquid state with the other elements in the alloy and that these clusters will act as nucleation centers for the crystalline phase and decrease the GFA of the alloy. The application of electromagnetic vibration destroys these clusters and consequently eliminates the opportunity for the crystalline phase to nucleate. Increasing the magnitude of the magnetic flux density and/or applying these vibrations for a longer time helps in the destruction of these clusters. On the other hand, by increasing the rest time between melting and water spraying, the molten alloy has enough time to reform the clusters, and therefore the GFA of the alloy is decreased. Thus, the authors concluded that by controlling the process parameters, this could be a very effective method to produce metallic glasses, and that larger BMGs could be obtained by this process under similar cooling conditions.

4.7 Bulk Metallic Glass Composites

Composites based on BMGs exhibit much better mechanical properties, especially enhanced plasticity, over the monolithic materials, and this aspect will be discussed in some detail later in Chapter 8. Because of this special attribute, there has been a lot of activity in recent years on synthesizing and characterizing BMG composites. Qiao et al. [72] recently presented a review on metallic glass matrix composites. Since these aspects will be covered in detail at a later stage, we will focus here only on the different methods to *synthesize* the BMG composites.

The second (reinforcement) phase in the BMG composites has been a crystalline phase, and its volume fraction has been different depending on the desired properties. To achieve very high plasticity, the volume fraction was chosen to be as high as 80% [73–75]. But, the composites studied did not always contain that large amount of crystalline phase reinforcement. The BMG composites have been designated *in situ* or *ex situ* composites depending on the way that these have been obtained. In the *in situ* composites, the second phase precipitates out of the metallic glass either during casting or subsequent processing of the fully glassy alloy. Accordingly, the interface between the glassy matrix and the crystalline reinforcement is very clean and strong. On the other hand, in the *ex situ* method, the reinforcement phase is added separately during the casting/processing of the alloy and stays "as is" without much interaction with the matrix. Consequently, the interface between the matrix and the reinforcement may not be very strong. Further, the volume fraction of the reinforcement phase is smaller in the *in situ* method and could be much higher in the *ex situ* method.

4.7.1 *In Situ* Composites

The *in situ* composites are usually produced by adjusting the chemical composition of the alloy. When the alloy composition is chosen in such a way that it does not correspond exactly to the actual glass-forming composition, then the product of solidification will not be a homogeneous glassy phase. Instead, a crystalline phase will coexist with the glassy phase. This method has been adopted in a number of cases to produce the *in situ* composites [76–80].

In the *in situ* method, the alloy is melted and cast directly into the mold. If the composition deviates substantially from the glass-forming composition range, the second (crystalline) phase forms and its volume fraction is determined by the extent of deviation from the glass-forming composition range. The shape of the crystalline phase is usually dendritic since it forms directly from the melt. Instead of directly casting from the liquid state, if the alloy is homogenized in the mushy (liquid + solid) region and then cast, the crystalline phase obtained will have a spherical shape [79], and this is expected to further improve the mechanical properties.

If the second phase is obtained during subsequent processing of the fully solidified casting, the microstructure of the phase can be controlled. The grain size and shape of the second phase can be different. For example, if the glassy alloy is annealed at a low temperature, the grain size of the crystalline phase will be of nanometric dimensions [81]. Thus, different possibilities exist to obtain the desired size and shape and volume fraction of the reinforcement phase through this route.

Gao et al. [82] tried to develop metalloid-free Fe-based BMG composites. They started with an Fe–Co–La–Ce–Al–Cu alloy and, on cooling, the authors noted that the melt phase separated into Fe-rich and Fe-depleted liquids. On further cooling, the Fe-rich liquid produced a mixture of body-centered cubic (bcc)-Fe(Co,Al) solid solution and $Ce(Fe,Co)_2$ intermetallic, while the remaining Fe-lean Co–La–Ce–Al–Cu liquid transformed into a glass. The authors also noted that the final microstructure of the composite was strongly dependent on the cooling rate; excessively fast cooling reduced the formation of the glassy matrix.

Sha et al. [83] investigated the effect of casting temperature on the microstructure and mechanical properties of a $Ti_{45.7}Zr_{33}Ni_3Cu_{5.8}Be_{12.5}$ composite prepared by copper mold casting. They observed that on casting from a high temperature, the solidification rate achieved was lower and also that the oxygen content was higher. Consequently, the alloy was partially crystallized. Additionally, the lower solidification rate facilitated ripening of the crystalline dendrites and changed the geometric scale in the microstructure. The partial crystallization in the matrix as well as variation of the morphology of the dendrites resulted in increased strength and decreased plastic deformation.

4.7.2 *Ex Situ* Composites

This method of producing the BMG composites has been very important since one could introduce a very large volume fraction of the second crystalline phase. The types of reinforcements used have been pure metals (tungsten, molybdenum, tantalum, nickel, copper, and titanium), alloys (1080 steel, stainless steel, and brass), and nonmetallics (SiC, diamond, and graphite).

Long and continuous fibers, short fibers, and particulates have been used. Particulates and short fibers have been directly added to the melt and composites have been produced. But, for long and continuous fibers, the melt infiltration technique has been most commonly used.

Figure 4.14 shows a schematic of the melt infiltration casting apparatus used by Dandliker et al. [73] to produce the Zr-based alloy BMG composites. The wires were first degreased and cleaned, and this reinforcement material was placed in the sealed end of a 7 mm inner-diameter quartz-glass tube. The tube was necked about 1 cm above the reinforcement, and then BMG alloy ingot pieces were placed in the tube above the neck. The constriction minimized premature contact between the wires and the alloy, and avoided reactions between the melt and the reinforcement. The open end of the quartz tube was then connected to a three-way switching valve. The tube could be evacuated and also flushed with argon. The cycle of evacuation and flushing with argon was repeated a few times to remove any residual gaseous impurities. In the last cycle, the tube was left under vacuum to minimize entrapment of gases in the composite sample to be formed. The sample tube was then heated to the desired temperature to melt the BMG alloy, and a positive pressure was applied above the melt for about 30 min to allow infiltration of the molten matrix material into the reinforcement. The sample was then quickly removed from the furnace and quenched in brine. The temperatures up to which the tube was heated, the amount, and the duration of positive pressure application will be different for different combinations of matrix and reinforcements. Special precautions may also have to be taken in some cases, depending on the materials involved.

4.8 Mechanical Alloying

MA has been another popular technique to synthesize amorphous phases in a number of alloy systems [9,10]. However, a major difference between MA and the techniques described so far is that while a bulk glassy alloy (either monolithic or composite) is produced in one single step using the methods described, MA produces the amorphous phase in a powder form,

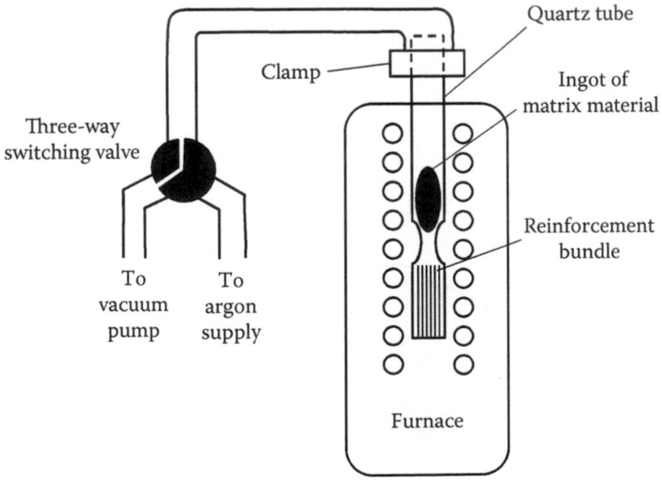

FIGURE 4.14
Schematic of the melt infiltration casting technique to produce *ex situ* BMG composites. (Reprinted from Dandliker, R.B. et al., *J. Mater. Res.*, 13, 2896, 1998. With permission.)

and this needs to be consolidated by some of the conventional or innovative methods through the application of pressure and/or temperature. But, one of the biggest advantages of the MA method is that this technique could be used to easily produce amorphous phases in those systems where conventional melting and casting methods prove difficult or impossible.

In this process, a mixture of the blended elemental powders is loaded, under inert atmosphere conditions, into the milling container along with the grinding medium (usually stainless steel or tungsten carbide or other hard material). This container is then placed inside the mill and the whole mass is violently agitated for the desired length of time. Depending on the type of mill used, the powder is subjected to shear, impact, or other types of mechanical forces.

As depicted in Figure 4.15a, the powder particles get trapped between two grinding balls and go through a process akin to rolling. But, there could be compressive, shear, and impact forces exerted on the surfaces of the powder particles. Under the action of

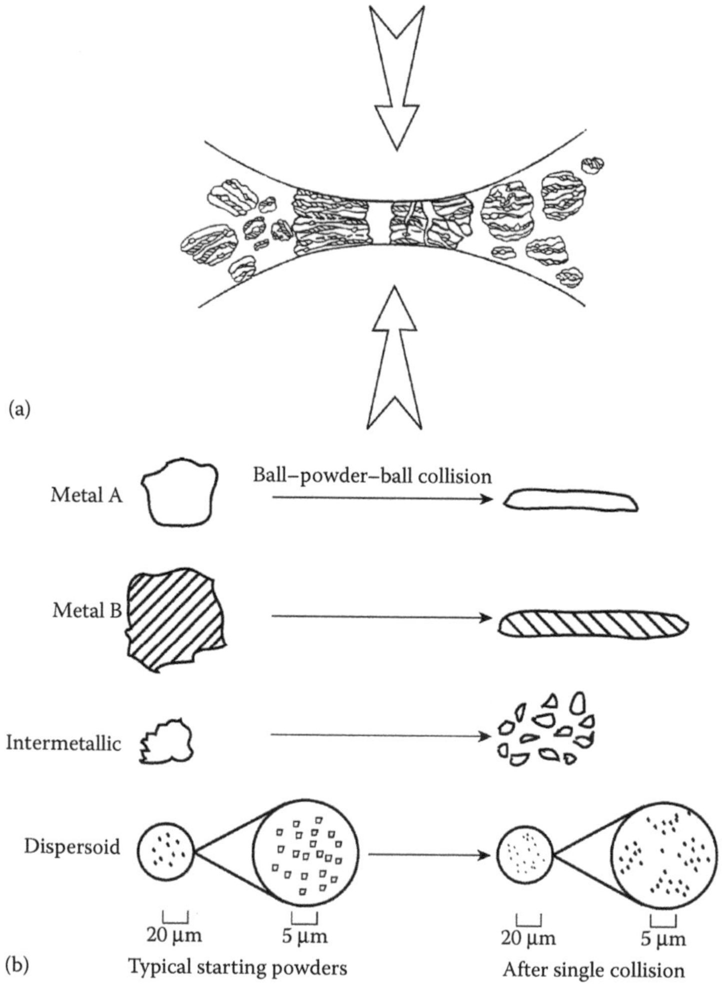

FIGURE 4.15
(a) Ball–powder–ball collision of powder mixture during MA and (b) deformation characteristics of representative constituents of starting powders in MA. Note that the ductile metal powders (metals A and B) get flattened, while the brittle intermetallic and dispersoid particles get fragmented into smaller particles.

TABLE 4.3
Some Selected BMG Alloy Compositions Produced in a Glassy Condition by MA

Alloy Composition	Mill	Ball-to-Powder Ratio	Time for Amorphization (h)	References
$Fe_{72}Al_5Ga_2C_6B_4P_{10}Si_1$	Planetary ball mill AGO-2U	10:1 or 20:1	8–12	[84]
$Fe_{60}Co_8Zr_{10}Mo_5W_2B_{15}$	SPEX mill	10:1	20	[85]
$Fe_{42}Ge_{28}Zr_{10}B_{20}$	SPEX mill	10:1	10	[86]
$Fe_{42}Ge_{28}Zr_{10}C_{10}B_{20}$	SPEX mill	10:1	8	[86]
$Mg_{65}Cu_{20}Y_{10}Ag_5 + ZrO_2$	Planetary mill	—	—	[87]
$Nb_{50}Zr_{10}Al_{10}Ni_{10}Cu_{20}$	Fritsch P5	14:1	200	[88]
$Ta_{55}Zr_{10}Ni_{10}Al_{10}Cu_{15}$	Tumbler mill	25:1	300	[89]
$Ti_{40.6}Zr_{9.4}Cu_{37.5}Ni_{9.4}Al_{3.1}$	Planetary ball mill QM-2SP20	12:1	20	[90]
$Ti_{60}Al_{15}Cu_{10}W_{10}Ni_5$	Tumbler mill	30:1	200	[91]
$V_{45}Zr_{20}Ni_{20}Cu_{10}Al_{2.5}Pd_{2.5}$	Tumbler mill	25:1	200	[92]
$W_{33}Co_{38}Fe_9C_{20}$	SPEX mill	5:1	60	[93]
$Zr_{52}Al_6Ni_8Cu_{14}W_{20}$	Tumbler mill	60:1	200	[94]

mechanical forces, the soft individual powder particles are flattened into a pancake shape, and if the powder particle is brittle, it becomes comminuted and the particle size gets smaller (Figure 4.15b). Due to the random process, the flat soft particles get stacked and form a layered structure. On continued milling, these layered structures become convoluted. This process of convolution continues with milling time and the layer spacing is significantly reduced. Additionally, the milled powder particles contain a high density of crystal defects (e.g., dislocations, stacking faults, and grain boundaries). Thus, the combination of a small particle size, reduced diffusion distances across the lamellar structure, fresh surfaces and interfaces, coupled with a slight rise in temperature, all contribute to increased atomic diffusivity and, therefore, alloying occurs. The kinetics of microstructural refinement depend on the mechanical properties of the powder, the type of mill used, the ball-to-powder weight ratio, and the temperature at which milling is carried out. Depending on the nature of the constituent elements present and also the proportion of the elements, different types of alloy phases are produced. These include amorphous phases in a large number of alloy systems.

Amorphous alloy powders of multicomponent BMG compositions have also been produced by MA in several alloy systems including those based on Fe, Mg, and Zr. Some of these are listed in Table 4.3. These powders can be conveniently consolidated into bulk in the supercooled liquid region. A clear advantage of this approach to synthesize monolithic BMGs is that the size limitations imposed by the solidification-processing methods will not apply here.

However, MA has a clear advantage in producing BMG composites. A very large number of oxide-, carbide-, and other ceramic-dispersed composites have been produced by MA. Because of the heavy deformation involved and the kneading that takes place, the ceramic phase gets uniformly dispersed in the metal matrix. In fact, it should be realized that the technique of MA was developed to satisfy an industrial necessity of producing oxide-dispersion strengthened nickel-based superalloys. Further, it has been demonstrated that using MA, one should be able to produce a composite containing a very high volume fraction of even nanometer-sized ceramic particles [95].

4.9 Bulk Metallic Glass Foams

We have so far seen how BMGs of different alloys could be produced by the solidification and non-solidification methods. In all these cases, the glass produced is fully dense (as dense as it could be in the glassy state) without any porosity (except for the free volume present). But, metallic foams are now receiving increased attention from materials scientists and engineers as one category of new industrially important materials. This is because they are known to have interesting combination of properties such as high stiffness in conjunction with very low specific weight, high gas permeability combined with high thermal conductivity, high mechanical energy absorption, and good acoustic damping [96–98].

Metallic foams can be classified into closed-cell, partially open-cell, and open-cell types [96]. Closed-cell-type metal foams have spatially separated pores and are useful for structural applications such as lightweight construction and energy absorption. On the other hand, open-cell-type metal foams have interconnected pores and are useful as functional materials for applications such as electrodes, catalyst support, fluid filters, and biomedical materials. Since the properties of these foams, especially the strength and modulus of elasticity, can be tailored by controlling the volume fraction as well as the structure of pores [99], these materials can be used as biomedical implants. This is because their structure allows bone tissue in-growth leading to the establishment of stable fixation with the surrounding tissues.

Banhart has briefly described the different routes available to manufacture metal foams [100]. BMGs have high strength and good corrosion resistance, but they possess very limited ductility in tension, and near-zero plastic strain; even in compression, it is typically <2%. The ductility of BMGs has been enhanced by producing composites through the dispersion of a ductile crystalline phase; but this reduces the strength of the composite. Therefore, it has been suggested that by producing BMGs as foams, their ductility could be increased without sacrificing the strength. This is because the struts within a foam can have cross sections small enough to impart high plasticity in bending, even when the loading is uniaxial. Therefore, several attempts have been made to produce BMG foams.

The production of metallic glassy foams is challenging to say the least. First, foams are unstable structures. Second, since the kinetics of foam expansion and collapse scale with the viscosity, the kinetics of foaming in BMGs are very sluggish (due to the high viscosity of glass-forming melts). But, this could be used to our advantage in controlling the foam homogeneity, bubble size distribution, and volume fraction of the pores. Third, to extract heat sufficiently fast to avoid crystallization, glassy foam size is restricted in one dimension. Last, techniques that introduce heterogeneous nucleation sites to improve uniformity of pore distribution will also facilitate crystallization. Therefore, this method cannot be used while producing metallic glass foams.

The possibility of producing metallic glassy foams was theoretically investigated by Apfel and Qiu [101]. They suggested that when a melt seeded with droplets of a volatile liquid is rapidly decompressed, the droplets vaporize explosively, taking their latent heat of vaporization from the melt, and therefore homogeneously cooling and expanding it. But, the actual process to produce amorphous metallic foam was first demonstrated by Schroers et al. [102]. They used hydrated B_2O_3 to create gas bubbles in the liquid melt of $Pd_{43}Ni_{10}Cu_{27}P_{20}$, which is known to be a very good glass former with a critical cooling rate for glass formation of as low as 0.1 K s^{-1}. The use of B_2O_3 in this case is acceptable since this is used to flux the Pd- and Pt-based alloys to remove impurities from the melt. Water vapor is released during the decomposition of the hydrated B_2O_3, which is entrapped by the melt. These bubbles expand when the pressure is decreased, resulting in a low-density closed-cell foam. They obtained bubbles varying in diameter between 0.1 and 1 mm and reported a very uniform bubble distribution. Densities as low as 1.4 g cm^{-3} were obtained, corresponding to a bubble volume fraction of 84%.

Brothers et al. [103 –105] also produced syntactic metallic glass foams (foams whose low density is achieved by incorporating hollow particulates into the alloy). Since Zr is very reactive, it cannot tolerate the presence of impurities such as hydrogen, oxygen, and boron. The metal Zr is especially sensitive to oxygen, even in small quantities. Therefore, the authors [103] used an alternative foaming method to produce foams in the $Zr_{57}Nb_5Cu_{15.4}Ni_{12.6}Al_{10}$ alloy. They used hollow carbon microspheres with a diameter of 25–50 μm and a wall thickness of 1–10 μm. A bed of these microspheres with 5 mm diameter and 8 mm height was placed into the sealed end of a stainless steel tube. The whole crucible assembly was then heated to 1250 K under a vacuum of 3×10^{-5} Torr, and a prealloyed charge of $Zr_{57}Nb_5Cu_{15.4}Ni_{12.6}Al_{10}$ was then lowered into the hot zone of the crucible and allowed to melt for 3 min. This melt was then infiltrated into the microsphere bed using 153 kPa of high-purity argon gas. After infiltration for 45 s, the sample was quenched by immersing the tube in an agitated brine solution. The investigators noted that the infiltration was uniform only in the lowest 3 mm at the bottom of the bed. The foam density was 3.4 ± 0.2 g cm^{-3}, corresponding to a relative density of about 50%. It was also reported that the glassy foam of 5 mm diameter showed no measurable loss in thermal stability as compared with the bulk glassy alloy.

Brothers and Dunand [104] also produced metallic glass foams by infiltrating the molten $Zr_{57}Nb_5Cu_{15.4}Ni_{12.6}Al_{10}$ alloy into a sintered BaF_2 pattern followed by rapid quenching. BaF_2 was leached out in a bath of 2 N nitric acid. These authors showed that the foams had a density of 1.52 g cm^{-3} (corresponding to 78% porosity) with a uniform pore size of 212–250 μm. During compression, these foams showed a strain of approximately 50%, with the sample remaining intact after unloading. This increased ductility was achieved through the bending of approximately 1 mm struts.

The Inoue group has also produced BMG foams, but by different methods. In one method, Wada and Inoue [106] mixed solid NaCl and liquid $Pd_{42.5}Cu_{30}Ni_{7.5}P_{20}$ alloy in a silica tube and water-quenched this mixture. The salt was later leached out by warm water at 353 K. By maintaining a volume fraction of 1:7 between the Pd-alloy and the salt, they were able to obtain a porosity of about 65% in the alloy. The pores in the foam had a polyhedral shape with sizes of 125–250 μm, corresponding to the shape and size of the salt crystals. Entrapment of pressurized hydrogen was another method used by them [107,108]. In this method, the $Pd_{42.5}Cu_{30}Ni_{7.5}P_{20}$ alloy was placed inside a quartz crucible, the crucible was evacuated, and the alloy heated to 853 K, about 50 K higher than the melting temperature of the alloy. Hydrogen gas at 1.5 MPa was then introduced into the crucible and the molten alloy was annealed for 10 min to dissolve the hydrogen into the melt. Subsequently, the crucible was evacuated and quenched immediately in water. Porous rods 7 mm in diameter and 50 mm in length were produced by this method. The volume fraction and pore size were controlled by changing the temperature of the molten alloy before quenching, the hydrogen pressure, and the duration of annealing. Figure 4.16 shows the transverse cross section and longitudinal cross section of a porous $Pd_{42.5}Cu_{30}Ni_{7.5}P_{20}$ alloy rod produced by melting the alloy for 10 min at 813 K under a hydrogen pressure of 1.5 MPa and then water quenching. It is clear from these micrographs that the alloy rod contains a high volume fraction of the pores and that there is no difference in the morphology of the pores between the two cross sections. Scanning electron micrographs of the etched transverse cross section of the porous $Pd_{42.5}Cu_{30}Ni_{7.5}P_{20}$ alloy both at low and high magnifications are

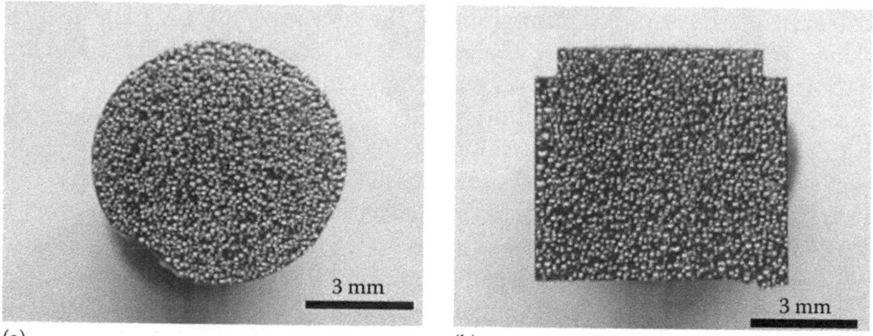

(a) (b)

FIGURE 4.16
(a) Transverse cross section and (b) longitudinal cross section of a 7 mm diameter porous $Pd_{42.5}Cu_{30}Ni_{7.5}P_{20}$ alloy rod produced by melting the alloy for 10 min at 813 k under a hydrogen pressure of 1.5 MPa and then water quenching. Note the uniformity of pore size in both the cross sections. (Reprinted from Wada, T. and Inoue, A., *Mater. Trans.*, 45, 2761, 2004. With permission.)

presented in Figure 4.17. It may be noted that the alloy is fully glassy, and contrast due to the presence of a crystalline phase is not seen even at high magnifications.

Xie et al. [109] used a clever method to fabricate porous BMG specimens. They first produced $Zr_{55}Cu_{30}Al_{10}Ni_5$ glassy powder by high-pressure gas atomization and then consolidated this powder through a spark plasma sintering process. The bulk samples had dimensions of 20 mm diameter and 5 mm height. The porosity in the bulk samples was controlled by varying the loading pressure and sintering temperature. For example, by sintering the gas-atomized powder for 10 min at 623 K and 30 MPa, and 613 K and 20 MPa pressure, they were able to obtain samples with 4.7% and 33.5% porosity, respectively.

Demetriou et al. [110] employed a similar method to obtain highly porous metallic glass foams. In this method, the glassy powder is combined with a blowing agent and

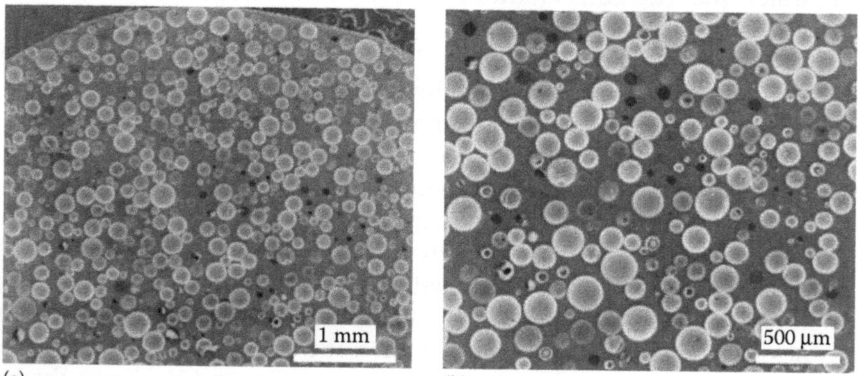

(a) (b)

FIGURE 4.17
Scanning electron micrographs of the transverse cross section of the porous glassy $Pd_{42.5}Cu_{30}Ni_{7.5}P_{20}$ alloy rod quenched from 833 K under a hydrogen pressure of 1.5 MPa. The specimen was etched in concentrated H_2SO_4 solution for 3 h to reveal any contrast due to the presence of a crystalline phase. (a) Low-magnification and (b) high-magnification micrographs. Contrast due to the presence of a crystalline phase is not seen even in the high magnification micrograph. (Reprinted from Wada, T. and Inoue, A., *Mater. Trans.*, 45, 2761, 2004. With permission.)

the mixture is consolidated at a temperature within the supercooled liquid region of the alloy, but below the decomposition point of the blowing agent. This produces the foam precursor. But, when this precursor is subsequently heated to a temperature, again in the supercooled liquid region of the glass, but above the decomposition point of the blowing agent, the precursor expands, and thus, a metallic glass foam is formed. In their study, the authors used the $Pd_{43}Ni_{10}Cu_{27}P_{20}$ glass, crushed it into powder, and mixed it with the particulates of the blowing agent, magnesium carbonate, and *n*-hydrate ($MgCO_3 \cdot nH_2O$). In the final product, they had obtained as much as 86% porosity. The foams inherited the high strength of the parent metallic glass and were able to deform heavily toward full densification absorbing high amounts of energy.

Chen et al. [111] produced interpenetrating phase composites of a Zr-based BMG and porous SiC. In this process, they preheated the $Zr_{41.25}Ti_{13.75}Cu_{12.5}Ni_{10}Be_{22.5}$ BMG and the porous SiC preform to a temperature above the liquidus temperature of the alloy in an evacuated quartz tube and then forced the viscous melt into the SiC preform by pressure infiltration. The whole system was then quickly quenched in supersaturated brine. The advantage of this method is that one could control the volume fraction of the desired reinforcement [112] and that these composites display a three-dimensional interconnected net structure [113]. The microstructure of the composite showed that both the glassy phase and the SiC phase were three-dimensionally interconnected. The mechanical behavior of the composite was different depending on the volume fraction of the SiC phase (ranging from 51% to 82%).

Cox et al. [114] were able to produce composites with aligned and elongated pores. They had warm-extruded an $Hf_{44.5}Cu_{27}Ni_{13.5}Ti_5Al_{10}$ BMG/crystalline W powder blend to densify it using equal channel angular extrusion. After extrusion, the W phase was dissolved by electrochemical means, resulting in an interconnected network of about 60% aligned, elongated pores within the amorphous Hf-based matrix. This composite exhibited ductility in compression, but at lower strengths than similar Zr-based foams due to incomplete bonding between the Hf-based amorphous powders.

4.10 Concluding Remarks

A number of different techniques have been described here to produce BMGs in different sizes and shapes. These include regular rod shapes, sheets, wedge shapes, and cone shapes. The later individual chapters also contain some details of preparation of BMGs. Since the critical cooling rate required to form the glassy phase in these alloys is relatively low, it is possible to use relatively simple techniques to obtain the BMGs. However, when the technique includes a conducting substrate on which the alloy is melted, it appears inevitable that heterogeneous nucleation of a crystalline phase occurs. But, it has been possible to obtain fully glass samples using methods such as high-pressure die casting, water quenching, copper mold casting, or suction/squeeze casting.

Naturally the cooling rates obtained differ in these different methods. But, the cooling rates achieved in most of the methods are above the critical cooling rate for the formation of the bulk glassy phases in different alloy systems, and therefore these techniques could be used to synthesize BMG alloys. Techniques to synthesize BMG composites and foams have also been described along with non-solidification methods to produce BMG alloys.

References

1. Suryanarayana, C. (1991). Rapid solidification. In *Processing of Metals*. Vol. 15 of *Materials Science and Technology: A Comprehensive Treatment*, ed. R.W. Cahn, pp. 57–110. Weinheim, Germany: VCH.
2. Hagiwara, M. and A. Inoue (1993). Production techniques of alloy wires by rapid solidification. In *Rapidly Solidified Alloys: Processes, Structures, Properties, Applications*, ed. H.H. Liebermann, pp. 139–155. New York: Marcel Dekker.
3. Liebermann, H.H. (1983). Sample preparation: Methods and process characterization. In *Amorphous Metallic Alloys*, ed. F.E. Luborsky, pp. 26–41. London, UK: Butterworths.
4. Anantharaman, T.R. and C. Suryanarayana (1987). *Rapidly Solidified Metals: A Technological Overview*, pp. 25–97. Zürich, Switzerland: Trans Tech Publications.
5. Klement, Jr., W., R.H. Willens, and P. Duwez (1960). Non-crystalline structure in solidified gold–silicon alloys. *Nature* 187: 869–870.
6. Duwez, P. and R.H. Willens (1963). Rapid quenching of liquid alloys. *Trans. Metall. Soc. AIME* 227: 362–365.
7. Boettinger, W.J. and J.H. Perepezko (1993). Fundamentals of solidification at high rates. In *Rapidly Solidified Alloys: Processes, Structures, Properties, Applications*, ed. H.H. Liebermann, pp. 17–78. New York: Marcel Dekker.
8. Suryanarayana, C., ed. (1999). *Non-Equilibrium Processing of Materials*. Oxford, UK: Elsevier.
9. Suryanarayana, C. (2001). Mechanical alloying and milling. *Prog. Mater. Sci.* 46: 1–184.
10. Suryanarayana, C. (2004). *Mechanical Alloying and Milling*. New York: Marcel-Dekker.
11. Lavernia, E.J. and Y. Wu (1996). *Spray Atomization and Deposition*. Chichester, UK: John Wiley & Sons.
12. Sharma, S.M. and S.K. Sikka (1996). Pressure induced amorphization of materials. *Prog. Mater. Sci.* 40: 1–77.
13. Turnbull, D. (1981). Metastable structures in metallurgy. *Metall. Trans. A* 12: 695–708.
14. Jacobson, L.A. and J. McKittrick (1994). Rapid solidification processing. *Mater. Sci. Eng.* R11: 355–408.
15. Strange, E.H. and C.A. Pim (1908). Process of manufacturing thin sheets, foil, strips, or ribbons of zinc, lead, or other metal or alloy. US Patent 905,758, December 1, 1908.
16. Pond, R.B. (1958). Metallic filaments and method of making same. US Patent 2,825,108, March 4, 1958.
17. Liebermann, H.H. (1980). The dependence of the geometry of glassy alloy ribbons on the chill block melt-spinning process parameters. *Mater. Sci. Eng.* 43: 203–210.
18. Chen, H.S. (1974). Thermodynamic considerations on the formation and stability of metallic glasses. *Acta Metall.* 22: 1505–1511.
19. Inoue, A., N. Nishiyama, and H.M. Kimura (1997). Preparation and thermal stability of bulk amorphous $Pd_{40}Cu_{30}Ni_{10}P_{20}$ alloy cylinder of 72 mm in diameter. *Mater. Trans., JIM* 38: 179–183.
20. Nishiyama, N., K. Takenaka, H. Miura, N. Saidoh, Y. Zeng, and A. Inoue (2012). The world's biggest glassy alloy ever made. *Intermetallics* 30: 19–24.
21. Drehman, A.J., A.L. Greer, and D. Turnbull (1982). Bulk formation of a metallic glass: $Pd_{40}Ni_{40}P_{20}$. *Appl. Phys. Lett.* 41: 716–717.
22. Kui, H.W., A.L. Greer, and D. Turnbull (1984). Formation of bulk metallic glass by fluxing. *Appl. Phys. Lett.* 45: 615–616.
23. Bie, L., Q. Li, D. Cao, H.X. Li, J. Zhang, C.T. Chang, and Y.F. Sun (2016). Preparation and properties of quaternary CoMoPB bulk metallic glasses. *Intermetallics* 71: 7–11.
24. Shechtman, D., I. Blech, D. Gratias, and J.W. Cahn (1984). Metallic phase with long-range orientational order and no translational symmetry. *Phys. Rev. Lett.* 53: 1951–1953.
25. Inoue, A., K. Kita, T. Zhang, and T. Masumoto (1989). An amorphous $La_{55}Al_{25}Ni_{20}$ alloy prepared by water quenching. *Mater. Trans., JIM* 30: 722–725.

26. Inoue, A., T. Nakamura, N. Nishiyama, and T. Masumoto (1992). Mg–Cu–Y bulk amorphous alloys with high tensile strength produced by a high-pressure die casting method. *Mater Trans., JIM* 33: 937–945.
27. Peker, A. and W.L. Johnson (1993). A highly processable metallic glass: $Zr_{41.2}Ti_{13.8}Cu_{12.5}Ni_{10.0}Be_{22.5}$. *Appl. Phys. Lett.* 63: 2342–2344.
28. Inoue, A., N. Nishiyama, and T. Matsuda (1996). Preparation of bulk glassy $Pd_{40}Ni_{10}Cu_{30}P_{20}$ alloy of 40 mm in diameter by water quenching. *Mater. Trans., JIM* 37: 181–184.
29. Takeuchi, A. and A. Inoue (2005). Classification of bulk metallic glasses by atomic size difference, heat of mixing and period of constituent elements and its application to characterization of the main alloying element. *Mater. Trans.* 46: 2817–2829.
30. Lin, X.H., W.L. Johnson, and W.K. Rhim (1997). Effect of oxygen impurity on crystallization of an undercooled bulk glass forming Zr–Ti–Cu–Ni–Al alloy. *Mater. Trans., JIM* 38: 473–477.
31. Eckert, J., N. Mattern, M. Zinkevitch, and M. Seidel (1998). Crystallization behavior and phase formation in Zr–Al–Cu-Ni metallic glass containing oxygen. *Mater. Trans., JIM* 39: 623–632.
32. de Oliveira, M.F., W.J. Botta, M.J. Kaufman, and C.S. Kiminami (2002). Phases formed during crystallization of $Zr_{55}Al_{10}Ni_5Cu_{30}$ metallic glass containing oxygen. *J. Non-Cryst. Solids* 304: 51–55.
33. Gebert, A., J. Eckert, and L. Schultz (1998). Effect of oxygen on phase formation and thermal stability of slowly cooled $Zr_{65}Al_{7.5}Cu_{17.5}Ni_{10}$ metallic glass. *Acta Mater.* 46: 5475–5482.
34. Gebert, A., J. Eckert, H.D. Bauer, and L. Schultz (1998). Characteristics of slowly cooled Zr–Al–Cu–Ni bulk samples with different oxygen content. *Mater. Sci. Forum* 269–272: 797–806.
35. Chen, C., Y.Y. Cheng, and T. Zhang (2016). Synthesis of impurity-insensitive Zr-based bulk metallic glass. *J. Non-Cryst. Solids* 439: 1–5.
36. Murty, B.S., D.H. Ping, K. Hono, and A. Inoue (2000). Influence of oxygen on the crystallization behavior of $Zr_{65}Cu_{27.5}Al_{7.5}$ and $Zr_{66.7}Cu_{33.3}$ metallic glasses. *Acta Mater.* 48: 3985–3996.
37. Sordelet, D.J., X. Yang, E.A. Rozhkova, M.F. Besser, and M.J. Kramer (2004). Influence of oxygen content in phase selection during quenching of $Zr_{80}Pt_{20}$ melt spun ribbons. *Intermetallics* 12: 1211–1217.
38. Scudino, S., J. Eckert, H. Breitzke, K. Lüders, and L. Schultz (2007). Influence of oxygen on the devitrification of Zr–Ti–Nb–Cu–Ni–Al metallic glasses. *Mater. Sci. Eng. A* 449–451: 493–496.
39. Kündig, A.A., D. Lepori, A.J. Perry, S. Rossmann, A. Blatter, A. Dommann, and P.J. Uggowitzer (2002). Influence of low oxygen contents and alloy refinement on the glass forming ability of $Zr_{52.5}Cu_{17.9}Ni_{14.6}Al_{10}Ti_5$. *Mater. Trans.* 43: 3206–3210.
40. Inoue, A., T. Zhang, N. Nishiyama, K. Ohba, and T. Masumoto (1993). Preparation of 16 mm diameter rod of amorphous $Zr_{65}Al_{7.5}Ni_{10}Cu_{17.5}$ alloy. *Mater. Trans., JIM* 34: 1234–1237.
41. Shen, T.D., Y. He, and R.B. Schwarz (1999). Bulk amorphous Pd–Ni–Fe–P alloys: Preparation and characterization. *J. Mater. Res.* 14: 2107–2115.
42. Amiya, K. and A. Inoue (2001). Preparation of bulk glassy $Mg_{65}Y_{10}Cu_{15}Ag_5Pd_5$ alloy of 12 mm in diameter by water quenching. *Mater. Trans.* 42: 543–545.
43. Guo, F., S.J. Poon, and G.J. Shiflet (2003). Metallic glass ingots based on yttrium. *Appl. Phys. Lett.* 83: 2575–2577.
44. Schroers, J. and W.L. Johnson (2004). Highly processable bulk metallic glass-forming alloys in the Pt–Co–Ni–Cu–P system. *Appl. Phys. Lett.* 84: 3666–3668.
45. Inoue, A., T. Nakamura, T. Sugita, T. Zhang, and T. Masumoto (1993). Bulky La–Al–TM (TM = transition metal) amorphous alloys with high tensile strength produced by a high-pressure die casting method. *Mater. Trans., JIM* 34: 351–358.
46. Inoue, A., T. Nakamura, N. Nishiyama, T. Sugita, and T. Masumoto (1993). Bulky amorphous alloys produced by a high-pressure die casting process. *Key Eng. Mater.* 81–83: 147–152.
47. Nishiyama, N. and A. Inoue (1996). Glass-forming ability of bulk $Pd_{40}Ni_{10}Cu_{30}P_{20}$ alloy. *Mater. Trans., JIM* 37: 1531–1539.
48. Park, E.S. and D.H. Kim (2004). Formation of Ca–Mg–Zn bulk glassy alloy by casting into cone-shaped copper mold. *J. Mater. Res.* 19: 685–688.

49. Inoue, A., Y. Shinohara, Y. Yokoyama, and T. Masumoto (1995). Solidification analyses of bulky $Zr_{60}Al_{10}Ni_{10}Cu_{15}Pd_5$ glass produced by casting into wedge-shape copper mold. *Mater. Trans., JIM* 36: 1276–1281.
50. Ma, H., E. Ma, and J. Xu (2003). A new $Mg_{65}Cu_{7.5}Ni_{7.5}Zn_5Ag_5Y_{10}$ bulk metallic glass with strong glass-forming ability. *J. Mater. Res.* 18: 2288–2291.
51. Amiya, K. and A. Inoue (2002). Formation and thermal stability of Ca–Mg–Ag–Cu bulk glassy alloys. *Mater. Trans.* 43: 2578–2581.
52. Park, E.S. and D.H. Kim (2005). Formation of Mg–Cu–Ni–Ag–Zn–Y–Gd bulk glassy alloy by casting into cone-shaped copper mold in air atmosphere. *J. Mater. Res.* 20: 1465–1469.
53. Yokoyama, Y., E. Mund, A. Inoue, and L. Schultz (2007). Production of $Zr_{55}Cu_{30}Ni_5Al_{10}$ glassy alloy rod of 30 mm in diameter by a cap-cast technique. *Mater. Trans.* 48: 3190–3192.
54. Inoue, A. and T. Zhang (1995). Fabrication of bulky Zr-based glassy alloys by suction casting into copper mold. *Mater. Trans., JIM* 36: 1184–1187.
55. Figueroa, I.A., P.A. Carroll, H.A. Davies, H. Jones, and I. Todd (2007). Preparation of Cu-based bulk metallic glasses by suction casting. In *SP-07, Proceedings of the Fifth Decennial International Conference on Solidification Processing*, Sheffield, UK, July 2007, pp. 479–482.
56. Gu, X., L.Q. Xing, and T.C. Hufnagel (2002). Glass-forming ability and crystallization of bulk metallic glass $(Hf_xZr_{1-x})_{52.5}Cu_{17.9}Ni_{14.6}Al_{10}Ti_5$. *J. Non-Cryst. Solids* 311: 77–82.
57. Wall, J.J., C. Fan, P.K. Liaw, C.T. Liu, and H. Choo (2006). A combined drop/suction-casting machine for the manufacture of bulk-metallic-glass materials. *Rev. Sci. Instrum.* 77: 033902-1–033902-4.
58. Shen, J., Q. Chen, J. Sun, H. Fan, and G. Wang (2005). Exceptionally high glass-forming ability of an FeCoCrMoCBY alloy. *Appl. Phys. Lett.* 86: 151907-1–151907-3.
59. Lynch, R.F., R.P. Olley, and P.C.J. Gallagher (1975). Squeeze casting of brass and bronze. *Trans. AFS* 83: 561–568.
60. Zhang, T. and A. Inoue (1998). Mechanical properties of Zr–Ti–Al–Ni–Cu bulk amorphous sheets prepared by squeeze casting. *Mater. Trans., JIM* 39: 1230–1237.
61. Kang, H.G., E.S. Park, W.T. Kim, D.H. Kim, and H.K. Cho (2000). Fabrication of bulk Mg–Cu–Ag–Y glassy alloy by squeeze casting. *Mater. Trans., JIM* 41: 846–849.
62. Flemings, M.C. (1974). *Solidification Processing*. New York: McGraw-Hill.
63. Abbaschian, R., L. Abbaschian, and R.E. Reed-Hill (2009). *Physical Metallurgy Principles*, 4th edn. Stamford, CT: CENGAGE Learning.
64. Inoue, A. (1998). *Bulk Amorphous Alloys: Preparation and Fundamental Characteristics*, p. 33. Uetikon-Zurich, Switzerland: Trans Tech Publications.
65. Kurz, W. and D.J. Fisher (1989). *Fundamentals of Solidification*. Aedermannsdorf, Switzerland: Trans Tech Publications.
66. Pfann, W.G. (1965). *Zone Melting*. New York: John Wiley & Sons.
67. Inoue, A., Y. Yokoyama, Y. Shinohara, and T. Masumoto (1994). Preparation of bulky Zr-based amorphous alloys by a zone melting method. *Mater. Trans., JIM* 35: 923–926.
68. Inoue, A. (1995). Slowly-cooled bulk amorphous alloys. *Mater. Sci. Forum* 179–181: 691–700.
69. Yokoyama, Y. and A. Inoue (1995). Solidification condition of bulk glassy $Zr_{60}Al_{10}Ni_{10}Cu_{15}Pd_5$ alloy by unidirectional arc melting. *Mater. Trans., JIM* 36: 1398–1402.
70. Zhang, T. and A. Inoue (1994). Tohoku University, Sendai, Japan, unpublished research.
71. Tamura, T., K. Amiya, R.S. Rachmat, Y. Mizutani, and K. Miwa (2005). Electromagnetic vibration process for producing bulk metallic glasses. *Nat. Mater.* 4: 289–292.
72. Qiao, J.W., H.L. Jia, and P.K. Liaw (2016). Metallic glass matrix composites. *Mater. Sci. Eng. R* 100: 1–69.
73. Dandliker, R.B., R.D. Conner, and W.L. Johnson (1998). Melt infiltration casting of bulk metallic-glass matrix composites. *J. Mater. Res.* 13: 2896–2901.
74. Xue, Y.F., H.N. Cai, L. Wang, F.C. Wang, and H.F. Zhang (2007). Strength-improved Zr-based metallic glass/porous tungsten phase composite by hydrostatic extrusion. *Appl. Phys. Lett.* 90: 081901-1–081901-3.

75. Choi-Yim, H., S.B. Lee, and R.D. Conner (2008). Mechanical behavior of Mo and Ta wire-reinforced bulk metallic glass composites. *Scr. Mater.* 58: 763–766.
76. Lee, M.L., Y. Li, and C.A. Schuh (2004). Effect of a controlled volume fraction of dendritic phases on tensile and compressive ductility in La-based metallic glass matrix composites. *Acta Mater.* 52: 4121–4131.
77. Shen, B.L., H. Men, and A. Inoue (2006). Fe-based bulk glassy alloy composite containing in-situ formed α-(Fe,Co) and $(Fe,Co)_{23}B_6$ microcrystalline grains. *Appl. Phys. Lett.* 89: 101915-1–101915-3.
78. Qin, C.L., W. Zhang, K. Asami, H.M. Kimura, X.M. Wang, and A. Inoue (2006). A novel Cu-based BMG composite with high corrosion resistance and excellent mechanical properties. *Acta Mater.* 54: 3713–3719.
79. Sun, G.Y., G. Chen, and G.L. Chen (2007). Comparison of microstructures and properties of Zr-based bulk metallic glass composites with dendritic and spherical bcc phase precipitates. *Intermetallics* 15: 632–634.
80. Chen, L.Y., Z.D. Fu, W. Zeng, G.Q. Zhang, Y.W. Zeng, G.L. Xu, S.L. Zhang, and J.Z. Jiang (2007). Ultrahigh strength binary Ni–Nb bulk glassy alloy composite with good ductility. *J. Alloys Compd.* 443: 105–108.
81. Fan, C., L. Kecskes, T. Jiao, H. Choo, A. Inoue, and P.K. Liaw (2006). Shear-band deformation in amorphous alloys and composites. *Mater. Trans.* 47: 817–821.
82. Gao, J.E., Z.P. Chen, Q. Du, H.X. Li, Y. Wu, H. Wang, X.J. Liu, and Z.P. Lu (2013). Fe-based bulk metallic glass composites without any metalloid elements. *Acta Mater.* 61: 3214–3223.
83. Sha, P.F., Z.W. Zhu, H.M. Fu, H. Li, A.M. Wang, H.W. Zhang, H.F. Zhnag, and Z.Q. Hu (2014). Effects of casting temperature on the microstructure and mechanical properties of the TiZr-based bulk metallic glass matrix composite. *Mater. Sci. Eng. A* 589: 182–188.
84. Chueva, T.R., N.P. Dyakonova, V.V. Molokanov, and T.A. Sviridova (2007). Bulk amorphous alloy $Fe_{72}Al_5Ga_2C_6B_4P_{10}Si_1$ produced by mechanical alloying. *J. Alloys Compd.* 434–435: 327–332.
85. Pail, U., S.J. Hong, and C. Suryanarayana (2005). An unusual phase transformation during mechanical alloying of an Fe-based bulk metallic glass composition. *J. Alloys Compd.* 389: 121–126.
86. Sharma, S. and C. Suryanarayana (2007). Mechanical crystallization of Fe-based amorphous alloys. *J. Appl. Phys.* 102: 083544-1–083544-7.
87. Chang, L.J., G.R. Fang, J.S.C. Jang, I.S. Lee, J.C. Huang, C.Y.A. Tsao, and J.L. Lou (2007). Mechanical properties of the hot-pressed amorphous $Mg_{65}Cu_{20}Y_{10}Ag_5$/nanoZrO_2 composite alloy. *Mater. Trans.* 48: 1797–1801.
88. Sherif El-Eskandarany, M., M. Matsushita, and A. Inoue (2001). Phase transformations of ball-milled $Nb_{50}Zr_{10}Al_{10}Ni_{10}Cu_{20}$ powders and the effect of annealing. *J. Alloys Compd.* 329: 239–252.
89. Sherif El-Eskandarany, M., W. Zhang, and A. Inoue (2002). Mechanically induced solid-state reaction for synthesizing new multicomponent $Ta_{55}Zr_{10}Ni_{10}Al_{10}Cu_{15}$ glassy alloy powders with extremely wide supercooled liquid region. *Mater. Trans.* 43: 1422–1424.
90. Liu, L.H., C. Yang, Y.G. Yao, F. Wang, W.W. Zhang, Y. Long, an Y.Y. Li (2015). Densification mechanism of Ti-based metallic glass powders during spark plasma sintering process. *Intermetallics* 66: 1–7.
91. Sherif El-Eskandarany, M. and A. Inoue (2006). Synthesis of new bulk metallic glassy $Ti_{60}Al_{15}Cu_{10}W_{10}Ni_5$ alloy by hot-pressing the mechanically alloyed powders at the supercooled liquid region. *Metall. Mater. Trans.* 37A: 2231–2238.
92. Sherif El-Eskandarany, M. and A. Inoue (2002). New $V_{45}Zr_{20}Ni_{20}Cu_{10}Al_{2.5}Pd_{2.5}$ glassy alloy powder with wide supercooled liquid region. *Mater. Trans.* 43: 770–772.
93. Cho, K. and C.A. Schuh (2015). W-based amorphous phase stable to high temperatures. *Acta Mater.* 85: 331–342.
94. Sherif El-Eskandarany, M. and A. Inoue (2006). Hot pressing and characterizations of mechanically alloyed $Zr_{52}Al_6Ni_8Cu_{14}W_{20}$ glassy powders. *J. Mater. Res.* 21: 976–987.
95. Prabhu, B., C. Suryanarayana, L. An, and R. Vaidyanathan (2006). Synthesis and characterization of high volume fraction Al–Al_2O_3 nanocomposite powders by high-energy milling. *Mater. Sci. Eng. A* 425: 192–200.

96. Banhart, J. (2001). Manufacture, characterisation and application of cellular metals and metal foams. *Prog. Mater. Sci.* 46: 559–632.
97. Ashby, M.F., A. Evans, N.A. Fleck, L.J. Gibson, J.W. Hutchinson, and H.N.G. Wadley (2000). *Metal Foams: A Design Guide*. Boston, MA: Butterworth-Heinemann.
98. Gibson, L.J. and M.F. Ashby (1997). *Cellular Solids: Structure and Properties*, 2nd edn. Cambridge, UK: Cambridge University Press.
99. Nakajima, H., T. Ikeda, and S.K. Hyun (2004). Fabrication of lotus-type porous metals and their physical properties. *Adv. Eng. Mater.* 6: 377–384.
100. Banhart, J. (2000). Manufacturing routes for metallic foams. *JOM* 52(12): 22–27.
101. Apfel, R.E. and N. Qiu (1996). Principle of dynamic decompression and cooling for materials processing. *J. Mater. Res.* 11: 2916–2920.
102. Schroers, J., C. Veazey, and W.L. Johnson (2003). Amorphous metallic foam. *Appl. Phys. Lett.* 82: 370–372.
103. Brothers, A.H. and D.C. Dunand (2004). Syntactic bulk metallic glass foam. *Appl. Phys. Lett.* 84: 1108–1110.
104. Brothers, A.H. and D.C. Dunand (2005). Ductile bulk metallic glass foams. *Adv. Mater.* 17: 484–486.
105. Brothers, A.H., R. Scheunemann, J.D. DeFouw, and D.C. Dunand (2005). Processing and structure of open-celled amorphous metal foams. *Scr. Mater.* 52: 335–339.
106. Wada, T. and A. Inoue (2003). Fabrication, thermal stability and mechanical properties of porous bulk glassy Pd–Cu–Ni–P alloys. *Mater. Trans.* 44: 2228–2231.
107. Wada, T. and A. Inoue (2004). Formation of porous Pd-based bulk glassy alloys by a high hydrogen pressure melting-water quenching method and their mechanical properties. *Mater. Trans.* 45: 2761–2765.
108. Wada, T., A. Inoue, and A.L. Greer (2005). Enhancement of room-temperature plasticity in a bulk metallic glass by finely dispersed porosity. *Appl. Phys. Lett.* 86: 251907-1–251907-3.
109. Xie, G.Q., W. Zhang, D.V. Louzguine-Luzgin, H.M. Kimura, and A. Inoue (2006). Fabrication of porous Zr–Cu–Al–Ni bulk metallic glass by spark plasma sintering process. *Scr. Mater.* 55: 687–690.
110. Demetriou, M.D., J.P. Schramm, C. Veazey, W.L. Johnson, J.C. Hanan, and N.B. Phelps (2007). High porosity metallic glass foam: A powder metallurgy route. *Appl. Phys. Lett.* 91: 161903-1–161903-3.
111. Chen, Y.L., A.M. Wang, H.M. Fu, Z.W. Zhu, H.F. Zhang, Z.Q. Hu, L. Wang, and H.W. Cheng (2011). Preparation, microstructure and deformation behavior of Zr-based metallic glass/porous SiC interpenetrating phase composites. *Mater. Sci. Eng. A* 530: 15–20.
112. Zhang, H.F., A.M. Wang, H. Li, W.S. Sun, B.Z. Ding, Z.Q. Hu, H.N. Cai, L. Wang, and W. Li (2006). Quasi-static compressive property of metallic glass/porous tungsten bi-continuous phase composite. *J. Mater. Res.* 21: 1351–1354.
113. Sun, Y., H.F. Zhang, A.M. Wang, H.M. Fu, Z.Q. Hu, C.E. Wen, and P.D. Hodgson (2009). Mg-based metallic glass/titanium interpenetrating phase composite with high mechanical performance. *Appl. Phys. Lett.* 95: 171910-1–171910-3.
114. Cox, M.E., L.J. Kecskes, S.N. Muthaudhu, and D.C., Dunand (2012). Amorphous Hf-based foams with aligned, elongated pores. *Mater. Sci. Eng. A* 533: 124–127.

5

Crystallization Behavior

5.1 Introduction

Metallic glasses, whether produced in the form of ribbons by the melt-spinning technique at high solidification rates or in bulk rod form by conventional solidification methods at relatively slow solidification rates, are in a high energy (metastable) state. Consequently, they lower their energy by transforming into the crystalline state, a process referred to as *crystallization* or *devitrification*. The crystallization of metallic glasses is expected to take place at or above the crystallization temperature, T_x, which is measured with the help of a differential scanning calorimeter (DSC) or a differential thermal analyzer (DTA), during continuous heating of the glassy ribbon/alloy at a constant heating rate. Note, however, that T_x is not a thermodynamic parameter like the melting temperature of a metal. It is a function of the heating rate employed, and the higher the heating rate, the higher is the T_x. Therefore, given sufficient time, the crystallization of metallic glasses can occur at temperatures lower than T_x also.

The crystallization studies of alloys are important from both scientific and technological points of view. The study of the crystallization behavior of metallic glasses is very interesting from a scientific point of view. Since crystallization of metallic glasses occurs by a nucleation and growth process, it offers an opportunity to study the growth of crystals into an isotropic medium. Further, this process also offers a chance to test the classical nucleation and growth theories at large undercoolings.

From a technological point of view, the crystallization temperature of metallic glasses provides a real upper limit to the safe use of metallic glasses without loss of their interesting combination of properties. The important and interesting properties of metallic glasses are lost as a result of crystallization. For example, the magnetic behavior of metallic glasses is different after crystallization occurs. Additionally, the improved corrosion resistance of metallic glasses has been found to deteriorate upon crystallization. Further, metallic glass ribbons were found to lose their bend ductility upon crystallization. But, it should be realized that T_x cannot be taken as an indicator of the *safe operating temperature* of a metallic glass. For example, even though many Fe-based metallic glasses have T_x in the region of 400°C, their maximum long-term operating temperatures are only of the order of 150°C [1].

Studies on the crystallization behavior of metallic glasses also provide opportunities to study the kinetics of crystallization and also the micromechanisms of crystallization. Such studies will provide a clear understanding of the way that the metallic glass transforms into the crystalline state and offer a means to impede or control the crystallization behavior. In other words, one could tailor the microstructure to obtain a glass + nanocrystal or an ultrafine-grained composite, or a completely crystalline material of different grain sizes by controlling the time and temperature of crystallization [2]. It has been reported in recent years that it is possible to obtain composites of a fine crystalline phase dispersed in the glassy matrix by the partial crystallization of glass. Such composites have been shown to exhibit interesting mechanical properties. For example, the strength could be substantially increased to very high values [3]. Ductility could be introduced into the bulk metallic glasses (BMGs) by introducing fine crystalline phase particles [4–6]. Further, it should be possible to produce a fully nanocrystalline material by the controlled crystallization of metallic glassy alloys of suitable compositions. As a result, the mechanical and other properties could be optimized through a proper understanding of the crystallization behavior of metallic glasses. It may not be possible to obtain such novel and unique microstructures by any other means or in other types of materials unless one starts with a fully glassy material.

Any properties of the material that change with the crystallization of the metallic glass may be used to monitor the crystallization behavior. These include electrical resistivity, saturation magnetization, magnetic coercivity, and elastic modulus, among others. When these properties are followed as a function of temperature during the heating of the metallic glass, there is a sharp and discontinuous change at the crystallization temperature. However, such methods are indirect, and therefore, caution should be exercised in using them to determine the kinetics of crystallization. Further, the detection limit of the presence of a crystalline phase by such methods depends on the resolution limit of the technique; it is not usually high for these indirect methods.

The crystallization behavior of melt-spun metallic glass ribbons has been studied most extensively using x-ray diffraction (XRD), transmission electron microscopy (TEM), and DSC methods. Atom probe tomography has been used in recent years to study the crystallization behavior of metallic glasses by obtaining atomic-scale information on the clustering and segregation of solute atoms [7,8]. Other techniques have also been employed to obtain specific information. From the findings of these investigations on the crystallization of melt-spun metallic glasses, some excellent reviews have been produced by Köster and Herold [9], Scott [10], and Ranganathan and Suryanarayana [11].

5.2 Methodology

The crystallization behavior of metallic glassy alloys (ribbons or bulk rod samples) is most commonly studied by first determining the characteristic temperatures at which the glassy phase transforms to the crystalline state (mostly in the case of ribbons), or mainly to the supercooled liquid state (in the case of BMGs) before crystallization starts. This is done by heating the sample containing the glassy phase, at a constant rate, usually between 5 and 40 K min^{-1} (even though lower and higher heating rates have been employed in a few cases) to a high temperature in a DSC or a DTA. In the DSC method, the glassy sample (a few mg in weight) and an inert reference sample of similar weight are subjected to identical thermal programs and the heat flow between the two is monitored. The heat evolved or absorbed is plotted on the y-axis and the temperature on the x-axis. Conventionally, a reaction is termed *endothermic* if heat is absorbed and *exothermic* if heat is evolved. While DSC is the most commonly used method to determine the glass transition (T_g) and crystallization (T_x) temperatures, the DTA is more commonly used to determine the melting (solidus and liquidus) temperatures. This is because the DTA can be operated to higher temperatures than the DSC.

5.2.1 Transformation Temperatures

A typical DSC plot from a BMG sample is shown in Figure 5.1. In this plot, one can notice three important transformations defined by T_g, the glass transition temperature, T_x, the crystallization temperature, and T_m, the melting temperature. The glass continues to be in the glassy state until T_g, the presence of which is identified by a change in the slope of the base line. An increase in the heat capacity is noted at T_g. Even though crystallization does not occur below T_g, the as-quenched glass undergoes structural relaxation. Crystallization is indicated by a large exotherm, from which a dynamic crystallization temperature can be defined. From this peak, one could identify either the crystallization onset temperature, T_x or the peak temperature, T_p. The number of crystallization events (and consequently the crystallization temperatures) observed depends on the number of stages in which the glassy phase transforms into the crystalline state. Thus, if equilibrium phases are formed in two stages, the DSC plots show two exothermic peaks representing two crystallization temperatures. If eutectic crystallization occurs (where more than one phase is formed simultaneously), only one crystallization peak is present in the DSC plot. Because of the presence of a large number of components in the BMGs, a number of different phases may form at different stages of heating. Consequently, the number of exothermic peaks (and crystallization temperatures) observed during the heating of the BMGs can be large, and as many as four exothermic peaks have been reported in $(Hf_xZr_{1-x})_{52.5}Cu_{17.9}Ni_{14.6}Al_{10}Ti_5$ glassy alloys for $x = 0$, 1/3, 1/2, and 2/3 [12].

The last transformation corresponds to the melting of the sample and is denoted by T_m, which is endothermic in nature. The temperature at which this peak occurs corresponds to the melting of the alloy. It is important to remember that most of the alloys, unless they correspond exactly to the eutectic composition, have a range of melting temperatures. Consequently, one needs to identify two temperatures, namely, T_s, the solidus temperature (at which melting begins) and T_ℓ, the liquidus temperature (at which melting is completed). The solidus temperature has also been occasionally referred to as the *onset*

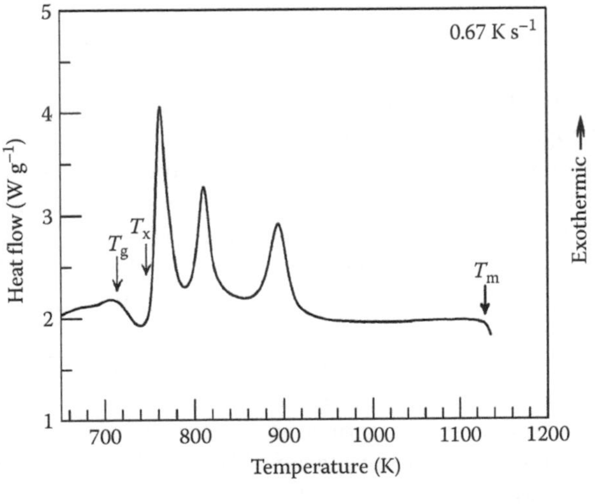

FIGURE 5.1
Schematic of a typical differential scanning calorimeter (DSC) curve obtained on heating a BMG alloy from room temperature to high temperatures at a constant heating rate of 40 K min^{-1}. Note that the curve displays three important temperatures—the glass transition temperature, T_g, the crystallization temperature, T_x, and the melting temperature, T_m. In some cases, there may be more than one crystallization temperature, depending on the number of stages in which the glass or the supercooled liquid transforms into the crystalline phase(s).

melting temperature. The measurement of T_ℓ, along with T_g, in the DSC/DTA plots, helps in determining the reduced glass transition temperature ($T_{rg} = T_g/T_\ell$), which is frequently used as a measure of the glass-forming alloy (GFA) of an alloy system (see Chapter 3 for details). Sometimes, researchers have used the term T_m to indicate the melting point of the alloy and some others have used this designation to indicate the solidus temperature. Also, at other times, the T_s has been used to indicate the melting temperature of the alloy. But, one should determine the T_ℓ temperature, especially if one is interested in correlating the GFA of the alloy by calculating T_{rg}. This is all the more important because Lu et al. [13] have clearly shown that T_{rg} given by T_g/T_ℓ showed a better correlation with critical cooling rate or maximum section thickness than T_{rg} given by T_g/T_m.

There were doubts in the minds of researchers, especially in the early years of research on melt-spun metallic glasses, whether the amorphous alloys produced by rapid solidification processing (RSP) were truly glasses. This was because a glass transition temperature, a hallmark of true glasses, was not detected in many cases. Even though the presence of T_g is absolute proof for the glassy nature of the material, the majority of the melt-spun metallic glass ribbons did not exhibit T_g in their DSC plots. The reason for this is that the difference between T_g and T_x for melt-spun ribbons is usually so small that one cannot clearly detect the presence of T_g in all cases. Only a few alloys, such as Au–Ge–Si, show the glass transition temperature [14]. In contrast, almost all of the BMGs exhibit T_g. But, there are also a few exceptions to this, and these BMGs do not show a clear T_g. For example, Nd–Fe–Al [15] and Pr–Fe–Al [16] glassy alloys do not exhibit any T_g, even though they have a high GFA and glassy rods with a diameter of >10 mm can be produced.

The temperature interval between T_g and T_x is referred to as ΔT_x (= $T_x - T_g$) or the *width of the supercooled liquid region*. The value of ΔT_x can be different depending on the alloy system in which the glass has formed and is usually in the range of 40–90 K. However, the smallest value of ΔT_x reported in BMG compositions is 12 K in a 1 mm diameter rod of $Ca_{66.4}Al_{33.6}$ alloy [17], and values larger than 120 K have also been reported. For example, ΔT_x of about 127 K was reported in a 16 mm diameter rod of $Zr_{65}Al_{7.5}Ni_{10}Cu_{17.5}$ glass [18,19]. The largest ΔT_x reported in the literature so far is 131 K in a $Pd_{43}Ni_{10}Cu_{27}P_{20}$ BMG alloy [20]. The width of the supercooled liquid region (ΔT_x) is an indicator of the stability of the supercooled liquid, and therefore, its resistance to the onset of crystallization. As pointed out in Chapter 3, there is a tendency for the GFA of the alloy to increase with an increasing value of ΔT_x, even though not in every case.

Table 5.1 lists the transformation temperatures for some BMG samples. At least one representative sample from each of the base metal categories was chosen, just for demonstration purposes. The T_g values were found to vary from as low as 348 K for an $Au_{55}Cu_{25}Si_{20}$ glassy alloy [21] to as high as 910 K for a $Co_{43}Fe_{20}Ta_{5.5}B_{31.5}$ glassy alloy [24]. Similarly, the T_x values ranged from about 383 K for an $Au_{55}Cu_{25}Si_{20}$ glassy alloy [21] to as high as 982 K for a $Co_{43}Fe_{20}Ta_{5.5}B_{31.5}$ glassy alloy [24].

The transformation temperatures for some of the melt-spun ribbons of the same composition are also listed in Table 5.1 for comparison. In fact, most commonly, a large quantity of the alloy is prepared, a small amount is used to synthesize bulk glassy rods, and another portion to synthesize melt-spun glassy ribbons. When the transformation temperatures of these glassy products are measured, it is noted that there is *no difference* between the melt-spun ribbons and bulk alloy rods. Therefore, usually, the transformation temperatures are measured on the melt-spun ribbons (of compositions identical to the rod samples) and these temperatures are used for heat treatment purposes to study structural relaxation, crystallization behavior, and so on, of the bulk rod samples. Also, there is no difference in the transformation temperatures of the glassy rods of different diameters.

TABLE 5.1

Transformation Temperatures Determined for Some Typical BMG and Melt-Spun Glassy Alloys

Composition	Glass-Forming Technique	T_g (K)	T_x (K)	ΔT_x ($T_x - T_g$) (K)	Heating Rate (K s^{-1})[a]	References
$Au_{55}Cu_{25}Si_{20}$	Cu mold casting	348	383	35	0.33	[21]
$Ca_{66.4}Al_{33.6}$	Cu mold casting	528	540	12	0.33	[17]
$Ca_{60}Al_{30}Ag_{10}$	Cu mold casting	483	531	48	0.33	[17]
$Ca_{58}Al_{32}Mg_{10}$	Cu mold casting	513	539	26	0.33	[17]
$Ca_{65}Mg_{15}Zn_{20}$	Cone-shaped Cu mold casting	374	412	38	0.67	[22]
$Ce_{60}Al_{10}Ni_{10}Cu_{20}$	Suction casting	374	441	67	0.167	[23]
$Co_{43}Fe_{20}Ta_{5.5}B_{31.5}$	Cu mold casting	910	982	72	0.67	[24]
$Cu_{50}Zr_{50}$	Cu mold casting	675	732	57	0.67	[4]
$Cu_{50}Zr_{50}$	Melt spinning	686	744	58	0.67	[4]
$Cu_{60}Zr_{30}Ti_{10}$	Cu mold casting (2.5 mm dia rod)	714	758	44	0.67	[25]
$Cu_{60}Zr_{30}Ti_{10}$	Melt spinning	711	754	43	0.67	[25]
$Cu_{46}Zr_{42}Al_7Y_5$	Injection casting into Cu mold (10 mm dia rod)	672	772	100	0.33	[26]
$Fe_{64}Mo_{14}C_{15}B_7$	Injection casting into Cu mold (2.5 mm dia rod)	793	843	50	0.33	[27]
$Hf_{52.5}Cu_{17.9}Ni_{14.6}Al_{10}Ti_5$	Suction casting	767	820	53	0.167	[12]
$La_{55}Al_{25}Ni_{10}Cu_{10}$	High-pressure die casting (9 mm dia rod)	460	527	67	0.67	[28]
$La_{55}Al_{25}Ni_{10}Cu_{10}$	Melt spinning	460	550	90	0.67	[29]
$Mg_{65}Cu_{7.5}Ni_{7.5}Zn_5Ag_5Y_{10}$	Melt spinning	430	459	29	0.67	[30]
$Ni_{60.25}Nb_{39.75}$	Injection casting into Cu mold (1 mm dia rod)	891	923	32	0.33	[31]
$Ni_{62}Nb_{33}Zr_5$	Injection molding (3 mm dia rod)	877	917	40	0.33	[32]
$Pd_{79}Ag_{4.5}Si_{16.5}$	Splat cooling	640	672	32	0.33	[33]
$Pd_{78}Ag_{5.5}Si_{16.5}$	Dropping a molten droplet onto metal substrate	642	683	41	0.33	[33]
$Pd_{80}Au_{3.5}Si_{16.5}$	Splat cooling	644	675	31	0.33	[33]
$Pd_{78}Au_4Si_{18}$	Roller quenching	656	696	40	0.67	[34]
$Pd_{43}Cu_{27}Ni_{10}P_{20}$ (fluxed)	Water quenching	585	716	131	0.67	[20]

(Continued)

TABLE 5.1 (CONTINUED)

Transformation Temperatures Determined for Some Typical BMG and Melt-Spun Glassy Alloys

Composition	Glass-Forming Technique	T_g (K)	T_x (K)	ΔT_x ($T_x - T_g$) (K)	Heating Rate (K s^{-1})[a]	References
$Pd_{43}Cu_{27}Ni_{10}P_{20}$ (foam)	Water quenching	577	667	90	0.33	[35]
$Pd_{40}Cu_{30}Ni_{10}P_{20}$ (unfluxed)	Melt spinning	572	663	91	0.33	[36]
$Pd_{40}Cu_{30}Ni_{10}P_{20}$ (fluxed)	Melt spinning	572	670	98	0.33	[36]
$Pd_{77.5}Cu_6Si_{16.5}$	Dropping a molten droplet onto metal substrate	646	686	40	0.33	[33]
$Pd_{40}Ni_{40}P_{20}$	Centrifugal spinning	583	650	67	0.33	[37]
$Pd_{40}Ni_{40}P_{20}$ (fluxed)	Water quenching (7 mm dia rod)	576	678	102	0.33	[38]
$Pd_{40}Ni_{40}P_{20}$ (unfluxed)	Melt spinning	580	643	63	0.33	[36]
$Pd_{40}Ni_{40}P_{20}$ (fluxed)	Melt spinning	590	671	91	0.33	[36]
$Pd_{81}Si_{19}$ (fluxed)	Air cooling	638	696	58	0.33	[39]
$Pd_{81}Si_{19}$ (fluxed)	Melt spinning	633	675	42	0.33	[39]
$Pd_{80}Si_{20}$	Splat cooling	655	667	12	0.33	[33]
$Pr_{60}Cu_{20}Ni_{10}Al_{10}$	Suction casting	409	452	43	0.167	[40]
$Pt_{57.5}Cu_{14.7}Ni_{5.3}P_{22.5}$ (fluxed)	Water quenching (16 mm dia rod)	508	606	98	0.33	[41]
$Sm_{56}Al_{22}Ni_{22}$	Suction casting	544	582	38	0.33	[42]
$Ti_{32}Hf_{18}Ni_{35}Cu_{15}$	Planar flow casting	722	766	44	0.167	[43]
$Ti_{50}Ni_{24}Cu_{20}B_1Si_2Sn_3$	Cu mold casting	726	800	74	0.67	[44]
$Ti_{40}Zr_{25}Ni_2Cu_{13}Be_{20}$	Cu mold casting	599	644	45	0.33	[45]
$Y_{36}Sc_{20}Al_{24}Co_{20}$	—	645	760	115	0.67	[46]
$Zr_{65}Al_{7.5}Ni_{10}Cu_{17.5}$	Water quenching (16 mm dia rod)	625	750	125	0.67	[19]
$Zr_{65}Al_{7.5}Ni_{10}Cu_{17.5}$	Melt spinning	622	749	127	0.67	[19]
$Zr_{41.2}Ti_{13.8}Cu_{12.5}Ni_{10}Be_{22.5}$	Cu mold casting	625	705	80	0.33	[47]

[a] 0.167 K s^{-1} = 10 K min^{-1}; 0.33 K s^{-1} = 20 K min^{-1}; and 0.67 K s^{-1} = 40 K min^{-1}.

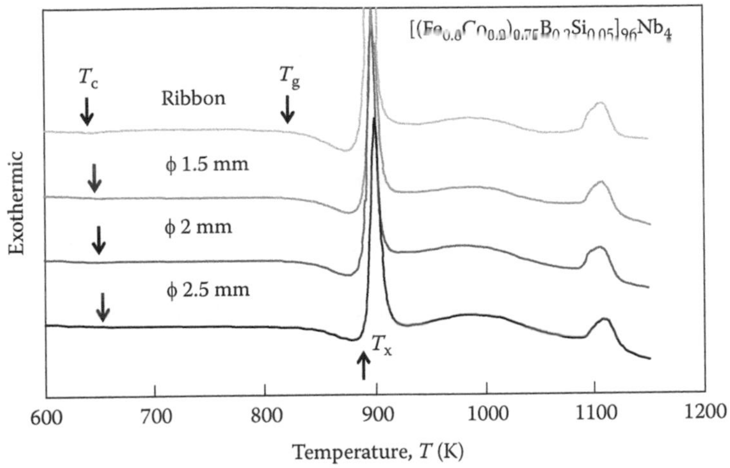

FIGURE 5.2
DSC curves of bulk glassy $[(Fe_{0.8}Co_{0.2})_{75}B_{20}Si_5]_{96}Nb_4$ alloy of different diameters (1.5, 2.0, and 2.5 mm) and melt-spun ribbon of the same composition. These curves clearly demonstrate that the transformation temperatures are identical for all the samples and that the transformation temperatures do not depend upon the diameter of the rod or the thickness of the ribbon. (Reprinted from Inoue, A. et al., *Acta Mater.*, 52, 4093, 2004. With permission.)

Figure 5.2 shows the DSC curves of glassy $[(Fe_{0.8}Co_{0.2})_{75}B_{20}Si_5]_{96}Nb_4$ alloy rods of different diameters and also of the melt-spun ribbon of the same composition [48]. It may be noted that the transformation temperatures are identical for all the samples. Similar observations have been made repeatedly by a number of researchers and in different alloy systems.

The glass transition temperature, T_g, is a kinetic parameter, and its value depends on the cooling rate at which the glass is formed (and also on the heating rate at which the glassy sample is reheated). It was also noted that T_g was lower when the glass had formed at lower cooling rates. Therefore, it would be possible to assume that the T_g for the melt-spun ribbon and BMG rod would be different. But, this is not true, because T_g, the temperature at which the glass is formed, is estimated during the cooling of the molten alloy. On the other hand, T_g is usually measured experimentally during the heating of the glassy alloy that has already formed. Once the glass is heated from room temperature to higher temperatures, it is structurally relaxed and, therefore, it does not matter how the glass had initially formed. Accordingly, both types of glasses will have the same T_g and T_x temperatures, when measured at the same heating rate. That is, there is no difference between the T_g values of glasses prepared by RSP or slow solidification methods.

Occasionally, there are reports in the literature where small differences have been reported between the transformation temperatures of the melt-spun ribbons and BMG alloy rods, even though the composition is identical for both the glasses. For example, Inoue et al. [28] measured the transformation temperatures for bulk glassy rods of the $La_{55}Al_{25}Ni_{10}Cu_{10}$ alloy and reported that the T_g and T_x values for this alloy were 460 and 527 K, respectively. These values were also shown to be independent of the diameter of the glassy rod from 1 to 9 mm. However, they reported the T_g and T_x temperatures for the melt-spun ribbon as 460 and 550 K, respectively [29]. These values suggest that while the T_g value was identical for the two types of glasses (ribbons and rods), the T_x value for the rod was lower by 23 K. The reduction of T_x for the BMG sample (by about 4%) was explained on the basis that perhaps it was difficult to protect the molten alloy against oxidation during the casting of the BMG samples. This brings out an important point—the melt purity is a vital consideration during the synthesis of metallic glassy alloys, especially in alloys based on reactive metals such as Zr and Ti. Therefore, the differences in T_g and T_x are not intrinsic, but due to extraneous factors such as oxidation or other environmental effects.

Determination of T_x helps in identifying the temperatures at which heat treatments can be carried out to determine the kinetics of crystallization and also the nature of the crystallization product(s). Once these temperatures are determined, the alloy specimen is heated to a temperature just above the transformation temperature, quenched from that temperature, and the sample is examined for the nature of the phases present, their morphology, volume fraction, size, and size distribution using techniques such as XRD, TEM, and others.

The heating rates are usually slow, ranging from about 5 K min^{-1} to about 40 K min^{-1}, even though much slower and much faster heating rates are occasionally employed for special reasons. It is important to realize that too slow a heating rate requires much longer experimentation time, and too high a heating rate could result in some of the transformation temperatures being missed.

5.2.2 Activation Energy for Crystallization

A transformation in a material during heating or cooling is indicated by a deflection or peak in the DSC/DTA curve. If the reaction is temperature dependent, that is, if it possesses an activation energy, the position of the peak varies with the heating rate employed, assuming that all other experimental conditions are constant. It has been noted that the transformation temperatures are higher at faster heating rates. For example, Figure 5.3 shows the variation of the first and second crystallization temperatures during heating of a $Ti_{40}Zr_{25}Ni_8Cu_9Be_{18}$

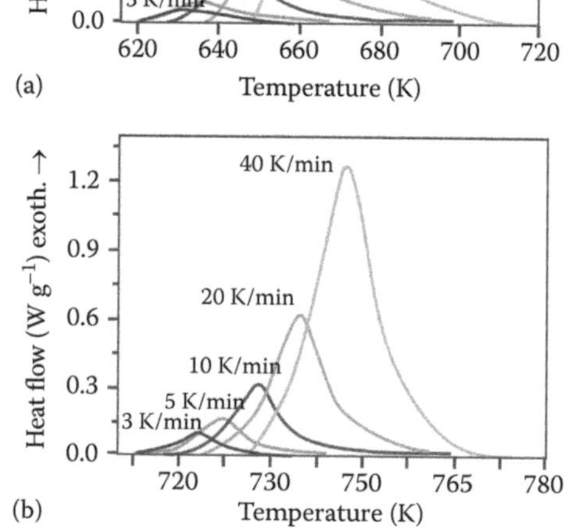

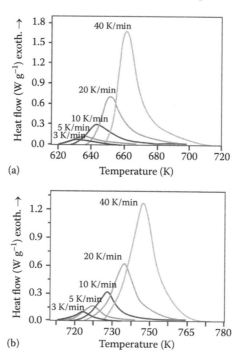

FIGURE 5.3
The non-isothermal DSC curves of (a) the first crystallization stage and (b) the second crystallization stage in $Ti_{40}Zr_{25}Ni_8Cu_9Be_{18}$ BMG after base line correction with different heating rates. (Reprinted from Hu, X., et al., *J. Non-Cryst. Solids*, 432, 254, 2016. With permission.)

BMG alloy at heating rates ranging from 3 to 40 K min^{-1} [49]. The increase in the transformation temperatures as a function of the increasing heating rate may be clearly noted. It was also shown that the peak temperature, T_p, corresponds to the temperature at which the reaction rate is a maximum. Using this concept of variation of T_x (or T_p) with heating rate (β), the activation energy for the crystallization of metallic glasses has been determined.

There are two important methods by which the activation energy for crystallization could be determined. One method is attributed to Kissinger [50,51] and the other to Ozawa [52]. In the Kissinger method, the temperature corresponding to the peak of the crystallization event (exothermic peak) is supposed to be determined. However , researchers have sometimes used the temperature corresponding to the start of the crystallization event, T_x. Whichever temperature is chosen, it is determined at different heating rates, and the activation energy for crystallization is determined using the equation:

$$\ln\left(\frac{\beta}{T_p{}^2}\right)=\left(-\frac{Q}{RT_p}\right)+A \tag{5.1}$$

where:

- A is a constant
- R is the universal gas constant

Thus, by plotting $\ln\left(\beta/T_p^2\right)$ against $1/T_p$, one obtains a straight line whose slope is $-Q/R$, from which the activation energy for the transformation, Q, can be calculated (Figure 5.4).

On the other hand, in the Ozawa method [52], log (β/T_x) is plotted against $1/T_x$ when a straight line is obtained. The activation energy for crystallization can then be determined from the slope of this straight line. There is disagreement in the literature about the merits and demerits of these two methods in determining the activation energies, even though the Kissinger method has been more commonly employed. In the majority of cases, however, both methods yield similar results.

With the combination of DTA/DSC and XRD/TEM techniques, it is possible to obtain full information about the transformation temperatures, the number of stages in which the transformation is occurring, details about the product(s) of each individual transformation

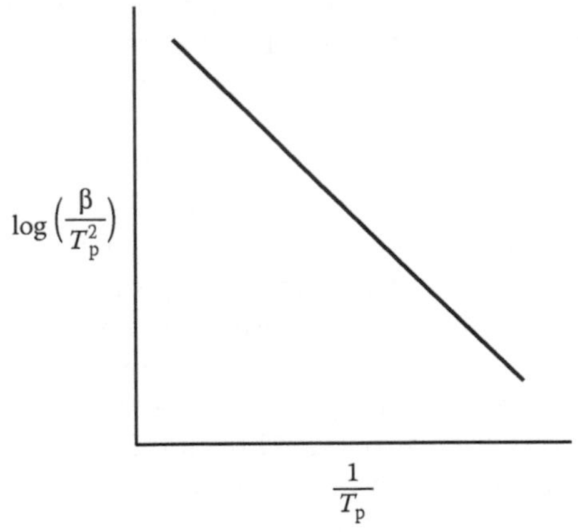

FIGURE 5.4
Kissinger plot in which $\ln(\beta/T_p^2)$ is plotted against $1/T_p$, when a straight line is obtained. The activation energy for crystallization can be calculated from the slope of this straight line.

(crystal structure, microstructure, and chemical composition), and the activation energy (and also the atomic mechanism) for the transformation. The greatest advantage of the Kissinger method is that one can get the required data during continuous heating in a DSC, which is far more convenient than time-consuming isothermal analysis studies.

The Kissinger method may not be useful in all studies of decomposition. For example, metallic glasses may decompose by nucleation, growth, or a combination of both processes. In such cases, the decomposition is seldom described by first [53,54].

In both methods, the metallic glass samples are heated continuously to high temperatures and at different rates. These are referred to in the literature as non-isothermal experiments. It is also possible to conduct these experiments isothermally, that is, the sample is maintained at a constant temperature and the transformation is followed as a function of time at that temperature. By recording the isothermal DSC scans at different temperatures, it is possible to study the kinetics of transformation by conducting the Johnson–Mehl–Avrami (JMA) analysis. For example, during the crystallization of a glassy phase, this is done by calculating the volume fraction of the crystalline phase formed, $x(t)$, at time t using the equation:

$$x(t) = 1 - \exp\left(-kt^n\right) \tag{5.2}$$

where:

$x(t)$ is the volume fraction of the glass that has transformed at time t

k is a temperature-sensitive factor [$k = k_0 \exp(-Q/RT)$, where k_0 is a constant]

n is an exponent (often referred to as the kinetic or Avrami exponent) that reflects the nucleation rate and/or the growth mechanism

$x(t)$ is obtained by measuring the area under the peak of the isothermal DSC curve at different times and dividing it by the total area. Equation 5.2 can also be written as

$$\ell n\left[-\ell n\left\{1 - x(t)\right\}\right] = \ell n\, k + n \ell n(t) \tag{5.3}$$

Thus, by plotting $\ell n[-\ell n\{1 - x(t)\}]$ against $\ell n(t)$, one obtains a straight line of slope n. Such analyses have been done in some studies [49,55–59]. In such cases, when first-order reaction kinetics are not obeyed, analysis of the data by the Johnson–Mehl–Avrami–Kolmogorov (JMAK) method will more accurately provide information about the nature of the transformation.

Since the crystallization of metallic glasses involves both nucleation and growth, such analysis gives the activation energy for the total transformation. That is, $Q = Q_n + Q_g$, where Q_n and Q_g represent the individual activation energies for the nucleation and growth stages of the transformation, respectively. It is possible to evaluate Q_n and Q_g separately, thereby ascertaining information about the mechanism of transformation. Ranganathan and Heimendahl [60] have developed formulae relating Q, Q_n, and Q_g. They showed that, by plotting the logarithm of nucleation rate against $1/T$, the value of Q_n could be obtained from the slope of the straight line. Similarly, the value of Q_g could be obtained from the slope of the straight line obtained by plotting the logarithm of the growth rate against $1/T$. They showed that the agreement between the theory and experiments on metallic glasses was very satisfactory.

The Avrami exponent, n, can be used to characterize the nucleation and growth behaviors with the increasing volume fraction of the crystalline phase. Following Ranganathan and Heimendahl [60], n can be expressed as

$$n = a + bc \tag{5.4}$$

where:

- a is the nucleation index
- b is the growth index
- c is the dimensionality of the growth

In cases where nucleation need not occur (e.g., when preexisting nuclei are present), $a = 0$, and only growth need occur for the transformation to proceed. Further, $a = 1$ for constant nucleation rate, $a > 1$ for increasing nucleation rate with time, and $0 < a < 1$ for decreasing nucleation rate with time. In addition, $b = 0.5$ for diffusion-controlled growth and 1 for interface-controlled growth. Finally, $c = 1$, 2, or 3 for one-, two-, or three-dimensional growth of particles. The value of n generally varies between 1.5 and 4. Thus, for linear growth,

$$Q = \left(\frac{aQ_n + bQ_g}{a + b} \right) \tag{5.5}$$

with $a + b = n$, and for parabolic growth

$$Q = \left(\frac{aQ_n + 0.5bQ_g}{a + 0.5b} \right) \tag{5.6}$$

with $a + 0.5b = n$.

5.2.3 Structural Details

Structural details of the crystallization products are usually determined by microscopy and diffraction methods. While XRD methods are common, less expensive, and the equipment is readily available in most laboratories, electron (and sometimes neutron) diffraction methods are used for specialized cases. For example, neutron diffraction could be used when one needs to obtain information from a larger volume of the specimen, or when the sample contains elements that are too close to each other in the periodic table. On the other hand, when the specimen contains crystalline phases with small dimensions, TEM is an ideal tool for obtaining information on the size, shape, and distribution of the crystalline phase. One can determine the size and morphology of the phases present in the crystallized sample in the imaging mode and the crystal structure features from the diffraction mode. Additionally, the chemical composition of the different phases can be easily determined from the energy dispersive spectroscopy (EDS) methods available in most electron microscopes. Thus, using TEM methods, it is possible to obtain full information about the number and nature of phases, their morphology, crystal structure, and composition. It is important to remember, however, that TEM studies should be conducted on samples that have been thinned after heat treatments and not as *in situ* studies on thin samples that have been heat treated. In the latter case, the occurrence of surface nucleation and diffusion could yield misleading results. As mentioned earlier, atom probe tomography methods have been used recently to obtain information about the distribution of phases and elements on an atomic scale [7,8]. But, it should be noted that the most appropriate technique depends on the desired structural information.

Glassy phases give rise to broad and diffuse halos in their diffraction patterns. It is desirable to confirm the formation of the glassy phase using direct TEM techniques.

Transmission electron micrographs from glassy phases do not show any diffraction contrast, and their electron diffraction patterns show broad and diffuse halos. Differentiation between a "truly" glassy (i.e., without translational symmetry) and micro- or nanocrystalline structure (i.e., an assembly of randomly oriented fragments of a bulk crystalline phase) has not been easy on the basis of diffraction studies alone; considerable confusion exists in the literature. In diffraction experiments on an "amorphous" structure, the intensity but not the phase of the scattered radiation is measured. Fourier inversion of these data can yield only the radial distribution function of the structure, which cannot uniquely specify the atomic positions. To determine the structure, the experimentally determined radial distribution function must be compared with the radial distribution functions calculated from the structural models being considered [61].

The occurrence of an amorphous phase is generally inferred by observing the presence of broad and diffuse peaks in the XRD patterns. It should be noted, however, that XRD patterns present only an average picture. Thus, by observing the broad x-ray peaks alone, it is not possible to distinguish between materials that are (1) truly glassy, (2) extremely fine-grained, or (3) a material in which a small volume fraction of fine crystals is embedded in a glassy matrix. Hence, in recent years, it has been the practice to designate these identifications as "x-ray amorphous," indicating that the conclusions were arrived at only on the basis of the observation of a broad peak (or a couple of diffuse peaks) in the XRD pattern. There have been several examples of a phase being reported as "glassy" on the basis of XRD studies alone. However, based on supplementary investigations by neutron diffraction and/or TEM techniques, which have specific advantages and/or a higher resolution capability than the XRD technique, it could be unambiguously confirmed whether the phase produced was truly glassy or not.

Neutron diffraction techniques have the advantage of detecting lighter atoms in the presence of heavy atoms, whereas the scattering intensity of the lighter atoms may be completely masked by that of the heavy atoms in techniques such as XRD or electron diffraction. Additionally, the technique of neutron diffraction has the ability to distinguish between neighboring elements in the periodic table. XRD techniques are unsuitable for this because neighboring elements will have their atomic scattering factors very close to each other. Consequently, the difference in their scattering factors, and hence the intensity, will be too small to be detected. Thus, it is desirable that XRD observations are confirmed by other techniques as well. For example, TEM studies can confirm the lack of contrast in the micrographs for a truly glassy phase.

Figure 5.5 shows transmission electron micrographs and the electron diffraction pattern from a BMG alloy. Figure 5.5a shows a (low-resolution) TEM micrograph, which is characterized by a complete lack of contrast. The presence of a crystalline phase would provide a diffraction contrast; therefore, using TEM images it is possible to determine if the BMG alloy produced is homogeneous or not. Figure 5.5b shows an electron diffraction pattern recorded from a fully glassy alloy. The presence of a very diffuse and broad halo suggests that the alloy is glassy. Figure 5.5c shows a high-resolution TEM (HRTEM) micrograph showing so-called salt-and-pepper contrast. Even at this high resolution, one does not see any evidence of crystallinity in the sample. The presence of well-defined fringe pattern in the micrograph is indicative of the presence of crystallites in the sample. It has been reported that, occasionally, the as-quenched glass contains a few regions exhibiting medium-range order (MRO) in the sample. This will be manifested in the form of isolated regions of a few nanometers in size (typically less than about 5 nm) showing a fringe-like pattern. Nanobeam diffraction patterns from alloys containing MRO zones or very fine crystalline particles give rise to well-defined diffraction spots.

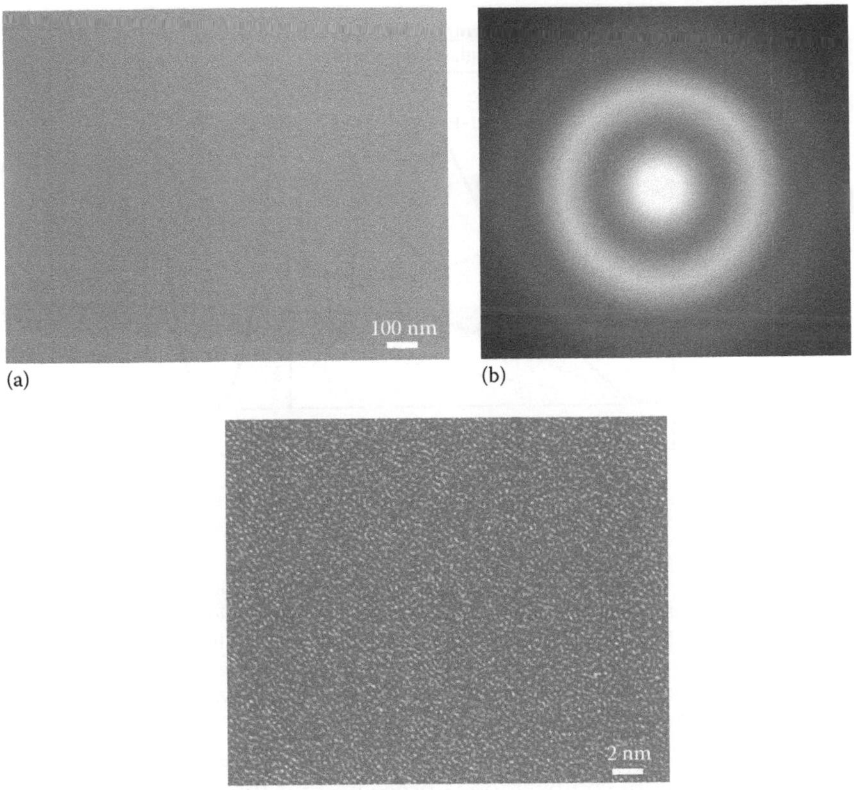

(a) (b) (c)

FIGURE 5.5
Transmission electron micrographs and electron diffraction pattern from $Zr_{55}Cu_{30}Al_{10}Ni_5$ glassy alloy. (a) Low-resolution TEM micrograph showing the absence of any contrast, (b) electron diffraction pattern showing the presence of a broad and diffuse halo, and (c) high-resolution TEM micrograph showing the typical salt-and-pepper contrast. All these features are typical of a fully glassy phase.

5.3 Crystallization Modes in Melt-Spun Ribbons

Crystallization of metallic glasses occurs by nucleation and growth processes. The driving force for crystallization is the difference in the free energy between the glassy phase and the corresponding crystalline phase(s) of the same composition. The crystalline phases formed as a result of crystallization may be either the equilibrium phases or, as is frequently noted, some metastable phases may also form in the initial stages of decomposition and these may eventually transform to equilibrium phases. Therefore, it should be possible to understand the crystallization behavior of metallic glasses with reference to a hypothetical free energy versus composition diagram [62].

Figure 5.6 shows the hypothetical free energy versus composition diagram for the glass and different crystalline phases in the Fe-rich Fe–B system. (Even though this diagram is specifically drawn for the Fe–B system, similar diagrams could be constructed for other alloy systems that form glassy phases. The differences will be essentially with respect to the extent of solid solubilities and the number of stable and metastable intermetallic phases and the compositions at which they are located in the phase diagram.) In this diagram, the

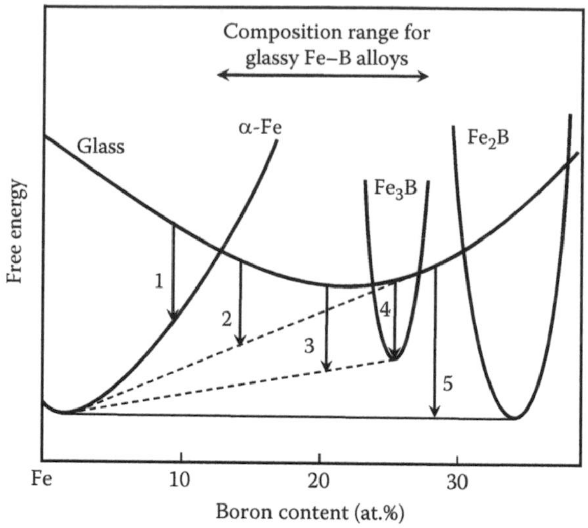

FIGURE 5.6
Hypothetical free energy vs. composition diagram for the Fe-rich Fe–B alloy system. The variation of free energy with composition is represented for the equilibrium α-Fe solid solution, the Fe_2B and metastable Fe_3B phases, and the glassy phase. The use of the common tangent approach will help in determining the compositions of the individual phases. The solid common tangent line represents the stable equilibrium between α-Fe and Fe_2B phases, while the dotted common tangent lines represent the metastable equilibrium between α-Fe and Fe_3B phases and α-Fe and glassy phases.

variation of free energy with composition is presented for the glassy phase, the Fe-rich α-Fe solid solution, the equilibrium Fe_2B phase, and the frequently detected metastable Fe_3B phase. The stable equilibrium between the α-Fe and Fe_2B phases is indicated by the solid common tangent line and the metastable equilibrium between the α-Fe and glassy phases or α-Fe and Fe_3B phases is denoted by the dashed common tangent lines.

Depending on the composition of the glassy alloy, the metallic glass could crystallize into the stable equilibrium phases in one of the following three ways.

5.3.1 Polymorphous Crystallization

In this mode of transformation, the glassy phase will transform into a single crystalline phase without any change in composition. (This transformation is designated polymorphous, as it is similar to what occurs in a pure metal or a compound when the crystal structure changes on varying the temperature and/or pressure. In the case of a glass, however, it is transformation from the noncrystalline to the crystalline state.) The polymorphous crystallization will occur only in composition ranges where the glassy phase had formed at a composition corresponding to either a stable or metastable crystalline solid solution or an intermetallic phase. With reference to Figure 5.6, polymorphous crystallization is possible when the glassy phase transforms into the α-Fe solid solution (Reaction 1), Fe_3B phase (Reaction 4), or Fe_2B phase. Since a glassy phase was reported to form only in the composition range between 12 and 27 at.% B in the Fe–B system by RSP methods and was not obtained in an Fe-33 at.% B alloy, the last reaction (glass → Fe_2B crystalline phase) will not be possible. Since Fe_3B (formed by Reaction 4) is a metastable phase, it will subsequently transform into a mixture of the equilibrium α-Fe and Fe_2B phases on further annealing. Similarly, the supersaturated α-Fe (formed by Reaction 1) will also transform to the equilibrium constitution on further annealing. During polymorphous crystallization, the growth of crystals is linear with time, and the growth rate has an Arrhenius dependence on temperature. Such transformations have been least common among the glass → crystal transformations. Figure 5.7a represents a bright-field TEM micrograph showing polymorphous crystallization in a $Ti_{50}Ni_{25}Cu_{25}$ BMG alloy.

5.3.2 Eutectic Crystallization

In this type of crystallization, the glassy phase transforms simultaneously into two (or more) crystalline phases by a discontinuous reaction. For example, through Reaction 3 in Figure 5.6, the glassy phase can form a mixture of α-Fe and Fe_3B, or through Reaction 5 it can form a mixture of α-Fe and Fe_2B phases. This mode of crystallization has the largest driving force and can occur in the whole concentration range between the two stable or metastable phases. (Even though the whole transformation takes place in the solid state and, therefore, it should be more appropriately called a *eutectoid* crystallization, the term *eutectic* has come to stay, presumably because the starting material [the glass] is more liquid like.)

Like polymorphous crystallization, the eutectic crystallization is a discontinuous reaction; the overall composition of the crystal and the glass are the same. The composition of the glassy matrix remains unchanged until the glass–crystal interface sweeps past it.

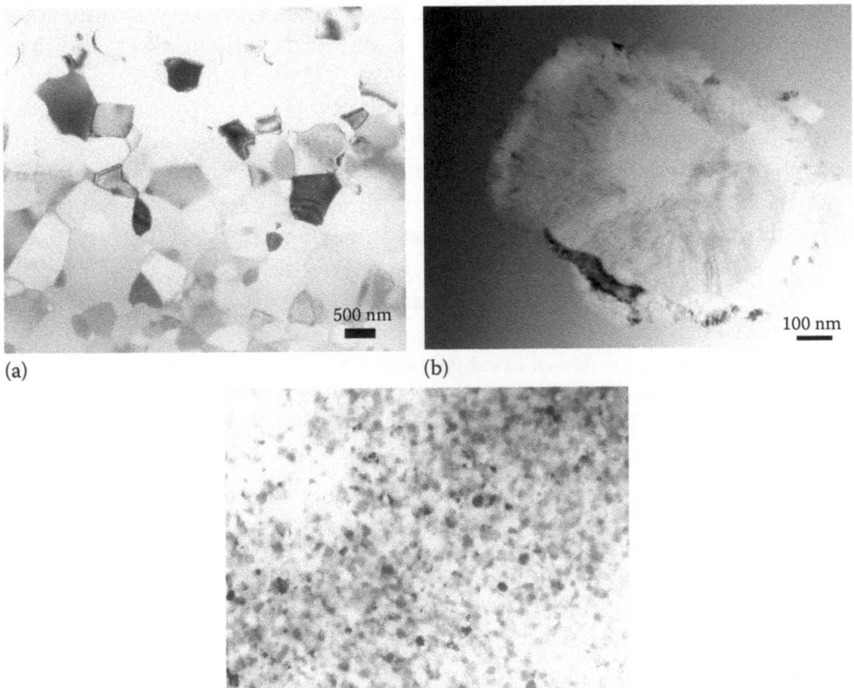

FIGURE 5.7
Transmission electron micrographs showing the microstructures obtained after polymorphous, eutectic, or primary crystallization of BMG alloys. (a) Polymorphous crystallization in a $Ti_{50}Ni_{25}Cu_{25}$ BMG alloy on annealing for 28 min at 709 K. (b) Eutectic crystallization in a $Zr_{62.5}Cu_{22.5}Al_{10}Fe_5$ glassy alloy annealed for 10 min at 713 K. (c) Primary crystallization in a $Ti_{50}Ni_{20}Cu_{23}Sn_7$ alloy on heating the glass in a DSC at 40 K min^{-1} up to 787 K.

For such reactions, the crystal growth rate is independent of time until hard impingement with another crystal occurs.

Figure 5.7b shows a bright-field TEM micrograph from a $Zr_{62.5}Cu_{22.5}Al_{10}Fe_5$ BMG alloy annealed for 10 min at 713 K, when eutectic crystallization had occurred.

5.3.3 Primary Crystallization

In this mode, a supersaturated solid solution, for example, α-Fe, forms first from the glassy phase, indicated by Reaction 2. Since the concentration of the solute in the α-Fe phase is lower than that in the glassy phase, the solute (boron) atoms are rejected into the glassy phase and, consequently, the remaining glassy phase becomes enriched in B until further crystallization is stopped. At this stage, a metastable equilibrium is established between α-Fe and the glassy Fe–B phase with the new composition. This B-enriched Fe–B glassy phase can transform later or at higher temperatures by one of the mechanisms described previously. For example, if the B concentration is close to 33 at.%, then a polymorphous crystallization event could occur, resulting in the formation of Fe_2B. Alternatively, if the B concentration in the new glassy phase is 25 at.%, then polymorphous crystallization to Fe_3B could occur. On the other hand, if the B concentration is different from these two values, then the glassy phase could crystallize in a eutectic mode.

Primary crystallization has been observed to be the main mode of transformation in many metallic glasses. The crystalline phase formed can be either a terminal solid solution or an intermediate phase. For example, in the Ti-based metallic glasses, the first phase to precipitate out during primary crystallization is the β-Ti (bcc) phase. Similarly, in Fe-based alloys, the α-Fe (bcc) phase precipitates out during primary crystallization. The morphology of the primary crystals is highly dependent on composition, and ranges from spherical to highly dendritic. Further, the growth rate of the primary crystals depends on their morphology. In the absence of interfacial instabilities, the growth is parabolic with time, that is, the mean radius of the crystals increases linearly with the square root of time.

Figure 5.7c presents a bright-field TEM micrograph obtained on heating a $Ti_{50}Ni_{20}Cu_{23}Sn_7$ BMG alloy in a DSC at a heating rate of 40 K min^{-1} up to 787 K. Note the distribution of fine crystallites in a glassy matrix.

Even though the different modes of crystallization have been described with respect to the Fe–B system, *all* metallic glasses crystallize according to one or more of these three types of reactions. The nature of the reaction depends on the composition of the alloy, the thermodynamic driving force for crystallization, and also the activation barrier to, and therefore kinetics of, each reaction. The details of growth rates and the morphologies of phases obtained in the different modes of crystallization have been reviewed earlier and can be found in Refs. [9,10].

5.4 Differences in Crystallization Behavior between Melt-Spun Ribbons and Bulk Metallic Glasses

BMGs are also metastable in nature. An important difference between the melt-spun metallic glassy ribbons and BMGs is that the critical cooling rates required for the formation of glassy phases are different in the two cases. While melt-spun metallic glasses are usually obtained by solidifying their metallic melts at a fast rate ($>10^5$–10^6 K s^{-1}), BMGs are synthesized at solidification rates slower than about 10^2–10^3 K s^{-1}. A consequence of this difference is that the melt-spun metallic glasses are much thinner in cross section and BMGs have a larger section thickness. Both the lower critical cooling rate and larger section thickness in BMGs are a result of the presence of a large number of components in the bulk glass-forming alloy. Consequently, the melt-spun metallic glass ribbons solidified at higher cooling rates are farther from equilibrium than the BMGs. This larger departure from equilibrium is manifested in a larger decrease in density and higher energy stored in the melt-spun ribbons. The extent of departure from equilibrium can be established with the help of DSC methods. Accordingly, DSC is an ideal tool to determine the basic differences in the energy stored in the melt-spun glassy ribbons and BMGs. One would also expect that, due to the larger departure from equilibrium, the kinetics of crystallization in melt-spun glassy ribbons would be faster than that in BMGs. But, as will be shown later, this is not necessarily true.

It is clear from Figure 5.1 that BMGs exhibit a glass transition temperature, T_g, in their DSC curves. The significance of this temperature is that once heated above it, the glass becomes a supercooled liquid (but still exists in the form of a "solid"). At this stage, there is no difference in the "structure" between the BMG and the melt-spun metallic glass that was obtained directly by rapidly solidifying the metallic melt, except that the extent of structural relaxation is different in the two glasses. In other words, the nature of the glass above T_g is the same irrespective of whether we had started with the melt-spun glassy ribbon or slowly solidified BMG. Therefore, once the BMG has been heated above T_g, its crystallization behavior will be identical to that of melt-spun metallic glassy ribbons (assuming that both glasses have the same chemical composition). This is why there have not been many detailed studies on the crystallization behavior of BMGs. Instead, researchers have been preparing melt-spun metallic glassy ribbons of the same composition as the BMG, and studying their crystallization behavior, with the understanding that the processes occurring in the two different samples are identical. It may be noted that it is also easier to prepare thin foil specimens for TEM investigations from melt-spun ribbons than from BMG samples.

Illeková et al. [63] produced metallic glasses in the $Zr_{55}Ni_{25}Al_{20}$ system by two different methods. One was by planar flow casting (30 μm thick ribbons at a solidification rate of about 10^5 K s^{-1}) and the other by quenching in flowing water (9 mm diameter rod samples at a solidification rate of about 10^2 K s^{-1}). A comparison of their thermodynamic states measured through DSC, as listed in Table 5.2, shows that the enthalpy of structural relaxation

TABLE 5.2
Comparison of the Thermal Properties of Rapidly Solidified and Bulk Metallic Glass Samples

Composition	Technique	Section Thickness	T_g (K)	T_x (K)	ΔH_{relax} (J g^{-1})	$\Delta H_{first\ exotherm}$ (J g^{-1})	ΔH_{cryst} (J g^{-1})	References
$Zr_{55}Ni_{25}Al_{20}$	Planar flow casting	30 μm thick ribbon	805	820	−34.79	—	−71.43	[63]
$Zr_{55}Ni_{25}Al_{20}$	Water quenching	9 mm dia rod	738	795	−10.32	—	−75.90	[63]
$Mg_{65}Cu_{15}Y_{10}Ag_{10}$	Melt spinning	45 μm thick ribbon	428	469	—	42.3	129.3	[64]
$Mg_{65}Cu_{15}Y_{10}Ag_{10}$	Injection casting	6 mm dia rod	428	469	—	36.7	120.8	[64]
$Mg_{65}Cu_{15}Y_{10}Ag_{10}$	Squeeze casting	10 mm dia rod	428	469	—	41.5	124	[64]

(ΔH_{relax}) was lower for the BMG sample. By measuring the full width at half maximum (FWHM) of the first diffuse peak in the XRD patterns of the slowly cooled and rapidly quenched specimens, the authors noted that the size of the coherently diffracting domains in both cases was approximately the same, namely, 1.35 nm. Thus, they attributed the difference in the glassy structures between the ribbon and rod samples to the higher degree of short-range order (SRO) within these domains in the ribbon sample. From these observations, it was concluded that the ribbon and bulk samples represented the same glassy states, differing mainly in the degree of departure from thermodynamic equilibrium. The characteristic SRO decreased approaching the "ideal" glassy state in the bulk sample.

It is also interesting that, in the foregoing investigation, the authors reported the T_g and T_x temperatures to be different in the ribbon and rod samples. This is, however, contrary to the observations reported by several authors and in different alloys that the T_g and T_x temperatures are independent of the section thickness of the glassy specimens.

A somewhat different result was reported by Kang et al. [64], who investigated the glassy phase in the $Mg_{65}Cu_{15}Y_{10}Ag_{10}$ alloy system obtained in three different forms and by three different methods—45 μm thick ribbons by melt spinning, 6 mm diameter rods by injection casting, and 10 mm diameter rods by the squeeze-casting method. They noted that neither the heat of crystallization for the first exothermic peak nor the total heat of crystallization had followed any trend with the cooling rate. For example, the heat of crystallization was highest for the melt-spun ribbon and least for the injection mold-cast 6 mm rod. It was in between these two values for the squeeze-cast 10 mm diameter rod. A similar trend was followed for the heat of crystallization based on the first exothermic peak as well (Table 5.2). If we ignore the small differences in these values, which are only about 7% between the melt-spun ribbon and the squeeze-cast bulk rod, one can assume that these values are almost the same. In fact, these values are expected to be the same. This is because, as mentioned earlier, once the glassy phase is heated to a temperature above T_g, the glass is completely relaxed, and it becomes the supercooled liquid. In this condition, the initial state from which it came to this stage is unimportant. Therefore, irrespective of the initial condition of the glassy sample, the heat of crystallization should be the same. But, the structural relaxation behavior is expected to be different because this happens to occur below T_g. We will see more about this later in the chapter.

Qin et al. [65] studied the crystallization behavior of the $Ni_{45}Ti_{23}Zr_{15}Si_5Pd_{12}$ metallic glass in both 50 μm thick ribbon and 1.5 mm diameter rod forms. They noted that the apparent activation energy for the onset of crystallization, determined using the Kissinger method, was slightly lower for the ribbon sample than for the rod sample. Since the activation energy for the onset of crystallization is related to the nucleation process [66], it was concluded that the nucleation process was easier in the ribbon sample, probably because of the large surface area available. A higher activation energy for the crystallization peak was also noted for the ribbon sample, suggesting that the growth process was more difficult. They also noted that the incubation time for crystallization was shorter for the ribbon samples.

It has recently been reported [67] that the kinetics of crystallization could be different between the melt-spun ribbons and the BMG rods. By conducting careful TEM analysis of the phases formed and the kinetics of transformation in samples isothermally transformed, these authors showed that even though the phases formed were the same in both melt-spun ribbons and BMG rods, the kinetics of crystallization at any given temperature were slower in the BMG alloy than in the ribbon sample.

Since BMGs contain a large number of components, the number of crystalline phases present under equilibrium conditions could also be quite large. Consequently, the XRD patterns will be quite complex and, therefore, it would not be easy to differentiate between the different crystalline phases and identify them unambiguously. Further, many of the intermetallic phases formed in these multicomponent alloy systems are quite complex in their crystal structures. Therefore, additional care needs to be taken for a clear and unmistakable identification of the different phases present in a fully crystallized BMG specimen. One solution to this problem would be to conduct detailed TEM studies. By observing the microstructural features and recording selected area diffraction (SAD) patterns from the fine features (using the nanobeam method, if the features are extremely fine), one could determine the crystal structure details unambiguously. Additionally, the determination of the composition of the individual phases with the help of the EDS method, assists in clearly identifying the individual phases. Thus, TEM investigations provide information about the microstructure, crystal structure, and chemical composition of the individual phases present in the samples.

Let us now look into the details of the crystallization behavior of BMGs. It should be cautioned, from the beginning, that a lot of information about the thermodynamics and kinetics of crystallization was gathered from extensive studies of the crystallization behavior of melt-spun metallic glass ribbons during the 1980s and 1990s. Information about the nature of the crystallization products, activation energies for crystallization, and the mechanisms of transformation is available, for example, in Refs. [9,10]. Consequently, there have not been many detailed investigations on the crystallization behavior of BMGs. But, before we discuss the crystallization behavior of BMGs, let us discuss the effects of the nature and amount of solute elements on the transformation temperatures of metallic glasses.

5.5 Thermal Stability of Metallic Glasses

BMGs generally have a high thermal stability, as indicated by the presence of a wide supercooled liquid region, ΔT_x. Values for ΔT_x have also been listed for some BMG alloy compositions in Table 5.1. Attempts have been made to explain the thermal stability of metallic glasses in terms of their transformation temperatures. The glass transition (T_g) and crystallization (T_x) temperatures of $Zr_{65}Al_xCu_{35-x}$ glassy samples (the subscripts represent the composition in atomic percentage) are presented in Figure 5.8 [68]. It should be noted that while T_g increases continuously and almost linearly with the Al content (but with a small increase at 7.5 at.% Al), the T_x value increases, shows a maximum at 7.5 at.% Al, and then

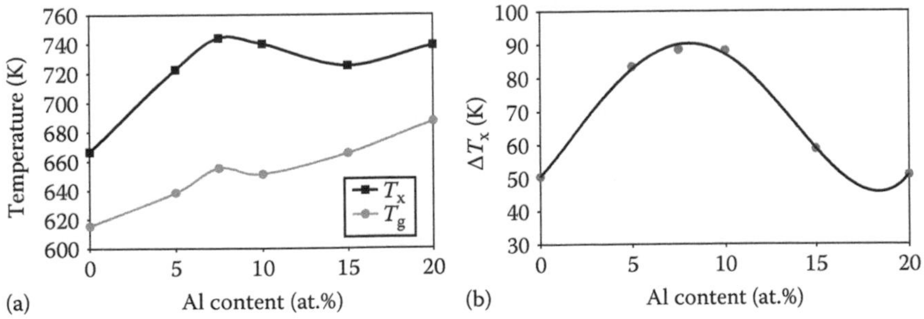

FIGURE 5.8
Variation of (a) T_g and T_x temperatures and (b) the width of the supercooled liquid region ΔT_x (=$T_x - T_g$), with Al content in the $Zr_{65}Al_xCu_{35-x}$ glassy alloys. (Reprinted from Inoue, A. et al., *Mater. Sci. Eng. A*, 178, 255, 1994. With permission.)

decreases with further increase in the Al content. This maximum value of T_x at $x = 7.5$ at.% Al suggests the presence of a maximum in ΔT_x at this composition. That is, crystallization is most retarded when the Al content in the alloy is 7.5 at.%. Reasons for this retardation of crystallization were sought by conducting isothermal annealing studies on the binary $Zr_{67}Cu_{33}$ and ternary $Zr_{65}Al_{7.5}Cu_{27.5}$ and $Zr_{65}Al_{20}Cu_{15}$ alloys. A single exothermic peak was observed in all the alloys and the incubation time (τ) for the precipitation of crystalline phases was also measured.

It was observed that, irrespective of the annealing temperature, crystallization in the binary alloy always occurred by a polymorphous mode through the formation of the Zr_2Cu phase, and in the ternary alloys by a eutectic mode through the simultaneous precipitation of $Zr_2(Cu,Al)$ and ZrAl phases [68]. An Arrhenius plot of the incubation time ($\ln \tau$) versus the reciprocal of the isothermal annealing temperature for the three alloys studied is presented in Figure 5.9. The τ values shifted from the linear variation to higher temperatures, above T_g, only for the ternary $Zr_{65}Al_{7.5}Cu_{27.5}$ alloy, and not for the others, suggesting that precipitation of the crystalline phases from the supercooled liquid was suppressed only in this ternary alloy and not in the others. A similar deviation was also noted in the variation of the viscosity, η, with temperature calculated from the equation:

$$\tau \propto \frac{T\eta}{(T_\ell - T)^2} \tag{5.7}$$

where:

T_ℓ is the liquidus temperature of the alloy
T is the temperature at which measurements were made [69]

From these variations, it could be concluded that the supercooled liquid region in the $Zr_{65}Al_{7.5}Cu_{27.5}$ alloy possesses highly dense randomly packed structures since it contains elements with significantly different atomic sizes, which also possess attractive atomic bonding among them. This leads to an increase in the viscosity of the alloy and consequently it

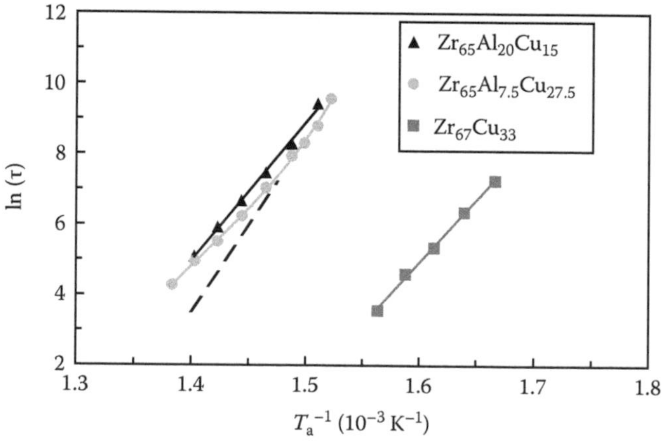

FIGURE 5.9
Arrhenius plot of the incubation time, τ, for the precipitation of crystalline phases in the binary $Zr_{67}Cu_{33}$, and ternary $Zr_{65}Al_{7.5}Cu_{27.5}$ and $Zr_{65}Al_{20}Cu_{15}$ alloys. Note the deviation of τ to the positive side of the linear variation (to higher temperatures) only for the ternary $Zr_{65}Al_{7.5}Cu_{27.5}$ alloy, signifying the delayed crystallization in the alloy with 7.5 at.% Al. Such a deviation is not observed for the other alloys. (Reprinted from Inoue, A. et al., *Mater. Sci. Eng. A*, 178, 255, 1994. With permission.)

is difficult to have long-range diffusion, which is necessary for solute redistribution to form crystalline phases. Thus, the increased viscosity and the difficulty in forming crystalline phases, due to the presence of dense randomly packed structures, explain the increased width of the supercooled liquid region and the thermal stability of this alloy.

Peak temperatures for the nucleation and growth reactions of the crystalline phases forming from the glassy alloy were also evaluated for the binary $Zr_{67}Cu_{33}$ and ternary $Zr_{65}Al_{7.5}Cu_{27.5}$ alloys. Both glassy alloys were pre-annealed for 60 s at different annealing temperatures, T_a. The glassy samples were heated to the annealing temperature, T_a, at a heating rate of 0.17 K s^{-1} (10 K min^{-1}), annealed there for 60 s (1 min), and then cooled to room temperature. The peak crystallization temperatures, T_p, of the exothermic reaction were then determined from the DSC plots obtained using a heating rate of 0.17 K s^{-1} (10 K min^{-1}). The difference in the reciprocals of the peak temperatures between the pre-annealed (T_p) and as-quenched $\left(T_p^o\right)$ samples was plotted against the annealing temperature and the result is shown in Figure 5.10. It may be noted that the difference in the reciprocal temperatures shows a peak only for the ternary alloy, which exhibits a large supercooled liquid region of 90 K, and consequently a high GFA [70]. This peak corresponds to a temperature of 688 K. Such a peak is not observed for the binary alloy, with a relatively smaller value of ΔT_x of only 50 K. A peak temperature in such a plot was earlier reported [71] to correspond to the peak temperature of the nucleation event of a crystalline phase formation.

The crystallization temperatures for the binary and ternary alloys in Figure 5.10 were measured at a very high heating rate of 5.33 K s^{-1} (320 K min^{-1}) and, thus, these temperatures could be considered to correspond to the maximum growth rates, that is, the growth temperatures. Thus, it can be clearly seen from Figure 5.10 that the binary $Zr_{67}Cu_{33}$ alloy crystallizes at a temperature just above the maximum temperature of 670 K. That is, the difference between the temperatures corresponding to the maximum nucleation rate and maximum growth rate is very small. Therefore, crystallization of the glassy alloy becomes easy. On the other hand, for the ternary $Zr_{65}Al_{7.5}Cu_{27.5}$ alloy, the difference between the maximum nucleation and maximum growth temperatures is as much as 143 K, resulting in enhanced resistance to crystallization. From these observations, it was concluded that the large ΔT_x value, and hence the high thermal stability, for the ternary $Zr_{65}Al_{7.5}Cu_{27.5}$ alloy results from the difficulty in both forming and growing crystalline nuclei.

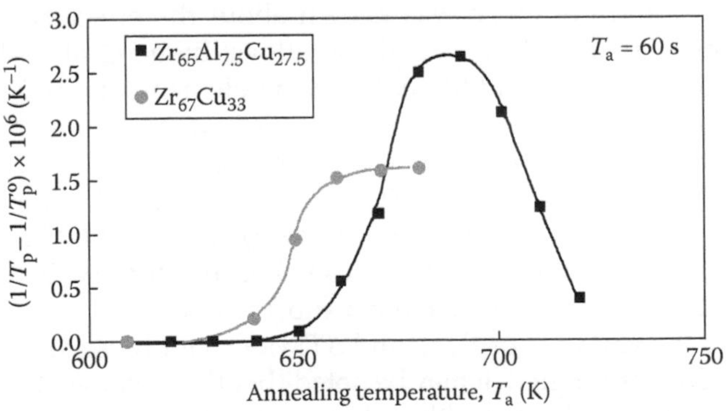

FIGURE 5.10

Variation of $\left(\frac{1}{T_p}-\frac{1}{T_p^0}\right)$ as a function of the pre-annealing temperature, T_a for the binary $Zr_{67}Cu_{33}$ and ternary $Zr_{65}Al_{7.5}Cu_{27.5}$ alloys. Note that the difference in the peak temperatures between the as-quenched and pre-annealed samples shows a peak only for the ternary alloy and not for the binary alloy.

From the nucleation and growth behavior of these two glassy alloys, it is expected that an increase in the heating rate will not significantly increase the grain size of the crystalline phase in the binary $Zr_{67}Cu_{33}$ alloy, whereas the grain size should be considerably larger in the ternary $Zr_{65}Al_{7.5}Cu_{27.5}$ alloy, due to the presence of fewer nuclei. This has been confirmed experimentally [72].

5.6 Crystallization Temperatures and Their Compositional Dependence

As mentioned earlier, the transformation temperatures of glasses—glass transition (T_g), crystallization (T_x), and liquidus (T_ℓ) temperatures—are measured routinely to determine their thermal stability and the temperatures at which relaxation and crystallization behavior can be studied, and also to conduct heat treatments to observe microstructural variations.

Unlike the melting temperature of a metal, the crystallization temperature (T_x) of a metallic glass is not a thermodynamic parameter; it is a kinetic value. Given sufficient time, the metallic glasses will transform into the crystalline stable phase(s) at any temperature. But, for all practical purposes, metallic glasses can be considered as "stable" indefinitely at room temperature. T_x is not a well-defined temperature, since it depends upon the rate at which the glass is heated to determine the crystallization temperature. Further, it is also dependent upon the thermal history of the glass, the way the glass was prepared, and other parameters. Despite this, researchers have been reporting the crystallization temperatures of metallic glasses in the literature. It is also important to remember that the T_x value measured this way is not the safe temperature up to which the glass could be used without any transformation occurring. But, practically, it is only used to estimate the thermal stability of the glass.

It is easy to realize that the higher the T_x, the higher the thermal stability of the glass. Further, it is easier to form the glass if its T_g is also high. Therefore, Inoue [73] has identified the difference between T_x and T_g, ΔT_x, as a measure of the stability of the glass produced. Accordingly, there have been many investigations to determine these temperatures.

It was shown earlier that the T_g of ternary $Zr_{65}Al_xCu_{35-x}$ glassy alloys increases with increasing Al content (Figure 5.8). Since Zr-based glassy alloys happen to be an important group of BMGs, and not much was known about the effect of additional alloying elements on the thermal stability of the glassy alloys, Inoue et al. [74] conducted a systematic study of the influence of additional alloying elements in the $Zr_{65}Al_{10}Cu_{15}Ni_{10}$ system, which exhibited a ΔT_x value of 107 K. The alloying additions (M) chosen were Ti, Hf, V, Nb, Cr, Mo, Fe, Co, Pd, and Ag. They were added to the alloy system in the pattern $Zr_{65-x}Al_{10}Cu_{15}Ni_{10}M_x$ (for M = Ti and Hf), $(Zr_{0.65}Al_{0.1}Cu_{0.15}Ni_{0.1})_{100-x}M_x$ (for M = V, Nb, Cr, and Mo), $Zr_{65}Al_{10}Cu_{15-x}Ni_{10}M_x$ (for M = Fe, Co, Pd, and Ag), and $Zr_{65}Al_{10}Cu_{15}Ni_{10-x}M_x$ (where M = Fe, Co, Pd, or Ag). This combination was chosen because these substitution modes gave rise to large ΔT_x values in comparison to other substitutions.

Figure 5.11 shows the variation of T_g, T_x, and ΔT_x values for these four groups of substitutions. From Figure 5.11a through d, it may be noted that the T_g value increases with increasing amounts of the alloying element, although the magnitude of the increase was relatively small when Ti or Hf was added. The T_x values always decreased, with the caveat that the magnitude of the decrease was again dependent on the type of solute added. The decrease was gradual and small in most cases, except when Ti was added; the decrease in T_x was substantial in this case. The only exception to this trend was an increase in the T_x value when Hf was added to the $Zr_{65-x}Al_{10}Cu_{15}Ni_{10}M_x$ alloy. As a result of these opposing trends

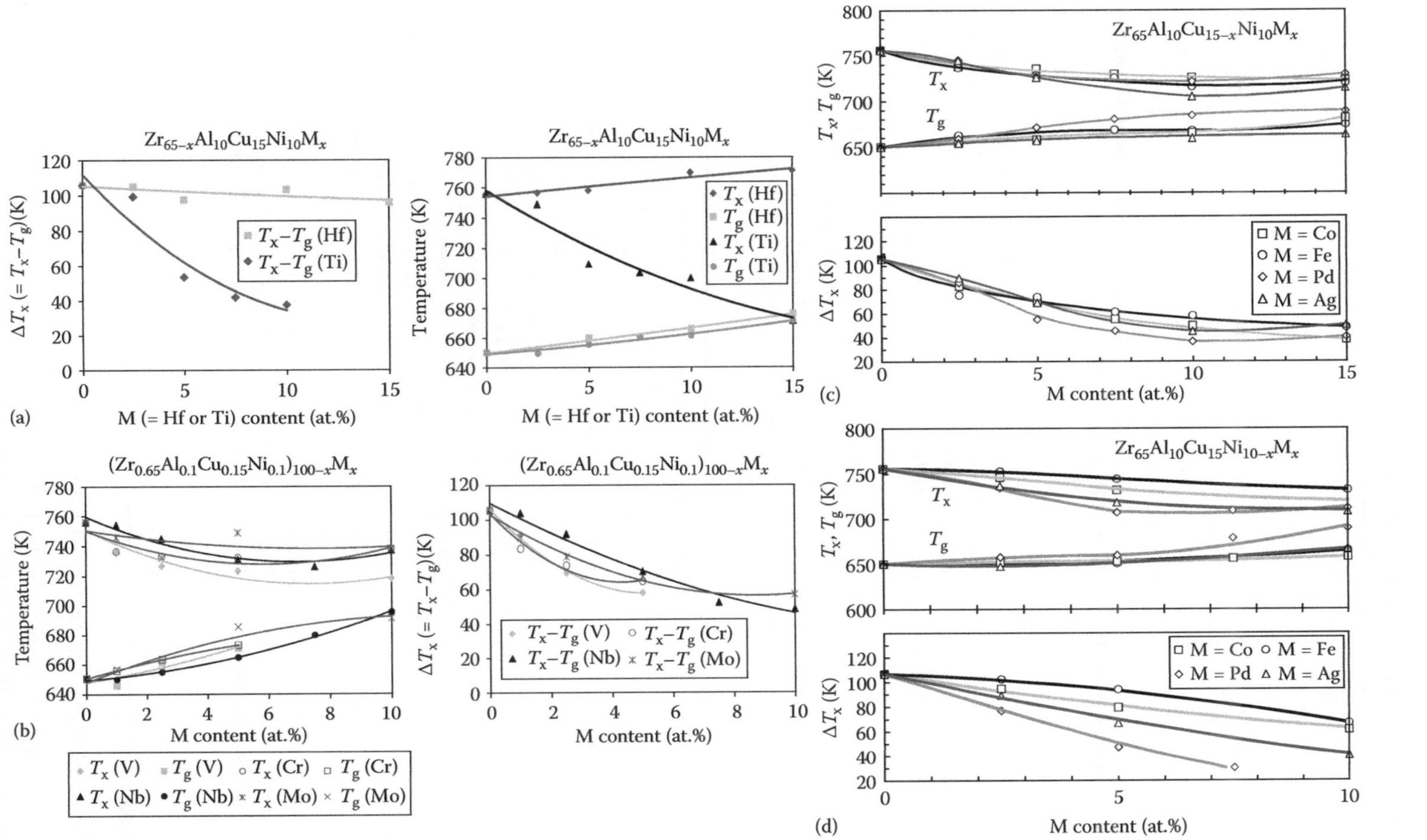

FIGURE 5.11
Variation of T_g, T_x, and ΔT_x with M content for melt-spun glassy alloys of the four groups. (a) $Zr_{65-x}Al_{10}Cu_{15}Ni_{10}M_x$ (where M = Ti and Hf), (b) $(Zr_{0.65}Al_{0.1}Cu_{0.15}Ni_{0.1})_{100-x}M_x$ (where M = V, Nb, Cr, and Mo), (c) $Zr_{65}Al_{10}Cu_{15-x}Ni_{10}M_x$ (where M = Fe, Co, Pd, and Ag), and (d) $Zr_{65}Al_{10}Cu_{15}Ni_{10-x}M_x$ (where M = Fe, Co, Pd, and Ag). The T_g and T_x of 0.67 K s^{-1} (40 K min^{-1}). (Reprinted from Inoue, A. et al., *Mater. Trans., JIM*, 36, 1420, 1995. With permission.)

in T_x and T_g with the alloying additions, the ΔT_x values always decreased with an increasing amount of the alloying addition. The decrease in the ΔT_x values for these alloy systems can be attributed to a significant decrease in T_x rather than a slight increase in the T_g value.

According to the Inoue criteria [73], a large value of ΔT_x is directly related to an increased GFA of the alloy, even though some exceptions have been noted, as described in Chapter 3. The increased GFA has been explained on the basis of the existence of dense randomly packed structures in the liquid state. It is easy to realize that the occurrence of such densely packed structures requires strong atomic bonding between the constituent elements. However, based on the heats of mixing [75], it can be seen that, in the case of the alloying additions considered here, V, Nb, Cr, Mo, Fe, and Co are highly repulsive to Cu atoms, while Pd and Ag with Cu, and Ti and Hf with Zr are only weakly repulsive. Thus, none of the elements exhibit strong attractive bonds with any of the constituents of the base alloy system (Zr, Cu, Al, and Ni) and, consequently, it is difficult to form densely packed random structures. That is why the width of the supercooled liquid region was found to decrease with increasing amounts of the alloying elements. On the other hand, if alloying elements are chosen that have a strong attractive force with the constituents of the base alloy system, then it is possible to form dense random structures and, as a result, increase the width of the supercooled liquid region. Thus, the bonding nature between the alloying elements and the constituents in the base alloy could be a useful guide in selecting alloying elements to increase the width of the supercooled liquid region and, eventually, the GFA.

Zhou et al. [76] investigated the effect of Ag addition on the thermal stability and crystallization behavior of $Zr_{70-x}Cu_{12.5}Ni_{10}Al_{7.5}Ag_x$ in the range $x = 0$–16 at.%, and noted that both T_g and T_x increased with increasing Ag content. They reported that while the ΔT_x was 72 K in the Ag-free alloy, it increased to 110 K when the Ag content was 16 at.%. They also noted that when the Ag content was low (0–10 at.%), crystallization took place in two or three stages with primary precipitation of the icosahedral (*i*) phase. But, when the Ag content was increased to 12–16 at.%, crystallization occurred in a single stage with the formation of Zr_2Cu and/or Zr_2Ni.

The factors that affect the value of T_x in metallic glasses are not clear. A number of different possibilities have been proposed, the atomic size being chief among them. It is suggested that the T_x value is higher when the atomic size of the alloying element in the Fe–Ni-based metal–metalloid-type metallic glasses is larger than that of Fe or Ni. Exactly the opposite result was reported in actinide-transition metal–metal-type metallic glasses. A reduction in free volume and diffusivity were suggested as reasons for a change in the T_x value. Other factors such as electron-to-atom ratio and differences in electronegativity were also considered to explain the variation of T_x with solute content. A brief summary of these effects on melt-spun ribbons was given by Scott [10]. No such systematic investigations have been carried out on BMG alloys.

The compositional ranges in which glassy phases have been obtained and the variation of T_x and ΔT_x in several ternary and higher-order alloy systems based on La, Mg, Zr, and Pd were discussed earlier by Inoue [77] and so they will not be repeated here.

5.7 Annealing of Bulk Metallic Glasses

When a BMG (or for that matter, any metallic glass) is annealed (heated to higher temperatures, maintained at that temperature for a given length of time and cooled slowly down to room temperature), a number of different reactions can take place in the glass.

With increasing annealing temperature, the glass may exhibit structural relaxation, phase separation, and then crystallization. While structural relaxation takes place at temperatures below T_g, phase separation in the glassy alloy can take place at temperatures around T_g (either below T_g, at T_g, or above T_g). On the other hand, crystallization generally takes place at temperatures above T_x. All these reactions may not take place in all the alloy systems. In fact, the observation of phase separation has not been very frequent, as will be shown later in the chapter. The nature of the transformation of the glassy phase depends on the temperature at which the glass is annealed and the types of constituents present in the alloy system. For example, if annealing is carried out at a high temperature, above T_x, crystallization will occur and it will not be possible to study the structural relaxation or phase separation behavior. On the other hand, if annealing is carried out at a very low temperature, it will be possible to study the structural relaxation in the glass, but its crystallization may take an unusually long time, and not occur on the timescale where experimental observations can be made. Figure 5.12 shows schematically the different pathways for the transformations that can take place in a metallic glassy alloy on heating.

It is clear from Figure 5.12 that the as-quenched metallic glass will start relaxing on being heated to higher temperatures to a more "stable" or "ideal" glass. As will be explained, this process of structural relaxation will be complete by the time the glass reaches the T_g temperature. Therefore, the relaxed glass is the starting point for all the transformations to be considered. This relaxed glass will transform first into the supercooled liquid state at T_g,

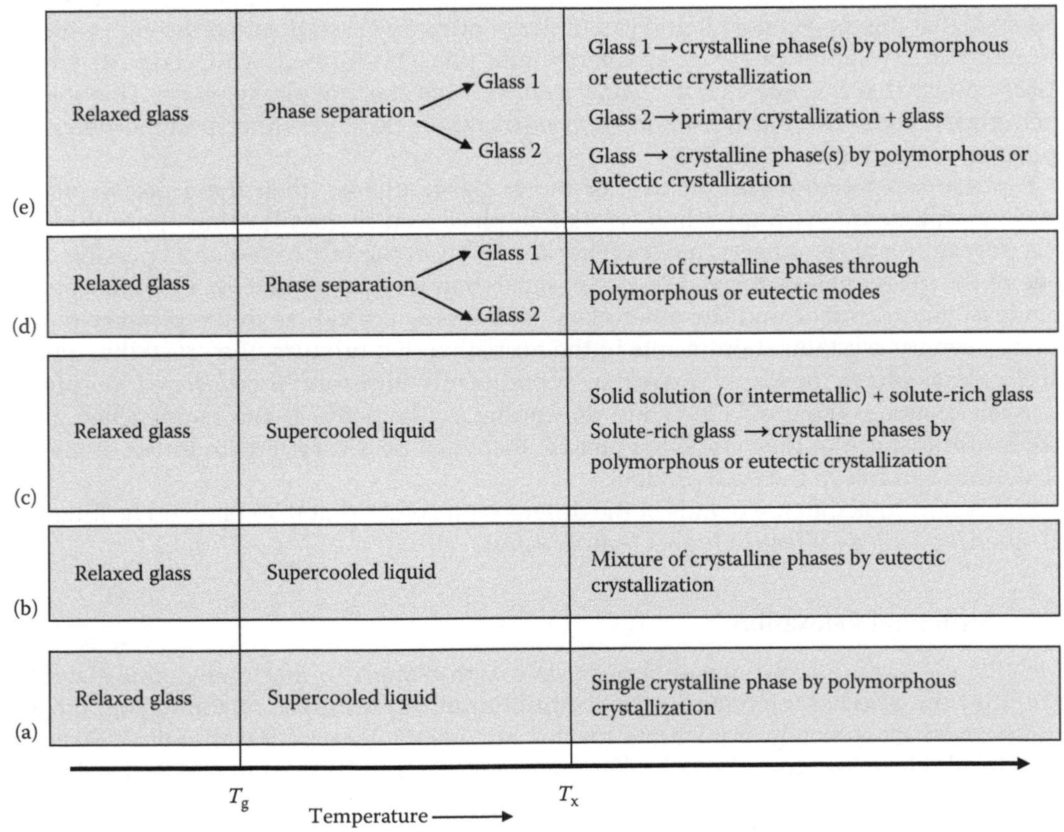

FIGURE 5.12
Different pathways for a metallic glass to crystallize into the equilibrium phases.

provided that one is able to differentiate between T_g and T_x. This is the case in all BMGs, and is also true in some melt-spun ribbon samples, notably those based on Pd and Pt [33,34].

The supercooled liquid can undergo phase separation into two different glassy phases with two different compositions. The phenomenon of phase separation is manifested by the presence of two T_gs in the DSC curve of the glassy alloy and also by the appearance of an interconnected structure in the electron micrographs—usually one of the phases appears lighter in contrast than the other. These two different phases will have widely differing compositions, and these have been measured in recent times with the help of a three-dimensional atom probe (3DAP) technique. It is also important to realize that phase separation does not occur in all metallic glasses. In fact, it is rare and has not been observed very frequently.

Depending on whether we are starting with the supercooled liquid or the phase-separated mixture of two glasses, further transformation to the crystalline phases may take place in different ways. We will consider these in two different categories—the first starting with the supercooled liquid and the second starting with the phase-separated glassy phase mixture.

Assuming that phase separation does not occur, the supercooled liquid, on further heating, may directly transform into a single crystalline phase through a polymorphous mode (Figure 5.12a), although this is very rare and has not been frequently reported in the literature so far. More frequently, the supercooled liquid will transform into a mixture of crystalline phases through the eutectic mode of crystallization (Figure 5.12b). It is also possible that the supercooled liquid can undergo primary crystallization through which a solvent-rich crystalline phase (or an intermetallic phase) is formed, which coexists with a glassy phase that is solute-rich in comparison with the starting glassy phase. This solute-rich glassy phase can further undergo crystallization through either polymorphous or eutectic modes (Figure 5.12c).

If phase separation occurs in two (or more) glassy phases, then these glassy phases may crystallize into two (or more) crystalline phases on further heating either through the polymorphous or eutectic modes (Figure 5.12d). On the other hand, it is possible that one of the glassy phases formed on phase separation may crystallize in a polymorphous mode or eutectic mode and the other glassy phase may crystallize in the primary mode. Since primary crystallization results in the formation of a mixture of a crystalline phase and a glassy phase, the glassy phase may subsequently transform through polymorphous or eutectic modes (Figure 5.12e). Thus, depending on the pathway the as-quenched glass takes to transform to the crystalline phases, there can be a very large number of microstructural variants in the final product.

Let us now look into the details of the different processes that take place on heating the as-quenched glassy phase to higher temperatures.

5.7.1 Structural Relaxation

Metallic glasses are in a nonequilibrium state with respect to thermodynamic stability. But, they are also not in configurational equilibrium. Accordingly, on annealing, the as-synthesized glass slowly transforms toward an "ideal" glass of lower energy through structural relaxation. To put this in the proper perspective, let us compare an ideal crystalline material with a real crystalline material. The real crystalline material contains a variety of defects, which include vacancies, dislocations, grain boundaries, stacking faults, and so on. Thus, a real material can be considered as an ideal material containing a variety of defects, whose character can be identified and well quantified. The as-synthesized glass can be similarly considered as an "ideal" or "perfect" glass containing defects. However, the nature of defects present in the glass is presently unknown, and all the defects put together are considered as free volume. Thus, the structural relaxation of a quenched glass results in the annihilation of these defects leading to the formation of an ideal glass.

In comparison to the melt-spun glassy ribbons, BMGs exhibit large ΔT_x values. This temperature range provides an opportunity to quantitatively examine the amount of free volume present in the glass as a function of temperature. It is also possible to investigate whether the glass will achieve a fully relaxed state as a result of annealing it in the supercooled liquid state. Most importantly, it allows the glass (the supercooled liquid) to be investigated near T_g without allowing crystallization to take place.

When a glass is annealed in the supercooled liquid region, the amount of free volume varies with the annealing temperature and reaches an equilibrium value typical of that temperature. The mechanism(s) by which structural relaxation occurs in glasses is through the annihilation of "defects" or free volume, or a recombination of the defects of opposing character, or by changes in both topological and chemical SRO [78].

Let us now see what we mean by chemical short-range order (CSRO) and topological short-range order (TSRO). Since most metallic glasses are alloys, and therefore contain at least two components, it is possible that a truly random distribution of the constituent atoms is not obtained even by RSP methods. This means that the chemical environment (chemical composition) around the constituent atoms is different from the average composition. Such a variation in chemical composition from point to point is referred to as CSRO. On the other hand, when the atomic configuration is different due to the collective motion of groups of atoms, the glass is then said to possess TSRO. In a fully relaxed glass that has attained the "equilibrium" free volume concentration, it is possible to separate the effects due to CSRO and TSRO.

Metallic glasses, especially those that have been obtained at high solidification rates by RSP methods, contain a large frozen-in structural disorder and, therefore, possess high atomic diffusivities and can undergo structural relaxation even at low temperatures. For example, Pd–Si, Fe–B, and Zr–Cu binary glassy ribbons have been reported to undergo structural relaxation just above room temperature; several hundred degrees below T_g. But more commonly, structural relaxation is investigated in two temperature regimes below T_g, which are clearly distinguishable from each other. Thus, the relaxation processes are studied either in the low temperature (sub-sub-T_g, i.e., $T_g - 200\text{ K} < T_a < T_g - 100\text{ K}$) or high-temperature (sub-T_g, i.e., $T_a \geq T_g - 100\text{ K}$) regimes, where T_a is the annealing temperature.

Structural relaxation in metallic glasses can be achieved by a low-temperature annealing process, which does not cause crystallization. But, significant changes in physical properties (and mechanical properties in some cases) have been reported. For example, relaxed glasses exhibit a decreased specific heat, reduced diffusivity, reduced magnetic anisotropy, increased elastic constants (by about 7%), significantly increased viscosity (by more than five orders of magnitude), and loss of (bend) ductility in some, in addition to changes in electrical resistivity (by about 2%), Curie temperature (by as much as 40 K), enthalpy (by about 200–300 cal mol^{-1}), superconductivity, and several other structure-sensitive properties [78]. Another important property that changes on structural relaxation is the density. A small increase in density (about 0.5% for melt-spun ribbons [79] and a smaller value of about 0.1%–0.15% for BMG alloys [80,81]) was also reported to occur due to structural relaxation in metallic glasses. Table 5.3 lists the differences in the densities of as-solidified and structurally relaxed metallic glasses based on Pd and Zr.

Structural relaxation in metallic glasses has been investigated using DSC, XRD, electrical resistivity, magnetic measurements, creep experiments, internal friction measurements using a torsion pendulum, Mössbauer spectroscopy, and dynamic mechanical analysis

TABLE 5.3

Changes in the Bulk Densities, ρ (g cm^{-3}) of Metallic Glassy Alloys in the As-Solidified and Structurally Relaxed Conditions

Alloy Composition	Synthesis Method	Rod Diameter (mm)/ Ribbon Thickness	$\rho_{as\text{-}solidified}$	$\rho_{relaxed}$	$\Delta\rho_{relaxed}$ (%)	References
$Pd_{77.5}Cu_6Si_{16.5}$	Melt spinning	30 μm thick ribbon	10.46	10.51	0.48	[79]
$Pd_{40}Cu_{30}Ni_{10}P_{20}$	Melt spinning	40 μm thick ribbon	9.318	9.337	0.2	[80]
$Pd_{77.5}Cu_6Si_{16.5}$	Water quenching	2 mm dia rod	10.48	10.51	0.29	[79]
$Pd_{40}Cu_{30}Ni_{10}P_{20}$	Cu mold casting	5 mm dia rod	9.27	9.28	0.11	[81]
$Zr_{55}Cu_{30}Al_{10}Ni_5$	Cu mold casting	5 mm dia rod	6.82	6.83	0.15	[81]

Note: $\Delta\rho_{relaxed} = \dfrac{\rho_{relaxed} - \rho_{as\text{-}solidified}}{\rho_{as\text{-}solidified}}$.

methods. Of these, electrical resistivity measurements and DSC have been the most popular techniques, while Mössbauer spectroscopy methods have been used to determine the atomic environments. It has been reported that the contribution of CSRO processes toward the electrical resistance is relatively small and that it is TSRO that essentially determines the resistance curve [82]. Egami [83] has summarized the use of diffraction techniques (x-ray, neutron, and electron scattering methods) to study structural relaxation in metallic glasses.

Dmowski et al. [84] investigated the structural relaxation behavior of a $Zr_{52.5}Cu_{17.9}Ni_{14.6}Al_{10.0}Ti_{5.0}$ BMG alloy through a combination of hardness measurement, DSC, and time-of-flight neutron diffraction techniques. The authors annealed the as-quenched BMG sample at 630 K (T_g and T_x for this glass were reported to be ~670 and ~720 K, respectively) for different times and also at and above T_x. It was noted that the hardness of the as-quenched sample increased quickly on annealing at 630 K and remained constant for up to 40 min of annealing. Another small rise in hardness was noted at 60 min of annealing and also after the sample had crystallized. DSC scans of these samples confirmed structural relaxation, indicated by an exothermic relaxation peak (loss of enthalpy). By measuring the area under the peak, the authors concluded that most of the structural relaxation was completed within 20 min of annealing at 630 K.

By analyzing the atomic pair distribution functions (PDF), the authors directly examined the structural relaxation behavior in these glasses and suggested that structural relaxation takes place in two stages. The process of structural relaxation was inferred from the sharpening of the PDF peaks, without their positions having shifted. By observing the profiles of the left and right shoulders of the main peak, they suggested that the free volume was not annihilated at short durations of annealing. Instead, they interpreted the increase in peak height and the sharpening of left and right shoulders as changes in topological order that did not involve significant atomic transport, which could occur on annealing at higher temperatures. Thus, the first stage of relaxation was suggested to be related to the elimination of short and long inter atomic distances and the second stage to the local chemical reordering in the glassy phase. The observed chemical ordering on annealing the samples for longer times was suggested as a possible reason for the observed phase separation and nanocrystallization after annealing at higher temperatures.

A common procedure that is followed to investigate structural relaxation in metallic glasses is as follows. The glassy alloy is first equilibrated for a given length of time at a temperature above T_g in the supercooled liquid region. This is considered as the reference sample since no more structural relaxation takes place in this sample on subsequent annealing. The time is chosen to be about $10\tau_r$, where τ_r is the relaxation time, defined as $\tau_r = \Delta T_g/\beta$, where ΔT_g is the temperature interval, defined as the temperature range between the onset and the end of the endothermic glass transition event, and β is the heating rate. This is because, at $10\tau_r$, the samples are expected to be fully relaxed [85]. (However, a fully relaxed structure was not obtained even after annealing a melt-spun $Pd_{40}Cu_{30}Ni_9Fe_1P_{20}$ glassy ribbon for 35 days (3×10^6 s) at 540 K, which is only 34 K below the T_g value for this alloy [82].) The variation of C_p for this reference sample is then plotted as a function of temperature. Other samples are then annealed at different temperatures below T_g for the same length of time as the reference sample, and the variation of their C_p with temperature is again determined and plotted for comparison purposes.

Figure 5.13 shows the variation of the specific heat, C_p, with temperature for a $Zr_{65}Al_{7.5}Cu_{27.5}$ glassy alloy annealed at different temperatures, T_a, for an annealing period of t_a = 12 h [86]. On heating the as-quenched sample, the specific heat $C_{p,q}$ begins to decrease, indicative of structural relaxation at about 380 K, and reaches a minimum value at a temperature of about 575 K. With further increases in temperature, $C_{p,q}$ increases gradually to about 620 K, then rapidly in the glass transition range from 630 to 675 K, reaching a maximum value of 47.4 J mol^{-1} K^{-1} for the supercooled liquid at around 680 K. It then decreases rapidly because of crystallization at 731 K. The variation of C_p for the pre-annealed samples shows a $C_{p,a}(T)$ behavior, which closely follows the C_p curve of the reference sample, $C_{p,s}$, (specimen annealed at 690 K) up to each T_a and then exhibits an excess specific heat relative to the reference sample before merging with that of the as-quenched sample in the supercooled liquid region above 642 K.

The main features of Figure 5.13 may be summarized as follows:

1. The sample annealed at T_g (reported as 646 K for this alloy) shows an excess endothermic specific heat beginning at T_a, implying that the C_p value in the temperature range above T_a is dependent on the thermal history of the sample and consists of a configurational contribution as well as the contribution from purely thermal vibrations.

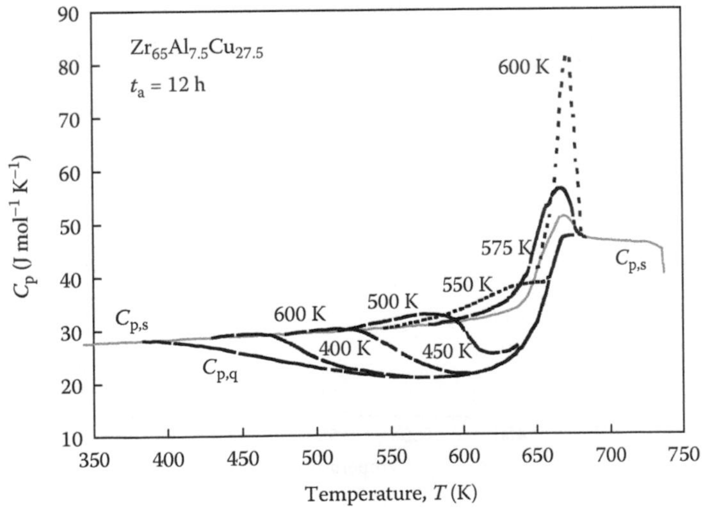

FIGURE 5.13
The variation of specific heat, C_p, with annealing temperature, T_a, for a glassy $Zr_{65}Al_{7.5}Cu_{27.5}$ BMG alloy annealed for 12 h at different temperatures from 400 to 620 K. The solid line represents the variation of C_p for the reference sample annealed for 12 h at 690 K. (Reprinted from Inoue, A. et al., *J. Non-Cryst. Solids*, 150, 396, 1992. With permission.)

2. The magnitude of the endothermic peak increases rapidly at annealing temperatures just below T_g.
3. The excess endothermic peak is recoverable, while the exothermic broad peak is irrecoverable.

Thus, the $C_{p,a}(T)$ curves for the annealed samples reflect a combination of the recoverable endothermic and irrecoverable exothermic reactions. Chen [87] suggested that the excess endothermic peak occurs because of a relaxed atomic configuration, achieved by the rearrangement of atoms on annealing, which is more stable (but still metastable) at temperatures above T_a. Thus, the excess endothermic reaction reflects the atomic configuration achieved by annealing and, hence, we can obtain information on structural relaxation during annealing by examining the change in the excess endothermic peak with T_a and t_a.

The temperature dependence of the difference in C_p between the annealed and the reference states [$\Delta C_{p,endo} = C_{p,a}(T) - C_{p,s}(T)$] for the $La_{55}Al_{25}Ni_{20}$ and $Zr_{65}Al_{7.5}Cu_{27.5}$ glassy

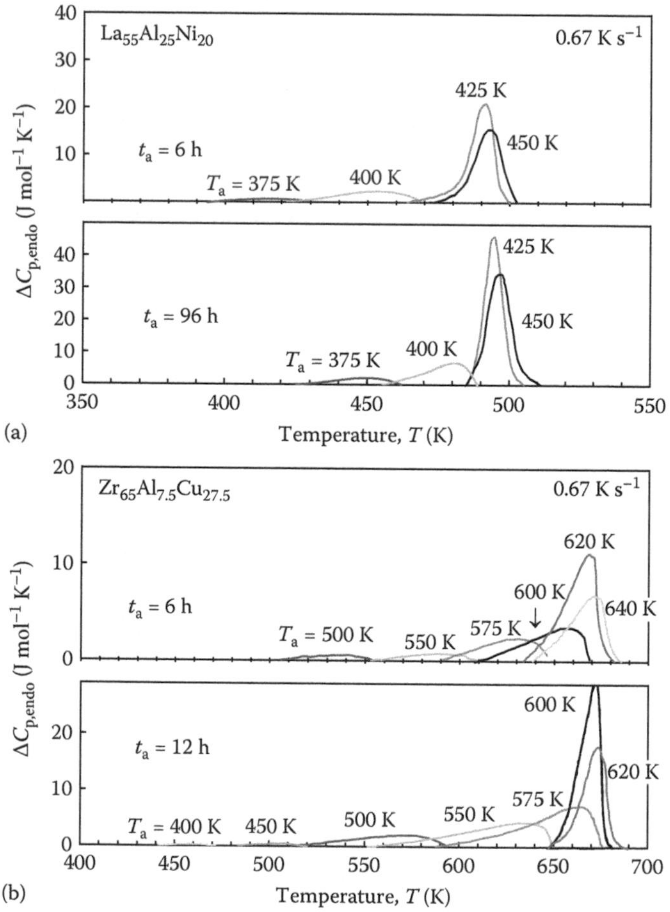

FIGURE 5.14
The differential specific heat, $\Delta C_p(T)$, between the reference and annealed samples for the glassy (a) $La_{55}Al_{25}Ni_{20}$ and (b) $Zr_{65}Al_{7.5}Cu_{27.5}$ alloys annealed for 6 and 96 h for the $La_{55}Al_{25}Ni_{20}$ alloy and for 1 and 12 h in the case of $Zr_{65}Al_{7.5}Cu_{27.5}$ alloy at different temperatures. The samples have been heated in a DSC at 0.67 K s^{-1} (40 K min^{-1}). (Reprinted from Inoue, A. et al., *J. Non-Cryst. Solids*, 150, 396, 1992. With permission.)

alloys annealed for different times at different temperatures is shown in Figure 5.14 [86]. With increasing T_a, the maximum in $\Delta C_{p,endo}$ for the two alloys exhibits an initial gradual increase, followed by a rapid increase at temperatures just below T_g, and then a rapid decrease above T_g. The rapid increase in the maximum of $\Delta C_{p,endo}$ as a function of T_a is interpreted as corresponding to the glass transition phenomenon. Similarly, the rapid decrease in the maximum of $\Delta C_{p,endo}$ above T_g is interpreted to be due to the achievement of internal equilibrium resulting from the very short relaxation times in the supercooled liquid region. Although the change in the maximum in $\Delta C_{p,endo}$ as a function of T_a is similar to that for the metal–metal-type Zr–Cu and Zr–Ni glassy alloys [88], it is significantly different from the two-stage change that has been observed for the metal–metalloid-type glassy alloys containing more than two types of metallic elements [89]. The appearance of a two-stage relaxation process was suggested to be due to the difference in the relaxation times between the metal–metal atom pairs with weaker bonding, and the metal–metalloid atom pairs with stronger bonding. Since no splitting of the maximum in $\Delta C_{p,endo}$ into two stages is seen as a function of T_a for the La–Al–Ni and Zr–Al–Cu amorphous alloys, it was concluded that the relaxation times were nearly the same for the La–Al, Al–Ni, and La–Ni atom pairs and also for the Zr–Al, Al–Cu, and Zr–Cu atom pairs, which have large negative enthalpies of mixing. This result suggests that there is no appreciable difference in the attractive bonding between the atomic pairs and that the constituent atoms in these glassy alloys are in an optimum bonding state. The optimum bonding state prevents easy atomic movement, even in the supercooled liquid, and suppresses the nucleation and growth of a crystalline phase, leading to the appearance of a very wide supercooled liquid region.

The mechanical properties of metallic glasses (including the BMGs) are affected by the magnitude of free volume present in them [90–92]. Hence, it becomes important to be able to quantitatively determine the free volume present in the glass to relate the magnitude of free volume to the changes in mechanical properties. Accordingly, Launey et al. [93] used the DSC technique to quantify the free volume changes in a $Zr_{44}Ti_{11}Ni_{10}Cu_{10}Be_{25}$ glassy alloy with structural relaxation. Assuming that the change in enthalpy is entirely due to structural changes in the glassy state and not in the crystal and that the average free volume per atom ($=V_f/V_m$, where V_f is the free volume and V_m is the atomic volume) is proportional to the change in enthalpy:

$$\frac{V_f}{V_m} = C\Delta H \tag{5.8}$$

where C is a constant. The proportionality constant, C, is determined by first calculating V_f using the Grest and Cohen model [94]:

$$V_f = \frac{k}{2s_0}\left(T - T_0 + \sqrt{(T - T_0)^2 + \frac{4V_a s_0}{k}T}\right) \tag{5.9}$$

where k is the Boltzmann constant. The appropriate fit parameters for the alloy were reported to be $bV_m s_0/k = 4933$ K with $b = 0.105$, $4V_a s_0/k = 162$ K, $T_0 = 672$ K. V_m for this alloy has been reported to be 1.67×10^{-29} m^3 near the liquidus temperature. Thus, by calculating V_f from Equation 5.9, V_f/V_m can be calculated. Further, by matching the curves of the average free volume from Equation 5.9 and enthalpy from Equation 5.8 with temperature, the constant C in Equation 5.8 can be determined. The Grest and Cohen model is capable of estimating the average free volume in the supercooled liquid at substantially lower temperatures than is possible by relaxation experiments. It is also important to remember that even though the free volume varies locally from one place in the sample to another, the globally averaged macroscopic free volume can be determined from the DSC measurements.

5.7.2 Glass Transition

The variation of C_p with temperature, described in Figure 5.13, is typical of all metallic glasses. The C_p value slowly increases with increasing temperature, and then begins to decrease, suggesting the onset of irreversible structural relaxation. With further increases in temperature, the C_p value reaches a minimum and then increases rapidly in the glass transition range. The C_p value reaches a maximum for the supercooled liquid, and then starts to decrease first gradually and then rapidly once crystallization has set in. These observations suggest that the transition from the glassy (g) state to the supercooled liquid (scl) condition is accompanied by a large increase in the specific heat, $\Delta C_{p,g \rightarrow scl}$. The magnitude of this increase is different for different types of glasses; the values for some of the glasses are listed in Table 5.4.

The $\Delta C_{p,s}$ of the reference sample consists of configurational contributions as well as those arising from purely thermal vibrations. Because of the linear variation of C_p of the reference sample with temperature, one could obtain the specific heat, purely due to thermal vibrations, $C_{p,v}$, by extrapolating the C_p values in the low-temperature region. For example, for the $Zr_{60}Ni_{25}Al_{15}$ glassy alloy the relationship is

$$C_{p,v} = 23.9 + 1.29 \times 10^{-2}(T - 340) \quad \text{at } 340 \le T \le 620 \text{ K} \tag{5.10}$$

Similarly, the equilibrium specific heat of the supercooled liquid, $C_{p,scl}$, including the vibrational and configurational-specific heat, can be expressed by the equation:

$$C_{p,e} = 33.7 - 2.59 \times 10^{-2}(T - 735) \quad \text{at } 735 \le T \le 760 \text{ K} \tag{5.11}$$

The difference between these two values at the temperature of interest will give us the $\Delta C_{p,g \rightarrow scl}$ value. It was suggested that there is a tendency for the value of $\Delta C_{p,g \rightarrow scl}$ to scale with ΔT_x and ΔH_x, the enthalpy of crystallization [77], and results on Zr-based alloy glasses support this (Table 5.4). However, it does not appear to be true when all the results are taken into consideration.

TABLE 5.4

Increase in Specific Heat from the As-Quenched Glassy (g) State to the Supercooled Liquid (scl) Condition, $\Delta C_{p,g \rightarrow scl}$ for Different Metallic Glasses Synthesized by Melt Spinning, and Measured at a Heating Rate of 0.67 K s^{-1} (40 K min^{-1})

Composition	$\Delta C_{p,g \rightarrow scl}$ (J mol^{-1} K^{-1})	ΔT_x (K)	References
$La_{55}Al_{20}Cu_{25}$	11.5	59	[95]
$La_{55}Al_{25}Ni_{20}$	14.0	69	[96]
$Mg_{50}Ni_{30}La_{20}$	17.4	58	[97]
$Zr_{60}Al_{15}Ni_{25}$	6.25	77	[98]
$Zr_{65}Cu_{27.5}Al_{7.5}$	—	88	[86]
$Zr_{65}Cu_{17.5}Ni_{10}Al_{7.5}$	14.5	127	[18]

Note: ΔT_x represents the width of the supercooled liquid region.

The $\Delta C_{p,g \rightarrow scl}$ values for the Zr-based metallic glasses are considerably smaller than those of Pd–Ni–P and Pt–Ni–P glasses. Even though the reasons for this difference are not clearly known at the moment, it is possible that it is related to (1) the higher packing fraction of atoms in the glassy Zr alloys, which require a lower cooling rate to form the glassy structure; (2) the possibility of the atomic configuration in the glassy and supercooled liquid structures being similar; and (3) the higher T_g values in comparison to those of La-, Mg-, Pd-, and Pt-based glassy alloys.

5.7.3 Phase Separation

Phase separation is the process in which a homogeneous glassy phase of a given composition is separated into two different glassy phases of different compositions. (Even though a more accurate description of this process would be to refer to it as *glassy* [or amorphous] phase separation, in the following paragraphs we will simply use the expression "phase separation" to denote that it is the glassy phase that is separating into two glassy phases.) Phase separation occurring by a nucleation and growth process or by spinodal decomposition without any barrier to the nucleation process is well known to occur in oxide glasses [99]. Phase separation was also reported to occur in metallic glassy alloy ribbons of $Pd_{74}Au_8Si_{18}$ [34], $(Pd_{0.5}Ni_{0.5})_{81}P_{19}$ [37], and $Zr_{36}Ti_{24}Be_{40}$ [100], among other alloy systems. The phase separation in melt-spun ribbons was noted mostly during the reheating of the glassy alloy ribbon specimen to higher temperatures, even though it was reported to occur in the as-quenched condition in the $Zr_{36}Ti_{24}Be_{40}$ alloy. Evidence for phase separation in these systems was gathered from small-angle x-ray scattering (SAXS), TEM, and DSC techniques. The presence of two glass transition temperatures in the DSC plot was considered a clear indication of the existence of two glassy phases in the sample. Recently, Luizguine-Luzgin et al. [101] showed that the clear existence of two different slopes within the glass transition region indicated the occurrence of two separate glass transition processes. It was suggested that this phenomenon was likely related to the different diffusion coefficients of the alloying elements in the $Au_{49}Cu_{26.9}Ag_{5.5}Pd_{2.3}Si_{16.3}$ alloy studied. In TEM micrographs, phase separation is manifested by the presence of multiphase light and dark structures with a characteristic interconnected morphology, often seen in ceramic glasses or isolated droplet structures. Kim et al. [102] have recently reviewed the current status in this area of research.

The study of phase separation in metallic glasses is important for two reasons. First, it will provide us with an opportunity to synthesize different types of composites (glass + glass or glass + crystals) with widely differing mechanical properties and performance. By annealing the metallic glasses at different temperatures and for different times, it is possible to obtain a variety of microstructures and, therefore, different properties. Second, it is of scientific interest to be able to study the decomposition behavior of glass over a wide temperature range (in the supercooled liquid range). This is possible only in BMGs because they exhibit a wide supercooled liquid region.

At this stage, it is important to differentiate between two clearly different situations. One is that the as-solidified glassy rods (especially in the case of binary BMGs) contain very fine crystals, often with nanometer dimensions, dispersed in a glassy matrix. The presence of such tiny crystals could be due to the relatively low GFA of the alloy. That is, if the critical cooling rate for glass formation has not been exceeded throughout the cross section of the sample, then a crystalline phase will be present in the sample. To check whether this is the reason for the presence of these tiny crystals, rods of smaller and smaller diameter could be cast to check whether they are fully glassy. It is also possible that the as-solidified samples sometimes contain crystalline particles on the surface as, for example, in arc melted ingots [74]. If this is the situation, then it may be due to heterogeneous nucleation occurring on the surface of the sample, or on those parts that are in contact with a metallic substrate. Also, during heating of the glassy sample to higher temperatures, primary crystallization could occur with the formation of a solid solution phase in a glassy matrix with a composition that is solvent rich. In this case, it is the process of precipitation that results in the formation of a glass + crystal composite. In all these situations, it is not correct to refer to the process that occurs as (glassy) phase separation: even though two different phases are present, they are not two different *glassy* phases.

The second situation is when two different glassy phases are formed in the supercooled melt either during the process of solidification or during the reheating of a homogeneous glassy phase. In these situations, it is most appropriate to refer to the process that occurs as *(glassy) phase separation*.

It is well known that phase separation occurs in alloy systems whose phase diagrams feature a miscibility gap between two phases that are thermodynamically stable. Such a situation arises when the two constituent elements in the binary alloy have a zero or positive heat of mixing. (The heats of mixing for different binary combinations of metals have been calculated by Miedema et al. and are available in Refs. [75,103].) Examples of such miscibility gaps may be found in binary alloy systems such as Cu–Ni, Cu–Rh, Au–Pt, Cr–Mo, and As–Sb [104].

Figure 5.15a shows the schematic of a hypothetical binary phase diagram featuring a miscibility gap; the corresponding variation of free energy (G) with composition at a temperature T_2 is shown in Figure 5.15b. Note that the free energy curve shows a convex upward shape with two minima, corresponding to compositions C_1 and C_2. When an alloy with a composition between C_1 and C_2, say C_0, is solution-treated at a temperature, T_1, and then quenched to a lower temperature, T_2, into the miscibility gap region, it is in a metastable high-energy state and, therefore, can lower its free energy by decomposing into two phases. The lowest energy state is obtained when the metastable solid solution decomposes into two phases with the compositions C_1 and C_2. However, if an alloy with the composition C_0' is quenched from temperature T_1 to temperature T_2, the alloy has a free energy G_0'. This alloy is now in an unstable situation, since any small changes in composition will result in a decrease in the free energy of the system. This will be true for any alloy with a composition from C_1' to C_2'. Such spontaneous transformation is referred to as *spinodal transformation*. Thus, the decomposition of the supersaturated solid solution can occur either by a nucleation and growth process (which requires a nucleation barrier to be overcome) or by a spinodal process (which does not have any nucleation barrier). The spinodal points in the free energy versus composition curve (and the miscibility gap) correspond to those compositions at which $\partial^2G/\partial C^2 = 0$, and these are marked as C_1' and C_2'. That is, at a temperature T_2, if the alloy has a composition between C_1 and C_1', or between C_2' and C_2, it will decompose by a nucleation and growth process. On the other hand, if the alloy composition lies between C_1' and C_2', it will decompose spinodally. The formation of spinodal structures in metallic alloys in the solid state has been known for a long time [105]. The process of phase separation in a glassy alloy can also be understood in a similar manner if the high-temperature phase in the hypothetical phase diagram is assumed to be a glassy phase and the low-temperature "stable" phases are also glassy phases with different compositions.

But, let us recall that it has been frequently emphasized in the literature that a negative heat of mixing among the constituent elements in the alloy is a prerequisite for the formation of BMGs [73]. Therefore, these two criteria seem to be apparently contradictory to each

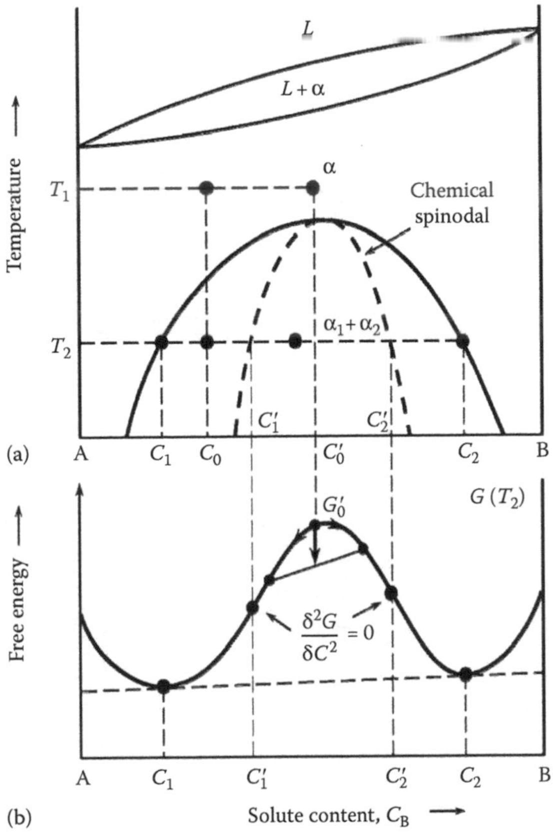

FIGURE 5.15
(a) Typical phase diagram showing a miscibility gap in the solid state. (b) The corresponding free energy vs. composition diagram featuring two minima. Phase separation is possible in such an alloy system either by a nucleation and growth process or by a spinodal decomposition process.

other; that is, a large negative heat of mixing is required to synthesize a BMG, but a positive heat of mixing is a prerequisite to observe phase separation in the glass that is produced. Thus, it appears to be impossible to imagine that one would observe phase separation in BMGs. But, as will be shown later, phase separation in BMGs has been reported to occur in alloy systems even when the constituent elements have a negative heat of mixing. But, it is much easier to observe phase separation when the alloy contains at least two elements that have a positive heat of mixing. Table 5.5 summarizes the results of phase separation reported in both melt-spun glassy ribbons and BMG alloys. Let us now look at the details of the reports of phase separation in some of the alloy systems and the types of microstructures that could be developed.

$Zr_{60}Al_{15}Ni_{25}$, with a ΔT_x of 77 K and a T_{rg} of 0.64, is a good GFA [98]. $Y_{60}Al_{15}Ni_{25}$ is also a reasonably good glass former, but with a ΔT_x of only 22 K [124]. Since Zr and Y are immiscible in the solid state (they have a positive heat of mixing of +35 kJ mol^{-1}), it was decided to combine these two alloy systems and to determine the GFA and also if any unusual phenomena occur. By solidification, $Zr_{60-x}Y_xAl_{15}Ni_{25}$ alloys (with x = 15, 21, 27, and 30) were found to be glassy, as determined from their XRD patterns [124,125]. But, an unusual behavior was also noted. This quaternary glassy alloy exhibited two glass transition temperatures and two supercooled liquid regions before the completion of crystallization.

TABLE 5.5

Alloy Systems Showing Phase Separation in the Glassy State

Alloy Composition	Synthesis Method	Characterization Method(s)	Compositions of the Two Glassy Phases	Comments	References
$Ag_{20}Cu_{48}Zr_{32}$	Melt spinning	TEM	—	—	[106]
$Cu_{43}Zr_{43}Al_7Ag_7$	Cu mold casting	TEM and 3DAP	$Cu_{40.7}Zr_{46.8}Al_{8.0}Ag_{4.5}$ and $Cu_{36.8}Zr_{43.5}Al_{7.0}Ag_{12.7}$	Phase separation due to unusually high plastic strain	[107]
$Cu_{46}Zr_{22}Y_{25}Al_7$	Melt spinning	DSC and TEM	$Cu_{35.7}Zr_{12.8}Y_{44.3}Al_{7.2}$ and $Cu_{53.4}Zr_{31.8}Y_{8.3}Al_{6.5}$	—	[108]
$La_{27.5}Zr_{27.5}Al_{25}Cu_{10}Ni_{10}$	Melt spinning	SEM and TEM	$La_{5.0}Zr_{51.4}Cu_{5.4}Ni_{13.2}Al_{25}$ and $La_{43.4}Zr_{10.9}Cu_{14.4}Ni_{8.2}Al_{22.1}$	—	[109]
$Nd_{60-x}Zr_xAl_{10}Co_{30}$ ($6 \leq x \leq 40$)	Melt spinning	DSC and TEM	—	—	[110]
$Ni_{70}Nb_{15}Y_{15}$	Melt spinning	DSC, TEM, and SAXS	—	—	[111]
$Ni_{66}Nb_{17}Y_{17}$	Melt spinning	DSC, TEM, and SAXS	—	—	[111]
$Ni_{58.5}Nb_{20.25}Y_{21.25}$	Melt spinning	DSC, SEM, TEM, and SAXS	$Ni_{59}Nb_{16}Y_{25}$ and $Ni_{57}Nb_{28}Y_{15}$ by SEM and $Ni_{53}Nb_{42}Y_5$ and $Ni_{60}Nb_{10}Y_{30}$ by TEM	Two T_gs were not observed	[111, 112]
$Ni_{54}Nb_{23}Y_{23}$	Melt spinning	DSC, TEM, and SAXS	$Ni_{50}Nb_{44}Y_6$ and $Ni_{58}Nb_7Y_{35}$	—	[111]
$Ni_{61}Zr_{28-x}Nb_7Al_4Ta_x$ ($x = 0, 2, 4, 6, 8$)	Melt spinning	—	—	No evidence of phase separation	[113]
$Pd_{80}Au_{3.5}Si_{16.5}$	Roller quenching	DSC and SAXS	—	Apparent phase separation	[34]
$Pd_{78}Au_6Si_{16}$	Splat cooling	DSC and TEM	Segregation into (Pd–Au)-rich and Si-rich glassy phases	No clear identification of the phases	[33]
$Pd_{40.5}Ni_{40.5}P_{19}$	Centrifugal spinning	DSC	—	Two T_gs were observed only after the original glassy sample was heated beyond the first exothermic peak, then cooled quickly and reheated	[37]
$Pd_{80}Si_{20}$	Splat cooling	DSC and TEM	Pd-rich particles embedded in a Si-rich matrix	No clear identification of the phases	[33]
$Ti_{28}Y_{28}Al_{24}Co_{20}$	Melt spinning	XRD and TEM	$Y_{40.4}Ti_{14.7}Al_{21.9}Co_{23}$ and $Ti_{45.6}Y_{11.6}Al_{26.7}Co_{16.1}$	No clear T_g in DSC	[114]

(Continued)

TABLE 5.5 (CONTINUED)

Alloy Systems Showing Phase Separation in the Glassy State

Alloy Composition	Synthesis Method	Characterization Method(s)	Compositions of the Two Glassy Phases	Comments	References
$Ti_{56-x}Y_xAl_{22}Co_{22}$ (x = 11, 20, or 28)	Melt spinning	TEM	$Y_{44.5}Ti_{8.8}Al_{36.9}Co_{9.8}$ and $Ti_{47.2}Y_{2.1}Al_{19.9}Co_{30.8}$. These compositions depend on the initial composition of the alloy	—	[115]
$Zr_{63.8}Ni_{16.2}Cu_{15}Al_5$	Cu mold casting	—	$Zr_{68.5}Cu_{8.1}Ni_{21.3}Al_{2.1}$ and $Zr_{62.4}Cu_{16.7}Ni_{14.6}Al_{6.3}$	Noted 30% plastic strain during compression at room temperature	[116]
$Zr_{36}Ti_{24}Be_{40}$	Melt spinning	DSC and TEM	—	Two T_gs were reported. Nagahama et al. [117] concluded that this alloy crystallized in a eutectic mode and that there was no phase separation	[100]
$Zr_{52.5}Ti_5Cu_{17.9}Ni_{14.6}Al_{10}$ (Vit 105)	Cu mold casting and melt spinning	SANS and TEM	—	Phase separation? Kajiwara et al. [119] suggested primary crystallization	[118]
$Zr_{41.2}Ti_{13.8}Cu_{12.5}Ni_{10.0}Be_{22.5}$ (Vit 1)	Water quenching	DSC, SANS, TEM and APFIM	Zr-rich and Be-rich phases	—	[120–122]
$Zr_{28}Y_{28}Al_{22}Co_{22}$	Melt spinning	Dynamic mechanical analysis and TEM	$Y_{30.9}Zr_{26.0}Al_{24.8}Co_{18.3}$ and $Zr_{36.4}Y_{15.8}Al_{28.8}Co_{19.0}$	Phase separation observed during heating of a homogeneous glassy phase	[123]
$Zr_{60-x}Y_xAl_{15}Ni_{25}$ (x = 15, 27, and 45)	Melt spinning	DSC	—	Two supercooled liquid regions	[124]

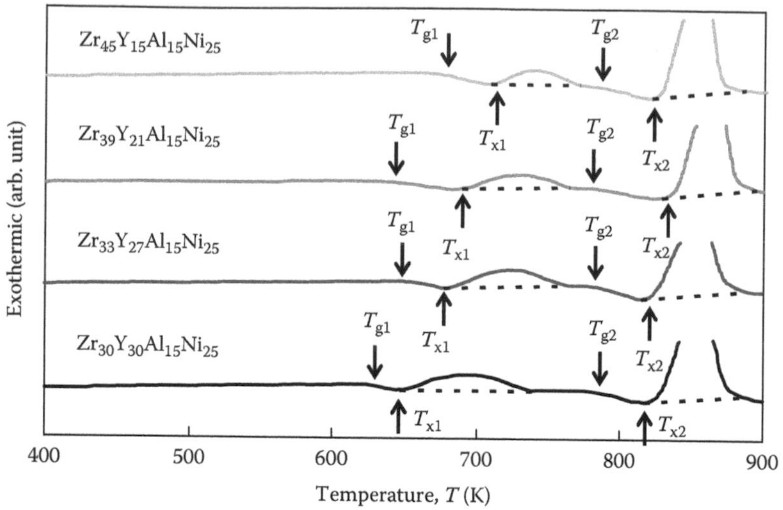

FIGURE 5.16
DSC curves of the glassy $Zr_{60-x}Y_xAl_{15}Ni_{25}$ (x = 15, 21, 27 and 30) alloys obtained at a heating rate of 0.67 K s^{-1} (40 K min^{-1}). Note the presence of two T_gs and two T_xs in all the alloys studied. (Reprinted from Inoue, A. et al., *Mater. Sci. Eng. A*, 179/180, 346, 1994. With permission.)

Figure 5.16 shows the DSC curves of the $Zr_{60-x}Y_xAl_{15}Ni_{25}$ (x = 15, 21, 27, and 30) glassy alloys recorded at a heating rate of 0.67 K s^{-1} (40 K min^{-1}). As marked by T_{g1}, T_{x1}, T_{g2}, and T_{x2}, these glassy alloys exhibited sequential phase transitions: the first-stage glass transition at T_{g1}, followed by the first-stage exothermic crystallization reaction at T_{x1}, the second-stage glass transition at T_{g2}, and then the second-stage crystallization at T_{x2}. Although T_{g1} and T_{x1} tended to decrease with increasing Y content, T_{g2} and T_{x2} remained almost constant. Since the T_g and T_x values were 626 and 648 K, respectively, for the $Y_{60}Al_{15}Ni_{25}$ alloy, and 699 and 766 K, respectively, for the $Zr_{60}Al_{15}Ni_{25}$ alloy, it was concluded that the first-stage reactions of glass transition and crystallization resulted from the Y-rich Y–Al–Ni glassy phase, and that the second-stage reactions were due to the Zr-rich Zr–Al–Ni glassy phase. The distinct split of the glass transition and crystallization reactions into two stages seems to originate from the immiscibility between Zr and Y. The temperature intervals of the two supercooled liquid regions, ΔT_{x1} and ΔT_{x2}, were examined as a function of Y content. ΔT_{x1} decreased monotonically from 26 to 19 K with increasing Y content, while ΔT_{x2} was in the range 31–40 K, with a maximum value around 25 at.% Y. It was therefore concluded that the stability of the supercooled liquid against the completion of crystallization was maximum in the vicinity of $Zr_{35}Y_{25}Al_{15}Ni_{25}$.

In order to evaluate the thermal stability of the supercooled liquid against the completion of crystallization, the $Zr_{33}Y_{27}Al_{15}Ni_{25}$ glassy alloy was chosen for a detailed study. When this glassy alloy was pre-annealed for 30 s at 773 K, the DSC curve showed that the ΔT_x value increased significantly from 40 to 104 K by the pre-annealing treatment (Figure 5.17). This result clearly indicates that the glassy phase, remaining after the Y-rich Y–Al–Ni components were removed in the form of a crystalline phase, has a much higher thermal stability than the as-quenched $Zr_{33}Y_{27}Al_{15}Ni_{25}$ glassy phase. This is even more surprising considering that this glassy phase coexisted with a crystalline Y–Al–Ni phase, which may provide heterogeneous nucleation sites. Therefore, further investigations were carried out on this remaining glassy phase.

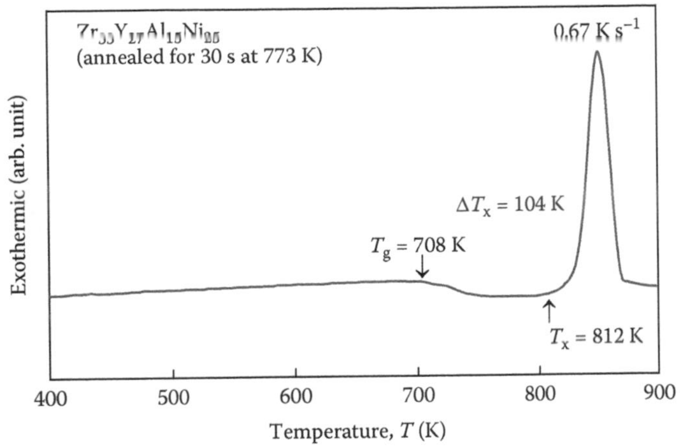

FIGURE 5.17
DSC curve of the glassy $Zr_{33}Y_{27}Al_{15}Ni_{25}$ alloy pre-annealed for 30 s at 773 K. The width of the supercooled liquid region, ΔT_x (=$T_x - T_g$), has now increased to 104 K from 40 K in the as-solidified condition. (Reprinted from Inoue, A. et al., *Mater. Sci. Eng. A*, 179/180, 346, 1994. With permission.)

XRD patterns from the pre-annealed glassy phase continued to show the presence of broad peaks indicative of the presence of the glassy phase; however, the peak widths were smaller than in the as-quenched condition. TEM investigations showed a featureless modulated contrast without any appreciable periodicity in the as-quenched sample, indicating that the as-quenched sample consisted of a homogeneous glassy phase. On the other hand, TEM micrographs of the pre-annealed samples showed that the pre-annealing treatment had generated a distinct modulation contrast with a size of about 3–5 nm. Additionally, the SAD patterns showed that the first broad peak had split into two rings. Coupled with the DSC data, these observations suggested that the modulated contrast observed in the TEM micrographs was due to the presence of the crystalline Y-rich Y–Ni–Al phase. That is, the homogeneous glassy phase in the as-quenched state had transformed into a mixed structure consisting of the Zr-rich Zr–Al–Ni glassy phase and the Y-rich Y–Al–Ni crystalline phase.

To confirm this hypothesis, HRTEM images of the $Zr_{33}Y_{27}Al_{15}Ni_{25}$ glassy alloy in both the as-quenched and pre-annealed conditions were recorded. A featureless modulated contrast without any appreciable periodicity was seen in the as-quenched sample (Figure 5.18a), indicating that the as-quenched sample consisted of a homogeneous glassy phase. However, in the pre-annealed sample, one could see a homogeneous distribution of very fine spherical particles with a distinct periodic contrast embedded in the glassy matrix. The size of these spherical particles with periodic contrast was as small as about 3 nm. This result clearly indicated that the pre-annealed alloy had a mixed structure consisting of nanoscale Y-rich crystalline particles embedded in the Zr-rich glassy phase. It could therefore be concluded that the small first-stage exothermic reaction at T_{x1} was due to precipitation of the nanoscale crystalline phase. The crystalline particles were too small for the crystal structure to be identified, due to the limited resolution of the TEM instruments at that time.

It is now possible to understand the reason for the significant extension of the supercooled liquid region for the remaining glassy phase after the precipitation of the nanoscale crystalline particles. From the DSC data and the TEM micrographs, it could be inferred that the nanoscale crystalline particles were mainly composed of the Y-rich Y–Al–Ni alloy

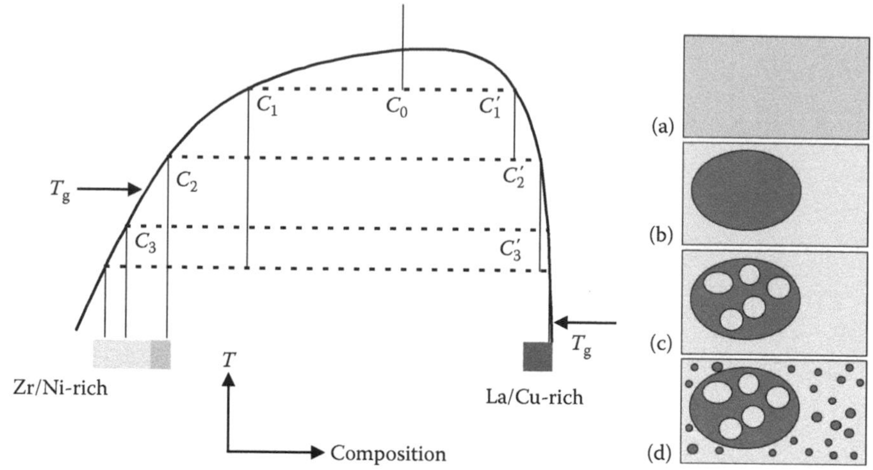

FIGURE 5.18
Schematic of the miscibility gap and the sequence of phase formation during cooling in the La–Zr–Al–Cu–Ni system. The positions of letters (a)–(d) in the diagram on the left correspond to the schematic microstructures (a)–(d) on the right. (Reprinted from Kündig, A.A. et al., *Acta Mater.*, 52, 2441, 2004. With permission.)

phase, which had been precipitated homogeneously by the primary mode in the glassy matrix. This result implies that Y-rich Y–Al–Ni units have been distributed homogeneously in the as-quenched glassy phase. It is therefore reasonable to presume that the remaining glassy phase is composed mainly of Zr-rich Zr–Al–Ni components. Therefore, if the exact composition of the remaining glassy phase is determined, it should be possible to produce a glassy alloy with a ΔT_x of 104 K as a single homogeneous glassy phase.

Thus, even though the DSC plots indicate the presence of two T_gs, it is not conclusive that (glassy) phase separation has occurred in this glassy alloy. This is because the second phase observed is crystalline, and it could only be indirectly inferred that the origin of the Y-rich crystalline phase was a Y-rich glassy phase.

Sugiyama et al. [126] used the anomalous SAXS method to investigate the origin of structural inhomogeneity in the glassy $Zr_{33}Y_{27}Al_{15}Ni_{25}$ alloy annealed for 300 s (5 min) at 773 K, a temperature higher than the first exothermic reaction temperature. By observing the anomalous SAXS intensity as a function of energy near the absorption edges of Zr, Y, and Ni, the authors concluded that the glassy alloy annealed at 773 K for 5 min possibly contained a Y-rich phase corresponding to the composition $Y_{60}Al_{15}Ni_{25}$. This observation suggested that the homogeneous glassy alloy underwent phase separation before the formation of the Zr_5Al_3 phase on annealing the glassy alloy for more than 15 min at 773 K.

Let us now look at some of the other alloy systems that have been shown to exhibit the phenomenon of phase separation. As mentioned earlier, phase separation was reported to occur in $Zr_{36}Ti_{24}Be_{40}$ glassy ribbons produced by melt spinning. The occurrence of phase separation was inferred from the presence of two T_gs in the DSC curve and the formation of an interconnected microstructure as observed by TEM [100]. Nagahama et al. [117] also investigated the microstructure of melt-spun ribbons of the $Zr_{36}Ti_{24}Be_{40}$ glassy alloy by advanced analytical techniques in the as-quenched condition and also during early crystallization by TEM and 3DAP techniques. They prepared thin foil specimens of the glassy alloy for TEM both by electropolishing and Ar-ion milling methods (to avoid the effect of artifacts introduced by the electropolishing technique) and concluded that the contrast attributed to phase separation by Tanner and Ray [100] was due to pitting corrosion from the overcurrent during the electropolishing process. From the 3DAP elemental maps, they concluded that there was no detectable fluctuation in the chemical composition in the sample and that an excellent match was obtained with the random distribution of the constituent elements. These authors concluded that, on annealing the glassy sample at 440°C, it crystallized in a eutectic mode by the formation of β-(Ti,Zr) solid solution and Be_2(Ti,Zr) phases.

Johnson and coworkers [120–122] investigated the decomposition behavior of $Zr_{41.2}Ti_{13.8}Cu_{12.5}Ni_{10}Be_{22.5}$ glassy alloy using different techniques such as small angle neutron scattering (SANS), TEM, DSC, and atom probe field ion microscopy (AP-FIM). They noted that the as-solidified glassy alloy showed both bright and dark areas. By conducting atom probe analysis, the authors identified the dark areas to be Be-rich and the bright areas to be Zr-rich. The concentration of Ti, Cu, and Ni was about the same in both areas [120]. Because of the change in composition, the T_g and T_x were expected to be different for the two regions. Therefore, by conducting SANS investigations on the glassy alloy annealed at 621, 631, and 641 K, the authors predicted that the formation of spatially periodically arranged Cu–Ti-rich nanocrystals would be preceded by a modulated chemical decomposition process. From these observations, they concluded that the homogeneous glassy alloy decomposed during cooling in the liquid state to a two-phase mixture of Be-rich and Zr-rich glassy regions with a typical length scale of tens of nanometers. Since the Zr-rich glassy phase had a lower crystallization temperature, the crystallization of the reheated glassy alloy was expected to start in the Zr-rich regions of the microstructure. Accordingly, on reheating the phase-separated alloy into the supercooled liquid region, primary crystallization (of a Zr-rich phase) occurred. The microstructure consisted of a high number density of crystals embedded in a glassy matrix.

Martin et al. [127] also investigated the decomposition behavior of the $Zr_{41.2}Ti_{13.8}Cu_{12.5}Ni_{10}Be_{22.5}$ alloy using 3DAP, TEM, and SAXS techniques. By annealing the glassy alloy at a temperature slightly above the T_g, they observed that an icosahedral (*i*) phase precipitated from a single homogeneous glassy phase, and this was followed by the precipitation of Be_2Zr and $CuZr_2$ phases. Only a uniform, featureless, glass-like contrast was observed before the *i*-phase particles were detected. The Ti-rich and Be-depleted regions that appeared in the early stage of annealing were explained to have formed due to the partitioning of alloying elements accompanied by the crystallization reaction. These authors did not see any evidence for prior decomposition in the glassy phase.

Phase separation was reported to occur in the $Pd_{74}Si_{18}Au_8$ system [33]. However, by employing a combination of techniques, such as SAXS, HRTEM, and 3DAP, it was shown [128] that the as-quenched glassy phase crystallized near T_g by the primary crystallization process through the formation of an Au-enriched Pd-based fcc solid solution and that at higher temperatures, the remaining glassy phase crystallized in a polymorphous mode by forming the Pd_3Si phase. Similarly, the reported phase separation in the $Zr_{52.5}Cu_{17.9}Ni_{14.6}Al_{10}Ti_5$ glassy alloy [118] was also explained to be due to the primary crystallization of the $NiTi_2$-type big cube phase [119]. Using advanced electron microscopy techniques, Hirotsu et al. [129] reported nanoscale phase separation in melt-spun ribbons of $Fe_{70}Nb_{10}B_{20}$ and $Fe_{73.5}Nb_3Si_{13.5}Cu_1B_9$ and in a copper mold cast $Zr_{71}Cu_{13}Ni_{10}Ti_3Al_3$ alloy.

On centrifugal spinning of $Pd_{40.5}Ni_{40.5}P_{19}$ alloy melts, a glassy phase was found to form. DSC curves of this glassy phase indicated the presence of two exothermic peaks, the first corresponding to the formation of an fcc phase (Pd-based solid solution) and the second due to the formation of a phosphide phase. When the samples were first scanned to the end of the first exothermic peak and then cooled quickly to lower temperatures, the second thermogram showed double glass transitions, suggesting phase separation in these glasses [37]. Madge et al. [130] investigated the structure of this glassy alloy using the energy-filtered TEM technique. They did not observe any compositional inhomogeneities and concluded that the second T_g was not attributable to phase separation in the material. Even though the origin of the T_g-like signals was not clear, they suggested that these could be related to changes in the SRO in the supercooled liquid.

It is known that phase separation in alloy systems is facilitated when the constituent elements in an alloy system have a zero or positive heat of mixing between them. Similar to the case of the Zr–Y–Al–Ni system [125], Kündig et al. [109] also combined two good GFAs and showed that phase separation is possible in such an alloy system during cooling in the supercooled liquid state.

La–Al–Cu–Ni [28] and Zr–Al–Cu–Ni [19] are two well-known good bulk GFA systems. That is, the enthalpies of mixing of Zr with Al, Cu, and Ni, and also of La with Al, Cu, and Ni, are negative. But, the enthalpy of mixing between La and Zr is positive (+74 kJ mol^{-1}). Therefore, by combining these two systems and forming an La–Zr–Al–Cu–Ni quinary alloy, Kündig et al. [109] produced BMG alloys of different compositions. A typical alloy investigated had the composition $La_{27.5}Zr_{27.5}Al_{25}Cu_{10}Ni_{10}$. Similar to the $Zr_{60-x}Y_xAl_{15}Ni_{25}$ (x = 15, 21, 27, and 30) glassy alloys [120,121], this alloy also exhibited a DSC plot that could be interpreted as a superimposition of two separate DSC traces—one for the La-based metallic glass and the other for the Zr-based metallic glass. Thus, it could be concluded that this alloy also exhibited phase separation during quenching from the melt. Scanning electron microscopy (SEM) examination showed that the phases exhibited rounded boundaries as expected from a phase-separated liquid. Both the phases had an upper size limit of about 20 μm. The second phase formed in a spherical shape inside the spheres of the first phase and, therefore, there was no characteristic size for the two phases, but their length scales varied from a few nanometers to a few tens of micrometers. By conducting EDS investigations, it was noted that the two phases were La–Cu-rich and Zr–Ni-rich glassy phases. The Zr–Ni-rich phase had a composition corresponding to $Zr_{50}La_7Cu_{3.6}Ni_{22.4}Al_{17}$. The two glassy phases had different T_g temperatures, and several self-similar generations of spheres in spheres were observed using SEM and TEM techniques. The presence of the two glassy phases was confirmed using a variety of experimental techniques including SEM, TEM, and SAXS. This was perhaps the first *direct* evidence for phase separation in a metallic glass during cooling from the liquid state.

Figure 5.18 shows a schematic of the miscibility gap and the sequence of phase separation in the La–Zr–Al–Cu–Ni BMG alloy during cooling. Since the homogeneous glassy alloy phase separates into Zr–Ni-rich and La–Cu-rich phases, the two ends of the miscibility gap have been marked accordingly. When the homogeneous melt of composition C_0 is cooled from high temperatures into the miscibility gap region, the liquid is undercooled. Immediately after the liquid is undercooled to a temperature T, below the critical temperature for phase separation, the homogeneous glassy structure does not undergo any phase separation. The microstructure of the alloy without undergoing any phase separation is depicted in Figure 5.18a. However, since the glass is unstable, and is sufficiently undercooled, it tends to get separated into Zr–Ni-rich and La–Cu-rich phases with the compositions C_1 and C_1', respectively. This microstructure is represented in Figure 5.18b. The area showing dark contrast represents the La–Cu-rich phase and the area with the gray contrast represents the Zr–Ni-rich phase. If the alloy is now further cooled down to a temperature low enough for one of the newly formed phases (C_1') to separate again into compositions C_2 and C_2', while the other phase (with the composition C_1) remains as is, the resulting microstructure is represented in Figure 5.18c. Such a phase separation process continues until the alloy is cooled to the T_gs of the individual glassy phases, below which there is no more transformation and the microstructure is preserved. In this sense, it is very similar to a situation where a solid phase undergoes a eutectoid reaction, and one of the newly formed solid phases transforms by another eutectoid reaction, and so on. But, since the transformation in the present case is taking place in the glassy state, the phase formed has a spherical shape to minimize the surface energy. Further, the phase is also forming in an isotropic medium. This is why the final microstructure will consist of spheres in spheres of decreasing dimensions.

It is also important to note that the compositions of the Zr–Ni-rich and La–Cu-rich phases are not fixed, but vary according to the slope of the miscibility gap curve on either side. Accordingly, the size of the spheres and their compositions will be different depending on the temperature at which the phase has formed. The two phases that form in the early stages of the cooling sequence will be larger in size and have compositions that are closer to the initial composition. On the other hand, the phases that have formed at a later stage of cooling will have smaller sizes and compositions that are far from the original alloy composition. Since the viscosity is low in the undercooled liquid, equilibrium shapes with the lowest interfacial energy, that is, spheres will be produced due to the availability of sufficient time.

Figure 5.19 shows two scanning electron micrographs of the as-solidified $La_{27.5}Zr_{27.5}Al_{25}Cu_{10}Ni_{10}$ glassy alloy showing the spherical particles of the glassy phases of different sizes. Since the microstructure is composed of several self-similar generations of spheres, it can be considered a surface fractal with a dimension of 2.6.

Figure 5.20 shows a transmission electron micrograph and the corresponding SAD patterns showing the presence of two glassy phases in the $Nd_{30}Zr_{30}Co_{30}Al_{10}$ alloy [110].

Phase separation in the as-cast $Cu_{43}Zr_{43}Al_7Ag_7$ [107], $Ni_{58.5}Nb_{202.25}Y_{21.25}$ [111,112], $Ag_{20}Cu_{48}Zr_{32}$ [106], and $Y_{28}Ti_{28}Al_{24}Co_{20}$ [114] alloys was also reported. In each of these systems, two of the elements had a positive heat of mixing, for example, Cu–Ag (+5 kJ mol^{-1}), Nb–Y (+30 kJ mol^{-1}), and Ti–Y (+58 kJ mol^{-1}). Using the computational thermodynamic approach, Du et al. [116] identified the chemical composition of an alloy system exhibiting the two-liquid miscibility region in the Zr–Cu–Ni–Al alloy system. Since this alloy is expected to show phase separation on solidification, the authors produced 2 mm diameter rods of $Zr_{63.8}Ni_{16.2}Cu_{15}Al_5$ glassy alloy by the copper mold casting method. TEM examination showed the presence of two glassy phases in the as-quenched alloy. Coupled with the EDS method, they identified the darker-appearing phase to be Ni-enriched ($Zr_{68.5}Cu_{8.1}Ni_{21.3}Al_{2.1}$) and the lighter matrix to be Cu-enriched ($Zr_{62.4}Cu_{16.7}Ni_{14.6}Al_{6.3}$). The important characteristic of this phase-separated glassy alloy is that it exhibited remarkable plasticity in compression exceeding 30% strain to failure.

Even though BMGs are expected to be good glass formers and, therefore, the constituent elements should have a negative heat of mixing, it is still theoretically possible to have phase separation in such systems, usually those containing three or more elements. According to Meijering [131,132], a ternary alloy phase can decompose into two phases with different compositions even when the enthalpy of mixing is negative. This is possible when the enthalpy of mixing for one of the three possible binary alloy systems is significantly more negative than the others. This suggests that a miscibility gap could be present in a ternary (or higher-order) BMG alloy system (even when all the constituent elements have a negative enthalpy of mixing). In other words, phase separation is possible even in an alloy with good GFA. But, there have not been any reports on experimental observations of phase separation in BMG systems that contain elements that have only negative heat of mixing.

Phase separation was observed in Pd–Si and Pd–M–Si (M = Cu, Ag, or Au) alloys by Chen and Turnbull [33] even though the heats of mixing of the pure liquid metals were negative.

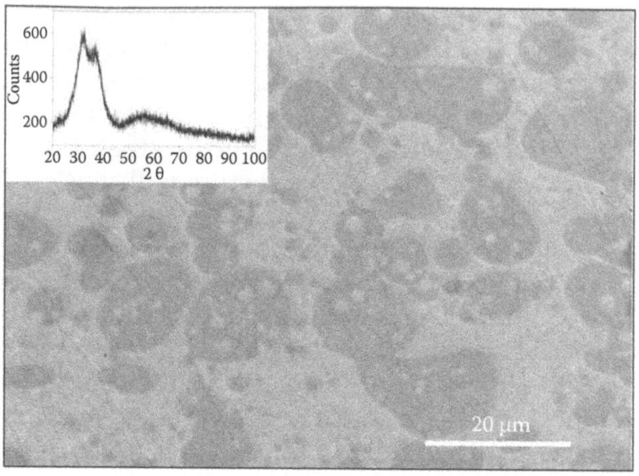

(a)

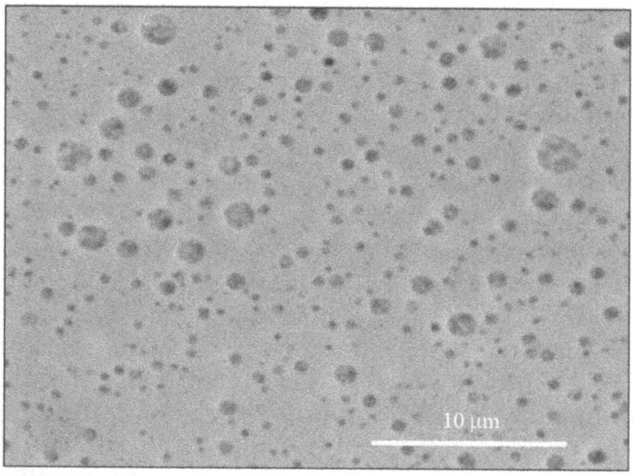

(b)

FIGURE 5.19
(a) Scanning electron micrograph of the cross section of a melt-spun ribbon of $La_{27.5}Zr_{27.5}Cu_{10}Ni_{10}Al_{25}$ and the XRD pattern taken from the wheel side of the ribbon (inset). (b) SEM image from the La-rich spheres in a 100 μm sphere of Zr-rich phase in a wedge of $La_{27.5}Zr_{27.5}Cu_{10}Ni_{10}Al_{25}$ alloy at 1 mm sample thickness. (Reprinted from Kündig, A.A. et al., *Acta Mater.*, 52, 2441, 2004. With permission.)

They agreed that this was contrary to the predictions of the regular solution model, but suggested that it may be accounted for qualitatively by supposing that the homogeneous system lowers its energy further by splitting into two liquid phases, each with a high degree of unique local order. Abe et al. [133] calculated the liquid phase miscibility gaps in ternary glass-forming systems using the sub-regular solution model. They noted that liquid miscibility gaps existed at low temperatures in most of the reported ternary bulk GFA systems, even though the enthalpies of mixing of the constituent elements were all negative. But, these miscibility gaps did not overlap with the glass-forming compositions, except in the Cu–Ti–Zr system. The main conclusion of this study was that the GFA of the system would be low at compositions showing phase separation.

An alloy quenched from the high-temperature single-phase region into the miscibility gap can decompose into the equilibrium phases either by nucleation and growth or spinodal processes. If the transformation occurs by the nucleation and growth processes, the

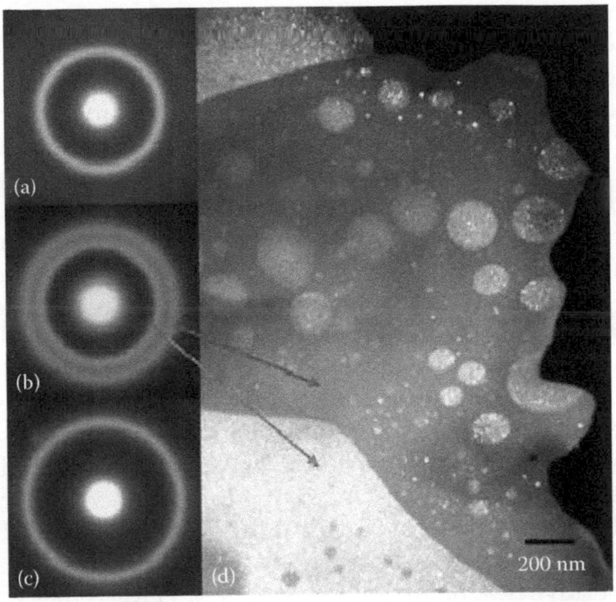

FIGURE 5.20
Selected area diffraction patterns and dark-field electron micrograph of BMG alloys. (a) $Nd_{60}Al_{10}Co_{30}$, (b) $Nd_{30}Zr_{30}Co_{30}Al_{10}$, (c) $Zr_{60}Al_{10}Co_{30}$, and (d) is the dark-field electron micrograph of the two-phase glassy $Nd_{30}Zr_{30}Co_{30}Al_{10}$ alloy recorded with an objective aperture on the inner halo ring as shown in (b). (Reprinted from Park, E.S. et al., *Scr. Mater.*, 56, 197, 2007. With permission.)

interface between the constituent phases is expected to be sharp. On the other hand, if the transformation occurs by the spinodal process, the interface is diffuse at the beginning but becomes sharp with increasing time. Eventually, of course, it will be difficult to differentiate between the microstructures formed by the two processes, solely based on the nature of the interfaces. An important point to remember also is that the size of the phases in the spinodal process depends on the spinodal wavelength, which in turn can be controlled by the solidification rate during the formation of the glassy phase. Figure 5.21 shows two

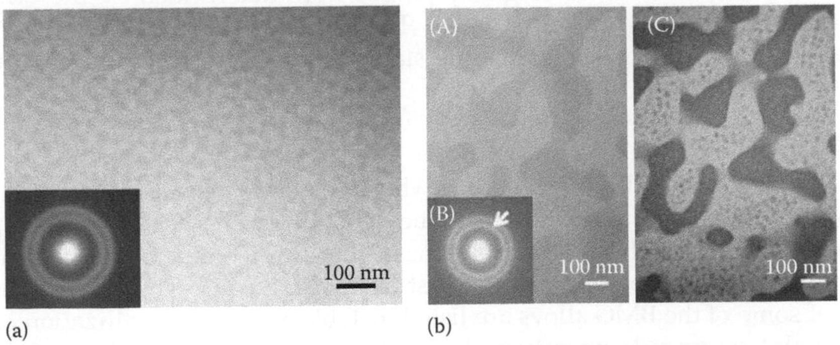

FIGURE 5.21
Transmission electron micrographs from rapidly solidified $Y_{28}Ti_{28}Al_{24}Co_{20}$ alloy showing that the size of the glassy phase is dependent upon the solidification rate experienced by the ribbon. (a) Microstructure on the wheel side of the ribbon of about 25 nm size and (b) microstructure from the region near the air side of the ribbon of about 250 nm size. (A) Bright-field micrograph, (B) corresponding selected area diffraction pattern, and (C) dark-field image obtained using the inner diffuse halo ring marked in (B). (Reprinted from Park, B.J. et al., *Appl. Phys. Lett.*, 85, 6353, 2004. With permission.)

bright-field TEM micrographs; one from the air side of the ribbon and the other from the wheel side. Because the ribbon on the wheel side would have solidified faster, the electron micrograph (Figure 5.21a) shows the glassy phase to be about 25 nm in size, while that on the air side (Figure 5.21b) shows the phase to be about 250 nm in size [114]. Therefore, it is possible to tailor the sizes of the glassy phases by varying the solidification rate during cooling.

It was mentioned earlier that phase separation in metallic glasses is usually manifested by the presence of two T_gs in the DSC plot. However, such clear T_gs may not always be observed. The measurement of the viscosity of the glass as a function of temperature can conclusively prove whether phase separation has occurred in the glassy alloy, irrespective of whether two T_gs are observed in the DSC plot. Inoue et al. [134] noted that the $Zr_{33}Y_{27}Ni_{25}Al_{15}$ glass exhibited two T_gs in the DSC plot and so concluded that phase separation had occurred in this alloy. By measuring the viscosity of the glass as a function of temperature, they observed that the viscosity started increasing rapidly just above the T_g, and that this was followed by a small decrease up to T_x. Measurements at higher temperatures showed one more set of such variations, indicating that another T_g followed by another T_x were also present. In the case of an $Nd_{30}Zr_{30}Co_{30}Al_{10}$ alloy glass, the DSC plot did not show the presence of two T_gs [110]. But, this glass was expected to show phase separation since Nd and Zr have a positive heat of mixing (+10 kJ mol^{-1}). TEM studies have also confirmed phase separation. By measuring the viscosity as a function of temperature, and noting that the viscosity showed a significant decrease in two different temperature regimes, these authors were able to conclude that this glass had phase separated into Nd-rich and Zr-rich glasses.

Most of the reports of phase separation have been concerned with a homogeneous glass phase separating into two glasses with different compositions. However, Park [135] has shown that by combining elements that possess positive heats of mixing, it should be possible to separate a homogeneous glass into three different glassy phases. The $Ti_{24}Y_{18}La_{18}Al_{22}Co_{18}$ BMG alloy contains elements that have zero or positive heats of mixing (Ti–Y: 58 kJ mol^{-1}; La–Ti: 20 kJ mol^{-1}; Y–La: 0 kJ mol^{-1}). Accordingly, the XRD pattern of the rapidly solidified $Ti_{24}Y_{18}La_{18}Al_{22}Co_{18}$ BMG alloy showed three diffuse peaks located at ~29.5°, ~32.1°, and ~41.2°, corresponding to the Y-rich, La-rich, and Ti-rich glassy phases, respectively. The compositions of the three different glassy phases were determined by the EDS method to be $Y_{40.5}La_{30.2}Ti_{5.7}Al_{31.7}Co_{1.9}$ (Y-rich), $La_{48.4}Y_{27.9}Ti_{14.5}Al_{16.5}Co_{2.7}$ (La-rich), and $Ti_{53.5}Y_{1.6}La_{0.6}Al_{7.5}Co_{36.8}$ (Ti-rich). TEM studies helped in understanding the microstructural evolution.

5.7.4 Crystallization

The crystallization of metallic glasses starts when the glassy alloy is heated to T_x or above. This T_x is a kinetic temperature, and its value depends on the heating rate employed to determine it. The higher the heating rate, the higher the T_x value, identified in the DSC curve as the starting temperature of the first exothermic peak. The crystallization temperatures of some of the BMG alloys are listed in Table 5.1. But, crystallization of metallic glasses can also occur at temperatures lower than T_x, for example, above T_g, that is, in the supercooled liquid region of the glass, or, in principle, at temperatures even below T_g. Theoretically, it is possible for a metallic glass to crystallize even at room temperature, but for all practical purposes, most of the metallic glasses can be considered "stable," that is, they do not undergo any transformation at room temperature. But, the kinetics of crystallization at low temperatures will be so slow that it will be impractical (and almost impossible) to study the crystallization behavior in a reasonable amount of time.

Some of the glassy alloys exhibit more than one exothermic peak in their DSC plots. Each exothermic peak corresponds to the heat evolved due to the formation of a crystalline phase. Therefore, to identify the crystalline phase that has formed at each exothermic peak, the glassy alloy is continuously heated to a temperature just beyond the exothermic peak, quenched from that temperature, and subjected to XRD and/or TEM studies to determine the nature (crystal structure and lattice parameters) of the phase and its microstructure. As mentioned earlier, the presence of a single peak in the DSC plot may also correspond to the eutectic crystallization taking place in the alloy resulting in the formation of more than one phase.

Isothermal crystallization studies were conducted on melt-spun $Zr_{65}Al_{7.5}Cu_{27.5}$ glassy ribbons at different temperatures, T_a, between 663 and 723 K, in the supercooled liquid region. These temperatures are above the glass transition temperature, T_g, of 646 K but below the crystallization temperature, T_x, of 731 K. The variation of the fraction crystallized, X, as a function of time, t_a, at each temperature showed a typical sigmoid curve in the temperature range investigated and a linear relationship was observed in the Johnson–Mehl–Avrami (JMA) plots when $\ell n[-\ell n\{1 - x(t)\}]$ was plotted against $\ell n(t_a)$. In Figure 5.22, the Avrami exponent (n) calculated from the slope of the linear JMA plots is plotted as a function of the annealing temperature. It may be noted that the n value is not constant, but increases continuously from 3.0 to 3.7 with increasing T_a.

It was also noted that the nature of the crystalline phase produced on long-time annealing was different depending on whether the annealing temperature was below or above T_g. For example, annealing for 2.5 h at 613 K (below T_g) resulted in the formation of a bct Zr_2Cu phase, while annealing for 20 min at 713 K (above T_g) led to the formation of bct ZrAl phase. This behavior was independent of the actual temperature, as long as the temperature was maintained below (i.e., in the glassy state) or above T_g (i.e., in the supercooled liquid state). Since the crystalline phase formed is the same and independent of the annealing temperature, but the Avrami exponent, n, changes continuously, it was concluded that the mechanism of crystallization was different in the two temperature regimes. In the glassy state, the number of crystalline nuclei is constant, while in the supercooled liquid state, the nucleation rate is constant. In both cases, the crystal nuclei grow through an interfacial reaction-controlling process. Further, the n value of near 4 at temperatures above T_g suggested that the nucleation rate had a distinct temperature dependence.

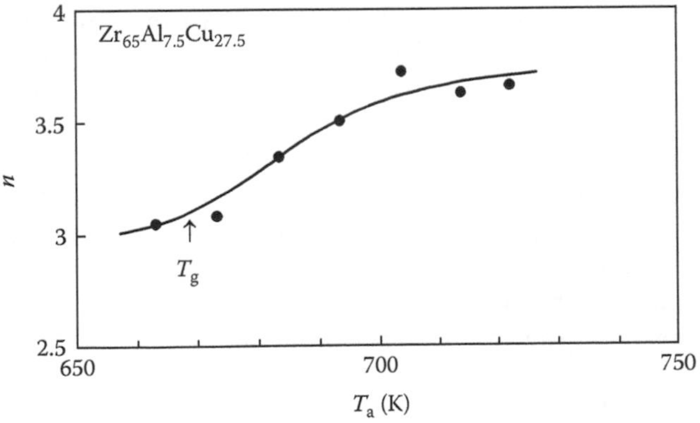

FIGURE 5.22
Variation of the Avrami exponent, n, with annealing temperature, T_a, during the isothermal annealing of glassy $Zr_{65}Al_{7.5}Cu_{27.5}$ alloy.

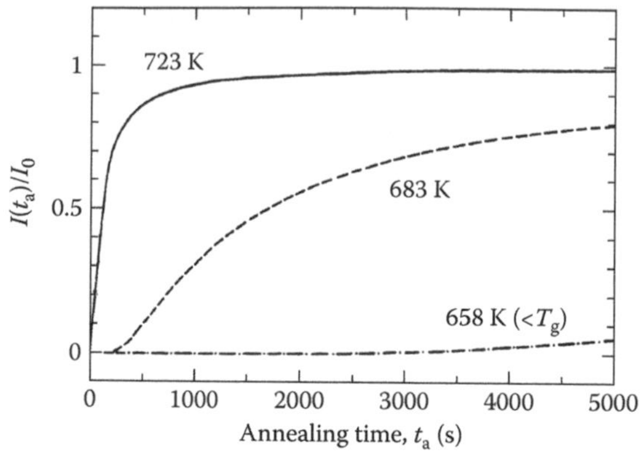

FIGURE 5.23
Variation of the reduced homogeneous nucleation rate, $I(t_a)/I_0$ evaluated from the incubation time for the precipitation of the $Zr_2(Cu,Al)$ phase with annealing time, t_a, for the glassy $Zr_{65}Al_{7.5}Cu_{27.5}$ alloy annealed at 658, 683, and 723 K.

The nucleation rate is a function of both t_a and T_a. For isothermal annealing, the nucleation rate, $I(t_a)$, at any temperature can be expressed as a function of annealing time, t_a, according to the equation

$$I(t_a) = I_0 \exp\left(-\frac{\tau}{t_a}\right) \tag{5.12}$$

where:
- I_0 is the steady-state homogeneous nucleation rate
- τ is the incubation time

From the experimentally determined values of the incubation time, τ, the variation of $I(t_a)/I_0$ at different temperatures both below and above T_g is plotted against t_a in Figure 5.23. The nucleation rate increases after an incubation time and then becomes saturated. It may also be noted that the increase in $I(t_a)/I_0$ becomes significant in the temperature range above T_g.

By plotting the variation of the incubation time as a function of the reciprocal of the annealing temperature, and noting its deviation from linearity for the $Zr_{65}Al_{7.5}Cu_{27.5}$ glassy alloy, it was noted that nucleation occurred through a non-Arrhenius-type thermal activation process. The activation energy was calculated to change from 400 kJ mol^{-1} in the glassy solid to 260 kJ mol^{-1} in the supercooled liquid state for the nucleation of the Zr_2Cu and ZrAl phases, and from 370 kJ mol^{-1} in the glassy solid to 230 kJ mol^{-1} in the supercooled liquid state for crystallization. This difference in the crystallization kinetics was thought to originate from the differences in the atomic mobility and/or viscosity in the two different conditions of the alloy.

Since the nucleation and growth of the crystalline phases are found to obey a non-Arrhenius behavior, it is expected that the crystallization behavior is strongly affected by the ΔT_x value. To test this hypothesis, the crystallization behavior of binary Zr–Cu and ternary Zr–Cu–Al glassy alloys with different ΔT_x values was investigated. Figure 5.24a shows the DSC plot of the $Zr_{65}Al_{7.5}Cu_{27.5}$ glassy alloy heated at a rate of 0.67 K s^{-1} (40 K min^{-1}). The T_g and T_x values are indicated in Figure 5.24a. The value of the peak crystallization

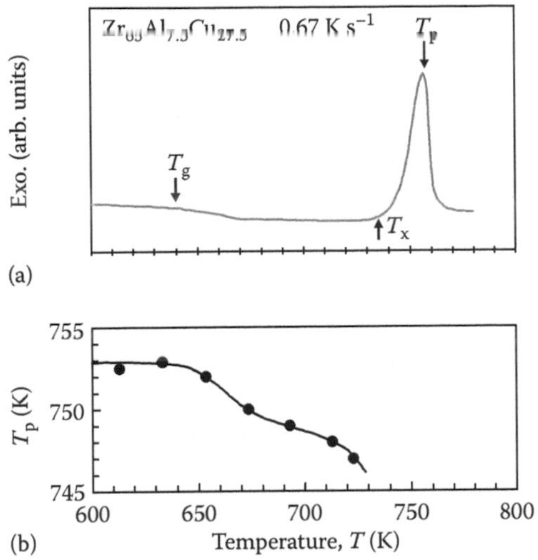

FIGURE 5.24
(a) DSC plot of the $Zr_{65}Al_{7.5}Cu_{27.5}$ glassy alloy continuously heated at a rate of 0.67 K s^{-1} (40 K min^{-1}). The T_g, T_x, and T_p values are indicated. (b) Variation of the peak temperature, T_p, in the exothermic reaction due to crystallization with heating temperature, T_a.

temperature, T_p, is also noted. By continuously heating the glassy alloy to a temperature, T_a, and cooling from that temperature down to room temperature, the DSC curve is obtained again, and the T_p values noted. Figure 5.24b shows the variation of T_p as a function of the annealing temperature, T_a. It may be noted that the value of T_p remained unchanged from the as-quenched value as long as $T_a < T_g$. But, in the supercooled liquid region, that is, when $T_a > T_g$, the peak temperature, T_p, decreased in two stages, attributed to increased nucleation rate resulting from a rapid decrease in viscosity. A similar trend was observed for the binary $Zr_{67}Cu_{33}$ glassy alloy, while for the Zr–Fe alloy that did not exhibit a glass transition, the T_p decreased sharply just below the crystallization temperature, T_x. This result clearly indicates that the peak temperatures for nucleation and growth are well separated for glassy alloys exhibiting a significant width of the supercooled liquid region (ΔT_x), whereas for glassy alloys that do not show a T_g, these two peak temperatures overlap. This difference is expected to reflect in the nature of the crystallized product obtained from the supercooled liquid region. Accordingly, the crystallized structure was examined as a function of the heating rate.

The activation energy for crystallization was evaluated from the Kissinger plot as 230 kJ mol^{-1} for $Zr_{65}Al_{7.5}Cu_{27.5}$ with $\Delta T_x = 90$ K, 360 kJ mol^{-1} for $Zr_{67}Cu_{33}$ with $\Delta T_x = 50$ K, and 390 kJ mol^{-1} for $Zr_{67}Fe_{33}$ without a glass transition. The grain size of the $Zr_2(Cu,Al)$ phase of the $Zr_{65}Al_{7.5}Cu_{27.5}$ alloy heated at a rate of 2.7 K s^{-1} (160 K min^{-1}) was about five times as large as that of the sample heated at a rate of 0.17 K s^{-1} (10 K min^{-1}). This result suggests that the contribution to the growth rate increases with increasing heating rate as compared with the contribution to the nucleation rate. However, no significant change in the grain size of the Zr_2Cu phase with heating rate was seen for the $Zr_{67}Cu_{33}$ alloy with the much smaller ΔT_x value. Thus, the increase in the contribution to the growth rate with increasing heating rate becomes significant at temperatures just below T_x in the wide supercooled liquid region. This result also supports the belief that the peak temperatures for the nucleation and growth reactions are well separated for the Zr–Al–Cu glassy alloy with a wide supercooled liquid region.

The crystallized $Zr_{65}Al_{7.5}Cu_{27.5}$ alloy consists of Zr_2Cu and ZrAl phases. However, the significant changes in the *n* value and the activation energies for the nucleation and crystallization of the Zr_2Cu phase allow us to expect the morphology of the Zr_2Cu phase to be dependent on the annealing temperature. Bright-field electron micrographs of the $Zr_{65}Al_{7.5}Cu_{27.5}$ glassy alloy annealed for 1020 s (17 min) at 693 K (above T_g) and for 780 ks (9 days) at 613 K (below T_g) showed significant differences in the morphology, size, and distribution of the Zr_2Cu phase in these samples. The Zr_2Cu phase precipitated from the supercooled liquid has a dendritic morphology with a preferential growth direction, indicating that the redistribution of the constituent elements at the liquid–solid interface was necessary for the growth of the crystalline phase. However, the Zr_2Cu phase, which precipitated from the glassy solid, has a nearly spherical morphology with a rather smooth interface, suggesting that the growth of the Zr_2Cu phase took place in the absence of significant redistribution of the constituent elements at the interface between Zr_2Cu and the glassy phases. Further, the distribution of the Zr_2Cu phase appears to be more homogeneous when formed from the glassy solid.

The preferential growth direction of the bct Zr_2Cu phase is [110] and the facet plane lies along the (1 0 0) plane, expected for dendritic precipitates with a bct structure. The result also confirms that the bct Zr_2Cu phase precipitates with a typical dendritic mode, when formed from the supercooled liquid.

The structure and morphology of the crystallized phases also appear to be a function of the Al content in the alloy. Accordingly, when the crystallized structures in the binary $Zr_{67}Cu_{33}$, and ternary $Zr_{65}Al_{7.5}Cu_{27.5}$ and $Zr_{65}Al_{20}Cu_{15}$ alloys were investigated by TEM methods, it was noted that the Zr_2Cu phase in the binary Zr–Cu alloy had a smooth interface, and growth appeared to have occurred in a polymorphous mode without the necessity for redistribution of the solute atoms. On the other hand, the alloy containing 7.5 at.% Al grew with a dendritic morphology accompanied by significant ruggedness of the interface, suggesting that solute redistribution (particularly Al) occurred at the solid–liquid interface for the growth of the $Zr_2(Cu,Al)$ phase. A further increase in Al content to 20 at.% caused the formation of a much finer structure consisting of Zr_2Cu and Zr_2Al phases by simultaneous precipitation.

5.8 Effect of Environment

It is well known that the crystallization behavior of metallic glasses depends on the environment, and this has been shown to be true especially in the case of reactive glasses, for example, those based on Zr and Fe. By studying the isothermal crystallization behavior of $Fe_{77}Gd_3B_{20}$ melt-spun ribbons both in air and vacuum, Mihalca et al. [136] noted that, on annealing at 828 K, the volume fraction of the crystalline phase was 0.7 in air and only 0.4 in vacuum. This significant difference in the crystallization behavior was attributed to surface oxidation, which reduces thermal stability.

The isothermal crystallization kinetics of $Zr_{41}Ti_{14}Cu_{12.5}Ni_{10}Be_{22.5}$ were studied by Wong and Shek [137] under different air pressures. With increasing annealing time at 653 K, they noted that the times required for the appearance of sharp crystalline peaks in the XRD patterns were different and were 3, 8, and 20 min, corresponding to air pressure of 5, 260, and 760 Torr, respectively. These results suggest that the crystallization of the glassy phase started earlier with decreasing air pressure during annealing. X-ray photoelectron spectroscopy investigations on these samples showed that the sample surfaces were enriched in oxygen and that the oxygen penetrated deeper into the sample that was annealed at 5 Torr pressure. A similar result was also reported by Kiene et al. [138]. Oxygen is known to speed up the crystallization process by promoting extra heterogeneous sites for nucleation. When the samples were annealed under atmospheric pressure (760 Torr), a thick surface oxide layer formed, which prevented further ingress of oxygen into the sample. This delayed the formation of heterogeneous sites and consequently the amount of the crystalline phase. On the other hand, when the oxygen pressure was low (say 5 Torr), the thickness of the surface oxide layer was small and therefore more oxygen could diffuse inside and accelerate the crystallization process. Hence, it was concluded that the crystallization kinetics were faster at low partial pressure of oxygen.

If this is the mechanism by which the crystallization kinetics are altered, then the nature of the oxide film should also play an important role. It is known that the oxide layer formed on Al is impervious to oxygen, and that is why Al and its alloys are oxidation resistant. Therefore, if the same investigations were carried out on Al alloys, one should not observe any effect of the partial pressure of oxygen on the crystallization behavior. Unfortunately, however, we do not have any Al-based BMGs at present.

Huang et al. [139] studied the nature of the oxide layer formed during oxidation of both crystalline and glassy $Zr_{55}Cu_{30}Al_{10}Ni_5$ alloys. They reported that tetragonal (*t*) ZrO_2 was formed during oxidation of the BMG alloy while both *t*-ZrO_2 and monoclinic (*m*)-ZrO_2 were detected on its crystalline counterparts of the same chemical composition. It was also reported that crystallization induced changes in the coefficient of thermal expansion and Young's modulus of the oxide and the substrate. This resulted in a difference in the mechanical properties related to the interfacial and thermal stresses and led the formation of *m*-ZrO_2. Formation of *m*-ZrO_2 on the crystalline substrate was attributed to the inward diffusion of oxygen ions, while in the BMG alloy the oxidation was dominated by outward diffusion of metal ions and inward diffusion of oxygen ions.

de Oliveira et al. [140] investigated the crystallization behavior of melt-spun ribbons and copper mold cast 5 mm diameter rods in the $Zr_{55}Al_{10}Ni_5Cu_{30}$ system. While the melt-spun ribbons were fully glassy, the bulk rods contained about 20 vol.% of a crystalline phase, identified as a big-cube fcc oxide with Ti_2Ni-type structure and lattice parameters near those of Zr_4Cu_2O. Since a fully glassy phase was not obtained, the authors concluded that their alloys probably contained oxygen. This was inferred from the fact that the observed ΔT_x was only 60 K, smaller than the expected value of 90 K [141]. Formation of the big-cube structure could also be attributed to the presence of oxygen [142]. Furthermore, the nature of the crystalline phases formed was also different in the alloys containing oxygen.

Sordelet et al. [143] intentionally added oxygen to Zr–Pt alloys, by mixing the metals with ZrO_2 before melting, to investigate its effect on the structure of melt-spun $Zr_{80}Pt_{20}$ ribbons. The measured oxygen contents were in the range 184–4737 ppm by weight. They reported that ribbons containing <500 ppm of oxygen were fully crystalline and consisted predominantly of a metastable β-Zr phase. With increasing oxygen content, the ribbons were fully glassy (at 1053 ppm oxygen) and mixed glassy + quasicrystalline phases (at 1547 ppm oxygen). At the highest oxygen content of 4737 ppm, the ribbon contained glassy and nanocrystalline phases. From these studies, it becomes clear that the presence of oxygen in the alloys has a significant effect on the nature of the phases formed after quenching and also those formed on crystallization. Therefore, it is essential that the alloys are clean and devoid of impurities, especially in reactive metals. Otherwise, comparing results from different investigators becomes difficult.

Köster et al. [144–146] reported the formation of quasicrystalline phases during the crystallization of melt-spun glassy $Zr_{65}Cu_{17.5}Ni_{10}Al_{7.5}$ and $Zr_{69.5}Cu_{12}Ni_{11}Al_{7.5}$ alloys. The DSC plots of these glassy alloys exhibited the glass transition temperature (T_g) in addition to the exothermic crystallization temperatures (one at 734 K in the case of $Zr_{65}Cu_{17.5}Ni_{10}Al_{7.5}$ and two, at 695 and 740 K, in the case of the $Zr_{69.5}Cu_{12}Ni_{11}Al_{7.5}$ alloy). By conducting isothermal crystallization studies at temperatures close to T_g, for example, at 633 K for 1 h in the $Zr_{69.5}Cu_{12}Ni_{11}Al_{7.5}$ glassy alloy (T_g = 645 K), they observed the formation of quasicrystals by a primary crystallization mechanism. They determined the structure of the quasicrystals to be primitive icosahedral with a quasilattice constant of a = 0.253 nm. The two exothermic peaks in the case of the $Zr_{69.5}Cu_{12}Ni_{11}Al_{7.5}$ alloy were attributed to the formation of quasicrystals and the decomposition of the quasicrystals, respectively [146]. Since the formation of quasicrystals was by a primary crystallization mode and that the growth of the quasicrystals was time dependent ($r \propto t^{1/2}$), the authors concluded that solute diffusion was involved.

Quasicrystal formation was not reported during the decomposition of Zr-based glassy alloys in earlier investigations [147–149]. It was pointed out by Eckert et al. [142] that the quasicrystals formed only in the early stages of crystallization and that their formation was very sensitive to the oxygen concentration in the alloy. For example, at low oxygen concentrations of 0.2 at.% in the $Zr_{65}Cu_{17.5}Ni_{10}Al_{7.5}$ alloy, the decomposition products consisted of the quasicrystalline, tetragonal $Cu(Al,Ni)Zr_2$ and hexagonal Zr_6NiAl_2 phases. Increasing oxygen concentration (up to 0.8 at.%) led to the formation of quasicrystalline and metastable fcc-Zr_2Ni phases and these formed at lower temperatures and over shorter times. The quasicrystal phase was reported to have a quasilattice constant of a = 0.4846 nm. This observation highlights the importance of controlling the oxygen content in the alloy to control its crystallization behavior.

Chen et al. [150] studied the crystallization behavior of the $Zr_{65}Al_{7.5}Ni_{10}$ $Cu_{12.5}Ag_5$ glassy alloy in the supercooled liquid region and observed the formation of icosahedral quasicrystals through a systematic TEM diffraction analysis. It was concluded that the addition of Ag to the Zr-based alloy promoted the formation of the quasicrystalline phase. It has been reported that Zr–Al–Ni ternary alloys have a high GFA [98] and that they crystallize through a single exothermic reaction with the simultaneous precipitation of Zr_4Al_3 and Zr_2Ni phases [68]. However, when noble metals, such as Ag, Au, Pd, or Pt, were added to the Zr–Al–Ni alloys, the crystallization mechanism was changed. The glassy phase in these noble metal-containing alloys transformed into the crystalline state in a two-stage process. The annealing of the glassy alloy at the first exothermic peak resulted in the formation of a quasicrystalline phase. The second stage of crystallization led to the formation of Zr_4Al_3 + Zr_2Ni + Zr_2Pd (or Zr_2Au or ZrPt) from the quasicrystalline phase [151]. Alloys containing the quasicrystalline phases were also reported to be ductile. The formation of icosahedral quasicrystalline phases in these noble metal-containing alloys has been attributed to the positive heat of mixing of the noble metals with Ni.

Murty et al. [152] confirmed that a quasicrystalline phase had formed during the crystallization of a $Zr_{65-x}Cu_{27.5}Al_{7.5}O_x$ (with x = 0.14, 0.43, and 0.82) glassy alloy and that there was significant oxygen enrichment in the quasicrystalline phase, as revealed by their 3DAP studies. Since no quasicrystalline phase had formed in the alloy with an oxygen content 0.14 at.%, the authors concluded that the quasicrystalline phase in Zr-based glasses was stabilized by oxygen. It was also shown that while the oxygen-rich alloys crystallized through two stages (the initial precipitation of a quasicrystalline phase followed by the precipitation of the stable $Zr_2(Cu,Al)$ phase), the low-oxygen (0.14 at.%) alloy crystallized through a polymorphous mode. Chen et al. [153] also studied the oxygen redistribution in a $Zr_{65}Cu_{15}Al_{10}Pd_{10}$ glassy alloy annealed at 730 K for 1 h (T_g of the glass was 709 K). They reported that nanocrystals of metastable phases formed during crystallization and that there was significant oxygen enrichment (up to 4 at.%) in these metastable phases. On the other hand, there was virtually no oxygen in the remaining glassy phase. Thus, they were able to show that impurity oxygen promotes crystallization by forming metastable phases enriched in oxygen.

Different investigators have reported different quasilattice constants for the icosahedral quasicrystalline phases. While Köster et al. [144] reported the quasicrystals to be primitive with a quasilattice constant of 0.253 nm, Wollgarten et al. [154] also reported them to be primitive, but with a quasilattice constant of 0.753 nm. Even though the importance of this quasilattice constant is not clear at this stage, Jiang et al. [155] monitored the quasilattice constant during the glass → quasicrystalline phase transformation. By annealing $Zr_{65}Al_{7.5}Ni_{10}Cu_{7.5}Ag_{10}$ metallic glassy ribbons at 663 K (between the T_g of 628 K and the first T_x of 672 K), the authors noted that the average quasilattice constant for the primitive icosahedral phase decreased from 0.4843 nm at 15 min of annealing to about 0.482 nm at 90 min of annealing and then remained constant. The authors speculated that the quasicrystalline phase had formed initially over a composition range, and that during annealing there was a redistribution of atoms. Since the Zr atom is quite large, it was suggested that the Zr atoms diffused out of the quasicrystalline phase during the annealing process. Chemical analysis by EDS methods indicated that the average Zr content in the quasicrystalline phase was 62.1 at.% on annealing for 10 min and that it was 58.4 at.% on annealing for 60 min.

Since there has been confusion in the literature regarding whether the glass-to-quasicrystal formation is polymorphous [156] or primary [157] in nature, Liu et al. [158] undertook detailed investigations on a 2 mm diameter rod of $Zr_{65}Ni_{10}Cu_{7.5}Al_{7.5}Ag_{10}$ glassy alloy samples using TEM (including high-resolution) methods. The T_g and T_x for this glass were reported to be 631 and 669 K, respectively, measured at a heating rate of 5 K min^{-1}, and the oxygen content in the glass was 1200 ppm. They noted that the glass-to-quasicrystal transformation did involve atomic diffusion; therefore, it was concluded that the transformation was not polymorphous in nature. Further, they reported that this system underwent a series of interprocesses and followed the sequence glass → fcc Zr_2Ni → tetragonal Zr_2Ni → tetragonal Zr_2Ni with a domain structure → quasicrystal. It is important to note that the formation of quasicrystals took place after the completion of the first exothermic crystallization peak and that the quasicrystals did not form in the early stages. In fact, quite surprisingly, some metastable and stable phases had formed prior to the formation of the quasicrystalline phase.

5.9 Effect of Pressure

Metallic glasses were generally produced in a ribbon, wire, or powder form, due to the requirement for high critical cooling rates. Even though the situation has now changed with the production of BMGs, it was noted in Chapter 3 that there is a limit to the section thickness of BMGs in some cases. Hence, if one is interested in obtaining bulk samples for "real" applications, it becomes necessary to consolidate these ribbon, wire, or powder forms to full density by applying high pressures and temperatures. Some of the methods commonly used for such purposes are hot pressing, hot extrusion, hot isostatic pressing, and other recently developed methods such as spark plasma sintering. It is also important to remember that the properties of the glasses could be retained only when crystallization did not occur in them as a result of the application of pressure. Thus, it becomes useful to evaluate the stability of glasses when exposed to high pressures. This happens to be critical for optimizing the consolidation process parameters to obtain fully dense and bulk samples with the desired structure and properties. Studying the effect of pressure on the crystallization of metallic glasses is not only of scientific interest from a fundamental point of view, but will also be commercially useful to understand the deformation of these glasses in the supercooled liquid region.

The application of pressure during crystallization is known to affect the kinetics of transformation and also the nature of the crystallization product [159–161]. The application of pressure during crystallization of metallic glasses may lead to four different effects.

1. Since there is an increase in the density of the product on crystallization (the glassy alloys are about 1%–2% less dense than their crystalline counterparts), it is natural to expect that the application of pressure would reduce the free volume in the glassy phase and, therefore, it is expected that crystallization will be accelerated. Such a process could easily happen when the glass crystallizes by a polymorphous mode.
2. Due to the retarded mobility of atoms (diffusivity) under high pressures, atomic diffusion is reduced and therefore crystallization is retarded, as evidenced by the increase of crystallization temperatures. Since atomic diffusion is required for primary and eutectic-type crystallization modes, the application of pressure is expected to retard the crystallization of metallic glasses when the transformation takes place by any of these modes.
3. The relative Gibbs free energies of the glassy and other competing crystalline phases and also the activation barriers could be altered by the application of pressure. Consequently, metastable phases could form, the relative amounts of the different phases could be different, or alternately, different crystallization paths could be followed. The situation will be decided by the sign and magnitude of the variation of the crystallization temperature with pressure, that is, dT_x/dP.

   As an example, during the primary crystallization of Fe–B glassy alloys, α-Fe is formed at atmospheric pressure. However, when crystallization is conducted at pressures above 100 kbar, the metastable hcp ε-Fe phase was found to form [162]. Similarly, instead of the equilibrium tetragonal Nb_3Si phase, the cubic A15 Nb_3Si phase formed in the Nb–Si system during crystallization at high pressures in the glassy Nb–Si alloys [163]. Again, in the case of the crystallization of the $Zr_{41}Ti_{14}Cu_{12.5}Ni_{10}Be_{22.5}$ glassy alloy, Yang et al. [164] reported that the primary crystallized phase was the same at all pressures studied, but the subsequent phase-formation sequence was different at different temperatures.
4. The last effect of the application of pressure to metallic alloys is that amorphization could occur, that is, pressure-induced amorphization takes place [165–167]. For example, Wang et al. [165] reported that by cooling $Zr_{41}Ti_{14}Cu_{12.5}Ni_{10}Be_{22.5}$ liquid at a high pressure of 6 GPa, they were able to obtain a high-density glassy alloy that had a structure and properties different from the low-density glassy alloy obtained by water quenching the melt.

A few investigations have been carried out on the effects of high pressure on the crystallization of metallic glasses and BMGs, as listed in Table 5.6.

TABLE 5.6

Effect of Pressure in Increasing the Crystallization Temperature of Bulk Metallic Glasses

Composition	Pressure Range Used (GPa)	Rate of Increase of T_x (K GPa^{-1})	References
$Al_{89}La_6Ni_5$	0–4	Decrease at a rate of 50 between 0 and 1 GPa and then increase at a rate of 25	[168]
$Fe_{72}P_{11}C_6Al_5B_4Ga_2$	0–2.4	30 (T_x dropped at higher pressures between 2.4 and 3.2)	[161]
$Mg_{60}Cu_{30}Y_{10}$	0–4	16	[169]
$Pd_{40}Ni_{40}P_{20}$	0–4.2	11	[170]
$Pd_{40}Cu_{30}Ni_{10}P_{20}$	0–4	11	[171]
$Zr_{66.7}Pd_{33.3}$	0–4	22	[149]
$Zr_{70}Pd_{30}$	0–3	11 ± 3 for quasicrystalline phase 9 ± 4 for intermetallic phase	[172]
$Zr_{65}Al_{7.5}Ni_{10}Cu_{7.5}Ag_{10}$	0–4.2	9.4 for T_{x1} No change for T_{x2}	[173]
$Zr_{48}Nb_8Cu_{14}Ni_{12}Be_{18}$	0–4.4	9.5	[174]
$Zr_{41.2}Ti_{13.8}Cu_{12.5}Ni_{10}Be_{22.5}$	0–3	19	[175]
$Zr_{46.8}Ti_{8.2}Cu_{7.5}Ni_{10}Be_{27.5}$	0–4.2	1.7	[176]
$Zr_{41}Ti_{14}Cu_{12.5}Ni_{10}Be_{22.5}$	0.5–6.5	12.8 (a sudden drop occurred at 5.6 GPa)	[164]

An important observation that can be made is that the crystallization temperature of the glassy alloy generally increases with increasing pressure, noting that the rate of increase is different for different alloy systems, and ranged from as low as 9 K GPa^{-1} in a $Zr_{48}Nb_8Cu_{14}Ni_{12}Be_{18}$ glass to as high as 30 K GPa^{-1} in an $Fe_{72}P_{11}C_6Al_5B_4Ga_2$ glass. There have also been instances where the T_x decreased with increasing pressure and either increased or remained almost constant with further application of pressure. Figure 5.25 shows the possible variations of T_x with pressure.

The effect of pressure on the change of crystallization temperature can be rationalized in the following way. Crystallization can be considered as a process involving the nucleation of a crystal with a size larger than the critical size and its subsequent growth. The rate of nucleation, I, can be represented by the equation:

$$I = I_0 \exp\left(-\frac{\Delta G^* + \Delta G^d}{RT}\right) \tag{5.13}$$

where:

- I_0 is a constant
- ΔG^* is the thermodynamic activation barrier, that is, free energy required to form the critical nucleus
- ΔG^d is the activation energy for diffusion (to transport atoms across the interface)
- R is the universal gas constant
- T is the temperature

The sum of $\Delta G^* + \Delta G^d = \Delta G$ is the total energy required for the nucleation process, and this value is responsible for the change of the crystallization temperature. ΔG^* can be expressed as

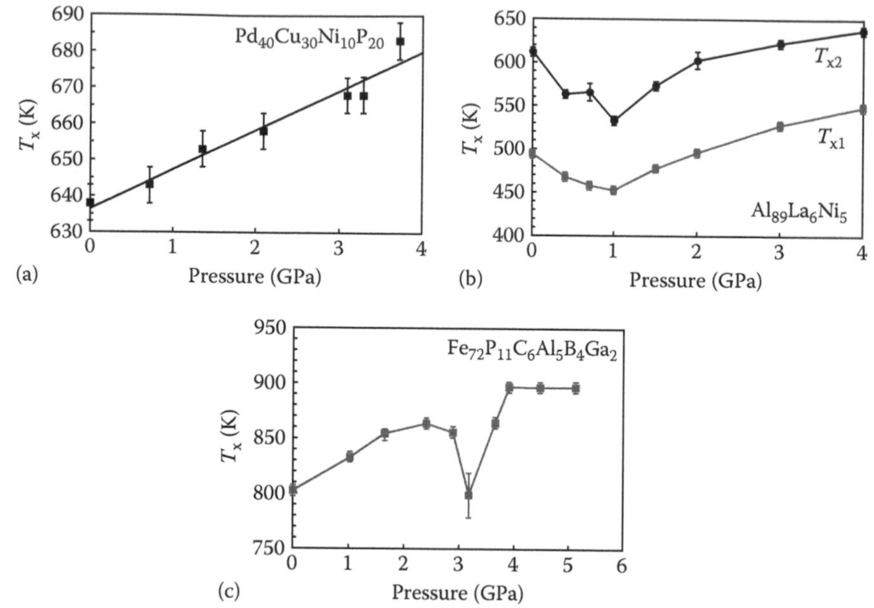

FIGURE 5.25
Variation of T_x with pressure in bulk metallic glassy alloys. Note that the T_x usually increases with increasing pressure, although there are cases where either a decrease or no change has also been observed. Three typical examples are shown in (a) $Pd_{40}Cu_{30}Ni_{10}P_{20}$ glass (Reprinted from Jiang, J.Z., et al., *Europhys. Lett.*, 54, 182, 2001. With permission.), (b) $Al_{89}La_6Ni_5$ glass (Reprinted from Zhuang, Y.X., et al., *Appl. Phys. Lett.*, 77, 4133, 2000. With permission.), and (c) $Fe_{72}P_{11}C_6Al_5B_4Ga_2$ glass. (Reprinted from Jiang, J.Z. et al., *J. Appl. Phys.*, 87, 2664, 2000. With permission.)

$$\Delta G^* = \frac{16\pi\sigma^3}{3\Delta G_v^2} = \frac{16\pi\sigma^3}{3(G_c - G_a)^2} \tag{5.14}$$

where:

- σ is the interfacial energy between the amorphous and crystalline phases
- G_c and G_a are the Gibbs free energies of the crystalline and amorphous phases, respectively

At a given temperature and pressure, ΔG^* can be expressed as

$$(\Delta G^*)_{P,T} = \frac{16\pi\sigma^3 (V_c)^2}{3\left[P(V_a - V_c) - \Delta G^{a\to c} + E\right]^2} \tag{5.15}$$

where:

- V_c and V_a are the molar volumes of the crystalline and amorphous phases, respectively
- $\Delta G^{a\to c}$ $= G_c - G_a$
- E is the elastic energy induced by the volume change when the phase transformed from the amorphous to the crystalline state

Assuming a negligible pressure dependence of $\Delta G^{a\to c}$, E, and σ, we can see that ΔG^* decreases with increasing pressure and, therefore, crystallization is favored. By combining the two equations, and assuming that σ is a constant, the variation of G with pressure, P, at a constant temperature, T, can be expressed as

$$\left(\frac{\partial G}{\partial P}\right)_T = -\frac{32\pi\sigma^3}{3\left(\Delta G^{a\to c}\right)^3}\left(V_c - V_a\right) + \left(\frac{\partial \Delta G^d}{\partial P}\right)_T \tag{5.16}$$

Since the crystalline phase is more dense than the corresponding amorphous phase, $V_c < V_a$, and since it is more stable than the amorphous phase, $\Delta G^{a\to c}$ is negative. Consequently, the first term in Equation 5.16 is always negative and therefore crystallization will be promoted. On the other hand, the variation of the activation energy increases with pressure (the atomic mobility is reduced with pressure and hence the diffusivity is reduced), and therefore, the second term in Equation 5.16 is always positive and, consequently, crystallization is retarded. Thus, whether increasing pressure promotes or retards crystallization is determined by the magnitudes of the two terms in Equation 5.16.

When the glassy alloy crystallizes in a polymorphous mode, atomic redistribution is not involved; thus, the second term becomes negligible. Therefore, increasing pressure always promotes crystallization. For the eutectic and primary crystallization, atomic rearrangement is necessary and, therefore, the increase of ΔG^d with pressure retards the crystallization process. Thus, the sum of the two terms is likely to be a minimum at some value of the pressure and this could explain the minimum observed in the T_x versus P plots in the $Al_{89}La_6Ni_5$ [168] and $Fe_{72}P_{11}C_6Al_5B_4Ga_2$ [169] glassy alloy systems.

5.10 Effect of Magnetic Field

There have been some investigations on the crystallization behavior of BMGs under the influence of a magnetic field. Miglierini et al. [177] studied the crystallization kinetics of $Fe_{90}Zr_7B_3$ melt-spun ribbons during magnetic annealing using nuclear forward scattering (NFS) of synchrotron radiation. They observed that when the ribbons were isothermally annealed at 753 K under a weak magnetic field of B_{ex} = 0.652 T, rapid crystallization was observed in comparison with zero-field conditions. This effect was attributed to energetic perturbations of magnetic interactions in comparison with the thermal energy. This had enhanced the nucleation centers. Using these techniques, the authors were able to characterize not only the initial and final stages of the structure, but also the intermediate transition states.

Zhuang et al. [178] studied the crystallization behavior of $Fe_{84}B_{10}C_6$ melt-spun ribbons under a high magnetic field of 12 T. It was observed that the crystallization kinetics during annealing were faster under the magnetic field and that the strongest influence was at temperatures close to T_x. The volume fraction and the number of precipitated α-Fe crystals annealed under the 12 T magnetic field were much larger than those in the alloys annealed at the same temperature without the magnetic field. The magnetic field, however, did not have any effect on the average grain size of the α phase. It has been suggested that both the reduction of ΔG^*, the activation barrier for forming a critical nucleus size, and the enhancement of atomic diffusion caused by the high magnetic field were responsible for the above observations.

5.11 Effect of Ion Irradiation

Ion irradiation is expected to improve the atomic diffusion and, consequently, enhance the crystallization of metallic glasses [179]. Since metallic glasses are expected to be used in a variety of environments, and virtually nothing is known about the effects of ion irradiation on metallic glasses, some investigations have been carried out. Chen et al. [180] studied the effect of 7 MeV Xe^{26+} ions on the crystallization behavior of a 3 mm diameter $Zr_{50.7}Al_{12.3}Cu_{28}Ni_9$ BMG alloy. Needle-like precipitates with a length of 100–200 nm and a width of 5–10 nm were observed along with some particles with a diameter of 10–50 nm. Based on interplanar spacings calculated from the electron diffraction patterns, these primary crystals were identified as belonging to the $Cu_{10}Zr_7$ phase. Since formation of the $Ni_{10}Zr_7$ phase is also a possibility, and the structures and lattice parameters of these two phases are very close to each other, the authors suggested the phase formed could be $(Ni_xCu_{1-x})_{10}Zr_7$. The formation of the needle-like precipitates was explained on the basis of anisotropic atomic diffusion along a direction normal to the (3 1 1) plane of the $(Ni_xCu_{1-x})_{10}Zr_7$ phase.

Myers et al. [181] studied the nanocrystallization behavior of melt-spun $Ni_{52.5}Nb_{10}Zr_{15}Ti_{15}Pt_{7.5}$ metallic glass irradiated with 1 MeV Ni+ ions to a fluence of 1×10^{16} cm^{-2} and a beam current of 160 nA cm^{-2}. Nanocrystals, showing band-like features, were observed within the precipitates after irradiation. Even though it is tempting to conclude that the nanocrystals formed as a result of ion irradiation, the authors conducted multiscale modeling combined with Monte-Carlo damage simulation and three-dimensional finite element analysis. Although the metallic glass had melted locally within the damage cascades, the subsequent solidification had a quenching rate (~10^{13} K s^{-1}) many orders of magnitude higher than the critical cooling rate (10^2 K s^{-1}) required for glass formation. Thus, the formation of the nanocrystals could not have been due to irradiation. Their formation was explained on the basis of structural changes caused by an enhanced system energy, due to displacement creation, which lead to enhanced atomic mobility.

5.12 Miscellaneous

Pradeep et al. [8] conducted atom-probe tomography studies on nanostructure formation in an $Fe_{73.5}Si_{15.5}Cu_1Nb_3B_7$ ribbon. They noted that nanocrystals of α-Fe(Si) formed during annealing by primary crystallization. By varying the annealing time in rapid annealing (4–10 s) or normal annealing (30–60 min), they did not observe any difference in the nanostructure formed or its chemical composition. Thus, they concluded that the annealing conditions did not have any effect on the initial phase formation. However, during the final stages of crystallization, rapid annealing resulted in the formation of ~30% smaller α-Fe(Si) nanocrystals with very high number densities of the order of ~10^{24} m^{-3} compared with conventional annealing. This was explained on the basis of the shorter processing time during rapid annealing at higher annealing temperatures, when growth is restricted.

Zhou et al. [182] investigated the crystallization behavior of $Ti_{40}Zr_{25}Ni_8Cu_9Be_{18}$ BMG alloys obtained by solidifying the melt from different temperatures. They observed that the first crystallization product to form from the metallic glass was dependent on the melt temperature from which the melt was solidified to form the glass. Thus, an icosahedral phase formed as the first crystallization product when the melt was solidified from a relatively low temperature of 1123 K. On the other hand, when the melt was solidified from a high temperature of above 1573 K, the first crystalline product was a mixture of α-Ti + icosahedral phase + an unknown phase. This difference was attributed to the different SROs present at different melt temperatures. It was also shown that metallic glasses with greater icosahedral SRO exhibit higher plasticity.

Kim et al. [183] had earlier demonstrated that a Zr-based BMG alloy crystallized at room temperature during nanoindentation using a Berkovich indenter and that the nanocrystals were 10–40 nm in size. It was suggested that the increased diffusivity due to local dilatation in the shear bands was responsible for the nanocrystallization observed. However, not all investigators have observed this phenomenon [184,185]. Some doubts were also expressed as to whether the nanocrystals had formed during specimen preparation for the TEM examination. With a view to resolving this ambiguity, Yoo et al. [186] undertook a detailed study of nanoindentation-induced nanocrystallization in a $Zr_{52.5}Cu_{17.9}Ni_{14.6}Al_{10}Ti_5$ (Vit 105) alloy. They used two different three-sided pyramidal indenters (Berkovich and cube-corner) to induce different levels of strains. Using TEM techniques, they noted that a larger amount of nanocrystals was formed at higher strains, suggesting that crystallization was induced by indentation-induced plastic deformation. Additionally, the strain field underneath the indenter was correlated with the amount of nanocrystallization using cross-sectional TEM analysis, thus unambiguously confirming that nanocrystallization had occurred by nanoindentation. It would be desirable, however, to quantitatively determine the critical strain conditions necessary for nanocrystallization under different indenters.

5.13 Concluding Remarks

The transformation of BMGs to their equilibrium state, that is, their crystallization behavior, has been discussed. It has been pointed out that only a few detailed studies of the crystallization behavior of BMGs have been undertaken and that most researchers have been investigating the crystallization behavior of melt-spun ribbons of the same composition as the BMGs. Even though the departure from equilibrium is expected to be different between the BMGs and the melt-spun metallic glasses, the crystallization behavior (nature of the crystalline phases and the kinetics of their formation) has been found to be the same in both cases. Further, the types of transformations occurring in thin metallic glass ribbons (even with a low GFA) and BMGs (with a high GFA) have also been found to be the same. This is because the state of the BMG after it is heated to a temperature beyond the glass transition temperature is the same as that of the melt-spun glass, though without exhibiting the glass transition temperature.

On annealing a BMG, it has been shown that the glass first undergoes structural relaxation and then phase separation (though not in every case) followed by crystallization. Similar to the case of melt-spun metallic glasses, polymorphous, eutectic, and primary crystallization modes have been observed in BMGs also. In contrast to the situation in melt-spun ribbons, the question of phase separation has been discussed at some length in view of the large number of reports. The effect of environmental (oxygen content) and external variables, for example, pressure, magnetic field, ion irradiation, and nanoindentation on the crystallization behavior of BMGs has also been studied in detail.

References

1. Luborsky, F.E. (1977). Perspective on application of amorphous alloys in magnetic devices. In *Amorphous Magnetism II*, eds. R.A. Levy. and R. Hasegawa, pp. 345–368. New York: Plenum Press.
2. Lu, K. (1996). Nanocrystalline metals crystallized from amorphous solids: Nanocrystallization, structure, and properties. *Mater. Sci. Eng. Rep.* R16: 161–221.
3. Inoue, A. (1998). Amorphous, nanoquasicrystalline and nanocrystalline alloys in Al-based systems. *Prog. Mater. Sci.* 43: 365–520.
4. Inoue, A., W. Zhang, T. Tsurui, A.R. Yavari, and A.L. Greer (2005). Unusual room-temperature compressive plasticity in nanocrystal-toughened bulk copper-zirconium glass. *Philos. Mag. Lett.* 85: 221–229.
5. Hoffmaan, D.C., J.-Y. Suh, A. Wiest, G. Duan, M.-L. Lind, M.D. Demetriou, and W.L. Johnson (2008). Designing metallic glass matrix composites with high toughness and tensile ductility. *Nature* 451: 1085–1090.
6. Ma, E. and J. Ding (2016). Tailoring structural inhomogeneities in metallic glasses to enable tensile ductility at room temperature. *Mater. Today* 19: 568–579.
7. Laws, K.J., D.W. Saxey, W.R. McKenzie, R.W.K. Marceau, B. Gun, S.P. Ringer, and M. Ferry (2012). Analysis of dynamic segregation and crystallization in $Mg_{65}Cu_{25}Y_{10}$ bulk metallic glass using atom probe tomography. *Mater. Sci. Eng. A* 556: 558–566.
8. Pradeep, K.G., G. Herzer, P. Choi, and D. Raabe (2014). Atom probe tomography study of ultrahigh nanocrystallization rates in FeSiNbBCu soft magnetic amorphous alloys on rapid annealing. *Acta Mater.* 68: 295–309.
9. Köster, U. and U. Herold (1981). Crystallization of metallic glasses. In *Glassy Metals I*, eds. H.-J. Güntherodt and H. Beck, pp. 225–259. Berlin, Germany: Springer-Verlag.
10. Scott, M.G. (1983). Crystallization. In *Amorphous Metallic Alloys*, ed. F.E. Luborsky, pp. 144–168. London, UK: Butterworths.
11. Ranganathan, S. and C. Suryanarayana (1985). Amorphous to crystalline phase transformations. *Mater. Sci. Forum* 3: 173–185.
12. Gu, X., L.Q. Xing, and T.C. Hufnagel (2003). Glass-forming ability and crystallization of bulk metallic glass $(Hf_xZr_{1-x})_{52.5}Cu_{17.9}Ni_{14.6}Al_{10}Ti_5$. *J. Non-Cryst. Solids* 311: 77–82.
13. Lu, Z.P., Y. Li, and S.C. Ng (2000). Reduced glass transition temperature and glass forming ability of bulk glass forming alloys. *J. Non-Cryst. Solids* 270: 103–114.
14. Chen, H.S. and D. Turnbull (1967). Thermal evidence of a glass transition in gold-silicon-germanium alloy. *Appl. Phys. Lett.* 10: 284–286.
15. Inoue, A., T. Zhang, W. Zhang, and A. Takeuchi (1996). Bulk Nd–Fe–Al amorphous alloys with hard magnetic properties. *Mater. Trans., JIM* 37: 99–108.
16. Inoue, A., T. Zhang, and A. Takeuchi (1996). Preparation of bulk Pr–Fe–Al amorphous alloys and characterization of their hard magnetic properties. *Mater. Trans., JIM* 37: 1731–1740.
17. Guo, F.Q., S.J. Poon, and G.J. Shiflet (2004). CaAl-based bulk metallic glasses with high thermal stability. *Appl. Phys. Lett.* 84: 37–39.
18. Zhang, T., A. Inoue, and T. Masumoto (1991). Amorphous Zr–Al–TM (TM=Co, Ni, Cu) alloys with significant supercooled liquid region of over 100 K. *Mater. Trans., JIM* 32: 1005–1110.
19. Inoue, A., T. Zhang, N. Nishiyama, K. Ohba, and T. Masumoto (1993). Preparation of 16 mm diameter rod of amorphous $Zr_{65}Al_{7.5}Ni_{10}Cu_{17.5}$ alloy. *Mater. Trans., JIM* 34: 1234–1237.
20. Lu, I.-R., G. Wilde, G.P. Görler, and R. Willnecker (1999). Thermodynamic properties of Pd-based glass-forming alloys. *J. Non-Cryst. Solids* 250–252: 577–581.
21. Schroers, J., B. Lohwongwatana, W.L. Johnson, and A. Peker (2005). Gold based bulk metallic glass. *Appl. Phys. Lett.* 87: 061912-1–061912-3.
22. Park, E.S. and D.H. Kim (2004). Formation of Ca–Mg–Zn bulk glassy alloy by casting into cone-shaped copper mold. *J. Mater. Res.* 19: 685–688.

23. Zhang, B., M.X. Pan, D.Q. Zhao, and W.H. Wang (2004). "Soft" bulk metallic glasses based on cerium. *Appl. Phys. Lett.* 85: 61–63.
24. Inoue, A., B.L. Shen, H. Koshiba, H. Kato, and A.R. Yavari (2003). Cobalt-based bulk glassy alloy with ultrahigh strength and soft magnetic properties. *Nat. Mater.* 2: 661–663.
25. Jiang, J.Z., J. Saida, H. Kato, T. Oshuna, and A. Inoue (2003). Is $Cu_{60}Ti_{10}Zr_{30}$ a bulk glass-forming alloy? *Appl. Phys. Lett.* 82: 4041–4043.
26. Xu, D., G. Duan, and W.L. Johnson (2004). Unusual glass-forming ability of bulk amorphous alloys based on ordinary metal copper. *Phys. Rev. Lett.* 92: 245504-1–245504-4.
27. Ponnambalam, V., S.J. Poon, G.J. Shiflet, V.M. Keppens, R. Taylor, and G. Petculescu (2003). Synthesis of iron-based bulk metallic glasses as nonferromagnetic amorphous steel alloys. *Appl. Phys. Lett.* 83: 1131–1133.
28. Inoue, A., N. Nakamura, T. Sugita, T. Zhang, and T. Masumoto (1993). Bulky La–Al–TM (TM=Transition Metal) amorphous alloys with high tensile strength produced by a high-pressure die casting method. *Mater. Trans., JIM* 34: 351–358.
29. Zhang, T., A. Inoue, and T. Masumoto. Unpublished research [quoted in Ref. 25].
30. Park, E.S., J.S. Kyeong, and D.H. Kim (2007). Enhanced glass forming ability and plasticity in Mg-based bulk metallic glasses. *Mater. Sci. Eng. A* 449–451: 225–229.
31. Chen, L.Y., Z.D. Fu, W. Zeng, Z.Q. Zhang, Y.W. Zeng, G.L. Xu, S.L. Zhang, and J.Z. Jiang (2007). Ultrahigh strength binary Ni–Nb bulk glassy alloy composite with good ductility. *J. Alloys Compd.* 443: 105–108.
32. Chen, L.Y., H.T. Hu, G.Q. Zhang, and J.Z. Jiang (2007). Catching the Ni-based ternary metallic glasses with critical diameter up to 3 mm in Ni–Nb–Zr system. *J. Alloys Compd.* 443: 109–113.
33. Chen, H.S. and D. Turnbull (1969). Formation, stability and structure of palladium–silicon based alloy glasses. *Acta Metall.* 17: 1021–1031.
34. Chou, C.P.P. and D. Turnbull (1975). Transformation behavior of Pd–Au–Si metallic glasses. *J. Non-Cryst. Solids* 17: 169–188.
35. Schroers, J., C. Veazey, and W.L. Johnson (2003). Amorphous metallic foam. *Appl. Phys. Lett.* 82: 370–372.
36. Inoue, A., N. Nishiyama, and H.M. Kimura (1997). Preparation and thermal stability of bulk amorphous $Pd_{40}Cu_{30}Ni_{10}P_{20}$ alloy cylinder of 72 mm in diameter. *Mater. Trans., JIM* 38: 179–183.
37. Chen, H.S. (1976). Glass temperature, formation and stability of Fe, Co, Ni, Pd and Pt based glasses. *Mater. Sci. Eng.* 23: 151–154.
38. He, Y., R.B. Schwarz, and J.I. Archuleta (1996). Bulk glass formation in the Pd–Ni–P system. *Appl. Phys. Lett.* 69: 1861–1863.
39. Yao, K.F. and F. Ruan (2005). Pd–Si binary bulk metallic glass prepared at low cooling rate. *Chin. Phys. Lett.* 22: 1481–1483.
40. Zhao, Z.F., Z. Zhang, P. Wen, M.X. Pan, D.Q. Zhao, W.H. Wang, and W.L. Wang (2003). A highly glass-forming alloy with low glass transition temperature. *Appl. Phys. Lett.* 82: 4699–4701.
41. Schroers, J. and W.L. Johnson (2004). Highly processable bulk metallic glass-forming alloys in the Pt–Co–Ni–Cu–P system. *Appl. Phys. Lett.* 84: 3666–3668.
42. Wu, J., Q. Wang, J.B. Qiang, F. Chen, C. Dong, Y.M. Wang, and C.H. Shek (2007). Sm-based Sm–Al–Ni ternary bulk metallic glasses. *J. Mater. Res.* 22: 573–577.
43. Pasko, A., V. Kolomytsev, P. Vermaut, F. Prima, R. Portier, P. Ochin, and A. Sezonenko (2007). Crystallization of the amorphous phase and martensitic transformations in multicomponent (Ti,Hf,Zr)(Ni,Cu)-based alloys. *J. Non-Crsyt. Solids* 353: 3062–3068.
44. Zhang, T. and A. Inoue (2001). Ti-based amorphous alloys with a large supercooled liquid region. *Mater. Sci. Eng. A* 304–306: 771–774.
45. Guo, F.Q., H.J. Wang, S.J. Poon, and G.J. Shiflet (2005). Ductile titanium-based glassy alloy ingots. *Appl. Phys. Lett.* 86: 091907-1–091907-3.
46. Guo, F.Q. and S.J. Poon (2003). Metallic glass ingots based on yttrium. *Appl. Phys. Lett.* 83: 2575–2577.
47. Peker, A. and W.L. Johnson (1993). A highly processable metallic glass: $Zr_{41.2}Ti_{13.8}Cu_{12.5}Ni_{10.0}Be_{22.5}$. *Appl. Phys. Lett.* 63: 2342–2344.

48. Inoue, A., B.L. Shen, and C.T. Chang (2004). Super-high strength of over 4000 MPa for Fe-based bulk glassy alloys in $[(Fe_{1-x}Co_x)_{0.75}B_{0.2}Si_{0.05}]_{96}Nb_4$ system. *Acta Mater.* 52: 4093–4099.
49. Hu, X., J. Qiao, J.M. Pelletier, and Y. Yao (2016). Evaluation of thermal stability and isochronal crystallization kinetics in the $Ti_{40}Zr_{25}Ni_8Cu_9Be_{18}$ bulk metallic glass. *J. Non-Cryst. Solids* 432: 254–264.
50. Kissinger, H.E. (1957). Reaction kinetics in differential thermal analysis. *Anal. Chem.* 29: 1702–1706.
51. Kissinger, H.E. (1956). Variation of peak temperature with heating rate in differential thermal analysis. *J. Res. Nat. Bureau of Standards* 57: 217–221.
52. Ozawa, T. (1970). Kinetic analysis of derivative curves in thermal analysis. *J. Therm. Anal.* 2: 301–324.
53. Greer, A.L. (1982). Crystallisation kinetics of $Fe_{80}B_{20}$ glass. *Acta Metall.* 30: 171–192.
54. Henderson, D.W. (1979). Thermal analysis of non-isothermal crystallization kinetics in glass forming liquids. *J. Non-Cryst. Solids* 30: 301–315.
55. Duman, N., M. Vedat Akdeniz, and A.O. Mekhrabov (2013). Magnetic monitoring approach to nanocrystallization kinetics in Fe-based bulk amorphous alloy. *Intermetallics* 43: 152–161.
56. Cui, J., J.S. Li, J. Wang, H.C. Kou, J.C. Qiao, S. Gravier, and J.J. Blandin (2014). Crystallization kinetics of $Cu_{38}Zr_{46}Ag_8Al_8$ bulk metallic glass in different heating conditions. *J. Non-Cryst. Solids* 404: 7–12.
57. Haratian, S. and M. Haddad-Sabzevar (2015). Thermal stability and non-isothermal crystallization kinetics of $Ti_{41.5}Cu_{42.5}Ni_{7.5}Zr_{2.5}Hf_5Si_1$ bulk metallic glass. *J. Non-Cryst. Solids* 429: 164–170.
58. Li, Y.H., C. Yang, L.M. Kang, H.D. Zhao, S.G. Qu, X.Q. Li, W.W. Zhang, and Y.Y. Li (2016). Non-isothermal and isothermal crystallization kinetics and their effect on microstructure of sintered and crystallized TiNbZrTASi bulk alloys. *J. Non-Cryst. Solids* 432: 440–452.
59. Zheng, Z., G. Zhao, L. Xu, A.F. Wang, and B. Yan (2016). Influence of Ni addition on nanocrystallization kinetics of FeCo-based amorphous alloys. *J. Non-Cryst. Solids* 434: 23–27.
60. Ranganathan, S. and M.v. Heimendahl (1981). The three activation energies with isothermal transformations: Applications to metallic glasses. *J. Mater. Sci.* 16: 2401–2404.
61. Waseda, Y. (1980). *The Structure of Non-Crystalline Materials.* New York: McGraw-Hill.
62. Köster, U. and P. Weiss (1975). Crystallization and decomposition of amorphous silicon-aluminium films. *J. Non-Cryst. Solids* 17: 359–368.
63. Illeková, E., M. Jergel, P. Duhaj, and A. Inoue (1997). The relation between the bulk and ribbon $Zr_{55}Ni_{25}Al_{20}$ metallic glasses. *Mater. Sci. Eng. A* 226–228: 388–392.
64. Kang, H.G., E.S. Park, W.T. Kim, D.H. Kim, and H.K. Cho (2000). Fabrication of bulk Mg–Cu–Ag–Y glassy alloy by squeeze casting. *Mater. Trans., JIM* 41: 846–849.
65. Qin, F.X., H.F. Zhang, B.Z. Ding, and Z.Q. Hu (2004). Nanocrystallization kinetics of Ni-based bulk amorphous alloy. *Intermetallics* 12: 1197–1203.
66. Wang, H.R., Y.L. Guo, G.H. Min, X.D. Hui, and Y.F. Ye (2003). Primary crystallization in rapidly solidified $Zr_{70}Cu_{20}Ni_{10}$ alloy from a supercooled liquid region. *Phys. Lett. A* 314: 81–87.
67. Louzguine-Luzgin, D.V., G.Q. Xie, S. Li, Q.S. Zhang, W. Zhang, C. Suryanarayana, and A. Inoue (2009). Glass-forming ability and differences in the crystallization behavior of ribbons and rods of $Cu_{36}Zr_{48}Al_8Ag_8$ bulk glass-forming alloy. *J. Mater. Res.* 24: 1886–1895.
68. Inoue, A., D. Kawase, A.P. Tsai, T. Zhang, and T. Masumoto (1994). Stability and transformation to crystalline phases of amorphous Zr–Al–Cu alloys with significant supercooled liquid region. *Mater. Sci. Eng. A* 178: 255–263.
69. Thompson, C.V., A.L. Greer, and F. Spaepen (1983). Crystal nucleation in amorphous $(Au_{100-y}Cu_y)_{77}Si_9Ge_{14}$ alloys. *Acta Metall. Mater.* 31: 1883–1894.
70. Inoue, A. (1995). Slowly-cooled bulk amorphous alloys. *Mater. Sci. Forum* 179–181: 691–700.
71. Zhou, X.L. and M. Yamane (1988). Effect of heat treatment for nucleation on the crystallization of $MgO–Al_2O_3–SiO_2$ glass containing TiO_2. *J. Ceram. Soc. Jpn.* 96: 152–158.
72. Kawase, D. (1993). Master's thesis, Institute for Materials Research, Tohoku University, Sendai, Japan.

73. Inoue, A. (2000). Stabilization of metallic supercooled liquid and bulk amorphous alloys. *Acta Mater.* 48: 279–306.
74. Inoue, A., T. Shibata, and T. Zhang (1995). Effect of additional elements on glass transition behavior and glass formation tendency of Zr–Al–Cu–Ni alloys. *Mater. Trans., JIM* 36: 1420–1426.
75. de Boer, F.R., R. Boom, W.C.M. Mattens, A.R. Miedema, and A.K. Niessen (1989). *Cohesion in Metals*. Amsterdam, the Netherlands: North-Holland Physics Pub.
76. Zhou, W., J. Hou, Z. Zhang, and J. Li (2015). Effect of Ag content on thermal stability and crystallization behavior of Zr-Cu-Ni-Al-Ag bulk metallic glass. *J. Non-Cryst. Solids* 411: 132–136.
77. Inoue, A. (1988). *Bulk Amorphous Alloys: Preparation and Fundamental Characteristics*. Uetikon-Zurich, Switzerland: Trans Tech Publications.
78. Chen, H.S. (1983). Structural relaxation in metallic glasses. In *Amorphous Metallic Alloys*, ed. F.E. Luborsky, pp. 169–186. London, UK: Butterworths.
79. Chen, H.S. (1978). The influence of structural relaxation on the density and Young's modulus of metallic glasses. *J. Appl. Phys.* 49: 3289–3291.
80. Haruyama, O., H.M. Kimura, N. Nishiyama, A. Inoue, and J.I. Arai (2002). Electrical resistivity and Mössbauer studies for the structural relaxation process in Pd–Cu–Ni–P glasses. *Mater. Trans.* 43: 1931–1936.
81. Inoue, A., T. Negishi, H.M. Kimura, T. Zhang, and A.R. Yavari (1998). High packing density of Zr- and Pd-based bulk amorphous alloys. *Mater. Trans., JIM* 39: 318–321.
82. Haruyama, O., M. Tando, H.M. Kimura, N. Nishiyama, and A. Inoue (2004). Structural relaxation in Pd–Cu–Ni–P metallic glasses. *Mater. Sci. Eng. A* 375–377: 292–296.
83. Egami, T. (1981). Structural relaxation in metallic glasses. *Ann. N.Y. Acad. Sci.* 371: 238–251.
84. Dmowski, W., C. Fan, M.L. Morrison, P.K. Liaw, and T. Egami (2007). Structural changes in bulk metallic glass after annealing below the glass-transition temperature. *Mater. Sci. Eng. A* 471: 125–129.
85. Busch, R. and W.L. Johnson (1998). The kinetic glass transition of the $Zr_{46.75}Ti_{8.25}Cu_{7.5}Ni_{10}Be_{27.5}$ bulk metallic glass former-supercooled liquids on a long time scale. *Appl. Phys. Lett.* 72: 2695–2697.
86. Inoue, A., T. Zhang, and T. Masumoto (1992). The structural relaxation and glass transition of La–Al–Ni and Zr–Al–Cu amorphous alloys with a significant supercooled liquid region. *J. Non-Cryst. Solids* 150: 396–400.
87. Chen, H.S. (1981). On mechanisms of structural relaxation in a $Pd_{48}Ni_{32}P_{20}$ glass. *J. Non-Cryst. Solids* 46: 289–305.
88. Inoue, A., T. Masumoto, and H.S. Chen (1985). Enthalpy relaxation behaviour of metal metal (Zr–Cu) amorphous alloys upon annealing. *J. Mater. Sci.* 20: 4057–4068.
89. Inoue, A., H.S. Chen, and T. Masumoto (1985). Two-stage enthalpy relaxation behaviour of $(Fe_{0.5}Ni_{0.5})_{83}P_{17}$ and $(Fe_{0.5}Ni_{0.5})_{83}B_{17}$ amorphous alloys upon annealing. *J. Mater. Sci.* 20: 2417–2438.
90. Hess, P.A. and R.H. Dauskardt (2004). Mechanisms of elevated temperature fatigue crack growth in Zr–Ti–Cu–Ni–Be bulk metallic glass. *Acta Mater.* 52: 3525–3533.
91. Murali, P. and U. Ramamurty (2005). Embrittlement of a bulk metallic glass due to sub-T_g annealing. *Acta Mater.* 53: 1467–1478.
92. Launey, M.E., R. Busch, and J.J. Kruzic (2006). Influence of structural relaxation on the fatigue behavior of a $Zr_{41.25}Ti_{13.75}Ni_{10}Cu_{12.5}Be_{22.5}$ bulk amorphous alloy. *Scr. Mater.* 54: 483–487.
93. Launey, M.E., J.J. Kruzic, C. Li, and R. Busch (2007). Quantification of free volume differences in a $Zr_{44}Ti_{11}Ni_{10}Cu_{10}Be_{25}$ bulk amorphous alloy. *Appl. Phys. Lett.* 91: 051913-1–051913-3.
94. Grest, G.S. and M.H. Cohen (1981). Liquids, glasses, and glass transition: A free-volume approach. *Adv. Chem. Phys.* 48: 455–525.
95. Inoue, A., H. Yamaguchi, T. Zhang, and T. Masumoto (1990). Al–La–Cu amorphous alloys with a wide supercooled liquid region. *Mater. Trans., JIM* 31: 104–109.
96. Inoue, A., T. Zhang, and T. Masumoto (1989). Al–La–Ni amorphous alloys with a wide supercooled liquid region. *Mater. Trans., JIM* 30: 965–972.
97. Inoue, A., M. Kohinata, A.P. Tsai, and T. Masumoto (1989). Mg–Ni–La amorphous alloys with a wide supercooled liquid region. *Mater. Trans., JIM* 30: 378–381.

98. Inoue, A., T. Zhang, and T. Masumoto (1990). Zr–Al–Ni amorphous alloys with high glass transition temperature and significant supercooled liquid region. *Mater. Trans., JIM* 31: 177–183.
99. Hilliard, J.E. (1970). Spinodal decomposition. In *Phase Transformations*, ed. H.I. Aaronson, pp. 497–560. Materials Park, OH: ASM International.
100. Tanner, L.E. and R. Ray (1980). Phase separation in Zr–Ti–Be metallic glasses. *Scr. Metall.* 14: 657–662.
101. Luizguine-Luzgin, D.V., I. Seki, S.V. Ketov, L.V. Luizguine-Luzgin, V.I. Polkin, N. Chen, H. Fecht, A.N. Vasiliev, and H. Kawaji (2015). Glass-transition process in an Au-based metallic glass, *J. Non-Cryst. Solids* 419: 12–15.
102. Kim, D.H., W.T. Kim, E.S. Park, N. Mattern, and J. Eckert (22013). Phase separation in metallic glasses. *Prog. Mater. Sci.* 58: 1103–1172.
103. Miedema, A.R., F.R. de Boer, and R. Boom (1977). Model predictions for the enthalpy of formation of transition metal alloys. *CALPHAD* 1: 341–359.
104. Massalski, T.B. (ed.) (1992). *Binary Alloy Phase Diagrams*. Materials Park, OH: ASM International.
105. Porter, D.A., K.E. Easterling, and M.Y. Sherif (2009). *Phase Transformations in Metals and Alloys*, 3rd edn. Boca Raton, FL: CRC Press.
106. Kündig, A.A., M. Ohnuma, T. Ohkubo, T. Abe, and K. Hono (2006). Glass formation and phase separation in the Ag–Cu–Zr system. *Scr. Mater.* 55: 449–452.
107. Oh, J.C., T. Ohkubo, Y.C. Kim, E. Fleury, and K. Hono (2005). Phase separation in $Cu_{43}Zr_{43}Al_7Ag_7$ bulk metallic glass. *Scr. Mater.* 53: 165–169.
108. Park, E.S. and D.H. Kim (2006). Phase separation and enhancement of plasticity in Cu–Zr–Al–Y bulk metallic glasses. *Acta Mater.* 54: 2597–2604.
109. Kündig, A.A., M. Ohnuma, D.H. Ping, T. Ohkubo, and K. Hono (2004). *In-situ* formed two-phase metallic glass with surface fractal microstructure. *Acta Mater.* 52: 2441–2448.
110. Park, E.S., E.Y. Jeong, J.-K. Lee, J.C. Bae, A.R. Kwon, A. Gebert, L. Schultz, H.J. Chang, and D.H. Kim (2007). *In situ* formation of two glassy phases in the Nd–Zr–Al–Co alloy system. *Scr. Mater.* 56: 197–200.
111. Mattern, N., T. Gemming, G. Goerigk, and J. Eckert (2007). Phase separation in amorphous Ni–Nb–Y alloys. *Scr. Mater.* 57: 29–32.
112. Mattern, N., U. Kühn, A. Gebert, T. Gemming, M. Zinkevich, H. Wendrock, and L. Schultz (2005). Microstructure and thermal behavior of two-phase amorphous Ni–Nb–Y alloy. *Scr. Mater.* 53: 271–274.
113. Na, J.H., S.W. Sohn, W.T. Kim, and D.H. Kim (2007). Two-step-like anomalous glass transition behavior in Ni–Zr–Nb–Al–Ta metallic glass alloys. *Scr. Mater.* 57: 225–228.
114. Park, B.J., H.J. Chang, D.H. Kim, and W.T. Kim (2004). *In situ* formation of two amorphous phases by liquid phase separation in Y–Ti–Al–Co alloy. *Appl. Phys. Lett.* 85: 6353–6355.
115. Park, B.J., H.J. Chang, D.H. Kim, W.T. Kim, K. Chattopadhyay, T.A. Abinandanan, and S. Bhattacharyya (2006). Phase separating bulk metallic glass: A hierarchical composite. *Phys. Rev. Lett.* 96: 245503-1–245503-4.
116. Du, X.H., J.C. Huang, K.C. Hsieh, Y.H. Lai, H.M. Chen, J.S.C. Jang, and P.K. Liaw (2007). Two-glassy-phase bulk metallic glass with remarkable plasticity. *Appl. Phys. Lett.* 91: 131901-1–131901-3.
117. Nagahama, D., T. Ohkubo, and K. Hono (2003). Crystallization of $Ti_{36}Zr_{24}Be_{40}$ metallic glass. *Scr. Mater.* 49: 729–734.
118. Pekarskaya, E., J.F. Löffler, and W.L. Johnson (2003). Microstructural studies of crystallization of a Zr-based bulk metallic glass. *Acta Mater.* 51: 4045–4057.
119. Kajiwara, K., M. Ohnuma, T. Ohkubo, D.H. Ping, and K. Hono (2004). APFIM/TEM/SAXS studies of early stage crystallization of a $Zr_{52.5}Cu_{17.9}Ni_{14.6}Al_{10}Ti_5$ metallic glass. *Mater. Sci. Eng. A* 375–377: 738–743.
120. Busch, R., S. Schneider, A. Peker, and W.L. Johnson (1995). Decomposition and primary crystallization in undercooled $Zr_{41.2}Ti_{13.8}Cu_{12.5}Ni_{10.0}Be_{22.5}$ melts. *Appl. Phys. Lett.* 67: 1544–1546.
121. Schneider, S., P. Thiyagarajan, and W.L. Johnson (1996). Formation of nanocrystals based on decomposition in the amorphous $Zr_{41.2}Ti_{13.8}Cu_{12.5}Ni_{10}Be_{22.5}$ alloy. *Appl. Phys. Lett.* 68: 493–495.

122. Löffler, J.F. and W.L. Johnson (2000). Model for decomposition and nanocrystallization of deeply undercooled $Zr_{41.2}Ti_{13.8}Cu_{12.5}Ni_{10}Be_{22.5}$. *Appl. Phys. Lett.* 76: 3394–3396.
123. Park, B.J., S.W. Sohn, D.H. Kim, H.T. Jeong, and W.T. Kim (2008). Solid-state phase separation in Zr–Y–Al–Co metallic glass. *J. Mater. Res.* 23: 828–832.
124. Zhang, T., A. Inoue, S. Chen, and T. Masumoto (1992). Amorphous $(Zr–Y)_{60}Al_{15}Ni_{25}$ alloys with two supercooled liquid regions. *Mater. Trans., JIM* 33: 143–145.
125. Inoue, A., S. Chen, and T. Masumoto (1994). Zr–Y base amorphous alloys with two glass transitions and two supercooled liquid regions. *Mater. Sci. Eng. A* 179/180: 346–350.
126. Sugiyama, K., A.H. Shinohara, Y. Waseda, S. Chen, and A. Inoue (1994). Anomalous SAXS study on structural inhomogeneity in amorphous $Zr_{33}Y_{27}Al_{15}Ni_{25}$ alloy. *Mater. Trans., JIM* 35: 481–484.
127. Martin, I., T. Ohkubo, M. Ohnuma, B. Deconihout, and K. Hono (2004). Nanocrystallization of $Zr_{41.2}Ti_{13.8}Cu_{12.5}Ni_{10.0}Be_{22.5}$ metallic glass. *Acta Mater.* 52: 4427–4435.
128. Kajiwara, K., M. Ohnuma, D.H. Ping, O. Haruyama, and K. Hono (2002). Nanocrystallization of $Pd_{74}Si_{18}Au_{8}$ metallic glass. *Intermetallics* 10: 1053–1060.
129. Hirotsu, Y., T. Hanada, T. Ohkubo, A. Makino, Y. Yoshizawa, and T.G. Nieh (2004). Nanoscale phase separation in metallic glasses studied by advanced electron microscopy techniques. *Intermetallics* 12: 1081–1088.
130. Madge, S.V., H. Rösner, and G. Wilde (2005). Transformations in supercooled $Pd_{40.5}Ni_{40.5}P_{19}$. *Scr. Mater.* 53: 1147–1151.
131. Meijering, J.L. (1950). Segregation in regular ternary solutions. 1. *Philips Res. Rep.* 5: 333–356.
132. Meijering, J.L. (1951). Segregation in regular ternary solutions. 2. *Philips Res. Rep.* 6: 183–210.
133. Abe, T., M. Shimono, K. Hashimoto, K. Hono, and H. Onodera (2006). Phase separation and glass-forming abilities of ternary alloys. *Scr. Mater.* 55: 421–424.
134. Inoue, A., S.J. Chen, E. Oikawa, T. Zhang, and T. Masumoto (1993). Elastic and viscous behavior of an amorphous $Zr_{33}Y_{27}Al_{15}Ni_{25}$ alloy with a two-stage glass transition. *Mater. Lett.* 16: 108–112.
135. Park, B.J. (2007). Phase separation in metallic glasses, PhD thesis, Yonsei University, Seoul, South Korea.
136. Mihalca, I., A. Ercuta, I. Zaharie, A. Jianu, V. Kuncser, and G. Filoti (1996). Annealing effects in amorphous $Fe_{77}Gd_{3}B_{20}$ ribbons. *J. Magn. Magn. Mater.* 157–158: 161–162.
137. Wong, C.H. and C.H. Shek (2004). Difference in crystallization kinetics of $Zr_{41}Ti_{14}Cu_{12.5}Ni_{10}Be_{22.5}$ bulk metallic glass under different oxidizing environments. *Intermetallics* 12: 1257–1259.
138. Kiene, M., T. Strunskus, G. Hasse, and F. Faupel (1999). Oxide formation in the bulk metallic glass $Zr_{46.75}Ti_{8.25}Cu_{7.5}Ni_{10}Be_{27.5}$. In *Bulk Metallic Glasses, MRS Symposium Proceedings*, eds. W.L. Johnson, A. Inoue, and C.T. Liu, Vol. 554, pp. 167–172. Warrendale, PA: Materials Research Society.
139. Huang, D., L. Huang, B. Wang, V. Ji, and T. Zhang (2012). The relationship between t-ZrO_2 stability and the crystallization of a Zr-based bulk metallic glass during oxidation. *Intermetallics* 31: 21–25.
140. de Oliveira, M.F., W.J. Botta F., M.J. Kaufman, and C.S. Kiminami (2002). Phases formed during crystallization of $Zr_{55}Al_{10}Ni_{5}Cu_{30}$ metallic glass containing oxygen. *J. Non-Cryst. Solids* 304: 51–55.
141. Inoue, A. and T. Zhang (1996). Fabrication of bulk glassy $Zr_{55}Al_{10}Ni_{5}Cu_{30}$ alloy of 30 mm in diameter by a suction casting method. *Mater. Trans., JIM* 37: 185–187.
142. Eckert, J., N. Mattern, M. Zinkevitch, and M. Seidel (1998). Crystallization behavior and phase formation in Zr–Al–Cu-Ni metallic glass containing oxygen. *Mater. Trans., JIM* 39: 623–632.
143. Sordelet, D.J., X.Y. Yang, E.A. Rozhkova, M.F. Besser, and M.J. Kramer (2003). Oxygen-stabilized glass formation in $Zr_{80}Pt_{20}$ melt-spun ribbons. *Appl. Phys. Lett.* 83: 69–71.
144. Köster, U., J. Meinhardt, S. Roos, and H. Liebertz (1996). Formation of quasicrystals in bulk glass forming Zr–Cu–Ni–Al alloys. *Appl. Phys. Lett.* 69: 179–181.
145. Köster, U., J. Meinhardt, S. Roos, and R. Rüdiger (1996). Influence of oxygen contents on nanocrystallization of $Co_{33}Zr_{67}$ and $Zr_{65}Cu_{17.5}Ni_{10}Al_{7.5}$ alloys. *Mater. Sci. Forum* 225–227: 311–316.

146. Köster, U., J. Meinhardt, S. Roos, and R. Busch (1997). Formation of quasicrystals in bulk glass forming Zr–Cu–Ni–Al alloys. *Mater. Sci. Eng. A* 226–228: 995–998.
147. Kawase, D., A.P. Tsai, A. Inoue, and T. Masumoto (1993). Crystallization on supercooled liquid in metallic Zr–Cu–Al glasses. *Appl. Phys. Lett.* 62: 137–139.
148. Frankwicz, P.S., R. Ram, and H.-J. Fecht (1996). Enhanced microhardness in $Zr_{65.0}Al_{7.5}Ni_{10.0}Cu_{17.5}$ amorphous rods on coprecipitation of nanocrystallites through supersaturated intermediate solid phase particles. *Appl. Phys. Lett.* 68: 2825–2827.
149. Jiang, J.Z., K. Saksi, J. Saida, A. Inoue, H. Franz, K. Messel, and C. Lathe (2002). Evidence of polymorphous amorphous-to-quasicrystalline phase transformation in $Zr_{66.7}Pd_{33.3}$ metallic glass. *Appl. Phys. Lett.* 80: 781–783.
150. Chen, M.W., T. Zhang, A. Inoue, A. Sakai, and T. Sakurai (1999). Quasicrystals in a partially devitrified $Zr_{65}Al_{7.5}Ni_{10}Cu_{12.5}Ag_5$ bulk metallic glass. *Appl. Phys. Lett.* 75: 1697–1699.
151. Inoue, A., J. Saida, M. Matsushita, and T. Sakurai (2000). Formation of an icosahedral quasicrystalline phase in $Zr_{65}Al_{7.5}Ni_{10}M_{17.5}$ (M=Pd, Au or Pt) alloys. *Mater. Trans., JIM* 41: 362–365.
152. Murty, B.S., D.H. Ping, K. Hono, and A. Inoue (2000). Direct evidence for oxygen stabilization of icosahedral phase during crystallization of $Zr_{65}Cu_{27.5}Al_{7.5}$ metallic glass. *Appl. Phys. Lett.* 76: 55–57.
153. Chen, M.W., A. Inoue, T. Sakurai, D.H. Ping, and K. Hono (1999). Impurity oxygen redistribution in a nanocrystallized $Zr_{65}Cr_{15}Al_{10}Pd_{10}$ metallic glass. *Appl. Phys. Lett.* 74: 812–814.
154. Wollgarten, M., S. Mechler, E. Davidov, N. Wanderka, and M.-P. Macht (2004). Decomposition and crystallization of $Pd_{40}Cu_{30}Ni_{10}P_{20}$ and $Zr_{46.8}Ti_{8.2}Cu_{7.5}Ni_{10}Be_{27.5}$ metallic glasses. *Intermetallics* 12: 1251–1255.
155. Jiang, J.Z., A.R. Rasmussen, C.H. Jensen, Y. Lin, and P.L. Hansen (2002). Change of quasilattice constant during amorphous-to-quasicrystalline phase transformation in $Zr_{65}Al_{7.5}Ni_{10}Cu_{7.5}Ag_{10}$ metallic glass. *Appl. Phys. Lett.* 80: 2090–2092.
156. Inoue, A., T. Zhang, M.W. Chen, T. Sakurai, J. Saida, and M. Matsushita (2000). Formation and properties of Zr-based bulk quasicrystalline alloys with high strength and good ductility. *J. Mater. Res.* 15: 2195–2208.
157. Lee, J.K., G. Choi, D.H. Kim, and W.T. Kim (2000). Formation of icosahedral phase from amorphous $Zr_{65}Al_{7.5}Cu_{12.5}Ni_{10}Ag_5$ alloys. *Appl. Phys. Lett.* 77: 978–980.
158. Liu, L., K.C. Chan, and G.K.H. Pang (2004). The microprocesses of the quasicrystalline transformation in $Zr_{65}Ni_{10}Cu_{7.5}Al_{7.5}Ag_{10}$ bulk metallic glass. *Appl. Phys. Lett.* 85: 2788–2790.
159. Wang, W.K., H. Iwasaki, C. Suryanarayana, and T. Masumoto (1983). Crystallization characteristics of an amorphous $Ti_{80}Si_{20}$ alloy at high pressures. *J. Mater. Sci.* 18: 3765–3772.
160. Ye, F. and K. Lu (1999). Pressure effect on crystallization kinetics of an Al–La–Ni amorphous alloy. *Acta Mater.* 47: 2449–2454.
161. Jiang, J.Z., J.S. Olsen, L. Gerward, S. Abdali, J. Eckert, N. Schlorke-de Boer, L. Schultz, J. Truckenbrodt, and P.X. Shi (2000). Pressure effect on crystallization of metallic glass $Fe_{72}P_{11}C_6Al_5B_4Ga_2$ alloy with wide supercooled liquid region. *J. Appl. Phys.* 87: 2664–2666.
162. Ogawa, Y., K. Nunogaki, S. Endo, M. Kiritani, and F.E. Fujita (1982). Crystallization of amorphous Fe–B alloys under pressure. In *Proceedings of Fourth International Conference on Rapidly Quenched Metals (RQIV)*, eds. T. Masumoto and K. Suzuki, pp. 675–678. Sendai, Japan: Japan Institute of Metals.
163. Suryanarayana, C., W.K. Wang, H. Iwasaki, and T. Masumoto (1980). High-pressure synthesis of A15 Nb_3Si phase from amorphous Nb–Si alloys. *Solid State Commun.* 34: 861–863.
164. Yang, C., W.K. Wang, R.P. Liu, Z.J. Zhan, L.L. Sun, J. Zhang, J.Z. Jiang, L. Yang, and C. Lathe (2006). Crystallization of $Zr_{41}Ti_{14}Cu_{12.5}Ni_{10}Be_{22.5}$ bulk metallic glass under high pressure examined by *in situ* synchrotron radiation x-ray diffraction. *J. Appl. Phys.* 99: 023525-1–023525-4.
165. Wang, W.H., R.J. Wang, D.Y. Dai, D.Q. Zhao, M.X. Pan, and Y.S. Yao (2001). Pressure-induced amorphization of ZrTiCuNiBe bulk glass-forming alloy. *Appl. Phys. Lett.* 79: 1106–1108.
166. Jiang, J.Z. (2002). Comment on "Pressure-induced amorphization of ZrTiCuNiBe bulk glass-forming alloy" [*Appl. Phys. Lett.* 79, 1106 (2001)]. *Appl. Phys. Lett.* 80: 700.

167. Wang, W.H., R.J. Wang, D.Y. Dai, D.Q. Zhao, M.X. Pan, and Y.S. Yao (2002). Response to "Comment on 'Pressure-induced amorphization of ZrTiCuNiBe bulk glass-forming alloy'" [*Appl. Phys. Lett.* 80, 700 (2002)]. *Appl. Phys. Lett.* 80: 701.
168. Zhuang, Y.X., J.Z. Jiang, T.J. Zhou, H. Rasmussen, L. Geward, M. Mezouar, W. Crichton, and A. Inoue (2000). Pressure effects on $Al_{89}La_6Ni_5$ amorphous alloy crystallization. *Appl. Phys. Lett.* 77: 4133–4135.
169. Lenderoth, S., N. Pryds, M. Eldrup, A.S. Pedersen, M. Ohnuma, T.-J. Zhou, L. Gerward, J.Z. Jiang, and C. Lathe (2001). Bulk Mg–Cu–Y–Al alloys in the amorphous, supercooled and crystalline states. In *Supercooled Liquid, Bulk Glassy, and Nanocrystalline States of Alloys*, eds. A. Inoue, A.R. Yavari, W.L. Johnson, and R.H. Dauskardt, Vol. 644, pp. L4.1.1–L4.1.12. Warrendale, PA: Materials Research Society.
170. Jiang, J.Z., K. Saksi, N. Nishiyama, and A. Inoue (2002). Crystallization in $Pd_{40}Ni_{40}P_{20}$ glass. *J. Appl. Phys.* 92: 3651–3656.
171. Jiang, J.Z., Y.X. Zhuang, H. Rasmussen, N. Nishiyama, A. Inoue, and C. Lathe (2001). Crystallization of $Pd_{40}Cu_{30}Ni_{10}P_{20}$ bulk glass under pressure. *Europhys. Lett.* 54: 182–186.
172. Jiang, J.Z., S. Jeppesen, J. Saida, and C. Lathe (2004). Pressure effect on crystallization temperature in $Zr_{70}Pd_{30}$ metallic glass. *J. Appl. Phys.* 95: 4651–4654.
173. Jiang, J.Z., Y.X. Zhuang, H. Rasmussen, J. Saida, and A. Inoue (2001). Formation of quasicrystals and amorphous-to-quasicrystalline phase transformation kinetics in $Zr_{65}Al_{7.5}Ni_{10}Cu_{7.5}Ag_{10}$ metallic glass under pressure. *Phys. Rev. B* 64: 094208-1–094208-10.
174. Zhuang, Y.X., L. Gerward, J.Z. Jiang, J.S. Olsen, Y. Zhang, and W.H. Wang (2001). Crystallization of bulk $Zr_{48}Nb_8Cu_{14}Ni_{12}Be_{18}$ metallic glass. In *Supercooled Liquid, Bulk Glassy, and Nanocrystalline States of Alloys*, eds. A. Inoue, A.R. Yavari, W.L. Johnson, and R.H. Dauskardt, Vol. 644, pp. L5.2.1–L5.2.6. Warrendale, PA: Materials Research Society.
175. Jiang, J.Z., T.J. Zhou, H. Rasmussen, U. Kuhn, J. Eckert, and C. Lathe (2000). Crystallization in $Zr_{41.2}Ti_{13.8}Cu_{12.5}Ni_{10}Be_{22.5}$ bulk metallic glass under pressure. *Appl. Phys. Lett.* 77: 3553–3555.
176. Jiang, J.Z., L. Gerwood, and Y.S. Xu (2002). Pressure effect on crystallization kinetics in $Zr_{46.8}Ti_{8.2}Cu_{7.5}Ni_{10}Be_{27.5}$ bulk glass. *Appl. Phys. Lett.* 81: 4347–4349.
177. Miglierini, M., V. Procházka, R. Rüffler, and R. Zbořil (2015). *In situ* crystallization of metallic glasses during magnetic field annealing. *Acta Mater.* 91 (2015) 50–56.
178. Zhuang, Y.X., W.B. Wang, B.T. Han, Z.M. Wang, and P.F. Xing (2016). Crystallization behavior of $Fe_{84}B_{10}C_6$ amorphous alloy under high magnetic field. *J. Non-Cryst. Solids* 432: 200–207.
179. Carter, J., E.G. Fu, M. Martin, G. Xie, X. Zhang, Y.Q. Wang, D. Wijesundera, X.M. Wang, W.-K. Chu, and L. Shao (2009). Effects of Cu ion irradiation in $Cu_{50}Zr_{45}Ti_5$ metallic glass. *Scr. Mater.* 61: 265–268.
180. Chen, H.C., L. Yan, R.D. Liu, M.B. Tang, G. Wang, H.F. Huang, Y. Hai, and X.T. Zhou (2014). Anisotropic nanocrystallization of a Zr-based metallic glass induced by Xe ion irradiation. *Intermetallics* 52: 15–19.
181. Myers, M., E.G. Fu, M. Myers, H. Wang, G. Xie, X. Wang. W.-K. Chu, and L. Shao (2010). An experimental and modeling study on the role of damage cascade formation in nanocrystallization of ion-irradiated $Ni_{52.5}Nb_{10}Zr_{15}Ti_{15}Pt_{7.5}$ metallic glass. *Scr. Mater.* 63: 1045–1048.
182. Zhou, X., H.C. Kou, J. Wang, J. Li, and L. Zhou (2012). Crystallization and compressive behaviors of $Ti_{40}Zr_{25}Ni_8Cu_9Be_{18}$ BMG cast from different liquid states. *Intermetallics* 28: 45–50.
183. Kim, J.J., Y. Choi, S. Suresh, and A.S. Argon (2002). Nanocrystallization during nanoindentation of a bulk amorphous metal alloy at room temperature. *Science* 295: 654–657.
184. Schuh, C.A. and T.G. Nieh (2004). A survey of instrumented indentation studies on metallic glasses. *J. Mater. Res.* 19: 46–57.
185. Fornell, J., E. Rossinyol, S. Suriñach, M.D. Baró, W.H. Li, and J. Sort (2010). Enhanced mechanical properties in a Zr-based metallic glass caused by deformation-induced nanocrystallization. *Scr. Mater.* 62: 13–16.
186. Yoo, B.-, I.-C. Choi, Y.-J. Kim, J.-Y. Suh, U. Ramamurthy, and J.-I. Jang (2012). Further evidence for room temperature, indentation-induced nanocrystallization in a bulk metallic glass. *Mater. Sci. Eng. A* 545: 225–228.

6

Physical Properties

6.1 Introduction

For the successful exploitation of bulk metallic glasses (BMGs), it is necessary to characterize them for their structure, thermal stability, and the different properties. The structure and thermal stability of BMGs were discussed in Chapter 5. The chemical, mechanical, and magnetic properties of BMGs will be discussed in Chapters 7, 8, and 9, respectively. The different physical properties of BMGs such as density, specific heat, viscosity, electrical resistivity, thermal expansion, and diffusivity will be discussed in this chapter. A measurement of these physical properties will greatly aid in understanding the structural relaxation and the crystallization processes occurring in these metastable materials and their effect on properties.

6.2 Density

It has been known for a long time that metallic glasses are less dense than their crystalline counterparts. The densities of both metallic glassy ribbons and BMG alloys have been measured mostly by using the Archimedes principle. The working fluids used are different, and include *n*-tridecane $[CH_3(CH_2)_{11}CH_3]$ because it has a low vapor pressure and low surface tension at room temperature, CCl_4, buromethane, dodecane ($C_{12}H_{26}$), and distilled water. Although different estimates were made by different research groups, it is now well accepted that metallic glassy ribbons produced by rapid solidification processing (RSP) methods are about 2%–3% less dense than their crystalline counterparts [1]. This conclusion was reached by measuring the density of the alloy ribbons both in the glassy condition and after fully crystallizing them. For example, Masumoto et al. [2] reported that the density of a $Pd_{80}Si_{20}$ alloy increased with annealing time when the melt-spun glassy ribbon was isothermally annealed at 200°C. The large density difference between the glassy and the crystalline conditions was attributed to the presence of free volume retained in the glassy ribbons. On annealing the glassy ribbons, this free volume gets annihilated, leading to an increase in the density.

BMGs are synthesized at relatively lower solidification rates than melt-spun ribbons and, therefore, the free volume content in the BMG alloys is expected to be lower than in melt-spun ribbons. It has also been suggested that under appropriate conditions, for example when the three empirical rules for BMG formation as suggested by Inoue [3] are satisfied, the liquid possesses a higher degree of dense random-packed structures. Additionally, the short-range atomic configurations are also expected to be different from those in the corresponding crystalline alloys. These points suggest that the supercooled liquid in a multicomponent alloy system exhibits new atomic configurations that have not been realized in any other kind of metallic alloy. Thus, the formation of unique atomic configurations leads to a change in the density of BMGs. Accordingly, it has been reported that the density of the BMG alloys is about 0.5% lower than that of the crystalline counterpart. Since the density of a material is often a reflection of the packing state of the constituent elements, measurement of density values can provide an understanding of the random atomic configurations in BMGs.

Frequently, the relative change of density $\Delta\rho$, of the alloy is reported in the literature, which is defined as

$$\%\Delta\rho = \frac{\rho_{\text{crystallized}} - \rho_{\text{as-cast}}}{\rho_{\text{as-cast}}} \times 100 \tag{6.1}$$

where ρ is the density, and subscripts crystallized and as-cast represent the fully crystallized and the as-cast glassy conditions of the sample, respectively.

The relative change of density between the structurally relaxed and the as-cast glassy conditions has also been occasionally reported. But, this value is usually much smaller than the relative change of density between the crystallized and as-cast conditions.

Table 6.1 lists the densities of some of the BMGs that have been reported in the literature. A majority of the BMG samples listed were obtained by the copper mold or squeeze-casting technique, except the Pd–Ni–Fe–P BMG alloys that were obtained by water quenching. As anticipated, the densities of the glassy alloys are lower than those in the crystallized state, typically by about 0.5%, but the differences are occasionally as large as 1%. The water-quenched Pd–Ni–Fe–P samples, however, appear to show even larger differences in the density between the glassy and the crystallized states [9]. No reasons were given for this larger difference. Therefore, it is not clear whether the technique used to synthesize the glassy alloy (since the cooling rate will be different) will also have an effect on the density of the glass.

Louzguine-Luzgin et al. [13] have obtained BMG samples of $Pd_{40}Ni_{40}Si_5P_{15}$ composition in the form of a long pyramid by conventional copper mold casting. They measured the different properties of the BMG sample, including the density, and noted that the general density difference between the as-solidified and annealed sample was only about 0.38%, which appears to be a very low value for metallic glasses.

It is well known that the cooling rates obtained in different techniques are different and, therefore, it is expected that the free volume will be different in samples produced by different methods. Since it is the free volume that determines the density of the material, it is possible that the density of samples obtained by different techniques would be different. However, the differences in the densities between BMGs produced by different methods are expected to be small. It is also possible that this difference in density (and free volume) may lead to differences in the kinetics of the crystallization and the mechanical behavior of the glassy alloys prepared by different synthesis methods.

Hu et al. [8] solidified $Pd_{40}Cu_{30}Ni_{10}P_{20}$ alloy melts in 2 mm diameter quartz capillaries at cooling rates varying from 0.63 to 75.4 K s^{-1}. One alloy was also chill cast, where the cooling rate was estimated to be about 500 K s^{-1}. Since the alloys solidified at rates <1.98 K s^{-1} contained crystalline phases and those above this value were fully glassy, it was concluded that the critical cooling rate for glass formation in this system was 1.98 K s^{-1}. The densities

TABLE 6.1

Relative Densities (ρ) of the As-Cast and Crystallized BMG Alloys

Alloy Composition	Synthesis Method	Rod Diameter (mm)	$\rho_{as\text{-}cast}$ (g cm^{-3})	$\rho_{crystallized}$ (g cm^{-3})	%$\Delta\rho_{crystallized}$[a]	References
$Ca_{65}Mg_{15}Zn_{20}$	C	V	2.140	2.147	0.33	[4]
$Mg_{65}Cu_{7.5}Ni_{7.5}Ag_5Zn_5Y_5Gd_5$	C	V	3.572	3.607	0.98	[5]
$Pd_{40}Cu_{30}Ni_{10}P_{20}$	C	5	9.27	9.31	0.43	[6,7]
$Pd_{40}Cu_{30}Ni_{10}P_{20}$	W	2	9.270	9.341	0.76	[8]
$Pd_{40}Ni_{40}P_{20}$	W	7	9.390	9.412	0.23	[9]
$Pd_{40}Ni_{38}Fe_2P_{20}$	W	7	9.353	9.423	0.75	[9]
$Pd_{40}Ni_{35}Fe_5P_{20}$	W	7	9.281	9.390	1.17	[9]
$Pd_{40}Ni_{33.5}Fe_{6.5}P_{20}$	W	7	9.259	9.371	1.21	[9]
$Pd_{40}Ni_{32.5}Fe_{7.5}P_{20}$	W	7	9.246	9.359	1.22	[9]
$Pd_{40}Ni_{31.5}Fe_{8.5}P_{20}$	W	7	9.236	9.350	1.23	[9]
$Pd_{40}Ni_{30}Fe_{10}P_{20}$	W	7	9.203	9.326	1.34	[9]
$Pd_{40}Ni_{27.5}Fe_{12.5}P_{20}$	W	7	9.166	9.262	1.05	[9]
$Pd_{40}Ni_{25}Fe_{15}P_{20}$	W	7	9.141	9.198	0.62	[9]
$Pd_{40}Ni_{22.5}Fe_{17.5}P_{20}$	W	7	9.107	9.180	0.80	[9]
$Pd_{40}Ni_{20}Fe_{20}P_{20}$	W	7	9.062	9.123	0.67	[9]
$Zr_{62.5}Al_{7.5}Ni_{10}Cu_{20}$	S	T	6.77	6.80	0.44	[10]
$Zr_{60}Cu_{30}Al_{10}$	C	5	6.72	6.74	0.30	[6,7]
$Zr_{60}Cu_{25}Al_{15}$			6.36	6.38	0.32	[7]
$Zr_{60}Cu_{20}Al_{10}Ni_{10}$	S	T	6.68	6.71	0.45	[10]
$Zr_{60}Cu_{20}Al_{10}Ti_2Ni_8$	C	3	6.64	6.66	0.30	[11]
$Zr_{57.5}Cu_{20}Al_{12.5}Ni_{10}$	S	T	6.60	6.63	0.46	[10]
$Zr_{57.5}Cu_{20}Al_{10}Ti_{2.5}Ni_{10}$	S	T	6.65	6.68	0.45	[10]
$Zr_{57.5}Cu_{20}Al_{7.5}Ti_5Ni_{10}$	S	T	6.68	6.72	0.60	[10]
$Zr_{55}Cu_{30}Al_{10}Ni_5$	C	5	6.82	6.85	0.44	[6,7]
$Zr_{55}Cu_{20}Al_{15}Ni_{10}$	S	T	6.51	6.54	0.46	[10]
$Zr_{55}Cu_{20}Ti_{2.5}Al_{12.5}Ni_{10}$	S	T	6.57	6.60	0.46	[10]

(Continued)

TABLE 6.1 (CONTINUED)

Relative Densities (ρ) of the As-Cast and Crystallized BMG Alloys

Alloy Composition	Synthesis Method	Rod Diameter (mm)	$\rho_{as\text{-}cast}$ (g cm^{-3})	$\rho_{crystallized}$ (g cm^{-3})	$\%\Delta\rho_{crystallized}$[a]	References
$Zr_{55}Cu_{20}Ti_5Al_{10}Ni_{10}$	S	T	6.62	6.64	0.30	[7,10]
$Zr_{52.5}Cu_{20}Ti_{2.5}Al_{15}Ni_{10}$	S	T	6.49	6.52	0.46	[10]
$Zr_{52.5}Cu_{20}Ti_5Al_{12.5}Ni_{10}$	S	T	6.52	6.55	0.46	[7,10]
$Zr_{52.5}Cu_{20}Ti_{7.5}Al_{10}Ni_{10}$	S	T	6.47	6.50	0.46	[10]
$Zr_{52}Ti_5Cu_{18}Ni_{15}Al_{10}$	C	3	6.60	6.77	2.58	[11]
$Zr_{50}Cu_{20}Ti_{7.5}Al_{12.5}Ni_{10}$	S	T	6.48	6.51	0.46	[10]
$Zr_{46.7}Ti_{8.3}Cu_{7.5}Ni_{10}Be_{27.5}$	C	10			0.80	[12]

Note: C, Cu mold casting; W, water quenching; S, squeeze casting; V, cone-shaped (15 mm at the top and 6 mm at the bottom); T, 2.5 mm thick sheet.

[a] $\%\Delta\rho_{crystallized} = \frac{\rho_{crystallized} - \rho_{as\text{-}cast}}{\rho_{as\text{-}cast}} \times 100.$

of the BMGs were measured by the Archimedes principle as a function of the cooling rate and these density values are plotted as a function of the cooling rate in Figure 6.1. The density of the fully crystallized alloy has been reported to be 9.31 [6,7] or 9.341 g cm^{-3} [8]. The trend of variation of density with the cooling rate is as expected. For example, it may be noted from Figure 6.1 that the density of the alloy decreased very rapidly when the alloy was solidified at relatively slow cooling rates, suggesting that the density is highly sensitive to the cooling rate, especially when the cooling rates are low. In other words, even though the difference in the density between the BMG alloy solidified at very slow cooling rates and the crystalline alloy is very small, its sensitivity to the cooling rate is very high. But, when the liquid alloy is solidified at higher rates, the density difference between the as-solidified and the crystallized states is larger, suggesting that the free volume retained in the liquid as a result of faster solidification is higher. But, at cooling rates higher than about 50–75 K s^{-1} the density is much lower than that of the crystalline alloy, and also seems to be almost independent of the cooling rate. This observation suggests that the constituent atoms have sufficient time to rearrange themselves into dense atomic configurations when the melt is solidified at relatively low cooling rates, and that the higher the density of the BMG (and closer to that of the crystalline alloy), the slower the cooling rate. By comparing the density of the BMG alloy with that of the fully crystalline alloy, the authors noted that the difference between these two states is only about 0.6%. Compared with the density changes between the melt-spun glassy ribbons and the fully crystallized alloys, where the differences were reported to be as large as 1%–2% [1,2], these values are much smaller.

The small differences in the densities between the crystalline and the glassy alloys clearly indicate that BMG alloys have a higher degree of dense random-packing state of the constituent atoms as compared with those of melt-spun ribbons, which require higher cooling rates for glass formation. This result is consistent with the trend obtained from the structural analyses. The dense random packing of atoms results in a decreased atomic mobility and an increased viscosity. Further, the frequency with which atoms are trapped at the liquid/solid interface decreases in the higher degree of dense random-packing state, leading to a decrease in the nucleation frequency of the crystalline phase. These changes in the atomic mobility and the viscosity can be reasonably expected to lead to an increase in the reduced glass transition temperature, $T_{rg} = T_g/T_m$, and also to a decrease in the growth rate of the crystalline phase.

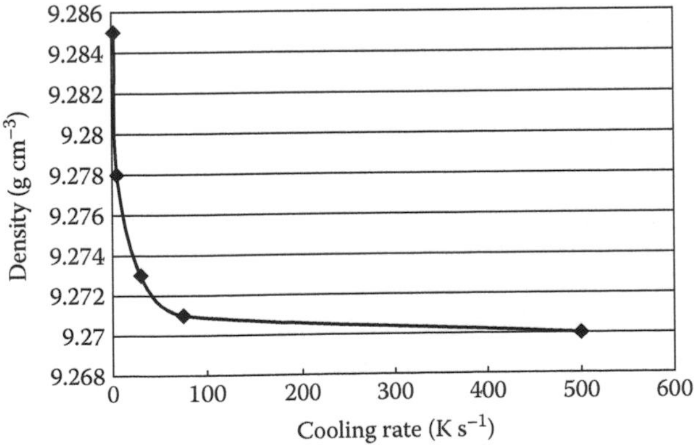

FIGURE 6.1
Density as a function of the cooling rate for the $Pd_{40}Cu_{30}Ni_{10}P_{20}$ BMG alloy.

The relative change of density of BMGs has also been measured on annealing the as-cast BMG alloy at different temperatures. Figure 6.2 shows the densities of $Pd_{40}Cu_{30}Ni_{10}P_{20}$ and $Zr_{55}Cu_{30}Al_{10}Ni_5$ BMG alloys isochronally annealed for 15 min as a function of the annealing temperature, T_a. The densities of these alloys in the as-cast glassy condition were measured to be 9.27 and 6.82 g cm^{-3}, respectively. It may be noted that the density slowly increases with the annealing temperature, as a result of the annihilation of the excess free volume, and this trend continues up to the glass transition temperature, T_g. But, the density of the BMG is much higher when the glass is annealed at temperatures higher than the crystallization temperature, T_x. This is because the glass crystallizes under these conditions and approaches equilibrium. Therefore, the difference in density between the annealed glass and the equilibrium value is reduced. Similar results have been reported for other BMG alloys as well [12].

The densities of $Pd_{40}Cu_{30}Ni_{10}P_{20}$ and $Zr_{55}Cu_{30}Al_{10}Ni_5$ glassy alloys were also measured in the structurally fully relaxed condition [6–8]. It was noted that the difference in the densities between the structurally relaxed and the fully glassy states was about 0.11% [6,7] or 0.26% [8] for the $Pd_{40}Cu_{30}Ni_{10}P_{20}$ glass, 0.15% for the $Zr_{55}Cu_{30}Al_{10}Ni_5$ glass [6,7], and 0.1% for the $Zr_{46.7}Ti_{8.3}Cu_{7.5}Ni_{10}Be_{27.5}$ glass [12]. Since the formation of equilibrium crystalline phases as the glassy alloy is progressively annealed at higher temperatures involves structural relaxation and crystallization of the metallic glass, it can be inferred that the total difference in density between the as-solidified and the fully crystallized condition is due to the combined effects of structural relaxation and crystallization of the glassy phase. Since BMGs are produced at much lower cooling rates than the melt-spun ribbons, it is only natural that the BMG alloys contain a much lower fraction of free volume and therefore their density is higher, but the difference in density between the as-solidified and crystallized conditions is lower in the BMGs than in the melt-spun ribbons [13,14].

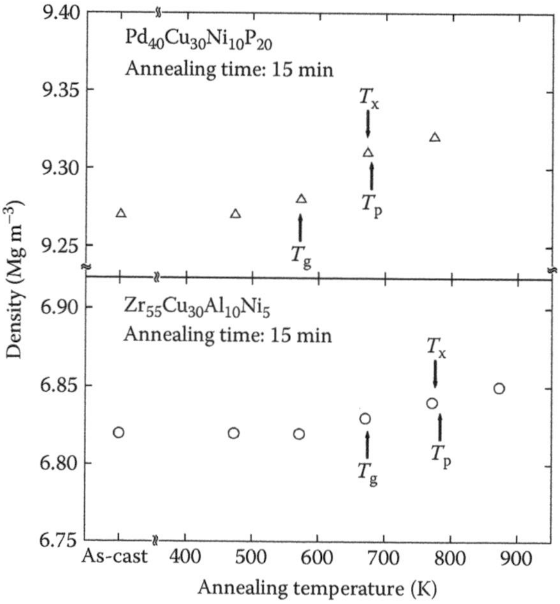

FIGURE 6.2
Change of density with annealing temperature for $Pd_{40}Cu_{30}Ni_{10}P_{20}$ and $Zr_{55}Cu_{30}Al_{10}Ni_5$ BMG alloys. Annealing time at each temperature was fixed at 15 min.

An ideal glass is expected to form at T_K (the Kauzmann temperature), at which the free volume or the configurational entropy disappears. Experimentally, it is observed that T_K is slightly lower than the glass transition temperature, T_g. Therefore, the density of such an "ideal" glass, ρ_i, can be estimated from the relationship:

$$\rho_i = \rho_g (1 + \Delta\beta \Delta T) \tag{6.2}$$

where:

ρ_g is the density of the fully relaxed glassy alloy
$\Delta\beta = \beta^L - \beta^g$ is the difference between the volumetric thermal expansion coefficients of the supercooled liquid and the glassy conditions
$\Delta T = T_g - T_K$

These ideal glasses seem to have a density about 0.5% higher than the fully relaxed glass (or the crystalline value).

Harms et al. [15] reported that the density of the $Pd_{40}Cu_{30}Ni_{10}P_{20}$ BMG alloy was reduced on plastically deforming it. It was suggested that plastic deformation reduced the short-range order (SRO) and increased the free volume and, consequently, the density had decreased. Annealing, on the other hand, had the opposite effect. Liu et al. [16] investigated the effect of pressure on the density of $Zr_{46}Cu_{37.6}Ag_{8.4}Al_8$ BMG and noted that the density of the as-solidified sample (7.184 g cm^{-3}) increased smoothly by 5.2% in the pressure range from ambient pressure to 6.3 GPa. After release of pressure, the density of the recovered sample was 7.231 g cm^{-3}, which is 0.7% denser than the starting specimen.

The free-volume model of Turnbull and Cohen [17] has been widely used to explain the properties of metallic glasses. BMG alloys provide an excellent opportunity to study the structural relaxation behavior in the sub-T_g range. This has been done, for example, through density measurements. Structural relaxation occurs on annealing the glassy alloy below T_g and this involves the annihilation (and/or the creation) of free volume. These processes have been most efficiently studied by measuring the density very accurately [12,18–20]. Through such measurements, it was reported that structural relaxation in some of the BMG alloys takes place in two steps, and that the density changes of the glassy alloy are directly connected with changes in the free volume. Russew and Sommer [18] established a quantitative relationship between the density changes and the length changes in metallic glass samples as

$$\Delta\rho = \rho(t) - \rho_0 = -\frac{\left[3W\Delta L_f(t)/L_0\right]}{L_0^3\left[1+3\Delta L_f(t)/L_0\right]} \tag{6.3}$$

where:

$\rho(t)$ is the density as a function of time
ρ_0 is the initial sample density
W is the sample weight
ΔL_f is the change in length of the sample
L_0 is the initial edge length

They measured the relative density changes, $\Delta\rho/\rho_0$, as a function of annealing time at different temperatures and related these to the non-isothermal length changes of the glassy alloy ribbons. Both the experimental results (density changes and length changes) could be described using the same set of free-volume model parameters. Figure 6.3 shows the relative density changes, $\Delta\rho/\rho_0$, of the $Pd_{40}Cu_{30}Ni_{10}P_{20}$ BMG alloy as a function of annealing time at different temperatures, ranging from 500 to 593 K. It may be noted that the density increases with an increasing annealing time and is expected to reach a saturation value once equilibrium is established. From the data, it is clear that while the samples annealed at 573 and 593 K showed saturation values for annealing times of about 2.5×10^3 and 6×10^2 s, respectively, the samples annealed at lower temperatures showed a monotonic increase and did not exhibit any tendency toward saturation in the times investigated. The plateau data could be assigned to the quasi-equilibrium density value at these temperatures.

It is useful to remember one important point while trying to study structural relaxation through density measurements. Structural relaxation is associated with the annihilation of excess free volume. It has been reported that the distribution of the quenched-in free volume was non-uniform in a rod-shaped $Zr_{50}Cu_{40}Al_{10}$ BMG alloy. The free volume was more at the surface of the rod than in the interior, mainly because of the higher cooling rate experienced by the sample at the surface [20]. But, density is the bulk property of the alloy and, therefore, relaxation studies using density measurements should be conducted only after the quenched glass is homogenized by pre-annealing treatments. This should, of course, ensure that the total amount of free volume has not changed by this treatment.

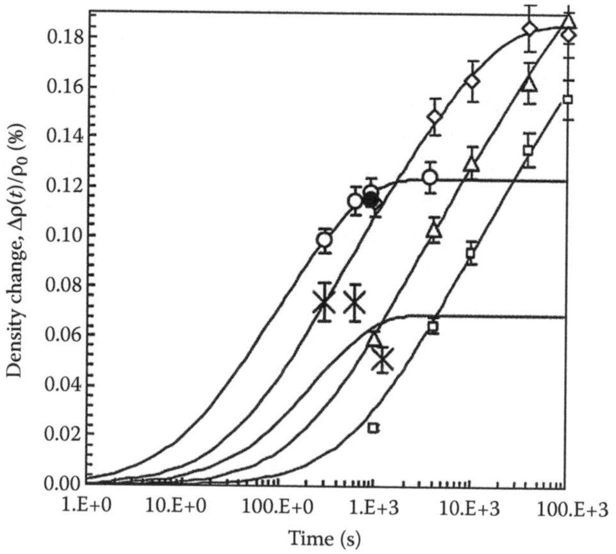

FIGURE 6.3
Relative change of density, $\Delta\rho/\rho_0$ (%), as a function of the isothermal annealing time at different temperatures in a $Pd_{40}Cu_{30}Ni_{10}P_{20}$ BMG alloy. The continuous lines are the fitting curves calculated using Equation 6.3 and are drawn to aid visualization. The different symbols represent the temperatures at which annealing was carried out. (□) 500 K, (△) 520 K, (◇) 550 K, (○) 573 K, and (×) 593 K. The symbol (●) denotes the value of $\Delta\rho/\rho_0$ (in %) (≈0.11%) obtained by Inoue for the same BMG alloy after annealing at 573 K for 15 min. (Reprinted from Russew, K. and Sommer, F., *J. Non-Cryst. Solids*, 319, 289, 2003. With permission.)

6.3 Thermal Expansion

Measurement of thermal expansion in metallic glassy alloy samples, just like density, can provide useful information about the structural relaxation behavior of glasses. The coefficient of thermal expansion (CTE) has been measured in the fully glassy state, the structurally relaxed glassy state, the supercooled liquid state (SLS), and in the crystallized condition. A dilatometer is typically used to measure the CTE, although other techniques have also been used. For example, Yavari et al. [21–23] measured the change in the wave vector and obtained information about the structural relaxation of metallic glasses using the equation:

$$Q_{\text{max}} = \frac{4\pi \sin\theta}{\lambda} \tag{6.4}$$

where:
- θ is the angular position corresponding to the maximum of the diffracted intensity
- λ is the wavelength of the radiation used

The wave vector is related to the mean interatomic distance in the first coordination shell. By measuring the change in the Q_{max} vector as a function of temperature, the volume coefficient of the thermal expansion was determined using the relationship

$$\left[\frac{Q_{\text{max}}(T^0)}{Q_{\text{max}}(T)}\right]^3 = \frac{V(T)}{V(T^0)} = 1 + \alpha_{\text{th}}(T - T^0) \tag{6.5}$$

where:
- T^0 is a reference temperature
- α_{th} is the volume coefficient of the thermal expansion

It is important to remember that the CTE values in these samples are measured usually under a small tensile load. Accordingly, the CTE value may be overestimated, since it contains the contributions due to creep in addition to the true thermal expansion. From such observations, it has been reported that the CTE values, at least in the SLS, may be overestimated by about one order of magnitude. The true value of CTE may be obtained by performing these experiments at different stress levels and then extrapolating the observed CTE values to a zero stress value.

The CTE values measured for different BMG alloys are listed in Table 6.2. A few general comments can be made from these values. The CTE values of samples in the fully glassy state are typically in the range $10–15 \times 10^{-6}$ K^{-1}, except in the case of Sm-based BMGs, where it was reported to be about 125×10^{-6} K^{-1}. The CTE values in the glassy state are usually larger than those of the crystalline pure metals on which they are based [24]. One exception appears to be Cu, where the glassy alloy exhibits a smaller CTE than the crystalline Cu metal. In comparison to the as-quenched glassy state, the CTE value of the glass in the structurally relaxed condition is higher. The CTE values are much higher in the supercooled liquid condition and lower in the crystallized condition, even lower than in the glassy state.

A typical plot of thermal expansion versus temperature is shown in Figure 6.4. The glassy material expands on heating to higher temperatures, at a particular rate of CTE,

TABLE 6.2

Linear Coefficients of Thermal Expansion (CTE) Measured for Different BMG Alloys

Alloy Composition (at.%)	Temperature Range (K)	Condition of the Alloy	Linear CTE ($\times 10^{-6}$ K^{-1})	References
Pure metal Cu	300	Crystalline	16.5	[24]
$Cu_{55}Hf_{25}Ti_{15}Pd_5$	300–600	Glassy (before structural relaxation)	12	[25]
	650–750	After structural relaxation	14	[23,25]
	—	Supercooled liquid	42	[25]
	300–530	Glassy	15 (TMA)	[25]
$Cu_{47}Ti_{33}Zr_{11}Ni_8Si_1$	Up to 373	Glassy	13.6	[26]
$Cu_{60}Zr_{30}Ti_{10}$	Up to 373	Glassy	10.9	[26]
$Cu_{55}Zr_{30}Ti_{10}Ni_5$	300–600	Glassy (before structural relaxation)	12.7	[25]
$Cu_{55}Zr_{30}Ti_{10}Ni_5$	300–600	Glassy (after structural relaxation)	13.7	[25]
$Cu_{55}Zr_{30}Ti_{10}Ni_5$	300–600	Glassy	13 (TMA)	[25]
$Cu_{55}Zr_{30}Ti_{10}Ni_5$	—	Supercooled liquid	30	[25]
Pure metal Mg	300	Crystalline	26.1	[24]
$Mg_{65}Cu_{25}Tb_{10}$	350–487	Glassy	44	[27]
Pure metal Pd	300	Crystalline	11.2	[24]
$Pd_{43}Cu_{27}Ni_{10}P_{20}$	—	Glassy	12.3	[28]
	—	Supercooled liquid	30	[28]
	—	Liquid	30	[29]
	—	Glassy	17	[29]
	—	Crystalline	14.7	[29]
$Pd_{40}Cu_{30}Ni_{10}P_{20}$	323–523	Relaxed glassy	17	[30]
	—	Supercooled liquid	27	[30]
	323–673	Crystallized	14.2	[30]
	873–1223	Liquid	39.9	[30]
$Pd_{40}Cu_{30}Ni_{10}P_{20}$	—	Glassy	16.5	[8]
	—	Supercooled liquid	34.8	[8]
$Pd_{40}Cu_{30}Ni_{10}P_{20}$	—	Glassy	18.6	[18]
$Pd_{40}Cu_{30}Ni_{10}P_{20}$	300–500	Glassy	8	[31]
$Pd_{40}Cu_{30}Ni_{10}P_{20}$	602–613	Supercooled liquid	26,000	[31]
$Pd_{40}Ni_{40}P_{20}$	—	Glassy	13–17	[28]
	—	Supercooled liquid	30	[28]
Pure metal Sm	300	Crystalline	10.4	[24]
$Sm_{55}Al_{25}Cu_{10}Co_{10}$	up to 450	Glassy	125	[32]
Pure metal Zr	300	Crystalline	5.78	[24]
$Zr_{65}Al_{10}Cu_{15}Ni_{10}$	Up to 553	Glassy	12.2	[22]
$Zr_{60}Ni_{10}Cu_{20}Al_{10}$	Up to 650	Glassy	10	[22]
$Zr_{60}Ti_2Ni_8Cu_{20}Al_{10}$	—	Glassy	11	[11]
	—	Supercooled liquid	57	[11]
	—	Crystalline	12	[11]
$Zr_{52}Ti_5Ni_{15}Cu_{18}Al_{10}$	—	Glassy	12	[11]
	—	Crystalline	14	[11]
$Zr_{41}Ti_{14}Cu_{12.5}Ni_{10}Be_{22.5}$	Up to 473	Glassy	9.26	[33]
	—	Glassy	8.51	[34]
	—	Crystalline	9.24	[33]

Note: TMA, thermomechanical analysis.

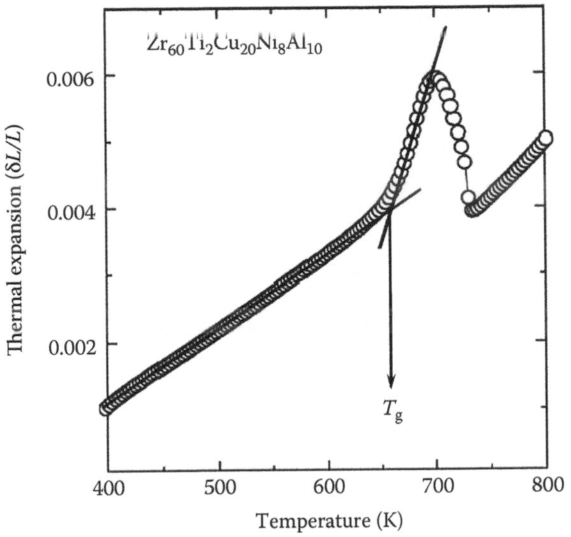

FIGURE 6.4
Variation of thermal expansion with temperature in a $Zr_{60}Ti_2Cu_{20}Ni_8Al_{10}$ BMG alloy. (Reprinted from Mattern, N. et al., *Mater. Sci. Eng. A*, 375, 351, 2004. With permission.)

nearly up to the glass transition temperature, T_g. But, when the specimen approaches T_g, structural relaxation takes place and the CTE value increases more rapidly. In those cases where a stress is applied to the specimen during measurement, the actual temperature to which the thermal expansion is linear may be a little lower. But, once the specimen enters the supercooled liquid region, that is the temperature region between T_g and T_x, the CTE is much larger than in the as-quenched glassy condition or after structural relaxation. In the crystallized condition, the CTE is again reduced and its value is quite similar to that of the glassy alloy.

As mentioned earlier, the glass prepared by quenching the melt (whether at slow rates or faster rates) contains an excess quenched-in free volume and its amount increases with increasing solidification rate. On reheating the glassy sample to a temperature below T_g, structural relaxation occurs through chemical short-range order (CSRO) and topological short-range order (TSRO). The CSRO involves slight changes in the local chemical environment and is sometimes reversible. On the other hand, the TSRO involves a long-range rearrangement of all the atoms and is an irreversible process. That means the process of structural relaxation is associated with the redistribution and the reduction of excess quenched-in free volume, resulting in a decrease in volume, and consequently an increase in the density. This will also have an effect on the thermal expansion of the material.

The relative length change during annealing of a BMG sample is directly related to the annihilation of excess quenched-in free volume and, therefore, this occurs due to TSRO. The disappearance of free volume will lead to a decrease in the length of the sample, and therefore in the CTE as well. This is a nonreversible process. But, when the glassy sample is heated to temperatures below T_g, where structural relaxation also occurs, but through CSRO, the process is reversible. That is, the same value of CTE is obtained during thermal cycling by cooling and through subsequent reheating, but only up to temperatures just below T_g.

The CTE value appears to be independent of the heating rate at which the sample is heated to high temperatures [26]. Figure 6.5a shows the variation of the thermal

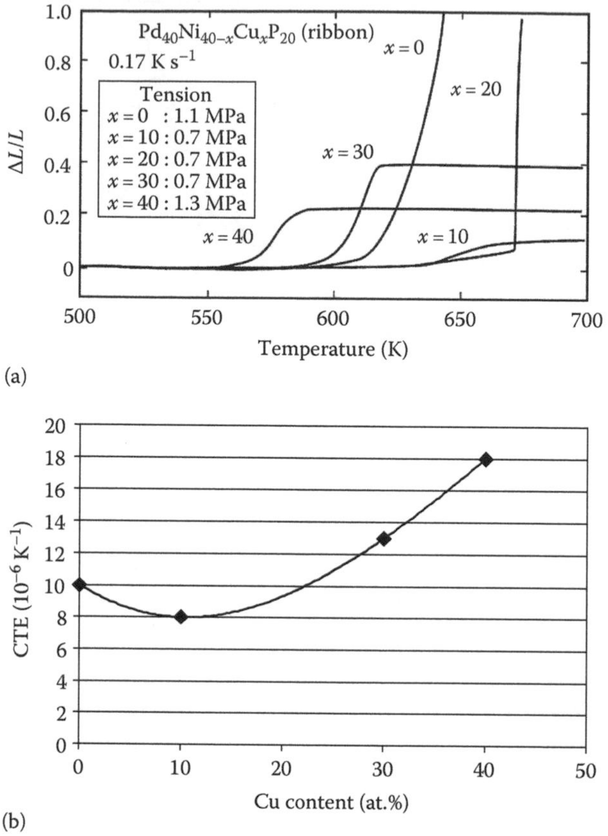

FIGURE 6.5
(a) Variation of thermal dilatation, $\Delta L/L$, as a function of temperature for different Cu contents in the melt-spun $Pd_{40}Ni_{40-x}Cu_xP_{20}$ ribbons. While $\Delta L/L$ remains almost unchanged in the glassy state, it increases with increasing temperature in the supercooled liquid state and stops after the sample is crystallized. (b) Variation of CTE as a function of Cu content in the melt-spun $Pd_{40}Ni_{40-x}Cu_xP_{20}$ glassy alloys with x = 0, 10, 30, and 40 at.% Cu. A tensile stress of 0.7–1.3 MPa was applied during the test.

dilatation ($\Delta L/L$) as a function of temperature for different Cu contents in the melt-spun $Pd_{40}Ni_{40-x}Cu_xP_{20}$ glassy alloy ribbons. No appreciable change in $\Delta L/L$ is seen in the temperature range below T_g, indicating that the alloy has an extremely low CTE in the glassy state. With a further increase in temperature, the alloy begins to elongate, and this is followed by a rapid increase in elongation in the supercooled liquid range. But, once the sample reaches the crystallization temperature range (600–615 K), elongation stops. This trend is observed in most BMG alloys, with a few exceptions. Figure 6.5b shows the variation of CTE as a function of the Cu content in the $Pd_{40}Ni_{40-x}Cu_xP_{20}$ glassy alloy ribbons. The CTE decreases with increasing Cu content up to about 10 at.% Cu and then increases with increasing Cu content.

The thermal expansion behavior of Sm-based BMGs and Cu-based BMGs appears to be quite different above T_g. At temperatures higher than T_g, the CTE for the Cu-based BMG alloy increases initially and then a slight contraction follows. A similar behavior is seen in some Zr- and Pd-based BMGs as well. But, in the case of Sm-based BMGs, no increase in CTE is observed above T_g, in the supercooled liquid region. This apparent anomaly has been explained on the basis of the following differences between the two glasses [32].

The change in CTE, above T_g, can be expressed as a combination of two parts, that is,

$$\Delta\alpha = \Delta\alpha_1 + \Delta\alpha_2 \tag{6.6}$$

Here, $\Delta\alpha_1$ is due to the transition from the glassy to the SLS and is always positive. When a compressive load is used during the length-change measurement, $\Delta\alpha_2$ is negative. Depending on the relative magnitudes of $\Delta\alpha_1$ and $\Delta\alpha_2$, the $\Delta\alpha$ value may be positive or negative. In the case of Sm-based BMGs, the specific amount of contraction was found to be an order of magnitude larger than in the case of Cu-based BMGs under the action of the compressive load. In other words, $\Delta\alpha_2$ is much more negative than $\Delta\alpha_1$. That is why the Sm-based BMGs exhibited a decrease in CTE in the supercooled liquid region. Whether this difference in CTE between the Cu-based and the Sm-based BMGs is real can easily be resolved by measuring the CTE values under a tensile load and seeing whether the behavior is any different.

6.4 Diffusion

Studies of diffusion are very important in the field of metallic glasses—both ribbons and BMGs. It is scientifically interesting to determine whether the mechanism of diffusion in metallic glasses is different from that in crystalline materials. More specifically, it will be important to decide whether the single-atom jumps observed in crystalline materials occur in metallic glasses as well. Increasing evidence suggests they do not, particularly in the supercooled liquid region. Further, since metallic glasses are nonequilibrium materials, they transform to the equilibrium constitution on annealing them at sufficiently high temperatures and/or for a long time. During annealing, depending on the temperature at which it is done, the glass undergoes structural relaxation, phase separation, and crystallization. All these processes involve diffusion, and therefore the thermal stability of the metallic glasses is determined by their diffusion behavior. Thus, a study of the diffusion behavior in metallic glasses is important both scientifically and technologically.

Diffusion studies were conducted earlier on melt-spun metallic glass ribbons [35,36], but mostly in the glassy state, at temperatures below the crystallization temperature, T_x. The width of the supercooled liquid region ($\Delta T_x = T_x - T_g$) is very small in these melt-spun metallic glasses, if observed at all, and therefore it was not possible to study the diffusion behavior in the supercooled liquid condition in these materials. But, the advent of BMGs, which show a significantly large ΔT_x range, has changed the situation significantly. Since ΔT_x in BMGs is a few tens of degrees wide, or in some cases even over 100 K, detailed diffusion studies have been conducted in the supercooled liquid region in BMGs. Such temperature regimes were previously not accessible. A number of studies are available now, and the results up to 2003 have been summarized in a succinct review [37].

Generally, radiotracer techniques have been used to study the diffusion behavior in BMG alloys due to their high sensitivity and also because self-diffusion can be very conveniently investigated. The procedure for these studies is rather simple. A thin layer of the radiotracer atoms is deposited on a well-polished flat surface of the sample and then an isothermal diffusion anneal is performed at the desired temperature and for the required length of time. The concentration/depth profile is then obtained by sectioning the sample and determining the number of tracer atoms. The diffusion condition is equivalent to an infinitely thin source diffusing into a semi-infinite cylinder, and therefore the concentration profile with distance can be described by the thin film solution of Fick's second law of diffusion as

$$C(x,t) = \frac{M}{\sqrt{\pi D t}} \exp\left(-\frac{x^2}{4Dt}\right) \tag{6.7}$$

where:

$C(x, t)$	is the tracer concentration measured at time t
M	is the initial amount of the tracer at the surface
D	is the diffusion coefficient
x	is the penetration depth

Thus, by plotting $\ell n\, C(x, t)$ against x^2, a straight line with the slope $-1/4Dt$ is obtained, from which the value of D is determined.

The diffusion behavior of melt-spun glassy ribbons is known to be sensitive to structural relaxation. Accordingly, in the as-quenched ribbon (i.e., in the unrelaxed sample), the diffusivity decreases as a function of time due to structural relaxation and annealing out of excess free volume. The diffusivity reaches a constant value only when the sample is fully relaxed. Such a situation is not obtained in the BMG alloy samples due to the relatively slow solidification rates at which these glassy alloys are produced. Therefore, whenever diffusion studies are carried out on thin ribbon specimens, they are given a pre-annealing relaxation treatment before the diffusion measurements are made.

Figure 6.6 shows the penetration profiles for the diffusion of ^{63}Ni in the glassy and the supercooled liquid phases, respectively, of the $Zr_{55}Al_{10}Ni_{10}Cu_{25}$BMG alloy [38]. All the diffusion profiles are Gaussian without a serious surface hold-up or noticeable non-Gaussian tails. The very first data points are affected by surface effects and so are not taken into account. Data points at large penetration depths are influenced by the background activity and therefore they are also neglected. By analyzing the penetration profiles using a least squares fitting to Equation 6.7, the D values are calculated at different temperatures. Figure 6.7 shows the temperature dependence of diffusivity of Ni in the $Zr_{55}Al_{10}Ni_{10}Cu_{25}$ BMG alloy.

The important feature to be noted in Figure 6.7 is the nonlinearity in the whole temperature range investigated. The data, however, can be divided into two subsets, each described by a different Arrhenius equation. The first region is for temperatures below the "kink" temperature, and the other is for temperatures above the "kink." This kink temperature, which represents the change in the temperature dependence of diffusivity, is frequently identified as the glass transition temperature, T_g. That is, the Arrhenius behavior above T_g represents the diffusion behavior in the SLS, and that below T_g, the behavior in the glassy state. Accordingly, the pre-exponential factor D_0 and the activation energy for diffusion are different in these two regions. The diffusivity in the supercooled liquid phase is much higher than that obtained by extrapolation from the low-temperature data in the glassy phase. This behavior is in contrast to melt-spun ribbons that show a single straight line described by a unique Arrhenius equation. This is

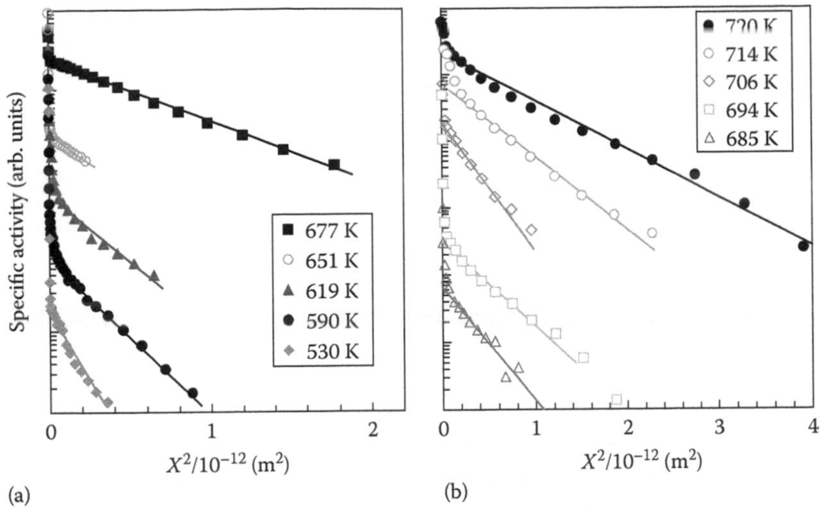

FIGURE 6.6
Depth penetration profiles for diffusion of ^{63}Ni in the glassy phase (a) and in the supercooled liquid condition (b) for the $Zr_{55}Al_{10}Ni_{10}Cu_{25}$ BMG alloy.

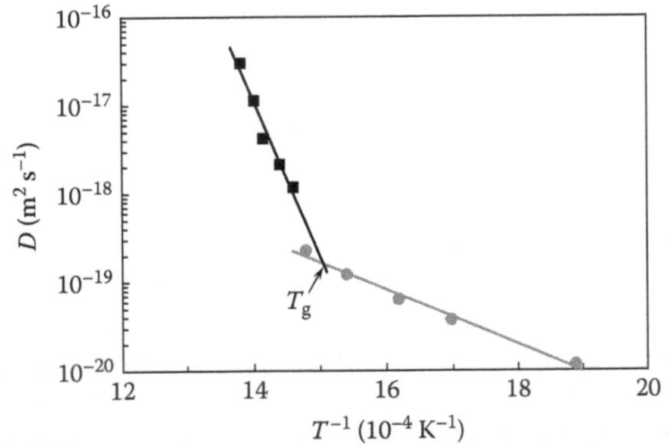

FIGURE 6.7
Temperature dependence of diffusivity of ^{63}Ni in the $Zr_{55}Al_{10}Ni_{10}Cu_{25}$ BMG alloy.

essentially because the supercooled region is not easily accessible in the case of melt-spun glassy ribbons.

Fitting the data to two different Arrhenius equations, the temperature dependence of diffusivity of Ni in the $Zr_{55}Al_{10}Ni_{10}Cu_{25}$ BMG alloy in the glassy and supercooled states are expressed as

$$D_{\text{glass}} = 6.8 \times 10^{-15} \exp\left(\frac{-59 \pm 2\ \text{kJ mol}^{-1}}{RT}\right) \text{m}^2\,\text{s}^{-1} \tag{6.8}$$

and

$$D_{\text{supercooled}} = 4.5\times10^{9}\exp\left(\frac{-363\pm32\text{ kJ mol}^{-1}}{RT}\right)\text{m}^2\,\text{s}^{-1} \tag{6.9}$$

The diffusion data available for different species in different BMG alloys are summarized in Table 6.3.

It is clear from Table 6.3 that both the pre-exponential factor, D_0, and the effective activation energy, Q, for BMG alloys in the high-temperature region (supercooled liquid region) are much higher than those in the low-temperature region (glassy state). These differences indicate that the operative mechanisms are different in these two regions. It has been suggested that single atom hopping is effective in explaining the diffusion behavior in the glassy state (like in the crystalline state). But, in the SLS, diffusion is envisaged as a medium-assisted, highly collective hopping process, in which groups of atoms (typically about 10) perform thermally activated transitions into new configurations [49]. That is, the diffusion mechanism at high temperatures is cooperative and at lower temperatures it is noncooperative.

This suggestion of different diffusion mechanisms has been confirmed by measuring the mass dependence of diffusion described by the isotope effect. The isotope effect parameter, E, is defined as

$$E_{\alpha,\beta} \equiv \frac{(D_\alpha / D_\beta)-1}{\sqrt{m_\beta / m_\alpha}-1} \tag{6.10}$$

where D and m represent the diffusivity and the mass of the isotope pair α and β, respectively.

In ordinary liquids, diffusion is governed by viscous flow, which can be described by uncorrelated binary collisions. Since all atoms contribute continually to the viscous flow, the $1/\sqrt{m}$ dependence of the diffusivity yields $E = 1$. On the other hand, a highly collective jump process leads to a very small isotope effect since the contribution of the individual mass is diluted [50]. Since E has been found to be very small (ranging from 0 to 0.1) in the case of multicomponent BMG alloys, it is suggested that the diffusion mechanism in the SLS is completely different from the uncorrelated single-atom jumps in the ordinary viscous flow [44].

As already mentioned, both the Q and D_0 values are much higher, typically by several orders of magnitude, in the supercooled liquid region, than in the glassy condition. The small values in the glassy state suggest a single jump motion of the atoms, while the larger values in the SLS might reflect a highly cooperative motion of a number of atoms. Distinct changes were observed in the apparent activation energies for Be diffusion at high and low temperatures, with a crossover at the laboratory glass transition temperature. Geyer et al. [42] suggested that this could be attributed to a change in the diffusion behavior. At low temperatures, in an essentially solid environment, diffusion occurs through hopping of small Be atoms, whereas at high temperatures, in the SLS, Be transport occurs through a superposition of single atomic jumps and a cooperative motion of neighbor atoms that support diffusion by increasing the frequency of critical free volume fluctuations. From nuclear magnetic resonance experiments, it was concluded that the hopping of Be atoms extends into the SLS, but is less dominant there than the cooperative motion of

TABLE 6.3

Details of Diffusion Data Obtained for BMG Alloys

		High Temperature		Low Temperature			
Alloy	**Tracer**	D_0 ($m^2\ s^{-1}$)	Q (eV)	D_0 ($m^2\ s^{-1}$)	Q (eV)	**"Kink" Temperature (K)**	**References**
$Pd_{43}Cu_{27}Ni_{10}P_{20}$	Co	1.6×10^{10}	3.2	—	—	—	[39]
$Pd_{40}Cu_{30}Ni_{10}P_{20}$	Co	1.3×10^{9}	3.2	—	—	—	[39]
$Pd_{40}Cu_{30}Ni_{10}P_{20}$	Ni	—	—	3.3×10^{7}	3.1	—	[40]
$Pd_{40}Cu_{30}Ni_{10}P_{20}$	Pd	—	—	9.4×10^{12}	3.45	—	[40]
$Zr_{55}Al_{10}Ni_{10}Cu_{25}$	Ni	—	—	3.0×10^{-12}	0.95	—	[40]
$Zr_{55}Al_{10}Ni_{10}Cu_{25}$	Ni	4.5×10^{9}	3.76	6.8×10^{-15}	0.61	—	[14]
$Zr_{46.75}Ti_{8.25}Cu_{7.5}Ni_{10}Be_{27.5}$ (Vitreloy 4)	Al	2.2×10^{12}	4.1	—	—	—	[41]
	B	2.84×10^{2}	2.59	1.89×10^{-4}	1.87	585	[41]
	Be	1.7×10^{-3}	1.9	8×10^{-10}	1.1	625	[42,43]
	Co	3.05×10^{5}	3.0	7.3×10^{-4}	2.01	575	[41]
	Co	1.5×10^{8}	3.4	—	—	—	[44]
	^{57}Co	6×10^{6}	3.2	—	—	—	[45]
	Fe	2.76×10^{6}	3.16	1.98×10^{-3}	2.11	575	[41]
	Hf	—	3.85	—	—	—	[46]
	^{63}Ni	4.32×10^{3}	2.76	—	—	580	[47]
$Zr_{41}Ti_{16.5}Cu_{12.5}Ni_{10}Be_{20}$	B	2.11×10^{3}	2.77	1.19×10^{-4}	1.87	—	[48]
$Zr_{41}Ti_{16.5}Cu_{12.5}Ni_{10}Be_{20}$	Fe	4.58×10^{7}	3.44	4.80×10^{-6}	1.85	—	[48]
$Zr_{41}Ti_{8.7}Cu_{12.5}Ni_{10}Be_{27.8}$	B	1.24×10^{1}	2.50	3.14×10^{-2}	2.18	—	[48]
$Zr_{41}Ti_{8.7}Cu_{12.5}Ni_{10}Be_{27.8}$	Fe	7.24×10^{5}	3.17	2.05×10^{-2}	2.27	—	[48]
$Zr_{41}Ti_{14}Cu_{12.5}Ni_{10}Be_{22.5}$	B	1.22×10^{5}	2.97	1.73×10^{-5}	1.80	593	[41,48]
	Be	1.1×10^{17}	4.47	1.8×10^{-11}	1.05	625	[42]
	Fe	2.26×10^{5}	3.48	1.08×10^{-5}	1.91	593	[41]

Note: D_0, pre-exponential factor; Q, effective activation energy.

the atoms [51]. In contrast to the diffusion of the small Be atom, no crossover was observed for the diffusion of larger-size atoms such as Al, Co, and Ni, nor for some other elements [52,53]. Diffusion data were also used to explain the nature of the glass transition by the mode-coupling theory [54].

Combining the measurements on viscosity and diffusion, Masuhr et al. [55] defined a characteristic jump time for atomic diffusion as

$$\tau_D = \frac{l^2}{6D} \tag{6.11}$$

where:
- l is the atomic diameter
- D is the diffusion coefficient

and compared it with the structural relaxation time:

$$\tau_\eta = \frac{\eta}{G_\eta} \tag{6.12}$$

where:
- η is the viscosity
- G_η is the high-frequency shear modulus

They observed that the temperature dependence of the Al self-diffusion compares well with the temperature dependence of the equilibrium viscosity ($\tau_{D,Al} = \tau_\eta/14$). In contrast, the self-diffusion coefficients of Be and Ni differ by three orders of magnitude at low temperatures, but tend to merge at higher temperatures. The authors concluded that the mobilities of small- and medium-size atoms, which have small characteristic jump times, decouple at low temperatures from the cooperative liquid-like mobility determined by the viscosity. With increasing temperature, the timescale for shear becomes comparable to the hopping times until, eventually, a cooperative, liquid-like motion predominates at high temperatures.

Geyer et al. [42,43] tried to explain the temperature dependence of diffusivity in the whole temperature range (covering both the glassy solid and the SLS) with a single Arrhenius-type equation by taking into account the temperature-dependent entropy change per atom, $\Delta S^{SLS}(T)$, due to the glass transition. Using this approach, the diffusivity in the SLS, $D^{SLS}(T)$, was expressed as

$$D^{SLS}(T) = D_0^{SLS} \exp\left(-\frac{\Delta H_M^{SLS}}{k_B T}\right) \tag{6.13}$$

where:

$$D_0^{SLS} = D_0^{glass} \exp\left(\frac{N \Delta S^{SLS}}{k_B}\right) \tag{6.14}$$

The change of entropy per atom, ΔS^{SLS}, in the supercooled liquid (above T_g) can be expressed as

$$\Delta S^{\mathrm{SLS}} = \frac{\Delta C_{\mathrm{p}}(T_{\mathrm{g}})}{N_{\mathrm{A}}}\left(\frac{T - T_{\mathrm{g}}}{T_{\mathrm{g}}}\right) \tag{6.15}$$

In Equation 6.15, the symbols N_A and C_p represent Avogadro's number and the specific heat, respectively. Since $\Delta C_p(T_g)/N_A$ has been experimentally found to be $2.9k_B$, where k_B is the Boltzmann constant, Equation 6.13 can be rewritten as

$$D^{\mathrm{SLS}}(T) = D_0^{\mathrm{glass}} \exp\left(-\frac{\Delta H_{\mathrm{M}}^{\mathrm{SLS}}}{k_{\mathrm{B}}T}\right) \exp\left(2.9N \frac{T - T_{\mathrm{g}}}{T_{\mathrm{g}}}\right) \tag{6.16}$$

Since changes of electronic and vibrational contributions to ΔS^{SLS} are negligible near T_g, ΔS^{SLS} is mainly a configurational ("communal") entropy term. In Equation 6.16, the product of the pre-exponential factor and the first exponential factor describes the temperature dependence of diffusivity in the glassy region. The second exponential factor contains the term N, which represents the number of the nearest and possibly the next-nearest neighbors that have a liquid-like behavior and participate in the cooperative jumps during diffusion in the supercooled state. The enhanced configurational entropy of this liquid-like cell of N atoms is accounted for by adding the ΔS^{SLS} term to the migration entropy. The value of N is obtained by fitting the experimental diffusivity data above T_g using Equation 6.16. Thus, the authors [42,43] were able to interpret the diffusion data in both the glassy and the SLSs in terms of a single atom jump diffusion mechanism.

Bartsch et al. [56] evaluated the codiffusion behavior of ^{32}P and ^{57}Co tracers in a $Pd_{43}Cu_{27}Ni_{10}P_{20}$ BMG alloy and noted that P diffused slower than the Co atom; the diffusivity of ^{32}P was found to be about 15% smaller than that of ^{57}Co. This situation was found to be true not only in the supercooled melt but also in the glassy state. Measurements of diffusivities of ^{32}P, ^{63}Ni, and Cu in a similar alloy were performed earlier [57,58], and these authors found that the diffusivity of ^{32}P was slower by about two orders of magnitude. This was explained on the basis that P forms strong covalent bonds with Ni and Cu and thus slows down its diffusivity. But, it was also noted that there was a difference of nearly two orders of magnitude between the diffusivity of ^{63}Ni measured by Bartsch et al. [56] and that of ^{63}Ni measured by Nakajima et al. [57]. Therefore, the situation is still not clear.

6.5 Electrical Resistivity

Due to their intrinsic disordered structure, metallic glasses exhibit an electrical resistivity of about two orders of magnitude higher than their crystalline counterparts. The two typical characteristics of metallic glasses in general are (1) a relatively high electrical resistivity of >1 μΩm and (2) a small and sometimes negative temperature coefficient of resistivity (TCR). Since there is no essential change in the spatial atomic configuration during glass formation, a liquid and a glass belong structurally and thermodynamically to the same phase. Accordingly, both the liquid and the glass of the same composition show similar magnitudes of TCR.

A measurement of electrical resistivity has been used as a means to study the glass transition behavior and the relaxation behavior of metallic glasses. The electrical resistivity of melt-spun Au–Ge–Si and Pd–Au–Si metallic glasses was measured by Chen and Turnbull [59,60]. They observed that the Au–Ge–Si glassy alloys showed a negative TCR, while the Pd–Au–Si alloys showed a positive TCR. Even though the slope of the electrical resistivity as a function of temperature increased after the glass transition, this was attributed to the precipitation of nanocrystalline phases in the supercooled liquid region.

There have been only a limited number of investigations on the electrical resistivity of the recently developed BMG alloys. The measurements were made mostly on Pd-based and Zr-based alloys and the available results are summarized in Table 6.4.

Figure 6.8 shows the variation of the normalized electrical resistance, $R(T)/R_0$, where $R(T)$ represents the resistance at temperature T, and R_0 is the resistance at 300 K, for the melt-spun $Zr_{60}Al_{15}Ni_{25}$ glassy alloy [67]. It may be noted that the electrical resistance decreases with increasing temperature from 300 K up to T_g; this is attributed to the negative TCR. But, the plot shows a point of inflection around 450 K, where the electrical resistance starts to slowly increase with increasing temperature, although with a smaller (but still negative) slope. Structural relaxation begins to occur at this temperature. However, once the sample has reached a temperature higher than T_g, that is, in the SLS, the alloy exhibits a positive

TABLE 6.4

Electrical Resistivity of BMG Alloys

Alloy	$\rho_{300}\mu$ (m)	$[dT\rho_{300}/dT]/\rho$ (×10^{-5})	$[d\rho_{sls}/dT]/\rho$ (×10^{-5})	TCR (×10^{-5} K^{-1})	References
$Au_{81}Ge_{11}Si_8$	—	—	—	−20	[60]
$Fe_{41}Co_7Cr_{15}Mo_{14}C_{15}B_6Y_2$	—	—	—	18.5 (g)	[61]
				25.2 (c)	
$Pd_{82}Si_{18}$	—	—	—	15 (unrelaxed glass)	[60]
	—	—	—	40 (relaxed glass)	[60]
$Pd_{81}Au_4Si_{15}$	0.9	—	—	15	[60]
$Pd_{76}Cu_6Si_{18}$	0.81	—	—	—	[62]
$Pd_{40}Cu_3Ni_{37}P_{20}$	2.33	—	—	−11	[62]
$Pd_{40}Cu_3Ni_{37}P_{20}$	1.37	7.3	—	—	[63]
$Pd_{40}Cu_{30}Ni_{10}P_{20}$	—	−8.61	−41.7	—	[64]
$Pd_{40}Ni_{40}P_{20}$	1.29	12.8	24.2	—	[63]
$Pd_{43}Ni_{37}P_{20}$	1.41	3.9	23.3	—	[63]
$Zr_{60}Ti_2Cu_{20}Ni_8Al_{10}$	1.4 (g)		−7 (g)	—	[11]
	—	—	−13 (sls)	—	[11]
	0.73 (c)	—	+85 (c)	—	[11]
$Zr_{55}Al_{10}Cu_{35}$	2.56	—	—	−6.8	[65]
$Zr_{55}Al_{10}Cu_{30}Ni_5$	2.60	—	—	−10.9	[65]
$Zr_{55}Al_{10}Cu_{25}Ni_{10}$	2.53	—	—	−15.9	[65]
$Zr_{67}Co_{33}$	1.80	—	—	−11.5	[66]
$Zr_{52}Ti_5Cu_{18}Ni_{15}Al_{10}$	1.6 (g)	—	−9 (g)	—	[11]
	—	—	−20 (sls)	—	[11]
	1.1 (c)		+10 (c)		[11]

Note: g, glassy state; sls, supercooled liquid state; c, crystalline state.

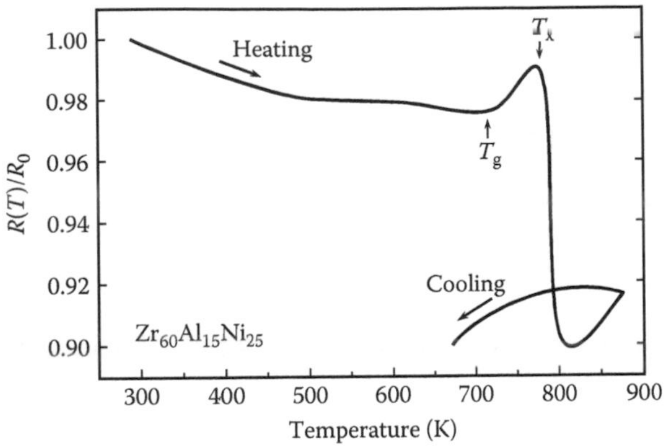

FIGURE 6.8

Variation of reduced electrical resistance with increasing annealing temperature for the melt-spun $Zr_{60}Al_{15}Ni_{25}$ glassy alloy. (Reprinted from Haruyama, O. et al., *Mater. Trans., JIM*, 37, 1741, 1996. With permission.)

TCR and the variation of the resistivity is almost linear with temperature, up to the crystallization temperature. Once crystallization sets in, the electrical resistance drops rapidly. Other Pd- and Cu-based BMG glassy alloys also show a similar behavior except that the specifics of TCR, T_g, and T_x are different for different alloy systems [62,68–70].

Lu et al. [61] investigated the transport properties of an $Fe_{48-x}Co_xCr_{15}Mo_{14}C_{15}B_6Y_2$ BMG alloy as a function of the Co content. It was reported that, as expected, the electrical resistivity of the glassy alloy was higher than that of its crystalline counterpart. They also observed that the electrical resistivity increased approximately linearly with increasing temperature until a critical temperature was reached, showing a positive TCR. After the critical temperature, the resistivity gradually decreased in three stages corresponding to the three crystallization stages (Figure 6.9).

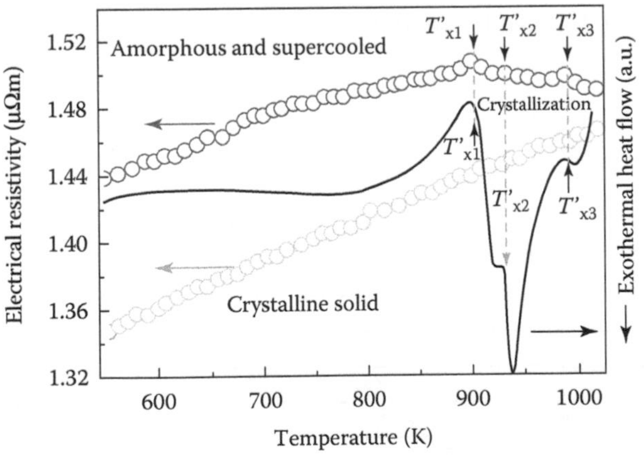

FIGURE 6.9

Variation of electrical resistivity with temperature for the $Fe_{43}Co_5Cr_{15}Mo_{14}C_{15}B_6Y_2$ BMG alloy and its crystalline counterpart. The electrical resistivity is compared with the DTA at the same heating rate of 80 K min^{-1}. Note that the different stages in the electrical resistivity and DSC plots match very well. (Reprinted from Lu, Y.Z. et al., *Intermetallics* 30, 144, 2012. With permission.)

Orbanic and Kokanvic [66] studied the electrical resistivity behavior of $Zr_{67}Co_{33}$ metallic glass ribbons obtained by melt spinning as a function of temperature. They noted that this alloy showed an increase of the room temperature electrical resistivity due to annealing at temperatures around the first crystallization temperature. This was explained on the basis of increased disorder caused by the nucleation of cubic Zr_2Co nanocrystalline phase.

The reduced electrical resistivity in $Zr_{60}Ti_2Cu_{20}Ni_8Al_{10}$ and $Zr_{52}Ti_5Cu_{18}Ni_{15}Al_{10}$ BMG alloys was shown to decrease with an increasing temperature in the glassy region. The variation was slow to start with and showed a kink near T_g. By measuring the electrical resistivity as a function of temperature, Mattern et al. [11] noted that a plot of resistivity against temperature showed a clear change in the slope at T_g (see, e.g., figure 3 in Ref. [11]). A further increase in temperature reduced the resistivity more rapidly in the supercooled region, and much more rapidly after crystallization [62,68–70]. A similar drop in resistivity was obtained at the other crystallization events too. The number of steps in the decrease of the reduced electrical resistivity has been related to the number of crystallization events, as determined from the differential scanning calorimeter (DSC) curves. From the strong evidence presented in a number of cases, it has been concluded that the variation in electrical resistivity after the glass transition is an inherent change in the electron transport property of the alloy. That is, this change in electrical resistance at the glass transition is real, and is not, as interpreted earlier, associated with the formation of nanocrystalline phases. This has been confirmed using transmission electron microscopy studies [62].

Figure 6.10 shows a plot of the TCR of $Pd_{40}Ni_{40-x}Cu_xP_{20}$ BMG alloys as a function of the Cu content. It may be noted from this plot that the TCR is negative at low Cu contents and it slowly becomes positive, passing through a zero value, as the Cu content is increased [31]. Thus, it should be possible to obtain a zero TCR by fine-tuning the Cu content in the BMG alloy. In this context, it is interesting to note that the $Pd_{40}Cu_{30}Ni_{10}P_{20}$ alloy has the highest glass-forming ability (GFA) in the Pd–Cu–Ni–P alloy system. It is also the same composition at which TCR ≈ 0. Therefore, it will be instructive to investigate if the GFA of alloys is in any way related to the zero TCR. A similar situation of TCR going from a negative to a positive value was reported to occur as the P content was reduced in the melt-spun ribbons of the Pd–Ni–P alloy system [71]. Even though all the alloy compositions for x = 15–27 at.% P were glassy, the TCR was close to zero at x = 24 at.% P. Detailed investigations are not available regarding the GFA of these alloys.

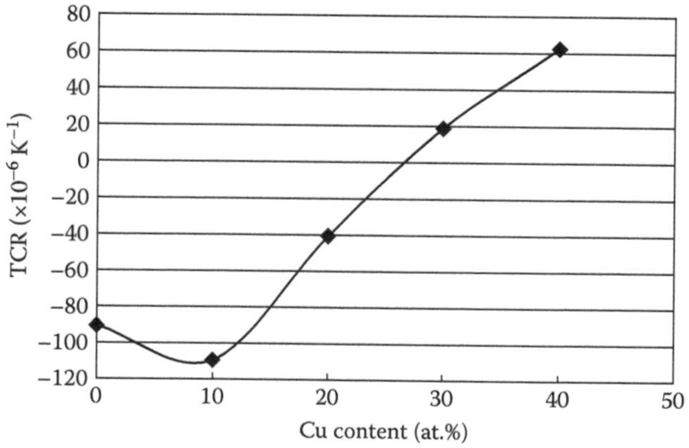

FIGURE 6.10
Variation of TCR with Cu content in a melt-spun $Pd_{40}Ni_{40-x}Cu_xP_{20}$ glassy alloy with x = 0, 10, 20, 30, and 40 at.% Cu.

In most cases, the electrical resistance is measured on heating the glassy samples from room temperature in the glassy state to the SLS, and sometimes into the liquid state [68,69]. Measurement of electrical resistivity at low temperatures has been reported only in a few Zr-based BMG alloys. Okai et al. [65,72] measured the electrical resistivity of $Zr_{55}Cu_{30-x}Al_{10}Ni_5Nb_x$ (x = 0, 1, 3, and 5 at.%) BMG alloys at temperatures from 300 K down to about 2 K. They noted that the electrical resistivity in these alloys increased slowly with decreasing temperature and that the resistivity dropped precipitously at 2.8 K in the $Zr_{55}Cu_{30}Al_{10}Ni_5$ alloy, suggesting that the electrical resistivity characteristics of the $Zr_{55}Al_{10}Cu_{30}Ni_5$ alloy are similar to those of glassy superconducting alloys [65]. An addition of 1–5 at.% of Nb (an element which has a positive heat of mixing with Zr) to the alloy resulted in the precipitation of the nanocrystalline superconducting phases, for example Zr_2Cu and Zr_2Ni, in the glassy matrix. These superconducting phases have a superconducting transition temperature of 2.3–2.9 K [72].

The crystallization behavior of $Zr_{41}Ti_{14}Cu_{12.5}Ni_{10}Be_{22.5}$ [73] and $Cu_{43}Zr_{43}Al_7Ag_7$ [70] BMG alloys was analyzed using electrical resistivity measurements. The $Zr_{41}Ti_{14}Cu_{12.5}Ni_{10}Be_{22.5}$ BMG alloy was isothermally annealed at 390°C (663 K). The measurement of the electrical resistivity of this alloy, in which the resistivity decreased with time, showed that there were two distinct stages of resistivity reduction with annealing time, suggesting that crystallization will be complete in two stages. This result is consistent with the results of DSC, which also showed two crystallization peaks. The results of the electrical resistivity were analyzed for the degree of crystallization, $x(t)$, using the relationship:

$$x(t) = \frac{\rho_0 - \rho(t)}{\rho_0 - \rho(\infty)} \tag{6.17}$$

where ρ_0 and $\rho(\infty)$ are the electrical resistivities at the beginning (t = 0) and end of the crystallization, respectively. The results were analyzed using the Johnson–Mehl–Avrami–Kolmogorov (JMAK) approach:

$$x(t) = 1 - \exp(-kt^n) \tag{6.18}$$

where:
- k is a kinetic factor
- n is the Avrami exponent

Since the crystallization took place in two stages, the two values of n were evaluated, and from these the mechanism of the growth of precipitates was inferred.

Since some applications of BMGs require good electrical conductivity, and the two traditional approaches of compositional modification and the reduction of defects have proven unsuccessful, Wang et al. [74] developed a composite of α-brass ($Cu_{80}Zn_{20}$) dispersed in an $Ni_{59}Zr_{20}Ti_{16}Si_2Sn_3$ BMG matrix. A microscopic examination showed that the brass phase was uniformly distributed in both the longitudinal and transverse cross sections. While the room temperature resistivity of the monolithic BMG was measured as 1.75×10^{-6} Ωm, the composite had a much lower room temperature resistivity of 1.63×10^{-7} Ωm in the longitudinal direction. The resistivity in the transverse direction was about three times higher than this, suggesting that the conductivity is anisotropic and that it is sensitive to the morphology of the high conducting phase.

6.6 Specific Heat

Like the other physical properties discussed so far, the specific heat of BMG alloys can also provide valuable information about structural relaxation. The specific heat is generally measured using a DSC. The sample is first heated to a certain temperature at a constant heating rate and held there isothermally for the desired length of time. The heat flux at this temperature is

$$\frac{dQ}{dt} = \dot{Q} = \left(\frac{\partial Q}{\partial t}\right)_{\dot{T}\neq 0} - \left(\frac{\partial Q}{\partial t}\right)_{\dot{T}=0} - C\cdot\frac{dT}{dt} \tag{6.19}$$

where:

$(\partial Q/\partial t)_{T\neq 0}$ represents the power required to heat the sample and the container with an overall heat capacity of C and to hold it at a certain temperature

$(\partial Q/\partial t)_{\dot{T}=0}$ is the power needed to keep the temperature just constant

The absolute specific heat capacity of the sample can be determined from a knowledge of the heat capacity of the empty sample container and of a standard sapphire sample, by repeating the previous experiment every 10 or 20 K or so, on (1) the metal sample in the sample pan, (2) a sapphire standard in the sample pan, and (3) the sample pan by itself. The absolute specific heat capacity of the sample is then calculated using the relationship:

$$C_p(T)_{\text{sample}} = \frac{\dot{Q}_{\text{sample}} - \dot{Q}_{\text{pan}}}{\dot{Q}_{\text{sapphire}} - \dot{Q}_{\text{pan}}}\cdot\frac{m_{\text{sapphire}}\cdot\mu_{\text{sample}}}{m_{\text{sample}}\cdot\mu_{\text{sapphire}}}\cdot C_p(T)_{\text{sapphire}} \tag{6.20}$$

where m and μ represent the mass and the mole mass, respectively. This has been done on some Mg- [75], Cu- [76], and Zr-based [76] BMG alloy systems.

The specific heat of the supercooled liquid can be described by the equation:

$$C_p(\text{Liquid}) = 3R + aT + bT^{-2} \tag{6.21}$$

and that of the crystal well above the Debye temperature as

$$C_p(\text{Crystal}) = 3R + cT + dT^2 \tag{6.22}$$

where:

R is the universal gas constant

a, b, c, and d are the fitting constants

By calculating the difference in the specific heats between the liquid and the crystalline states, other thermodynamic parameters, such as enthalpy and entropy, can be calculated using the equations:

$$\Delta H^{\ell-x}(T) = \Delta H_f - \int_T^{T_f} \Delta C_p^{\ell-x}(T')\,dT' \tag{6.23}$$

and

$$\Delta S^{\ell-x}(T) = \Delta S_f - \int_T^{T_f} \frac{\Delta C_p^{\ell-x}(T')}{T'} dT' \quad (6.24)$$

where:

ΔH_f and ΔS_f	represent the enthalpy and the entropy of fusion, respectively
T_f	is the temperature at which the Gibbs free energies of the liquid and crystal are equal
ℓ and x	represent the liquid and crystal, respectively

ΔS_f can be calculated as $\Delta S_f = \Delta H_f/T_f$. From these parameters, the Gibbs free energy difference between the crystal and the liquid phases can be calculated using the relationship:

$$\Delta G^{\ell-x}(T) = \Delta H^{\ell-x}(T) - T \cdot \Delta S^{\ell-x}(T) \quad (6.25)$$

A knowledge of the free energy difference between the crystal and the liquid phases can provide information about the GFA of alloys. In general, the lower the $\Delta G^{\ell-x}(T)$ value, the better the GFA. It has also been noted that the smaller the entropy of fusion, the higher the GFA. The free energy difference between the crystal and the liquid phases is the driving force for crystallization in the case of polymorphous crystallization of metallic glasses. But, in those cases where crystallization takes place by eutectic or primary modes, this value is the lower limit of the thermodynamic driving force for crystallization.

Kauzmann temperature, T_K, is the temperature at which the entropies of the liquid and the crystal are equal [77]. Since the entropies of the liquid and the crystal are measured using the calorimetric method as described, it is possible to determine T_K using these results [75,76,78]. By extrapolating the entropy of the supercooled liquid to low temperatures and identifying the temperature at which the entropy of the liquid is the same as that of the crystal, one can determine T_K. The significance of T_K is that it is the lowest temperature at which a supercooled liquid can exist. Below this temperature, the supercooled liquid has to either spontaneously crystallize or form a glass.

Nishiyama et al. [78] noted that during continuous cooling of the liquid in a DSC to low temperatures, an abrupt change in the specific heat was observed at the glass transition temperature, T_g.

As mentioned at the beginning of this section, the measurement of specific heat can greatly aid in understanding the structural relaxation behavior of metallic glasses. Figure 6.11 shows the thermograms of the $Zr_{60}Al_{10}Co_3Ni_9Cu_{18}$ glassy samples in three different conditions—in melt-spun ribbon of about 20 μm thickness, a 5 mm diameter rod, and a 7 mm diameter rod [79]. Here, the symbol $C_{p,q}$ represents the specific heat of the sample in the as-quenched condition and $C_{p,s}$ denotes the specific heat of the structurally relaxed sample (that was earlier heated to the supercooled liquid region at 700 K). The general trend of the variation of the specific heat with temperature as the as-quenched sample is heated to high temperatures may be described as follows.

With reference to Figure 6.11, note that the value of $C_{p,q}$ of the BMG samples increases slowly with increasing temperature up to about 510–520 K and then begins to decrease due to structural relaxation. The $C_{p,q}$ shows a minimum at about 580 K, and this is followed by a rapid increase due to the glass transition, when the glass reaches a fully relaxed supercooled state. On continued increase of temperature, one notices a gradual decrease in C_p

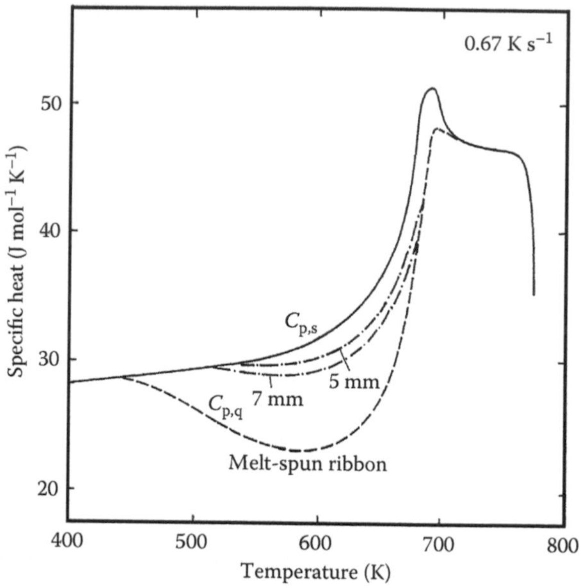

FIGURE 6.11
Variation of specific heat as a function of temperature for the $Zr_{60}Al_{10}Co_3Ni_9Cu_{18}$ glassy samples in three different forms—a melt-spun ribbon, a 5 mm diameter rod, and a 7 mm diameter rod. (Reprinted from Inoue, A. et al., *Mater. Trans., JIM*, 36, 391, 1995. With permission.)

in the supercooled liquid region and then a rapid decrease due to crystallization. In comparison, $C_{p,s}$ increases continuously in the temperature range below T_g and then steeply in the glass transition temperature range. No appreciable difference in the specific heat between $C_{p,q}$ and $C_{p,s}$ is seen in the SLS, indicating that the supercooled liquid is in an internal equilibrium state and that the previous thermal history of the sample has completely disappeared. The temperature dependence of $C_{p,s}$ for the glassy solid and the supercooled liquid can be expressed as

$$C_{p,s} = 30.5 + 8 \times 10^{10} (T - 570) + 0.04 (T - 570) \tag{6.26}$$

for the glassy solid in the temperature range 570–690 K, and

$$C_{p,s} = 51.2 + 0.05 (T - 690) - (T - 690)^{0.5} \tag{6.27}$$

for the supercooled liquid in the temperature range 690–760 K. Additionally, the heat of the structural relaxation (ΔH_r) caused by continuous heating can be determined using the equation:

$$\Delta H_r = \int \Delta C_p \cdot dT \tag{6.28}$$

where $\Delta C_p = C_{p,s} - C_{p,q}$. The ΔH_r values decreased from the melt-spun ribbon to increasing diameters of the BMG alloys. For example, the ΔH_r value for the cylindrical sample with a 5 mm diameter is 147 J mol^{-1}, while it is 683 J mol^{-1} for the melt-spun ribbon, suggesting that the cylindrical sample has a much more relaxed structure in comparison to the melt-spun ribbon. This is directly related to the much lower cooling rate experienced by the large cylinder during the casting process.

The usefulness of the specific heat in studying the structural relaxation behavior of metallic glasses, both melt-spun ribbons and cast large-diameter rods, was discussed in Chapter 5.

6.7 Viscosity

Viscosity is perhaps the most important physical property of the liquid alloy that determines its GFA and flow behavior. Several studies have been conducted to evaluate the viscosity of alloys as a function of temperature, the shear rate, the alloy constituents and the composition. It was explained in Chapter 2 that if a liquid could be undercooled down to a temperature below the glass transition temperature T_g, without allowing it to crystallize, then a glassy phase is formed. The formation of the glassy phase is related to a significant increase in the viscosity of the undercooled melt at T_g. Whereas the viscosity of simple alloys is typically about 10^{-3} Pa s at the melting point, it increases to about 10^{12} Pa s at T_g, a value that is taken as a measure to define T_g. That is, any liquid that attains a viscosity of 10^{12} Pa s without crystallization will be considered a glass, irrespective of the alloy system and composition.

Depending on the rate of change of viscosity with temperature, liquids have been traditionally classified into two categories—strong and fragile [80]. In the case of a strong liquid, the viscosity is high near the melting point and increases gradually with decreasing temperature following an Arrhenius behavior. In a fragile liquid, on the other hand, the viscosity is low at the melting temperature, increases slowly with decreasing temperature in a non-Arrhenius fashion, and rises suddenly near the glass transition temperature, T_g. We will come back to this aspect again later in this section.

It is observed during isothermal annealing experiments that the apparent viscosity increases gradually with the annealing time, t, and reaches an almost constant value after long annealing times. This value will be referred to as the *equilibrium viscosity*. The equilibrium viscosity value is different at different temperatures. The change in viscosity occurs because of the relaxation processes that take place in the glassy phase to reach the "equilibrium" supercooled region. The equilibrium viscosity is obtained by fitting the apparent viscosity data using a stretched exponential function:

$$\log \eta(t,T) = \log \eta_g + \log \frac{\eta_{eq}(T)}{\eta_g}\left[1-\exp\left\{1-\left(\frac{t}{\tau}\right)^{\beta}\right\}\right] \tag{6.29}$$

where:

$\eta(t, T)$ is the measured apparent viscosity
η_g is the apparent viscosity for the as-cast glassy alloy
$\eta_{eq}(T)$ is the equilibrium viscosity at temperature T
τ is the relaxation time for the shear viscous flow
β is the stretched exponent

The viscosity of liquid alloys has been measured by different methods. Since viscosity is defined as a measure of the resistance to shear flow, it is possible to determine the viscosity, η, by measuring the flow stress at different strain rates and using the relation:

$$\eta = \frac{\sigma_{\text{flow}}}{\dot{\gamma}} \tag{6.30}$$

where:

σ_{flow} represents the experimentally determined equilibrium flow stress
$\dot{\gamma}$ is the strain rate

The viscosity as a function of strain rate can be described using the equation:

$$\eta = A \times \dot{\gamma}^{n-1} \tag{6.31}$$

where n represents the shear thinning coefficient, which can be determined by fitting Equation 6.31 to the viscosity versus the shear rate data. If $n < 1$, the liquid exhibits a reduced viscosity, with an increasing shear rate or non-Newtonian behavior. If the viscosity is independent of the shear rate, that is, when $n = 1$, the liquid is said to exhibit Newtonian flow. It has been shown that the value of n increases with increasing temperature, suggesting increasingly Newtonian behavior with increasing temperature.

Another way to determine the viscosity is the three-point beam-bending method, in which a load is applied to the center of a beam of uniform cross section supported on both ends, and by measuring the deflection of the center of the beam with time [81,82]. The viscosity is calculated using the equation:

$$\eta = \frac{gL^3}{144I_c} \nu \left(M + \frac{5\rho AL}{8} \right) \tag{6.32}$$

where:

g is the gravitational constant (9.8 m s^{-2})
I_c is the cross section moment of inertia (m^4)
ν is the midpoint deflection rate (m s^{-1})
M is the applied load (kg)
ρ is the density of the glass (kg m^{-3})
A is the cross-sectional area (m^2)
L is the support span (m)

Other methods such as parallel plate rheometry [83], decay of resonant oscillations [84], and length changes of the glassy ribbons [85], among others, have also been used. A measurement of viscosity at very high temperatures has been difficult because of the possible interaction of the liquid melt with the container. But, these problems have now been resolved and several investigators have measured the viscosity in the liquid state as a function of temperature.

The variation of viscosity with temperature of a $Pd_{40}Cu_{30}Ni_{10}P_{20}$ BMG alloy at a heating rate of 20 K min^{-1} and a compressive stress of 20 kPa is shown in Figure 6.12a. The viscosity remains almost constant up to near T_g, at which temperature it starts to decrease, slowly

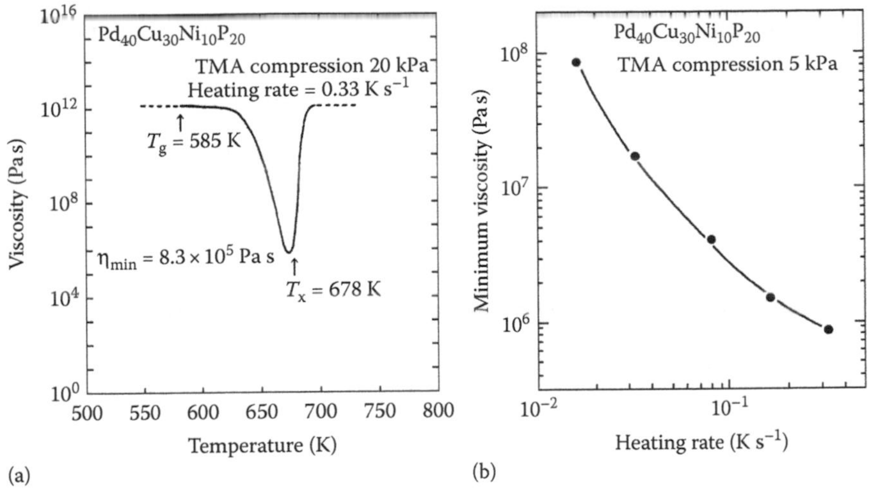

FIGURE 6.12
(a) Variation of viscosity with temperature for the $Pd_{40}Cu_{30}Ni_{10}P_{20}$ BMG alloy at a heating rate of 20 K min^{-1} under a compressive stress of 20 kPa. (b) Change in the minimum viscosity (η_{min}) with the heating rate for the $Pd_{40}Cu_{30}Ni_{10}P_{20}$ BMG alloy.

at first and then very rapidly reaching a minimum. The minimum temperature (of 675 K in the present case) corresponds to the onset temperature of crystallization. After crystallization, the viscosity increases steeply with increasing temperature and then reaches a steady-state value on completion of crystallization. It has been suggested [86] that the absolute value of η_{min} is an indication of the thermal stability of the supercooled liquid. The value of 8.3×10^5 Pa s for the $Pd_{40}Cu_{30}Ni_{10}P_{20}$ BMG alloy is two to three orders of magnitude lower than that (about 10^8 Pa s) for the $Pd_{77.5}Cu_6Si_{16.5}$ alloy [1], suggesting that the supercooled liquid of the Pd–Cu–Ni–P alloy has a higher resistance to crystallization than the Pd–Cu–Si alloy. This is an indication that the Pd–Cu–Ni–P glass is appropriate for secondary working through viscous flow in the supercooled liquid region. Further, it has been noted that the viscosity decreases with an increasing heating rate employed, as shown in Figure 6.12b. It has also been suggested that a measurement of viscosity at high heating rates is one effective way of reducing the oxygen contamination from the measuring environment [87].

It is known that the onset temperature of crystallization depends on the heating rate. When the secondary working is carried out in the SLS, the heating rate is an important factor to determine the holding time before the onset of crystallization. Even though η_{min} decreases continuously with increasing heating rate, the rate of decrease slows at higher heating rates, and this suggests the existence of a lower limit of η_{min} even at very high heating rates. The lower η_{min} value in the higher heating rate range indicates the possibility of a secondary working at a low applied stress, while it must be realized that the working time before crystallization decreases. Thus, the secondary working using viscous flow of the supercooled liquid must be done under conditions where the low viscosity, leading to easy deformation, and the short time available for crystallization must be satisfied simultaneously.

The variation of viscosity with temperature is of great importance in understanding the formation of metallic glasses, realizing that the higher the viscosity at the melting temperature, the easier it is for the glass to form. The variation of viscosity with temperature in the supercooled liquid region is usually expressed by the Vogel–Fulcher–Tammann (VFT) equation (even though Angell [80] prefers to call this the *VTF equation*, since it was Tammann who had provided a physical significance for this equation):

$$\eta = \eta_0 \exp\left[\frac{B}{T - T_0}\right] \tag{6.33}$$

where:

T_0 is referred to as the *VFT temperature* (or *fictive temperature*), and represents the temperature at which the relaxation time and thus the barrier to viscous flow would become infinite

η_0 is the high temperature limit of viscosity

B is a fitting parameter

The value of T_0 is found to lie near, but below, T_g. The value of η_0 can be expressed as

$$\eta_0 = \frac{N_A \times h}{V} \tag{6.34}$$

where:

N_A represents Avogadro's number

h is Planck's constant

V is the molar volume [88]

Angell [80] has modified Equation 6.33 as

$$\eta = \eta_0 \exp\left[\frac{D^* T_0}{T - T_0}\right] \tag{6.35}$$

where the parameter D^*, referred to as the fragility parameter, controls how closely the system obeys the Arrhenius behavior ($D^* = \infty$).

Liquids with very large values of D^* are known as *strong liquids* and those with small values as *fragile liquids*. Strong liquids show Arrhenius behavior for the temperature dependence of viscosity, have a high viscosity at the melting point, and form stable glasses. On the other hand, fragile liquids have a low viscosity at the melting temperature, but the viscosity rises sharply as the liquid approaches the glass transition temperature. Such liquids are marginal glass formers and if they form glasses, they are unstable with respect to crystallization. That is, the GFA of the alloy increases with the fragility parameter. Accordingly, the value of D^* is of the order of 2 for the most fragile liquids, while it is 100 for the strongest glass formers, such as SiO_2. Figure 6.13 shows a typical Angell plot [89]. Senkov [90] has recently analyzed the available data on the fragility parameter (D^*), the critical cooling rate for glass formation (R_c), and the reduced glass transition temperature (T_{rg}), and came up with a new GFA parameter. Table 6.5 summarizes the available data on the D^* values and also the T_0 temperatures for different BMG alloys.

Lu et al. [82] measured the viscosity of $Fe_{48-x}Co_xCr_{15}Mo_{14}C_{15}B_6Y_2$ BMG alloys using a non-isothermal three-point creep bending test and noted that the viscosity first increased and then decreased with increasing Co content, yielding the largest value at 7 at.% Co. In this

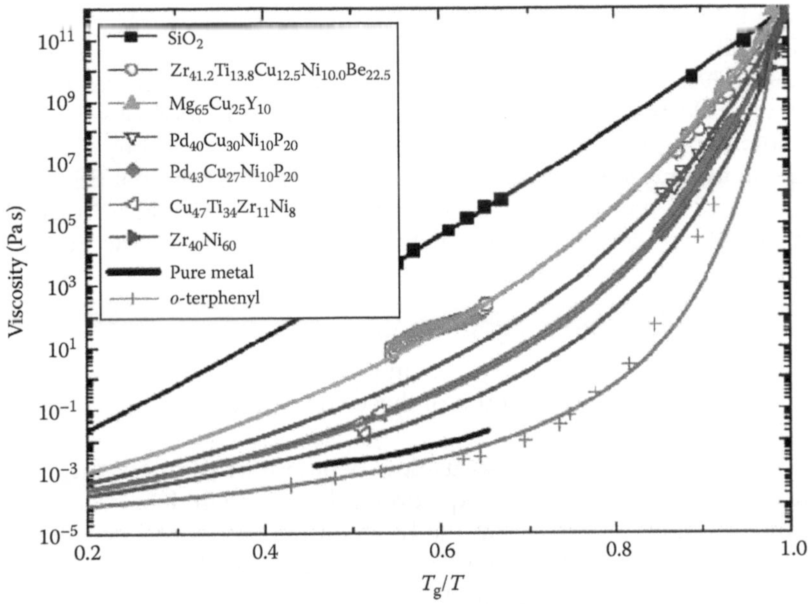

FIGURE 6.13
Viscosity of glass-forming liquids as a function of inverse temperature, normalized by the glass transition temperature, T_g. The metallic alloys are intermediate between strong network-forming liquids, such as SiO_2, and fragile liquids such as *o*-terphenyl. (Reprinted from Busch, R. et al., *MRS Bulletin*, 32, 620, 2007. With permission.)

TABLE 6.5
Fragility Parameter, D^*, and T_0 Temperatures for Some of the BMG Alloys in the Equation $\eta = \eta_0 \exp[(D^* T_0)/(T - T_0)]$

Alloy	D^*	T_0 (K)	References
$Au_{77.8}Ge_{13.8}Si_{8.4}$	8.4	—	[93]
$Cu_{47}Ti_{34}Zr_{11}Ni_8$	12	500	[93]
$Mg_{65}Cu_{25}Y_{10}$	22.1	260	[75]
$Ni_{65}Nb_{35}$	6.2	784	[94]
$Ni_{59.5}Nb_{40.5}$	5.6	800	[84]
$Ni_{60}Nb_{35}Sn_5$	11.0	670	[94]
$Ni_{59.35}Nb_{34.45}Sn_{6.2}$	11.0	670	[94]
$Ni_{57}Fe_3Nb_{35}Sn_5$	16.4	591	[94]
$Ni_{60}(Nb_{0.4}Ta_{0.6})_{34}Sn_6$	19.0	581	[94]
$Pd_{40}Ni_{40}P_{20}$	18.1	390	[95]
$Pd_{48}Ni_{32}P_{20}$	16.6	392	[90,93]
$Pd_{77.5}Cu_6Si_{16.5}$	11.1	493	[90,93]
$Zr_{65}Al_{7.5}Cu_{17.5}Ni_{10}$	13.8	—	[75]
$Zr_{40}Ni_{60}$	7.9	—	[75]
$Zr_{41.2}Ti_{13.8}Cu_{12.5}Ni_{10}Be_{22.5}$	23.8	390	[84]
$Zr_{41.2}Ti_{13.8}Cu_{12.5}Ni_{10}Be_{22.5}$	18.5	412.5	[96]
$Zr_{58.5}Cu_{15.6}Ni_{12.8}Al_{10.3}Nb_{2.8}$	21.0	—	[97]
$Zr_{57}Cu_{15.4}Ni_{12.6}Al_{10}Nb_5$	11.3	524.7	[84,97]
$Zr_{52.5}Cu_{17.9}Ni_{14.6}Al_{10}Ti_5$	11.6	521	[84]
$Zr_{46.75}Ti_{8.25}Cu_{7.5}Ni_{10}Be_{27.5}$	22.7	372	[98]

context, it will be useful to recall that while the Co-free Fe–Cr–Mo–C–B–Y alloy could be made glassy at a maximum diameter of 9 mm, this could be increased to 16 mm by adding 7 at.% Co [91]. Lu et al. [82] explained that this observation was due to this alloy having the lowest electrical conductivity, leading to the formation of strong covalent-like metal–metalloid bonds, which suppress the atomic viscous flow. They also noted that the average positron lifetime was the shortest at 7 at.% Co, suggesting that this alloy possesses the smallest average free volume. In other words, the alloy with 7 at.% Co is more densely packed than the other alloys and consequently this alloy exhibits the highest GFA. This observation is also consistent with the previous x-ray photoelectron spectroscopy (XPS) and Raman studies [92].

It is interesting to note in this context that the value of D^* increases with an increase in the number of components in the alloy system [94]. Therefore, this is like an indirect validation of the Inoue rule that a minimum of three constituents are required for bulk glass formation [3].

Fan et al. [99] have analyzed the viscous flow behavior in a $Pd_{43}Ni_{10}Cu_{27}P_{20}$ BMG glass-forming liquid and noted that the VFT equation did not fit their data over a wide temperature range. Instead, they reported that the free-volume model of Cohen and Grest [100] was better able to describe the behavior. Fan et al. [99] have also analyzed their viscosity data on the basis of an earlier cluster model proposed by them [101]. In this cluster model, a supercooled liquid contains some clusters. Liquid regions that have clusters will have a higher activation energy for viscous flow in comparison to regions that do not have clusters. The liquid above the liquidus temperature will not have any clusters, and its viscosity follows Arrhenius behavior. On the other hand, the liquid below the liquidus temperature (i.e., in the supercooled region) will contain clusters, and when these form on a large scale, glass transition sets in. The flow behavior in this region can be described by an Arrhenius behavior on which a fragility term is superimposed. (This fragility term is different from the Angell fragility term.) Fan et al. [99] were able to satisfactorily explain their viscosity data using either the free volume or the cluster model.

Using viscosity as the main parameter, Takeuchi and Inoue [102] estimated the three most important parameters to explain the GFA of alloys. They had estimated the critical cooling rate for glass formation, R_c; the reduced glass transition temperature, $T_{rg} = T_g/T_l$; and the width of the supercooled liquid region, $\Delta T_x = T_x - T_g$. While R_c and T_{rg} were estimated on the basis of the homogeneous nucleation and growth theory by the construction of a *T–T–T* diagram, the ΔT_x parameter was estimated from the free volume theory. By estimating these parameters for Fe-, Pd-, Pt-, Zr-, and Mg-based systems, the authors summarized their data in R_c versus T_{rg} and R_c versus ΔT_x diagrams and showed good qualitative agreement with the experimental data.

References

1. Chen, H.S. (1980). Glassy metals. *Rep. Prog. Phys.* 43: 353–432.
2. Masumoto, T., H.M. Kimura, A. Inoue, and Y. Waseda (1976). Structural stability of amorphous metals. *Mater. Sci. Eng.* 23: 141–144.
3. Inoue, A. (2000). Stabilization of metallic supercooled liquid and bulk amorphous alloys. *Acta Mater.* 48: 297–306.

4. Park, E.S. and D.H. Kim (2004). Formation of Ca–Mg–Zn bulk glassy alloy by casting into cone-shaped copper mold. *J. Mater. Res.* 19: 685–688.
5. Park, E.S. and D.H. Kim (2005). Formation of Mg–Cu–Ni–Ag–Zn–Y–Gd bulk glassy alloy by casting into cone-shaped copper mold in air atmosphere. *J. Mater. Res.* 20: 1465–1469.
6. Inoue, A., T. Negishi, H.M. Kimura, T. Zhang, and A.R. Yavari (1998). High packing density of Zr- and Pd-based bulk amorphous alloys. *Mater. Trans., JIM* 39: 318–321.
7. Yavari, A.R. and A. Inoue (1999). Volume effects in bulk metallic glass formation. In *Bulk Metallic Glasses*, eds. W.L. Johnson, A. Inoue, and C.T. Liu, Vol. 554, pp. 21–30. Warrendale, PA: Materials Research Society.
8. Hu, X., S.C. Ng, Y.P. Feng, and Y. Li (2001). Cooling-rate dependence of the density of $Pd_{40}Ni_{10}Cu_{30}P_{20}$ bulk metallic glass. *Phys. Rev. B* 64: 172201-1–172201-4.
9. Shen, T.D., Y. He, and R.B. Schwarz (1999). Bulk amorphous Pd–Ni–Fe–P alloys: Preparation and characterization. *J. Mater. Res.* 14: 2107–2115.
10. Zhang, T. and A. Inoue (1998). Mechanical properties of Zr–Ti–Al–Ni–Cu bulk amorphous sheets prepared by squeeze casting. *Mater. Trans., JIM* 39: 1230–1237.
11. Mattern, N., U. Kühn, H. Hermann, S. Roth, H. Vinzelberg, and J. Eckert (2004). Thermal behavior and glass transition of Zr-based bulk metallic glasses. *Mater. Sci. Eng. A* 375–377: 351–354.
12. Nagel, C., K. Rätzke, E. Schmidtke, J. Wolff, U. Geyer, and F. Faupel (1998). Free-volume changes in the bulk metallic glass $Zr_{46.7}Ti_{8.3}Cu_{7.5}Ni_{10}Be_{27.5}$ and the undercooled liquid. *Phys. Rev. B* 57: 10224–10227.
13. Louzguine-Luzgin, D.V., N. Chen, V.Yu. Zadorozhny, I. Seki, and A. Inoue (2013). $Pd_{40}Ni_{40}Si_5P_{15}$ bulk metallic glass properties variation as a function of sample thickness. *Intermetallics* 33: 67–72.
14. Inoue, A. (1998). *Bulk Amorphous Alloys: Preparation and Fundamental Characteristics*. Zürich, Switzerland: Trans Tech Publications.
15. Harms, U., O. Jin, and R.B. Schwarz (2003). Effects of plastic deformation on the elastic modulus and density of bulk amorphous $Pd_{40}Ni_{10}Cu_{30}P_{20}$. *J. Non-Cryst. Solids* 317: 200–205.
16. Liu, W., Q.S. Zeng, Q.K. Jiang, L.P. Wang, and B.S. Li (2011). Density and elasticity of $Zr_{46}Cu_{37.6}Ag_{8.4}Al_8$ bulk metallic glass at high pressure. *Scr. Mater.* 65: 497–500.
17. Turnbull, D. and M.H. Cohen (1961). Free-volume model of the amorphous phase: Glass transition. *J. Chem. Phys.* 34: 120–125.
18. Russew, K. and F. Sommer (2003). Length and density changes of amorphous $Pd_{40}Cu_{30}Ni_{10}P_{20}$ alloys due to structural relaxation. *J. Non-Cryst. Solids* 319: 289–296.
19. Haruyama, O. (2007). Thermodynamic approach to free volume kinetics during isothermal relaxation in bulk Pd–Cu–Ni–P_{20} glasses. *Intermetallics* 15: 659–662.
20. Haruyama, O., Y. Yokoyama, and A. Inoue (2007). Precise measurement of density in the isothermal relaxation processes of $Pd_{42.5}Cu_{30}Ni_{7.5}P_{20}$ and $Zr_{50}Cu_{40}Al_{10}$ glasses. *Mater. Trans.* 48: 1708–1710.
21. Yavari, A.R., N. Nikolov, N. Nishiyama, T. Zhang, A. Inoue, J.L. Uriarte, and G. Heunen (2004). The glass transition of bulk metallic glasses studied by real-time diffraction in transmission using high-energy synchrotron radiation. *Mater. Sci. Eng. A* 375–377: 709–712.
22. Hajlaoui, K., T. Benameur, G. Vaughan, and A.R. Yavari (2004). Thermal expansion and indentation-induced free volume in Zr-based metallic glasses measured by real-time diffraction using synchrotron radiation. *Scr. Mater.* 51: 843–848.
23. Louzguine-Luzgin, D., A. Inoue, A.R. Yavari, and G. Vaughan (2006). Thermal expansion of a glassy alloy studied using a real-space pair distribution function. *Appl. Phys. Lett.* 88: 121926-1–121926-3.
24. Martienssen, W. and H. Warlimont (2005). *Springer Handbook of Condensed Matter and Materials Data*. Berlin, Germany: Springer.
25. Louzguine, D.V., A.R. Yavari, K. Ota, G. Vaughan, and A. Inoue (2005). Synchrotron X-radiation diffraction studies of thermal expansion, free volume change and glass transition phenomenon in Cu-based glassy and nanocomposite alloys on heating. *J. Non-Cryst. Solids* 351: 1639–1645.

26. Lin, T., X.F. Bian, and J. Jiang (2006). Relation between calculated Lennard-Jones potential and thermal stability of Cu-based bulk metallic glasses. *Phys. Lett. A* 353: 497–499.
27. Li, G., Y.C. Li, Z.K. Jiang, T. Xu, L. Huang, J. Liu, T. Zhang, and R.P. Liu (2009). Elasticity, thermal expansion and compressive behavior of $Mg_{65}Cu_{25}Tb_{10}$ bulk metallic glass. *J. Non-Cryst. Solids* 355: 521–524.
28. Lu, I.R., G.P. Görler, H.-J. Fecht, and R. Willnecker (2000). Investigation of specific heat and thermal expansion in the glass-transition regime of Pd-based metallic glasses. *J. Non-Cryst. Solids* 274: 294–300.
29. Lu, I.R., G.P. Görler, H.-J. Fecht, and R. Willnecker (2002). Investigation of specific volume of glass-forming Pd–Cu–Ni–P alloy in the liquid, vitreous and crystalline state. *J. Non-Cryst. Solids* 312–314: 547–551.
30. Nishiyama, N., M. Horino, and A. Inoue (2000). Thermal expansion and specific volume of $Pd_{40}Cu_{30}Ni_{10}P_{20}$ alloy in various states. *Mater. Trans., JIM* 41: 1432–1434.
31. Kimura, H.M., A. Inoue, N. Nishiyama, K. Sasamori, O. Haruyama, and T. Masumoto (1997). Thermal, mechanical, and physical properties of supercooled liquid in Pd–Cu–Ni–P amorphous alloy. *Sci. Rep. RITU* A43: 101–106.
32. Guo, J., X.F. Bian, T. Lin, Y. Zhao, T.B. Li, B. Zhang, and B. Sun (2007). Formation and interesting thermal expansion behavior of novel Sm-based bulk metallic glasses. *Intermetallics* 15: 929–933.
33. Jing, Q., R.P. Liu, G. Li, and W.K. Wang (2003). Thermal expansion behavior and structure relaxation of ZrTiCuNiBe bulk amorphous alloy. *Scr. Mater.* 49: 111–115.
34. Shek, C.H. and G.M. Lin (2003). Dilatometric measurements and calculation of effective pair potentials for $Zr_{41}Ti_{14}Cu_{12.5}Ni_{10}Be_{22.5}$ bulk metallic glass. *Mater. Lett.* 57: 1229–1232.
35. Cantor, B. and R.W. Cahn (1983). Atomic diffusion in amorphous alloys. In *Amorphous Metallic Alloys*, ed. F.E. Luborsky, pp. 487–505. London, UK: Butterworths.
36. Horváth, J. (1990). Diffusion in amorphous alloys. In *Diffusion in Solid Metals and Alloys*, ed. H. Meher, and Landolt Börnstein, New Series, Vol. 26, pp. 437–470. Berlin, Germany: Springer.
37. Faupel, F., W. Frank, M.-P. Macht, H. Mehrer, V. Naundorf, K. Rätzke, H.R. Schober, S.K. Sharma, and H. Teicher (2003). Diffusion in metallic glasses and supercooled melts. *Rev. Mod. Phys.* 75: 237–280.
38. Nonaka, K., Y. Kimura, K. Yamauchi, H. Nakajima, T. Zhang, A. Inoue, and T. Masumoto (1997). Diffusion in $Zr_{55}Al_{10}Ni_{10}Cu_{25}$ amorphous alloy with a wide supercooled liquid region. *Defect Diff. Forum* 143–147: 837–842.
39. Zöllmer, V., K. Rätzke, and F. Faupel (2003). Diffusion and isotope effect in bulk-metallic glass-forming Pd–Cu–Ni–P alloys from the glass to the equilibrium melt. *J. Mater. Res.* 18: 2688–2696.
40. Nakajima, H., T. Kojima, T. Nonaka, T. Zhang, A. Inoue, and N. Nishiyama (2001). Self-diffusion in $Zr_{55}Al_{10}Ni_{10}Cu_{25}$ and $Pd_{40}Cu_{30}Ni_{10}P_{20}$ bulk metallic glass. In *Supercooled Liquid, Bulk Glassy, and Nanocrystalline States of Alloys*, eds. A. Inoue, A.R. Yavari, W.L. Johnson, and R.H. Dauskardt, Vol. 644, pp. L.2.2.1–L.2.2.10. Warrendale, PA: Materials Research Society.
41. Fielitz, P., M.-P. Macht, V. Naundorf, and G. Frohberg (1999). Diffusion in ZrTiCuNiBe bulk glasses at temperatures around the glass transition. *J. Non-Cryst. Solids* 250–252: 674–678.
42. Geyer, U., S. Schneider, W.L. Johnson, Y. Qiu, T.A. Tombrello, and M.-P. Macht (1995). Atomic diffusion in the supercooled liquid and glassy states of the $Zr_{41.2}Ti_{13.8}Cu_{12.5}Ni_{10}Be_{22.5}$ alloy. *Phys. Rev. Lett.* 75: 2364–2367.
43. Geyer, U., W.L. Johnson, S. Schneider, Y. Qiu, T.A. Tombrello, and M.-P. Macht (1996). Small atom diffusion and breakdown of the Stokes–Einstein relation in the supercooled liquid state of the $Zr_{46.7}Ti_{8.3}Cu_{7.5}Ni_{10}Be_{27.5}$ alloy. *Appl. Phys. Lett.* 69: 2492–2494.
44. Ehmler, H., A. Heesemann, K. Rätzke, and F. Faupel (1998). Mass dependence of diffusion in a supercooled metallic melt. *Phys. Rev. Lett.* 80: 4919–4922.
45. Ehmler, H., K. Rätzke, and F. Faupel (1999). Isotope effect of diffusion in the supercooled liquid state of bulk metallic glasses. *J. Non-Cryst. Solids* 250–252: 684–688.
46. Zumkley, TH., M.-P. Macht, V. Naundorf, J. Rusing, and G. Frohberg (2000). Impurity diffusion in ZrTiCuNiBe bulk metallic glasses. *J. Metastable Nanocryst. Mater.* 8: 135–139.

47. Knorr, K., M.-P. Macht, K. Freitag, and H. Mehrer (1999). Self-diffusion in the amorphous and supercooled liquid state of the bulk metallic glass $Zr_{46.75}Ti_{8.25}Cu_{7.5}Ni_{10}Be_{27.5}$. *J. Non-Cryst. Solids* 250–252: 669–673.
48. Macht, M.-P., V. Naundorf, P. Fielitz, J. Rüsing, Th. Zumkley, and G. Frohberg (2001). Dependence of diffusion on the alloy composition in ZrTiCuNiBe bulk glasses. *Mater. Sci. Eng. A* 304–306: 646–649.
49. Sjögren, L. (1990). Temperature dependence of viscosity near the glass transition. *Z. Phys. B* 79: 5–13.
50. Faupel, F., P.W. Hüppe, and K. Rätzke (1990). Pressure dependence and isotope effect of self-diffusion in a metallic glass. *Phys. Rev. Lett.* 65: 1219–1222.
51. Tang, X.P., U. Geyer, R. Busch, W.L. Johnson, and Y. Wu (1999). Diffusion mechanisms in metallic supercooled liquids and glasses. *Nature* 402: 160–162.
52. Budke, E., P. Fielitz, M.-P. Macht, V. Naundorf, and G. Frohberg (1997). Impurity diffusion in Zr–Ti–Cu–Ni–Be bulk glasses. *Defect Diff. Forum* 143: 825–830.
53. Wenwer, F., K. Knorr, M.-P. Macht, and H. Mehrer (1997). Ni tracer diffusion in the bulk metallic glasses $Zr_{41}Ti_{14}Cu_{12.5}Ni_{10}Be_{22.5}$ and $Zr_{65}Cu_{17.5}Ni_{10}Al_{7.5}$. *Defect Diff. Forum* 143: 831–835.
54. Faupel, F., H. Ehmler, C. Nagel, and K. Rätzke (2001). Does the diffusion mechanism change at the caloric glass transition of bulk metallic glasses? *Defect Diff. Forum* 194: 821–826.
55. Masuhr, A., T.A. Waniuk, R. Busch, and W.L. Johnson (1999). Time scales for viscous flow, atomic transport, and crystallization in the liquid and supercooled liquid states of $Zr_{41.2}Ti_{13.8}Cu_{12.5}Ni_{10}Be_{22.5}$. *Phys. Rev. Lett.* 82: 2290–2293.
56. Bartsch, A., K. Rätzke, F. Faupel, and A. Meyer (2006). Codiffusion of ^{32}P and ^{57}Co on glass-forming $Pd_{43}Cu_{27}Ni_{10}P_{20}$ alloy and its relation to viscosity. *Appl. Phys. Lett.* 89: 121917-1–121917-3.
57. Nakajima, H., T. Kojima, Y. Zumkley, N. Nishiyama, and A. Inoue (1999). Diffusion of nickel in $Pd_{40}Cu_{30}Ni_{10}P_{20}$ metallic glass. In *Proceedings of International Conference on Solid-Solid Phase Transformations (JIMIC-3)*, eds. M. Koiwa, K. Otsuka, and T. Miyazaki, Vol. 12, pp. 441–444, Kyoto, Japan.
58. Yamazaki, Y., T. Nihei, J. Koike, and T. Ohtsuki (2005). Self-diffusion of P in Pd–Cu–Ni–P bulk metallic glass. In *Proceedings of First International Conference on Diffusion in Solids and Liquids*, eds. A. Ochsner, J. Gracio, and F. Barlat, Part II, pp. 831–834, Aveiro, Portugal.
59. Chen, H.S., and D. Turnbull (1969). Formation, stability and structure of Pd–Si-based alloy glasses. *Acta Metall.* 17: 1021–1031.
60. Chen, H.S. (1980). The influence of structure on electrical resistivities of Pd–Au–Si and Au–Ge–Si glass forming alloys. *Solid State Commun.* 33: 915–919.
61. Liu, Y.Z., Y.J. Huang, X.S. Wei, and J. Shen (2012). Close correlation between transport properties and glass-forming ability of an FeCoCrMoCBY alloy system. *Intermetallics* 30: 144–147.
62. Haruyama, O., H.M. Kimura, N. Nishiyama, and A. Inoue (1999). Change in electron transport property after glass transition in several Pd-based metallic glasses. *J. Non-Cryst. Solids* 250–252: 781–785.
63. Haruyama, O., N. Annoshita, H.M. Kimura, N. Nishiyama, and A. Inoue (2002). Anomalous behavior of electrical resistivity in glass transition region of a bulk $Pd_{40}Ni_{40}P_{20}$ metallic glass. *J. Non-Cryst. Solids* 312–314: 552–556.
64. Haruyama, O., H.M. Kimura, A. Inoue, and N. Nishiyama (2000). Change in electrical resistivity associated with the glass transition in a continuously cooled $Pd_{40}Cu_{30}Ni_{10}P_{20}$ melt. *Appl. Phys. Lett.* 76: 2026–2028.
65. Okai, D., T. Fukami, T. Yamasaki, T. Zhang, and A. Inoue (2004). Temperature dependence of heat capacity and electrical resistivity of Zr-based bulk glassy alloys. *Mater. Sci. Eng. A* 375–377: 364–367.
66. Orbanić, F. and I. Kokanović (2015). Impact of quenched disorder and crystallization on electrical resistivity in $Zr_{67}Co_{33}$ metallic glass. *J. Non-Cryst. Solids* 428: 31–35.
67. Haruyama, O., H.M. Kimura, and A. Inoue (1996). Thermal stability of Zr-based glassy alloys examined by electrical resistance measurement. *Mater. Trans., JIM* 37: 1741–1747.

68. Haruyama, O., H.M. Kimura, N. Nishiyama, and A. Inoue (2001). Behavior of electrical resistivity through glass transition in $Pd_{40}Cu_{30}Ni_{10}P_{20}$ metallic glass. *Mater. Sci. Eng. A* 304–306: 740–742.
69. Haruyama, O., N. Annoshita, N. Nishiyama, H.M. Kimura, and A. Inoue (2004). Electrical resistivity behavior in Pd–Cu–Ni–P Metallic glasses and liquids. *Mater. Sci. Eng. A* 375–377: 288–291.
70. Ji, Y.S., S.J. Chung, M.-R. Ok, K.T. Hong, J.-Y. Suh, J.W. Byeon, J.-K. Yoon, K.H. Lee, and K.S. Lee (2007). Analysis on the phase transition behavior of Cu base bulk metallic glass by electrical resistivity measurement. *Mater. Sci. Eng. A* 449–451: 521–525.
71. Boucher, B.Y. (1972). Influence of phosphorus on the electrical properties of Pd–Ni–P amorphous alloys. *J. Non-Cryst. Solids* 7: 277–284.
72. Okai, D., A. Nanbu, T. Fukami, T. Yamasaki, T. Zhang, Y. Yokoyama, G. Motoyama, Y. Oda, H.M. Kimura, and A. Inoue (2004). Temperature dependence of electrical resistivity of Zr–Cu–Al–Ni–Nb bulk metallic glasses below 300 K. *Mater. Sci. Eng. A* 449–451: 548–551.
73. Chung, S.J., K.T. Hong, M.-R. Ok, J.-K. Yoon, G.-H. Kim, Y.S. Ji, B.S. Seong, and K.S. Lee (2005). Analysis of the crystallization of $Zr_{41}Ti_{14}Cu_{12.5}Ni_{10}Be_{22.5}$ bulk metallic glass using electrical resistivity measurement. *Scr. Mater.* 53: 223–228.
74. Wang, K., T. Fujita, M.W. Chen, T.G. Nieh, H. Okada, K. Koyama, W. Zhang, and A. Inoue (2007). Electrical conductivity of a bulk metallic glass composite. *Appl. Phys. Lett.* 91: 154101-1–154101-3.
75. Busch, R., W. Liu, and W.L. Johnson (1998). Thermodynamics and kinetics of the $Mg_{65}Cu_{25}Y_{10}$ bulk metallic glass forming liquid. *J. Appl. Phys.* 83: 4134–4141.
76. Glade, S.C., R. Busch, D.S. Lee, W.L. Johnson, R.K. Wunderlich, and H.-J. Fecht (2000). Thermodynamics of $Cu_{47}Ti_{34}Zr_{11}Ni_8$, $Zr_{52.5}Cu_{17.9}Ni_{14.6}Al_{10}Ti_5$, and $Zr_{57}Cu_{15.4}Ni_{12.6}Al_{10}Nb_5$ bulk metallic glass forming alloys. *J. Appl. Phys.* 87: 7242–7248.
77. Kauzmann, W. (1948). The nature of the glassy state and the behavior of liquids at low temperatures. *Chem. Rev.* 43: 219–256.
78. Nishiyama, N., M. Horino, O. Haruyama, and A. Inoue (2001). Abrupt change in heat capacity of supercooled Pd–Cu–Ni–P melt during continuous cooling. *Mater. Sci. Eng. A* 304–306: 683–686.
79. Inoue, A., T. Zhang, and T. Masumoto (1995). Preparation of bulky amorphous Zr–Al–Co–Ni–Cu Alloys by copper mold casting and their thermal and mechanical properties. *Mater. Trans., JIM* 36: 391–398.
80. Angell, C.A. (1995). Formation of glasses from liquids and biopolymers. *Science* 267: 1924–1935.
81. Hagy, H.E. (1963). Experimental evaluation of beam-bending method of determining glass viscosities in the range 10^8 to 10^{15} Poises. *J. Am. Ceram. Soc.* 46: 93–97.
82. Lu, Y.Z., Y.J. Huang, J. Shen, X. Lu, Z.X. Qin, and Z.H. Zhang (2014). Effect of Co addition on the shear viscosity of Fe-based bulk metallic glasses. *J. Non-Cryst. Solids* 403: 62–66.
83. Diennes, G.J. and H.F. Klemm (1946). Theory and application of the parallel plate plastometer. *J. Appl. Phys.* 17: 458–471.
84. Mukherjee, S., J. Schroers, Z. Zhou, W.L. Johnson, and W.-K. Rhim (2004). Viscosity and specific volume of bulk metallic glass-forming alloys and their correlation with glass-forming ability. *Acta Mater.* 52: 3689–3695.
85. Vlasák, G., P. Švec, and P. Duhaj (2001). Application of isochronal dilatation measurements for determination of viscosity of amorphous alloys. *Mater. Sci. Eng. A* 304–306: 472–475.
86. Taub, A.I. (1985). Measurement of the flow of amorphous alloys near T_g using continuous heating. In *Rapidly Quenched Metals V*, eds. S. Steeb, and H. Warlimont, pp. 1365–1368. Amsterdam, the Netherlands: Elsevier.
87. Maeda, S., T. Yamasaki, Y. Yokoyama, D. Okai, T. Fukami, H.M. Kimura, and A. Inoue (2007). Viscosity measurements of $Zr_{55}Cu_{30}Al_{10}Ni_5$ and $Zr_{50}Cu_{40-x}Al_{10}Pd_x$ (x = 0, 3, and 7 at.%) supercooled liquid alloys by using a penetration viscometer. *Mater. Sci. Eng. A* 449–451: 203–206.
88. Nemilov, S.V. (1995). Correlation of crystallization character of glass melts with the temperature dependence of their viscosity and the degree of spatial structural connectiveness. *Glass Phys. Chem.* 21: 91–96.

89. Busch, R., J. Schroers, and W.H. Wang (2007). Thermodynamics and kinetics of bulk metallic glass. *MRS Bulletin* 32: 620–623.
90. Senkov, O.N. (2007). Correlation between fragility and glass-forming ability of metallic alloys. *Phys. Rev. B* 76: 104202-1–104202-6.
91. Shen, J., Q.J. Chen, J.F. Sun, H.B. Fan, and G. Wang (2005). Exceptionally high glass-forming ability of an FeCoCrMoCBY alloy. *Appl. Phys. Lett.* 86: 151907-1–151907-3.
92. Lu, Y.Z., Y.J. Huang, and J. Shen (2011). The electronic structure origin for ultrahigh glass-forming ability of the FeCoCrMoCBY alloy system. *J. Appl. Phys.* 110: 033720-1–033720-4.
93. Glade, S.C. and W.L. Johnson (2000). Viscous flow of the $Cu_{47}Ti_{34}Zr_{11}Ni_8$ glass forming alloy. *J. Appl. Phys.* 87: 7249–7251.
94. Shadowspeaker, L. and R. Busch (2004). On the fragility of Nb–Ni-based and Zr-based bulk metallic glasses. *Appl. Phys. Lett.* 85: 2508–2510.
95. Kawamura, Y., and A. Inoue (2000). Newtonian viscosity of supercooled liquid in a $Pd_{40}Ni_{40}P_{20}$ metallic glass. *Appl. Phys. Lett.* 77: 1114–1116.
96. Waniuk, T.A., R. Busch, A. Masuhr, and W.L. Johnson (1998). Equilibrium viscosity of the $Zr_{41.2}Ti_{13.8}Cu_{12.5}Ni_{10}Be_{22.5}$ bulk metallic glass-forming liquid and viscous flow during relaxation, phase separation, and primary crystallization. *Acta Mater.* 46: 5229–5236.
97. Evenson, Z., S. Raedersdorf, I. Gallino, and R. Busch (2010). Equilibrium viscosity of Zr–Cu–Ni–Al–Nb bulk metallic glasses. *Scr. Mater.* 63: 573–576.
98. Busch, R., E. Bakke, and W.L. Johnson (1998). Viscosity of the supercooled liquid and relaxation at the glass transition of the $Zr_{46.75}Ti_{8.25}Cu_{7.5}Ni_{10}Be_{27.5}$ bulk metallic glass forming alloy. *Acta Mater.* 46: 4725–4732.
99. Fan, G.J., H.-J. Fecht, and E.J. Lavernia (2004). Viscous flow of the $Pd_{43}Ni_{10}Cu_{27}P_{20}$ bulk metallic glass-forming liquid. *Appl. Phys. Lett.* 84: 487–489.
100. Cohen, M.H. and G.S. Grest (1979). Liquid-glass transition, a free-volume approach. *Phys. Rev. B* 20: 1077–1098.
101. Fan, G.J. and H.-J. Fecht (2002). A cluster model for the viscous flow of glass-forming liquids. *J. Chem. Phys.* 116: 5002–5006.
102. Takeuchi, A. and A. Inoue (2004). Calculations of dominant factors of glass-forming ability for metallic glasses from viscosity. *Mater. Sci. Eng. A* 375–377: 449–454.

7

Corrosion Behavior

7.1 Introduction

Bulk metallic glasses (BMGs) are relatively new materials with many potential applications in diverse engineering areas. However, for effective use of these novel materials with an interesting combination of properties, it is necessary to fully characterize them on the basis of their physical, magnetic, mechanical, and chemical properties. While a great deal of effort is being spent on characterizing their mechanical (Chapter 8) and magnetic (Chapter 9) properties, only a limited amount of information is available on their corrosion behavior.

The corrosion behavior of BMGs becomes important when these materials need to be used in aggressive and hostile environments (high temperatures, oxidizing atmospheres, and corrosive media). Knowledge of the corrosion behavior becomes crucial when these BMGs are considered for biomedical applications and for decorative applications, or when surface appearance assumes importance. Hence, it is imperative to evaluate the corrosion behavior of these alloys in different environments.

The corrosion behavior of metallic glassy alloy ribbons (about 20–50 μm in thickness) produced by the rapid solidification processing (RSP) method of melt spinning was first evaluated in 1974 [1,2]. It was reported that Cr-containing Fe-based $Fe_{80-x}Cr_xP_{13}C_7$ glassy ribbons exhibited much higher corrosion resistance than the crystalline Fe–Cr alloys. Figure 7.1a shows that while the crystalline Fe–Cr alloys corroded at a rate of about 0.5–1 mm year^{-1}, the glassy Fe–Cr–P–C alloy did not show any measurable corrosion under identical conditions of exposure in 1 N NaCl solution at 30°C (303 K). Another important observation made was that the minimum amount of Cr required to achieve this corrosion resistance was only 8 at.%, much less in the glassy state than that required (>12 at.%) in the crystalline state. Figure 7.1b shows that the glassy alloy did not exhibit any measurable weight change with exposure to varying concentrations of HCl (from 0.01 to 1 N) for 1 week at 373 K. On the other hand, the corrosion rate of the crystalline 18–8 austenitic stainless steel (Fe–18 wt.% Cr–8 wt.% Ni) increased from 10^{-3} mm year^{-1} in 0.01 N HCl to over 10 mm year^{-1} in 1 N HCl solution; severe pitting corrosion occurred in the range 0.5–1 N HCl in the crystalline alloy.

X-ray photoelectron spectroscopy (XPS) studies indicated that the passive film on the glassy alloy consisted mainly of hydrated chromium oxyhydroxide, which is also a common major constituent of passive films on crystalline stainless steels. But, an important difference is that the Cr content in the passive layer of the glassy alloy was much higher than that on a conventional austenitic stainless steel. Thus, it was concluded that the extremely high corrosion resistance of the glassy alloy was only partially attributed to the formation of a protective hydrated chromium oxyhydroxide film [3], and that other factors also play an important role.

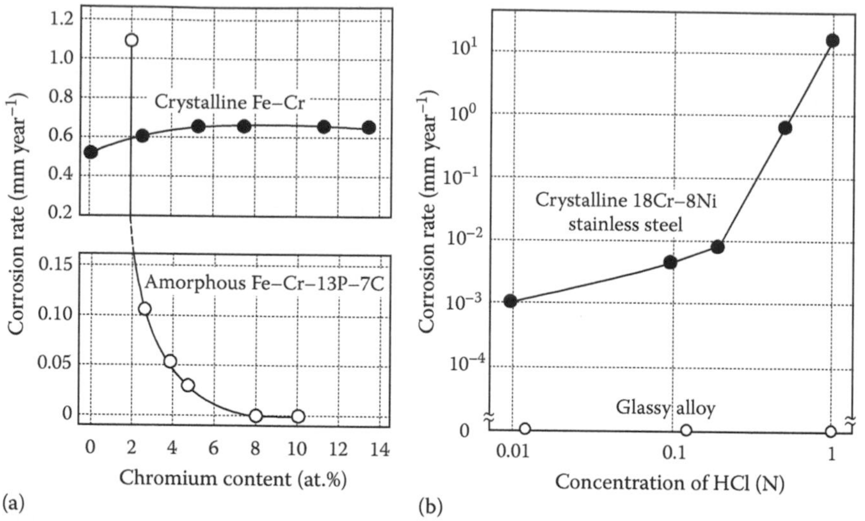

FIGURE 7.1
Variation of corrosion rate as a function of (a) Cr content and (b) concentration of HCl in crystalline Fe–Cr alloys and rapidly solidified glassy $Fe_{80-x}Cr_xP_{13}C_7$ alloy. Note that while the crystalline alloy exhibits a corrosion rate of about 0.5–1 mm year^{-1} at all Cr contents, the glassy alloy does not show any measurable corrosion rate when the Cr content is more than about 8 at.%. Further, the corrosion rate increases with an increasing concentration of HCl in the crystalline alloy, but there is no measurable effect on the glassy alloy.

Metallic glasses are in a high-energy state since they have been solidified at rates much faster than the equilibrium solidification rates. Consequently, they are expected to exhibit higher corrosion rates. But, the superior corrosion resistance of the glassy alloys, in comparison with the crystalline alloys of a similar composition, has been attributed to the following factors:

1. Since metallic glassy alloy ribbons are produced by RSP methods, there is not sufficient time for appreciable solid-state diffusion to take place and solute partitioning to occur. Consequently, the glass exhibits chemical homogeneity.
2. Second, the glassy phase does not contain any crystal defects, such as grain boundaries and dislocations, or second-phase precipitates, which could act as galvanic cells to initiate localized corrosion.
3. Third, since the passive films form uniformly on the glassy alloy surfaces, lower amounts of passivating alloying elements are needed than in a crystalline alloy to achieve a similar stability of the passive film.

High corrosion resistance has been reported in a number of other rapidly solidified melt-spun glassy ribbons [4–7].

In contrast to the thin metallic glassy alloys obtained by RSP methods (at solidification rates of about 10^5–10^6 K s^{-1}), BMGs synthesized at relatively slow solidification rates of about 1–100 K s^{-1} may not always show chemical homogeneity, especially in marginal glass-forming alloys. For example, it has been occasionally noted that small quantities of crystalline second phases are present in such alloys [8,9]. Secondary crystalline phases could also form as a result of heterogeneous nucleation occurring on impurities in the melt or on mold walls. A typical example is the presence of oxide particles when Zr-based BMGs are produced in the presence of oxygen in the environment [10,11].

BMGs have been synthesized in a number of alloy systems based on Au, Ca, Ce, Co, Cu, Fe, La, Mg, Nb, Nd, Ni, Pd, Pr, Pt, Sm, Ti, Y, and Zr. But, since it is unlikely that all these glassy alloys will be used in a corrosive or aggressive environment, the corrosion behavior of all these alloys has not been investigated. Results from different research groups are available on the corrosion behavior of BMGs based on Cu, Fe, Mg, Ni, Ti, Zr, and a few others. It is also important to remember at this stage that extensive information is not available on all the alloy systems and that the theories and mechanisms developed for one alloy system are not necessarily applicable to other alloy types. Scully et al. [12] and Green et al. [13] have reviewed the corrosion behavior of a number of BMG alloys.

We will now discuss the corrosion behavior of different BMG alloy systems based on Cu, Fe, Mg, Ni, Ti, and Zr. Even though there may be some overlap in the nature of results obtained in different alloy systems, it will be useful to discuss them system-wise, because the mechanisms and theories could be different in different alloy systems. But, before going into a detailed discussion of the results in different alloy systems, let us first understand the basic terminology used in the literature and the methods used to determine the corrosion behavior of alloys.

7.2 Terminology and Methodology

The corrosion behavior of alloys has been mostly evaluated using weight-loss and/or electrochemical methods. In the weight-loss method, the alloy is exposed to the corrosive medium at the given temperature for a predetermined length of time. Since corrosion involves a transfer of mass from the surface of the specimen to the environment by physical, chemical, or electrochemical means, the weight of the sample is measured before and after exposure to the corrosive medium. Knowing the weight loss and the cross-sectional surface area of the specimen, the thickness of the surface layer corresponding to the weight loss is calculated. The corrosion rate is then reported in units of micrometers or millimeters per year (μm $year^{-1}$ or mm $year^{-1}$). Generally speaking, it will be very difficult to calculate the thickness accurately, due to uncertainties in the weight-change measurements. Therefore, even if one does not observe any change in the weight of the specimen before and after exposure to the corrosive medium, the corrosion rate is usually reported as less than 1 μm $year^{-1}$. This is because it represents the reproducibility limit for measurements.

Another common technique to evaluate the corrosion behavior of alloys is to use potentiodynamic polarization curves. A typical cyclic anodic polarization curve is shown in Figure 7.2, which plots the potential as a function of the current density [14]. Such curves are obtained by measuring the current while varying the potential of the material using a potentiostat. The current density is obtained by dividing the current with the specimen cross-sectional area. The potential is always measured with reference to a saturated calomel electrode (SCE), and the current density is plotted on a logarithmic scale. The process is referred to as *anodic polarization* when the potential is above E_{corr} and *cathodic polarization* if it is below E_{corr}.

A few important parameters can be defined with reference to Figure 7.2. The specimen is first immersed in the corrosive solution for about 20 min, when the open-circuit potential becomes steady. The potential scan is then started below the corrosion potential,

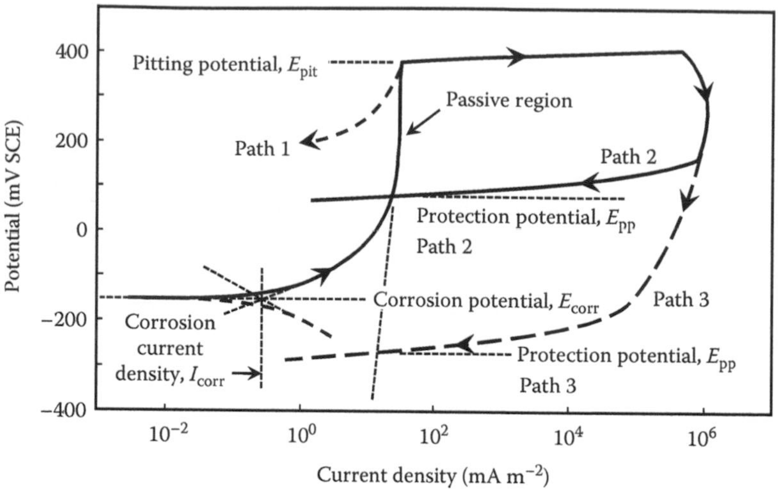

FIGURE 7.2
Schematic representative cyclic polarization curve of a material that is susceptible to pitting corrosion. (Reprinted from Peter, W.H. et al., *Intermetallics*, 10, 1157, 2002. With permission.)

E_{corr}, which corresponds to the potential at which the current density approaches zero. During the test, the current density first decreases, reaches the lowest value, and then starts to increase again. This lowest value is taken as E_{corr}. During the process, when the current density rises, it remains constant, at about 10 mA m^{-2}, over a range of potentials, suggesting that the alloy has passivated. That is, the material is protected from high corrosion rates through the formation of a thin oxide/hydroxide film. This region is called the *passive region*, and this value of the current density is referred to as the *passive current density*. It is desirable that the passive current density is as low as possible. It may be noted that the lower the passive current density the higher is the corrosion resistance of the alloy. Additionally, the width of the passive region, that is, the voltage range in which the current density remains constant, is also important. The wider the region, the more corrosion resistant the alloy is. That is, lower current density and wider passive region are both indicators of high corrosion resistance of an alloy. With continued increase of the potential, the passive film breaks down and the current density suddenly increases. The potential at which this happens is called the *pitting potential*, E_{pit}.

Additional information about the ability of the material to repassivate (i.e., formation of the passive film again) can be obtained by conducting cyclic anodic polarization experiments. These experiments are similar to the experiments described previously except that the potential scan is in the reverse direction at a specified current density. The potential at which the current density returns to the passive value is known as the *repassivation potential* or *protection potential*, E_{pp}. Between E_{pit} and E_{pp}, pits initiate and propagate. However, in path "2," pits will not initiate at E_{corr}, the natural corrosion potential, and therefore the material will not undergo pitting corrosion under natural corrosion conditions. But, if path "3" is followed, when E_{pp} is below E_{corr}, the material will undergo pitting corrosion at surface flaws or after incubation time periods at E_{corr}. The values of pitting overpotential ($\eta_{pit} = E_{pit} - E_{corr}$) and protection overpotential ($\eta_{pp} = E_{pp} - E_{corr}$) are important parameters. These represent a measure of the ability of the material to resist pitting corrosion, including at surface flaws and after incubation times. That is, higher positive values of η_{pit} and η_{pp} are desirable for improved corrosion resistance of the materials.

The corrosion rate under natural corrosion conditions, that is, at E_{corr}, is related to the corrosion current density, i_{corr}, which must be determined by extrapolation methods. Most commonly, it is determined by an extrapolation of the anodic and cathodic portions of the polarization curves, as indicated in Figure 7.2. The corrosion current is proportional to the corrosion penetration rate (CPR) of the metal at the open circuit condition (i.e., external current = 0 and potential = E_{corr}).

CPR, in units of micrometers per year, is calculated using Faraday's law:

$$\text{CPR} = 0.327\frac{Mi_{\text{corr}}}{m\rho} \tag{7.1}$$

where:

M is the atomic fraction-weighted value of the atomic weight (g mol^{-1})
m is the ion valence (g)
ρ is the density (g cm^{-3})
i_{corr} is the corrosion current density (mA m^{-2})

The CPR values calculated from Faraday's law may differ slightly from the corrosion rates estimated from the weight-loss measurements mentioned earlier. Full details about the general corrosion behavior of metals and alloys may be found in standard textbooks on corrosion (see, e.g., Ref. [15]).

Even though cyclic anodic polarization experiments provide a lot of information, many investigators have been reporting weight-loss measurements to evaluate the corrosion behavior of glassy alloys, since these methods are relatively easy to perform.

As mentioned in Section 7.1, formation of a passive film on the surface of the glassy alloy is responsible for the improved corrosion resistance of the alloy. Further, the composition of the passive layer gives important information about the stability and protectiveness of the passive film. Therefore, to investigate the origin of this corrosion resistance, it becomes necessary to analyze the chemistry of the passive layer film and also the substrate alloy surface. Such an analysis will provide important information regarding the distribution of alloying elements in the passive layer film, and is done most conveniently by using XPS (also known as *electron spectroscopy for chemical analysis*, or ESCA). In this technique, low-energy radiation, typically Al K_α (E = 1.4866 keV), is used to obtain the yield of photoelectrons from each chemical element over a range of energies. This enables qualitative identification and quantitative evaluation of the elements present in the sample. An advantage of this technique is that it will be possible to identify the different electron states of the elements based on their binding energies. Measurement of the thickness of the surface film and a quantitative determination of the composition of the surface film and the underlying alloy surface are also possible. This can be done with the help of a previously proposed method using the integrated intensities of photoelectrons under the assumption of a three-layer model [16,17]. The three layers involved are the outermost contaminant hydrocarbon layer of uniform thickness, the passive surface film of uniform thickness, and the underlying alloy surface, which can be considered to be of infinite thickness for the purposes of XPS analysis.

Before actually conducting the corrosion tests, the specimens are mechanically polished in cyclohexane with silicon carbide paper up to grit 2000, degreased in acetone, washed in distilled water, and dried in air. After immersion in the corrosion medium for the desired length of time, the specimens are again washed in distilled water, dried, and then weighed to determine the weight loss.

7.3 Copper-Based Bulk Metallic Glasses

Copper-based alloys have been traditionally used for marine applications due to their excellent corrosion resistance. A number of Cu-based glassy alloys have been synthesized in recent years both in ribbon and bulk rod forms, with the maximum diameter of the BMG alloy reaching up to 25 mm in a $(Cu_{47}Zr_{42.5}Al_8)_{96}Y_4$ alloy [18]. It was reported in 2001 that Cu–Zr–Ti [19], Cu–Hf–Ti [20], and Cu–Zr–Ti–Hf [21] alloys could be produced in a fully glassy condition with a critical diameter of 4 mm. These alloys possessed excellent mechanical properties with a tensile fracture strength of about 2100 MPa and a compressive plastic strain of 0.8%–1.7%, which is much larger than the values reported for other BMG alloys. It was also reported that ternary Cu–Zr–Al glassy alloys with a critical diameter of 3 mm could be synthesized with a compressive fracture strength of 2200 MPa and a plastic strain of 0.2% [22]. The good mechanical properties of these alloys suggest that they could be used for structural applications. However, for a successful utilization, it also becomes important to evaluate their corrosion behavior in different media. Therefore, a study of the corrosion behavior of several Cu-based BMG alloys has been undertaken.

Binary Cu–Zr [23] and Cu–Hf [24] alloys have also been produced in a glassy condition as 1–2 mm rods. These as-cast binary alloys have been occasionally found to contain a few nanometer-sized crystallites embedded in the glassy matrix. Corrosion studies have not been performed on these binary BMG alloys.

The majority of the corrosion studies have been focused on ternary Cu–Zr–Al, Cu–Zr–Ti, and Cu–Hf–Ti BMG alloys and their derivatives with further additions of Nb, Mo, or Ta. Table 7.1 summarizes the alloy compositions, the conditions under which corrosion studies were conducted, corrosion rates reported, and the electrochemical properties, if available. Weight-loss measurements and electrochemical measurements were performed to determine the corrosion behavior of these glassy alloys and the XPS technique was used to determine the chemistry of the surface passive layer and the underlying substrate material. The tests have been generally conducted at room temperature (nominally 298 K), unless otherwise specified.

7.3.1 Influence of Composition

Among the three major groups of alloys investigated—Cu–Hf–Ti, Cu–Zr–Al, and Cu–Zr–Ti—the Cu–Hf–Ti alloy glasses corresponding to the composition $Cu_{60}Hf_{25}Ti_{15}$ seem to have the lowest corrosion rates (Table 7.1). But, a common observation made was that the corrosion resistance of these glassy alloys was substantially improved by alloying them with additional elements such as Mo, Nb, or Ta. Further, among these alloying additions, Nb seems to be the most effective, and therefore investigations have been focused on evaluating the effect of Nb content on the corrosion behavior of $Cu_{60}Hf_{25}Ti_{15}$, $Cu_{50}Zr_{45}Al_5$, and $Cu_{55}Zr_{40}Al_5$ glassy alloys. Figure 7.3 presents the average corrosion rates of $(Cu_{60}Hf_{25}Ti_{15})_{100-x}Nb_x$ BMG alloys as a function of the Nb content [28]. These alloys were immersed in 1 N HCl, 3 wt.% NaCl, and 1 N H_2SO_4 + 0.001 N NaCl solutions at 298 K, open to air for 1 week (168 h). It is clear from Figure 7.3 that, while the Nb-free alloy showed a high corrosion rate of 340 μm year^{-1} in 1 N HCl solution, the addition of Nb has significantly reduced the corrosion rate to 2.6 μm year^{-1} at 6 at.% Nb level and less than the detectable value (1 μm year^{-1}) at 8 at.% Nb [27,28].

The addition of 10 at.% Nb to $Cu_{60}Hf_{25}Ti_{15}$ produced a composite of Nb dendritic crystals dispersed in a glassy matrix (Figure 7.4) [25,34]. The volume fraction and the mean size of

TABLE 7.1

Alloy Compositions and Conditions under Which Corrosion Studies Were Conducted on Cu-Based BMG Alloys

Alloy Composition	Electrolyte	Immersion Time	Corrosion Rate (μm year⁻¹)	References
$Cu_{60}Hf_{25}Ti_{15}$	1 N HCl	168 h	340	[25]
$Cu_{60}Hf_{25}Ti_{15}$	1 N HCl	336 h	340	[26]
$Cu_{60}Hf_{25}Ti_{15}$	3 wt.% NaCl	336 h	100	[26]
$Cu_{60}Hf_{25}Ti_{15}$	1 N H_2SO_4	336 h	<1	[27,28]
$(Cu_{0.6}Hf_{0.25}Ti_{0.15})_{98}Nb_2$	1 N HCl	168 h	170	[27,28]
$(Cu_{0.6}Hf_{0.25}Ti_{0.15})_{96}Nb_4$	1 N HCl	168 h	76	[27,28]
$(Cu_{0.6}Hf_{0.25}Ti_{0.15})_{94}Nb_6$	1 N HCl	168 h	2.6	[27,28]
$(Cu_{0.6}Hf_{0.25}Ti_{0.15})_{92}Nb_8$	1 N HCl	168 h	<1	[27,28]
$(Cu_{0.6}Hf_{0.25}Ti_{0.15})_{90}Nb_{10}$	1 N HCl	168 h	11	[25]
$(Cu_{0.6}Hf_{0.25}Ti_{0.15})_{98}Nb_2$	1 N H_2SO_4	336 h	<1	[27]
$(Cu_{0.6}Hf_{0.25}Ti_{0.15})_{96}Nb_4$	1 N H_2SO_4	336 h	<1	[27]
$(Cu_{0.6}Hf_{0.25}Ti_{0.15})_{94}Nb_6$	1 N H_2SO_4	336 h	<1	[27]
$(Cu_{0.6}Hf_{0.25}Ti_{0.15})_{92}Nb_8$	1 N H_2SO_4	336 h	<1	[27]
$(Cu_{60}Hf_{25}Ti_{15})_{98}Mo_2$	3 wt.% NaCl	336 h	1	[26]
$(Cu_{60}Hf_{25}Ti_{15})_{98}Nb_2$	3 wt.% NaCl	336 h	1	[26]
$(Cu_{60}Hf_{25}Ti_{15})_{96}Nb_4$	3 wt.% NaCl	168 h	<1	[27]
$(Cu_{60}Hf_{25}Ti_{15})_{98}Ta_2$	3 wt.% NaCl	336 h	1	[26]
$Cu_{0.5}NiAlCoCrFeSi$	High-purity water at 561 K	336 h	<1	[29]
$Cu_{55}Zr_{40}Al_5$	1 N HCl	24 h	29,800	[30]
$Cu_{55}Zr_{40}Al_5$	1 N HCl	24 h	1,902	[31]
$Cu_{55}Zr_{40}Al_5$	3 wt.% NaCl	168 h	200	[30]
$Cu_{52.5}Zr_{42.5}Al_5$	1 N HCl	24 h	1,617	[31]
$Cu_{50}Zr_{45}Al_5$	1 N HCl	24 h	15,600	[30]
$Cu_{50}Zr_{45}Al_5$	1 N HCl	24 h	1,609	[31]
$Cu_{50}Zr_{45}Al_5$	3 wt.% NaCl	168 h	120	[30]
$Cu_{50}Zr_{45}Al_5$	0.5 N H_2SO_4	336 h	<1	[30]
$Cu_{54}Zr_{40}Al_5Nb_1$	1 N HCl	24 h	940	[31]
$Cu_{52}Zr_{40}Al_5Nb_3$	1 N HCl	24 h	245	[31]
$Cu_{50}Zr_{40}Al_5Nb_5$	1 N HCl	24 h	211	[31]
$Cu_{50}Zr_{40}Al_5Nb_5$	1 N HCl	24 h	120	[30]
$Cu_{50}Zr_{40}Al_5Nb_5$	0.5 N H_2SO_4	336 h	<1	[30]
$Cu_{49}Zr_{45}Al_5Nb_1$	1 N HCl	24 h	1,605	[31]
$Cu_{47}Zr_{45}Al_5Nb_3$	1 N HCl	24 h	339	[31]
$Cu_{45}Zr_{45}Al_5Nb_5$	1 N HCl	24 h	325	[31]
$Cu_{55}Zr_{40}Al_5$	3 wt.% NaCl	97 h	609	[31]
$Cu_{52.5}Zr_{42.5}Al_5$	3 wt.% NaCl	97 h	539	[31]
$Cu_{50}Zr_{45}Al_5$	3 wt.% NaCl	97 h	539	[31]
$Cu_{54}Zr_{40}Al_5Nb_1$	3 wt.% NaCl	97 h	545	[31]
$Cu_{52}Zr_{40}Al_5Nb_3$	3 wt.% NaCl	97 h	395	[31]
$Cu_{50}Zr_{40}Al_5Nb_5$	3 wt.% NaCl	97 h	84	[31]
$Cu_{49}Zr_{45}Al_5Nb_1$	3 wt.% NaCl	97 h	417	[31]
$Cu_{47}Zr_{45}Al_5Nb_3$	3 wt.% NaCl	97 h	488	[31]
$Cu_{45}Zr_{45}Al_5Nb_5$	3 wt.% NaCl	97 h	<1	[31]

(Continued)

TABLE 7.1 (CONTINUED)

Alloy Compositions and Conditions under Which Corrosion Studies Were Conducted on Cu-Based BMG Alloys

Alloy Composition	Electrolyte	Immersion Time	Corrosion Rate (μm year⁻¹)	References
$Cu_{55}Zr_{40}Al_5$	1 N H_2SO_4	168 h	5.7	[31]
$Cu_{52.5}Zr_{42.5}Al_5$	1 N H_2SO_4	168 h	3.6	[31]
$Cu_{50}Zr_{45}Al_5$	1 N H_2SO_4	168 h	3.2	[31]
$Cu_{54}Zr_{40}Al_5Nb_1$	1 N H_2SO_4	168 h	3.7	[31]
$Cu_{52}Zr_{40}Al_5Nb_3$	1 N H_2SO_4	168 h	1.5	[31]
$Cu_{50}Zr_{40}Al_5Nb_5$	1 N H_2SO_4	168 h	1.3	[31]
$Cu_{49}Zr_{45}Al_5Nb_1$	1 N H_2SO_4	168 h	1.2	[31]
$Cu_{47}Zr_{45}Al_5Nb_3$	1 N H_2SO_4	168 h	<1	[31]
$Cu_{45}Zr_{45}Al_5Nb_5$	1 N H_2SO_4	168 h	<1	[31]
$Cu_{60}Zr_{30}Ti_{10}$	1 N HCl	168 h	660	[28,32]
$Cu_{60}Zr_{30}Ti_{10}$	0.01 N HCl	168 h	320	[28,32]
$Cu_{60}Zr_{30}Ti_{10}$	3 wt.% NaCl	168 h	290	[28,32]
$Cu_{60}Zr_{30}Ti_{10}$	1 N H_2SO_4	336 h	26	[28,32]
$Cu_{60}Zr_{30}Ti_{10}$	1 N HNO_3	336 h	18	[28,32]
$(Cu_{0.6}Zr_{0.3}Ti_{0.1})_{99}Nb_1$	1 N HCl	168 h	350	[28,32]
$(Cu_{0.6}Zr_{0.3}Ti_{0.1})_{99}Mo_1$	1 N HCl	168 h	360	[28,32]
$(Cu_{0.6}Zr_{0.3}Ti_{0.1})_{99.8}Ta_{0.2}$	1 N HCl	168 h	410	[28,32]
$(Cu_{0.6}Zr_{0.3}Ti_{0.1})_{99}Nb_1$	0.01 N HCl	168 h	160	[28,32]
$(Cu_{0.6}Zr_{0.3}Ti_{0.1})_{99}Mo_1$	0.01 N HCl	168 h	170	[28,32]
$(Cu_{0.6}Zr_{0.3}Ti_{0.1})_{99.8}Ta_{0.2}$	0.01 N HCl	168 h	230	[28,32]
$(Cu_{0.6}Zr_{0.3}Ti_{0.1})_{99}Nb_1$	3 wt.% NaCl	168 h	120	[28,32]
$(Cu_{0.6}Zr_{0.3}Ti_{0.1})_{99}Mo_1$	3 wt.% NaCl	168 h	140	[28,32]
$(Cu_{0.6}Zr_{0.3}Ti_{0.1})_{99.8}Ta_{0.2}$	3 wt.% NaCl	168 h	200	[28,32]
$(Cu_{0.6}Zr_{0.3}Ti_{0.1})_{99}Nb_1$	1 N H_2SO_4	336 h	6.8	[28,32]
$(Cu_{0.6}Zr_{0.3}Ti_{0.1})_{99}Mo_1$	1 N H_2SO_4	336 h	8.2	[28,32]
$(Cu_{0.6}Zr_{0.3}Ti_{0.1})_{99.8}Ta_{0.2}$	1 N H_2SO_4	336 h	14	[28,32]
$(Cu_{0.6}Zr_{0.3}Ti_{0.1})_{99}Nb_1$	1 N HNO_3	336 h	1.7	[28,32]
$(Cu_{0.6}Zr_{0.3}Ti_{0.1})_{99}Mo_1$	1 N HNO_3	336 h	3.5	[28,32]
$(Cu_{0.6}Zr_{0.3}Ti_{0.1})_{99.8}Ta_{0.2}$	1 N HNO_3	336 h	7.6	[28,32]
$Cu_{47}Zr_{11}Ti_{34}Ni_8$	1 N H_2SO_4	336 h	15.3	[33]
$(Cu_{47}Zr_{11}Ti_{34}Ni_8)_{99}Mo_1$	1 N H_2SO_4	336 h	1.5	[33]
$(Cu_{47}Zr_{11}Ti_{34}Ni_8)_{98}Mo_2$	1 N H_2SO_4	336 h	0.9	[33]
$(Cu_{47}Zr_{11}Ti_{34}Ni_8)_{95}Mo_5$	1 N H_2SO_4	336 h	18.9	[33]
$Cu_{47}Zr_{11}Ti_{34}Ni_8$	1 N NaOH	672 h	2	[33]
$(Cu_{47}Zr_{11}Ti_{34}Ni_8)_{99}Mo_1$	1 N NaOH	672 h	0.8	[33]
$(Cu_{47}Zr_{11}Ti_{34}Ni_8)_{98}Mo_2$	1 N NaOH	672 h	0.3	[33]
$(Cu_{47}Zr_{11}Ti_{34}Ni_8)_{95}Mo_5$	1 N NaOH	672 h	4.5	[33]

the dendritic body-centered cubic phase have been estimated to be about 8% and 5–10 μm, respectively. Electron probe micro analysis showed that the chemical composition of these dendrites was 2.6 Cu, 5.4 Hf, 11.6 Ti, and the rest Nb (by atomic percent), suggesting that these dendrites were Nb-rich in composition. The remaining glassy phase was, as expected, Cu-rich, and had a composition of 22.9 Hf, 14.2 Ti, and 6.9 Nb, and the rest Cu

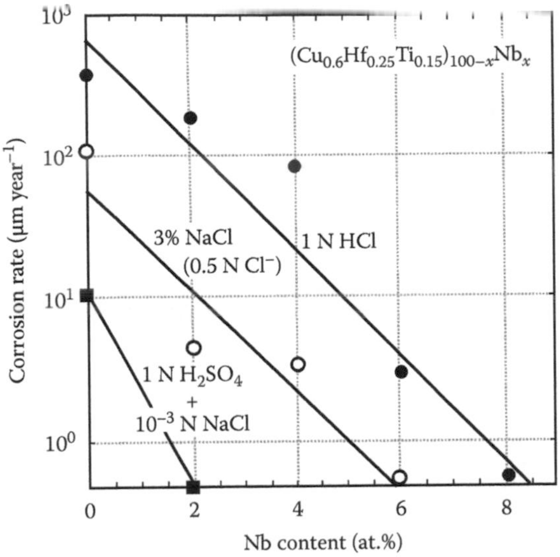

FIGURE 7.3
Corrosion rates of $(Cu_{0.6}Hf_{0.25}Ti_{0.15})_{100-x}Nb_x$ alloys (with $x = 0, 2, 4, 6$, and 8 at.% Nb) immersed in 1 N HCl, 3 wt.% NaCl, and 1 N H_2SO_4 + 0.001 N NaCl solutions for 168 h at 298 K open to air.

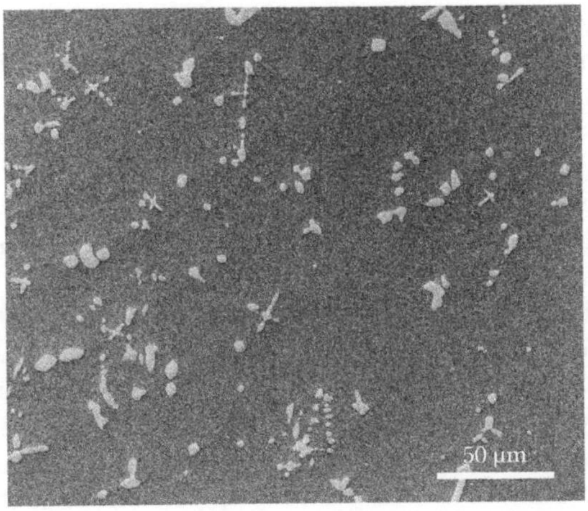

FIGURE 7.4
Scanning electron micrograph of the central region in the transverse cross section of the as-cast $(Cu_{0.6}Hf_{0.25}Ti_{0.15})_{90}Nb_{10}$ alloy rod with a diameter of 2 mm. (Reprinted from Qin, C.L. et al., *Acta Mater.*, 54, 3713, 2006. With permission.)

(by atomic percent). Since Cu and Nb exhibit a positive heat of mixing, they are immiscible with each other. Further, since Nb has a higher melting point than Cu, it may be inferred that the Nb-rich phase precipitated as the primary dendrite phase from the Cu-rich liquid. Despite being a composite alloy, its corrosion resistance was found to be significantly better than that of the base fully glassy $Cu_{60}Hf_{25}Ti_{15}$ alloy. Whereas the $Cu_{60}Hf_{25}Ti_{15}$ alloy showed a corrosion rate of 340 μm year^{-1} in 1 N HCl solution, the composite alloy showed a corrosion rate of only 11 μm year^{-1}, which is about an order of magnitude lower than that of the Nb-free alloy [25,34]. But, the lowest corrosion rate was observed at an Nb level of

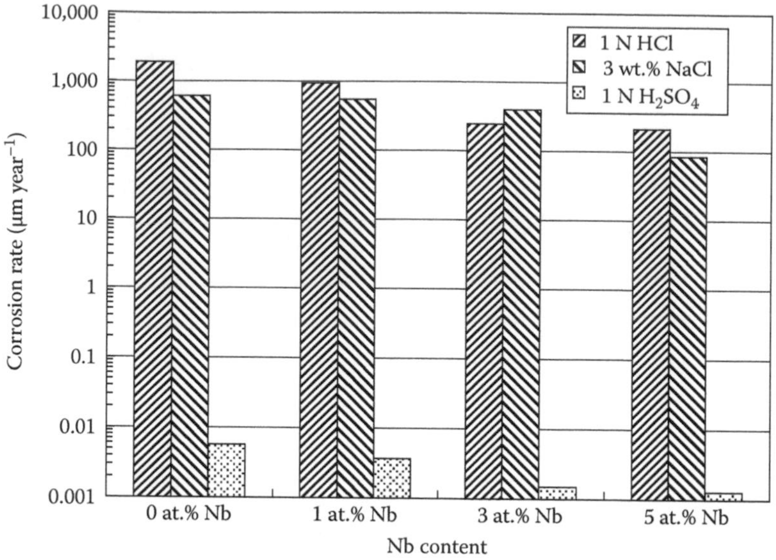

FIGURE 7.5
Effect of Nb addition on the corrosion resistance of $Cu_{50}Zr_{45}Al_5$ glassy alloys immersed in 1 N HCl solution for 24 h, 3 wt.% NaCl solution for 97 h, and 1 N H_2SO_4 solution for 168 h.

8 at.% [27,28]. By potentiodynamic polarization measurements, it was shown that while the Nb-free alloy immersed in 3 wt.% NaCl or 1 N H_2SO_4 + 0.001 N NaCl solutions did not show any passivation (the alloy dissolves quickly by slight anodic polarization), the Nb-containing alloy exhibited spontaneous passivation with low passive current density. It had, of course, suffered pitting corrosion.

Additions of Nb seem to have a similar beneficial effect in improving the corrosion resistance of other Cu-based BMGs, such as $Cu_{50}Zr_{45}Al_5$ and $Cu_{55}Zr_{40}Al_5$, in different acidic environments (Figure 7.5). Note that while the corrosion rate was very high (at least 1900 μm $year^{-1}$) in 1 N HCl solution in the Nb-free $Cu_{55}Zr_{40}Al_5$ alloy, the addition of 5 at.% Nb brought this value down by an order of magnitude to about 200 μm $year^{-1}$ [31].

Additions of Nb to $Cu_{50}Zr_{45}Al_5$ alloy have also been shown to improve the corrosion resistance significantly [30,31]. This effect was more obvious when the Nb content was greater than 3 at.%. The corrosion rate decreased to one-fifth of that of the base alloy in 1 N HCl solution when 5 at.% Nb was added [35]. The effect of the simultaneous addition of Nb and Ni to $Cu_{60}Zr_{30}Ti_{10}$ alloy on the corrosion behavior was investigated [36]. It was reported that the corrosion rate of the $Cu_{60}Zr_{30}Ti_{10}$ alloy in 1 N HCl was 660 μm $year^{-1}$ and decreased significantly by the addition of 5–7 at.% Ni or 2–4 at.% Nb. No weight loss in the sample was detected for the $(Cu_{0.6}Zr_{0.3}Ti_{0.1})_{89}Ni_5Nb_6$ alloy immersed in 1 N HCl. Similar remarkable improvement of corrosion resistance was also recognized in 3 wt.% NaCl solution. The addition of 5 at.% Ni or 5 at.% Nb caused a significant decrease in the corrosion rate from 290 μm $year^{-1}$ for the $Cu_{60}Zr_{30}Ti_{10}$ to 13 and 18 μm $year^{-1}$, respectively [36].

Improved corrosion resistance was also reported in the $Cu_{60}Zr_{30}Ti_{10}$ alloy, even though the corrosion rate of the Zr-containing Cu-based alloy was found to be higher than that of the Hf-containing alloy. The corrosion behavior of the $Cu_{60}Zr_{30}Ti_{10}$ alloy and also of the alloys modified with the addition of 1 at.% Nb, 1 at.% Mo, and 0.2 at.% Ta immersed in

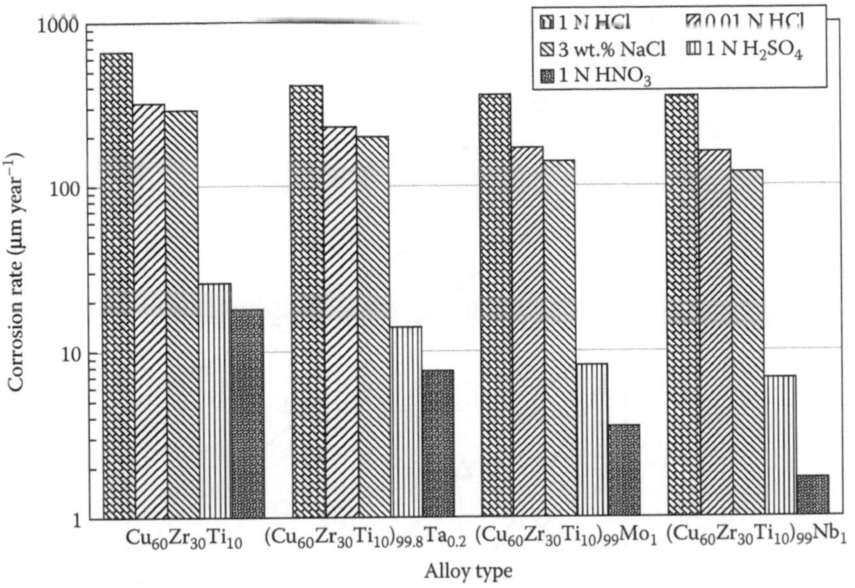

FIGURE 7.6
Effect of additional alloying elements (Ta, Mo, or Nb) on the corrosion behavior of $Cu_{60}Zr_{30}Ti_{10}$ BMG alloy in 1 N HCl, 0.01 N HCl, and 3 wt.% NaCl solutions after exposure for 168 h and in 1 N H_2SO_4 and 1 N HNO_3 solutions after exposure for 336 h.

different corrosive solutions have been investigated [32] and the trend of corrosion rates is presented in Figure 7.6.

The addition of alloying elements has substantially improved the corrosion resistance of the Cu-based BMG alloys. For example, the addition of 1 at.% Mo or Nb to the $Cu_{60}Zr_{30}Ti_{10}$ alloy reduced the corrosion rate by half in the chloride-containing solutions. Nb and Mo substitutions had a more drastic effect in 1 N H_2SO_4 or 1 N HNO_3 solutions. The corrosion rate decreased by about an order of magnitude when the alloy was exposed to 1 N HNO_3 solution. The variation of the corrosion rate of $Cu_{60}Zr_{30}Ti_{10}$ alloy as a function of Nb content in different test environments is shown in Figure 7.7 [28,32,37]. It is clear from Figure 7.7 that increasing the amounts of Nb significantly increased the corrosion resistance of the $Cu_{60}Zr_{30}Ti_{10}$ alloy, even in chloride-containing solutions. It may be instructive to check whether a higher concentration of Nb, say, for example, 10 at.%, will exhibit any further improvement in the corrosion resistance.

Zhang et al. [38] developed new types of Cu–Zr–based BMGs with a maximum diameter of up to 25 mm, and reported formation mechanism, thermal stability, mechanical properties, and corrosion resistance of the new $Cu_{42}Zr_{42}Ag_8Al_8$ BMG. Figure 7.8 shows the potentiodynamic polarization curves of the as-cast Cu–Zr–Ag–Al glassy rods with a diameter of 2 mm in 1 N H_2SO_4 solution, open to air at 298 K. The data for α–β industrial brass and SUS316L stainless steel are also shown for comparison. The Cu–Zr–Ag–Al glassy alloys are passivated at a significant low current density of the order of 10^{-2} A m^{-2}, followed by a slow rate of increase with increasing potential. In contrast, the current density of the α–β brass increases rapidly by slight anodic polarization, indicating high dissolution of the brass in the solution. Although the SUS316L is passivated until the transpassive dissolution of chromium occurs, its passive current density is about an order of magnitude higher than that of the Cu–Zr–Ag–Al glassy alloys. Accordingly, the newly

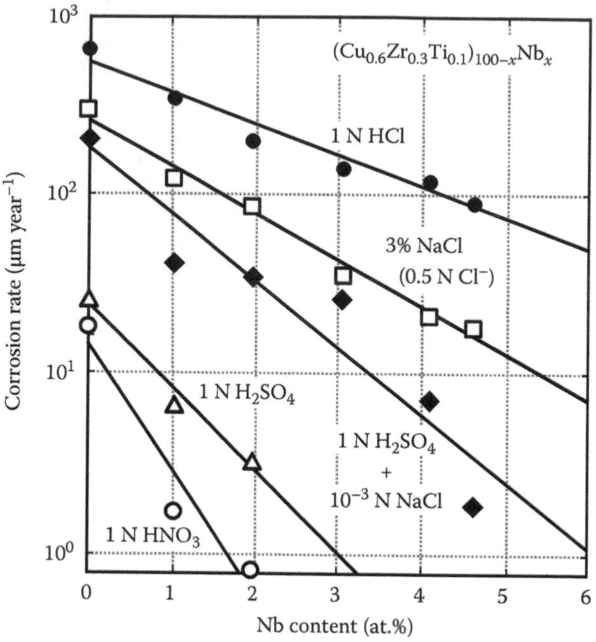

FIGURE 7.7
Variation of corrosion rate as a function of the Nb content in $(Cu_{0.6}Zr_{0.3}Ti_{0.1})_{100-x}Nb_x$ ($x = 0–4.6$ at.%) alloys in different chloride- and non-chloride-containing solutions (1 N HCl, 3% NaCl, 1 N H_2SO_4, 1 N H_2SO_4 + 0.001 N NaCl, and 1 N HNO_3 solutions).

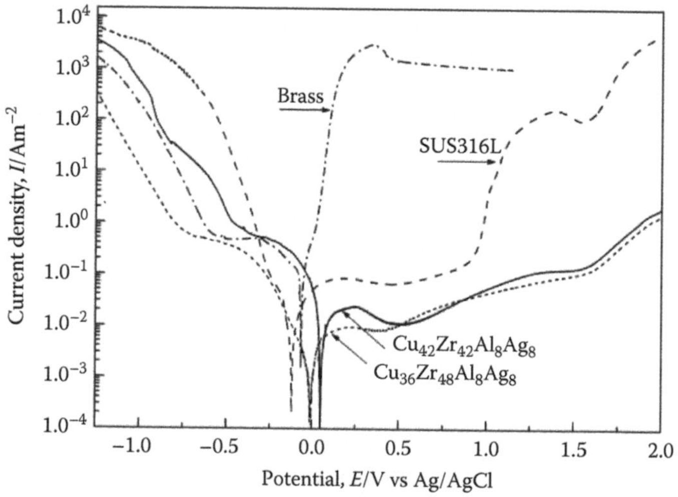

FIGURE 7.8
Potentiodynamic polarization curves of Cu–Zr–Ag–Al glassy alloys with a diameter of 2 mm in 1 N H_2SO_4 solution open to air at 298 K. The data for α–β brass alloy and SUS316L stainless steel are also shown for comparison.

developed Cu–Zr–Ag–Al BMGs exhibit excellent corrosion resistance in 1 N H_2SO_4 solution and their corrosion resistance is much better than those of the brass and SUS316L. The high corrosion resistance of the Cu–Zr–Ag–Al glassy alloys is attributed to their single glassy phase nature and the formation of Zr^{4+}- and Al^{3+}-enriched highly protective thin surface film in corrosion solutions.

7.3.2 Influence of Test Environment

The majority of the investigations in this category dealt with a change in the type and concentration of corrosive medium in which the alloys were immersed to evaluate their corrosion behavior. There is only one investigation where the corrosion behavior was studied at higher temperatures.

The corrosion behavior of a multicomponent $Cu_{0.5}NiAlCoCrFeSi$ alloy, containing mostly the glassy phase but some nanoscale precipitates, was studied at 288°C (561 K) in high-purity water simulating the boiling water reactor environment [29]. A small amount of 0.01 N sodium sulfate (Na_2SO_4) was added to the water, because pure water does not have sufficient conductivity for performing potential-controlled tests. It was reported that the BMG alloy exhibited a wide passive region, and that the passive current density was ~2×10^{-4} A cm^{-2}. A low weight loss of ~4.5 μg mm^{-2} was reported after immersing the alloy in deaerated water at 561 K for 12 weeks. This weight loss is much lower than the ~28 μg mm^{-2} value obtained for carbon steel under similar test conditions [29].

Looking through the results presented in Table 7.1, it is clear that the $Cu_{60}Zr_{30}Ti_{10}$ alloy corroded severely in chloride-containing solutions. Even when the concentration of HCl was reduced to 0.01 N, the corrosion rate was still substantial [32]. On the other hand, when the corrosive medium was changed to 1 N H_2SO_4 or 1 N HNO_3, the corrosion rates were about an order of magnitude lower than what they were in the chloride-containing environment. These results prove conclusively that the copper-based BMG alloys corrode severely in chloride-containing environments.

As discussed, Cu-based alloys fare very badly in the presence of chloride ions, with a corrosion rate of typically 100–300 μm $year^{-1}$. The corrosion rate is much smaller in media such as HNO_3 or H_2SO_4. Qin et al. [25,26] reported that the corrosion rate of $Cu_{60}Hf_{25}Ti_{15}$ BMG alloy was 340 μm $year^{-1}$ when exposed to 1 N HCl for 168–336 h at 298 K. The same alloy, however, had a corrosion rate of only 100 μm $year^{-1}$ when exposed to 3 wt.% NaCl solution. The corrosion resistance of the $Cu_{60}Hf_{25}Ti_{15}$ BMG alloy could be increased by the addition of alloying elements such as Mo, Nb, or Ta. Addition of these alloying elements at a level of 2 at.% reduced the corrosion rate significantly in H_2SO_4, HNO_3, and NaOH solutions to undetectable levels. Similarly, the Nb-free alloy had a corrosion rate of 100 μm $year^{-1}$ in NaCl solution, while the Nb-reinforced alloy did not show any measurable corrosion rate. Both the Nb-containing and Nb-free alloys did not show any measurable corrosion rate in 1 N H_2SO_4 solution.

7.3.3 Nature of the Passive Film

To understand the reasons for the improved corrosion resistance of the $Cu_{60}Hf_{25}Ti_{15}$ BMG alloy in the presence of Nb, XPS analysis was conducted on the surface film formed after exposure to air and also after immersion in different corrosive solutions. On the as-polished alloy, the atomic fractions of the elements Cu, Hf, Ti, and Nb in the surface film were found to be 41.5, 26.4, 23.1, and 9.0, respectively (as against the nominal values of 54, 22.5, 13.5, and 10.0). In the surface film formed after exposure to air, Hf and Ti were enriched, while Cu and Nb were deficient with respect to the alloy composition. This suggests that preferential oxidation of Hf and Ti took place in the air-formed film. Figure 7.9 shows the surface film composition for the $(Cu_{0.6}Hf_{0.25}Ti_{0.15})_{90}Nb_{10}$ alloy in the as-cast condition, after mechanical polishing and exposure to air, and after immersion in different chloride-containing solutions [25]. It was noted that the Cu content decreased significantly after immersion in 1 N HCl solution; the decrease was even more significant in 3 wt.% NaCl and

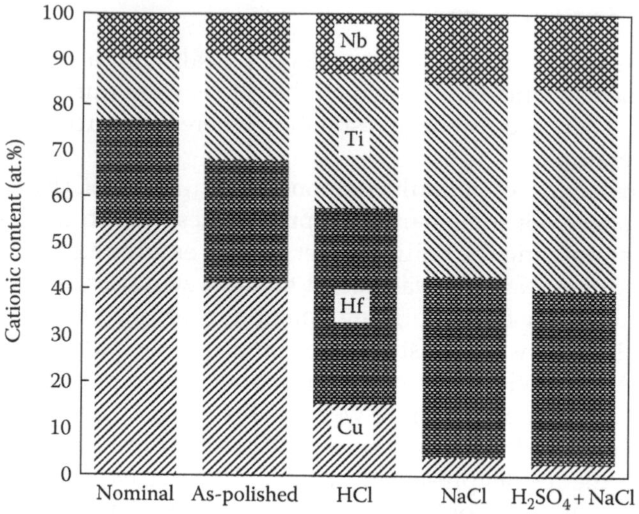

FIGURE 7.9
Surface film compositions for the $(Cu_{0.6}Hf_{0.25}Ti_{0.15})_{90}Nb_{10}$ alloy in the as-cast condition (nominal), as-polished condition, and after immersion in 1 N HCl, 3 wt.% NaCl, and 1 N H_2SO_4 + 0.01 N NaCl solutions open to air for 1 week after mechanical polishing. (Reprinted from Qin, C.L. et al., *Acta Mater.*, 54, 3713, 2006. With permission.)

1 N H_2SO_4 + 0.01 N NaCl solutions. At the same time, the concentrations of Ti and Hf were significantly higher, although the magnitudes are slightly different in different solutions. More significantly, the film is enriched with Nb.

Since Cu has the lowest corrosion resistance of all the elements present in the alloy, it easily dissolves in acidic and chloride-containing solutions. But Hf, Ti, and Nb exhibit excellent corrosion resistance in the solutions considered here. Consequently, a stable surface film cannot be formed in the Nb-free alloy. Therefore, the formation of a highly stable and protective surface film enriched with Hf, Ti, and Nb on the $(Cu_{0.6}Hf_{0.25}Ti_{0.15})_{90}Nb_{10}$ composite may be the reason for the improved corrosion resistance of the alloy.

7.3.4 Hierarchical Surface Structures

It has been shown recently that, by modifying the outer surface structure of BMGs through dealloying treatment using HF and HNO_3 solutions, the resulting BMGs can have nano-spherical, nano-disk, nano-rod, nano-cube, or nanoporous structures and exhibit unique chemical characteristics [39–41]. Work in this and related areas has been steadily increasing and can be considered as a new attractive engineering field for BMGs. Besides, the use of BMGs provides a bulk shape of ligament that cannot be obtained for ribbon-shaped alloys, possibly leading to more widespread application areas. Therefore, it will be useful to look at some typical data on the unique functional properties caused by surface structure modification of BMGs.

Qin et al. [42] synthesized hierarchical structured nanoporous copper (NPC)/BMG composite rods by one-pot chemical dealloying of $Cu_{50}Zr_{45}Al_5$ BMG rods in 0.05 M HF solution for 24 h. Figure 7.10a shows cross-sectional scanning electron microscopy (SEM) images of the NPC/BMG composite rod. The SEM images illustrate that the NPC/BMG composite rods exhibit the perfect combination of inner rod-shaped Cu–Zr–Al amorphous phase core and outer tube-shaped NPC layer with a thickness of 85 μm. The NPC tubular layer has a good interface combination with the BMG matrix. The dealloyed alloy exhibits a

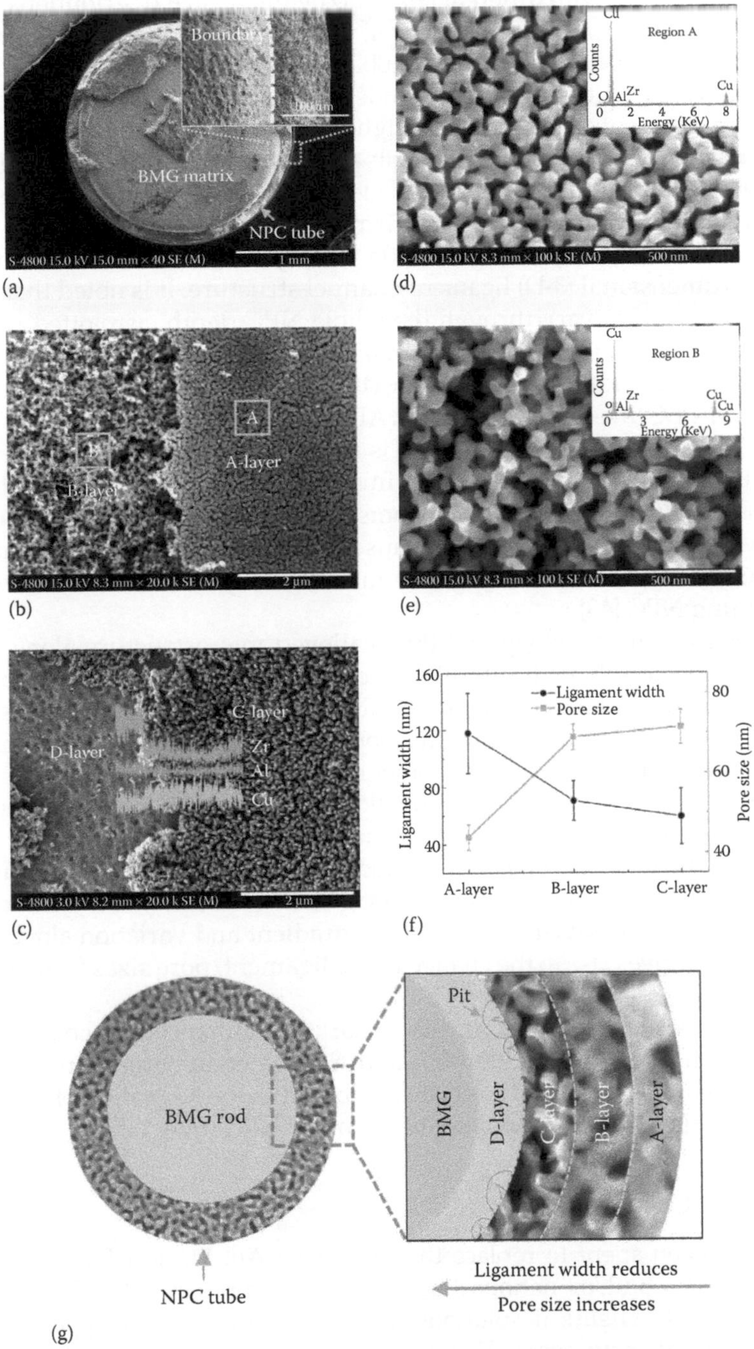

FIGURE 7.10
SEM images showing the surface microstructure of the dealloyed $Cu_{50}Zr_{45}Al_5$ BMG immersed in the 0.05 M HF solution for 24 h at 298 K. (a) the cross-sectional morphology of the NPC/BMG composite rod; (b) and (c) the four-layer hierarchical architectures after carefully peeling; (d) and (e) magnified SEM images with their corresponding EDS spectra from A and B regions, respectively; (f) changes in the ligament width and nanopore size with the NPC depth; (g) schematic illustration of the hierarchical architectures along the dealloying depth. (Reprinted from Qin, C.L. et al., *Intermetallics*, 77, 1, 2016. With permission.)

hierarchical architecture with different channel/ligament size distributions as a function of the dealloying depth (A-layer: outer layer, B-layer: beneath A-layer, C-layer: beneath B-layer, and D-layer: inner layer). Figure 7.10b and c shows the dealloyed microstructure along the depth for the $Cu_{50}Zr_{45}Al_5$ BMG rods as one moves into the interior of the rod. Figure 7.10d and e presents SEM images at higher magnifications with their corresponding energy-dispersive spectroscopy (EDS) results taken from regions A and B, respectively, in Figure 7.10b. From the EDS spectra, it is clear that the elemental compositions of the porous surfaces mainly consist of Cu, confirming that the porous surfaces covering the BMG matrix are NPC. The NPC (Figure 7.10b through e) exhibits an open, interpenetrating, and three-dimensional (3-D) ligament-channel structure. It is noted that the ligament width and nanopore size vary through the whole NPC depth, as plotted in Figure 7.10f. The nanopore size gradually increases from 45 nm at the A-layer to 72 nm at the C-layer, whereas the ligament width decreases from 118 nm at the A-layer to 60 nm at the C-layer. Besides, the SEM-EDS line-scan profile of Zr, Al, and Cu elements sweeping from the amorphous matrix (D-layer) to the NPC (C-layer) is inserted in Figure 7.10c. The results further confirm that the dealloying process results in selective dissolution of Al and Zr elements and the self-assembly of the residual Cu atoms. On the other hand, from Figure 7.10c, it is clearly observed that many corrosion pits caused from the constituent elements Zr and Al in the alloy are formed on the inner glassy matrix (D-layer), which is believed to be the origin for forming NPC [43].

The cross-sectional morphology and the dealloyed microstructure along the depth of the composite rod are schematically illustrated in Figure 7.10g. The possible formation mechanism of the hierarchically structured NPC tubular layer can be proposed from two aspects. First, unlike a thin film and thin ribbon precursor, the dealloying of a BMG rod with a diameter of 2 mm starts from the outer surface and gradually moves into the inner layer. This causes a large time interval of dealloying between the inner and outer layers of the BMG rod. Thus, the ligaments of NPC formed early at the outer part coarsen, whereas the fresh NPC in the inner part has just formed. Second, because the constituent elements (Cu, Zr, and Al) in HF solution exhibit different atomic diffusivity and chemical reactivity, the dealloying process results in a composition gradient and variation along the depth of the rod sample that gives rise to the change in the ligament/pore sizes from the outer layer to the inner layer.

These new composite rods demonstrate remarkable enhanced mechanical properties, with an ultra-high strength of 1500 MPa and a large compression strain of 2.9%. The increase in the compression strain has been attributed to the formation of the buffer deformation zone resulting from the existence of the outer NPC tube.

7.3.5 Nanophase Composites

Much effort has been spent to replace the expensive Au, Pt, and Pd catalysts by much cheaper nanoporous and/or nanoparticle composites prepared by dealloying suitable BMGs in appropriate chemical solutions. Besides, the composite-type nanoporous or nanoparticle materials have some advantages that cannot be obtained from single-phase monolithic nanoporous nanoparticle materials. We present here some examples showing the distinct advantages of the composite-type catalysts.

The production of nanoporous components can be done in two different ways. One is to use suitable multicomponent glassy alloys, where more than two elements leading to the composite are included. The other is to give the physical or chemical deposition of the second phase onto the nanoporous and/or nanoparticle material.

A typical Cu-based BMG is based on the ternary Cu–Hf–Al system [44]. Wang et al. [45] successfully synthesized monolithic NPC ribbons with good mechanical integrity and bendability by chemical dealloying the amorphous $Cu_{52.5}Hf_{40}Al_{7.5}$ alloy ribbon in 1.5M HF solution for 300s at 298 K. A ductile amorphous phase-containing interlayer with a certain thickness was successfully designed and fabricated in the dealloyed NPC ribbon, which ensures good bendability of the final dealloyed product. The SEM images in Figure 7.11a and b show an obvious interlayer consisting of Cu, Hf, and Al elements (confirmed by EDS analysis) with thicknesses of 250–400 nm between the NPC phases. It is seen that the residual amorphous phase is retained in the interlayer by stopping the dealloying treatment at a certain stage. Figure 7.11c shows a high-resolution transmission electron microscopy (HRTEM) image of the NPC ligamental structure. It is clear that the area marked between the dotted lines shows typical amorphous structure, as identified by the inserted fast Fourier transformation (FFT) pattern. Since the SEM images coupled with EDS analyses shown in Figure 7.11a and b exhibit a distinct interlayer containing Cu, Hf, and Al elements, it is reasonable to speculate that the amorphous structure in Figure 7.11c is the residual amorphous phase. The transmission electron microscopy (TEM) sample was gradiently thinned to expose the inner amorphous structure by two-sided ion milling treatment. Figure 7.11d shows an HRTEM image of a segment of the NPC ligament. In addition to the residual amorphous area, lattice fringes can be seen throughout the ligament. The interplanar distances measured from the adjacent lattice fringes marked in the IFFT image (Figure 7.11e) are 0.181 nm and 0.209 nm in two directions, which correspond to the (200) and (111) planes of face-centered cubic (fcc) Cu, respectively. Therefore, the dealloying products of the interlayer in the HF-treated $Cu_{52.5}Hf_{40}Al_{7.5}$ alloy are identified as NPC plus residual amorphous phase.

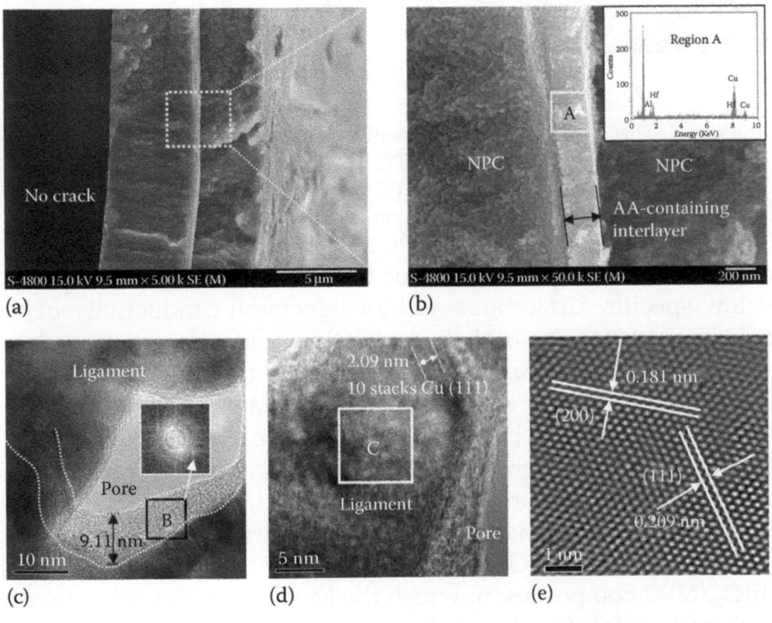

FIGURE 7.11
Sectional view of $Cu_{52.5}Hf_{40}Al_{7.5}$ ribbons subjected to dealloying in 0.5M HF for 300s at 298K. (a) and (b) are SEM images, the inserted image in (b) shows EDS analysis of Region A. (c) and (d) are HRTEM images of interlayer area in (b); the amorphous phase is identified from the FFT pattern of Region B. (e) IFFT pattern taken from Region C. (Reprinted from Wang, Z.F. et al., *Intermetallics*, 56, 48, 2015. With permission.)

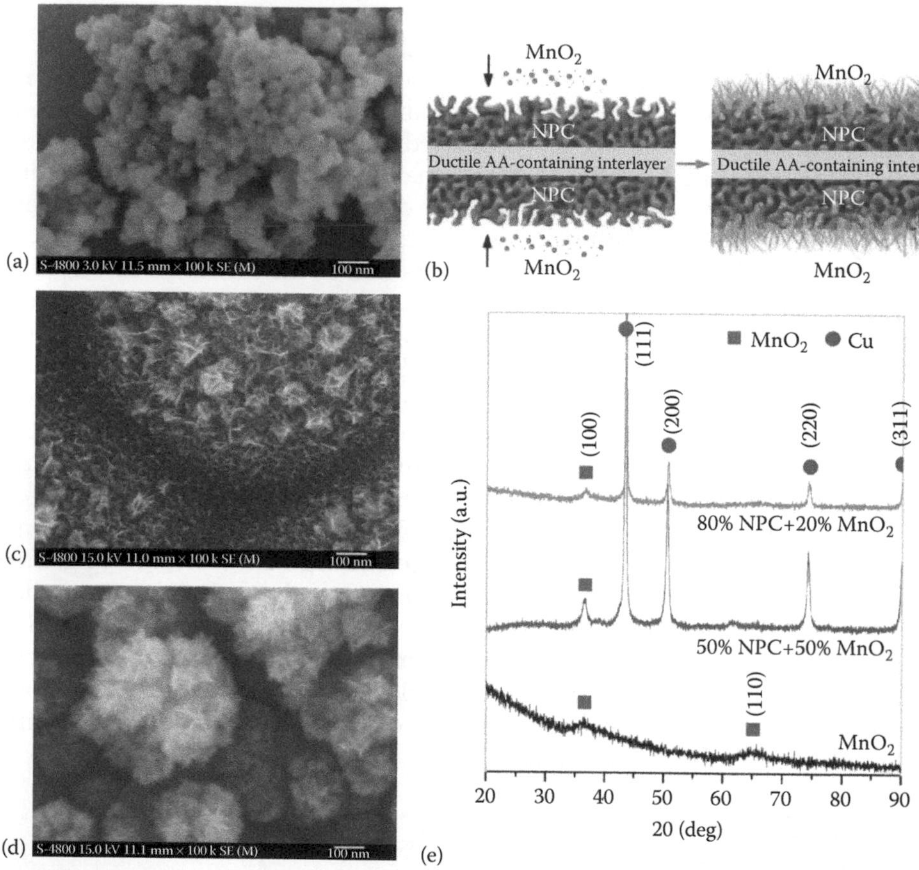

FIGURE 7.12
SEM images of the sample surfaces. (a) As-prepared MnO_2 powder. (b) Fabrication process of the MnO_2/NPC composites. (c) MnO_2/NPC composite with 80 wt.% NPC. (d) MnO_2/NPC composite with 50 wt.% NPC. (e) XRD patterns of as-prepared MnO_2 powders and MnO_2/NPC composites with different MnO_2 contents. (Reprinted from Wang, Z.F. et al., *Intermetallics*, 56, 48, 2015. With permission.)

MnO_2 has been a promising candidate for the next generation of supercapacitors [46]. However, the low specific surface area and low electrical conductivity of MnO_2 powder have restricted further extension of their application fields. Considering the high specific surface area and excellent electrical conductivity of nanoporous Cu, NPC-supported MnO_2 composite electrode could be a new type of electrode material. Most importantly, the ductile NPC ribbon could offer great benefit to a binder-free electrode substrate.

MnO_2 can be prepared through the classical chemical reaction between $KMnO_4$ and ethanol [47]. However, the resulting MnO_2 is composed of nano-sized globular particles, as shown in Figure 7.12a. One can easily observe the particle aggregation. Wang et al. [45] prepared an NPC-supported MnO_2 composite. Figure 7.12b illustrates the fabrication process of the MnO_2/NPC composites, in which the MnO_2 layer was directly deposited on a ductile NPC substrate. It is interesting to find in Figure 7.12c that the MnO_2 nanoflakes [46] are homogeneously deposited on the surface of the NPC substrate. The nanoflake morphology enables a much larger surface area of MnO_2, which is favorable for electrode/electrolyte interfacial contact. When the MnO_2 content is increased from 20 to 50 wt.%, the MnO_2 nanoflakes grow to the nanostructured flower-like morphology on the NPC

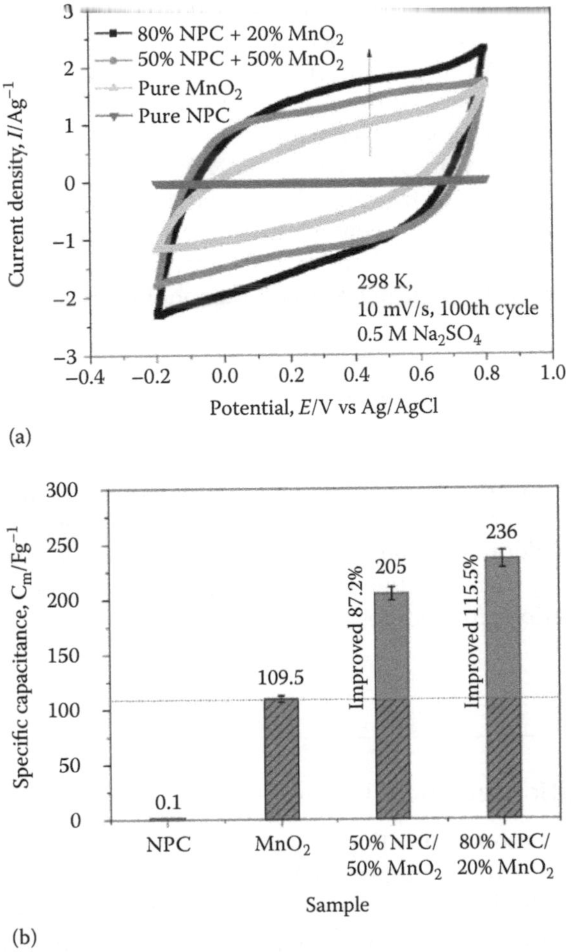

FIGURE 7.13
Current density-voltage curves of monolithic NPC, pure MnO_2, and MnO_2/NPC composites with different MnO_2 contents in 0.5M Na_2SO_4 solution. (b) Specific capacitance of monolithic NPC, pure MnO_2, and MnO_2/NPC composites with different MnO_2 contents. (Reprinted from Wang, Z.F. et al., *Intermetallics*, 56, 48, 2015. With permission.)

substrate (Figure 7.12d). This change in morphology suggests that the smaller-sized MnO_2 nanoflakes can be formed on NPC substrates for the composites with lower MnO_2 content. The NPC substrate remains stable during the fabrication process of MnO_2. According to the surface morphologies of MnO_2 with and without the NPC substrate (Figure 7.12a, c, and d), it can be seen that the NPC substrate provides 3-D nanoscale channels for the precipitation and growth of Mn cations, which greatly improves the specific surface area of MnO_2.

For NPC serving as a working electrode or an electrode substrate, it should be emphasized that the mechanical integrity of NPC is a key requirement. Wang et al. [45] used monolithic NPC ribbon with good mechanical integrity obtained from dealloyed Cu–Hf–Al BMG for electrochemical tests. Figure 7.13a shows the current density–voltage (C–V) curves of NPC, pure MnO_2, and MnO_2/NPC composites with different weight ratios in 0.5M Na_2SO_4 solution at 298 K. From their C–V curves, it is clearly seen that the capacitive current of the NPC in the voltage window is negligibly low. However, it is surprising that the signal and the closed area of C–V curves for the MnO_2/NPC composite electrode are remarkably enhanced as compared with those for the pure MnO_2 powders fabricated under similar conditions. Moreover, the current response increases greatly as the weight ratio of NPC to MnO_2 increases. Figure 7.13b shows the specific capacitance of monolithic NPC, pure MnO_2, and MnO_2/NPC composites with different weight ratios. It is noticed that the specific capacitance values of two MnO_2/NPC composites are more than 200 F g^{-1}, which are higher than those for traditional supercapacitor materials [48]. Thus, by using NPC substrate, a significant improvement in the specific capacitance of MnO_2 can be clearly seen. Combining the electrochemical performance of pure MnO_2 and MnO_2/NPC composites with their corresponding surface morphologies and distribution of MnO_2 (Figure 7.12a, c, and d), it is noted that NPC substrate with large specific surface area and excellent electrical conductivity can effectively improve the utilization of MnO_2 surface active sites and promote MnO_2 chemical reactions.

Wang et al. [45] further examined the effect of cycling on specific capacitance of MnO_2/NPC composites and reported that the specific capacitance of the composites exceeded 200 F g^{-1} even after 800 cycles. Thus, NPC substrate with large specific surface area and excellent electrical conductivity can effectively promote the morphological change of MnO_2 from globular to nanoflakes for larger specific surface areas, and improve the utilization of MnO_2 surface active sites, revealing a new electrochemical potential of NPC in future applications for composite electrodes.

7.4 Iron-Based Bulk Metallic Glasses

Ever since the first Fe-based BMG alloy was synthesized in 1995 in an Fe–Al–Ga–P–C–B system corresponding to the composition $Fe_{72}Al_5Ga_2P_{11}C_6B_4$ [49], a number of Fe-based alloys have been cast into the bulk glassy condition by the copper mold casting technique, with a diameter of a few millimeters. They possess excellent magnetic and mechanical properties, with the strength exceeding 4 GPa [50–52]. Fe-based BMG alloys investigated for their corrosion behavior generally contained Cr, Mo, C, and B in varying proportions. Cr has been shown to be essential in forming a passive layer and this is further facilitated by the addition of Mo. Both carbon and boron (to a combined total of approximately 20 at.%) were found to be necessary to achieve glass formation. By studying a number of alloys, the Cr and Mo contents were optimized to achieve the highest glass-forming ability. The alloy with the composition $Fe_{41}Co_7Cr_{15}Mo_{14}C_{15}B_6Y_2$ seems to have the highest glass-forming ability among the Fe-based alloys since it was possible to produce a fully glassy rod of 16 mm diameter in this alloy composition [53].

A typical Fe-based glassy alloy composition appears to be $Fe_{45}Cr_{16}Mo_{16}C_{18}B_5$. In some of the investigations, either additional metalloid elements, especially P, have been added, or Mo has been partially replaced with Ta or Nb. The actual compositions investigated include $Fe_{43}Cr_{16}Mo_{16}C_{10}B_5P_{10}$ [54,55], $Fe_{45}Cr_{16}Mo_{16-x}M_xC_{18}B_5$ (where M = Ta or Nb and $x = 0$ or 2 at.%) [56], $Fe_{50-x}Cr_{16}Mo_{16}C_{18}B_x$ (with $x = 3, 4, 6, 8$, and 10 at.%) [57], $Fe_{60-x}Cr_xMo_{15}C_{15}B_{10}$ (with $x = 0, 7.5, 15, 22.5$, and 30 at.%) [58], $Fe_{75-x-y}Cr_xMo_yC_{15}B_{10}$ (with $x = 0–30$ at.% and $y = 0–22.5$ at.%) [59], $Fe_{76-x}Cr_xMo_2Ga_2P_{10}C_4B_4Si_2$ (with $x = 0–6$ at.%) [60], $[\{(Fe_{0.6}Co_{0.4})_{0.75}B_{0.2}Si_{0.05}\}_{0.96}Nb_{0.04}]_{100-x}Cr_x$ (with $x = 0–4$) [61], $Fe_{49}Cr_{15.3}Mo_{15}Y_2C_{15}B_{3.4}N_{0.3}$ [62], $Fe_{50-x}Cr_{15}Mo_{14}C_{15}B_6Y_x$ [63], and $(Fe_{44.3}Cr_5Co_5Mo_{12.8}Mn_{11.2}C_{15.8}B_{5.9})_{98.5}Y_{1.5}$ [64], among others. A reasonably comprehensive list of alloy compositions of Fe-based BMGs, up to 2012, is available in Ref. [52]. The corrosion behavior of these glassy alloys was generally investigated by immersing the alloy samples in 3 wt.% NaCl or HCl (of different concentrations, usually 1, 6, and 12 N), by both weight-loss and electrochemical measurement techniques, and the compositions of the passive films and the underlying substrate materials were determined using XPS techniques.

7.4.1 Influence of Composition

All the glassy alloys investigated exhibited good corrosion resistance in concentrated HCl, with the measured corrosion rates as low as 1–10 μm year^{-1}. Figure 7.14a shows that the corrosion rate of the glassy $Fe_{50-x}Cr_{16}Mo_{16}C_{18}B_x$ alloy decreases with increasing amount of B in the alloy [57]. A similar decrease of corrosion rate with increasing Cr content was also noted in the $[\{(Fe_{0.6}Co_{0.4})_{0.75}B_{0.2}Si_{0.05}\}_{0.96}Nb_{0.04}]_{100-x}Cr_x$ alloy, when the corrosion rate decreased from 700 to 1.6 μm year^{-1} as the Cr content increased from 0 to 4 at.% (Figure 7.14b) [61]. The addition of Mo to Fe-based alloys has also been reported to improve their corrosion resistance in HCl solution, since it prevents the dissolution of Cr during passivation [65]. However, it has been noted that when the dissolution rate of Fe-based alloys is very high,

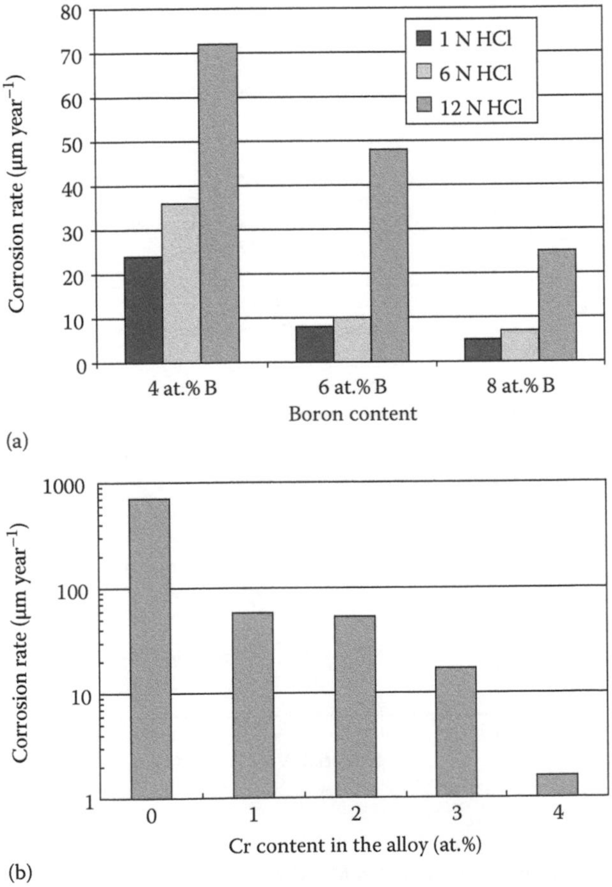

FIGURE 7.14
(a) Decreasing corrosion rate with increasing B content in an $Fe_{50-x}Cr_{16}Mo_{16}C_{18}B_x$ BMG alloy, and (b) decreasing corrosion rate with increasing Cr content in 0.5 N NaCl solution at 298 K open to air for 168 h for the $[\{(Fe_{0.6}Co_{0.4})_{0.75}B_{0.2}Si_{0.05}\}_{0.96}Nb_{0.04}]_{100-x}Cr_x$ alloy. (Reprinted from Long, Z.L. et al., *Intermetallics*, 15, 1453, 2007. With permission.)

as in the active region, Mo selectively remains in the alloy because the dissolution rate of Mo is slower than that of other constituents. Further, Mo has not been able to form its own passive film in the passive region of the alloys. Mo is also known to dissolve even at the lower potentials in the passive region of the alloys, indicating a lower stability of the passive film of Mo in comparison with a passive hydrated chromium or iron oxyhydroxide film. Consequently, excessive amounts of Mo added to replace Fe have been reported to be detrimental for the corrosion resistance of Fe-based glassy alloys [66].

The corrosion resistance of the BMG alloys has been shown to be higher due to the presence of P in the alloy. Figure 7.15a shows the potentiodynamic polarization curves of $Fe_{43}Cr_{16}Mo_{16}C_{15}B_{10}$ and $Fe_{43}Cr_{16}Mo_{16}C_{10}B_5P_{10}$ bulk glassy alloys in 1 N HCl at 298 K [55]. From these curves, it may be noted that both the alloys passivate spontaneously. However, the passive current density is about 10^{-1} A m^{-2} for the $Fe_{43}Cr_{16}Mo_{16}C_{15}B_{10}$ alloy, whereas it is approximately half (5×10^{-2} A m^{-2}) for the $Fe_{43}Cr_{16}Mo_{16}C_{10}B_5P_{10}$ alloy. The lower passive current density in the P-containing alloy clearly demonstrates that this alloy has a better corrosion resistance. Figure 7.15b shows that a similar improvement in the corrosion

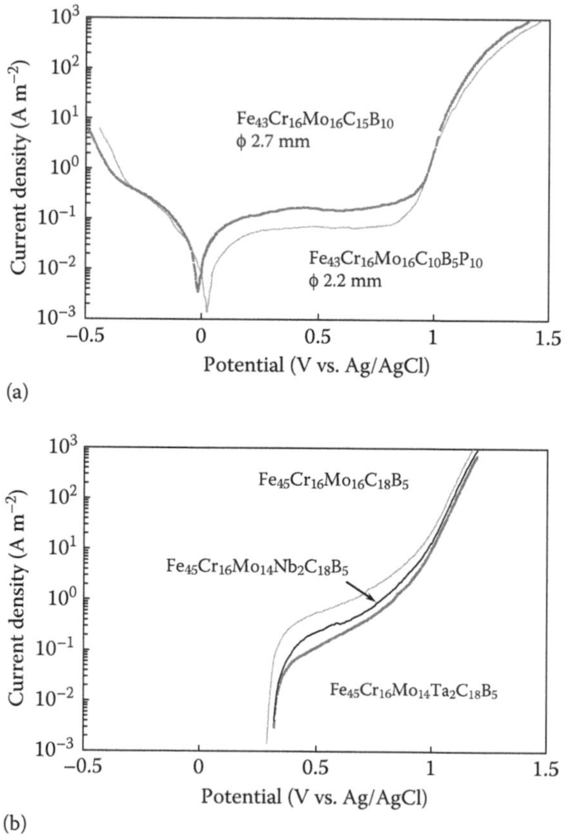

FIGURE 7.15
(a) Potentiodynamic polarization curves of $Fe_{43}Cr_{16}Mo_{16}C_{15}B_{10}$ and $Fe_{43}Cr_{16}Mo_{16}C_{10}B_5P_{10}$ bulk glassy alloys in 1 N HCl solution open to air at 298 K, showing that the passive current density for the P-containing alloy is lower. (Reprinted from Pang, S.J. et al., *Acta Mater.*, 50, 489, 2002. With permission.) (b) Anodic polarization curves of $Fe_{45}Cr_{16}Mo_{16}C_{18}B_5$ and $Fe_{45}Cr_{16}Mo_{14}M_2C_{18}B_5$ (where M = Nb or Ta) BMG alloys in 6 N HCl solution open to air at 298 K. (Reprinted from Pang, S.J. et al., *Mater. Trans.*, 42, 376, 2001. With permission.) Note the lower passive current density when Nb or Ta is present in the alloy.

resistance of $Fe_{43}Cr_{16}Mo_{16}C_{10}B_5$ and $Fe_{43}Cr_{16}Mo_{14}M_2C_{10}B_5$ (where M = Nb or Ta) BMG alloys in 6N HCl solution has also been obtained by substituting Mo with Nb or Ta [56].

The presence of N in the $Fe_{49}Cr_{15.3}Mo_{15}Y_2C_{15}B_{3.4}N_{0.3}$ BMG alloy demonstrated that the corrosion resistance of the N-containing alloy in concentrated HCl solution was at least an order of magnitude higher than that of the N-free alloy. XPS analysis indicated that the enrichment of Cr oxide and the presence of MoN nitride in the passive surface layer were responsible for the increased corrosion resistance [62].

Wang et al. [63] have investigated the effect of Y addition on the corrosion behavior of $Fe_{50-x}Cr_{15}Mo_{14}C_{15}B_6Y_x$ bulk glassy alloys in the range of x = 0–2 at.%. Using potentiodynamic polarization measurements, Mott–Schottky analysis, and XPS studies, the authors showed that i_{corr} increases with Y addition up to 0.5 at.% Y and then starts to decrease with further increase in the Y content. From this observation, they concluded that the stability of the passive film decreased sharply at 0.5 at.% Y addition. It was further shown that the corrosion resistance of the alloy is related to the semiconducting properties of the alloy, since Y doping induced changes in the structural defects in the passive films.

7.4.2 Influence of Test Environment

Different test environments have been explored to evaluate the corrosion behavior of Fe-based BMG alloys. These include the nature and concentration of the corrosive medium and the temperature at which corrosion was studied. The corrosive media investigated include HCl, NaCl, H_2SO_4, and Na_2SO_4.

It is natural to expect that corrosion becomes more severe as the concentration of the corrosive medium increases. Consequently, as shown in Figure 7.16a, the corrosion rate of the $Fe_{50-x}Cr_{16}Mo_{16}C_{18}B_x$ BMG alloy increased with increasing concentration of HCl, irrespective of the B content in the alloy [57]. It was also shown that pitting occurred on the surface of the alloy after immersion for 1 week in 12N HCl at room temperature, especially when the B content was only about 4 at.%; pitting did not occur at higher B levels. Such a phenomenon of pitting did not occur in 1N and 6N HCl solutions; instead they passivated spontaneously. Asami et al. [54] also reported that while the $Fe_{43}Cr_{16}Mo_{16}C_{10}B_5P_{10}$ BMG alloy passivated spontaneously in a 1N HCl solution, it did not have a stable passive region, with its current increasing drastically when anodically polarized in a 12N HCl solution.

A similar result of lowered passive current density in P-containing alloys was also reported at higher concentrations of HCl and also on partially replacing Mo with either Nb or Ta. Increasing B content in $Fe_{50-x}Cr_{16}Mo_{16}C_{18}B_x$ alloys was shown to have a similar effect (Figure 7.16b) [57].

Long et al. [67] reported that the corrosion resistance of $[(Fe_{0.6}Co_{0.4})_{0.75}B_{0.2}Si_{0.05}]_{96}Nb_4$ BMG alloys decreased with an increase in temperature in 0.1N and 0.5N H_2SO_4 solutions.

7.4.3 Influence of Structural Changes

The influence of structurally relaxing or crystallizing the BMG alloys on the structure and corrosion behavior of Fe-based BMG alloys has also been investigated. Pardo et al. [68,69] studied the effect of Cr content on the corrosion behavior of $Fe_{73.5}Si_{13.5}B_9Nb_3Cu_1$ BMG alloys in different concentrations of H_2SO_4 (1, 3, and 5N). They investigated the corrosion behavior of this alloy in three different conditions: (1) in the as-solidified fully glassy condition, (2) by annealing it for 1h at 813 K to obtain a nanocrystalline structure (10–15 nm grain size), and (3) by fully crystallizing the samples (to achieve a grain size of 0.1–1 μm) through annealing at 973 K for 1h. The corrosion resistance was higher with increasing

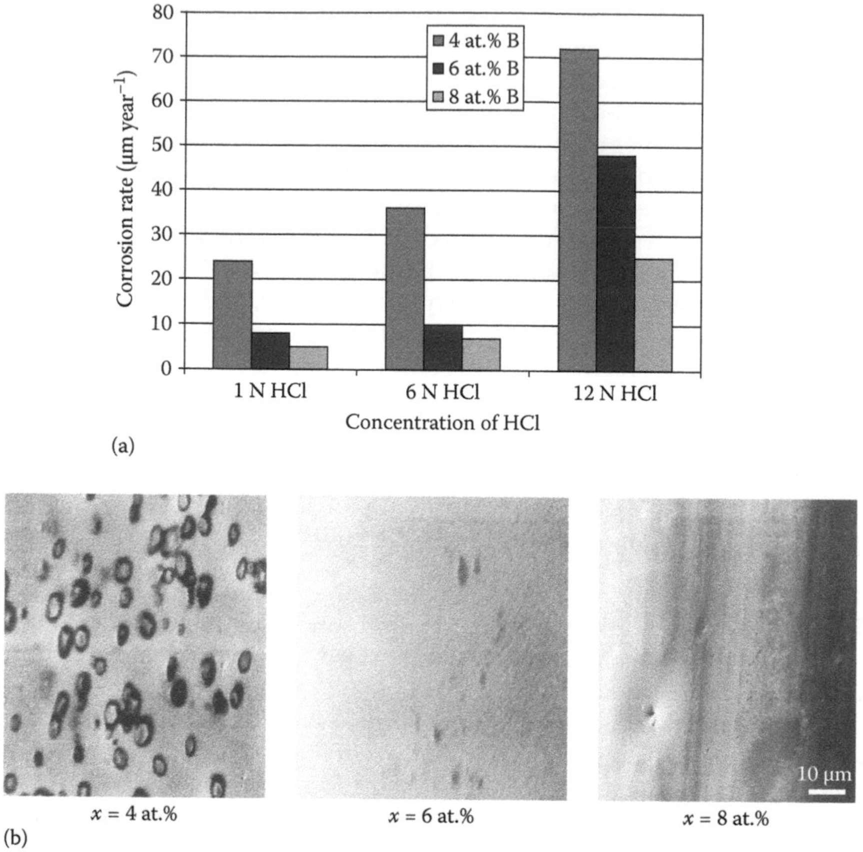

FIGURE 7.16
(a) Increasing corrosion rate with increasing concentration of HCl in an $Fe_{50-x}Cr_{16}Mo_{16}C_{18}B_x$ BMG alloy. (b) SEM micrographs showing that pitting occurs in a 12 N HCl solution, especially when the B content is less. Lower concentrations of HCl and higher concentrations of B prevented this from happening. (Reprinted from Pang, S.J. et al., *Corrosion Sci.*, 44, 1847, 2002. With permission.)

Cr content in the range studied (0–8 at.%). However, a minimum Cr concentration of 8 at.% was found necessary to generate a stable passive layer. Among all the conditions studied, the glassy structure showed the best corrosion resistance, followed by the nanocrystalline state. The fully crystallized alloy showed the least corrosion resistance. Figure 7.17 shows the weight loss of this alloy as a function of immersion time in 5 N H_2SO_4 solution, confirming this finding [68]. These authors had also investigated the corrosion behavior of a $Co_{73.5}Si_{13.5}B_9Nb_3Cu_1$ metallic glassy alloy, for which similar results of best corrosion resistance in the glassy state and least corrosion resistance in the fully crystallized condition were reported. They further noted that the corrosion resistance of the Co-based alloys was better than that of the Fe-based alloys [68].

Long et al. [67] also investigated the influence of structural changes on the corrosion behavior of an $[(Fe_{0.6}Co_{0.4})_{0.75}B_{0.2}Si_{0.05}]_{96}Nb_4$ BMG alloy in 0.1 N and 0.5 N H_2SO_4 solutions. They studied the alloy in three different conditions—in the as-cast glassy state, in a structurally relaxed state by annealing the glassy alloy at 573 or 773 K for 1 h, and in the fully crystallized condition by annealing the glassy alloy at 973 K for 1 h.

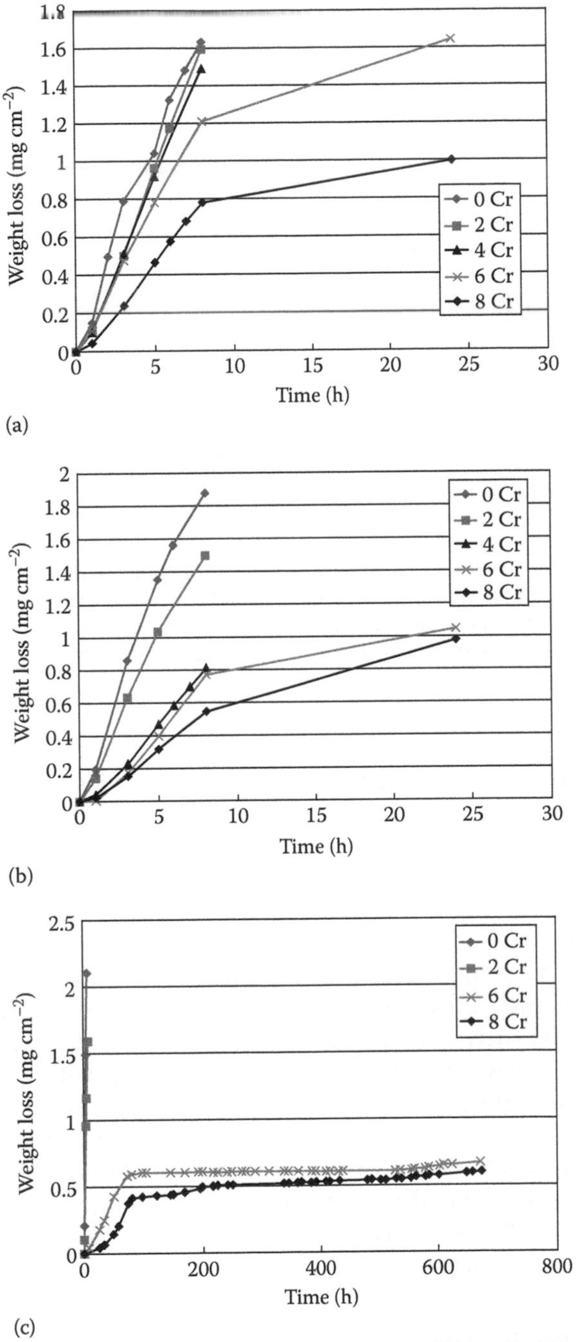

FIGURE 7.17
Weight loss as a function of immersion time in 5N H_2SO_4 for the $Fe_{73.5}Si_{13.5}B_9Nb_3Cu_1$ alloy in the (a) glassy, (b) nanocrystalline, and (c) fully crystallized conditions. The weight loss decreased with increasing Cr content in all cases. It may also be noted that the corrosion rate was the highest in the fully crystalline state, least in the glassy state, and intermediate in the nanocrystalline state. (Reprinted from Pardo, A. et al., *J. Non-Cryst. Solids*, 352, 3179, 2006. With permission.)

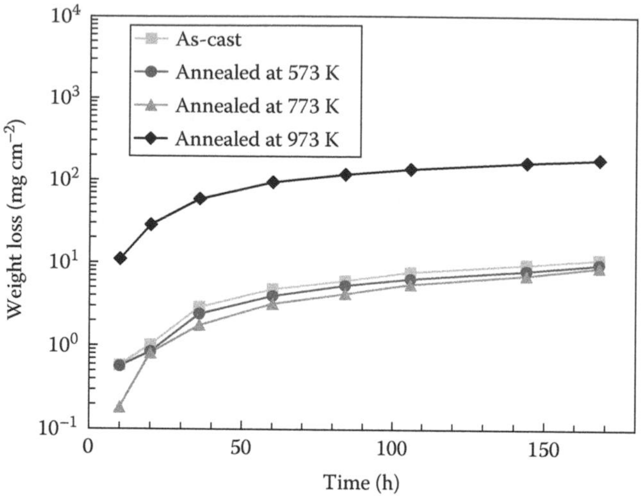

FIGURE 7.18
Weight loss of the $[(Fe_{0.6}Co_{0.4})_{0.75}B_{0.2}Si_{0.05}]_{96}Nb_4$ alloy in 0.1 N H_2SO_4 solution in three different conditions—as-cast glassy, structurally relaxed glassy, and fully crystallized conditions. The corrosion resistance was the lowest (the weight loss was maximum) for the fully crystallized alloy and the highest for the structurally relaxed glass. The as-cast glassy structure had an intermediate value. (Reprinted from Long, Z.L. et al., *Mater. Trans.*, 47, 2566, 2006. With permission.)

XRD patterns from these alloys confirmed that the samples annealed at 573 or 773 K remained glassy, but the glassy structure relaxed, and the alloy annealed at 973 K was fully crystallized. It was reported that the alloy had readily passivated in aerated H_2SO_4 solution in all the three different conditions; but the corrosion resistance of the alloy was different in the three different states. The fully crystallized alloy exhibited the highest corrosion rate, with a weight loss of one order of magnitude greater than in other conditions. The structurally relaxed glassy alloy showed the lowest corrosion rate and the as-cast glassy alloy showed an intermediate value (Figure 7.18). The highest corrosion rate in the fully crystallized alloy was attributed to the generation of galvanic effects between adjacent phases with different compositions. On the other hand, the structurally relaxed glassy alloy was relieved of all the solidification stresses, and so the glass reached a more "stable" structure, thereby causing a decrease in the reactivity of the constituent elements and increasing the chemical stability of the alloy. The smaller weight loss of the alloy annealed at 773 K suggests that annealing at the higher temperature resulted in a greater structural relaxation and consequently led to increased corrosion resistance.

7.4.4 Nature of the Passive Film

Since the increased corrosion resistance is attributed, at least in part, to the passive films formed in aggressive and corrosive environments, the XPS technique was used to characterize the passive films and also the substrate alloy surface. It is noted that the metallic elements exist in the oxidized state in the surface passive film and in the metallic state in the underlying alloy surface. The Fe spectrum is normally composed of peaks of iron species in metallic, Fe^{2+}, and Fe^{3+} states. The Cr spectrum consists of peaks of chromium species in metallic and Cr^{3+} states. The Mo spectrum is generally deconvoluted into peaks corresponding to metallic, Mo^{4+}, Mo^{5+}, and Mo^{6+} states. The C 1s peaks appear from carbon in the alloy, and the so-called contaminant carbon on the top surface of the specimen. The O 1s spectrum consists of peaks originating from oxygen in the metal–O–metal bond, metal–OH bond, and/or bound water vapor.

Assuming that the metallic state peaks arise from the underlying metal surface and the oxidized state peaks from the surface film, the XPS spectra are analyzed for the concentration of the different elements by deconvoluting and curve fitting. This is done both in the as-prepared condition and also after immersion in the corrosive medium. Figure 7.19 shows the metallic and cationic fractions of Fe, Cr, and Mo in an $Fe_{50-x}Cr_{16}Mo_{16}C_{18}B_x$ glassy alloy as a function of the B content in the as-cast condition and after immersion in the corrosive medium [57]. While the atomic fractions of the transition metals, that is, Fe, Cr, and Mo, were measured in the underlying substrate alloy for the cast condition (as-prepared, i.e., exposed to air after mechanical polishing), the cationic fraction was measured in the surface film after immersion in 1 N HCl at 298 K for 168 h (1 week). The superscripts "m" and "ox" represent the metallic and oxidized states, respectively. Comparing Figure 7.19a through d, between the two conditions, it can be seen that the Cr concentration in the surface film is apparently higher than that in the air-formed film.

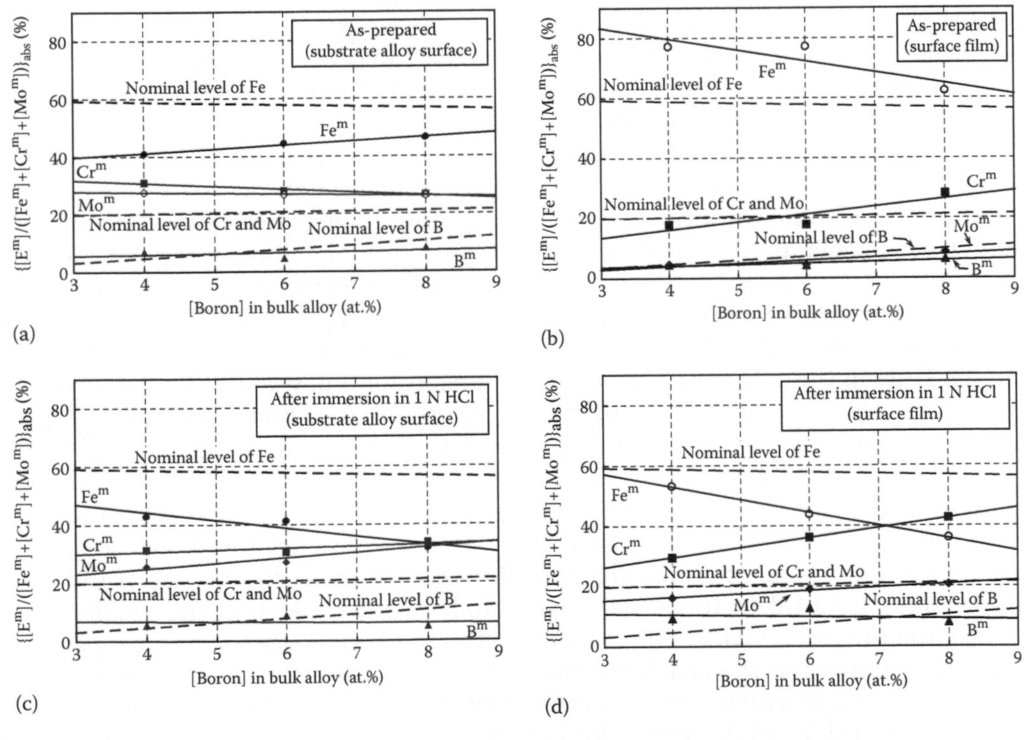

FIGURE 7.19
Atomic fractions of the transition metals Fe, Cr, and Mo in the underlying substrate alloy (a, b) and the cationic fraction in the surface film (c, d) as a function of boron content for the glassy $Fe_{50-x}Cr_{16}Mo_{16}C_{18}B_x$ alloys. (a, b) As-prepared, that is, exposed to air after mechanical polishing and (c, d) after immersion in 1 N HCl at 298 K for 168 h. The superscripts m and ox represent the metal surface and oxide surface, respectively. (Reprinted from Pang, S.J. et al., *Corrosion Sci.*, 44, 1847, 2002. With permission.)

This suggests that the passive film is enriched with Cr. It is also inferred that Fe from the air-formed film is preferentially dissolved back into the substrate and that the passive film is regrown with Cr enrichment. But, judging from the comparison between the compositions before and after immersion, the underlying alloy must have dissolved to some extent at the same time.

The thickness of the passive surface film is also an indicator of the corrosion resistance of the alloy. In general, a thick surface film is formed when the alloy is not resistant to the environment. On the other hand, the film is thin when the corrosion resistance of the alloy is higher. Accordingly, the thickness of the surface film formed on the bulk $Fe_{50-x}Cr_{16}Mo_{16}C_{18}B_x$ alloys after immersion in the acid ranged from 20 to 25 nm, which is almost the same as that of the air-formed film [57]. But, it does not mean that the air-formed surface film is responsible for the high corrosion resistance of the alloy. It is the Cr-enriched passive film that has been found to be responsible for the enhanced corrosion resistance.

7.4.5 Bioactivity

Fe-based BMGs can be formed in a Fe–B–Si–Nb system by copper mold casting [70]. It was noted that even though the corrosion resistance of Fe–Si–B alloys without glass transition was rather poor, the addition of small amounts of Nb were shown to increase the corrosion resistance, implying that the corrosion behavior can be controlled by the Nb content [55]. Considering that Fe–Si–B amorphous alloys are very inexpensive and do not contain any toxic elements, it is expected that Fe–Si–B–Nb BMGs can be used as bioactive materials. It is thus important to examine the formation tendency of a hydroxyapatite phase of Fe–Si–B and Fe–Si–B–Nb glassy alloys in Hank's solution.

Qin et al. [71] demonstrated, for the first time, that the $(Fe_{0.75}B_{0.15}Si_{0.1})_{100-x}Nb_x$ (x = 0, 1, and 3 at.%) metallic glasses without toxic and allergic elements exhibit excellent apatite-forming ability in simulated body fluids (SBF). Thus, these alloys are expected to be a new generation of biomaterials in stents and orthopedic implants. For the alloys without any surface treatment, spherical particles corresponding to octacalcium phosphate nucleate spontaneously and precipitate throughout the alloy surface after immersion for only 1 day, indicating that these alloys possess an unusually high bioactivity. During the subsequent in-vitro immersion for 3 days, SEM images shown in Figure 7.20 reveal the typical "cauliflower" morphology of bone-like hydroxyapatite with a Ca/P ratio of 1.65. In addition, it is surprising to find that the *in vitro* SBF immersion not only leads to the formation and growth of the apatite layer but also causes the progressive development of the underlying alloy substrate. Moreover, for the alloys immersed for 3–9 days, the substrate alloy just beneath the apatite layer consists of a hierarchical nano/macroporous structure through selective dissolution of the active components Fe and B in the surface. XPS analysis indicated that the apatite nucleation on the present alloys in SBF is attributed to the specific dissolution properties of the present alloys and the fast formation of Si–OH and Fe–OH or Nb–OH functional groups, followed by combination of these groups with Ca^{2+} and phosphate ions. The schematic illustration of the apatite formation process on the Fe–B–Si surface in an SBF is shown in Figure 7.21. Once the apatite nuclei are formed, they can grow spontaneously by consuming calcium and phosphate ions in the surrounding environment. As a result, the apatite nucleation on the Fe–B–Si base alloy in SBF was attributed to the fast formation of Si–OH and Fe–OH groups, followed by combination of these groups with Ca^{2+} and phosphate ions.

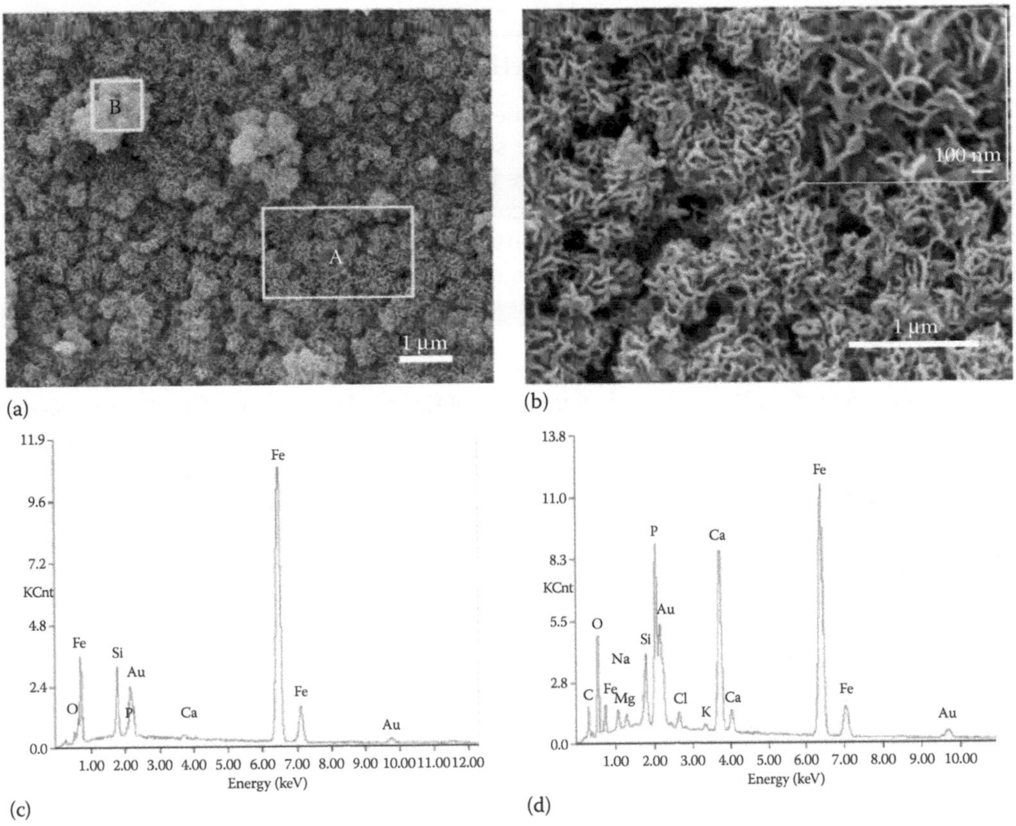

FIGURE 7.20
SEM images of (a, b) the peeled substrate alloy surface of $Fe_{75}B_{15}Si_{10}$ alloy immersed in SBF for 9 days at different magnifications, and (c, d) EDS spectra of the areas A and B, respectively, in (a).

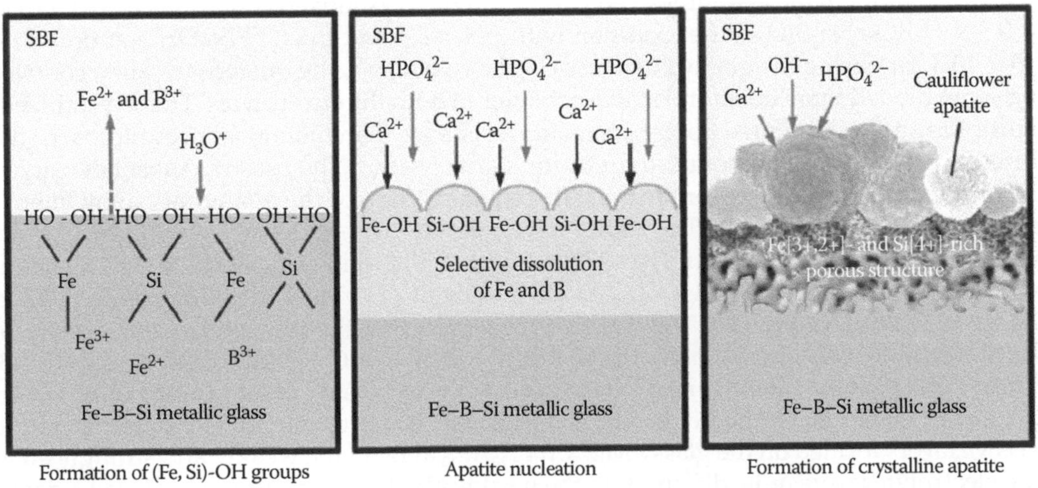

FIGURE 7.21
Schematic illustration of apatite formation process on the Fe–B–Si surface in SBF.

7.5 Magnesium-Based Bulk Glassy Alloys

The pure metal magnesium and its crystalline alloys are generally known to display poor corrosion resistance. In fact, even in rapidly solidified binary Mg alloys (which are crystalline), lack of sufficient corrosion resistance has been one of the major limitations in the commercial exploitation of these alloys. In contrast to the other alloy systems, the corrosion behavior of Mg-based BMG compositions has been investigated mostly on melt-spun ribbon samples. This has been done with the assumption and understanding that there is no difference in the corrosion behavior of melt-spun ribbons and slowly cast rod samples of the same composition as long as both the samples are fully glassy.

A majority of the corrosion behavior studies in Mg-based BMGs have been conducted on the ternary $Mg_{65}Cu_{25}Y_{10}$ and its derivative containing Ag, i.e., $Mg_{65}Cu_{15}Y_{10}Ag_{10}$. Gebert et al. [72] investigated the corrosion behavior of $Mg_{65}Cu_{25}Y_{10}$ in 0.1 N NaOH solution with pH = 13 at room temperature. The alloy was prepared in the form of melt-spun ribbons and also as 7 mm diameter rods by copper mold casting. The melt-spun ribbon was mostly glassy, but it contained a minor amount of Y_2O_3 embedded as small particles in the glassy matrix. On the other hand, the as-cast BMG alloy sample exhibited a very fine-grained crystalline structure consisting of Mg_2Cu, MgY, and other phases. The corrosion behavior of these alloys was compared with the behavior of pure Mg.

Significant differences were noted in the electrochemical behavior of these three materials. The melt-spun ribbon and the multiphase heterogeneous crystalline alloy showed significantly lower oxidation tendencies than pure magnesium, ascribed to the presence of Cu, a noble element. The lowest corrosion rate and lowest passive current density were observed for the "glassy" ribbon sample. These observations suggest that a highly protective surface film had formed on the "glassy" ribbon. The presence of structural and chemical heterogeneity, resulting in the setting-up of galvanic currents, was thought to be responsible for the higher corrosion rate and the worse passivation behavior of the crystalline alloy.

A comparison of the corrosion behavior between the glassy and crystalline alloys of the ternary $Mg_{65}Cu_{25}Y_{10}$ and the quaternary $Mg_{65}Cu_{15}Y_{10}Ag_{10}$ alloys was also reported [73]. These two alloys in the as-synthesized glassy and the crystallized conditions were studied in 0.3 N $H_3BO_3/Na_2B_4O_7$ buffer solution with pH = 8.4 and in 0.1 N NaOH solution with pH = 13. Comparing the glassy structures of the two alloys, the quaternary alloy containing Ag exhibited more noble and stable behavior in both the electrolytes. The (crystallized) multiphase alloys exhibited poorer corrosion resistance than their glassy counterparts, the difference being more pronounced for the quaternary alloy. The passive current density of the ternary alloy was lower for the crystallized alloy than for the quaternary alloy in both the electrolytes. But, for the glassy alloys, the passive current density was higher for the ternary alloy in both the electrolytes [73]. Gebert et al. [74] investigated the electrochemical behavior of the ternary $Mg_{65}Cu_{25}Y_{10}$ BMG alloy at pH values ranging from 6 to 13 in the 0.3 N $H_3BO_3/Na_2B_4O_7$ electrolyte. They observed that the passive current density of the glassy alloy was always lower in the ternary alloy at all pH values except at pH = 13. Further, the passive potential range decreased with decreasing pH and was not observed at all at pH = 5, where anodic dissolution processes were dominating. Additionally, the passive layers formed on the glassy $Mg_{65}Cu_{15}Y_{10}Ag_{10}$ alloy in the weakly and strongly alkaline electrolytes were quite different. In the weakly alkaline electrolyte, Cu played a dominant role in the passivation of the quaternary alloy along with Y. On the other hand, in the highly alkaline electrolyte, magnesium hydroxide, along with yttrium hydroxide and silver oxides formed the thin homogeneous passive layer. It was again confirmed that the glassy alloy had better corrosion resistance than the crystalline alloy.

Yao et al. [75] investigated the corrosion behavior of melt-spun ribbons of pure Mg, binary $Mg_{82}Ni_{18}$ and $Mg_{79}Cu_{21}$, and ternary $Mg_{65}Ni_{20}Nd_{15}$ and $Mg_{65}Cu_{25}Y_{10}$ alloys in 0.01 N NaCl (pH = 12) electrolyte. All the alloys in the melt-spun condition were reported to be glassy. Both the ternary alloys were shown to exhibit significantly higher corrosion resistance in strongly alkaline hydroxide electrolyte than pure Mg or the binary glassy alloys. Electrochemical investigations revealed that the anodically formed passive films containing NiO and Nd_2O_3 (in $Mg_{65}Ni_{20}Nd_{15}$) and CuO and Y_2O_3 (in $Mg_{65}Cu_{25}Y_{10}$), in addition to MgO, improved the corrosion resistance of the ternary alloys. The rare earth oxides were shown to be particularly useful. Hydroxydation of these oxides, however, led to the breakdown of passivity.

Magnesium in the crystalline state has been known to be a biodegradable material; but its corrosion is accompanied by hydrogen gas evolution, which is problematic in many biomedical applications. The scope for control of degradation and hydrogen gas evolution in crystalline Mg alloys by alloying is limited. On the other hand, the hydrogen gas evolution in a Zn-rich $Mg_{60+x}Zn_{35-x}Ca_5$ $(0 \leq x \leq 7)$ glass was reported to be significantly less [76]. It was also shown that above ~28 at.% Zn, a Zn- and oxygen-rich passivating layer formed on the alloy surface. Clinical trials with animals clearly showed that in addition to a reduction in hydrogen gas evolution, the glassy alloy also exhibited the same good tissue compatibility as the crystalline implants.

7.6 Nickel-Based Bulk Metallic Glasses

Nickel-based crystalline alloys are known to be very resistant to both general and localized corrosion, even in highly concentrated HCl solutions. Since metallic glasses are in general expected to be more resistant to corrosion than their crystalline counterparts, study of the corrosion behavior of Ni-based metallic glasses assumes importance. Further, it has been possible to produce both metal–metal- and metal–metalloid-type Ni-based alloys in the glassy state. However, the majority of these alloys have been produced in a ribbon form by the melt-spinning process and only a few of the Ni-based alloys have been produced in the BMG condition. The diameters of many of the fully glassy alloys have been only 1 or 2 mm, and the maximum diameter of the Ni-based BMG alloy reported so far has been 21 mm in $Ni_{50}Pd_{30}P_{20}$ [77]. Consequently, the majority of the corrosion studies on Ni-based glasses have been conducted on melt-spun ribbons of BMG-forming compositions and not many have examined the BMG rods. Comparative studies of the corrosion behavior of these alloys in the glassy condition and after crystallization have also been made.

An Ni–Cr–P–B alloy has been chosen as the basis and alloying additions of other metals, such as Mo, Nb, and Ta, have been made to it. Among the metal–metalloid-type alloys, the Ni-based alloys investigated for corrosion behavior include Ni–Cr–Mo–16 at.% P–4 at.% B [78], Ni–Cr–Ta–16 at.% P–4 at.% B [78], Ni–Cr–Nb–16 at.% P–4 at.% B (with Cr, Nb = 5 or 10 at.%) [79], $Ni_{55}Nb_{40-x}Ta_xP_5$ [80], $Ni_{70-x}Cr_5Ta_5Mo_xP_{16}B_4$ (x = 3 and 5 at.%) [81], and $(Ni_{60}Nb_{40-x}Ta_x)_{0.95}P_{0.05}$ [82]. Several metal–metal-type glassy alloys based on Ni were also studied for their corrosion behavior. These include $Ni_{59}Zr_{20}Ti_{16}Si_2Sn_3$ [83], $Ni_{53}Nb_{20}Ti_{10}Zr_8Co_6Cu_3$ [83], $Ni_{60-x}Co_xNb_{20}Ti_{10}Zr_{10}$ (with x = 0–20 at.%) [84], $Ni_{60}Nb_{20-x}Ta_xTi_{15}Zr_5$ (with x = 5, 15, and 20 at.%) [85], $Ni_{60}Ta_{40-x-y}Ti_xZr_y$ (with x = 0–25 at.% and y = 0–15 at.%) [86], $Ni_{60-x}Co_xNb_{20}Ti_{10}Zr_{10}$ [87,88], $Ni_{80-x}Pd_xP_{17}B_3$ (x = 10–30 at.%) [89,90] and $Ni_{80-x}Pd_xP_{16}B_4$ (x = 10–30 at.%) [89,90].

Since Ni-based glassy alloys are generally highly resistant to corrosion, their corrosion behavior has been investigated in concentrated acids such as 6 N HCl and 12 N HCl. The majority of the studies were conducted at 303 K and in electrolytes open to air.

7.6.1 Influence of Composition

Addition of some specific elements has been known to generally improve the corrosion resistance. Such investigations were conducted earlier on a number of melt-spun ribbon alloys (see, e.g., Ref. [91]). Similar results are also reported in the case of BMG alloys.

Addition of Ta to an Ni–Nb–P glassy alloy reduced the corrosion rate in a 6 N HCl solution and the corrosion rate of an $(Ni_{60}Nb_{40-x}Ta_x)_{95}P_5$ alloy with x = 5 was lower than the detectable value [80]. Glassy alloys with x > 20 at.% Ta have been reported to be immune to corrosion even in 12 N HCl at 303 K. It was further reported that during potentiodynamic polarization measurements, the curves showed wider passive regions and lower current densities with increasing Ta content in the alloy. Spontaneous passivation took place in alloys with high Ta content of 30 and 40 at.%. Figure 7.22 shows the corrosion rates of the ternary Ni–Nb–P and quaternary Ni–Nb–Ta–P alloys of 1 and 2 mm diameter and melt-spun ribbons measured in 6 N and 12 N HCl open to air at 303 K. The corrosion rate of the $Ni_{55}Nb_{40}P_5$ alloy of 2 mm diameter was about three orders of magnitude lower than that of Ni metal and is less than half of the Nb metal. It may also be noted that the corrosion rate decreases with Ta addition.

The corrosion behavior of Ni–Ta–P–B alloys with further additions of Cr and Mo was also investigated in 6 N HCl solution at 303 K open to air [78]. Potentiodynamic polarization measurements were made on 1 mm diameter fully glassy rods of $Ni_{75-x}Cr_xTa_5P_{16}B_4$

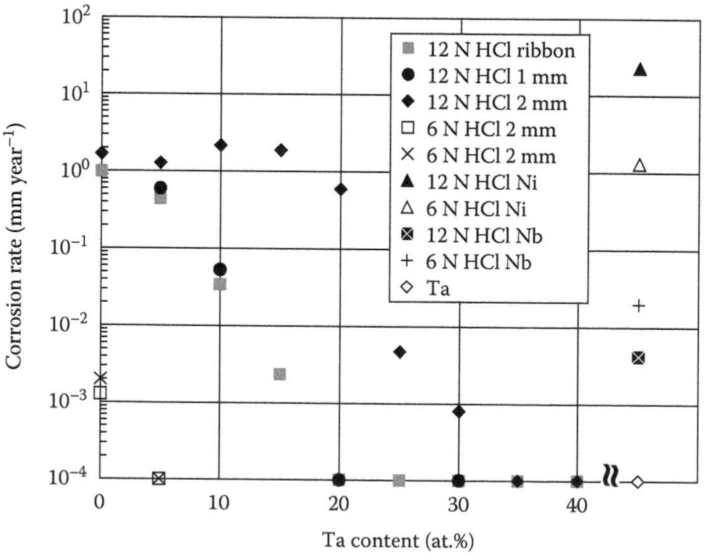

FIGURE 7.22
Corrosion rates of the bulk $Ni_{55}Nb_{40}P_5$ and $(Ni_{60}Nb_{40-x}Ta_x)_{95}P_5$ (x = 5–40 at.%) alloys of 1 and 2 mm diameter in 12 N HCl, open to air at 303 K, as a function of Ta content. The results of 6 N HCl open to air at 303 K are labeled in the figure. (Reprinted from Kawashima, A. et al., *Mater. Sci. Eng. A*, 304–306, 753, 2001. With permission.)

(with $x = 5$, 10, and 15 at.%). It was reported that all the alloys passivated spontaneously in 6 N HCl at 303 K. The alloy with 5 at.% Cr showed a passive current density of one order of magnitude higher than that exhibited by the melt-spun ribbon, attributed to the presence of crystalline 2 nm diameter Ni precipitates in the sample. However, on increasing the Cr content further, the behavior was identical to that of the melt-spun ribbon. The increased corrosion resistance was attributed to the enormous enrichment of Cr and Ta in the stable passive surface film with a formula of $(Cr,Ta)_xO_y(OH)_z$.

The cylindrical bulk samples of $Ni_{60}Pd_{20}P_{17}B_3$, $Ni_{65}Pd_{15}P_{17}B_3$ and $Ni_{70}Pd_{10}P_{16}B_4$ consist of a single glassy phase that was evident from a main halo peak without crystalline peaks in their x-ray diffraction (XRD) patterns in the diameter range up to at least 15 mm [89,92]. Figure 7.23 shows the average corrosion rates of the Ni–Pd–P–B rod samples in 1 N H_2SO_4 and 3 wt.% NaCl solutions for 168 h (7 days) open to air at 298 K [90]. In 1 N H_2SO_4 solution, the $Ni_{60}Pd_{20}P_{17}B_3$ alloys exhibited a low corrosion rate of 0.0058 mm year^{-1}. With increasing Ni content in the alloys up to 65 or 70 at.%, the corrosion rates of the $Ni_{65}Pd_{15}P_{17}B_3$ and $Ni_{70}Pd_{10}P_{16}B_4$ alloys increased to 0.039 and 0.072 mm year^{-1}, respectively. On the other hand, after immersion in neutral 3 wt.% NaCl, which contains about 0.5 N Cl^-, no weight loss was detected for the $Ni_{60}Pd_{20}P_{17}B_3$ glassy alloy, suggesting that the corrosion rate of the alloy was less than 1×10^{-3} μm year^{-1}. Meanwhile, it also appears that the average corrosion rates of the $Ni_{65}Pd_{15}P_{17}B_3$ and $Ni_{70}Pd_{10}P_{16}B_4$ increased with increasing Ni content in the alloys. Thus, in both solutions, their corrosion resistance decreased with increasing Ni content in the alloys. Among the present glassy alloys, the $Ni_{60}Pd_{20}P_{17}B_3$ glassy alloy demonstrates the highest corrosion resistance in the strong acidic and chloride-ion-containing solutions.

SEM micrographs of the specimens immersed in 3 wt.% NaCl solution for 7 days are shown in Figure 7.24 [90]. The $Ni_{60}Pd_{20}P_{17}B_3$ alloy still keeps the previous metallic luster, and almost no changes in its surface were seen before and after immersion (not shown here), indicating its high resistance to localized corrosion, in agreement with the

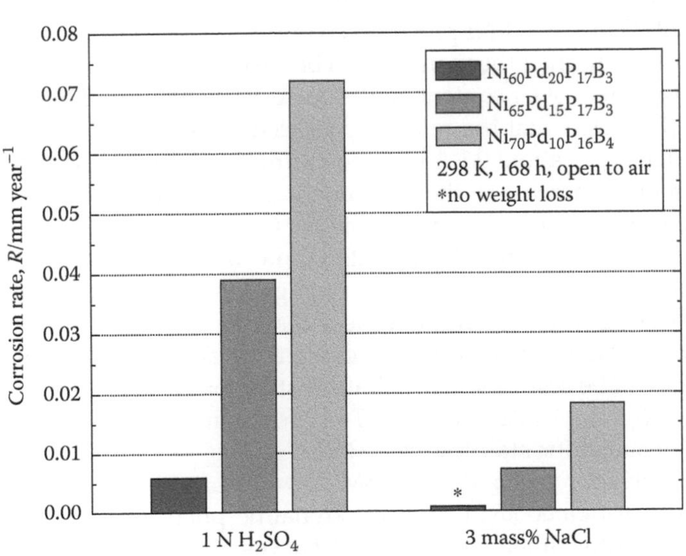

FIGURE 7.23
Average corrosion rates of the Ni–Pd–P–B bulk glassy alloys in 1 N H_2SO_4 and 3 wt.% NaCl solutions at 298 K open to air.

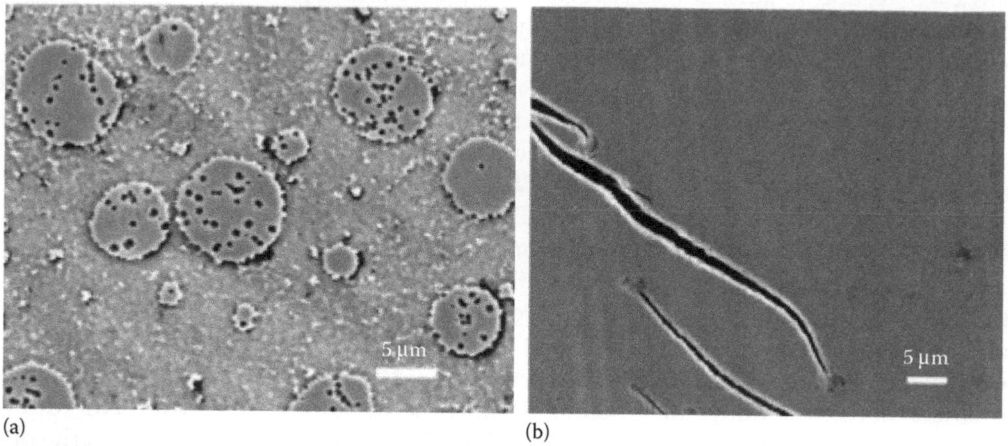

FIGURE 7.24
SEM micrographs of the surfaces of the (a) $Ni_{65}Pd_{15}P_{17}B_3$ and (b) $Ni_{70}Pd_{10}P_{16}B_4$ glassy rods after immersion in 3 wt.% NaCl solution for 7 days.

corrosion rate of the $Ni_{60}Pd_{20}P_{17}B_3$ alloy. The $Ni_{65}Pd_{15}P_{17}B_3$ alloy after immersion suffers uneven chloride-induced localized corrosion products (Figure 7.24a). When the Ni content in the alloy was increased to 70 at.%, several flaws were observed in the $Ni_{70}Pd_{10}P_{16}B_4$ alloy surface (Figure 7.24b), confirming that the alloy had suffered crevice corrosion in chloride-containing solution.

The corrosion behavior of melt-spun ribbons and 1 and 2 mm diameter BMG rods of $Ni_{55}Cr_{15}Mo_{10}P_{16}B_4$ composition was investigated through potentiodynamic polarization curves [78]. It was noted that, whereas the melt-spun ribbon and 1 mm diameter rod showed similar polarization curves, the 2 mm diameter rod showed a passive current density that was two orders of magnitude higher (Figure 7.25). This difference was attributed to the presence of 20 nm-sized Ni precipitates in the glassy matrix of the 2 mm diameter rod and their absence in the other samples (to be further discussed later).

Investigating the effect of Cr content on the corrosion behavior of Ni–Cr–Nb–P–B alloys, Habazaki et al. [79] noted that the relatively Cr-rich $Ni_{65}Cr_{10}Nb_5P_{16}B_4$ revealed the lowest passive current density of about 0.1 A m^{-2}, indicating that an increase in the Cr content was more effective than an increase in the Nb content in enhancing the stability of the passive film. A similar result was also obtained in binary Cr–Nb alloys [93].

Ni-based metal–metal-type BMG alloys also showed high corrosion resistance in H_2SO_4 and HCl corrosion media as suggested by negligible (and below the level of detection) weight loss, indicating a corrosion rate of less than 0.1 μm $year^{-1}$. All the alloys are also spontaneously passivated with low passive current densities of the order of 10^{-2} A m^{-2} in a wide passive region. The results suggest that substitution of Ta for Nb improves the corrosion resistance of $Ni_{60}Nb_{20}Ti_{15}Zr_5$ alloy [85]. The corrosion resistance of these alloys is much better than that of austenitic stainless steel 316L.

The corrosion behavior of Ni–Pd–P–B BMGs with large maximum diameter of centimeter class was investigated using potentiodynamic polarization measurements [89,90]. Figures 7.26 and 7.27 show anodic polarization curves of the Ni–Pd–P–B glassy rods in 1 N H_2SO_4 and 3 wt.% NaCl solutions, respectively, open to air at 298 K. In 1 N H_2SO_4 solution, the $Ni_{60}Pd_{20}P_{17}B_3$ and $Ni_{65}Pd_{15}P_{17}B_3$ glassy alloys passivated spontaneously with a significantly low current density of the order of 10^{-3} to 10^{-2} A m^{-2}. The alloy with 60 at.% Ni showed

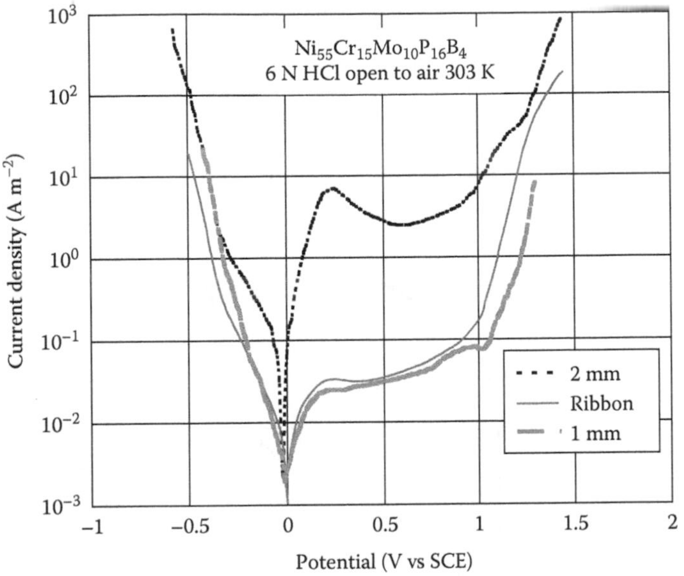

FIGURE 7.25
Potentiodynamic polarization curves of the bulk (1mm and 2mm diameter) and melt-spun ribbons of $Ni_{55}Cr_{15}Mo_{10}P_{16}B_4$ alloy specimens measured in 6N HCl solution open to air at 303 K. (Reprinted from Habazaki, H. et al., *Mater. Sci. Eng. A*, 304–306, 696, 2001. With permission.)

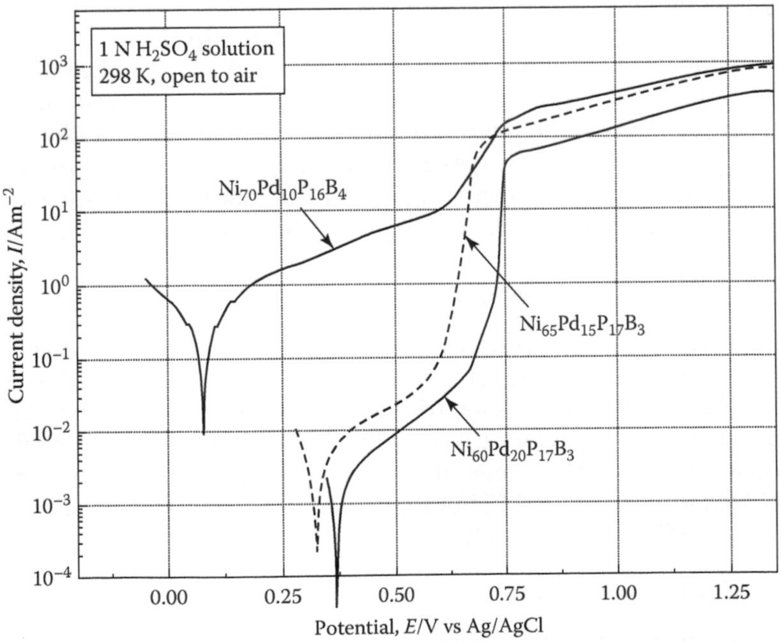

FIGURE 7.26
Anodic polarization curves of the Ni–Pd–P–B bulk glassy alloys in 1 N H_2SO_4 solution open to air at 298 K.

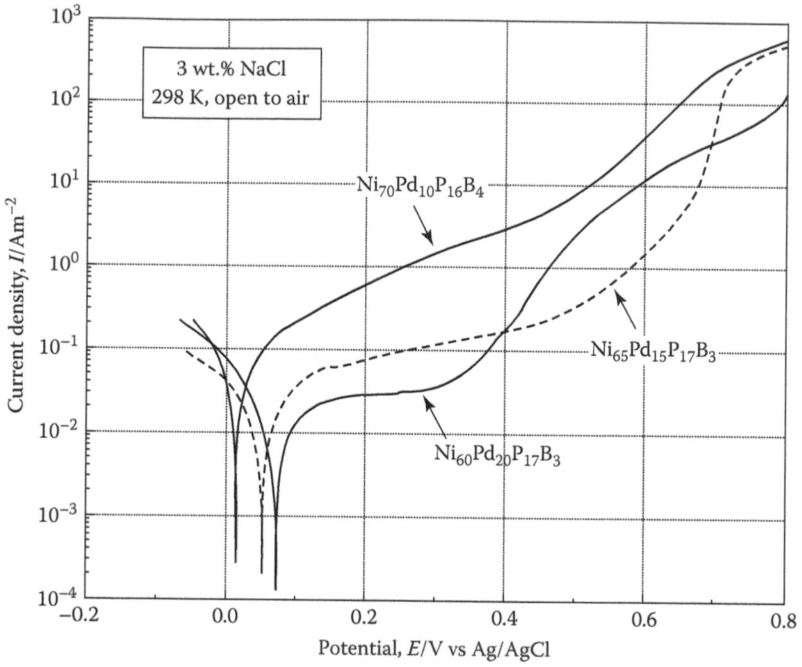

FIGURE 7.27
Anodic polarization curves of the Ni–Pd–P–B glassy alloys in 3 wt.% NaCl solution open to air at 298 K.

a lower passive current density at all potentials. The current density for the $Ni_{70}Pd_{10}P_{16}B_4$ alloy was much higher than that of the $Ni_{60}Pd_{20}P_{17}B_3$ and $Ni_{65}Pd_{15}P_{17}B_3$ glassy alloys. In addition, the open circuit potentials of the $Ni_{60}Pd_{20}P_{17}B_3$, $Ni_{65}Pd_{15}P_{17}B_3$, and $Ni_{70}Pd_{10}P_{16}B_4$ alloys were about 0.37, 0.33, and 0.077 mV (versus Ag/AgCl), respectively. It is clearly seen that the open-circuit potentials of the alloys were nobler with an increase in the Pd content. On the other hand, in 3 wt.% NaCl solution, spontaneous passivation took place for all the glassy alloys. Increasing the Ni content in the alloys resulted in an increase in the passive current densities of the present glassy alloys. As it will be shown later, the corrosion resistance of these glassy alloys decreases with increasing Ni content, due to a decrease in the Pd content, suggesting that Pd is the key element for the protection of alloys against corrosion.

7.6.2 Influence of Test Environment

As mentioned earlier, higher corrosion rates can be expected when the concentration of the acid is higher. Accordingly, the corrosion rate of the alloy in 12 N HCl is expected to be higher than in 6 N HCl electrolyte. Katagiri et al. [81] investigated the effect of concentration of HCl on the corrosion behavior of Ni–Cr–Ta–Mo–B glassy alloys through potentiodynamic polarization curves. They noted that the polarization curves were almost identical in 6 and 12 N HCl solutions at 303 K, but the damage to the specimen by transpassivation was milder in 6 N HCl than in 12 N HCl.

7.6.3 Nature of the Passive Film

Based on the corrosion rates and polarization curves of Ni–Pd–P–B BMGs, it was noted that the $Ni_{60}Pd_{20}P_{17}B_3$ BMG demonstrated high corrosion resistance in the H_2SO_4 and NaCl solutions. The reason for its high corrosion resistance can be understood by measuring the surface composition and knowing their chemical states. XPS analyses were performed on the $Ni_{60}Pd_{20}P_{17}B_3$ specimens as polished mechanically or immersed in 1 N H_2SO_4 solution for 7 days. The XPS spectra of the specimens over a wide binding energy region exhibited peaks of Ni 2p, Pd 3d, P 2p, P 2s, B 1s, O 1s, and C 1s. The C 1s spectrum showing a peak at around 285.0 eV came from a contaminant hydrocarbon layer covering the topmost surface of the specimen. The O 1s spectrum was composed of at least two overlapping peaks, which were assigned to OM and OH oxygen. The OM oxygen corresponds to O^{2-} ions in oxyhydroxide and/or oxide, and the OH oxygen is oxygen linked to proton in the film [94]. The XPS spectra of Cl and S arising from the solution species were less than the detectable level. The Ni $2p_{3/2}$ spectrum consisted of two peaks of oxidized state (Ni^{2+}) and metallic state (Ni^m), whose binding energies were approximately 856.0–856.2, and 852.5–852.7 eV, respectively. By comparison with the standard spectra of pure Pd metal, the Pd 3d peaks were confirmed to exhibit a pair of peaks at about 335.2–335.8 and 340.5–341.1 eV, which were identified as Pd metal (Pd^m). A doublet of the P 2p peak was assigned to be metallic phosphorus (P^0) at 129.7–130.3 eV and pentavalent phosphorus (P^{5+}) at 133.0–133.9 eV. Because B 1s and P 2s overlap each other, we cannot get B element content in the surface according to the B 1s peaks. Qin et al. [90] determined the Ni, Pd, and P contents in the surface. The spectrum peaks from alloy constituents were composed of peaks of oxidized states and metallic states; the oxidized states (OX) and metallic states (m) are assigned to signals from the oxidized layer and underlying alloy surface just beneath the oxidized layer, respectively.

After the integrated intensities of the peaks for individual species were obtained, the thickness and composition of the oxidized layer and the composition of the underlying alloy surface (metallic states) were determined quantitatively using a previously proposed method [93,95]. Figure 7.28 shows the cationic fraction of elements ($[M^{ox}]/\{Ni^{ox}\}$; $[Pd^{ox}]$; $[P^{ox}]$) in the oxidized layer and the atomic fraction of elements ($[M^m]/\{Ni^m\} + [Pd^m] + [P^m]$) in the underlying

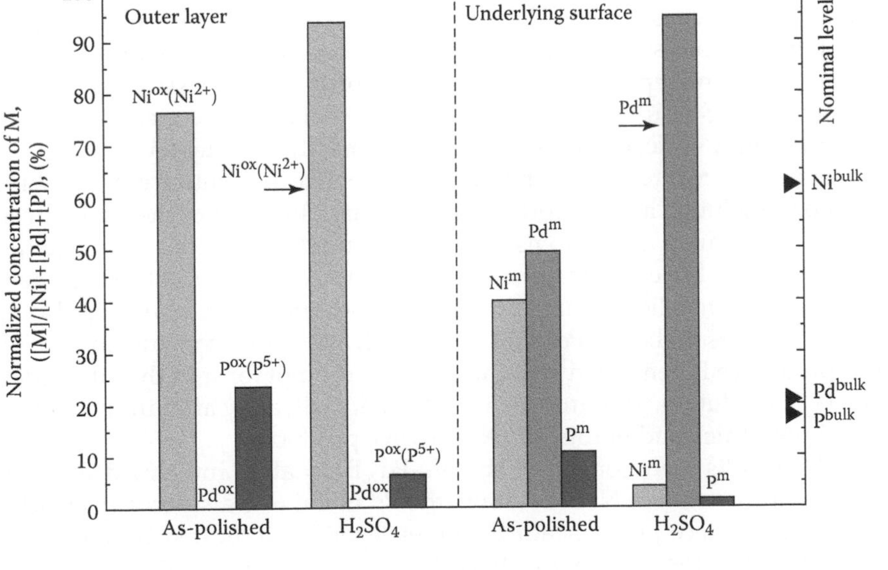

FIGURE 7.28
The cationic fraction of elements ($[M^{ox}]/[Ni^{ox}] + [Pd^{ox}] + [P^{ox}]$) in the oxidized layer and the atomic fraction of elements ($[M^m]/[Ni^m] + \{Pd^m\} + [P^m]$) in the underlying alloy surface just below the oxidized layer for the $Ni_{60}Pd_{20}P_{17}B_3$ alloy exposed to air and that immersed in 1 N H_2SO_4 solution open to air for 7 days after mechanical polishing.

alloy surface just below the oxidized layer for the $Ni_{60}Pd_{20}P_{17}B_3$ alloy exposed to air and a sample immersed in 1N H_2SO_4 solution open to air for 7 days after mechanical polishing. For air-exposed $Ni_{60}Pd_{20}P_{17}B_3$ samples after mechanical polishing, a large amount of Ni^{2+} and P^{5+} were detected in the outer oxidized surface. In addition, the chemical affinity of elemental Pd to oxygen was very low, and the presence of Pd was noted in the oxidized layer. In contrast, the content of Pd in the metallic state increased in the underlying alloy surface just beneath the oxidized layer, while those of metallic Ni and P elements decreased with respect to the alloy composition. The thickness of the outer oxidized surface was about 1.6 nm. Immersion in 1N H_2SO_4 solution for 7 days gave rise to the reconstruction of an alloy surface that was different from the air-formed surface. When the alloy was immersed in the solution, illustrated in Figure 7.28, Pd element in the oxidized state could not be found in the outer layer. In contrast, Ni^{2+} and P^{5+} cations were detected in the outer layer. On the other hand, Qin et al. [90] recognized that the noble metal Pd in the metallic state was remarkably concentrated in the underlying alloy surface, whereas the metallic Ni and P elements were significantly deficient in this surface. The outer layer after immersion became much thinner and was less than 0.8 nm. The formation of these distinct regions should imply a diffusion mechanism that comes from the interaction between the constituent elements Ni, P, and Pd and corrosive solutions. The diffusion of Ni and P occurred toward the outer surface, while Pd diffused a short distance to form a dense and protective underlying surface layer (also called the *sub-surface layer*). It is valuable to note that, generally, the high corrosion resistance of the alloy was attributed to the highly protective oxidized surface films of the alloys. However, the mechanism of corrosion resistance of the present alloy was quite different. During the immersion of the alloy in H_2SO_4 solution, Ni and P elements preferentially dissolved into the acidic solution. Some of the Ni^{2+} and P^{5+} cations from the acidic solution were also re-adsorbed onto the alloy surface together with contaminant C and hydrates in different forms to form an outer surface layer. The composition of the outer surface layer after immersion was similar to that of the air-formed outer surface except for the small thickness. Several researchers [91,96] reported that the amorphous Ni–P alloys exhibited low corrosion resistance in acidic solutions owing to the formation of thick porous surface films. The outer oxidized layer consisting of a large amount of Ni and P cations cannot protect the alloy against corrosion in the strong acidic solution. In contrast, from the XPS result in Figure 7.28, it is seen that the noble metal Pd in the metallic state was significantly enriched on the surface just beneath the outer layer.

The immersion process resulted in rapid initial selective dissolution of less noble Ni and P elements into the solution and the formation of the metallic Pd-enriched sub-surface structure by the self-assembly of Pd atoms at alloy/electrolyte interfaces. The metallic Pd-rich sub-surface layer just beneath the outer oxidized surface plays a vital role in the high corrosion resistance of the alloy against corrosion. In addition, Qin et al. [90] also measured the surface properties of the $Ni_{70}Pd_{10}P_{16}B_4$ BMG with a lower Pd content, and noted that the dense and protective Pd-enriched sub-surface layer did not form in the alloy after immersion in the H_2SO_4 solution. The surface of the $Ni_{70}Pd_{10}P_{16}B_4$ alloy after immersion for 7 days was covered with thick and porous black corrosive products.

Similar XPS results were found for the $Ni_{60}Pd_{20}P_{17}B_3$ alloy immersed in 3 wt.% NaCl solution. After immersion in NaCl solution, the Pd metal was also greatly enriched in the underlying alloy surface just beneath the outer layer. As indicated by the results of the immersion tests, the present alloy exhibited high corrosion resistance in H_2SO_4 and NaCl solutions, which in turn verifies that the Pd-enriched sub-surface layer was dense, compact, and stable. Therefore, it can be concluded that the high corrosion resistance is based on the formation of the Pd-enriched sub-surface layer with high protective quality and high uniformity.

7.6.4 Influence of Microstructure

Copper mold cast $Ni_{55}Nb_{40}P_5$ alloys of 1 and 2 mm diameter were found to be fully glassy as determined by XRD methods. But, on addition of Ta, the $(Ni_{60}Nb_{40-x}Ta_x)_{95}P_5$ alloys were found to be completely glassy only when the diameter of the cast rod was 1 mm. On increasing the diameter to 2 mm, the alloy was glassy only when Ta = 5, 30, and 35 at.%. At other compositions, the alloy consisted of crystalline phases such as Ni_3Nb, Ni_3Ta and nickel phosphides in the alloys containing 10–20 at.% Ta, and Ni_3Ta, Ta, and nickel phosphides in the alloy containing 40 at.% Ta [80].

For the quaternary alloy of 1 mm diameter, the corrosion rates of the BMG rod and of the melt-spun ribbon were identical. But, when the 2 mm diameter was tested, the corrosion rate of the BMG rod was much higher, with a value of about 1 mm $year^{-1}$. This value is somewhat similar to the Ta-free alloy or low-Ta-containing alloys. The inferior corrosion resistance of the 2 mm diameter BMG rod was attributed to the presence of crystalline inclusions. This observation suggests that a completely glassy phase exhibits superior corrosion resistance to the heterogeneous structure consisting of the glassy plus crystalline structure.

In addition to the fact that crystalline inclusions were found to be detrimental to the corrosion behavior of these alloys, the size of the crystalline inclusions also seems to be important. While investigating the corrosion behavior of Ni–Cr–Ta–Mo–P–B BMG alloys Katagiri et al. [81] reported that only $Ni_{67}Cr_5Ta_5Mo_3P_{16}B_4$ and $Ni_{65}Cr_5Ta_5Mo_5P_{16}B_4$ alloys were fully glassy as determined by XRD methods. All other alloys investigated contained minor amounts of crystalline phases, including an fcc Ni phase. The corrosion behavior of the alloys was found to be dependent upon the size of the crystalline inclusion. The corrosion resistance was better when the inclusions were about 2 nm in size, but worse when they were larger than about 5 nm in size. This has been explained on the basis of the ability of the passivating film to cover the inclusions. The presence of crystalline inclusions was also shown to increase the passive current density [78], as described earlier.

When precipitation of the fcc Ni phase occurs during solidification, the remaining glassy phase is enriched with alloying elements such as Cr, Ta, and Mo, which have high passivating efficiency. Consequently, the passive layer consisting of highly passivating elements can completely cover the Ni-phase inclusions, when they are small in size. But, when the size of the inclusions is large, the covering of this less pitting-resistant nanocrystalline phase by the passive layer is not possible. Consequently, the presence of nanocrystalline inclusions of about 5–20 nm in size was found to be detrimental to the passivating ability of the alloy. Such alloys show higher anodic current density and therefore poorer corrosion resistance [78].

In an interesting study, Habazaki et al. [79] compared the corrosion behavior of $Ni_{65}Cr_{10}Nb_5P_{16}B_4$ glassy alloys in the as-cast 1 mm diameter sample and another sample of 20 × 20 × 2 mm that was obtained by consolidating the gas-atomized glassy powder by sheath rolling at 708 K, in the supercooled liquid region. The corrosion resistance of the sheath-rolled specimen was inferior to that of the as-cast sample. After immersion in 6 N HCl, spherical pit-like attacks were observed on the surface of the sheath-rolled sample. Scanning electron micrographs of the fractured surfaces of the two samples were also different. While the as-cast sample showed the vein pattern typical of glassy alloys, the sheath-rolled specimen showed a particle-like fracture, possibly associated with incomplete consolidation of the original powders. This observation again confirms the importance of a homogeneous glassy phase to achieve increased corrosion resistance in the alloy.

Wang et al. [83] investigated the corrosion behavior of $Ni_{59}Zr_{20}Ti_{16}Si_2Sn_3$ and $Ni_{53}Nb_{20}Ti_{10}Zr_8Co_6Cu_3$ alloy ribbons both in the glassy state and after crystallizing them for 2 h at 903 and 973 K, respectively, in a vacuum. The samples were immersed in 1 N HCl for 1 week and their corrosion behavior was investigated through XPS. In contrast to the preceding observations, these authors noted no significant difference in the XPS spectra between the glassy alloys and their crystalline counterparts. Based on this observation, they concluded that the composition of the alloy rather than the microstructure determines the corrosion behavior of the alloy. But, much more convincing information is required to accept that microstructure does not play an important role in determining the corrosion behavior of an alloy.

By suitably designing a nickel-based BMG alloy starting from topological instability criterion, Wang and Wang [94] showed that an $Ni_{60}Nb_{35}Zr_5$ alloy had a high glass-forming ability (the alloy could be cast into 2 mm diameter rod with a supercooled liquid region of 56 K) and also high corrosion resistance.

7.7 Titanium-Based Bulk Metallic Glasses

Titanium alloys have been used in the industry for applications in the aerospace and biomedical sectors due to their high specific strength, low Young's modulus, high corrosion resistance, and good biocompatibility. A number of Ti-based alloys have been solidified into the glassy state—both in ribbon form and as BMG rods. The maximum diameter of the rod that can be produced in a fully glassy state depends on the constituents present in the alloy; Be-containing alloys seem to have a higher glass-forming ability. It is possible to produce a fully glassy rod of >14 mm diameter in a Ti alloy with the composition $Ti_{40}Zr_{25}Ni_2Cu_{13}Be_{20}$ [95]. More recently, this has been increased, by fine-tuning the composition, to >50 mm in $Ti_{32.8}Zr_{30.2}Cu_9Fe_{5.3}Be_{22.7}$ [97] and $(Ti_{36.1}Zr_{33.2}Ni_{5.8}Be_{24.9})_{100-x}Cu_x$ with $x = 5$ and 7 [98].

The specific Ti-based multicomponent alloys that have been investigated for their corrosion resistance studies include melt-spun $Ti_{47.5}Cu_{37.5}Ni_{7.5}Zr_{2.5}M_5$ (where M = Cu, Co, Nb, or Ta) [99], $Ti_{41.5}Zr_{2.5}Hf_5Cu_{42.5}Ni_{7.5}Si_1$ [100], ribbon and cylindrical rods of $Ti_{45}Zr_5Cu_{45}Ni_5$ [101], cylindrical rods of $Ti_{45}Zr_{10}Pd_{10}Cu_{31}Sn_4$ [102], $Ti_{47.5}Zr_{2.5+x}Cu_{37.5-x}Pd_{7.5}Sn_5$ (with $x = 0$, 5, and 7.5 at.%) [103], and 2.5 mm-thick plates of $Ti_{43.3}Zr_{21.7}Ni_{7.5}Be_{27.5}$ [104].

Ti-based glassy alloys have been shown to possess very high corrosion resistance. The measured corrosion rates for the 3 mm diameter fully glassy rods of $Ti_{45}Zr_5Cu_{45}Ni_5$ alloy showed extremely low values of 1, 3.7, and 10.7 μm $year^{-1}$ in 3 wt.% NaCl, 1 N H_2SO_4, and 1 N HCl solutions, respectively [101]. The alloys were passive in these solutions, but pitting corrosion occurred by anodic polarization at higher potentials for the alloys in chloride-containing solutions. Relatively low corrosion rates of 0.9, 1.5, and 3.1 μm $year^{-1}$ were also reported for the glassy $Ti_{41.5}Zr_{2.5}Hf_5Cu_{42.5}Ni_{7.5}Si_1$ alloy in 3 wt.% NaCl, 1 N HCl, and 1 N H_2SO_4 solutions, respectively [100]. These alloys had always revealed spontaneous passivation in the anodic polarization curve. The passive current densities were very low, at about 10^{-2} A m^{-2}, in 3 wt.% NaCl solution and even lower in the HCl solution. Pitting corrosion occurred by anodic polarization at higher potential in both the alloys in the chloride-containing solutions.

Alloying additions of M (= Co, Cu, Nb, or Ta) have been made to the $Ti_{47.5}Cu_{37.5}Ni_{7.5}Zr_{2.5}M_5$ alloy, and it was noted that, in comparison with the base glassy alloy, the modified glasses containing Co, Nb, or Ta exhibited much lower current densities and wider passive regions. This suggests that these alloying additions promote the formation of a more protective film on the surface of the Ti-based glasses and consequently enhance the corrosion resistance [99].

All the alloys listed in the preceding paragraphs contain Ni as an alloying element. The presence of Ni improves the glass-forming ability of the alloy and therefore it becomes easier to produce large-diameter glassy rods. But, Ni is one of the most commonly known allergens, causing allergy and Ni-hypersensitivity. Therefore, Ti-based metallic glasses aimed at biomedical applications should be developed without any Ni in them.

The corrosion resistance of $Ti_{47.5}Zr_{2.5+x}Cu_{37.5-x}Pd_{7.5}Sn_5$ (with $x = 0$, 5, and 7.5 at.%) glassy alloy rods was investigated in Hanks' solution (8.00 NaCl, 0.40 KCl, 0.35 $NaHCO_3$, 0.19 $CaCl_2{\cdot}2H_2O$, $0.09Na_2HPO_4{\cdot}7H_2O$, 0.2 $MgSO_4{\cdot}7H_2O$, 0.06 KH_2PO_4, and 1.00 glucose, all in units of g L^{-1}) with pH = 7.4 and at 310 K, close to the body temperature [103]. The tests were conducted by immersing the glassy alloy in this solution for 1 week. The potentiodynamic polarization curves of these alloys showed a passive region where the passive current density was low (Figure 7.29); it was lower than that of pure Ti, suggesting that the passive film formed on this alloy surface is more protective than that on pure Ti. The passive current density of the alloy with $x = 7.5$, that is, with a Zr content of 10 at.%, was only about 10^{-3} A m^{-2} and is much lower than that of the other two glassy alloys with $x = 0$ and 5. However, with increasing anodic potential, passivity breakdown occurred and the pitting potential increased with increasing Zr content. Figure 7.30 shows the optical micrographs of the $Ti_{47.5}$

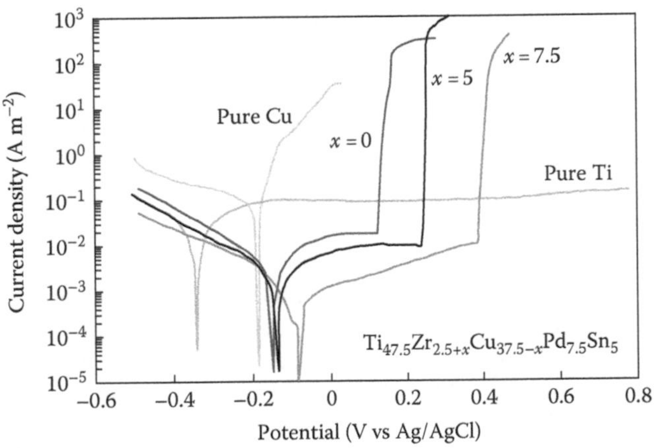

FIGURE 7.29
Anodic and cathodic polarization curves of the $Ti_{47.5}Zr_{2.5+x}Cu_{37.5-x}Pd_{7.5}Sn_5$ (with $x = 0$, 5, and 7.5 at.%) glassy alloy rods in Hanks' solution at 310 K. (Reprinted from Qin, F.X. et al., *Mater. Trans.*, 48, 515, 2007. With permission.)

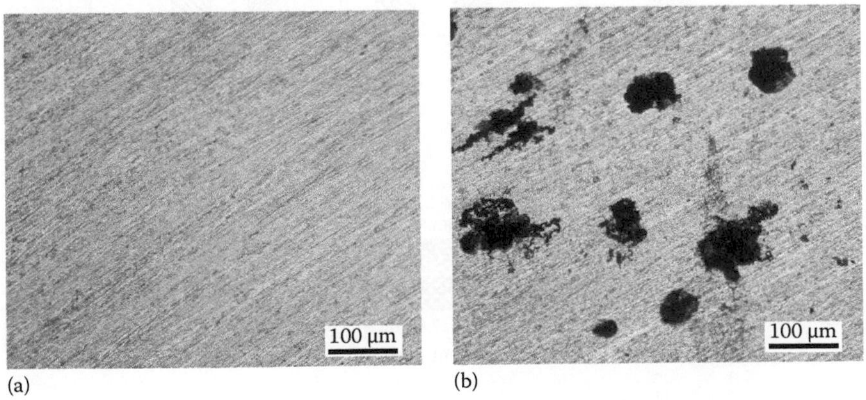

FIGURE 7.30
Optical micrographs of the surface of the $Ti_{47.5}Zr_{10}Cu_{30}Pd_{7.5}Sn_5$ BMG alloy (a) before and (b) after polarization in Hanks' solution at 310 K. Notice the formation of the pits after polarization in (b). (Reprinted from Qin, F.X. et al., *Mater. Trans.*, 48, 515, 2007. With permission.)

$Zr_{10}Cu_{30}Pd_{7.5}Sn_5$ alloy before and after potentiodynamic polarization testing, where pitting can be clearly observed after the polarization test [103].

The composition of the surface film of the $Ti_{47.5}Zr_{2.5+x}Cu_{37.5-x}Pd_{7.5}Sn_5$ (with $x = 0$, 5, and 7.5 at.%) glassy alloy rods after immersion for 1 week in Hanks' solution at 310 K was determined using XPS [103], and the results for the three different alloys are presented in Figure 7.31. Here, [M^{ox}] and [M^b] represent the concentration of the different elements (in at.%) in the oxide film and BMG substrate, respectively. The relative concentration, defined as the ratio of the cationic fraction in the surface film to the bulk concentration, that is, [M^{ox}]/[M^b], corresponds to the enrichment or deficiency of the element [M] in the surface film, when the ratio is >1 or <1, respectively. It can be clearly seen from Figure 7.31 that enrichment of Ti and Zr in the surface film is responsible for the improved corrosion resistance of these alloys. It was also reported that the anodically grown films on Ti-based metallic glasses contained mixed oxides of TiO_2, CuO, ZrO_2, as well as Nb_2O_5 or Ta_2O_5.

Further refinement of composition in this alloy series led to the development of the $Ti_{45}Zr_{10}Pd_{10}Cu_{31}Sn_4$ alloy [102]. It was reported that this Ni-free alloy showed a very high corrosion resistance in 1 N HCl solution. The corrosion rate was ~46 μm $year^{-1}$, an order of magnitude lower than that of austenitic 316 stainless steel (~280 μm $year^{-1}$). After immersion in 3 wt.% NaCl, 1 N H_2SO_4, or 1 N H_2SO_4 + 0.01 N NaCl solutions, the weight loss of this BMG alloy was undetectable, indicating a corrosion rate of <1 μm $year^{-1}$. From anodic polarization curves, it was noted that spontaneous passivation occurred and the alloy exhibited a significantly low passive current density and a wide passive potential region. No pitting corrosion was observed. The homogeneous single-phase nature, absence of crystalline defects, and formation of a uniform passive film that is able to separate the bulk of the alloy from the aggressive environment were found responsible for the high corrosion resistance of the alloy. Because of the absence of Ni in this highly corrosion-resistant alloy, it was suggested that this BMG alloy could be suitable for biomedical and implant applications.

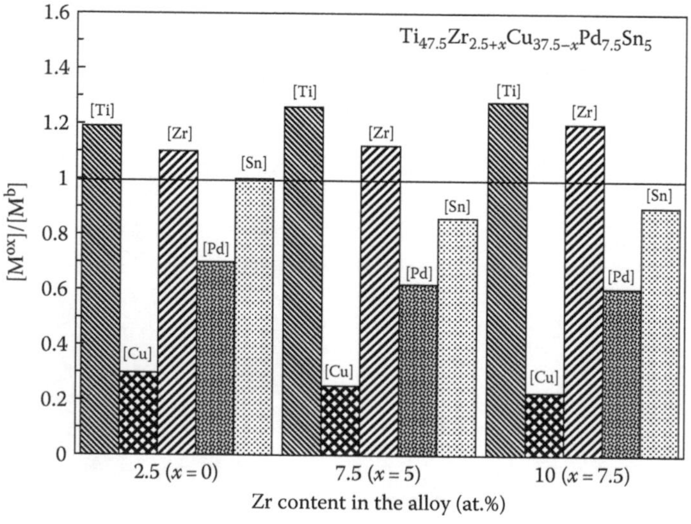

FIGURE 7.31
Ratio of cationic element M (Ti, Zr, Cu, Pd, and Sn) in the surface film of the $Ti_{47.5}Zr_{2.5+x}Cu_{37.5-x}Pd_{7.5}Sn_5$ (with $x =$ 0, 5, and 7.5 at.%) BMG alloys, that is, [M^{ox}]/[M^b] in Hanks' solution for 1 week at 310 K. The superscripts "ox" and "b" denote values in the "oxide" and "bulk," respectively. (Reprinted from Qin, F.X. et al., *Mater. Trans.*, 48, 515, 2007. With permission.)

Morrison et al. [104] investigated the corrosion behavior of a Ti-based BMG alloy with the composition $Ti_{43.3}Zr_{21.7}Ni_{7.5}Be_{27.5}$. While both Ti and Zr present in this alloy are most biocompatible, the alloy also contains Ni and Be, neither of which are supposed to be biocompatible. However, the authors argued that Ni is an important alloying element in many of the commonly used biomaterials, such as 316L stainless steel, and that the literature on the cytotoxicity of Be is less than conclusive. They studied the corrosion behavior of the LM-010 BMG alloy in a phosphate buffer saline (PBS) solution at 310 K. This electrolyte is intended to simulate surgical implant conditions similar to those found *in vivo*. The mean pH of the PBS electrolyte was 7.44, which is near that commonly found in the human body. Very low corrosion rates of 2.9 μm $year^{-1}$ were reported, and the alloy exhibited susceptibility to localized corrosion at elevated potentials of 206 mV (SCE).

7.8 Zirconium-Based Bulk Metallic Glasses

Zr-based BMGs were the first type of metallic glasses to be prepared as large diameter rods in 1993 [105]. Published results show that the largest diameter of the fully glassy Zr-based alloys has been continuously increasing and that the critical diameter for a fully glassy alloy has now been reported to be 73 mm in a $Zr_{46}Cu_{30.14}Ag_{8.36}Al_8Be_{7.5}$ composition, produced by the copper mold casting technique [106]. The Zr-based BMG alloys are characterized by high strength, large elastic strain, and low Young's modulus. Therefore, these alloys have already found some important applications in the industry [107,108]. Since Zr-based alloys are also biocompatible and therefore it is possible for these Zr-based BMG alloys to be used in biomedical applications, it is necessary to evaluate their corrosion behavior both for their general corrosion and also as replacements for transplants and other biomedical applications. Consequently, a number of investigations have been devoted to study the corrosion behavior of Zr-based BMGs in not only conventional electrolytes such as HCl, NaCl, H_2SO_4, but also in PBS, Eagle's Minimum Essential Medium (MEM), and Artificial Body Fluid (ABF) solutions.

Two major groups of alloys have been studied extensively for their corrosion behavior. One is the Zr–Cu–Al-type alloys and their derivatives developed by Inoue and his group [109] at Tohoku University in Japan, and the other is the Zr–Ti–Ni–Cu [110] and Zr–Ti–Ni–Cu–Be BMG alloys [105] developed by Johnson and his group at Caltech in the United States. Even though the corrosion resistance of the pure metal Zr itself is reasonably good, the corrosion resistance of Zr-based BMGs in non-chloride-containing solutions has been shown to be better than their crystalline counterparts; however, the corrosion resistance in halide-containing solutions has been inferior. Table 7.2 summarizes the results in this area.

One of the very early experiments conducted on the corrosion behavior of the $Zr_{41.2}Ti_{13.8}Cu_{12.5}Ni_{10}Be_{22.5}$ BMG was to compare the corrosion behavior in both the glassy and crystallized conditions. Based on the results of potentiodynamic polarization curves, Schroeder et al. [127] reported that the glassy alloy was only slightly more resistant to pitting corrosion in 0.5 N NaCl, and no more resistant to general corrosion in 0.5 N $NaClO_4$ solution, than the corresponding crystalline structure. Hence, they concluded that the homogeneous structure alone may not be sufficient to impart general or localized corrosion resistance to the alloy. On the other hand, Peter et al. [14] noted that even though both the glassy and crystalline states exhibited susceptibility to pitting corrosion in NaCl solutions, electrochemical results indicated that the glassy alloy was more resistant to the onset of pitting corrosion under natural corrosion conditions than the crystalline material.

TABLE 7.2
Alloy Compositions and Conditions under Which Corrosion Studies Were Conducted on Zr-Based BMG Alloys

Alloy Composition	Immersion Medium	Temp. (K)	Time	Corrosion Rate (μm year^{-1})	i_{corr} (A m^{-2})	η_{pit} (mV, SCE)	η_{pp} (mV)	References
$Zr_{55}Al_{20}Co_{25}$	3 wt.% NaCl	298	—	<1	<0.1	690	—	[111]
$Zr_{55}Al_{17.5}Co_{25}Nb_{2.5}$	3 wt.% NaCl	298	—	<1	<0.1	540	—	[111]
$Zr_{55}Al_{15}Co_{25}Nb_5$	3 wt.% NaCl	298	—	<1	<0.1	860	—	[111]
$Zr_{55}Al_{20}Co_{25}$	1 N H_2SO_4	298	—	<1	0.04–0.2	—	—	[111]
$Zr_{55}Al_{17.5}Co_{25}Nb_{2.5}$	1 N H_2SO_4	298	—	<1	0.04–0.2	—	—	[111]
$Zr_{55}Al_{15}Co_{25}Nb_5$	1 N H_2SO_4	298	—	<1	0.04–0.2	—	—	[111]
$Zr_{65}Al_{10}Ni_{10}Cu_{15}$	Hanks' solution	310	1 week	0.03	—	730	—	[112]
$Zr_{65}A_{17.5}Ni_{10}Cu_{17.5}$	Deaerated PBS	—	—	0.001	—	—	—	[113]
$Zr_{65}Al_{7.5}Ni_{10}Cu_{17.5}$	Deaerated PBS	—	—	0.002	—	—	—	[114]
$Zr_{65}Al_{7.5}Ni_{10}Cu_{17.5}$	Deaerated Hanks'	—	—	0.13	—	—	—	[112]
$Zr_{65}Al_{7.5}Ni_{10}Cu_{17.5}$	MEM	—	—	0.08	—	—	—	[112]
$Zr_{65}Al_{7.5}Ni_{10}Cu_{17.5}$	MEM + FBS	—	—	0.09	—	—	—	[112]
$Zr_{50}Cu_{40}Al_{10}$	0.6 N NaCl	—	—	—	0.75×10^{-2}	—	—	[115]
$Zr_{55}Cu_{30}Al_{10}Ni_5$	0.1 N Na_2SO_4	298	—	—	—	—	—	[116]
$Zr_{55}Cu_{30}Al_{10}Ni_5$	0.1 N NaOH	298	—	—	—	—	—	[116]
$Zr_{55}Cu_{30}Al_{10}Ni_5$	0.1 N Na_2SO_4	298	—	—	—	—	—	[117]
$Zr_{55}Cu_{30}Al_{10}Ni_5$	0.1 N Na_2SO_4	423	—	—	—	—	—	[117]
$Zr_{55}Cu_{30}Al_{10}Ni_5$	0.1 N Na_2SO_4	523	—	—	—	—	—	[117]
$Zr_{55}Cu_{30}Al_{10}Ni_5$	0.1 N Na_2SO_4	298	—	—	—	—	—	[117]
$Zr_{55}Cu_{30}Al_{10}Ni_5$	0.1 N Na_2SO_4	423	—	—	—	—	—	[117]
$Zr_{55}Cu_{30}Al_{10}Ni_5$	0.1 N Na_2SO_4	523	—	—	—	—	—	[117]
$Zr_{55}Cu_{30}Al_{10}Ni_5$ (as-prepared)	0.001 N NaCl	298	—	—	—	450	—	[117]
$Zr_{55}Cu_{30}Al_{10}Ni_5$ (as-prepared)	0.001 N NaCl	423	—	—	—	50	—	[117]
$Zr_{55}Cu_{30}Al_{10}Ni_5$ (as-prepared)	0.001 N NaCl	523	—	—	—	−100	—	[117]

(Continued)

TABLE 7.2 (CONTINUED)

Alloy Compositions and Conditions under Which Corrosion Studies Were Conducted on Zr-Based BMG Alloys

Alloy Composition	Immersion Medium	Temp. (K)	Time	Corrosion Rate (μm year^{-1})	i_{corr} (A m^{-2})	η_{pit} (mV, SCE)	η_{pp} (mV)	References
$Zr_{55}Cu_{30}Al_{10}Ni_5$ (pre-passivated)	0.001 N NaCl	298	—	—	—	750	—	[117]
$Zr_{55}Cu_{30}Al_{10}Ni_5$ (pre-passivated)	0.001 N NaCl	423	—	—	—	700	—	[117]
$Zr_{55}Cu_{30}Al_{10}Ni_5$ (pre-passivated)	0.001 N NaCl	523	—	—	—	100	—	[117]
$Zr_{60}Cu_{20}Al_{10}Ni_8Nb_2$	0.01 N NaCl	298	—	—	—	415	40	[118]
$Zr_{59}Cu_{20}Al_{10}Ni_8Nb_3$	0.01 N NaCl	298	—	—	—	625	125	[118]
$Zr_{57}Cu_{15.4}Al_{10}Ni_{12.6}Nb_5$	0.01 N NaCl	298	—	—	—	750	430	[118]
$Zr_{59}Cu_{20}Al_{10}Ni_8Ti_3$	0.01 N NaCl	298	—	—	—	625	50	[118]
$Zr_{65}Cu_{17.5}Ni_{10}Al_{7.5}$	ABF	310	20 days	—	—	600	—	[119]
$Zr_{63}Cu_{17.5}Ni_{10}Al_{7.5}Nb_2$	ABF	310	20 days	—	—	975	—	[119]
$Zr_{60}Cu_{17.5}Ni_{10}Al_{7.5}Nb_5$	ABF	310	20 days	—	—	1,430	—	[119]
$Zr_{65}Cu_{17.5}Ni_{10}Al_{7.5}$	3 wt.% NaCl	298	—	—	0.5–1.2 × 10^{-2}	125	—	[120]
$Zr_{64}Hf_1Cu_{17.5}Ni_{10}Al_{7.5}$	3 wt.% NaCl	298	—	—	0.5–1.2 × 10^{-2}	690	—	[120]
$Zr_{65}Cu_{17.5}Ni_{10}Al_{7.5}$	1 N H_2SO_4	298	—	—	—	—	—	[120]
$Zr_{64}Hf_1Cu_{17.5}Ni_{10}Al_{7.5}$	1 N H_2SO_4	298	—	—	—	—	—	[120]
$Zr_{65}Cu_{17.5}Ni_{10}Al_{7.5}$	1 N HCl	298	—	—	—	A	—	[120]
$Zr_{64}Hf_1Cu_{17.5}Ni_{10}Al_{7.5}$	1 N HCl	298	—	—	—	A	—	[120]
$Zr_{55}Al_{10}Cu_{30}Ni_5$	Deaerated PBS	—	—	9.9	5.43 × 10^{-3}	349	<0	[121]
$(Zr_{55}Al_{10}Cu_{30}Ni_5)_{99}Y_1$	Deaerated PBS	—	—	0.6	0.43 × 10^{-3}	46	<0	[121]
$Zr_{65}Cu_{17.5}Ni_{10}Al_{7.5}$	0.5 N H_2SO_4	—	—	—	7 × 10^{-2}	—	—	[122]
$Zr_{65}Cu_{17.5}Ni_{10}Al_{7.5}$	0.5 N HNO_3	—	—	—	4 × 10^{-2}	—	—	[122]
$Zr_{65}Cu_{17.5}Ni_{10}Al_{7.5}$	0.5 N HCl	—	—	240	2.1 × 10^{-1}	—	—	[122]
$Zr_{65}Cu_{17.5}Ni_{10}Al_{7.5}$	0.5 N NaOH	—	—	—	1.7 × 10^{-1}	—	—	[122]
$Zr_{52.5}Cu_{17.9}Ni_{14.6}Al_{10}Ti_5$	PBS	310	—	0.8	—	474	225	[123]
$Zr_{52.5}Cu_{17.9}Ni_{14.6}Al_{10}Ti_5$	0.01 N NaCl	298	—	—	—	500	165	[118]
$Zr_{52.5}Cu_{17.9}Ni_{14.6}Al_{10}Ti_5$	0.5 N H_2SO_4	298	—	—	—	225	—	[124]

(Continued)

TABLE 7.2 (CONTINUED)

Alloy Compositions and Conditions under Which Corrosion Studies Were Conducted on Zr-Based BMG Alloys

Alloy Composition	Immersion Medium	Temp. (K)	Time	Corrosion Rate (μm $year^{-1}$)	i_{corr} (A m^{-2})	η_{pit} (mV, SCE)	η_{pp} (mV)	References
$Zr_{52.5}Cu_{17.9}Ni_{14.6}Al_{10}Ti_5$	0.5 N NaCl	298	—	—	—	—	—	[124]
$Zr_{52.5}Cu_{17.9}Ni_{14.6}Al_{10}Ti_5$	3.5 wt.% NaCl	—	20 days	1.8 (525 K)	—	340	—	[125]
$Zr_{52.5}Cu_{17.9}Ni_{14.6}Al_{10}Ti_5$	1 N HNO_3	—	20 days	61 (525 K)	—	—	—	[125]
$Zr_{52.5}Cu_{17.9}Ni_{14.6}Al_{10}Ti_5$	1 N H_2SO_4	—	20 days	81.3 (525 K)	—	—	—	[125]
$Zr_{52.5}Cu_{17.9}Ni_{14.6}Al_{10}Ti_5$	1 N HCl	—	20 days	1270 (5253 K)	—	—	—	[122]
$Zr_{52.5}Cu_{17.9}Ni_{14.6}Al_{10}Ti_5$	0.6 N NaCl	—	—	1.3 (glassy)	1.2×10^{-3} (a)	320	40	[14]
$Zr_{52.5}Cu_{17.9}Ni_{14.6}Al_{10}Ti_5$	0.6 N NaCl	—	—	0.6 (cryst.)	0.5×10^{-3} (a)	—	—	[14]
$Zr_{52.5}Cu_{17.9}Ni_{14.6}Al_{10}Ti_5$	0.6 N NaCl	—	—	9 (glassy)	8×10^{-3} (b)	100	−20	[14]
$Zr_{52.5}Cu_{17.9}Ni_{14.6}Al_{10}Ti_5$	0.6 N NaCl	—	—	15 (cryst.)	14×10^{-3} (b)	—	—	[14]
$Zr_{52.5}Cu_{17.9}Ni_{14.6}Al_{10}Ti_5$	0.05 N Na_2SO_4	—	—	0.4	0.4×10^{-3}	—	—	[14]
$Zr_{52.5}Cu_{17.9}Ni_{14.6}Al_{10}Ti_5$	Deaerated PBS	—	—	0.8	—	—	—	[104]
$Zr_{52.5}Cu_{17.9}Ni_{14.6}Al_{10}Ti_5$	Deaerated PBS	—	—	0.8	—	474	225	[123]
$Zr_{46.75}Ti_{8.25}Cu_{7.5}Ni_{10}Be_{27.5}$	0.5 N H_2SO_4	—	—	—	0.1	—	—	[122]
$Zr_{46.75}Ti_{8.25}Cu_{7.5}Ni_{10}Be_{27.5}$	0.5 N HNO_3	—	—	—	5.5×10^{-2}	—	—	[122]
$Zr_{46.75}Ti_{8.25}Cu_{7.5}Ni_{10}Be_{27.5}$	0.5 N HCl	—	—	410	3.5×10^{-1}	—	—	[122]
$Zr_{46.75}Ti_{8.25}Cu_{7.5}Ni_{10}Be_{27.5}$	0.5 N NaOH	—	—	—	2.7×10^{-1}	—	—	[122]
$Zr_{41.2}Ti_{13.8}Cu_{12.5}Ni_{10}Be_{22.5}$	Deaerated PBS	295	—	0.8	—	478	215	[126]
$Zr_{41.2}Ti_{13.8}Cu_{12.5}Ni_{10}Be_{22.5}$	Deaerated PBS	310	—	1	—	410	184	[126]
$Zr_{41.2}Ti_{13.8}Cu_{12.5}Ni_{10}Be_{22.5}$	0.6 N NaCl	295	—	2.5	—	97	26	[126]
$Zr_{41.2}Ti_{13.8}Cu_{12.5}Ni_{10}Be_{22.5}$	0.6 N NaCl	310	—	1.3	—	142	68	[126]
$Zr_{41.2}Ti_{13.8}Cu_{12.5}Ni_{10}Be_{22.5}$	0.5 N NaCl	—	—	—	—	207 (glassy)	—	[127]
$Zr_{41.2}Ti_{13.8}Cu_{12.5}Ni_{10}Be_{22.5}$	0.5 N NaCl	—	—	—	—	111 (cryst.)	—	[127]
$Zr_{41.2}Ti_{13.8}Cu_{12.5}Ni_{10}Be_{22.5}$	0.5 N $NaClO_4$	—	—	—	—	215 (glassy)	—	[127]
$Zr_{41.2}Ti_{13.8}Cu_{12.5}Ni_{10}Be_{22.5}$	0.5 N $NaClO_4$	—	—	—	—	230 (cryst.)	—	[127]

Notes: SCE: saturated calomel electrode (reference); PBS: phosphate buffer saline electrolyte; MEM: Eagle's minimum essential medium; FBS: fetal bovine serum; ABF: artificial body fluid; A: active dissolution; (a): polished to 600 grit; (b): metallographic polish; cryst.: crystalline.

7.8.1 Influence of Composition, Test Environment, and Temperature

The addition of alloying elements has been shown to generally improve the corrosion resistance of Zr-based BMG alloys. The substitution of some of the Zr with corrosion-resistant elements such as Cr, Nb, Ta, or Ti has been found to improve the corrosion resistance. The addition of Cr is not encouraged, because Cr-substituted alloys showed lower glass-forming ability. For example, the substitution of Zr with Cr at 10 at.% failed to produce the alloy in the fully glassy state even by rapid quenching methods. On the other hand, substitution by Nb or Ta up to the 20 at.% level resulted in the formation of a fully glassy structure. Further, Nb substitution was found to be most effective for improving the corrosion resistance of the alloys.

Asami et al. [28] investigated the effect of substituting Zr with Nb on the corrosion resistance of $Zr_{60-x}Nb_xCu_{20}Al_{10}Ni_{10}$ BMG alloys. The corrosion rate of the Nb-free alloy in 1 N H_2SO_4 solution at 298 K open to air was about 0.2 μm $year^{-1}$. But, in a 1 N HCl solution, it was almost 100 μm $year^{-1}$. When Zr was substituted with 20 at.% Nb, the corrosion rate in HCl was reduced to less than 1 μm $year^{-1}$. The dynamic polarization curves of $Zr_{60-x}Nb_xCu_{20}Al_{10}Ni_{10}$ alloys with $x = 0–20$ at.% in a 1 N HCl solution at 298 K open to air are shown in Figure 7.32. It may be seen that the corrosion potential increases with increasing Nb content, suggesting an improvement in the corrosion resistance of the Nb-containing alloys. The sharp increase in the current density corresponds to the onset of pitting corrosion. The pitting potential also increased with increasing Nb content. Additionally, in contrast to the Nb-free alloy, where spontaneous passivation was not effectively observed, the Nb-containing alloys showed spontaneous passivation [28].

The addition of Nb or Ti to the base $Zr_{55}Cu_{30}Al_{10}Ni_5$ alloy has been shown to decrease the pitting susceptibility in 0.01 N NaCl solution [118]. However, the addition of Pd to $Zr_{55}Cu_{30}Al_{10}Ni_{5-x}Pd_x$ (with $x = 0$, 1, 3, and 5 at.%) showed a different result. It was reported that the Pd-free glassy alloy showed an active–passive transition by anodic polarization in 0.6 N NaCl solution and that the alloy was spontaneously passivated with a large passive region indicating high corrosion resistance. But, the alloy with the 5 at.% Pd addition

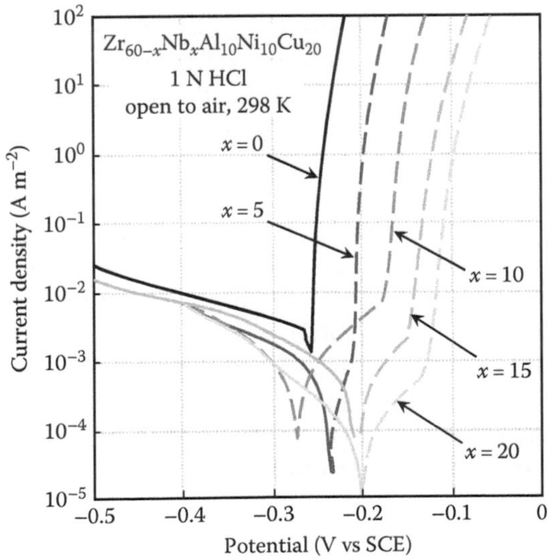

FIGURE 7.32
Dynamic polarization curves of $Zr_{60-x}Nb_xCu_{20}Al_{10}Ni_{10}$ (with x = 0–20 at.%) in 1 N HCl solution at 298 K, open to air.

showed a single active state with a limited passive region and higher current density, suggesting that the corrosion resistance decreased with the addition of 5 at.% Pd [128,129].

Pang et al. [111] investigated the corrosion behavior of 2.5–5 mm diameter rods of copper mold cast $Zr_{55}Al_{20-x}Co_{25}Nb_x$ (with x = 0, 2.5, and 5 at.%) BMGs. After immersion in 3 wt.% NaCl and 1 N H_2SO_4 solutions open to air for 1 week, no weight loss was observed in this alloy, suggesting a corrosion rate of less than 1 μm $year^{-1}$. As determined by potentiodynamic polarization curves, the alloys have spontaneously passivated in a wide passive region in these solutions. The passive current density was low, of the order of 10^{-1} A m^{-2}, before the occurrence of pitting, and the pitting potential increased significantly with increasing Nb content, indicating increased corrosion resistance. Thus, replacing part of the Al with Nb has been shown to be effective in improving the pitting corrosion resistance. XPS analysis showed that the air-formed films were enriched in Zr and Al, slightly deficient in Nb, and significantly deficient in Co with respect to the bulk alloy composition. A reverse trend was noted for the underlying alloy surface.

Quite surprisingly, Chieh et al. [125] reported that the glassy $Zr_{52.5}Cu_{17.9}Ni_{14.6}Al_{10}Ti_5$ alloy exhibited excellent corrosion resistance in 3.5 wt.% NaCl solution and good corrosion resistance in HNO_3 and H_2SO_4 solutions. The alloys were also reported to exhibit excellent pitting corrosion resistance. However, they had extremely poor corrosion resistance in HCl solution. This result, however, is with reference to the AISI 304L austenitic stainless steel and not the crystalline state of the Zr-based alloy.

Gebert et al. [117] investigated the effect of test temperature on the corrosion behavior of $Zr_{55}Cu_{30}Al_{10}Ni_5$ BMG alloy rods in 0.1 N Na_2SO_4 and 0.001 N NaCl electrolytes from room temperature up to 523 K. They observed an accelerated degradation of the cast BMG alloy samples under hot conditions up to 523 K in comparison with that observed at room temperature.

7.8.2 Influence of Microstructure

The corrosion behavior of both glassy and crystalline $Zr_{52.5}Cu_{17.9}Ni_{14.6}Al_{10}Ti_5$ alloys was also studied in 0.6 N NaCl, and that of the glassy alloy in 0.05 N Na_2SO_4 solution [14]. Both the glassy and crystalline alloys showed passive behavior with low corrosion rates (15 μm $year^{-1}$ or less) in the NaCl solution, but susceptibility toward pitting corrosion was noted. The glassy alloy was reported to be more resistant to the onset of pitting corrosion. In the 0.05 N Na_2SO_4 solution, however, the glassy alloy exhibited a very low corrosion rate (0.4 μm $year^{-1}$) and was also found to be immune to pitting corrosion.

An interesting comparison of 20–30 μm-thick melt-spun ribbons and 6 mm diameter copper mold cast rods of $Zr_{52}Ti_6Cu_{18}Ni_{14}Al_{10}$ composition was made with respect to their structure and corrosion behavior [130]. It was noted that the wheel-side surface of the ribbon contained a higher concentration of quenched-in defects (air pockets), whereas surface irregularities were noticed on the air-side surface. Fluctuation microscopy indicated that the BMG rods contained more medium-range order (MRO), especially in the center of the rod, in comparison to the melt-spun ribbon. Corrosion experiments were conducted across the cross section of the BMG rod in 0.025 N HCl and 0.25 N H_2SO_4 + 0.025 N NaCl solutions. It was observed that the corrosion resistance decreased with increasing MRO. But, the MRO was less detrimental to the corrosion behavior than the quenched-in defects and surface irregularities.

7.8.3 Polarization Studies

$Zr_{55}Cu_{30}Al_{10}Ni_5$ is another Zr-based alloy system in which it has been easy to produce large-diameter fully glassy rods. Gebert et al. [131] investigated the corrosion behavior of 3–5 mm diameter glassy rods synthesized by the copper mold casting technique in 0.1 N

Na_2SO_4 (pH = 8) solution. Potentiodynamic and potentiostatic polarization measurements revealed that the alloy formed strong protective surface layers by anodization, which is similar to that which happens in pure Zr. However, the barrier effect of the surface layers on the alloy is slightly lower than that of films on Zr. An important observation made was that these glassy alloys are susceptible to pitting corrosion, possibly due to the existence of micrometer-sized crystalline inclusions embedded in the glassy matrix. The chloride attack is possible at the interface between the crystalline and glassy phases [116].

The enrichment of the different elements in the passive layer seemed to be different when the BMG alloy was immersed in NaCl or H_2SO_4 solutions. Zr was further concentrated, while the Al content decreased in the spontaneously passivated films after immersion in both the solutions (3 wt.% NaCl and 1 N H_2SO_4). Further, the Nb content in the passive films on the alloys immersed in 1 N H_2SO_4 solution was higher, while in those immersed in the NaCl solution it remained almost the same as in the air-formed film, that is, slightly lower than the bulk alloy composition. These results confirm that the formation of passive films enriched in Zr and Al is responsible for the high corrosion resistance of the alloys. The addition of Nb is favorable for the alloys to form a protective surface film with higher chemical stability.

Cyclic anodic polarization studies were conducted on $Zr_{41.2}Ti_{13.8}Cu_{12.5}Ni_{10}Be_{22.5}$ alloys in their fully glassy and crystallized states [126]. Two different temperatures (295 and 310 K) and two different electrolytes (0.6 N NaCl and a PBS solution with a physiologically relevant dissolved oxygen content) were used. For all tested conditions, the alloy demonstrated passive behavior at the open-circuit potential with low corrosion rates of 1–3 μm $year^{-1}$. However, susceptibility to localized pitting corrosion was observed in all of the test conditions, with the caveat that the susceptibility was lower for the PBS electrolyte.

Huang et al. [121] investigated the bio-corrosion of $Zr_{55}Cu_{30}Al_{10}Ni_5$ BMG alloy with and without 1 at.% Y addition in a PBS electrolyte. It was noted that the alloy with 1 at.% Y addition exhibited lower current density and corrosion rates than the Y-free alloy. Further, the Y-containing alloy showed better resistance to the onset of pitting than the Y-free alloy. The general conclusion was that the addition of 1 at.% Y had enhanced the bio-corrosion resistance of the $Zr_{55}Cu_{30}Al_{10}Ni_5$ BMG alloy and that this was probably due to the Y addition having changed the structure and/or accelerated the formation of the passive film.

7.8.4 Pitting Corrosion

The addition of Hf to $Zr_{65}Cu_{17.5}Ni_{10}Al_{7.5}$ BMG alloys on their glass-forming ability and corrosion behavior has been investigated [120]. The alloy with 1 at.% Hf addition was found to be fully glassy and, consequently, the BMG with 1 at.% Hf addition exhibited a wider passive region and the highest pitting potential of 340 mV as compared with the Hf-free alloy or the BMG composite obtained when 2 at.% Hf was added. When a $Zr_{55}Cu_{30}Al_{10}Ni_5$ liquid alloy with 0.4 at.% oxygen was cooled rather slowly, micron-sized crystalline inclusions were noted to be present in them. The presence of such inclusions has also been shown to be responsible for chloride-induced pitting [131]. The presence of fine crystalline inclusions in the glassy matrix has, in general, been shown to lead to a deterioration in corrosion resistance [129].

The effect of Hf content on the surface morphology of the corroded $Zr_{65}Cu_{17.5}Ni_{10}Al_{7.5}$ alloy was studied by Liu et al. [120]. Figure 7.33 shows the corrosion morphologies of the alloys with 0, 1, and 2 at.% Hf as revealed by scanning electron microscopy. While the alloys with Hf = 0 and 1 at.% are fully glassy, the alloy with 2 at.% Hf showed the presence of dendritic crystals dispersed in a glassy matrix. In fact, the alloy with 1 at.% Hf addition appears to

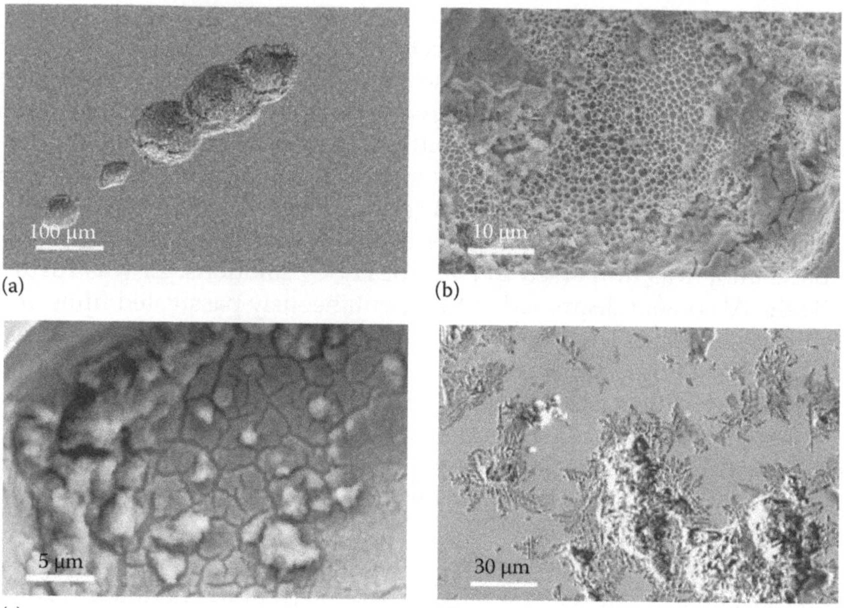

(a) (b) (c) (d)

FIGURE 7.33
Scanning electron micrographs of the corroded surfaces of the three $Zr_{65-x}Hf_xCu_{17.5}Ni_{10}Al_{7.5}$ ($x = 0$, 1, and 2) alloys. (a) General appearance of the Hf-free BMG indicating that pitting corrosion had occurred. (b) High-magnification view of a pit in the Hf-free alloy showing a "honeycomb" structure. (c) High-magnification view of a pit in the alloy with 1 at.% Hf showing that a protective film had formed. (d) Micrograph of the composite (alloy with 2 at.% Hf) showing a heavy break-off of the dendritic phase. (Reprinted from Liu, L. et al., *J. Alloys Compd.*, 399, 144, 2005. With permission.)

be "more glassy" as indicated by the very broad halo in the XRD pattern. In contrast, the halo in the Hf-free alloy was a little sharp. Figure 7.33a shows that a few pits with different sizes have formed on the Hf-free alloy and that these are randomly distributed on the surface. A similar situation was obtained in the alloy with 1 at.% Hf. But, when observed at higher magnifications, significant differences were noted between the two alloy surfaces. In the Hf-free alloy, as shown in Figure 7.33a and b "honeycomb-like" structures, with cells of about 2 μm in diameter, were observed inside the pits. On the other hand, in the alloy with 1 at.% Hf, Figure 7.33c shows that a protective film was present at the bottom of each pit. Since the alloy was exposed to heavy attack in HCl solution, the film was already ruptured. This observation suggests that the addition of Hf had promoted the formation of passive films. Figure 7.33d from the alloy with 2 at.% Hf clearly shows that pitting corrosion occurred mainly around the crystalline phase, and caused a scaling-off of dendrites.

A number of researchers have reported that chloride-induced pitting occurs in Zr-based BMGs. Resistance to pit initiation was low and the growth of the pits was quite pronounced. Coupled with these, it was also reported that re-formation of the passive film was not easy. All these findings were related to the heterogeneity in the microstructure of the Zr-based BMGs, which were shown to contain fine nanocrystalline particles distributed in the glassy matrix. Pitting was demonstrated to occur at the interface between the glassy and crystalline phases. The presence of such crystalline particles could be attributed to several factors. First, it is possible that the critical cooling rate for glass formation was not exceeded during the processing to achieve a fully glassy structure. This can be easily overcome by either increasing the cooling rate during processing or reducing the diameter of the as-cast rod. Second, it is possible that heterogeneous nucleation occurred during solidification as a result of contact between the melt and the substrate, resulting in the formation of crystalline phases at those sites. This could also be avoided by choosing the appropriate processing technique. Third, it is possible that impurities present in the melt, for example, oxide particles because of the high reactivity of the metal Zr or introduced from the gaseous atmosphere, could act as nucleation sites, leading to the formation of crystalline phases.

The mechanism suggested for chloride-induced pitting is as follows. Chloride ions can be preferentially adsorbed during anodic polarization at the interface between the glassy and crystalline phases, and this could lead to initiation of pitting in the Zr-based BMG. At this region, the passive film can be broken down even in solutions containing low chloride ion concentrations, and the pits grow preferentially into the reactive glassy matrix. Since the crystalline particles are unattacked and the glassy matrix gets pitted, the crystalline particles detach themselves from the surface, leaving holes behind. This process gets repeated when the chloride ions accumulate in the holes, and further pitting and its growth occurs.

This mechanism of chloride-induced pitting raises some doubts. For example, though it may be at the interface between the glassy and crystalline phases at which pitting occurs, there have been other alloy systems and compositions in which a heterogeneous structure is observed. And pitting corrosion has not been reported in those cases. If it is the reactivity of the glass that is responsible, then this should also occur in Ti- or Hf-based alloys, since these are also reactive metals. But, no such reports are available. A simple and possible solution to test the validity of this mechanism will be to produce a very clean Zr-based alloy (with the least amount of oxygen in the melt) of a diameter smaller than the maximum possible diameter. This ensures that the critical cooling rate for glass formation is exceeded. Since it is possible that surface nucleation could occur, the outer surface can be machined off, for example, using electro discharge machining methods, and the pitting behavior of the remaining sample can then be studied. With an alloy processed in this manner, pitting corrosion should not occur. An alternative experiment could be to conduct the pitting corrosion studies in very pure and also not-so-pure titanium alloys and compare the results.

7.9 Other Bulk Metallic Glassy Alloys

BMGs based on Ca, for example, $Ca_{57}Mg_{19}Cu_{24}$ and $Ca_{60}Mg_{20}Ag_{20}$ with 4 mm diameter [132] and $Ca_{60}Mg_{20}Ag_{10}Cu_{10}$ with 7 mm diameter [133] have been produced by copper mold casting. Subsequently, a number of other Ca-based BMGs including Ca–Mg–Zn, Ca–Mg–Cu, Ca–Mg–Al, and combinations of these alloys have also been produced [134]. Dahlman et al. [135] and Senkov et al. [134] reported results on the corrosion behavior of four different BMGs based on Ca in distilled water. The electrochemical behavior of these alloys was also investigated in 0.05 N Na_2SO_4 [136]. Two ternary alloys, *viz.*, $Ca_{65}Mg_{15}Zn_{20}$ and $Ca_{50}Mg_{20}Cu_{30}$, one quaternary $Ca_{55}Mg_{18}Zn_{11}Cu_{16}$ [136], and a quinary alloy $Ca_{55}Mg_{15}Al_{10}Zn_{15}Cu_5$ [134] were investigated and the corrosion behavior was studied using static aqueous immersion at room temperature. Both the ternary glassy alloys experienced destructive corrosion reactions. But, the quaternary and quinary alloys exhibited corrosion resistance, forming protective films 18–23 μm thick on the quaternary alloy and 7–11 μm thick on the quinary alloy. These results indicate that the corrosion resistance of Ca-based BMGs can be significantly improved by the addition of alloying elements such as Zn and Cu or Al. From electrochemical measurements, the corrosion rates were estimated as 5691 μm year^{-1} for the $Ca_{65}Mg_{15}Zn_{20}$, 311 μm year^{-1} for the $Ca_{55}Mg_{18}Zn_{11}Cu_{16}$, and 1503 μm year^{-1} for the $Ca_{50}Mg_{20}Cu_{30}$. By comparing these corrosion rates with those of Zr-based and Fe-based BMG alloys, the authors concluded that the electrochemical corrosion resistance of some of the Ca-based BMGs is comparable to some of the Fe-based BMGs and Mg-based crystalline alloys.

7.10 Concluding Remarks

All the investigations discussed have quite unambiguously confirmed that a chemically and microstructurally homogeneous glassy phase exhibits very good corrosion resistance. The corrosion resistance of the homogeneous glassy alloy is better than either an alloy that has been completely crystallized or a composite containing nanocrystalline particles dispersed in a glassy matrix. However, it was also noted that the effect of dispersion of nanometer-sized particles in a glassy matrix becomes significant, as far as corrosion resistance is concerned, only when the particles are reasonably large, say >20 nm or so. If these particles are extremely fine, say 2–3 nm, then they do not seem to affect the corrosion behavior adversely. If the corrosion resistance of the glassy alloy is not adequate, then the addition of a sufficient amount of alloying elements that are basically corrosion resistant, for example, Mo, Nb, or Ta, seems to improve the corrosion resistance. In these respects, the corrosion behavior of BMG alloys does not appear to be significantly different from that of melt-spun metallic glass ribbons.

References

1. Naka, M., K. Hashimoto, and T. Masumoto (1974). Corrosion resistivity of amorphous Fe alloys containing Cr. *J. Jpn. Inst. Metals* 38: 835–841 (in Japanese).
2. Naka, M., K. Hashimoto, and T. Masumoto (1976). High corrosion resistance of Cr-bearing amorphous Fe alloys in neutral and acidic solutions containing chloride. *Corrosion* 32: 146–152.
3. Asami, K., K. Hashimoto, T. Masumoto, and S. Shimodaira (1976). ESCA study of the passive film on an extremely corrosion-resistant amorphous iron alloy. *Corrosion Sci.* 16: 909–914.
4. Masumoto, T. and K. Hashimoto (1978). Chemical properties of amorphous metals. *Ann. Rev. Mater. Sci.* 8: 215–233.
5. Hashimoto, K. (1983). Chemical properties. In *Amorphous Metallic Alloys*, ed. F.E. Luborsky, pp. 471–486. London, UK: Butterworths.
6. Hashimoto, K. (1993). Chemical properties of rapidly solidified alloys. In *Rapidly Solidified Alloys: Processes, Structures, Properties, Applications*, ed. H.H. Liebermann, pp. 591–615. New York: Marcel Dekker.
7. Hashimoto, K. (2002). W.R. Whitney award lecture: In pursuit of new corrosion-resistant alloys. *Corrosion* 58: 715–722.

8. Inoue, A., W. Zhang, T. Tsurui, A.R. Yavari, and A.L. Greer (2005). Unusual room-temperature compressive plasticity in nanocrystal-toughened bulk copper-zirconium glass. *Philos. Mag. Lett.* 85: 221–229.
9. Louzguine, D.V. and A. Inoue (2003). Effect of Ni on stabilization of the supercooled liquid and devitrification of Cu–Zr–Ti bulk glassy alloys. *J. Non-Cryst. Solids* 325: 187–192.
10. Murty, B.S., D.H. Ping, K. Hono, and A. Inoue (2000). Direct evidence for oxygen stabilization of icosahedral phase during crystallization of $Zr_{65}Cu_{27.5}Al_{7.5}$ metallic glass. *Appl. Phys. Lett.* 76: 55–57.
11. Gebert, A., J. Eckert, and L. Schultz (1998). Effect of oxygen on phase formation and thermal stability of slowly cooled $Zr_{65}Al_{7.5}Cu_{17.5}Ni_{10}$ metallic glass. *Acta Mater.* 46: 5475–5482.
12. Scully, J.R., A. Gebert, and J.H. Payer (2007). Corrosion and related mechanical properties of bulk metallic glasses. *J. Mater. Res.* 22: 302–313.
13. Green, B.A., P.K. Liaw, and R.A. Buchanan (2008). Corrosion behavior. In *Bulk Metallic Glasses: An Overview*, eds. M.K. Miller and P.K. Liaw, pp. 205–234. New York: Springer.
14. Peter, W.H., R.A. Buchanan, C.T. Liu, P.K. Liaw, M.L. Morrison, J.A. Horton, C.A. Jr Carmichael, and J.L. Wright (2002). Localized corrosion behavior of a zirconium-based bulk metallic glass relative to its crystalline state. *Intermetallics* 10: 1157–1162.
15. Stansbury, E.E. and R.A. Buchanan (2000). *Principles and Prevention of Corrosion*. Materials Park, OH: ASM International.
16. Asami, K. and K. Hashimoto (1977). The x-ray photo-electron spectra of several oxides of iron and chromium. *Corrosion Sci.* 17: 559–570.
17. Asami, K. and K. Hashimoto (1984). An XPS study of the surfaces on Fe–Cr, Fe–Co and Fe–Ni alloys after mechanical polishing. *Corrosion Sci.* 24: 83–97.
18. Deng, L., B.W. Zhou, H.S. Yang, X. Jiang, B. Jiang, and X.G. Zhang (2015). Roles of minor rare-earth elements addition in formation and properties of Cu–Zr–Al bulk metallic glasses. *J. Alloys Compd.* 632: 429–434.
19. Inoue, A., W. Zhang, T. Zhang, and T. Kurosaka (2001). High-strength Cu-based bulk glassy alloys in Cu–Zr–Ti and Cu–Hf–Ti ternary systems. *Acta Mater.* 49: 2645–2652.
20. Inoue, A., W. Zhang, T. Zhang, and T. Kurosaka (2001). Formation and mechanical properties of Cu–Hf–Ti bulk glassy alloys. *J. Mater. Res.* 16: 2836–2844.
21. Inoue, A., W. Zhang, T. Zhang, and T. Kurosaka (2001). Cu-based bulk glassy alloys with good mechanical properties in Cu–Zr–Hf–Ti system. *Mater. Trans.* 42: 1805–1812.
22. Inoue, A. and W. Zhang (2002). Formation, thermal stability and mechanical properties of Cu–Zr–Al bulk glassy alloys. *Mater. Trans.* 43: 2921–2925.
23. Xu, D., B. Lohwongwatana, G. Duan, W.L. Johnson, and C. Garland (2004). Bulk metallic glass formation in binary Cu-rich alloy series – $Cu_{100-x}Zr_x$ (x = 34, 36, 38.2, 40 at.%) and mechanical properties of bulk $Cu_{64}Zr_{36}$ glass. *Acta Mater.* 52: 2621–2624.
24. Xia, L., D. Ding, S.T. Shan, and Y.D. Dong (2006). The glass forming ability of Cu-rich Cu–Hf binary alloys. *J. Phys. Condens. Matter* 18: 3543–3548.
25. Qin, C.L., W. Zhang, K. Asami, H. Kimura, X.M. Wang, and A. Inoue (2006). A novel Cu-based BMG composite with high corrosion resistance and excellent mechanical properties. *Acta Mater.* 54: 3713–3719.
26. Qin, C.L., K. Asami, T. Zhang, W. Zhang, and A. Inoue (2003). Effects of additional elements on the glass formation and corrosion behavior of bulk glassy Cu–Hf–Ti alloys. *Mater. Trans.* 44: 1042–1045.
27. Qin, C.L., W. Zhang, H. Kimura, K. Asami, N. Ohtsu, and A. Inoue (2005). Glass formation, corrosion behavior and mechanical properties of bulk glassy Cu–Hf–Ti–Nb alloys. *Acta Mater.* 53: 3903–3911.
28. Asami, K., H. Habazaki, A. Inoue, and K. Hashimoto (2005). Recent development of highly corrosion resistant bulk glassy alloys. *Mater. Sci. Forum* 502: 225–230.
29. Chen, Y.Y., U.T. Hong, J.W. Yeh, and H.C. Shih (2006). Selected corrosion behaviors of a $Cu_{0.5}NiAlCoCrFeSi$ bulk glassy alloy in 288 °C high-purity water. *Scr. Mater.* 54: 1997–2001.
30. Qin, C.L., W. Zhang, H. Kimura, K. Asami, and A. Inoue (2004). New Cu–Zr–Al–Nb bulk glassy alloys with high corrosion resistance. *Mater. Trans.* 45: 1958–1961.

31. Tam, M.K., S.J. Pang, and C.H. Shek (2006). Effects of niobium on thermal stability and corrosion behavior of glassy Cu–Zr–Al–Nb alloys. *J. Phys. Chem. Solids* 67: 762–766.
32. Asami, K., C.L. Qin, T. Zhang, and A. Inoue (2004). Effect of additional elements on the corrosion behavior of a Cu–Zr–Ti bulk metallic glass. *Mater. Sci. Eng. A* 375–377: 235–239.
33. Liu, L. and B. Liu (2006). Influence of the micro-addition of Mo on glass forming ability and corrosion resistance of Cu-based bulk metallic glasses. *Electrochim. Acta* 51: 3724–3730.
34. Qin, C.L., W. Zhang, K. Amiya, K. Asami, and A. Inoue (2007). Mechanical properties and corrosion behavior of $(Cu_{0.6}Hf_{0.25}Ti_{0.15})_{90}Nb_{10}$ bulk metallic glass composites. *Mater. Sci. Eng. A* 449–451: 230–234.
35. Tam, M.K., S.J. Pang, and C.H. Shek (2007). Corrosion behavior and glass-forming ability of Cu–Zr–Al–Nb alloys. *J. Non-Cryst. Solids* 353: 3596–3599.
36. Yamamoto, T., C.L. Qin, T. Zhang, K. Asami, and A. Inoue (2003). Formation, thermal stability, mechanical properties and corrosion resistance of Cu–Zr–Ti–Ni–Nb bulk glassy alloys. *Mater. Trans.* 44: 1147–1152.
37. Qin, C.L., K. Asami, T. Zhang, W. Zhang, and A. Inoue (2003). Corrosion behavior of Cu–Zr–Ti–Nb bulk glassy alloys. *Mater. Trans.* 44: 749–753.
38. Zhang, W., Q.S. Zhang, C.L. Qin, and A. Inoue (2008). Synthesis and properties of Cu–Zr–Ag–Al glassy alloys with high glass-forming ability. *Mater. Sci. Eng.* B148: 92–96.
39. Xu, W., S. Zhu, Y. Liang, Z. Li, Z. Cui, X. Yang, and A. Inoue (2015). Nanoporous CuS with excellent photocatalytic property. *Sci. Rep.* 5: 18125.
40. Xu, W., S.L. Zhu, Y. Liang, Z. Cui, X. Yang, and A. Inoue (2016). Synthesis of rutile-brookite TiO_2 by dealloying Ti–Cu amorphous alloy. *Mater. Res. Bull.* 73: 290–295.
41. Ren, H.G., W.C. Xu, S.L. Zhu, Z.D. Cui, X.J. Yang, and A. Inoue (2016). Synthesis and properties of nanoporous Ag_2S/CuS catalyst for hydrogen evolution reaction. *Electrochim. Acta* 190: 221–228.
42. Qin, C.L., C.Y. Wang, Q.F. Hu, Z.F. Wang, W.M. Zhao, and A. Inoue (2016). Hierarchical nanoporous metal/BMG composite rods with excellent mechanical properties. *Intermetallics* 77: 1–5.
43. Aburada, T., J.M. Fitz-Gerald, and J.R. Scully (2011). Synthesis of nanoporous copper by dealloying of Al–Cu–Mg amorphous alloys in acidic solution; the effect of nickel. *Corrosion Sci.* 53: 1627–1632.
44. Inoue, A. and W. Zhang (2003). Formation and mechanical properties of Cu–Hf–Al bulk glassy alloys with a large supercooled liquid region of over 90 K. *J. Mater. Res.* 18: 1435–1440.
45. Wang, Z.F., J.Y. Liu, C.L. Qin, L. Liu, W.M. Zhao, and A. Inoue (2015). Fabrication and new electrochemical properties of nanoporous Cu by dealloying amorphous Cu–Hf–Al alloys. *Intermetallics* 56: 48–55.
46. Dubal, D., R. Holze, and P. Kulal (2013). Enhanced supercapacitive performances of hierarchical porous nanostructure assembled from ultrathin MnO_2 nanoflakes, *J. Mater. Sci.* 48: 714–719.
47. Hu, Y., J. Wang, X.H. Jiang, Y.F. Zheng, and Z.X. Chen (2013). Facile chemical synthesis of nanoporous layered δ-MnO_2 thin film for high-performance flexible electrochemical capacitors, *Appl. Surf. Sci.* 271: 193–201.
48. Fan Y.F., X.D. Zhang, Y.S. Liu, Q. Cai, and J.M. Zhang (2013). One-pot hydrothermal synthesis of Mn_3O_4/graphene nanocomposite for supercapacitors. *Mater. Lett.* 95: 153–156.
49. Inoue, A. and J.S. Gook (1995). Fe-based ferromagnetic glassy alloys with wide supercooled liquid region. *Mater. Trans., JIM* 36: 1180–1183.
50. Inoue, A., A. Takeuchi, and T. Zhang (1998). Ferromagnetic bulk amorphous alloys. *Metall. Mater. Trans. A* 29A: 1779–1793.
51. Inoue, A., B.L. Shen, and H.M. Kimura (2004). Fundamental properties and applications of Fe-based bulk glassy alloys. *J. Metastable Nanocryst. Mater.* 20–21: 3–12.
52. Suryanarayana, C. and A. Inoue (2013). Iron-based bulk metallic glasses. *Internat. Mater. Rev.* 58: 131–166.
53. Shen, J., Q. Chen, J. Sun, H. Han, and G. Wang (2005). Exceptionally high glass-forming ability of an FeCoCrMoCBY alloy. *Appl. Phys. Lett.* 86: 151907-1–151907-3.

54. Asami, K., S.J. Pang, T. Zhang, and A. Inoue (2002). Preparation and corrosion resistance of Fe–Cr–Mo–C–B–P bulk glassy alloys. *J. Electrochem. Soc.* 149: B366–B369.
55. Pang, S.J., T. Zhang, K. Asami, and A. Inoue (2002). Synthesis of Fe–Cr–Mo–C–B–P bulk metallic glasses with high corrosion resistance. *Acta Mater.* 50: 489–497.
56. Pang, S.J., T. Zhang, K. Asami, and A. Inoue (2001). New Fe–Cr–Mo–(Nb,Ta)–C–B glassy alloys with high glass-forming ability and good corrosion resistance. *Mater. Trans.* 42: 376–379.
57. Pang, S.J., T. Zhang, K. Asami, and A. Inoue (2002). Bulk glassy Fe–Cr–Mo–C–B alloys with high corrosion resistance. *Corrosion Sci.* 44: 1847–1856.
58. Pang, S.J., T. Zhang, K. Asami, and A. Inoue (2002). Effects of chromium on the glass formation and corrosion behavior of bulk glassy Fe–Cr–Mo–C–B alloys. *Mater. Trans.* 43: 2137–2142.
59. Pang, S.J., T. Zhang, K. Asami, and A. Inoue (2002). Formation of bulk glassy $Fe_{75-x-y}Cr_xMo_yC_{15}B_{10}$ alloys and their corrosion behavior. *J. Mater. Res.* 17: 701–704.
60. Shen, B.L., M. Akiba, and A. Inoue (2006). Effect of Cr addition on the glass-forming ability, magnetic properties, and corrosion resistance in FeMoGaPCBSi bulk glassy alloys. *J. Appl. Phys.* 100: 043523-1–043523-5.
61. Long, Z.L., Y. Shao, X.H. Deng, Z.C. Zhang, Y. Jiang, P. Zhang, B.L. Shen, and A. Inoue (2007). Cr effects on magnetic and corrosion properties of Fe–Co–Si–B–Nb–Cr bulk glassy alloys with high glass-forming ability. *Intermetallics* 15: 1453–1458.
62. Jayaraj, J., K.B. Kim, H.S. Ahn, and E. Fleury (2007). Corrosion mechanism of N-containing Fe–Cr–Mo–Y–C–B bulk amorphous alloys in highly concentrated HCl solution. *Mater. Sci. Eng. A* 449–451: 517–520.
63. Wang, Z.M., J. Zhang, X.C. Chang, W.L. Hou, and J.Q. Wang (2008). Susceptibility of minor alloying to corrosion behavior in yttrium-containing bulk amorphous steel. *Intermetallics* 16: 1036–1039.
64. Gostin, F., U. Siegel, C. Mickel, S. Baunack, A. Gebert, and L. Schultz (2009). Corrosion behavior of the bulk glassy $(Fe_{44.3}Cr_5Co_5Mo_{12.8}Mn_{11.2}C_{15.8}B_{5.9})_{98.5}Y_{1.5}$ alloy. *J. Mater. Res.* 24: 1471–1479.
65. Tan, M.W., E. Akiyama, H. Habazaki, A. Kawashima, K. Asami, and K. Hashimoto (1996). The role of chromium and molybdenum in passivation of amorphous Fe–Cr–Mo–P–C alloys in deaerated 1 M HCl. *Corrosion Sci.* 38: 2137–2154.
66. Asami, K., M. Naka, K. Hashimoto, and T. Masumoto (1980). Effect of molybdenum on the anodic behavior of amorphous Fe–Cr–Mo–B alloys in hydrochloric acid. *J. Electrochem. Soc.* 127: 2130–2138.
67. Long, Z.L., B.L. Shen, Y. Shao, C.T. Chang, Y.Q. Zeng, and A. Inoue (2006). Corrosion behaviour of $[(Fe_{0.6}Co_{0.4})_{0.75}B_{0.2}Si_{0.05}]_{96}Nb_4$ bulk glassy alloy in sulphuric acid solutions. *Mater. Trans.* 47: 2566–2570.
68. Pardo, A., M.C. Merino, E. Otero, M.D. López, and A. M'hich (2006). Influence of Cr additions on corrosion resistance of Fe- and Co-based metallic glasses and nanocrystals in H_2SO_4. *J. Non-Cryst. Solids* 352: 3179–3190.
69. Pardo, A., E. Otero, M.C. Merino, M.D. López, M. Vázquez, and P. Agudo (2001). The influence of Cr addition on the corrosion resistance of $Fe_{73.5}Si_{13.5}B_9Nb_3Cu_1$ metallic glass in SO_2 contaminated environments. *Corrosion Sci.* 43: 689–705.
70. Amiya, K., A. Urata, N. Nishiyama, and A. Inoue (2004). Fe-B-Si-Nb bulk metallic glasses with high strength above 4000 MPa and distinct plastic elongation. *Mater. Trans.* 45: 1214–1218.
71. Qin, C.L., Q.F. Hu, Y.G. Li, Z.F. Wang, W.M. Zhao, D.V. Louzguine-Luzgin, and A. Inoue (2016). Novel bioactive Fe-based metallic glasses with excellent apatite-forming ability. *Mater. Sci. Eng. C* 69: 513–521.
72. Gebert, A., U. Wolff, A. John, and J. Eckert (2000). Corrosion behaviour of $Mg_{65}Y_{10}Cu_{25}$ metallic glass. *Scr. Mater.* 43: 279–283.
73. Subba Rao, R.V., U. Wolff, S. Baunack, J. Eckert, and A. Gebert (2003). Corrosion behaviour of the amorphous $Mg_{65}Y_{10}Cu_{15}Ag_{10}$ alloy. *Corrosion Sci.* 45: 817–832.
74. Gebert, A., R.V. Subba Rao, U. Wolff, S. Baunack, J. Eckert, and L. Schultz (2004). Corrosion behaviour of the $Mg_{65}Y_{10}Cu_{15}Ag_{10}$ bulk metallic glass. *Mater. Sci. Eng. A* 375–377: 280–284.

75. Yao, H.B., Y. Li, and A.T.S. Wee (2003). Corrosion behavior of melt-spun $Mg_{65}Ni_{20}Nd_{15}$ and $Mg_{65}Cu_{25}Y_{10}$ metallic glasses. *Electrochim. Acta* 48: 2641–2650.
76. Zberg, B., P.J. Uggowitzer, and J.F. Löffler (2009). MgZnCa glasses without clinically observable hydrogen evolution for biodegradable implants. *Nat. Mater.* 8: 887–891.
77. Zeng, Y., N. Nishiyama, T. Yamamoto, and A. Inoue (2009). Ni-rich bulk metallic glasses with high glass-forming ability and good metallic properties. *Mater. Trans.* 50: 2441–2445.
78. Habazaki, H., T. Sato, A. Kawashima, K. Asami, and K. Hashimoto (2001). Preparation of corrosion-resistant amorphous Ni–Cr–P–B bulk alloys containing molybdenum and tantalum. *Mater. Sci. Eng. A* 304–306: 696–700.
79. Habazaki, H., H. Ukai, K. Izumiya, and K. Hashimoto (2001). Corrosion behaviour of amorphous Ni–Cr–Nb–P–B bulk alloys in 6M HCl solution. *Mater. Sci. Eng. A* 318: 77–86.
80. Kawashima, A., H. Habazaki, and K. Hashimoto (2001). Highly corrosion-resistant Ni-based bulk amorphous alloys. *Mater. Sci. Eng. A* 304–306: 753–757.
81. Katagiri, H., S. Meguro, M. Yamasaki, H. Habazaki, T. Sato, A. Kawashima, K. Asami, and K. Hashimoto (2001). An attempt at preparation of corrosion-resistant bulk amorphous Ni–Cr–Ta–Mo–P–B alloys. *Corrosion Sci.* 43: 183–191.
82. Kawashima, A., T. Sato, N. Ohtsu, and K. Asami (2004). Characterization of surface of amorphous Ni–Nb–Ta–P alloys passivated in a 12 kmol/m^3 HCl solution. *Mater. Trans.* 45: 131–136.
83. Wang, A.P., X.C. Chang, W.L. Hou, and J.Q. Wang (2007). Corrosion behavior of Ni-based amorphous alloys and their crystalline counterparts. *Corrosion Sci.* 49: 2628–2635.
84. Pang, S.J., T. Zhang, K. Asami, and A. Inoue (2004). Bulk glassy Ni(Co–)Nb–Ti–Zr alloys with high corrosion resistance and high strength. *Mater. Sci. Eng. A* 375–377:368–371.
85. Qin, C.L., W. Zhang, H. Nakata, H.M. Kimura, K. Asami, and A. Inoue (2005). Effect of tantalum on corrosion resistance of Ni–Nb(–Ta)–Ti–Zr glassy alloys at high temperature. *Mater. Trans.* 46: 858–862.
86. Zhang, T., S.J. Pang, K. Asami, and A. Inoue (2003). Glassy Ni–Ta–Ti–Zr(–Co) alloys with high thermal stability and high corrosion resistance. *Mater. Trans.* 44: 2322–2325.
87. Pang, S.J., T. Zhang, K. Asami, and A. Inoue (2002). Formation of bulk glassy Ni–(Co–)Nb–Ti–Zr alloys with high corrosion resistance. *Mater. Trans.* 43: 1771–1773.
88. Pang, S.J., C.H. Shek, T. Zhang, K. Asami, and A. Inoue (2006). Corrosion behavior of glassy $Ni_{55}Co_5Nb_{20}Ti_{10}Zr_{10}$ alloy in 1 N HCl solution studied by potentiostatic polarization and XPS. *Corrosion Sci.* 48: 625–633.
89. Zeng, Y.Q., C.L. Qin, N. Nishiyama, and A. Inoue (2010). New nickel-based bulk metallic glasses with extremely high nickel content. *J. Alloys Compd.* 489: 80–83.
90. Qin, C.L., Y.Q. Zeng, D.V. Louzguine, N. Nishiyama, and A. Inoue (2010). Corrosion resistance and XPS studies of Ni-rich Ni–Pd–P–B bulk glassy alloys. *J. Alloys Compd.* 504S: S172–S175.
91. Lee, H.-J., E. Akiyama, H. Habazaki, A. Kawashima, K. Asami, and K. Hashimoto (1997). The roles of tantalum and phosphorus in the corrosion behavior of Ni–Ta–P alloys in 12 M HCl. *Corrosion Sci.* 39: 321–332.
92. Zeng, Y.Q., N. Nishiyama, T. Yamamoto, and A. Inoue (2009). Ni-rich bulk metallic glasses with high glass-forming ability and good metallic properties. *Mater. Trans.* 50: 2441–2445.
93. Li, X.-Y., E. Akiyama, H. Habazaki, A. Kawashima, K. Asami, and K. Hashimoto (1998). An XPS study of passive films on sputter-deposited Cr–Nb alloys in 12 M HCl solution. *Corrosion Sci.* 40: 821–838.
94. Wang, A.P. and J.Q. Wang (2007). A topological approach to design Ni-based bulk metallic glasses with high corrosion resistance. *J. Mater. Res.* 22: 1–4.
95. Guo, F., H.-J. Wang, and S.J. Poon (2005). Ductile titanium-based glassy alloy ingots. *Appl. Phys. Lett.* 86: 091907-1–0919017-3.
96. Kawashima, A., K. Asami, and K. Hashimoto (1985). Change in corrosion behavior of amorphous Ni–P alloys by alloying with chromium, molybdenum or tungsten. *J. Non-Cryst. Solids* 70: 69–83.

97. Zhang, L., M.Q. Tang, Z.W. Zhu, H.M. Fu, H.W. Zhang, A.M. Wang, H. Li, H.F. Zhang, and Z.Q. Hu (2015). Compressive plastic metallic glasses with exceptional glass-forming ability in the Ti–Zr–Cu–Fe–Be system. *J. Alloys Compd.* 638: 349–355.
98. Tang, M.Q., H.F. Zhang, Z.W. Zhu, H.M. Fu, A.M. Wang, H. Li, and Z.Q. Hu (2010). TiZr-base bulk metallic glass with over 50 mm in diameter. *J. Mater. Sci. Technol.* 26: 481–486.
99. Qin, F.X., X.M. Wang, A. Kawashima, S.L. Zhu, H.M. Kimura, and A. Inoue (2006). Corrosion behavior of Ti-based metallic glasses. *Mater. Trans.* 47: 1934–1937.
100. Pang, S.J., C.H. Shek, C.L. Ma, A. Inoue, and T. Zhang (2007). Corrosion behavior of a glassy Ti–Zr–Hf–Cu–Ni–Si alloy. *Mater. Sci. Eng. A* 449–451: 557–560.
101. Pang, S.J., H. Men, C.H. Shek, C.L. Ma, A. Inoue, and T. Zhang (2007). Formation, thermal stability and corrosion behavior of glassy $Ti_{45}Zr_5Cu_{45}Ni_5$ alloy. *Intermetallics* 15: 683–686.
102. Qin, C.L., J.J. Oak, N. Ohtsu, K. Asami, and A. Inoue (2007). XPS study on the surface films of a newly designed Ni-free Ti-based bulk metallic glass. *Acta Mater.* 55: 2057–2063.
103. Qin, F.X., X.M. Wang, S.L. Zhu, A. Kawashima, K. Asami, and A. Inoue (2007). Fabrication and corrosion property of novel Ti-based bulk glassy alloys without Ni. *Mater. Trans.* 48: 515–518.
104. Morrison, M.L., R.A. Buchanan, A. Peker, P.K. Liaw, and J.A. Horton (2007). Electrochemical behavior of a Ti-based bulk metallic glass. *J. Non-Cryst. Solids* 353: 2115–2124.
105. Peker, A. and W.L. Johnson (1993). A highly processable metallic glass: $Zr_{41.2}Ti_{13.8}Cu_{12.5}Ni_{10.0}Be_{22.5}$. *Appl. Phys. Lett.* 63: 2342–2344.
106. Lou, H.B., X.D. Wang, F. Xu, S.Q. Ding, Q.P. Cao, K. Hono, and J.Z. Jiang (2011). 73 mm-diameter bulk metallic glass rod by copper mold casting. *Appl. Phys. Lett.* 99: 051910-1–051910-3.
107. Inoue, A. and N. Nishiyama (2007). New bulk metallic glasses for applications as magnetic-sensing, chemical, and structural materials. *MRS Bull.* 32: 651–658.
108. Liquid metal technologies. http://www.liquidmetal.com.
109. Inoue, A., T. Zhang, N. Nishiyama, K. Ohba, and T. Masumoto (1993). Preparation of 16 mm diameter rod of amorphous $Zr_{65}Al_{7.5}Ni_{10}Cu_{17.5}$ alloy. *Mater. Trans., JIM* 34: 1234–1237.
110. Lin, X.H. and W.L. Johnson (1995). Formation of Ti–Zr–Cu–Ni bulk metallic glasses. *J. Appl. Phys.* 78: 6514–6519.
111. Pang, S.J., T. Zhang, K. Asami, and A. Inoue (2003). Formation, corrosion behavior, and mechanical properties of bulk glassy Zr–Al–Co–Nb alloys. *J. Mater. Res.* 18: 1652–1658.
112. Hiromoto, S., A.P. Tsai, M. Sumita, and T. Hanawa (2002). Polarization behavior of bulk Zr-base amorphous alloy immersed in cell culture medium. *Mater. Trans.* 43: 3112–3117.
113. Hiromoto, S., A.P. Tsai, M. Sumita, and T. Hanawa (2000). Effect of chloride ion on the anodic polarization behavior of the $Zr_{65}Al_{7.5}Ni_{10}Cu_{17.5}$ amorphous alloy in phosphate buffered solution. *Corrosion Sci.* 42: 1651–1660.
114. Hiromoto, S., A.P. Tsai, M. Sumita, and T. Hanawa (2000). Effects of surface-finishing and dissolved oxygen on the polarization behavior of $Zr_{65}Al_{7.5}Ni_{10}Cu_{17.5}$ amorphous alloy in phosphate buffered solution. *Corrosion Sci.* 42: 2167–2185.
115. Green, B.A., R.V. Steward, I. Kim, C.K. Choi, P.K. Liaw, K.D. Kihm, and Y. Yokoyama (2009). In situ observation of pitting corrosion of the $Zr_{50}Cu_{40}Al_{10}$ bulk metallic glass. *Intermetallics* 17: 568–571.
116. Gebert, A., K. Mummert, J. Eckert, L. Schultz, and A. Inoue (1997). Electrochemical investigations on the bulk glass forming $Zr_{55}Cu_{30}Al_{10}Ni_5$ alloy. *Mater. Corrosion* 48: 293–297.
117. Gebert, A., K. Buchholz, A.M. El-Aziz, and J. Eckert (2001). Hot water corrosion behaviour of Zr–Cu–Al–Ni bulk metallic glass. *Mater. Sci. Eng. A* 316: 60–65.
118. Raju, V.R., U. Kühn, U. Wolff, F. Schneider, J. Eckert, R. Reiche, and A. Gebert (2002). Corrosion behaviour of Zr-based bulk glass-forming alloys containing Nb or Ti. *Mater. Lett.* 57: 173–177.
119. Qiu, C.L., L. Liu, M. Sun, and S.M. Zhang (2005). The effect of Nb addition on mechanical properties, corrosion behavior, and metal-ion release of ZrAlCuNi bulk metallic glasses in artificial body fluid. *J. Biomed. Mater. Res.* 75A: 950–956.
120. Liu, L., C.L. Qiu, H. Zou, and K.C. Chan (2005). The effect of the microalloying of Hf on the corrosion behavior of ZrCuNiAl bulk metallic glass. *J. Alloys Compd.* 399: 144–148.

121. Huang, L., D.C. Qiao, B.A. Green, P.K. Liaw, J.F. Wang, and S.J. Pang (2009). Bio-corrosion study on zirconium-based bulk-metallic glasses. *Intermetallics* 17: 195–199.
122. Dhawan, A., S. Roychowdhury, P.K. De, and S.K. Sharma (2005). Potentiodynamic polarization studies on bulk amorphous alloys and $Zr_{46.75}Ti_{8.25}Cu_{7.5}Ni_{10}Be_{27.5}$ and $Zr_{65}Cu_{17.5}Ni_{10}Al_{7.5}$. *J. Non-Cryst. Solids* 351: 951–955.
123. Morrison, M.L., R.A. Buchanan, R.V. Leon, C.T. Liu, B.A. Green, P.K. Liaw, and J.A. Horton (2005). The electrochemical evaluation of a Zr-based bulk metallic glass in a phosphate-buffered saline electrolyte. *J. Biomed. Mater. Res.* 74A: 430–438.
124. He, G., Z. Bian, and G.L. Chen (2001). Corrosion behavior of a Zr-base bulk glassy alloy and its crystallized counterparts. *Mater. Trans.* 42: 1109–1111.
125. Chieh, T.C., J. Chu, C.T. Liu, and J.K. Wu (2003). Corrosion of $Zr_{52.5}Cu_{17.9}Ni_{14.6}Al_{10}Ti_5$ bulk metallic glasses in aqueous solutions. *Mater. Lett.* 57: 3022–3025.
126. Morrison, M.L., R.A. Buchanan, A. Peker, W.H. Peter, J.A. Horton, and P.K. Liaw (2004). Cyclic-anodic-polarization studies of a $Zr_{41.2}Ti_{13.8}Ni_{10}Cu_{12.5}Be_{22.5}$ bulk metallic glass. *Intermetallics* 12: 1177–1181.
127. Schroeder, V., C.J. Gilbert, and R.O. Ritchie (1998). Comparison of the corrosion behavior of a bulk amorphous metal, $Zr_{41.2}Ti_{13.8}Cu_{12.5}Ni_{10}Be_{22.5}$, with its crystallized form. *Scr. Mater.* 38: 1481–1485.
128. Qin, F.X., H.F. Zhang, P. Chen, F.F. Chen, D.C. Qiao, and Z.Q. Hu (2004). Corrosion behavior of bulk amorphous $Zr_{55}Al_{10}Cu_{30}Ni_{5-x}Pd_x$ alloys. *Mater. Lett.* 58: 1246–1250.
129. Qin, F.X., H.F. Zhang, Y.F. Deng, B.Z. Ding, and Z.Q. Hu (2004). Corrosion resistance of Zr-based bulk amorphous alloys containing Pd. *J. Alloys Compd.* 375: 318–323.
130. Vishwanadh, B., G.J. Abraham, Jagannath, S. Neogy, R.S. Dutta, and G.K. Dey (2009). Effect of structural defects, surface irregularities, and quenched-in defects on corrosion of Zr-based metallic glasses. *Metall. Mater. Trans. A* 40A: 1131–1141.
131. Gebert, A., K. Buchholz, A. Leonhard, K. Mummert, J. Eckert, and L. Schultz (1999). Investigations on the electrochemical behaviour of Zr-based bulk metallic glasses. *Mater. Sci. Eng. A* 267: 294–300.
132. Amiya, K. and A. Inoue (2002). Formation, thermal stability and mechanical properties of Ca-based bulk glassy alloys. *Mater. Trans.* 43: 81–84.
133. Amiya, K. and A. Inoue (2002). Formation and thermal stability of Ca–Mg–Ag–Cu bulk glassy alloys. *Mater. Trans.* 43: 2578–2581.
134. Senkov, O.N., D.B. Miracle, V. Keppens, and P.K. Liaw (2008). Development and characterization of low-density Ca-based bulk metallic glasses: An overview. *Metall. Mater. Trans. A* 39A: 1888–1900.
135. Dahlman, J., O.N. Senkov, J.M. Scott, and D.B. Miracle (2007). Corrosion properties of Ca-based bulk metallic glasses. *Mater. Trans.* 48: 1850–1854.
136. Morrison, M.L., R.A. Buchanan, O.N. Senkov, D.B. Miracle, and P.K. Liaw (2006). Electrochemical behavior of Ca-based bulk metallic glasses. *Metall. Mater. Trans. A* 37A: 1239–1245.

8

Mechanical Behavior

8.1 Introduction

The mechanical properties of materials play a very important role in their applications, and therefore their characterization has been of great value and critical to the exploitation of materials in human civilization. Metallic materials are traditionally crystalline in nature, and their mechanical behavior is determined essentially by the nature and density of dislocations, and their ability to move and interact with other dislocations and microstructural features. The presence of dislocations in crystals has also been noted to be the reason for their low strength (compared to the theoretical value) and their plastic deformation behavior. Since metallic glasses are noncrystalline in nature, and therefore do not contain any dislocations, they are expected to exhibit high theoretical strength, but not sufficient ductility.

Even though metallic glasses were synthesized in the form of splats as early as 1960 [1], the study of the mechanical behavior of these novel materials started only in the 1970s, with the first publication on a study of $Pd_{80}Si_{20}$ ribbons by Masumoto and Maddin [2]. Ribbons about 15–50 μm thick and 0.05–0.5 mm in width were obtained by rapidly quenching the molten alloy along the inside cylindrical surface of a rotating drum. Since then, numerous research papers have been published on glassy materials produced by improved solidification techniques that have yielded specimens that are better in their surface appearance and also have a very uniform cross section. Interesting and seminal contributions have been made regarding the deformation mechanisms, failure criteria, and the origins of strength, ductility, and toughness, and these have been well documented in some reviews (see, e.g., Refs. [3–7], to mention a few). It was shown that metallic glasses are very strong with the yield strength exceeding 3 GPa in some metal–metalloid systems. It was also shown that they exhibited very limited plastic strain (often less than about 0.5%) in tension and that inhomogeneous deformation occurred through the formation of shear bands. Even though many different theories were proposed to understand and explain the mechanical behavior of these thin ribbon specimens, the reliability of the results was poor due to the limited size of the samples and thus a poorly defined and ill-controlled stress state during mechanical testing. It was felt that the availability of "bulk" specimens would answer some of the puzzling questions most satisfactorily.

The successful synthesis of bulk metallic glasses (BMGs) in the late 1980s/early 1990s [8,9] changed the situation completely. There have been many different alloy compositions and a number of different alloy systems in which BMGs could be produced with a diameter of at least a few millimeters and in some cases a few centimeters. As a result of this development and the availability of advanced characterization techniques, research on the mechanical behavior of BMGs has been very active during the last few years. In stark contrast to the large number of reviews on the mechanical behavior of melt-spun glassy alloy ribbons, there has been a conspicuous absence of reviews on the *mechanical behavior of BMGs*, even though some general reviews have appeared in the literature. However, a very comprehensive and exhaustive review appeared in 2007 [10], and a few more subsequently [11–20].

8.2 Deformation Behavior

Metallic glasses, including BMGs, have a very high tensile strength. But, their Achilles' heel is their low room temperature ductility. The metallic glasses generally fail soon after yielding without showing any signs of a reasonable amount of plastic deformation. At high temperatures, however, the deformation behavior is quite different.

The deformation behavior of metallic glasses can be described as inhomogeneous at low temperatures and high stresses and strain rates, and as homogeneous at high temperatures and high strain rates. At low temperatures, lower than about $0.5\,T_g$, where T_g is the glass transition temperature, deformation is mostly concentrated in a few very thin "shear bands" that form approximately on the planes of maximum resolved shear stress. These planes are inclined close to 45° to the loading axis. This localized deformation is referred to as *inhomogeneous deformation*. It is the inhomogeneous deformation in the metallic glass that renders it mechanically unstable at high stresses. Consequently, it fails catastrophically.

On the other hand, at high temperatures, greater than about $0.5\,T_g$, metallic glasses undergo viscous flow, in which plastic strain is distributed continuously, but not necessarily equally, between different volume elements within the material. That is, each volume element of the specimen contributes to the strain. This type of deformation is referred to as *homogeneous deformation*.

When a BMG sample is subjected to a tensile test, the sample deforms elastically with a maximum elastic strain of about 2%, and then fractures catastrophically. That is, a typical stress–strain curve shows only the elastic portion. On the other hand, when a BMG sample is tested under compression, the stress–strain curve consists of elasticity followed by a small amount (up to 1%) of plastic strain. Serrated flow is usually observed at an early stage of compressive deformation. But, it will be shown later in the chapter that innovative methods of altering the chemical composition or introducing a crystalline phase into the glassy alloy have helped in achieving large amounts of ductility in these traditionally brittle materials.

8.2.1 Inhomogeneous Deformation

This type of deformation is characterized by the formation of shear bands, their rapid propagation, and the sudden fracture of the sample.

Metallic glasses have been shown to exhibit the phenomenon of strain softening. That is, an increase in strain makes the material softer and allows the material to be deformed at lower stresses and higher rates. In contrast, crystalline materials undergo strain hardening, that is, with increasing strain, the material becomes harder and therefore it is difficult to further deform the material. Shear band formation or shear localization has been considered a direct consequence of strain softening.

Shear softening and the formation of shear bands in metallic glasses have been attributed to a local decrease in the viscosity of the glass. A number of different reasons have been suggested for this phenomenon. These include local production of free volume due to flow dilatation, the local evolution of structural order due to the shear transformation zone (STZ) operations, redistribution of internal stresses associated with STZ operations, and local heating.

All the reasons suggested for inhomogeneous deformation through the formation of shear bands can be grouped under two hypotheses. The first hypothesis suggests that the viscosity in shear bands decreases during deformation due to the formation of free volume. This decreases the density of the glass and, consequently, its resistance to deformation. Spaepen [21] derived an expression for steady-state inhomogeneous flow in metallic glasses on the basis of a competition between the stress-driven creation and the diffusional annihilation of free volume. Argon [22] has demonstrated that flow localization occurs in a shear band in which the strain rate has been perturbed due to the creation of free volume. The model attributed to Spaepen was subsequently developed by Steif et al. [23], who derived an expression for the stress at which catastrophic softening due to free volume creation occurs during uniform shearing of a homogeneous body under constant applied strain rate.

The second hypothesis suggests that local adiabatic heating occurs in the shear bands, which leads to a decrease in the viscosity of the metallic glass by several orders of magnitude [24]. This adiabatic heating could lead to a substantial increase in the temperature to a level above the glass transition temperature or even beyond the melting temperature of the alloy. Experimental evidence is available for both the increase in free volume and rise in temperature in shear bands during deformation. The aspect of temperature rise in the shear bands will be revisited at a later stage. Schuh et al. [10] have examined the mechanism of local increase in the free volume or evolution of structural order in the glass.

When shear bands form and propagate, a sudden drop in load is noticed in the stress–strain plot. The surrounding material recovers elastically and arrests the shear-band propagation. When this process gets repeated, we obtain serrated flow. The free volume within the shear band increases during deformation, thereby decreasing its density and hence resistance to deformation. Figure 8.1 shows the compressive stress–strain curve for a $Zr_{40}Ti_{14}Ni_{10}Cu_{12}Be_{24}$ BMG alloy tested at a strain rate of 1×10^{-4} s^{-1} [25]. The formation of

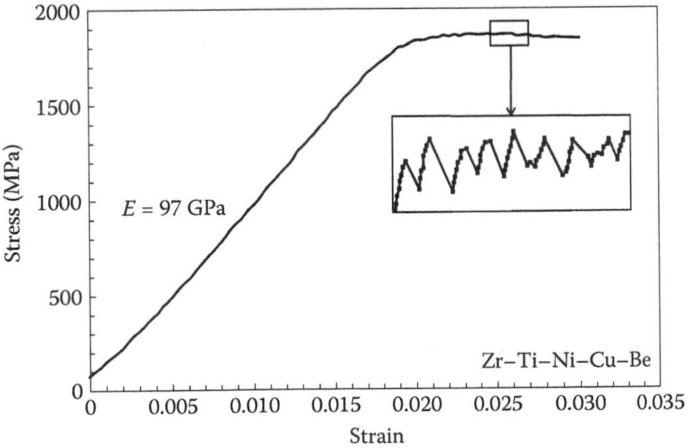

FIGURE 8.1
Compressive stress–strain curve for $Zr_{40}Ti_{14}Ni_{10}Cu_{12}Be_{24}$ BMG alloy tested at a strain rate of 1×10^{-4} s^{-1}. (Reprinted from Wright, W.J. et al., *Mater. Trans.*, 42, 642, 2001. With permission.)

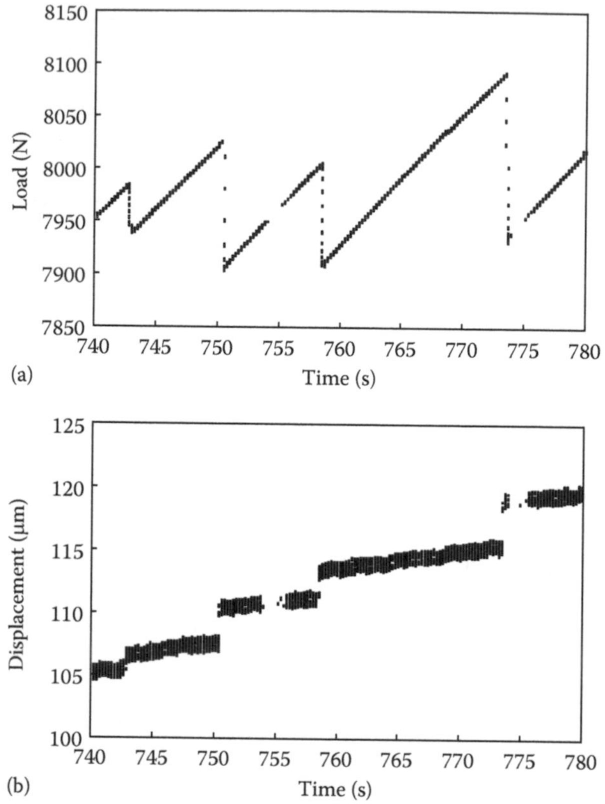

FIGURE 8.2
(a) Load as a function of time and (b) total displacement as a function of time in the serrated flow region of the $Zr_{40}Ti_{14}Ni_{10}Cu_{12}Be_{24}$ BMG alloy tested in uniaxial compression. (Reprinted from Wright, W.J. et al., *Mater. Trans.*, 42, 642, 2001. With permission.)

multiple serrations is worth noting. The magnitude of the load drop, the displacements (elastic, plastic, and total), and the time elapsed during unloading were analyzed for individual serrations. The load versus time and displacement versus time are plotted in Figure 8.2a and b, respectively. Using these data and the modulus of elasticity, and through some analyses, these authors were able to estimate the temperature rise in a shear band.

Serrated flow is also observed during nanoindentation, but only at slow loading rates. Activation of each individual shear band is associated with the occurrence of a discrete "pop-in" event (sudden rise in load). A single shear band can rapidly accommodate the deformation at slow loading rates. But, when the loading rate is high, there is not enough time to accommodate the strain and, consequently, multiple shear bands have to operate simultaneously, resulting in a smooth load–displacement curve. Therefore, the "pop-in" events are more pronounced at low rates of loading and their occurrence reduces with increasing loading rates. This situation is quite similar to the one during compression, where serrations are observed only at slow strain rates and not at high strain rates. Figure 8.3 shows the formation of "pop-in" events (i.e., serrations during nanoindentation) in a number of different BMGs. In all cases, it may be noted that serrations are observed only at low loading rates and not at high loading rates; a smooth curve is obtained at high loading rates [26]. The analysis of the experimental data reveals that there exists a critical applied

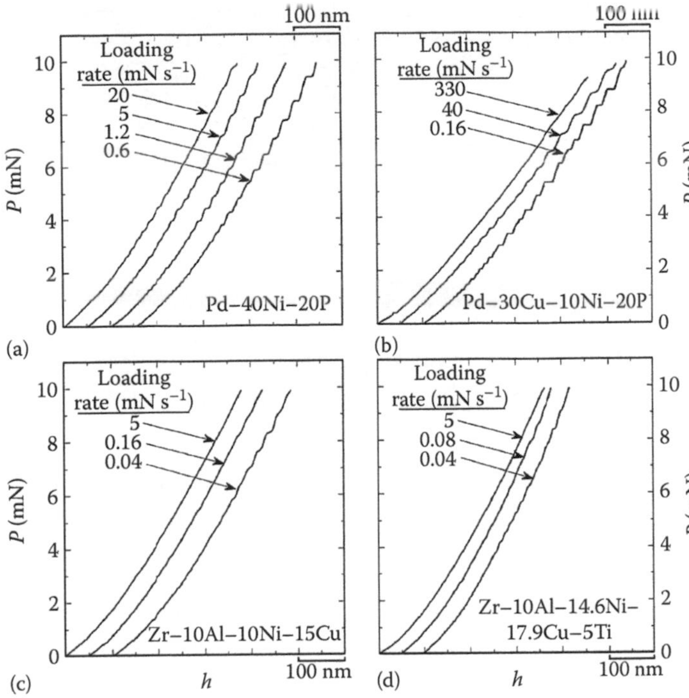

FIGURE 8.3
Typical load–displacement (*P*–*h*) curves measured on the loading portion of nanoindentation experiments, for four different BMGs investigated. (a) $Pd_{40}Ni_{40}P_{20}$, (b) $Pd_{40}Cu_{30}Ni_{10}P_{20}$, (c) $Zr_{65}Al_{10}Ni_{10}Cu_{15}$, and (d) $Zr_{52.5}Al_{10}Ni_{14.6}Cu_{17.9}Ti_5$. Curves are offset from the origin for clear viewing, and the rate of indentation loading is specified in each graph. (Reprinted from Schuh, C.A. and Nieh, T.G., *Acta Mater.*, 51, 87, 2003. With permission.)

strain rate above which serrated flow is completely suppressed. By separating the elastic and plastic contributions to deformation, the authors were able to show that, at sufficiently low indentation rates, plastic deformation occurs entirely in discrete events of isolated shear banding, while at higher rates, deformation is continuous, without any evidence of discrete events at any size scale.

Figure 8.4 shows a scanning electron micrograph of the surface of a specimen deformed under compression showing the presence of shear bands [27]. However, the nature of the shear band is not very clear, except that there are regions where deformation is concentrated. The role of shear bands in the deformation of metallic glasses will be discussed again in detail at a later stage.

8.2.2 Homogeneous Deformation

Homogeneous deformation in metallic glasses occurs at high temperatures, for example, $>0.7\,T_g$, and also in the supercooled liquid state. This deformation could be thought of as viscous flow of the supercooled liquid (it can result in significant plasticity) and is therefore of commercial importance. It is possible to achieve net-shape forming capability by working the metallic glass in this temperature regime [28–31]. The transition temperature between the inhomogeneous and homogeneous deformation can be thought of as a brittle-to-ductile transition, and is strongly dependent on the applied strain rate, suggesting that homogeneous deformation is associated with a rate process.

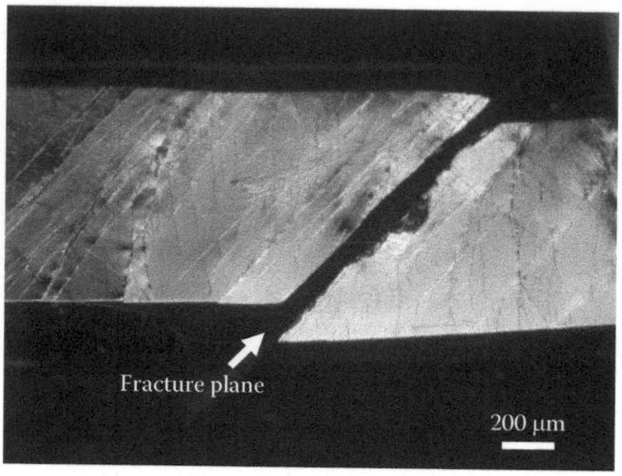

FIGURE 8.4
An optical micrograph showing the formation of multiple shear bands in an $Ni_{50}Pd_{30}P_{20}$ BMG specimen subjected to compression testing. (Reprinted from Wang, K. et al., *Acta Mater.*, 56, 2834, 2008. With permission.)

By conducting steady-state flow tests over a large range of stresses and strain rates, and in a number of different alloy systems at different temperatures near the glass transition, it has been observed that at high temperatures and low applied stresses, flow is Newtonian, that is, strain rate is proportional to stress. But, with increasing applied stresses, the stress sensitivity of deformation drops rapidly and the flow is non-Newtonian. The homogeneous deformation of metallic glasses can also be understood using the concept of STZ proposed by Argon [22].

Schuh et al. [10] have elegantly described the concept of an STZ. The STZ is essentially a local cluster of atoms (a few to perhaps up to 100) that undergoes an inelastic shear distortion from one relatively low-energy configuration to another low-energy configuration, crossing an activation barrier. In other words, an STZ may be considered as the basic shear unit, which consists of a free volume site with immediate adjacent atoms. These atomic-scale STZs collectively deform under an applied shear stress to produce macroscopic shear deformation. But, it should be realized that the STZ is not a structural defect in a glassy material in the way that a dislocation is a defect in a crystal. Instead, an STZ is an event defined in a local volume and not a feature of the glass. In other words, an STZ is defined by its transience, that is, an observer, inspecting the glass at a single instant in time, cannot identify the STZ.

Homogeneous deformation can be considered as a statistical superposition of many independent atomic-scale events with a characteristic size and energy scale. By fitting the rate equations to the experimental flow stress data, it was noted that the activation energy for homogeneous deformation was about 4.6 eV for a $Zr_{41.2}Ti_{13.8}Cu_{12.5}Ni_{10}Be_{22.5}$ BMG alloy, which represents an energy that is about 85 times kT_g, where k is the Boltzmann constant. The activation volume was calculated to be ~7.5×10^{-28} m^{-3}, roughly 40 times the Goldschmidt volume of the Zr atom, suggesting that the STZ unit contains about 20–30 atoms. Both these estimates are in the expected range.

The steady-state shear strain rate, $\dot{\gamma}$, in the Newtonian regime can be expressed as

$$\dot{\gamma} = \frac{\alpha_o \nu_o \gamma_o V}{kT} . \exp\left(-\frac{Q}{kT}\right)\tau \qquad \left(\tau \ll \frac{kT}{V}\right) \tag{8.1}$$

and that in the non-Newtonian regime as

$$\dot{\gamma}=\frac{1}{2}\alpha_o\nu_o\gamma_o.\exp\left(-\frac{Q-\tau V}{kT}\right)\quad\left(\tau\gg\frac{kT}{V}\right) \tag{8.2}$$

where:

- α_o incorporates numerical factors as well as the fraction of the material that is available to deform via the activated process
- ν_o is the attempt frequency, which is essentially the frequency of the fundamental mode vibration along the reaction pathway
- γ_o is the shear strain
- V is the product of the characteristic STZ volume and shear strain
- Q is the activation energy for the process
- k is the Boltzmann constant
- T is the absolute temperature
- τ is the applied shear stress

The stress sensitivity varies continuously between the low and high stress limits expressed by Equations 8.1 and 8.2.

Expressing Equations 8.1 and 8.2 as a general power law, we can include the stress sensitivity term, n, as

$$\dot{\gamma}=A\tau^n \tag{8.3}$$

where A is a temperature-dependent constant. As the strain rate increases, the value of n increases from unity (Equation 8.1) to extremely high values. Whereas the value of $n = 1$ is associated with stable flow (Newtonian behavior), higher values of n are associated with flow instabilities. It is possible to achieve large plastic strains in metallic glasses in the Newtonian range ($n = 1$). For example, Kawamura et al. [32] obtained elongations of up to 1260% in a $Pd_{40}Ni_{40}P_{20}$ BMG alloy at a strain rate of 1.7×10^{-1} s^{-1} at a temperature of 620 K, and under about 70 MPa. This temperature is about 1.07 times the T_g and 0.64 times the T_m of the BMG alloy (T_m is the melting point of the alloy). Large elongations have also been obtained in other alloy systems. Kim et al. [33] achieved an elongation of 1350% in a $Zr_{65}Al_{10}Ni_{10}Cu_{15}$ BMG alloy at 696 K and a strain rate of 2×10^{-2} s^{-1}. Wang et al. [34] reported elongations of 1623% at 676 K and 1.52×10^{-2} s^{-1} and 1624% at 656 K and 7.58×10^{-3} s^{-1} in a $Zr_{41.25}Ti_{13.75}Cu_{12.5}Ni_{10}Be_{22.5}$ BMG alloy. In contrast, it was noted that deformation with non-Newtonian rheology exhibited much reduced elongations. Nieh et al. [35,36] suggested that the transition from Newtonian to non-Newtonian behavior is associated with the precipitation of nanocrystals within the glassy matrix during high-temperature deformation. A number of investigations have confirmed that homogeneous flow is often associated with the partial crystallization of glass. In fact, crystallization is known to be normally accelerated by pressure (in tension, compression, and more complex loading conditions). The effect of pressure on the crystallization of metallic glass ribbons is well known [37,38].

When nanocrystallization occurs in a BMG, the material contains nanocrystals dispersed in a glassy matrix. The deformation rate of such a composite can then be expressed by the equation:

$$\dot{\gamma}_{\text{total}}=f_c\dot{\gamma}_c+(1-f_c)\dot{\gamma}_g \tag{8.4}$$

where:

$\dot{\gamma}_{total}$	is the total strain rate
$\dot{\gamma}_c$ and $\dot{\gamma}_g$	are the strain rates resulting from the crystalline and glassy phases, respectively
f_c	is the volume fraction of the crystalline phase in the composite

The plastic flow of the glassy matrix is described by Newtonian behavior, that is, $\dot{\gamma}_g = A\tau$, but the plastic flow of the nanocrystalline phase is described by another power law, with a nonlinear power-law dependence. Therefore, Equation 8.4 can be rewritten in the form:

$$\dot{\gamma}_{total} = (1 - f_c)A\tau + f_c B\tau^n \tag{8.5}$$

where A and B are material constants. Thus, when nanocrystals are present in the composite, the strain rate sensitivity of the composite is no longer unity.

8.3 Deformation Maps

Deformation maps were developed by Ashby and Frost [39,40] for crystalline materials. These maps help in delineating the different modes and mechanisms of plastic deformation of a material as a function of stress, temperature, and structure. The steady-state constitutive flow law describes each deformation mode using an equation of the type

$$\dot{\gamma} = f(\tau, T, \text{structure}) \tag{8.6}$$

where:

$\dot{\gamma}$	is the strain rate
τ	is the shear stress
T	is the temperature
structure	represents all the relevant structural parameters of the material

The steady-state condition implies that the structural parameters are uniquely determined by the external parameters—stress and temperature—and hence remain constant during the course of the flow.

Based on these concepts, Spaepen [21] introduced an empirical deformation map for metallic glasses, in which he calculated the boundary between homogeneous and inhomogeneous flow regions. Figure 8.5 represents a schematic deformation map of a metallic glass, in which the various modes of deformation are indicated. The stress is plotted on the y-axis on a logarithmic scale as a fraction of the shear stress, and temperature is plotted on the x-axis. The temperature region between T_x and T_m (where T_x and T_m represent the temperatures at which crystallization of the glass and melting of the crystallized material occur, respectively) is inaccessible for mechanical measurements since the glass crystallizes at T_x. But, reasonable extrapolations of the strain rate contours could be made through this region. Based on experimental observations, the map indicates that the homogeneous mode of deformation occurs at low stresses and high temperatures, and that the stress is a strong function of the strain rate, as indicated by the strain rate contours. On the other hand, inhomogeneous deformation occurs at high stress levels and low temperatures, and

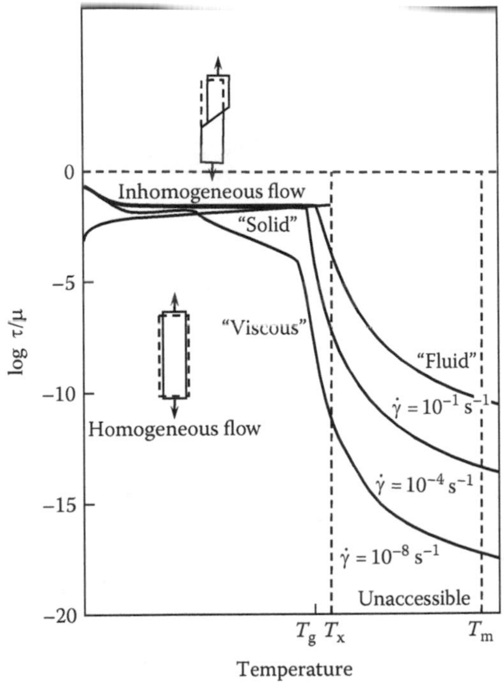

FIGURE 8.5
A schematic deformation map for a metallic glass with stress plotted on the *y*-axis and temperature on the *x*-axis. In this diagram, τ represents the shear stress, μ the shear modulus, *T* the temperature, T_g the glass transition temperature, T_x the crystallization temperature, and T_m the melting temperature. The stress–temperature regions where inhomogeneous and homogeneous flow occurs are indicated. (Reprinted from Spaepen, F., *Acta Metall.*, 25, 407, 1977. With permission.)

in this region, the stress is virtually insensitive to the strain rate. But, it will be shown later that the available results indicate that there is no consensus on this aspect. It is also important to note that, unlike melt-spun metallic glass ribbons, BMGs exhibit a wide supercooled liquid region, ΔT_x (= $T_x - T_g$). Therefore, detailed investigations have been carried out to study the deformation behavior of BMGs in this temperature regime [28,34,35].

It has been clearly shown by Schuh et al. [10] that both the modes of deformation (inhomogeneous and homogeneous) can be conveniently explained by using the concept of STZ. That is, one single mechanism is able to describe every aspect of glass deformation. Therefore, instead of "deformation mechanism map," these authors preferred to call this simply a *deformation map*.

Schuh et al. [10] have developed deformation maps for BMGs, very similar to the deformation mechanism map developed by Spaepen [21] using the results for melt-spun metallic glasses. The maps in both cases look somewhat similar except for the details. Figure 8.6 shows two complementary deformation maps. The first one plots the coordinates of stress and temperature and follows the form developed by Spaepen [21]. The second one plots the strain rate as a function of temperature, and follows the analysis suggested by Megusar et al. [41]. In the stress–temperature plot, strain rates are represented as a series of contours. Similarly, in the strain rate–temperature plot, stress is represented as a series of contours. In both the maps, the stress is represented as a fraction of the shear modulus, which allows for approximate generalizations to be made.

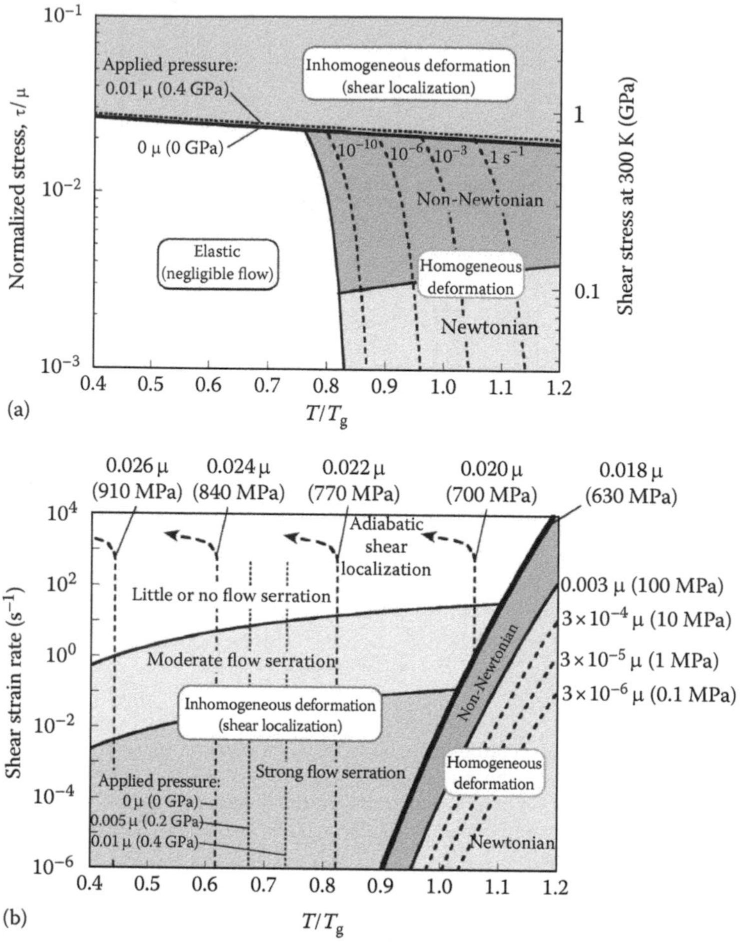

FIGURE 8.6
Deformation maps for metallic glasses in (a) stress–temperature and (b) strain rate–temperature axes. The main division on the map separates homogeneous deformation at high temperatures and low stresses/rates from the inhomogeneous flow (shear localization) that occurs at lower temperatures and higher stresses/rates. In the homogeneous regime, contours for steady-state flow are indicated, as is the transition from Newtonian to non-Newtonian flow. In the inhomogeneous regime, the effect of confining hydrostatic pressure is shown, and various degrees of flow serration are denoted in (b). The thick solid lines in both the maps indicate the transition from inhomogeneous to homogeneous flow. The absolute stress values shown are for the specific glass $Zr_{41.25}Ti_{13.75}Cu_{12.5}Ni_{10}Be_{22.5}$. (Reprinted from Schuh, C.A. et al., *Acta Mater.*, 55, 4067, 2007. With permission.)

Just like in the Spaepen map, here also the most important aspect is the stress–strain rate and/or temperature at which transition occurs from the homogeneous mode to inhomogeneous mode of deformation. Whereas Spaepen [21] considered this transition in terms of the free-volume model, Schuh et al. [10] examined this in terms of the STZ operation. The transition from inhomogeneous to homogeneous flow is indicated in both the maps of Figure 8.6 by a thick solid line. In the stress–temperature map, this line also directly indicates the operational strength of the glass prior to flow localization, which decreases slightly with temperature. In the strain rate–temperature map, the shape of the transition curve is roughly parallel to a family of iso-stress contours that can be drawn in the homogenous regime.

Further, within the homogeneous regime, subregions were marked with "elastic," "Newtonian," and "non-Newtonian." The homogeneous deformation can be neglected below a strain rate of 10^{-12} s^{-1}. The Newtonian–non-Newtonian transition is delineated at ~10^{-5} s^{-1}. Non-Newtonian flow is observed at strain rates of <10^{-5} s^{-1} and Newtonian flow above this value. However, it is important to note that at high enough shear rates, non-Newtonian flow as well as shear localization can occur at high temperatures—even in the supercooled liquid region. Schuh et al. [10] caution that these deformation maps do not consider the evolution of the glass structure during deformation and also that they lack information about their fracture behavior. Despite this, it is suggested that "these maps could be used as semiquantitative tools, for example, for rationalizing observed trends with applied rate or temperature, or for comparing observed mechanical responses for different glasses tested at a common absolute temperature (room temperature) but at different homologous temperatures (T/T_g)" [10].

Schuh et al. [42] conducted instrumented nanoindentation studies over four decades of indentation strain rate and temperatures on $Pd_{40}Ni_{40}P_{20}$ and $Mg_{65}Cu_{25}Gd_{10}$ BMG alloys, examining indentation strain rates at different temperatures from ambient to the glass transition. Based on their analysis, they had observed two distinct regimes of homogeneous flow. While the first regime, as expected, occurs with the onset of viscous flow at high temperatures and low rates, the second one occurs at high deformation rates well below the glass transition temperature. This arises when deformation rates exceed the characteristic rate for shear band nucleation, kinetically forcing strain distribution. The authors have labeled this as "Homogeneous II."

One would notice that Figure 8.6 does not show the Homogeneous II mode of deformation. The transition from the inhomogeneous to the homogeneous mode was shown to be rather sudden in Ref. [42]. But, in reality, the shear bands proliferate at higher rates also and the transition from the inhomogeneous to homogeneous mode is quite gradual and spread over a wide range of strain rates and temperatures. Therefore, instead of putting a single dividing line between the two, Schuh et al. [10] modified the map with the inclusion of strain rate contours.

Furukawa and Tanaka [43] have considered the fracture behavior of BMGs and suggested that the nonlinear behavior associated with their fracture is a consequence of the coupling between density fluctuations and deformation fields. The shear-induced enhancement of density fluctuations is self-amplified by the resulting enhancement of dynamic and elastic asymmetry between denser and less-dense regions.

8.4 Temperature Rise at Shear Bands

As mentioned earlier, inhomogeneous deformation at low temperatures and high stresses takes place through the formation and propagation of shear bands. Most of the plastic strain is localized in these narrow shear bands, which form approximately on the planes of maximum resolved shear stress. It has been specifically noted that in samples loaded in uniaxial tension (under plane stress conditions), failure occurs almost immediately after the formation of the first shear band. Because of this, metallic glasses exhibit virtually zero plastic strain before failure in tension. On the other hand, specimens loaded under constrained geometries such as uniaxial compression, bending, rolling, or under localized indentation (plane strain conditions) fail in elastic, perfectly plastic manner by the generation of multiple shear bands. Since the shear bands carry very large plastic strains, metallic glasses are ductile when deformed in compression or by rolling or indentation at room temperature. Consequently, catastrophic failure is avoided and the specimens exhibit some degree of plasticity.

It has been mentioned earlier that the inhomogeneous flow in metallic glasses appears to be related to a local decrease in the viscosity in shear bands. One of the reasons suggested for this was the local adiabatic heating that could lead to a substantial increase in the temperature.

The vein pattern seen on the fracture surfaces of metallic glass specimens and the localization of deformation are consistent with shear softening in the bands. It has also been shown in some cases that the temperature rise was sufficient to cause localized melting and that molten droplets were reported to be observed on shear lips of tensile-tested metallic glass samples [44,45]. But, the extent to which this softening is associated with local heating has remained controversial. The local temperature rise in the shear bands has been variously estimated to range from less than 0.1 to a few thousand degrees Kelvin [44–51]. This large discrepancy is mainly due to the difficulty of directly measuring the temperature in extremely small distances of the shear band widths (~10–20 nm) [52–56] and short timescales (~10^{-5} s) for shear band propagation [57,58].

Leamy et al. [24] proposed that shear banding in metallic glasses could occur due to adiabatic heating. But, this suggestion was criticized on the basis that the heat generated could be quickly dissipated by the large mass of metal surrounding the shear band and, consequently, the temperature rise would be minimal [4].

Since it is difficult to directly measure the actual temperature rise in a shear band, alternative and indirect methods have been employed. One solution to the problem was to actually calculate the temperature rise resulting from the shear. This can be done either by calculating the temperature rise on the entire shear band at once [25] or as a propagating shear front [25,49]. These authors assumed that the formation of each shear band is manifested in a single serration and that all of the work done in producing the shear band is dissipated as heat. By assuming that shear occurred simultaneously over the entire band, the predicted temperature increase was of the order of a few degrees Kelvin in $Zr_{40}Ti_{14}Ni_{10}Cu_{12}Be_{24}$ [25] and $Pd_{40}Ni_{40}P_{20}$ BMG alloys [49]. But, assuming that shear initiates at one point in the band and then propagates progressively across the sample, the temperature increases were estimated to be in the range of 90–120 K for the $Zr_{40}Ti_{14}Ni_{10}Cu_{12}Be_{24}$ BMG [25] and over 280 K for the $Pd_{40}Ni_{40}P_{20}$ BMG alloy [49]. In fact, the authors point out that since the magnitude of the load drop during failure was much larger than that during serrated flow, the temperature rise could be even higher. Consequently, the temperature rise during failure is sufficient even to melt the sample locally and this could explain why the fracture surface contained features that resembled resolidified droplets.

Assuming that the shear banding events are adiabatic (even though strictly speaking they are not, as we now know), Schuh et al. [10] calculated some of their properties and compared them with the experimental measurements.

Heat is generated in the shear band due to plastic work. The heat is also dissipated to its surroundings by thermal conduction. A steady-state condition is obtained when the rate at which heat is generated equals the rate at which heat is dissipated. By balancing these two terms, Bai and Dodd [59] developed an expression for the thickness (δ) of a fully developed adiabatic operating shear band as

$$\delta = \sqrt{\frac{\kappa \times \Delta T}{\chi \times \tau \times \dot{\gamma}}} \tag{8.7}$$

where:

- κ is the thermal conductivity
- ΔT is the temperature increase above ambient
- χ is the fraction of plastic work converted into heat
- τ is the shear stress
- $\dot{\gamma}$ is the shear strain rate

(While it is true that ΔT, τ, and $\dot{\gamma}$ vary through the thickness of the shear band, the authors have considered their values at the center of the band.) Further, comparison with shear bands in crystalline alloys suggests that this expression is accurate to within a factor of about 2 [59].

The fraction of plastic work converted to heat, χ, is generally estimated to be between 0.9 and 1.0 [60]. Therefore, it was assumed to be equal to 1.0 in the analysis by Schuh et al. [10]. It was also assumed that the thermal conductivity of the glass was known. Even though τ would be obtained from a constitutive relation for the flow stress in terms of $\dot{\gamma}$ and T, it was assumed that it was determined by the macroscopic yield stress in shear, and a typical value of $\tau \approx 1\,\text{GPa}$ was taken. Therefore, the other two parameters required to calculate the steady-state thickness of the adiabatic shear bands are ΔT and $\dot{\gamma}$. Even though different experimental methods were used to measure ΔT, Schuh et al. [10] calculated the shear band thickness, δ, as a function of ΔT. It is important to realize that ΔT needs to be substantially large to approach the glass transition temperature, T_g, where significant softening is expected to occur.

Determining $\dot{\gamma}$ is not straightforward, since the strain rate inside the operating shear band can be substantially higher than the macroscopic strain rate. However, in tests with macroscopic strain rates of $\sim 10^{-4}\ \text{s}^{-1}$, both Neuhauser [61] and Wright et al. [25] measured shear displacement rates $\dot{u} \approx 10^{-4}\ \text{m s}^{-1}$. The shear strain rate, $\dot{\gamma}$, was obtained from the relationship $\dot{\gamma} \approx u/\delta$, and substituting this value into Equation 8.7, one gets

$$\delta = \frac{\kappa \times \Delta T}{\chi \times \tau \times \dot{u}} \tag{8.8}$$

The shear band width, calculated using Equation 8.8, is plotted in Figure 8.7 as a function of ΔT, for various displacement rates, $\dot{u}$, in the operating shear band. The maximum displacement rate considered was $\dot{u} \approx 0.9c_t$, where c_t is the transverse wave speed.

It is clear from Figure 8.7 that a displacement rate of $\approx 10^2\ \text{m s}^{-1}$ is required to produce an adiabatic shear band consistent with the observed thickness and with the necessary ΔT to achieve significant softening. Further, this displacement rate is approximately six orders of magnitude greater than the displacement rates that are actually observed for quasi-static loading. That is, if shear bands in metallic glasses were primarily adiabatic, then either the displacement rates or the shear band thicknesses would be much greater than the experimental observations suggest. Thus, the authors concluded that it is unlikely that shear localization in metallic glasses is driven primarily by thermal (adiabatic) softening.

Some experimental measurements of temperature rise were also made. Yang et al. [51,62] used a high-speed infrared (IR) camera to capture the dynamic shear band evolution process in a $Zr_{52.5}Cu_{17.9}Ni_{14.6}Al_{10}Ti_5$ BMG alloy. They noted that the intense heat was conducted along the shear bands into the surrounding material, resulting in wider, cooler "hot bands" in the IR images. These hot bands were about 0.4 mm in width. Further, it was observed that the temperature was the highest at the initiation site and the lowest at the end. For all the "hot bands," the averaged maximum and mean temperature increases were observed

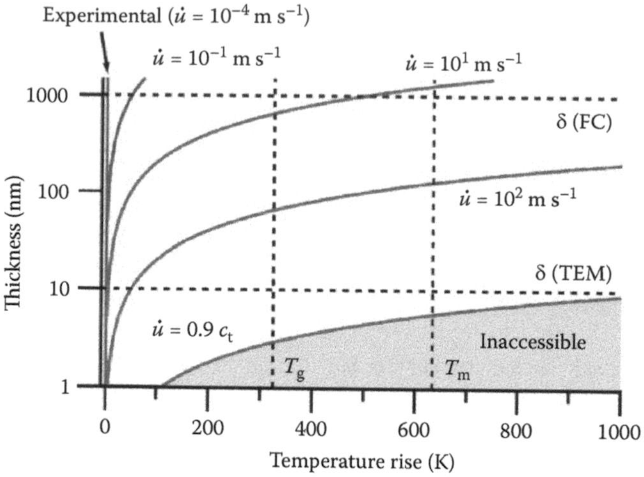

FIGURE 8.7
Shear band thicknesses as a function of the adiabatic temperature rise for various shear displacements on the operating shear band. The ranges of experimental observations for shear band thickness in metallic glasses, δ, are also included in the figure as dashed horizontal lines. For the purpose of illustration, the glass transition (T_g) and melting (T_m) temperatures are shown. (Reprinted from Schuh, C.A. et al., *Acta Mater.*, 55, 4067, 2007. With permission.)

to be ~0.77 and 0.25 K, respectively. Assuming that all the heat was originally generated in a shear band about 10 nm in width (a typical value), the measured temperature rise of 0.25 K in a single "hot band" will translate to an estimated temperature rise of 650 K at the initiation site.

Bruck et al. [47] measured the temperature rise during dynamic testing of the $Zr_{41.25}Ti_{13.75}Ni_{10}Cu_{12.5}Be_{22.5}$ BMG at strain rates of 10^2–10^3 s^{-1}. These measurements indicated that the temperature increase due to adiabatic heating occurred only after the onset of inhomogeneous deformation, and that the temperature may approach the melting point of the alloy within shear bands after the specimen has failed. The temperature rise measured was ~500 K using a single detector. Since only a single detector was used, the authors were not able to identify the detailed nature of thermal distribution near the shear band. A few other measurements were also reported [63,64].

It is known that on heating the metallic glass, the viscosity drops rapidly near T_g and, therefore, significant softening can occur in the metallic glass. This softening will result in rapid shear band propagation, eventually leading to catastrophic failure. Yang et al. [65] used the STZ model of Argon [22] to estimate the temperature rise in a shear band. By balancing the mechanical work and heat generation within an STZ unit, and also considering the collective STZ deformation and that the process is adiabatic, the temperature rise, ΔT, was calculated as

$$\Delta T = \frac{\chi\tau}{\rho C_p} \tag{8.9}$$

where:

- χ is the fraction of mechanical energy converted to heat (assumed to be 0.9 in this case)
- τ is the applied shear stress

ρ is the density
C_p is the specific heat of the BMG

The temperature rise at the time of fracture was calculated to be

$$\Delta T = \frac{\chi \sigma_f}{2\rho C_p} \tag{8.10}$$

where:
τ in Equation 8.9 is replaced by $\sigma_f/2$ in Equation 8.10
σ_f represents the nominal fracture strength during monotonic loading

In this derivation, it was assumed that the maximum shear plane was at an angle of 45° to the loading axis and that the slight strength asymmetry between tension and compression conditions was neglected. The temperature rise was calculated for different BMG alloys using Equation 8.10, and these values are also consistent with the thermographic measurements. Assuming that the shear bands have softened due to the temperature in the shear bands reaching the glass transition temperature, T_g, the authors assumed that ($T_o + \Delta T$) is at least equal to T_g, if not higher. Figure 8.8 shows the results of the temperature rise in the shear bands at the time of fracture plotted against T_g. A linear relationship is obtained between ΔT and T_g, suggesting that this relation could be a practical guide and useful in predicting T_g from the fracture strength, σ_f, of BMGs, and vice versa. An interesting conclusion of this observation was that the catastrophic failure of BMGs was caused by the sudden drop in viscosity inside the shear band as a result of local heating to a level close to the glass transition temperature.

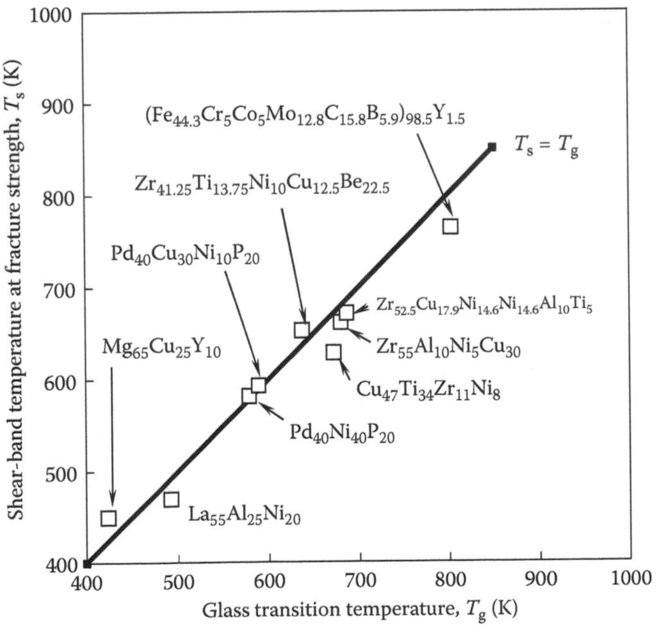

FIGURE 8.8
Temperature rise, ΔT, in the shear bands at the time of fracture for different BMG alloys plotted against the glass transition temperature, T_g. (Reprinted from Yang, B. et al., *J. Mater. Res.*, 21, 915, 2006. With permission.)

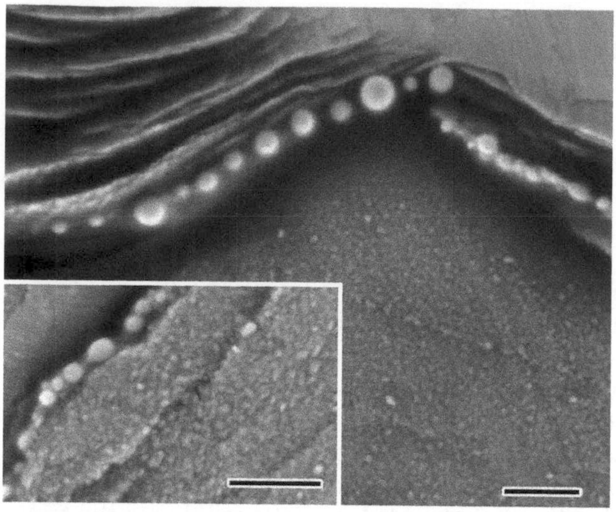

FIGURE 8.9
Scanning electron micrograph of the surface of $Zr_{41.2}Ti_{13.8}Cu_{12.5}Ni_{10}Be_{22.5}$ BMG, which was originally coated with a tin coating. During deformation, the "fusible coating" had melted near the shear bands. The round shape of the tin beads clearly suggests that the coating had melted due to the temperature rise as a result of deformation and had resolidified. The bar in the micrographs corresponds to 1 μm. (Reprinted from Lewandowski, J.J. and Greer, A.L., *Nat. Mater.*, 5, 15, 2006. With permission.)

Lewandowski and Greer [66] used a clever method to estimate the temperature rise in a shear band. Specimens of $Zr_{41.2}Ti_{13.8}Cu_{12.5}Ni_{10}Be_{22.5}$ BMG were deposited with a 50 nm thick tin coating by ultra-high vacuum magnetron sputtering. The double-notched samples were then subjected to four-point bend testing. The authors noted that the coating had melted during deformation at places where the shear bands had intersected the surface. This was inferred from scanning electron micrographs such as those shown in Figure 8.9, in which it was noted that the coating had formed into spherical beads at shear bands as a result of local melting and resolidification.

Such melting of the coating was not observed on Al alloy substrates where the deformation was not localized. Nor was it observed on glass slides at the line of fracture where there was intense local deformation in the tin sample, but no heat was produced in the glass. Thus, the authors concluded that the local melting was due to the heat produced near shear bands.

These observations have provided direct evidence of temperature increases of over 200 K in the operating shear bands (i.e., the actual temperature is about 500 K), at least near the surface (since tin has a melting temperature of 505 K and it had melted). Even though this fusible coating "detector" is limited to a single temperature per experiment (the melting point of the coating), it has excellent spatial (~100nm) and temporal (~30 ps) resolution. These should be compared with the values of ~11 μm and 1.4 ms during infrared imaging [51,62] or single infrared detector (100 μm and ~10 μs) [47]. To obtain information about the actual maximum temperature requires additional approximations or repeating the experiments with other metal coatings that have progressively higher melting temperatures.

The measurements of Lewandowski and Greer [66] suggest that the temperature rise in the shear bands, over a few nanoseconds, can be as high as a few thousand degrees Kelvin. Therefore, it is not surprising that there could be structural changes such as the formation of nanovoids or nanocrystals. It was further suggested by the authors that the shear bands were not fully adiabatic, and that even though local heating was not the origin of shear localization (nor the shear band thickness), temperature changes should be taken into account in analyzing shear band operation. Zhang and Greer [56] reached a similar conclusion based on the thinness of the shear bands and the time available for thermal conduction.

8.4.1 Nanocrystallization Near Shear Bands

Through transmission electron microscopy (TEM) studies, Chen et al. [67] were the first to show that nanocrystals were present within the shear bands of bent $Al_{90}Fe_5Gd_5$ glassy samples. The nanocrystals were face-centered cubic Al with a size of 7–10 nm and seemed to have formed as a result of local atomic rearrangements in the regions of high plastic strain. Jiang et al. [68,69] investigated the structure of shear bands in an $Al_{90}Fe_5Gd_5$ glassy alloy bent at −40°C. They detected the presence of a high density of nanocrystals within the shear bands. It was also reported that there was massive precipitation of nanocrystals at the fracture surface of the bent sample. Such precipitation, both in the shear bands and at the fracture surface, was attributed to plastic-deformation-assisted atomic transport. Similar experiments were extended to $Al_{90}Fe_5Gd_5$ glassy samples bent through 180°, where one could obtain areas deformed in either tension or compression. It was noted that nanocrystals were only observed in the shear bands in the compressive regions, but not in the tensile regions. Surprisingly, no such nanocrystals were observed in a 90° bent $Al_{86.8}Ni_{3.7}Y_{9.5}$ glassy ribbon even in compression. The difference between these two alloys was that the $Al_{90}Fe_5Gd_5$ glassy sample contained a low density of nanoscale defects, predominantly in the shear bands, near the boundary with the undeformed matrix. On the other hand, such defects (nanovoids) were distributed uniformly in the entire shear band of the $Al_{86.8}Ni_{3.7}Y_{9.5}$ glassy sample. Thus, it was concluded that nanocrystallization occurred at shear bands only in samples that contained fewer nanovoids, while no crystallization occurred if the shear bands had a large number of nanovoids.

Kim et al. [48] have also investigated nanocrystallization in a $Zr_{52.5}Cu_{17.9}Ni_{14.6}Al_{10}Ti_5$ BMG sample through a combination of nanoindentation, TEM, and atomic force microscopy (AFM) techniques. The nanocrystals formed around the indentation were identified to be the same as those formed during annealing without deformation at 783 K. These authors concluded that nanocrystallization in their samples was due to the radically enhanced atomic diffusional mobility inside actively deforming shear bands. Lee et al. [70] investigated the effect of brass reinforcement on the formation of nanocrystals at shear bands in a $Ni_{59}Zr_{20}Ti_{16}Si_2Sn_3$ BMG alloy. Based on a detailed TEM investigation on both the monolithic and the composite glassy alloy specimens, the authors noted that the volume fraction of nanocrystals was much higher near the brass particles than farther into the matrix. But, they did not observe the presence of nanocrystals in the monolithic alloy. Further, they noted that the structure of nanocrystals at the shear band was the same as previously observed after isothermal annealing of a monolithic glassy alloy. The authors had attributed nanocrystallization at the shear band to stress concentration by a geometrical effect of the reinforcement phase on the compressive loading conditions and the thermally activated processes.

Chen et al. [71] conducted a very detailed high-resolution TEM study of shear bands in a rapidly solidified melt-spun $Cu_{50}Zr_{50}$ metallic glass ribbon, which was found to be ductile, with a room temperature compressive strain of ~50% [72]. Plastic deformation was introduced into this ribbon by manually bending the ribbons to 180° and then straightening them out until approximately flat. TEM studies of the thinned samples showed the

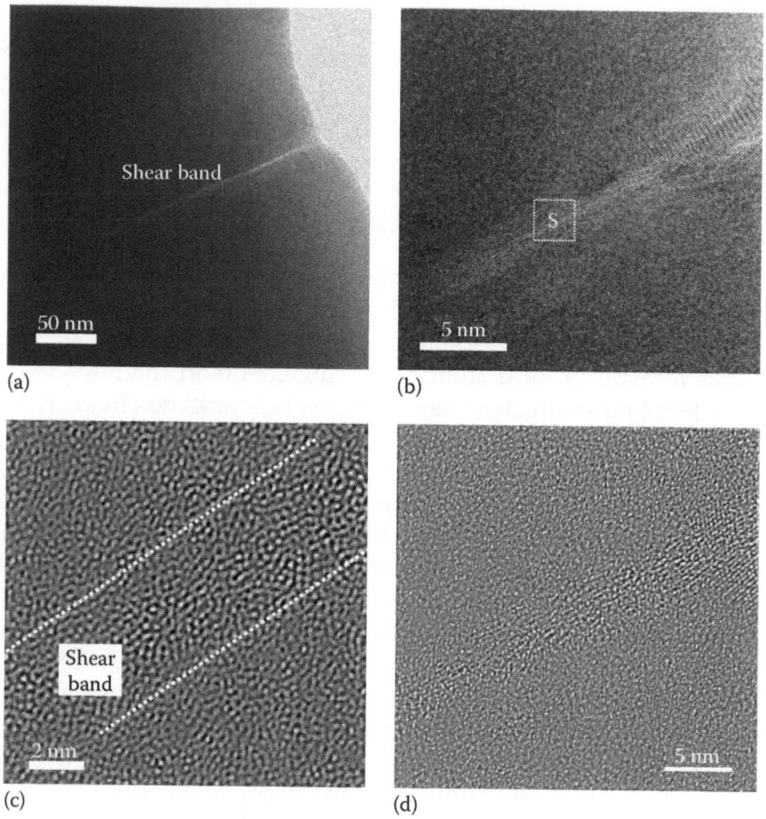

FIGURE 8.10
TEM observations of shear bands in a deformed $Cu_{50}Zr_{50}$ melt-spun ribbon. (a) Bright-field TEM image of a narrow shear band with lighter contrast across the thin region of the TEM specimen. (b) Low-magnification HRTEM image of a shear band with nanocrystalline precipitates inside the shear band. (c) HRTEM image of the shear band taken from the region S in (b). Slightly coarser maze clusters within the shear band can be seen. (d) Fourier-filtered HRTEM image showing that a shear band gradually becomes wider as it approaches a nanoparticle. (Reprinted from Chen, M.W. et al., *Phys. Rev. Lett.*, 96, 245502–1, 2006. With permission.)

presence of shear bands with a width of 2–20 nm and lengths of several micrometers. Figure 8.10a shows the presence of a thin shear band, 3–10 nm wide, in the thin region of the sample. The shear band loses its contrast in the thicker region of the sample. Even in a high-resolution image (Figure 8.10b), the shear bands appear lighter than the regions outside the bands, which may be due to the mass contrast difference produced by either smaller thickness or the lower density of the shear bands. A high-resolution image of the shear band (Figure 8.10c) shows that the maze structure is coarser inside the shear band than that outside the band. These structures were suggested to correspond to the STZs. An important observation is that while the as-spun ribbon did not contain any nanocrystals at all, a number of nanocrystallites, 5–20 nm in size, were observed inside the shear bands in the deformed specimen. The formation of such nanocrystals was attributed to the temperature rise and mass transport as a result of significant plastic flow within the narrow shear bands. Interestingly, the authors claimed that the deformation-induced nanocrystals had a crystal structure very different from that obtained by thermal annealing. These authors also noted a strong interaction between the shear bands and the nanocrystals. For example, it was noted that the width of the shear band increased from about 2–5 nm when it approached a nanocrystal (Figure 8.10d). The authors claimed that nanocrystals generated in the matrix produced strain hardening to compensate for the strain softening of shear bands and, consequently, suppressed catastrophic failure of the specimen.

8.5 Strength

One of the important attributes of metallic glasses (both in thin ribbon form and in bulk form) is their high strength, exceeding 1 GPa in most cases. Table 8.1 lists the mechanical properties of some glassy alloys. Some of the mechanical properties of the BMG composites are presented later in Section 8.9. It is clear from Table 8.1 that BMGs are very strong with very high yield strength, reaching values of up to about 5 GPa. The typical strain at yielding is about 2%. But, they suffer from lack of ductility in most cases, and the plastic strain is virtually zero. By clever design of alloy compositions or by addition of reinforcements, it has been possible to increase their ductility. But, let us first look at the strength aspect of BMGs.

The theoretical shear strength of a dislocation-free crystal is known to be of the order of $\mu/5$, where μ is the shear modulus of the crystal. Johnson and Samwer [102] have recently analyzed the room temperature elastic constants and compressive yield stresses of about 30 metallic glasses. Based on their analysis, they concluded that plastic yielding in metallic glasses at the glass transition temperature, T_g, can be described by an average elastic shear limit criterion:

$$\tau_y = \gamma_c \mu \tag{8.11}$$

where:

τ_y is the maximum resolved shear stress at yielding
γ_c is the critical shear strain
μ is the shear modulus of the unstressed glass

The yield strength of metallic glasses is thought to be determined by the cooperative shear motion of STZ, similar to the movement of dislocations in crystalline materials. Considering that shear deformation occurs through the movement of STZs, Johnson and Samwer [102] showed that the average shear limit for metallic glasses under compression is $\gamma_c = 0.0267 \pm 0.0020$. It was also shown that γ_c depends on T/T_g. Using a cooperative shear model, inspired by Frenkel's work and recent molecular dynamics simulations, these authors showed that

$$\tau_{c,T} = \mu\left[\gamma_{c_0} - \gamma_{c_1}\left(\frac{T}{T_g}\right)^{2/3}\right] \tag{8.12}$$

where:

$\tau_{c,T}$ is the compressive shear stress at temperature T
γ_{c_0} and γ_{c_1} are fitting constants, which are approximately universal constants with the values of 0.036 ± 0.002 and 0.016 ± 0.002, respectively

TABLE 8.1

Mechanical Properties of Some Metallic Glasses

Material	E (GPa)	Strain Rate (s^{-1})	σ_y (MPa)	σ_f (MPa)	ε_y (%)	ε_p (%)	References
$Co_{43}Fe_{20}Ta_{5.5}B_{31.5}$	268	5×10^{-4}	—	5185	2	—	[73]
$Cu_{60}Hf_{40}$	120	5×10^{-4}	—	2245	—	0.4	[74]
$Cu_{60}Hf_{25}Ti_{15}$	124	5×10^{-4}	2024	2088	—	1.6	[75]
$(Cu_{60}Hf_{25}Ti_{15})_{96}Nb_4$	130	5×10^{-4}	—	2405	—	2.8	[76]
$Cu_{47}Ti_{33}Zr_{11}Ni_6Sn_2Si_1$	—	—	1930	2250	—	—	[77]
$Cu_{50}Zr_{50}$	84	8×10^{-4}	1272	1794	1.7	6.2	[78]
$Cu_{50}Zr_{50}$	107	5×10^{-4}	—	1920	—	0.2	[74]
$Cu_{64}Zr_{36}$	92.3	8×10^{-4}	—	2000	2.2	—	[79]
$Cu_{42}Zr_{42}Ag_8Al_8$	104	5×10^{-4}	1844	1873	—	0.1	[80]
$Cu_{46}Zr_{46}Al_8$	—	—	—	1960	—	—	[81]
$Cu_{47.5}Zr_{47.5}Al_5$	87	8×10^{-4}	1547	2265	2.0	16	[78]
$Fe_{65}Mo_{14}C_{15}B_6$	195	1×10^{-4}	3400	3800	—	0.6	[82]
$Fe_{59}Cr_6Mo_{14}C_{15}B_6$	204	1×10^{-4}	3800	4400	—	0.8	[82]
$Fe_{60}Ni_{20}P_{13}C_7$	—	5×10^{-4}	—	2500	—	2.2	[83]
$(Fe_{0.9}Co_{0.1})_{64.5}Mo_{14}C_{15}B_6Er_{0.5}$	192	1×10^{-4}	3700	4100	—	0.55	[82]
$Fe_{71}Nb_6B_{23}$	—	1×10^{-4}	—	4850	—	1.6	[84]
$Fe_{72}Si_4B_{20}Nb_4$	200	—	—	4200	2.1	1.9	[85]
$Fe_{74}Mo_6P_{10}C_{7.5}B_{2.5}$	—	2.1×10^{-4}	3330	3400	—	2.2	[86]
$Fe_{80}P_{13}C_7$	—	5×10^{-4}	—	3100	—	1.5	[83]
$[(Fe_{0.6}Co_{0.4})_{0.75}B_{0.2}Si_{0.05}]_{96}Nb_4$	210	6×10^{-4}	4100	4250	2	2.25	[87]
$Fe_{49}Cr_{15}Mo_{14}C_{15}B_6Er_1$	220	4×10^{-4}	3750	4140	—	0.25	[88]
$Gd_{60}Co_{15}Al_{25}$	70	5×10^{-4}	—	1380	—	1.97	[89]
$Gd_{60}Ni_{15}Al_{25}$	64	5×10^{-4}	—	1280	—	2.01	[89]
$Mg_{70}Ca_5Zn_{25}$	—	—	566	—	—	—	[90]
$Mg_{75}Cu_{15}Gd_{10}$	50	5×10^{-4}	—	743	1.5	0	[91]
$Mg_{75}Cu_5Ni_{10}Gd_{10}$	54	5×10^{-4}	—	874	1.55	0.2	[91]
$Mg_{65}Cu_{7.5}Ni_{7.5}Zn_5Ag_5Y_{10}$	39	1×10^{-4}	—	490–650	1.7	0	[92]
$Ni_{53}Nb_{20}Ti_{10}Zr_8Co_6Cu_3$	140	—	2700	3100	2.7	—	[93]
$Ni_{61}Zr_{28}Nb_7Al_4$	—	1×10^{-4}	—	2620	—	—	[94]
$Ni_{61}Zr_{22}Nb_7Al_4Ta_6$	—	1×10^{-4}	—	3080	—	5	[94]
$Pd_{77.5}Cu_6Si_{16.5}$	—	4×10^{-4}	1476	1600	—	11.4	[95]
$Pd_{79}Cu_6Si_{10}P_5$	82	4.2×10^{-4}	1475	1575	—	3.5	[96]
$Pd_{40}Ni_{40}P_{20}$	96	1×10^{-4}	1700	—	—	—	[49]
$Pt_{57.5}Cu_{14.7}Ni_{5.3}P_{22.5}$	—	1×10^{-4}	1400	1470	2	20	[97]
$Ti_{41.5}Zr_{2.5}Hf_5Cu_{42.5}Ni_{7.5}Si_1$	103	3×10^{-4}		2080	—	—	[98]
$Ti_{41.5}Zr_{2.5}Hf_5Cu_{42.5}Ni_{7.5}Si_1$	95	3×10^{-4}		2040	—	0	[98]
$Zr_{55}Cu_{30}Al_{10}Ni_5$	—	—	1410	1420	—	—	[99]
$Zr_{41.25}Ti_{13.75}Cu_{12.5}Ni_{10}Be_{22.5}$	96	—	1900	1900	2	—	[100]
$Zr_{57}Nb_5Al_{10}Cu_{15.4}Ni_{12.6}$	86.7	—	1800	1800	2	—	[101]

Note: E: Young's modulus; σ_y: yield strength, σ_f: fracture strength; ε_y: elongation at yielding; ε_p: plastic elongation. All the tests were conducted under compression.

Just as the strength of a crystalline metal is determined by the Peierls stress—the internal frictional stress for dislocation motion—the fracture strength of a metallic glass is expected to be determined by the atomic bond strength [103]. This is because a glass does not contain any crystal defects. Therefore, it is logical to assume that the strength of a BMG is related to the physical parameters determined by the atomic cohesive energy, such as T_g, elastic modulus, and coefficient of thermal expansion. Therefore, many different attempts have been made to correlate the strength of BMGs to these parameters.

When the fracture strength of a number of metallic glasses is plotted against T_g, a general trend is—the strength increases with increasing values of T_g [104]. But, the data points are scattered between two bounds with slopes of 1.22 and 2.15 MPa K^{-1} (Figure 8.11a). Because of the significant amount of scatter, the authors concluded that T_g may not be the sole factor determining the strength of a BMG.

As already mentioned, plastic deformation of metallic glasses is facilitated by the movement of STZs. The glass transition is also similarly facilitated by the movement of a group of atoms. In this sense, both these processes are similar. But, while the thermal energy

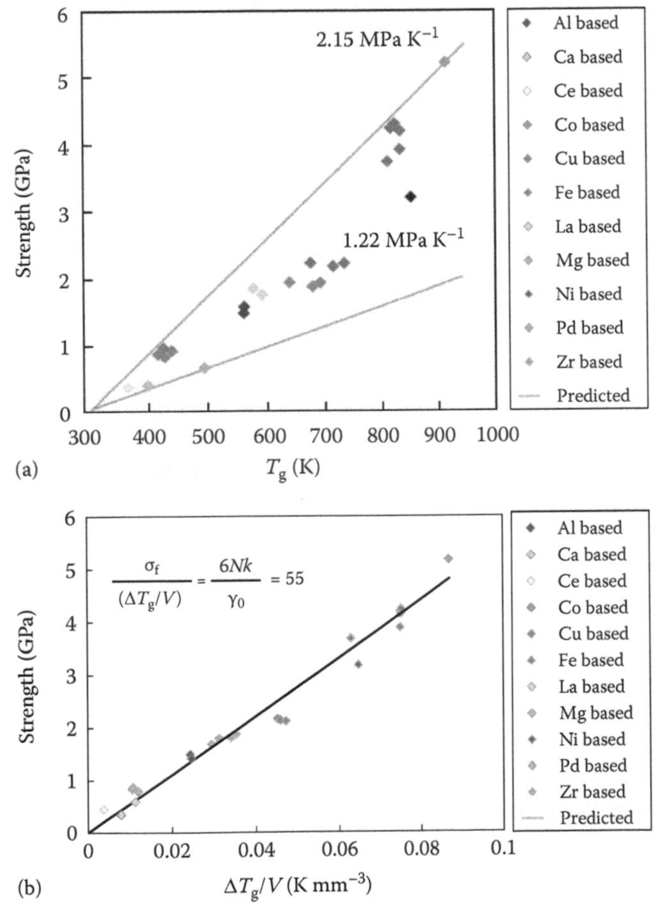

FIGURE 8.11
(a) Variation of strength with glass transition temperature, T_g, for a number of BMGs. (b) Relationship between the calculated fracture strength from a free-volume model and the ratio $\Delta T_g/V$ for a variety of BMGs. (Reprinted from Yang, B. et al., *Appl. Phys. Lett.*, 88, 221911–1, 2006. With permission.)

for glass transition is spread around the entire body of the specimen during heating, the mechanical energy of shearing is highly localized in shear bands. However, the initial and final states are the same in both cases and, consequently, the energy density is expected to be equal in both cases. Thus,

$$\tau_y \gamma_o = \int_{T_o}^{T_g} \rho C_p dT \tag{8.13}$$

where:

- τ_y is the maximum shear stress on yielding
- $\gamma_o = 1$ is the shear strain of the basic shear unit (STZ)
- ρ is the density of the material
- C_p is the specific heat
- T_o and T_g are the ambient and glass transition temperatures, respectively

Through mathematical manipulation and substituting the different variables, the authors arrived at an equation for the yield strength, σ_y, as

$$\sigma_y = \frac{6\rho_o Nk}{M\gamma_o}\left(T_g - T_o\right) = \frac{6Nk}{\gamma_o}\frac{\left(T_g - T_o\right)}{V} = 50\frac{\left(T_g - T_o\right)}{V} = 50\frac{\Delta T_g}{V} \tag{8.14}$$

where:

- N is the Avogadro number
- k is the Boltzmann constant
- ρ_o is the density at ambient temperature
- V is the molar volume
- M is the molar mass

Equation 8.14 suggests that the room temperature strength of BMGs is determined by the T_g and the molar volume. Since BMGs have limited plasticity, it is also assumed that the fracture strength is about 10% higher than the yield strength. Accordingly, Equation 8.14 is rewritten as

$$\sigma_f = 1.1\sigma_y = 55\frac{\Delta T_g}{V} = 55\frac{\rho_o}{M}\Delta T_g \tag{8.15}$$

A linear relationship is found between the calculated fracture strengths and the published experimental fracture strengths for 27 BMGs from 11 different alloy systems (Figure 8.11b). The slope of this plot (the ratio of strength over $\Delta T_g/V$) was found to match exactly with the theoretically predicted value of 6 Nk/γ_o.

Liu et al. [105] analyzed the strength data of 15 BMG alloys with a view to exploring the underlying physics of the linearity between strength versus $\Delta T_g/V$ in terms of fundamental thermodynamics principles. They plotted the room temperature yield shear strength of these BMGs as a function of $(T_g - T_o)/V$ and observed a distinct linear dependence, satisfying the equation:

$$\tau_y = 3R\frac{T_g - T_o}{\gamma_o \cdot V} \tag{8.16}$$

where:

- R is the gas constant
- T_o is room temperature
- γ_o is the critical local shear strain leading to the destabilization of local shearing events
- V is the molar volume

The authors also noted that the fitting slope is just equal to $3R$, the Dulong–Petit limit, while γ_o is equal to 1. By comparing Equation 8.16 with the empirical equation (Equation 8.15), one can easily find that the slope of 50 reported by Yang et al. [104] is actually the product of Dulong–Petit limit ($3R$) and a Schmid factor of two, both of which are invariable for BMGs.

On the basis of Equation 8.16, the authors also deduced the correlation between Young's modulus and T_g. Since $E = \sigma_y/\varepsilon_E$ and the room temperature elastic strain limit of BMGs is approximately 2%, and assuming that $\tau_y = \sigma_y/2$ and that $\gamma_o = 1$, Equation 8.16 can be rewritten as

$$E = 3R\frac{2}{\varepsilon_E}\frac{T_g - T_o}{V} \tag{8.17}$$

Figure 8.12a and b shows these relationships between room temperature yield shear strength and $(T_g - T_o)/V$, and Young's modulus and $(T_g - T_o)/V$, in which the solid lines are the plots representing the slopes of $3R$ and $3R(2/\varepsilon_E)$, respectively.

There have also been other correlations in the literature. For example, Inoue [106] noted that the fracture strength and hardness of BMGs scale with Young's modulus, as shown in Figure 8.13.

8.6 Ductility

Associated with the high strength of BMGs is their brittleness. Even though some BMGs show a small amount of plastic deformation, the majority of them fail soon after yielding. This is because metallic glasses do not, in general, show work-hardening characteristics. But, it should be pointed out in this context that there have been some reports recently mentioning "work hardenable" BMGs [78].

Ductility is an ambiguous and ill-defined term. As has been occasionally pointed out, ductility is like beauty. It is difficult to describe, and as the proverb goes, beauty is in the eyes of the beholder (and it is not absolute). However, if we define ductility in terms of plastic strain to failure, irrespective of its distribution, then metallic glasses are very ductile since most of the deformation is concentrated in the shear bands, where the plastic strain could be $\gg$100%. But, BMGs do not show any significant macroscopic ductility. Thus, metallic glasses showing plasticity (in the shear bands) cannot properly be termed globally *ductile,* though they are *malleable* (capable of plastic compression) and can be bent plastically [107].

Metallic glasses are very brittle in tension, with almost no plastic strain to failure. When BMGs are tested in compression, they seem to exhibit some amount of plastic strain before failure. The reasons for this difference may be found in the way the shear bands are nucleated and propagated. During a tensile test, crack initiation occurs almost immediately after the formation of the first shear band and, as a result, metallic glasses fail catastrophically

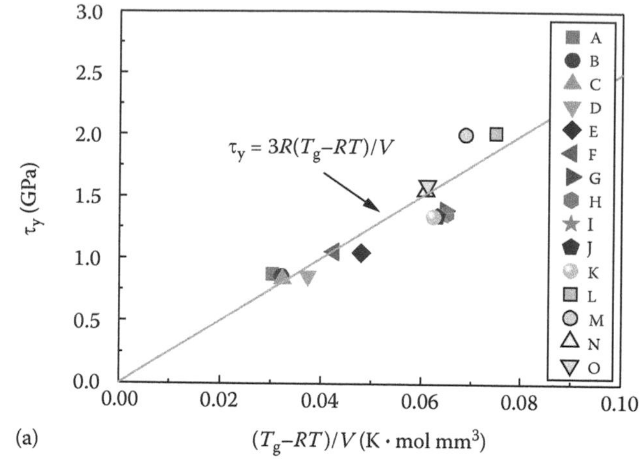

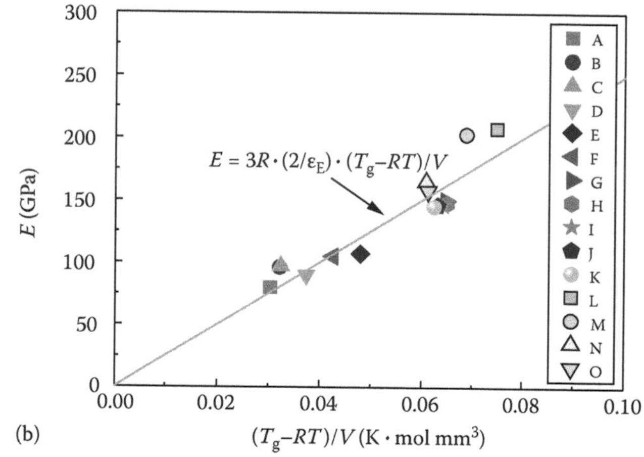

FIGURE 8.12
(a) Relationship between yield shear stress, τ_y, and glass transition temperature, T_g. The solid line is the plot of Equation 8.16, $\tau_y = 3R(T_g - RT)/V$, where R is the gas constant. (b) Relationship between Young's modulus, E, and glass transition temperature, T_g. The solid line is the plot of Equation 8.17, $E = 3R(2/\varepsilon_E)\ (T_g - RT)/V$, in which R is the gas constant and ε_E is the elastic limit in uniaxial compression. (Reprinted from Liu, Y.H. et al., *Phys. Rev. Lett.*, 103, 065504–1, 2009. With permission.)

and show essentially zero plastic strain prior to failure. But, specimens loaded under constrained geometries (e.g., compression) fail in an elastic, perfectly plastic manner by the generation of multiple shear bands, i.e., some plastic strain is observed before failure.

As will be discussed later, one way of overcoming this lack of ductility in BMGs is to introduce a crystalline phase into the BMG matrix to produce a composite. The crystalline phase may be obtained in two different ways. One is the *in situ* process, in which the crystalline phase forms during solidification (due to the fact that the alloy composition is not an exact glass-forming composition) or subsequent processing of the BMG, when partial crystallization of the glassy phase occurs. The other method is the *ex situ* process, in which the crystalline phase is added separately during melting and then subsequently cast into a composite. But in recent years, there have been some reports of significant achievement of plasticity in monolithic BMGs. Some typical examples are the plastic strains to failure of 16% in $Cu_{47.5}Zr_{47.5}Al_5$ [78] and 20% in $Pt_{57.5}Cu_{14.7}Ni_{5.3}P_{22.5}$ [97] BMG alloys.

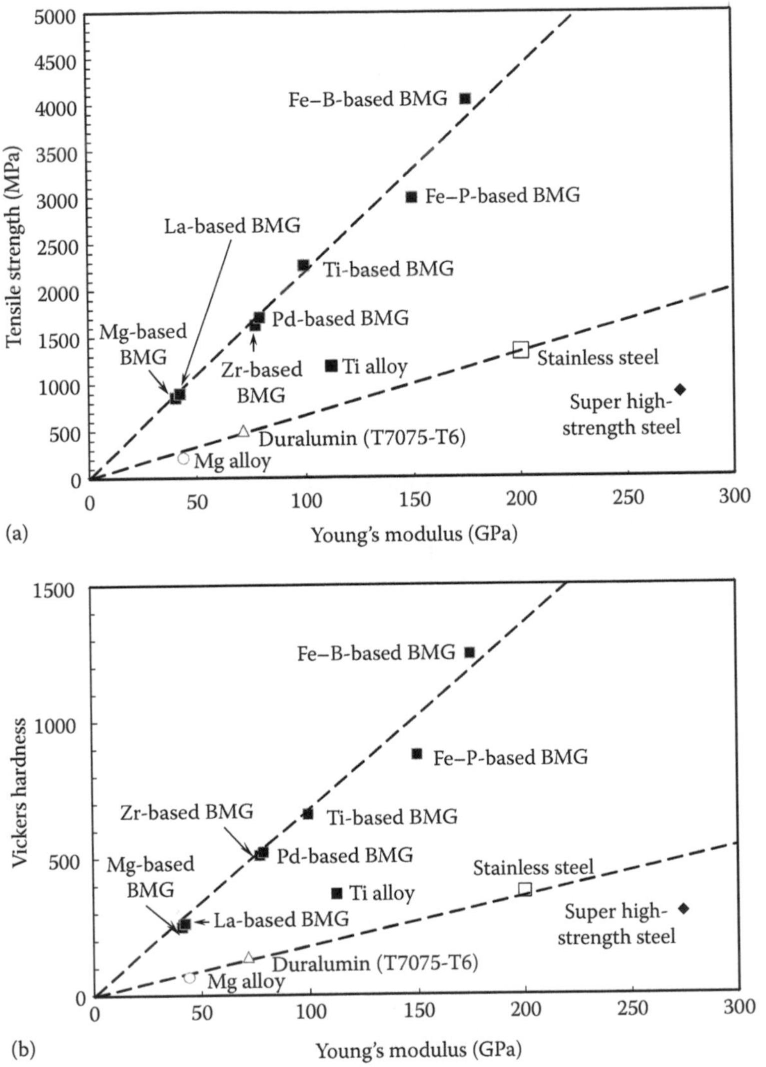

FIGURE 8.13
Relationship between (a) tensile strength and Young's modulus and (b) Vickers hardness and Young's modulus for some typical BMGs. The data for crystalline alloys are also shown for comparison. (Reprinted from Inoue, A., *Acta Mater.*, 48, 279, 2000. With permission.)

During an analysis of competition between plastic flow and brittle failure in metals, Pugh [108] concluded that materials with a high solidity index, S, which is defined as the ratio of the shear modulus, μ, to the bulk modulus, B, that is,

$$S = \frac{\mu}{B} \tag{8.18}$$

favored brittleness, and those with a low value behaved in a ductile manner. The data collected by Pugh provided only a qualitative ranking from ductile to brittle behavior as S increased. S has a value of zero for a liquid and reaches its maximum value of 1.3 in diamond. Kelly et al. [109] made the analysis rigorous and Cottrell [110] suggested a dividing point of 0.3, above which the material is brittle and below which it is ductile.

Lewandowski et al. [30,107] extended this concept to BMG alloys, using the fracture energy, G_c, which is the energy required to create two new fracture surfaces. For ideally brittle materials, $G_c = 2\gamma$, where γ is the surface energy per unit area. Under plane strain conditions,

$$G_c = \frac{K^2}{E(1-\nu)^2} \tag{8.19}$$

where:

K is the stress intensity
E is Young's modulus
ν is Poisson's ratio

The authors calculated the fracture energy, G_c, based on the elastic constants and the stress intensity, whenever it is not reported/available. By plotting the fracture energy, G_c against S (= μ/B), the authors noted that a clear correlation existed between G_c and μ/B. BMGs based on Pd, Zr, Cu, and Pt, which have a low value of μ/B and a fracture energy much in excess of 1 kJ m^{-2}, exhibited extensive shear banding and their fracture surfaces showed a vein pattern. Consequently, they could be considered ductile. On the other hand, BMGs based on Mg had a fracture energy of ~1 kJ m^{-2} approaching brittle behavior. The critical value of μ/B differentiating a brittle BMG from a ductile BMG appears to be in the range 0.41–0.43 (Figure 8.14a). BMGs with a μ/B value of over 0.41–0.43 are brittle.

For isotropic materials, the ratio μ/B can also be expressed in terms of Poisson's ratio, ν, as

$$S = \frac{\mu}{B} = \frac{3(1-2\nu)}{2(1+\nu)} \tag{8.20}$$

Therefore, one should be able to relate Poisson's ratio to the solidity index, noting that the higher the solidity index the lower is Poisson's ratio. Figure 8.14b shows the fracture energy plotted against Poisson's ratio, confirming that BMGs are ductile when Poisson's ratio is high. The critical Poisson's ratio above which the BMGs are ductile (or tough) appears to be 0.32–0.33 [107].

Chen et al. [111, p. 170] were the first to point out that "It is the high Poisson's ratio (ν) which is responsible for the ductile behavior of many metallic glasses. The decreasing ν with falling temperature, together with a relatively lower ν (<0.40) results in a rapid increase in the fracture strength and the brittle behavior of Fe-based glasses." Schroers and Johnson [97] also noted later that the plasticity of BMGs could be related to Poisson's ratio, based on their observation of 20% compressive plastic strain prior to fracture in their $Pt_{57.5}Cu_{14.7}Ni_{5.3}P_{22.5}$ BMG. They suggested that a high Poisson's ratio allows for shear collapse before the extensional instability of crack formation can occur. This causes the tip of a shear band to extend rather than to initiate a crack. This leads to the formation of multiple shear bands and the improvement of global ductility and very high fracture toughness. Gu et al. [112] also noted that large plastic strain (of about 0.8%) was obtained in their $Fe_{65-x}Mo_{14}C_{15}B_6Er_x$ (x = 0–2 at.%) BMG alloys only when x = 0 and 0.15. In the other BMGs with higher amounts of Er (with x = 1–2 at.% Er), ductility was limited, and almost

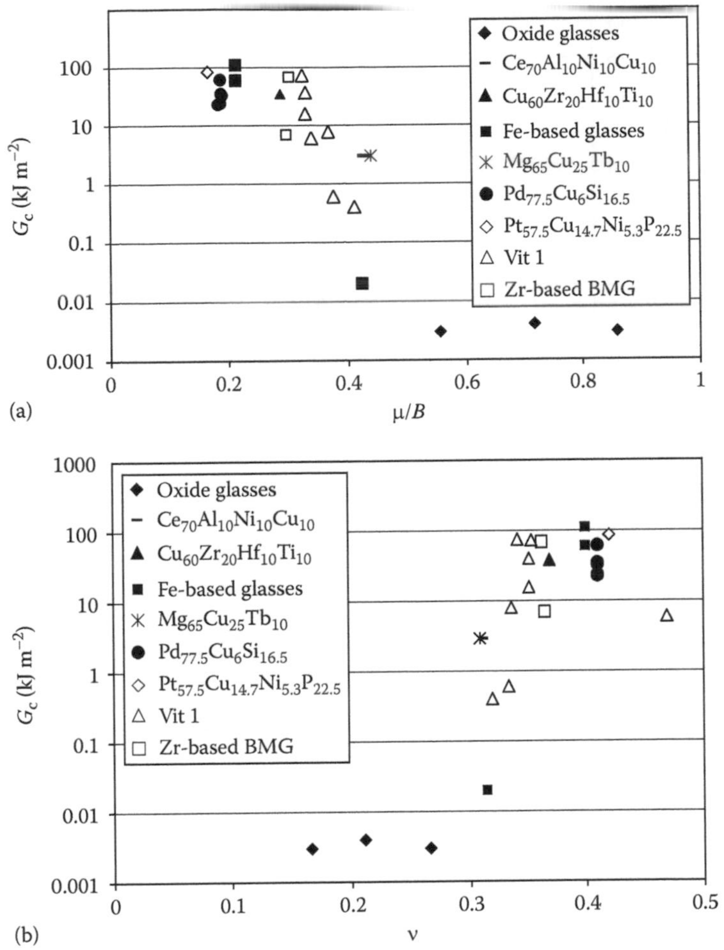

FIGURE 8.14
The correlation of fracture energy, G_c, with (a) the ratio μ/B and (b) Poisson's ratio, ν, for oxide glasses, melt-spun ribbons, and BMGs. The divide between tough and brittle regimes is in the range of 0.41–0.43 for μ/B and 0.32–0.33 for ν.

zero. The high ductility was related to the large value of Poisson's ratio at small values of x. This relationship was again successfully confirmed in Zr–Cu–Ni–Al BMGs [113]. By compositional changes, these authors produced these BMG alloys with a high Poisson's ratio. Compression testing revealed that superplastic elongations of up to 160% could be achieved at room temperature.

Yu and Bai [114] synthesized a number of $(Cu_{50}Zr_{50})_{100-x}Al_x$ BMG alloys with $x = 0$ and 4 to 10, and evaluated their elastic moduli and plastic elongations. Among all the alloys investigated, the $(Cu_{50}Zr_{50})_{95}Al_5$ alloy exhibited the maximum plastic strain of 16% (Figure 8.15a). The other alloys showed much lower plastic strains to failure. A similar trend was noted for the variation of Poisson's ratio with Al content in these BMG alloys (Figure 8.15b). Thus, a very good correlation could be established between Poisson's ratio and the plastic strain to failure in these BMG alloys.

Large amounts of ductility have been reported in other BMG alloys also. For example, Kawashima et al. [115] studied the mechanical properties of $Ni_{60}Pd_{20}P_{17}B_3$ BMG alloys at

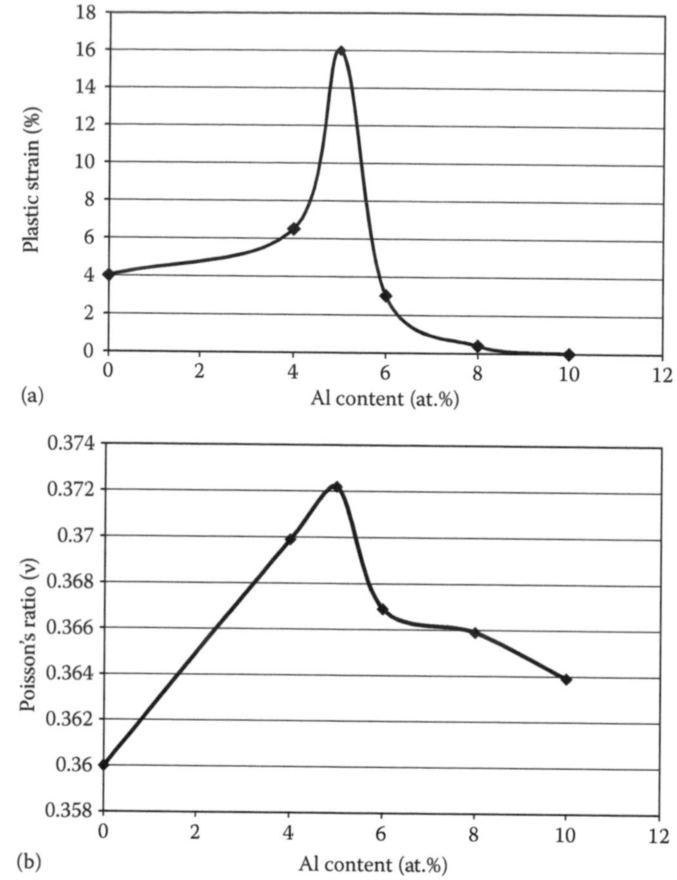

FIGURE 8.15
Variation of (a) plastic strain and (b) Poisson's ratio with Al content in the $(Cu_{50}Zr_{50})_{100-x}Al_x$ BMGs. It may be noted that the variation of Poisson's ratio with Al content mimics the variation of plastic strain. (Reprinted from Yu, P. and Bai, H.Y., *Mater. Sci. Eng. A*, 485, 1, 2008. With permission.)

and below room temperature down to liquid nitrogen temperature (77 K) at a strain rate of 5×10^{-4} s^{-1}. They noted that the strength of the glassy alloy increased with a decrease in temperature from about 2 GPa at room temperature to about 2.5 GPa at the liquid nitrogen temperature (Figure 8.16). But, the more interesting observation was that along with the high strength, the ductility of the glass was also higher with a decrease in temperature [115]. Whereas there was hardly any plastic strain at room temperature, the alloy exhibited about 20% plastic strain at liquid nitrogen temperature. The authors had analyzed the data and showed that the ratio of shear modulus to bulk modulus (μ/B) increased with a decrease in temperature.

Results of increased ductility have also been reported in a hypoeutectic $Zr_{70}Ni_{16}Cu_6Al_8$ glassy alloy where the authors noted a plastic tensile elongation of 1.7% [116]. In this case also, the alloy was shown to exhibit a large value of Poisson's ratio (0.39). Similarly, 3.1% plastic strain in compression in a $(Fe_{0.76}Si_{0.096}B_{0.084}P_{0.06})_{99.9}Cu_{0.1}$ alloy [117] and about 15% total strain in a $Zr_{62}Cu_{15.5}Al_{10}Ni_{12.5}$ alloy [118] were also reported.

It has also been reported that superplasticity could be achieved over a range of strain rates in the supercooled liquid region and use this attribute to form parts of complex

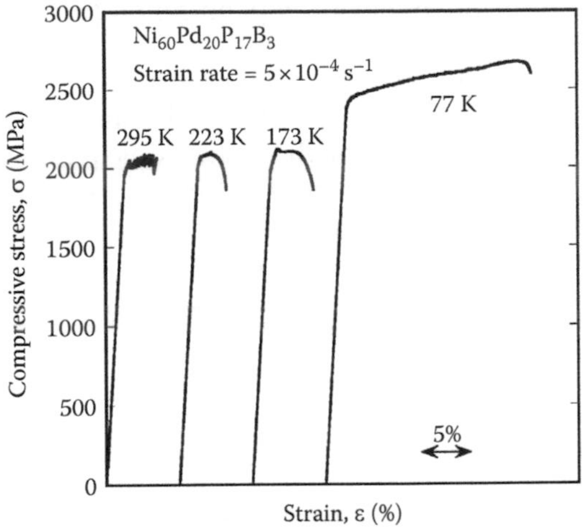

FIGURE 8.16
Stress–strain curves of $Ni_{60}Pd_{20}P_{17}B_3$ BMG alloy as a function of temperature down to liquid nitrogen temperature. Note the simultaneous increase in strength and ductility with decreasing temperature. (Reprinted from Kawashima, A. et al., *Mater. Sci. Eng. A*, 498, 475, 2008. With permission.)

design. Figure 8.17 shows the large elongations one could obtain by superplastically deforming a 3 mm diameter $La_{55}Al_{25}Ni_{20}$ BMG alloy at an initial strain rate of 1×10^{-1} s^{-1} at 500 K [119].

A very important point to remember while evaluating the ductility of metallic glasses (or for that matter any sample) is the specimen size. It has been clearly shown that, with a decrease in specimen size, the compressive plastic strain of a Zr-based $Zr_{52.5}Ni_{14.6}Al_{10}Cu_{17.9}Ti_5$ BMG increased from near zero to as high as 80% without failure [120]. This observation indicates that two samples of identical chemistry and structure

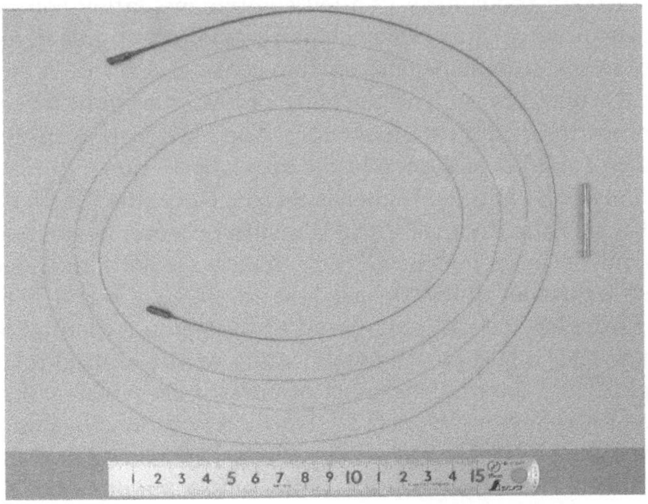

FIGURE 8.17
Extended elongation in a 3 mm diameter BMG rod of $La_{55}Al_{25}Ni_{20}$ alloy superplastically deformed in the supercooled liquid region at 500 K and a strain rate of 10^{-1} s^{-1}.

can show either ductile or brittle behavior depending on the size of the specimen used for the investigation.

The shear offset of two parts separated by the local shear band at room temperature is responsible for the plastic strain in metallic glasses. For failure to occur, the shear band has to propagate to a critical length. Thus, there should be a critical shear offset λ_c, above which the shear band becomes unstable and fracture occurs. It has also been noted that the plastic strain of metallic glasses at fracture increases linearly with increasing critical shear offset. Furthermore, if the sample size, w, is significantly larger than the critical shear offset, λ_c, then the plastic strain at fracture, ε_p, is given by

$$\varepsilon_p = \frac{\lambda_c \cos\theta}{2w} \tag{8.21}$$

where:

- θ is the shear angle between the shear plane and the loading direction (see Section 8.8.2 for further discussion)
- w is the specimen size (diameter or width)

Since the critical shear offset is constant for a given metallic glass, the plastic strain at failure increases with a decrease in specimen size. Wu et al. [120] also considered the situation when the specimen size is equal to or smaller than the critical shear offset.

8.6.1 Effect of Composition

It was shown in the previous section that the Poisson's ratio is the most important parameter to obtain a ductile BMG. A number of results have been presented in the literature on the relationship between the Poisson's ratio and ductility to show the importance of ν on the static and dynamic mechanical properties of BMGs.

The search for ductile BMGs in conjunction with high mechanical strength has been actively pursued through the control of alloy components, alloy compositions, preparation technique, subsequent treatments (such as thermal, mechanical, and working), irradiation, *in situ* and *ex situ* composite, and surface coating. A number of factors have been reported to affect the ductility and/or plasticity of BMGs, and all these can be classified into intrinsic, semi-intrinsic, and extrinsic types. The most important intrinsic parameter is the Poisson's ratio ($\nu \simeq B/G$), as shown in the previous section.

The higher the Poisson's ratio the higher is the ductility (the plasticity). For BMGs, the critical ν value, at which the ductile/brittle transition occurs, lies between 0.32 and 0.33, and there is a general tendency for the ductility to increase with increasing ν values above 0.33 [121]. There are a number of BMGs based on Pd, Pt, La, Zr, and Ti that exhibit high ν values of above 0.33, but Fe-, Co-, and Mg-based BMGs without distinct plasticity have low ν values of less than 0.32 [121]. As a systematic study on the mutual relationship between the ν value and plasticity, data on the most important Zr–Al–Ni–Cu quaternary BMGs have been reported by Yokoyama et al. [122]. Figure 8.18 shows the compositional dependence of ν value for Zr–Al–Ni–Cu BMGs. The ν value increased with increasing Zr content from 50 to 70 at.%. The $Zr_{70}Ni_{16}Cu_6Al_8$ BMG, with a high ν of 0.393 ± 0.003, exhibits extremely high plasticity and does not fracture under a uniaxial compressive deformation mode, as shown in Figure 8.19. A large number of shear bands are observed on the lateral surface of the deformed rod.

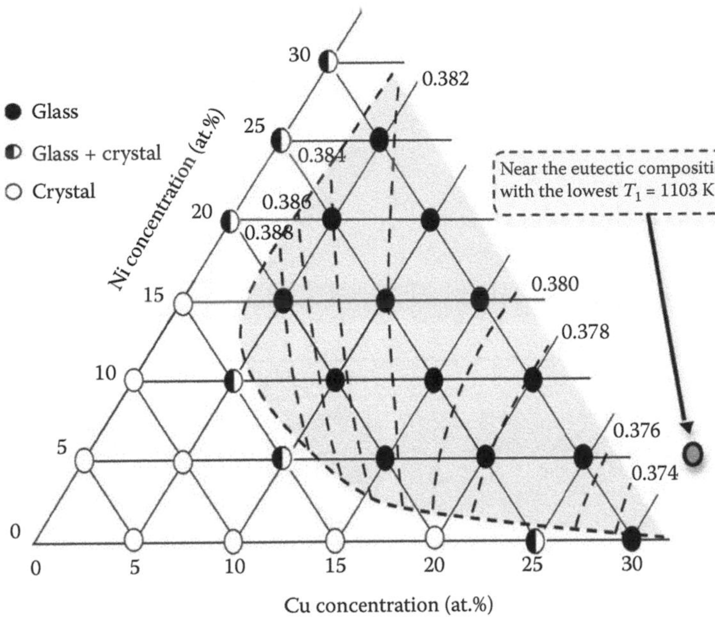

FIGURE 8.18
Compositional dependence of Poisson's ratio of hypoeutectic Zr–Ni–Cu–Al BMGs. (Reprinted from Yokoyama, Y. et al., *Philos. Mag. Lett.*, 89, 322, 2009. With permission.)

This highly plastic BMG also exhibits very high fracture toughness of about 110MPa $m^{-0.5}$ in conjunction with a significant plastic zone even in the front region of the pre-crack edge [123,124]. In addition, the high fracture toughness level of about 90 MPa $m^{-0.5}$ was maintained after annealing for 1.5 h at 671 K (~T_g). The 70% Zr alloy also exhibits high corrosion resistance with anodic current density in the passive state in 0.5 M NaCl solution that is much better than those for $Zr_{50}Al_{10}Ni_{10}Cu_{30}$ and $Zr_{50}Al_{10}Cu_{40}$ as well as pure Zr metal. Owing to the high corrosion resistance and high ductility, the 70% Zr alloy also exhibits high resistance to stress-corrosion cracking in NaCl solution [125].

It is thus clear that the Poisson's ratio is the most important parameter to determine the plasticity of BMGs and that this factor is strongly dependent on alloy component and composition. The results shown in Figure 8.19 clearly indicate that the selection of the host metal-rich alloy composition for multicomponent BMGs is important to get higher ν values, and consequently high plasticity.

8.6.2 Extrinsic Factors

A number of extrinsic factors have been identified that affect the ductility of BMGs. These include increase in stored energy by cold drawing, introduction of a large amount of free volume by increasing the cooling rate, warm deformation in the supercooled liquid region, increase in the degree of disordering by irradiation, introduction of residual stress by shot peening, achievement of rejuvenation by the introduction of free volume, and stored energy by cyclic thermal treatment. Thus, controlling the extrinsic parameters and having a high ν can ensure high plasticity in the BMGs. Although no experimental data are available on the compositional dependence of ν for metal–metalloid-type BMGs, one can

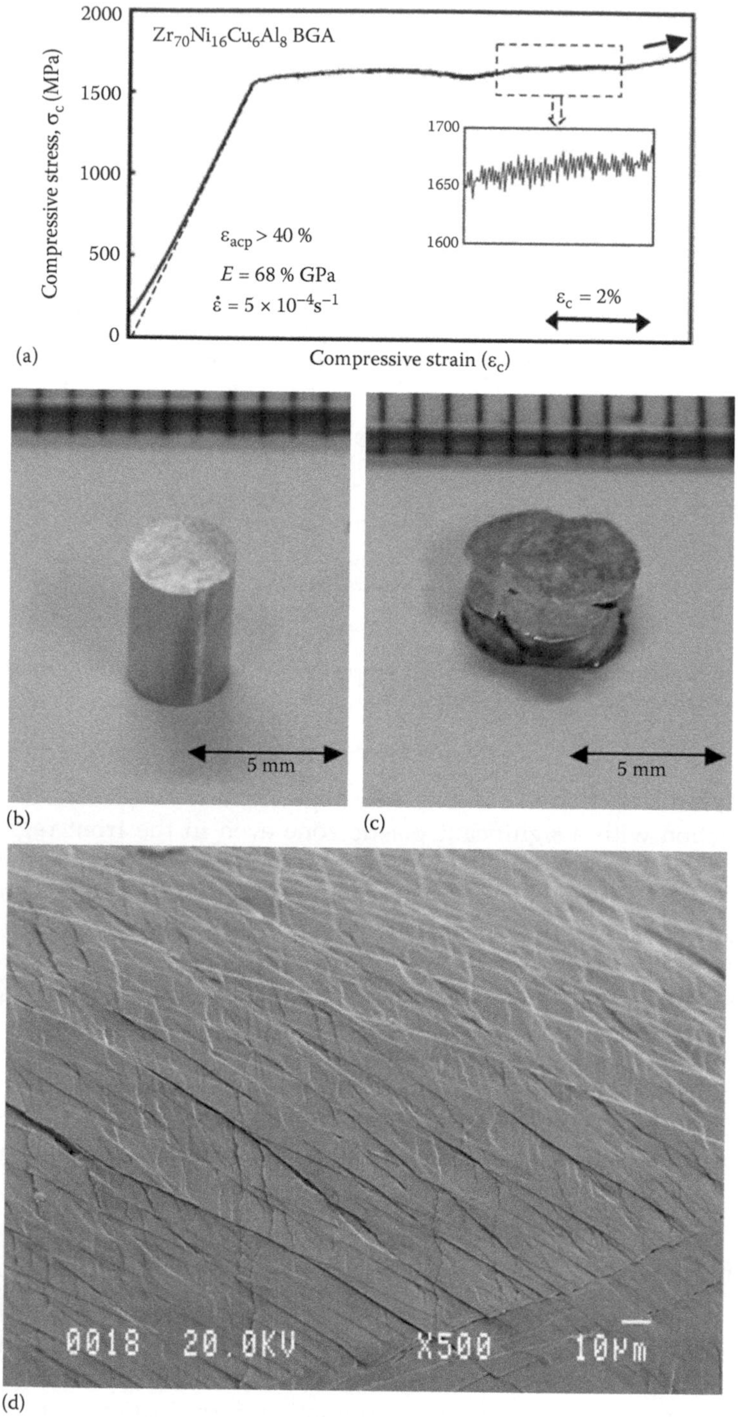

FIGURE 8.19
(a) Compressive stress–strain curve of hypoeutectic $Zr_{70}Ni_{16}Cu_6Al_8$ BMG and outer views of compression test specimen, (b) before, and (c) after compression test, and (d) SEM image after over 40% deformation. (Reprinted from Yokoyama, Y. et al., *Philos. Mag. Lett.*, 89, 322, 2009. With permission.)

recognize recent data on the importance of alloy components to get ductile metal–metalloid-type BMGs. Figure 8.20 shows the results of bending plasticity [126].

BMGs are formed in rather wide composition ranges in metal–metal and metal–metalloid-type multicomponent alloy systems. Although the plasticity of BMGs tends to increase with decreasing metalloid content for the metal–metalloid-type alloys, there is no information on the relationship between plasticity and alloy composition in metal–metal-type BMGs.

Recently, Zhao et al. [127] found an interesting general tendency in the ductile/brittle criterion for metal–metal-type BMGs. That is, BMGs with compositions at or near intermetallic compounds, in contrast to the ones at or near eutectics, were extremely ductile and also were insensitive to annealing-induced embrittlement. They selected two types of BMGs—one with compositions located at or near intermetallic compounds, namely, $Cu_{47}Zr_{47}Ti_6$ (C1), $Cu_{44}Zr_{52}Al_4$ (C2), and $Zr_{58}Cu_{15}Ni_{15}Al_{12}$ (C3), and the other with compositions lying at or near eutectic compositions, *viz.*, $Cu_{60}Zr_{30}Ti_{10}$ (E1) and $Cu_{49}Hf_{42}Al_9$ (E2). Optical micrographs of the cross sections of the five alloy ingots (E1, E2, C1, C2, and C3) prepared by arc melting showed near-equilibrium solidification microstructures. While the E1 and E2 alloys showed a mixed-phase structure, the other three alloys showed mostly the single-phase intermetallic compound phase. The differential scanning calorimetry data also corresponded to near-eutectic for E1, eutectic for E2, near-intermetallic compound for C1 and C2, and intermetallic compound for C3. X-ray diffraction patterns of the five alloys confirmed the presence of a glassy phase.

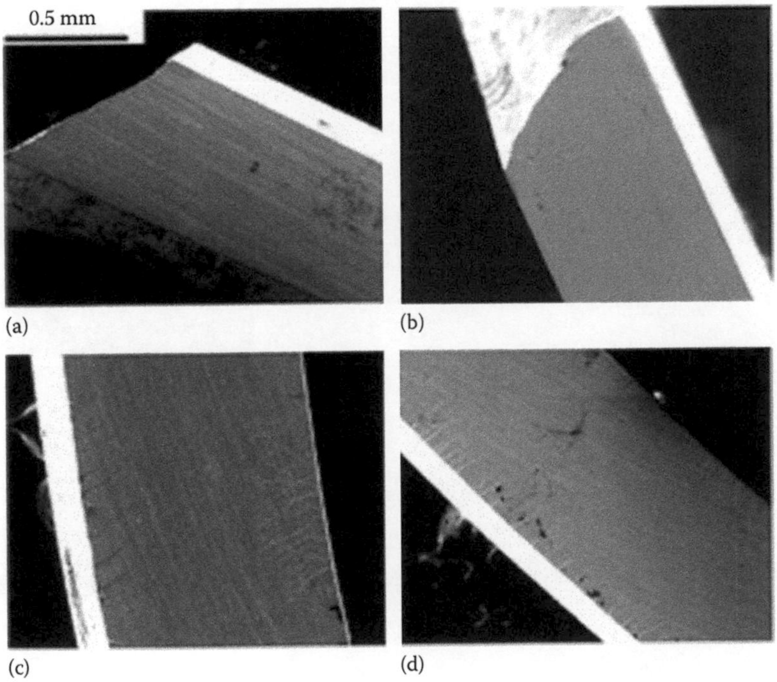

FIGURE 8.20
Scanning electron micrographs of as-cast BMG beams after bending or fracture. Fracture of (a) $Pd_{43}Cu_{27}Ni_{10}P_{20}$ and (b) $Pd_{36}Pt_{9.3}Cu_{25}Ni_{9.3}P_{20.4}$ without any plastic deformation. (c) $Pd_{27.5}Pt_{21}Cu_{22.4}Ni_{8.1}P_{21}$ and (d) $Pt_{57.5}Cu_{14.7}Ni_{5.3}P_{22.5}$ undergo significant plastic deformation, as reflected by multiple shear band formation on both the compression and tensile sides. (Reprinted from Kumar, G. et al., *Scr. Mater.*, 65, 585, 2011. With permission.)

Figure 8.21 shows the compressive stress–displacement curves of the glassy alloys in the as-cast and annealed (3h at T_g-20 K) states for the five alloys [127] It is notable that all the C-type alloys exhibited distinct plastic strains in the as-cast and annealed states (and there was no distinct change in the plastic strain with annealing), while no appreciable plastic strain was noted for the two E-type alloys in both the as-cast and annealed states. The good plasticity for the C-type alloys was confirmed from the displacement due to the generation and propagation of shear bands, as shown in Figure 8.21d and e.

The distinct difference in plasticity between the C-type and E-type BMGs has been presumed to originate from the difference in the homogeneity of alloy components [128]. Energy dispersive spectroscopy (EDS) measurements confirmed that the element distribution in the E-type alloys was rather inhomogeneous in the as-cast state and that it was worse after annealing. However, the deformation-induced heating in the local shear bands

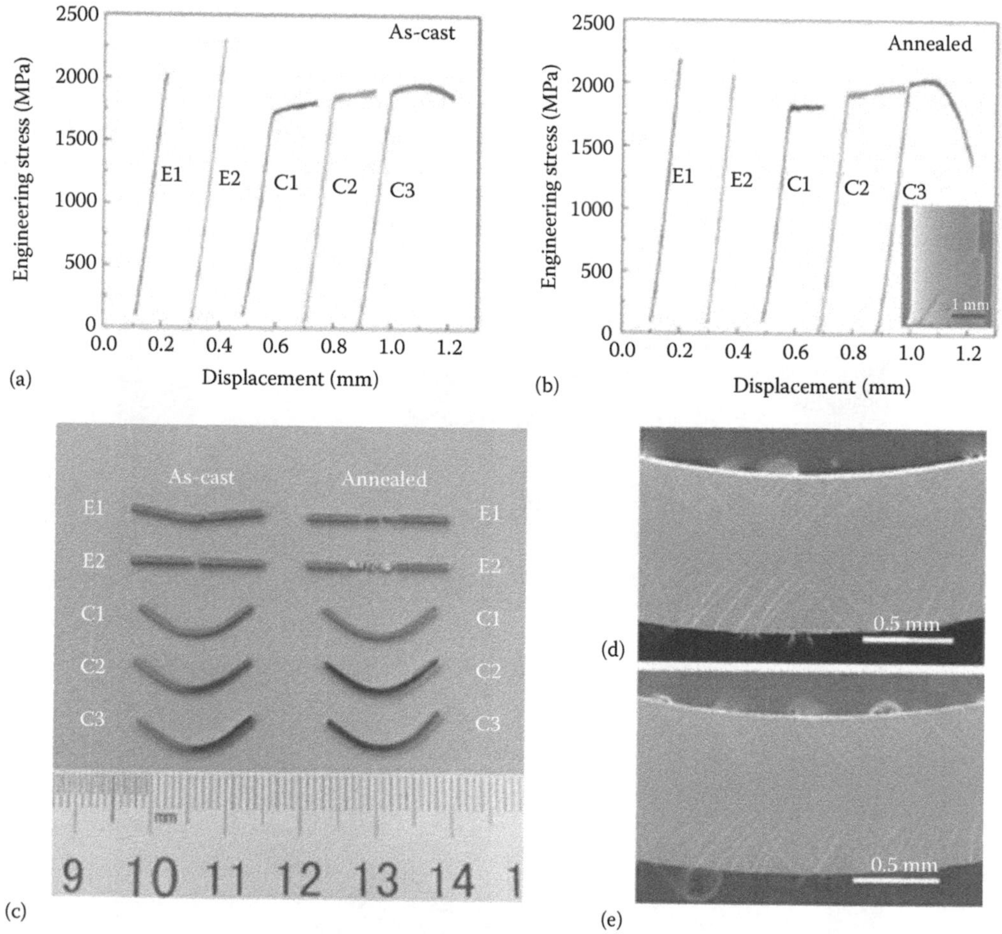

FIGURE 8.21
Deformation behavior in compressive and bending tests. (a, b) Compressive stress–displacement curves of the as-cast and annealed E1, E2, C1, C2, and C3 BMG rods, respectively. The inset of (b) shows that the deformation of the annealed C3 BMG was concentrated within a single dominative shear band. (c) Pictures of the outside of the as-cast and annealed E1, E2, C1, C2, and C3 plate samples bent over a mandrel with a 5 mm radius. (d, e) SEM micrographs of the as-cast and annealed C3 BMG plate samples after bending, respectively. (Reprinted from Zhao, Y.Y. et al., *Sci. Rep.*, 4, 5733, 2014. With permission.)

caused the change to a homogeneous state of element distribution, which corresponds to a higher internal energy state. The more unstable atomic configuration state causes easy generation and propagation of cracks in the shear band region, resulting in the absence of plastic strain. On the other hand, the constituent elements were distributed homogeneously in the C-type alloys and this remained unchanged even after annealing as well as in the deformation-induced shear bands. That is, the internal energy state remained unchanged. Consequently, the alloy homogeneity appears to be an important factor in deciding the ductility of BMGs, in addition to a high ν value. Thus, the difference in the near-eutectic and near-intermetallic compound compositions is shown schematically in Figure 8.22.

From this, it can be concluded that the homogeneous element distribution in the alloy seems to lead to a high stability state and this appears to be the reason for the high plasticity of the C-type BMGs in the as-cast, annealed, and locally deformed states. The high stability has also been confirmed from the distribution map data of the constituent elements. These data also demonstrate that no appreciable difference in the tendency in the plasticity between the C-type and the E-type alloys is recognized for the Cu-based and the Zr-based BMGs belonging to metal–metal-type alloys. Although high glass-forming ability (GFA) is obtained in the alloy composition range including the deep eutectic valley, it is important for the development of ductile BMGs to search for BMGs at intermetallic compound compositions where no phase transition occurs over the whole temperature range in the solid state. Considering that there are no established equilibrium phase diagrams for multicomponent alloy systems, the search should be conducted with the assistance of computer simulation and combinatorial search techniques.

8.6.3 Ductile Fe-Based Bulk Metallic Glasses

It is known that BMGs can be formed in a variety of systems based on transition metals (TM), lanthanides, and simple metals, and that the total number of BMG systems exceeds 500 types [129]. Among these systems, BMGs with good ductility (with a compressive plastic strain of >2%–3%) have been produced in almost all TM-based systems except for those based on Fe and Co. Fe-based BMGs have been commercialized as soft magnetic materials

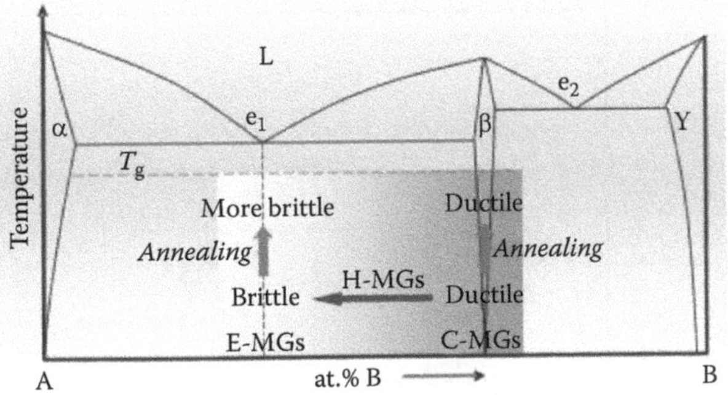

FIGURE 8.22
Schematic binary phase diagram showing the dependence of composition on the intrinsic plasticity or brittleness in metallic glasses with different compositions and thermal histories. Figure shows the glass-forming alloy system in which the glass-forming composition regions can extend from the eutectics to the corresponding intermetallic compounds. The light gray and dark gray regions represent the glass-forming composition regions of E-type BMGs are brittle before and after annealing while the C-type BMG compositions are ductile in the as-cast state and continue to be ductile after annealing. (Reprinted from Zhao, Y.Y. et al., *Sci. Rep.*, 4, 5733, 2014. With permission.)

in Fe–TM–metalloid systems and as structural coating materials in Fe–Cr–Mo–C–B-based systems, even though these Fe-based BMGs fracture without appreciable plastic strain. To further exploit them within their application fields and to explain the origin of their brittle nature, it is important to examine the possibility of synthesizing Fe-based BMGs with good ductility.

Very recently, Yang et al. [130] reported the formation of 2 mm diameter rods in the $Fe_{80-x}Ni_xP_{13}C_7$ system with an Ni content up to 30 at.% by water quenching. It was noted that the plastic strain increased from ~2% for the 0% Ni alloy to 22% for the 30% Ni alloy with a decrease in the yield strength from ~2900 to ~2100 MPa (Figure 8.23). It is notable that a large plastic strain above 5% is obtained in the wide Ni concentration range from 10% to 30% Ni. Thus, the compositional dependence data shown in Figure 8.23 indicate that Ni plays an important role in achieving high ductility. However, the addition of Ni was not always effective in enhancing the ductility, as no appreciable increase in the plastic strain was detected for $Fe_{72-x}Ni_xSi_4B_{20}Nb_4$ (x = 0, 7.2, 14.4, 21.6, 28.8) BMGs. This result suggests that the combination of Ni and metalloid elements (P and C) may be an important factor in the ductilization of Fe-based BMGs.

The mechanism underlying good compressive ductility was further examined by measuring the compositional dependence of binding energy among the constituent elements by x-ray spectroscopy and ultraviolet spectroscopy analyses [130]. As shown in Figure 8.24, the peak of the x-ray photoelectron spectroscopy (XPS) profile shifts continuously to the lower-energy side from 1.30 to 1.62 eV with increasing Ni content for the Fe–Ni–P–C BMGs, while there is no appreciable change in the binding energy with Ni content for the Fe–Ni–Nb–Si–B BMGs, where good compressive ductility was not obtained. Based on the close relationship between the systematic change in binding energy and the change in ductility, it was interpreted that Ni–Ni atomic pairs were present only in the ductile Fe–20% Ni–P–C glassy alloy, and not in the Fe–Ni–Nb–Si–B glassy alloys. Thus, it was concluded that the ductile nature of the Fe–Ni–P–C BMGs was due to the generation of Ni–Ni atomic bonding pairs with lower binding energy. However, there was no definite reason given for why the change in the binding energy with Ni content occurred only in the Fe–Ni–P–C BMGs, even though the bonding states of Fe–C and Ni–C atomic pairs are rather weak, as is evident

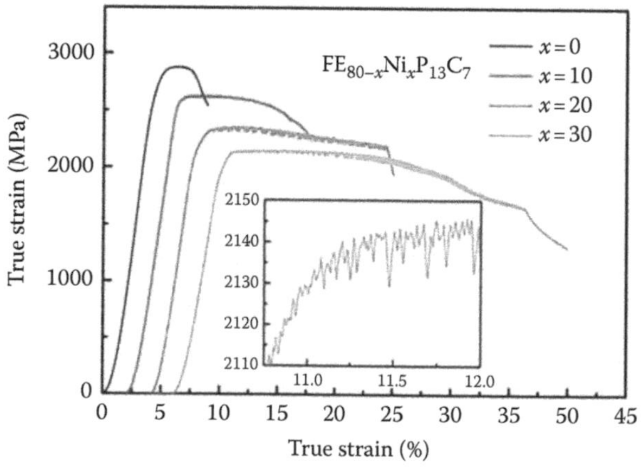

FIGURE 8.23
Compressive true stress–strain curves of $Fe_{80-x}Ni_xP_{13}C_7$ (x = 0, 10, 20, 30) BMGs at room temperature. The inset is the enlarged section of $Fe_{50}Ni_{30}P_{13}C_7$ BMG. (Reprinted from Yang, W. et al., *Sci. Rep.*, 4, 6233, 2014. With permission.)

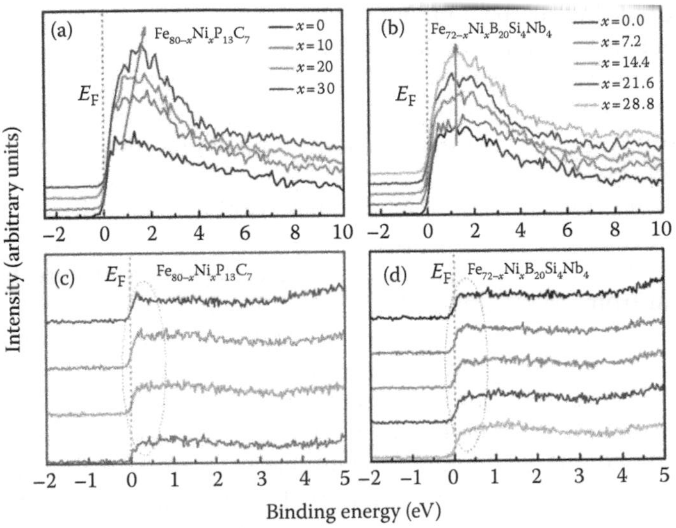

FIGURE 8.24
(a) XPS valence-band spectra for $Fe_{80-x}Ni_xP_{13}C_7$ BMGs. (b) XPS valence-band spectra for $Fe_{72-x}Ni_xB_{20}Si_4Nb_4$ BMGs. (c) Ultraviolet photoelectron spectroscopy valence-band spectra for $Fe_{80-x}Ni_xP_{13}C_7$ BMGs. (d) Ultraviolet photoelectron spectroscopy valence-band spectra for $Fe_{72-x}Ni_xB_{20}Si_4Nb_4$ BMGs. (Reprinted from Yang, W. et al., *Sci. Rep.*, 4, 6233, 2014. With permission.)

from their positive heats of mixing. The minor addition of an element with a positive heat of mixing may be the key factor to develop a ductile Fe-based BMG.

The first paper on ductile Fe-based amorphous alloy ribbons showing good soft magnetic properties after optimum annealing treatment reported Fe–B–P–Si and Fe–B–P–Si–C alloys with high Fe contents above 84 at.% [131,132]. These Fe-based amorphous alloys have a high saturation magnetization near 1.7 T, low coercive force below 10 A m^{-1}, high initial permeability above 10^4, and high effective permeability above 8000 at 1 kHz, in addition to good bend ductility (the ribbon could be bent through 180°). The saturation magnetization value is believed to be the highest of all the Fe-based glassy alloys reported to date. A similar ductile bending nature was also recognized in Fe–Cr–P–B–Si(–C), Fe–Mo–P–B–Si–(–C), and Fe–Cr–Nb–P–B–Si(–C) [133] glassy alloys with good soft magnetic properties. The good bending ductility for these Fe-rich amorphous alloys has been interpreted to be due to the generation of Fe–Fe atomic bonding pairs resulting from an increase in the Fe content above about 84 at.%. However, further increase in Fe content to above 86 at.% causes a decrease in the GFA and a lowering of the Curie temperature in conjunction with a decrease in saturation magnetization. It is therefore important to control the atomic configuration of the Fe-rich amorphous alloy so as not to reduce the Curie temperature, though achieving this control is very challenging.

8.7 Fatigue

Fatigue is a form of failure in structures subjected to dynamic and fluctuating stresses. Fatigue failures have been detected in different metallic and polymeric materials, especially in load-bearing components that have been subjected to repetitive stresses. Fatigue failure is catastrophic, occurring suddenly and without warning. There is very little gross plastic deformation associated with fatigue failure and so it is brittle-like. The mechanism of fatigue failure involves initiation of cracks, their propagation, and final fast fracture [134]. Studies on the fatigue behavior of BMGs are essentially focused on Zr-based alloys, even though results of a few investigations on other BMG alloys (Co-, Fe-, and Ti-based BMGs) are also reported.

For the successful exploitation of BMGs as structural materials, it is important to evaluate their fatigue behavior, especially when they are expected to experience alternating loads. However, BMGs have very high tensile strengths and, therefore, based on the conventional wisdom from crystalline materials, the BMGs are expected to show very high fatigue limits. But, this has not been realized. However, even though the propagation of fatigue cracks appears to be similar between crystalline materials and BMG alloys (striations are seen in both types of material), reasons for the susceptibility of BMGs to crack initiation are not known. Since BMG alloys are not crystalline and therefore they do not have grain boundaries and well-defined slip systems, it is difficult to visualize the sites for the initiation of cracks. It is also important to realize that fatigue lifetimes are dominated by the loading cycles needed to initiate damage rather than the cycles needed to propagate the damage, generally in the form of cracks, to critical size. Therefore, study of fatigue behavior of BMG alloys assumes importance. The fatigue behavior of Zr-based BMGs and their composites has been recently reviewed [12]. The general background to the topic of fatigue may be found in Ref. [135].

8.7.1 Methodology

Fatigue testing has been done in different ways. But, the most important requirement for the specimen is that its surface should be smooth and defect free. The presence of any imperfections on the surface act as initiation sites for fatigue cracks and, therefore, it is very important for the specimen to be free of these. Generally, the specimen is cut from the bulk alloy by an electro-discharge machining process and then the surface is polished to a 1200 SiC grit surface finish to ensure that the sides are parallel.

Different types of fatigue testing are conducted. They are conducted in bending, uniaxial, or rotating modes. The rotating-beam testing methods are most common, since this type of instrument allows fatigue tests to be run in torsion, combined bending and torsion, or biaxial bending at a constant frequency of cycling and under a constant amplitude loading. However, closed-loop servohydraulic testing machines are used nowadays, since specimens may be subjected to constant amplitude cycling with controlled loads, strains, or deflections, with the possibility to choose the amplitude, mean, and cyclic frequency. For tension–tension fatigue testing and button-head fatigue testing, specimens with a small radial sharp notch are used (see, e.g., Ref. [136]).

Bend testing methods also have been very commonly employed. Two possible approaches are the four-point bend test and the three-point bend test methods. In the four-point bend testing method, the specimen of square or rectangular cross section is supported at two points from the bottom and loaded from the top at two different outer points. The maximum stress, σ_{max}, is experienced by the specimen over the entire outermost tensile surface in the region between the two inner loading points, and is calculated using the equation:

$$\sigma_{\max} = \frac{3P(S_2 - S_1)}{2bh^2} \tag{8.22}$$

where:

- P is the applied load
- S_2 is the distance between the outer pins
- S_1 is the distance between the inner pins
- b is the specimen thickness
- h is the specimen height

Typically, the inner span length, S_1, is maintained at about 10 mm and the outer span length, S_2, at about 20 mm. This σ_{max} is used to obtain the stress versus fatigue cycles (S–N) curve.

In the three-point bend test, the specimen is supported at the center from the bottom and loaded at two points from the top. In this case, the maximum stress is experienced only at the center of the beam at the outermost tensile surface and is calculated using the equation:

$$\sigma_{max} = \frac{1.5PL}{bh^2} \tag{8.23}$$

where L is the span between the outer loading points and the other symbols have the same meaning as in Equation 8.22. Even though both methods have been used to evaluate the fatigue strengths and lives of BMG alloys, it is important to realize that the volume of material experiencing the maximum stress is much greater in the four-point bend test and, consequently, the fatigue lives are expected to be longer from the three-point bend tests.

A constant sinusoidal load is applied for each test, with the ratio R of the minimum to maximum stress ($\sigma_{min}/\sigma_{max}$) ranging from 0.1 to 0.3, and then the number of cycles to failure is determined. The tests are conducted at a frequency of 10–25 Hz. If the lifetime (number of cycles to failure) is very short, then a lower frequency is used. If the specimen does not fail after 2×10^7 cycles, the test is stopped and this is indicated with an arrow at the end of the plot. A large number of specimens are tested to get a reasonably reliable S–N plot.

The fractured surfaces of the test specimens are then examined in an optical or scanning electron microscope (SEM) to observe the microstructural features and obtain information about the origin and propagation of fatigue cracks, and the spacing of fatigue striations. The presence of second-phase particles that could act as nucleation sites for cracks can also be observed. The EDS attachment of the SEM can be used to obtain compositions of phases and also to identify those responsible for fatigue crack initiation.

8.7.2 High-Cycle Fatigue

There are only a few groups of researchers who have investigated the fatigue behavior of BMG alloys. The majority of the work reported so far has been on Zr-based BMG alloys. Sometimes, significant differences have been noted in the results reported by different groups even though apparently the alloy composition appears to have been almost the same. The differences have been attributed to specimen geometries and test procedures. There has also been some discussion in the open literature between two of these groups [137–139].

A number of studies have been conducted to evaluate the fatigue lives, fatigue strengths, and fatigue ratios of monolithic BMGs and their composites. Table 8.2 summarizes the available results. It may be noted that fatigue tests have been conducted on different types of alloys and also using different methods, not necessarily by the same group and on the

TABLE 8.2

Fatigue Limits and Fatigue Ratios of BMGs under Bending, Uniaxial, and Rotating Loads

Material	Fracture Strength (MPa)	Geometry (mm)	Loading Mode	Frequency (Hz)	*R* Ratio	Fatigue Limit (MPa)	Fatigue Ratio	References
Bending Loads								
$Cu_{47.5}Zr_{47.5}Al_5$	—	3 × 3 × 25	4PB	10	0.1	224	0.12	[140]
$Cu_{47.5}Zr_{38}Hf_{9.5}Al_5$ (composite)	—	3 × 3 × 25	4PB	10	0.1	378	0.23	[140]
$(Cu_{60}Zr_{30}Ti_{10})_{99}Sn_1$	1800	2.85 × 2.85 × 25	4PB	10	0.1	350	0.19	[141]
$(Cu_{60}Zr_{30}Ti_{10})_{99}Sn_1$	—	—	3PB	10	0.1	475	0.26	[141]
$Fe_{48}Cr_{15}Mo_{14}Er_2C_{15}B_6$	4000–4400	2.85 × 2.85 × 25	4PB	10	0.1	682	0.17	[142]
$Zr_{39.6}Ti_{33.9}Nb_{7.6}Cu_{6.4}Be_{12.5}$ (composite)	1210	3 × 3 × 50	4PB	25	0.1	340	0.28	[143]
$Zr_{41.2}Ti_{13.8}Cu_{12.5}Ni_{10}Be_{22.5}$	1900 (YS)[a]	3 × 3 × 50	4PB	25	0.1	76	0.04	[144,145]
$Zr_{41.2}Ti_{13.8}Cu_{12.5}Ni_{10}Be_{22.5}$	1900 (UTS)[b]	3 × 3 × 40	4PB	5	0.1	—	0.05	[146]
$Zr_{41.2}Ti_{13.8}Cu_{12.5}Ni_{10}Be_{22.5}$	1900 (UTS)	2 × 2 × 60	3PB	20	0.1	359	0.09	[147]
$Zr_{41.2}Ti_{13.8}Cu_{12.5}Ni_{10}Be_{22.5}$	1900 (UTS)	2 × 2 × 60	3PB	20	0.1	768	0.2	[147]
$Zr_{41.2}Ti_{13.8}Cu_{12.5}Ni_{10}Be_{22.5}$	1900 (YS)	7 × 38	C-T	—	—	—	—	[144]
$Zr_{41.2}Ti_{13.8}Cu_{12.5}Ni_{10}Be_{22.5}$ (liquidmetal)	1900 (UTS)	2 × 2 × 60	3PB	20	0.1	359	0.09	[147]
$Zr_{41.2}Ti_{13.8}Cu_{12.5}Ni_{10}Be_{22.5}$ (Howmet)	1900 (UTS)	2 × 2 × 60	3PB	20	0.1	768	0.2	[147]
$Zr_{44}Ti_{11}Ni_{10}Cu_{10}Be_{25}$	1900 (UTS)	2.3 × 2.0 × 85	4PB	5–20	0.3	195 (stress relief)	0.17	[148]
$Zr_{44}Ti_{11}Ni_{10}Cu_{10}Be_{25}$	1900 (UTS)	2.3 × 2.0 × 85	4PB	5–20	0.3	275 (10τ) less free vol	—	[148]
$(Zr_{58}Ni_{13.6}Cu_{18}Al_{10.4})_{99}Nb_1$	1700 (UTS)	2 × 2 × 25	4PB	10	0.1	559	0.33	[149]
$Zr_{52.5}Cu_{17.9}Ni_{14.6}Al_{10.0}Ti_{5.0}$	1700 (UTS)	3.5 × 3.5 × 30	4PB	10	0.1	425	0.25	[150]
$Zr_{47}Ti_{12.9}Nb_{2.8}Cu_{11}Ni_{9.6}Be_{16.7}$ + 25 vol.% $Zr_{71}Ti_{16.3}Nb_{10}Cu_{1.8}Ni_{0.9}$	1480	2.1–2.9 × 38.1	C-T	25	0.1	148	0.1	[151]
$Zr_{55}Cu_{30}Ni_5Al_{10}$	1560 (UTS)	2 × 20 × 50	Plate bend	40	0.1	410	0.26	[152]
$Zr_{39.6}Ti_{33.9}Nb_{7.6}Cu_{6.4}Be_{12.5}$ (composite)	1210	3 × 3 × 50	4PB	25	0.1	340	0.28	[143]
Uniaxial Loading								
$Cu_{60}Hf_{25}Ti_{15}$	2130	—	—	5–10	0.1	860	0.4	[153]
$Cu_{60}Zr_{30}Ti_{10}$	2000	—	—	5–10	0.1	980	0.49	[153]
$Ti_{41.5}Zr_{2.5}Hf_5Cu_{42.5}Ni_{7.5}Si_1$ (NC Comp)	2040	—	—	5–10	0.1	1610	0.79	[153]

(Continued)

TABLE 8.2 (CONTINUED)

Fatigue Limits and Fatigue Ratios of BMGs under Bending, Uniaxial, and Rotating Loads

Material	Fracture Strength (MPa)	Geometry (mm)	Loading Mode	Frequency (Hz)	*R* Ratio	Fatigue Limit (MPa)	Fatigue Ratio	References
$Zr_{41.2}Ti_{13.8}Cu_{12.5}Ni_{10}Be_{22.5}$	1850 (UTS)	—	T-T	10	0.1	703 (batch 59)	0.38	[154]
$Zr_{41.2}Ti_{13.8}Cu_{12.5}Ni_{10}Be_{22.5}$	1850 (UTS)	—	T-T	10	0.1	615 (batch 94)	0.33	[154]
$Zr_{41.2}Ti_{13.8}Cu_{12.5}Ni_{10}Be_{22.5}$ + $Zr_{47}Ti_{12.9}Nb_{2.8}Cu_{11}Ni_{9.6}Be_{16.7}$ (composite)	—	—	—	—	—	—	—	[155]
$Zr_{44}Ti_{11}Ni_{10}Cu_{10}Be_{25}$	1900 (UTS)	2.3 × 25.4	C-T	—	0.1	—	—	[148]
$Zr_{50}Cu_{40}Al_{10}$	1820 (UTS)	—	T-T	10	0.1	752	0.41	[156]
$Zr_{50}Cu_{40}Al_{10}$	—	—	—	—	—	752	0.41	[157]
$Zr_{50}Cu_{30}Ni_{10}Al_{10}$	—	—	—	—	—	865	0.46	[157]
$Zr_{50}Cu_{30}Ni_{10}Al_{10}$	1900 (UTS)	—	T-T	10	0.1	865	0.46	[156]
$Zr_{50}Cu_{37}Al_{10}Pd_{3}$	1900 (YS)	5.33 × 10.66	C-C	10	0.1	—	—	[158]
$Zr_{50}Cu_{37}Al_{10}Pd_{3}$	—	—	T-T	—	—	—	—	[158]
$Zr_{50}Cu_{37}Al_{10}Pd_{3}$	1900 (UTS)	—	T-T	10	0.1	983	0.52	[156]
$Zr_{55}Cu_{30}Ni_{5}Al_{10}$ (A)	—	5.4 × 3.8 × 20	SENB	57	0.33	—	—	[159]
$Zr_{55}Cu_{30}Ni_{5}Al_{10}$ (B)	—	—	SENB	25	0.33	—	—	[159]
$Zr_{55}Cu_{30}Ni_{5}Al_{10}$ (C)	—	B = 3.2; W = 15	C-T	—	—	—	—	[159]
$Zr_{59}Cu_{20}Al_{10}Ni_{8}Ti_{3}$	1580	6 × 3 × 1.5	T-T	1	0.1	No fatigue limit	1	[160]
$Zr_{52.5}Cu_{17.9}Ni_{14.6}Al_{10.0}Ti_{5.0}$	1700 (UTS)	—	Uniaxial	10	0.1	907 (in air)	0.53	[161]
$Zr_{52.5}Cu_{17.9}Ni_{14.6}Al_{10.0}Ti_{5.0}$	—	—	Uniaxial	10	0.1	907 (in air)	0.53	[136]
Rotating Beam								
$Pd_{40}Cu_{30}Ni_{10}P_{20}$	1700	Ø4	R	60	−1	340	0.2	[162]
$Zr_{50}Cu_{40}Al_{10}$	1820 (UTS)	Ø4	R	50	−1	250	—	[163]
$Zr_{50}Cu_{30}Ni_{10}Al_{10}$	1900 (UTS)	Ø4	R	50	−1	500	—	[163]
$Zr_{50}Cu_{40}Al_{10}$	—	Ø4	R	50	−1	250	—	[164]
$Zr_{50}Cu_{39}Al_{10}Pd_{1}$	1910	Ø4	R	50	−1	650	—	[164]
$Zr_{50}Cu_{38}Al_{10}Pd_{2}$	1910	Ø4	R	50	−1	700	—	[164]
$Zr_{50}Cu_{37}Al_{10}Pd_{3}$	1900	Ø4	R	50	−1	1050	—	[164]
$Zr_{50}Cu_{35}Al_{10}Pd_{5}$	1930	Ø4	R	50	−1	800	—	[164]
$Zr_{50}Cu_{33}Al_{10}Pd_{7}$	1950	Ø4	R	50	−1	550	—	[164]

Note: 4PB: four-point bend test; 3PB: three-point bend test; C-T: compact-tension specimen; SENB: single edge-notched beam; R: rotating beam; T-T: tensile-tensile; C-C: compressive-compressive.

[a] YS: yield strength.

[b] UTS: ultimate tensile strength.

TABLE 8.3
Fatigue Endurance Limits (at 10^7 Cycles) and Fatigue Ratios (Based on the Stress Amplitudes) of Selected Crystalline Alloys

Material	Yield Strength (MPa)	Tensile Strength (MPa)	Fatigue Endurance Limit (MPa)	Fatigue Ratio
300 M steel	1634	1930	428	0.22
Inconel 625 Ni alloy	509	918	217	0.24
Ti–6Al–4V	1014	1075	283	0.26
2090-T86 Al alloy	517	570	139	0.24
7075-T6 Al alloy	462	524	118	0.23

Source: Brandes, E.A. and Brook, G.B. (eds.), *Smithells Metals Reference Book*, 7th edn, Butterworth-Heinemann, London, UK, 1992.
Note: The yield and ultimate tensile strengths are also given. All the tests were performed in uniaxial condition and at an R value of 0.1.

same alloy. Selected fatigue data on some high-strength crystalline alloys are presented in Table 8.3 [165]. Typical S–N plots of some BMG alloys are shown in Figure 8.25. By comparing the fatigue results of BMGs with those from the crystalline alloys, the following points become clear.

Like crystalline alloys, BMG alloys also exhibit a fatigue limit. But, despite BMG alloys being much stronger than the crystalline alloys, they show much lower fatigue limits. Consequently, the fatigue ratio, defined as the ratio of fatigue limit to the ultimate tensile strength of the alloy (sometimes people use the yield strength instead of the ultimate tensile strength, when the material deforms mostly elastically), is expected to be quite high. The fatigue ratio is typically in the range 0.4–0.5 in the case of crystalline alloys. For BMG alloys, however, the ratio is much smaller, going down to a value of 0.04 (reported by Gilbert et al. [144] for $Zr_{41.2}Ti_{13.8}Cu_{12.5}Ni_{10}Be_{22.5}$). The values are higher in other cases, as reported by Liaw et al. [136,149,156,157,161]. These values are still lower than the values observed in crystalline alloys, and significantly lower than the anticipated values.

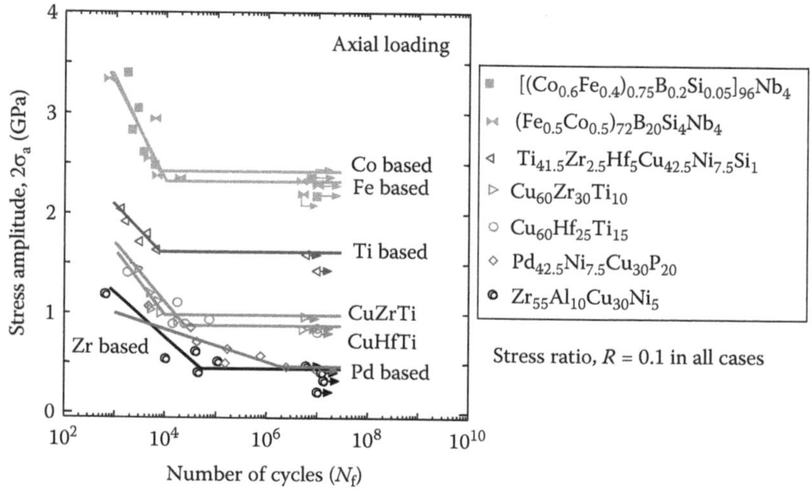

FIGURE 8.25
Typical S–N plots for some selected BMG alloys.

The reasons for this significant difference in the fatigue ratios between crystalline and BMG alloys are not clear. The group of Liaw et al. has mostly used notched cylindrical rods for evaluating fatigue behavior, in addition to four-point bend fatigue testing in some cases. Menzel and Dauskardt and Ritchie et al. have mostly used the four-point bend fatigue testing method. Menzel and Dauskardt [137] suggested that the stress state, and final fracture surfaces for the notched specimens, together with an incorrect stress concentration factor used in the work of Liaw et al. could account for the higher endurance limit reported by them. They also believe that Liaw et al. did not properly account for the location of the crack initiation site. Whereas the maximum tensile stress at the root of the notch was used in determining the endurance limit, regardless of the actual location of the fatigue initiation site, the actual fatigue initiation site was in fact far from the notch. Thus, it is possible that the tensile stress at the fracture initiation site in the cylindrical bars used by Liaw et al. was considerably less than the maximum tensile stress at the root of the notch used by Dauskardt's group. As a result, only a very small volume of the material is sampled by the highest stresses compared with the bend specimens. Therefore, the probability of a crack initiation site being present in this volume is low and, consequently, the fatigue endurance limit is higher. The situation is still not clear, and the reader is encouraged to consider the different arguments put forward by these two groups [138,139].

Another point of confusion in the literature is in reporting the fatigue ratio. The *S–N* plots are reported as stress amplitude against number of cycles. Thus, it is expected that the fatigue limit is determined as the stress amplitude below which the specimen does not fail irrespective of the number of cycles to which it is subjected. Sometimes, researchers have used the applied stress, rather than stress amplitude. In this case, the fatigue ratio will be twice the real value. One should take proper precautions to report the value correctly and also describe the way the fatigue ratio was determined.

Let us now discuss the effect of different test conditions on the fatigue behavior of BMG alloys.

8.7.2.1 Effect of Composition

A number of Zr-based BMG alloys have been investigated for their fatigue behavior, but only a few studies are available on BMG alloys based on other metals. Unfortunately, there have not been any investigations on systematically changing the composition in a regular fashion and identifying the trends, if any. From the available data, the following points may be noted.

Substituting 10 at.% Cu with Ni seems to improve the fatigue limit and fatigue ratio of $Zr_{50}Cu_{40}Al_{10}$ BMG alloys [157,163]. For example, the fatigue limit of the $Zr_{50}Cu_{40}Al_{10}$ alloy was 752 MPa, and it increased to 865 MPa in the $Zr_{50}Cu_{40}Ni_{10}Al_{10}$ alloy. Accordingly, the fatigue ratio increased from 0.41 to 0.46. It was suggested [163] that the difference in the fatigue strength could be attributed to the ease of oxidation of the fresh fatigue-fractured surface of the Ni-free alloy. The presence of Ni increases the ability to resist oxygen absorption, which results in longer fatigue life and higher fatigue endurance limit. But, decreasing the Ni content and increasing the Zr content had the opposite effect. The $Zr_{55}Cu_{30}Al_{10}Ni_5$ alloy had a fatigue limit of only 410 MPa. The tensile strengths of most of these alloys are about the same, 1.8–1.9 GPa, except for the $Zr_{55}Cu_{30}Al_{10}Ni_5$ alloy, which has a strength of only 1.56 GPa. The addition of 1 at.% Nb to a quaternary Zr-based BMG ($Zr_{58}Ni_{13.6}Cu_{18}Al_{10.4}$) also resulted in a lower fatigue limit [149]. But, it is not clear whether this is because of the Nb addition, the increase in Zr content, or both.

Yokoyama et al. [164] conducted a systematic study of the effect of Pd (and Ag, Pt, and Au) additions to a $Zr_{50}Cu_{40}Al_{10}$ BMG alloy to evaluate the fatigue strength enhancement. Of these additions, they noted that Pd was most effective. Figure 8.26a shows the variation of fatigue strength with Pd content in $Zr_{50}Cu_{40-x}Al_{10}Pd_x$ BMG alloys in the range $x = 0$–7 at.%. The fatigue strength increased with Pd content from 250 MPa at 0% Pd up to a peak value of 1050 MPa at the 3 at.% level. High-resolution TEM (HRTEM) studies showed that the alloy with 3 at.% Pd contained nanocrystalline regions, which could be responsible for increasing the fatigue limit by blocking shear bands and crack propagation [166]. The microstructural observations on alloys with other Pd contents are not available. Beyond this value of 3 at.%, the fatigue strength decreased with increasing Pd content. The authors had also calculated the volume change due to structural relaxation (corresponding to the excess free volume) in these alloys and found that the variation of volume change is also dependent on the Pd content, as shown in Figure 8.26b. It is interesting to note that the variation of fatigue strength and volume change with Pd content mimic each other and, as shown in Figure 8.26c, there is a linear relationship between the two.

Zr-based alloys are very reactive and, consequently, react with oxygen and other interstitial impurities. It was shown in Chapter 5 that the presence of oxygen in the Zr-based alloys is responsible for the formation of quasicrystalline phases. The effect of different amounts of oxygen on the fatigue behavior of Zr-based BMGs has also been investigated.

Two different samples of $Zr_{55}Cu_{30}Al_{10}Ni_5$ with different oxygen levels—1000 and 300 atomic ppm—were studied by Keryvin et al. [159]. The first sample, with the larger amount of oxygen, contained dendritic crystals of micron size dispersed in a glassy matrix. The second sample, prepared by a careful casting process, contained only 300 atomic ppm oxygen, and did not show any crystalline phases since the oxygen was dissolved in the matrix. It was

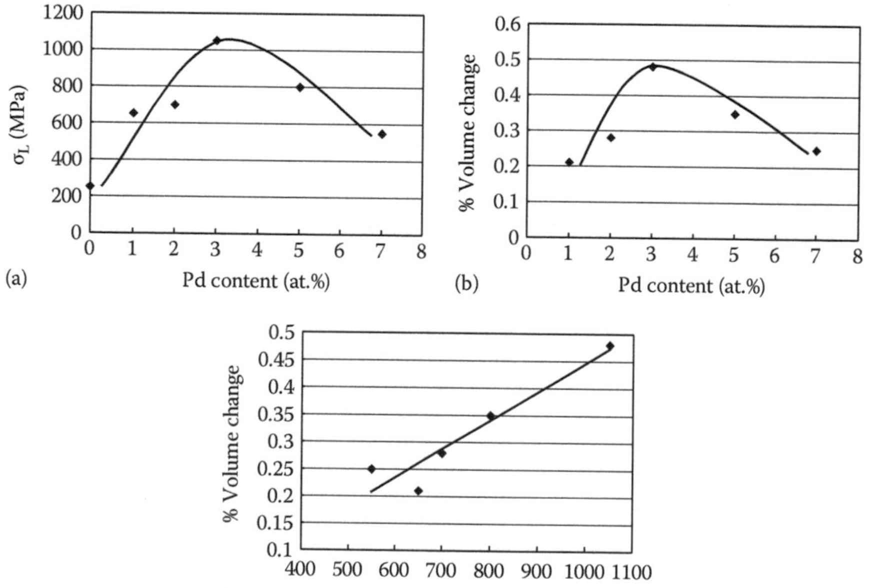

FIGURE 8.26
Variation of (a) fatigue strength, σ_L, and (b) volume change on relaxation with Pd addition to $Zr_{50}Cu_{40}Al_{10}$ BMG alloys. Note that both the curves peak at 3 at.% Pd. (c) Linear correlation between fatigue limit and volume change on relaxation in $Zr_{50}Cu_{40-x}Al_{10}Pd_x$ BMG alloys.

shown that the dendritic crystals caused embrittlement of the glassy alloy by allowing the cracks to propagate in a straight-line manner. On the other hand, in the sample in which oxygen was dissolved in the matrix, crack initiation was difficult, and it was not possible for the crack to pass straight through the glass. These results clearly show that oxygen, even in small amounts, has a major effect on the fatigue behavior of Zr-based BMG alloys.

Wang et al. [156] investigated the fatigue behavior of a $Zr_{41.2}Ti_{13.8}Cu_{12.5}Ni_{10}Be_{22.5}$ BMG alloy from two different batches (batch 59 and batch 94). The material from batch 59 contained less oxygen than that from batch 94, due to a better process control during the fabrication of batch 59. It was observed that the fatigue strengths of the two batches of alloys were different—703 MPa for batch 59 and 615 MPa for batch 94. It was suggested that this difference could be due to the presence of oxide inclusions in the specimen from batch 94, which aided in the easy initiation of cracks and, consequently, resulted in lower fatigue strengths.

It was also shown that the fatigue threshold (K_{th}) and the fatigue crack growth rates in the $Zr_{41.2}Ti_{13.8}Cu_{12.5}Ni_{10}Be_{22.5}$ BMG alloy were essentially identical in laboratory air and deionized water environments [167].

It is important to ponder a critical issue at this stage. Evaluating the effect of composition on the fatigue behavior of BMG alloys from the published literature is not easy. This is because researchers have been using different compositions, different methods to produce the glassy alloys, and different methods to evaluate the fatigue behavior. The "quality" of the samples, especially with reference to surface finish and melt cleanliness can seriously affect the fatigue life. Unfortunately, we do not have many data on alloys of different composition, processed from the same raw material, cast by the same clean process, and tested on a large number of specimens by the same fatigue method for these data to be considered reliable. For this to be achieved, we need large quantities of well-characterized glassy samples and long testing periods, which involves heavy expenditure. Until such results are available, it will be difficult to compare the results effectively, and to draw serious conclusions.

8.7.2.2 Effect of Environment

Some studies have been carried out to investigate the effect of the environment on the fatigue behavior of BMGs. Investigations were conducted in vacuum, laboratory air, aerated deionized water, and different corrosive environments (chloride solutions with different chloride ion concentrations and sulfate solutions), and phosphate-buffered saline (PBS) solution. These investigations were carried out only on Zr-based BMG alloys.

Wang et al. [166] measured the fatigue behavior of $Zr_{50}Cu_{40}Al_{10}$, $Zr_{50}Cu_{30}Al_{10}Ni_{10}$, and $Zr_{50}Cu_{37}Al_{10}Pd_3$ BMG alloys in both air and vacuum and did not find any significant difference in the fatigue lifetimes, especially at high stress levels, between the two environments. Peter et al. [136,161] also investigated the effect of air and vacuum on the fatigue behavior of a $Zr_{41.2}Ti_{13.8}Cu_{12.5}Ni_{10}Be_{22.5}$ BMG alloy and noted that the lifetimes were shorter in vacuum than in air (the relative humidity ranged from 18% to 61%, with an average value of 41%) for any given stress amplitude. One of the reasons for this deterioration in fatigue life could be that the hot tungsten filament of the ionization gauge dissociated the residual water vapor inside the vacuum chamber into atomic hydrogen and oxygen. Absorption of this hydrogen by the alloy could embrittle it, thus lowering the time for crack initiation and, consequently, reduce the fatigue life. No difference in fatigue life in air and vacuum was found when the ionization gauge was turned off.

The influence of different corrosive atmospheres on the fatigue behavior of Zr-based BMG alloys was also investigated. Greatly reduced fatigue strengths were noted in the presence of 0.6 N NaCl in a $Zr_{52.5}Cu_{17.9}Ni_{14.6}Al_{10}Ti_5$ alloy. While the fatigue endurance limit was 425 MPa in air, it was only 50 MPa in 0.6 N NaCl solution [168]. The fatigue properties of $Zr_{65}Cu_{15}Ni_{10}Al_{10}$ glassy alloy obtained by consolidating the amorphous powder were investigated in both air and PBS solution. No difference was found between the *S–N* curves of the alloy tested in air and PBS solution, both giving a fatigue strength of 150 MPa [169].

Early studies [170] indicated that the $Zr_{41.2}Ti_{13.8}Cu_{12.5}Ni_{10}Be_{22.5}$ BMG alloy was highly susceptible to stress corrosion cracking and stress corrosion fatigue in the presence of aqueous NaCl solution. These studies were later extended to investigate the fatigue behavior in a variety of environments [167]. The crack growth rates in 0.5 N aerated NaCl solution were 2–3 orders of magnitude faster than in air or deionized water. While the crack growth rates at a ΔK value of 1 MPa $m^{-0.5}$ were about 10^{-10} to 10^{-9} m per cycle in air or deionized water, they were about 10^{-8} to 10^{-7} m per cycle in 0.5 N NaCl solution. Further, compared with the fatigue threshold value (ΔK_{th}) of almost 3 MPa $m^{-0.5}$ in air, there was a significant reduction in the ΔK_{th} value to about 0.9 MPa $m^{-0.5}$ in the NaCl solution. Additionally, the fracture surfaces in the two conditions were also different. While the overall fatigue fracture surfaces are similar, with the three main regions of crack initiation, crack propagation, and final fast fracture, the surface of the sample in air was macroscopically uneven and that tested in the NaCl solution was smooth and optically reflective. Atomic force microscopy images showed that these smooth areas had roughnesses of below 100 nm and were separated by steps, 0.1–50 μm in height. Similar features were also noted by Kawashima et al. [171]. The effect of systematically varying the loading cycle, stress-intensity range, solution concentration, anion identity, solution deaeration, and bulk electrochemical potential was also studied. The obtained results suggest that crack growth mechanisms in air and in aqueous solutions are different [167,172]. Fatigue crack propagation behavior in air seems to be similar to that observed in polycrystalline metals and is associated with alternating blunting and resharpening of the crack tip, as evidenced by the presence of fatigue striations. In contrast, in NaCl solution, the mechanism seems to be driven by a stress-assisted anodic reaction at the crack tip.

Kawashima et al. [171] also reported that the fatigue crack growth behavior in their $Zr_{55}Cu_{30}Al_{10}Ni_5$ BMG alloy was almost the same both in air and deionized water. But, the crack growth rate in 0.5 N NaCl solution was about one to two orders of magnitude larger than that in air. In addition, they had studied the fatigue crack growth behavior in phosphate-citric acid (PCA)-buffered solutions containing different chloride ion concentrations. The interesting result was that while the crack growth rate increased with increasing chloride content, there was no effect if the PCA contained less than 0.1 N NaCl. Another interesting result was that the crack growth rates in PCA + NaCl were lower than those in NaCl of the same concentration. It is possible that the phosphate ions adsorbed at the crack tip will prevent the adsorption of the chloride ions, leading to a decreased growth rate.

8.7.2.3 Effect of Microstructure

It is generally expected that BMGs are fully glassy and homogeneous. But, the presence of a small volume fraction of nanocrystalline particles is occasionally noted in the samples (see, e.g., Refs. [149,158]). The presence of such nanocrystalline phases, in small quantities, may not be deleterious for the fatigue properties of glasses. Metallic glass specimens are extremely strong, but inherently brittle. One of the ways in which their plasticity could be increased is by incorporation of a discontinuous crystalline phase into the homogeneous glass matrix [100]. Such a composite will exhibit enhanced ductility and, consequently, the fatigue properties could also be enhanced. Following this approach, a few investigations have been carried out, and these aspects will be discussed in Section 8.9.

Flores et al. [151] produced a composite based on the commercially available Vitreloy 1 ($Zr_{41.2}Ti_{13.8}Cu_{12.5}Ni_{10}Be_{22.5}$) BMG. The composite consisted of about 25 vol.% of a crystalline phase with the composition $Zr_{71}Ti_{16.3}Nb_{10}Cu_{1.8}Ni_{0.9}$ in a glassy matrix of $Zr_{47}Ti_{12.9}Nb_{2.8}Cu_{11}Ni_{9.6}Be_{16.7}$. Since the crystalline phase had formed directly from the melt, it had a dendritic shape, and these dendrites were finely dispersed with a spacing of about 10 μm in the glassy matrix. Such a composite showed increased fatigue life. While the fatigue ratio for the monolithic alloy was as low as 0.04 [144], it was ~0.1 for the composite based on the composite ultimate tensile strength value of 1.48 GPa [173]. It was suggested that the presence of the crystalline phase allowed the residual surface stresses to be relieved and any surface strains to be accommodated, such that defects did not form. But, if the defects were present, they may have been blocked by the second phase.

Fujita et al. [153] reported a high fatigue strength of 1610 MPa for a $Ti_{41.5}Zr_{2.5}Hf_5Cu_{42.5}Ni_{7.5}Si_1$ composite, with a fatigue ratio of 0.79. These high values, even in comparison with the crystalline alloys, were explained on the basis that no micro-defects existed at the crack initiation sites and that the nanocrystals prevented the initiation and growth of the shear bands.

A surprisingly different result was, however, reported by Wang et al. [155]. They studied the fatigue behavior of a Zr-based BMG composite with the composition $Zr_{47}Ti_{12.9}Nb_{2.8}Ni_{9.6}Cu_{11}Be_{16.7}$ that contained precipitates of a crystalline dendritic Zr–Ti–Nb (β) phase dispersed in the glassy matrix. The observed fatigue strength was only 239 MPa with a fatigue ratio of 0.16. They commented that this value was much lower than that of the monolithic BMG with the composition $Zr_{41.2}Ti_{13.8}Cu_{12.5}Ni_{10}Be_{22.5}$, which had a composition identical to that of Vitreloy 1. The authors also noted that their results using round, tapered samples in uniaxial tension at a frequency of 10 Hz were consistent with the results of Flores et al. [151], who conducted a four-point bend test on rectangular specimens at a frequency of 25 Hz. Thus, they suggested that the test volume had little effect on the fatigue behavior of their BMG composite.

Two points are worth noting here. Firstly, Wang et al. [155] were comparing the fatigue strength of the composite (based on $Zr_{47}Ti_{12.9}Nb_{2.8}Cu_{11}Ni_{9.6}Be_{16.7}$, LM 002?) with that of the monolithic $Zr_{41.2}Ti_{13.8}Cu_{12.5}Ni_{10}Be_{22.5}$ to suggest that the composite had a lower fatigue strength. Instead, they should have compared the strength of the composite with that of the monolithic LM 002. (From the paper by Wang et al. [155], it is not clear what the alloy composition was for their investigation. The composition they provide for the LM 002 is the same as the composition of the *glassy* phase in the work of Flores et al. [151]. But, the microstructure of LM 002 shows the presence of a dendritic phase and therefore it can be assumed that they were dealing with a composite. But, the compositions of the glassy matrix and dendritic phases are not clear.) Secondly, Wang et al. [155] suggest that the test volume may not have any impact on the fatigue behavior (in the composite). This needs further support, because the major differences in the results between the groups of Liaw and Dauskardt have been traced mainly to the small test volumes of specimens used in the investigations of Liaw [137–139].

The stress range ($\Delta\sigma$) and the number of fatigue cycles (N_f) data of the *S–N* plots for different alloys were fitted, for engineering applications, using the equation for a straight line of the type:

$$\Delta\sigma(\text{MPa}) = A \times \log N_f - B \tag{8.24}$$

The values of *A* and *B* are listed in Table 8.4 for the different BMG alloys for which data are available.

TABLE 8.4

Stress-Range and Fatigue-Cycle Data for the Fatigue Behavior of BMG Alloys, Fitted According to the Equation: $\Delta\sigma$ (MPa) = $A \times \log N_f - B$

Material	$-A$ (MPa/Cycles)	B (MPa)	References
$(Cu_{60}Zr_{30}Ti_{10})_{99}Sn_1$	61.394	1186.6	[141]
$Fe_{48}Cr_{15}Mo_{14}Er_2C_{15}B_6$	4	790.86	[142]
$Zr_{41.2}Ti_{13.8}Cu_{12.5}Ni_{10}Be_{22.5}$	438.2	2666.5	[155]
$Zr_{41.2}Ti_{13.8}Cu_{12.5}Ni_{10}Be_{22.5}$ (batch 59)	558.64	5877.5	[154]
$Zr_{41.2}Ti_{13.8}Cu_{12.5}Ni_{10}Be_{22.5}$ (batch 94)	505.97	5076.8	[154]
$Zr_{41.2}Ti_{13.8}Cu_{12.5}Ni_{10}Be_{22.5}$	233.1	2052	[136]
$Zr_{47}Ti_{12.9}Nb_{2.8}Cu_{11}Ni_{9.6}Be_{16.7}$	336.3	2032.4	[155]
$Zr_{52.5}Cu_{17.9}Ni_{14.6}Al_{10}Ti_5$	61.7	1329.5	[150]
$Zr_{50}Cu_{40}Al_{10}$	971.5	4907.8	[157]
$Zr_{50}Cu_{30}Ni_{10}Al_{10}$	963.6	4928.9	[157]
$(Zr_{58}Ni_{13.6}Cu_{18}Al_{10.4})_{99}Nb_1$	56	1178	[148]
$Zr_{50}Al_{10}Cu_{37}Pd_3$	410.7 (compression–compression)	3584.5	[158]
$Zr_{50}Al_{10}Cu_{37}Pd_3$	369.4 (tension–tension)	2468.8	[158]

8.7.2.4 Effect of Testing Method

As mentioned earlier, different methods have been used to determine the fatigue behavior of BMG alloys. But, there have been only a limited number of investigations to compare the fatigue strengths of the same material by two different methods, and by the same group [141]. The fatigue behavior of a $(Cu_{60}Zr_{30}Ti_{10})_{99}Sn_1$ BMG alloy was tested by both four-point bend and three-point bend test methods and the results are shown in Figure 8.27. As expected, the fatigue lifetimes are higher for the three-point bend test condition than for the four-point bend test condition. The fatigue endurance limits, based on the applied stress range, are 475 and 350 MPa, respectively, for the three-point and four-point bend test methods. The larger test volume in the four-point bend test condition, in comparison with

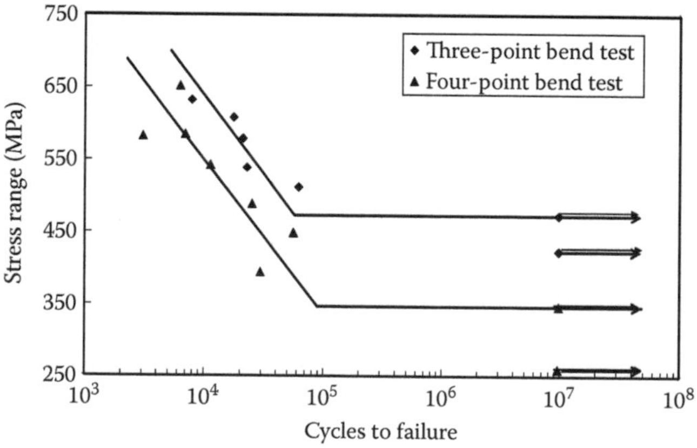

FIGURE 8.27
Cycles to failure, N_f, measured as a function of the applied stress range, σ_R $(=\sigma_{max} - \sigma_{min})$, during three-point and four-point bend fatigue tests on $(Cu_{60}Zr_{30}Ti_{10})_{99}Sn_1$ BMG alloy. The R ratio $(\sigma_{min}/\sigma_{max})$ was 0.1 and the frequency was 10 Hz in an air environment. (Reprinted from Freels, M. et al., *J. Mater. Res.*, 22, 374, 2007. With permission.)

the three-point bend test condition, could contain more critical defects, stress raisers, and free volume, which would enhance the possibilities for shear band formation, crack initiation, and therefore lower the fatigue endurance limit.

Morrison et al. [150] conducted fatigue tests on $Zr_{52.5}Cu_{17.9}Ni_{14.6}Al_{10}Ti_5$ using both four-point bend and uniaxial testing methods. They noted that there was large scatter in the data for the results from the four-point bend test, particularly below a stress amplitude of 600 MPa, as the stress amplitude approached the fatigue limit. The fatigue-endurance limit, in terms of the stress amplitude, was determined to be approximately 425 MPa, which is close to the value of 454 MPa determined from the uniaxial test method [136].

8.7.2.5 Effect of Structural Relaxation

Launey et al. [147] studied two different glassy alloy samples—one produced in 1994 as 4.8 mm thick sheet by Howmet Corporation, Whitehall, MI, and the other as 2.6 mm thick sheet by Liquidmetal Technologies, Lake Forest, CA, in 2004. Both had identical chemical compositions, *viz.*, $Zr_{41.25}Ti_{13.75}Ni_{10}Cu_{12.5}Be_{22.5}$, and both were found to be fully glassy. The fatigue strengths of the two samples were, however, quite different. The Liquidmetal sample showed a fatigue strength of 359 MPa and the Howmet sample 768 MPa. There was no difference between the two samples as regards their thermal properties. Both the specimens showed the same glass transition and crystallization temperatures and there was no difference in the hydrogen content. But, the Howmet sample showed a lower fracture toughness than the Liquidmetal alloy.

Generally speaking, a lower fracture toughness is expected to reduce the fatigue life of a material, but the present result runs contrary to this expectation, suggesting that the fatigue crack initiation and/or growth portions of the fatigue lifetimes were sufficiently longer to offset this effect. However, it was noted that the free volumes were different in the two cases, attributed to differences in the cooling rates during processing. It is possible that the 4.8 mm thick sheet of Howmet was cooled at a slower rate than the 2.6 mm thick sheet by Liquidmetal and, consequently, the Howmet alloy had a reduced free volume. Reduced free volume is known to result in longer fatigue lives and lower fracture toughness [174], and the present results confirm this.

Launey et al. [148] confirmed this hypothesis in a $Zr_{44}Ti_{11}Ni_{10}Cu_{10}Be_{25}$ BMG alloy by annealing the sample below the glass transition temperature, T_g. The free volume and residual stresses in the sample were altered by isothermally annealing the samples for the integral numbers of the relaxation time, τ, calculated on the basis of $\tau = \Delta T_g/\beta$, where β is the heating rate, and ΔT_g is the temperature interval between the onset and end of the endothermic glass transition events. The specimens annealed for a time of 10τ and for 2 min at 573 K (to relieve residual compressive stresses at the surface due to thermal tempering during processing) were tested using a four-point-bend fatigue test. While the specimen annealed for a time of 10τ showed a fatigue strength of 275 MPa, the stress-relieved sample showed a strength of 195 MPa. These results again show that a reduction in the free volume increases the fatigue strength of alloys.

From this general description, it is clear that the fatigue ratios of BMG alloys are significantly lower than those of the crystalline alloys. Whereas the ratios are typically in the range of 0.4–0.6 for crystalline alloys, they seem to be typically less than about 0.2, except for the results reported from Liaw's group, for the BMG alloys. In the alloys investigated by Liaw's group, the ratios reported ranged up to about 0.5. The reasons for the differences in the fatigue ratios have been discussed by Menzel and Dauskardt [137,139] and Wang et al. [138]. The possible causes for the differences include the mean stress, material quality, specimen geometry, chemical environment, temperature, cyclic frequency, residual stresses, and surface condition. Much has been said about these in the open literature. The real solution to this problem would be for the same research group to take samples of the same composition, cast by different techniques, and test them using different methods to see if the differences persist. This is going to be a tall order involving heavy expenditure and time. But, because of the importance of the issue, this may be a worthwhile thing to do. It should, however, be noted that rapid improvements have been occurring during the last several years with respect to improving the quality of BMG samples (better samples without many casting defects are being produced), increasing the sample size, and making available improved characterization tools that can be brought to bear on the problem.

Since the sample volume examined during fatigue testing has been an issue, the availability of high-quality and larger-size samples that are well characterized should be able to resolve the differences.

8.7.3 Fatigue Crack Growth Rate

Fatigue crack growth rate is an important aspect of studying the fatigue behavior of materials. The fatigue crack grows with each cycle and, therefore, the time for it to reach the critical size at which fracture occurs is determined by the rate at which it grows and also the fracture toughness of the material.

The fatigue crack growth rate is determined on fatigue pre-cracked compact-tension (C-T) specimens as a function of the stress intensity range, ΔK ($K_{max} - K_{min}$), where K represents the stress intensity. At very small values of ΔK, the crack growth rate is very small, or may be zero, until the curve starts approaching the fatigue threshold, ΔK_{th}. The significance of ΔK_{th} is that the crack does not grow, or grows at an extremely low rate, if the specimen is subjected to this stress intensity range. In the intermediate region of crack growth, the crack growth rate is determined from the Paris power law relationship:

$$\frac{da}{dN} = C(\Delta K)^m \tag{8.25}$$

where:

da/dN represents the crack growth rate (a is the crack length and N is the number of cycles)

C and m are constants, determined experimentally

These constants are different for different materials, and depend on the microstructure of the test material and test environment. At very large values of ΔK, the crack growth rate is again very high due to unstable crack growth just prior to final failure. The plots are made on a log–log scale. Thus, when log (da/dN) is plotted against log (ΔK), in the intermediate crack growth region, one obtains a straight line, the slope of which is m. This slope represents the rate at which the crack growth rate changes as a function of ΔK. The larger the value of m, the faster the crack growth rate increases. A desirable situation for long fatigue life is to achieve as small a value of m as possible.

The fatigue crack growth rate tests can be conducted under either increasing or decreasing ΔK conditions. During cycling, the crack length is continuously monitored using the unloading elastic compliance, obtained from a back-face strain gauge. It can also be measured using a high-resolution optical microscope mounted on a traveling stage. The fatigue threshold, ΔK_{th}, is determined during a decreasing ΔK test by allowing the stress intensity range to decrease until the crack growth rates are near 10^{-10} m^{-1} per cycle. ΔK_{th} is then defined as the highest ΔK value at which da/dN is less than 10^{-10} m^{-1} per cycle.

Compared with the number of published high-cycle fatigue studies to obtain the fatigue strengths and fatigue lives of BMG alloys, investigations on fatigue crack growth rates are limited. Gilbert et al. [144,145] conducted the fatigue crack growth rate studies on 7mm thick and 38mm wide compact-tension (C-T) specimens of $Zr_{41.2}Ti_{13.8}Cu_{12.5}Ni_{10}Be_{22.5}$ BMG alloy at a frequency of 25Hz and load ratio (minimum load/maximum load) of 0.1, and their results are shown in Figure 8.28. These authors were the first to report that the crack propagation behavior in the BMG alloy was not any different from that in ductile crystalline alloys either in terms of growth rate dependence on applied stress intensity range or the presence of striations on fatigue-fractured specimens. By fitting the growth rates in the mid-range to the Paris power law equation (Equation 8.25), they obtained a value of 2.7 for the exponent m. The exponent values for typical ductile crystalline metals are in the range 2–4. On the other hand, the m value is typically in the range 15–50 for brittle materials such as ceramics, for example, Al_2O_3 or oxide glasses. The fatigue crack growth threshold value (K_{th}) for the BMG alloy was also calculated as 3 MPa $m^{0.5}$, which was also comparable to the many high-strength steel and aluminum alloys. They noted that as the load ratio was increased to 0.5, cyclic crack growth rates were accelerated and fatigue thresholds were reduced. This trend is again similar to that in ductile crystalline metals at near-threshold growth rates.

When this BMG alloy was heat treated in a vacuum for 24h at 723K, the glassy phase was fully crystallized, containing a Laves phase with the $MgZn_2$-type structure, a phase with the $CuAl_2$-type structure, and at least one other unidentified phase. The fracture toughness of the crystallized glass was 1MPa $m^{0.5}$, a 50-fold reduction from that of the fully glassy alloy (55 MPa $m^{0.5}$) [175]. Consequently, the fully crystallized alloy behaved like a typical brittle material. Fatigue cracking was not observed and catastrophic failure occurred immediately after loading.

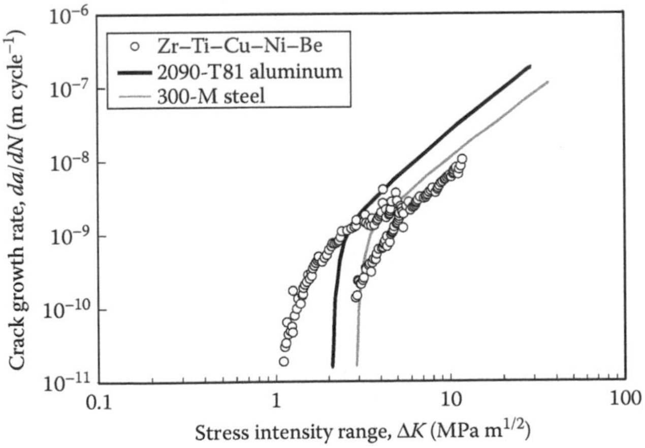

FIGURE 8.28
Fatigue crack growth rate, da/dN vs. stress intensity range, ΔK for the $Zr_{41.2}Ti_{13.8}Cu_{12.5}Ni_{10}Be_{22.5}$ BMG alloy. Data from a large number of samples are included. Also shown are the data from high-strength 300 M steel and an age-hardened 2090-T81 Al–Li alloy, whose ultimate tensile strengths are 2300 and 589 MPa, respectively. (Reprinted from Gilbert, C.J. et al., *Scr. Mater.*, 38, 537, 1998. With permission.)

Fatigue crack growth studies were also conducted on other Zr-based BMG alloys [146,148,151,176] and the available results for C, m, and ΔK_{th} are presented in Table 8.5. In all these cases, the m values are those typically observed in crystalline alloys (in the range 2–4) and the same is true with respect to the ΔK_{th} values.

The fatigue life of specimens is determined mostly by the fatigue crack growth rate and, therefore, crack initiation may not be important in BMG alloys [146]. Thus, it becomes necessary to understand the mechanisms and behavior of fatigue crack growth to be able to improve the fatigue lives of BMGs and BMG composites. Fatigue crack growth mechanisms in melt-spun ribbon specimens were explained on the basis of fractographic evidence. It was suggested that fatigue cracks propagate by fracturing ligaments between crack tip shear bands at higher growth rates. However, at lower growth rates, the cracks propagate within shear bands. But, due to the absence of these shear bands, especially at low to moderate growth rates, this mechanism may not be applicable to BMGs.

The fracture surfaces of BMGs after fatigue crack growth rate experiments seem to be quite different from those of crystalline materials. The surface roughness of BMG samples has been found to increase with increasing crack growth rate. The surfaces change from a mirror-like surface in the near-threshold fatigue crack growth region to a rough morphology exhibiting ridge-like features in the fast fracture process [145]. However, closer examination of the fatigue fracture surfaces, especially in the higher growth-rate region, shows a clear feature of classic fatigue striations parallel to the crack front, representing the cycle-by-cycle advance of the crack.

It is known that the mechanism for striation formation in ductile crystalline materials is due to repeated crack tip blunting and resharpening, and it has been proposed that a similar mechanism may be applicable in the case of BMGs as well [145]. The distance over which blunting causes a deviation in near-tip stress fields is proportional to the crack-tip-opening displacement (CTOD), d. If growth rates are dependent upon the blunting distance, then they may be shown to scale with the change in CTOD, Δd, that is,

$$\frac{da}{dN} \propto \Delta d \propto \frac{\Delta K^2}{\sigma_y E'} \tag{8.26}$$

where:

σ_y is the cyclic flow stress

$E' = E$ for plane stress and $E/(1-\nu^2)$ for plane strain condition, where E is Young's modulus and ν is Poisson's ratio

TABLE 8.5

Fatigue Crack Growth Properties of Zr-Based BMG Alloys

Material	C (m cycle^{-1}) (MPa m$^{0.5}$)$^{-m}$	m	ΔK_{th} (MPa m$^{0.5}$)	References
$Zr_{44}Ti_{11}Ni_{10}Cu_{10}Be_{25}$	1.8×10^{-10}	2.03	1.35–1.79	[148]
$Zr_{41.25}Ti_{13.75}Ni_{10}Cu_{12.5}Be_{22.5}$	5×10^{-10}	1.72	1.5–2.0	[146]
$Zr_{41.25}Ti_{13.75}Ni_{10}Cu_{12.5}Be_{22.5}$	1.6×10^{-11}	2.7	1–3	[145]
$Zr_{41.25}Ti_{13.75}Ni_{10}Cu_{12.5}Be_{22.5}$	1.7×10^{-13}	4.9	—	[145]
$Zr_{41.25}Ti_{13.75}Ni_{10}Cu_{12.5}Be_{22.5}$	6.3×10^{-12}	3.4	—	[145]
$Zr_{41.25}Ti_{13.75}Ni_{10}Cu_{12.5}Be_{22.5}$	1.5×10^{-11}	2.7	—	[175]
$Zr_{41.25}Ti_{13.75}Ni_{10}Cu_{12.5}Be_{22.5}$	2.3×10^{-10}	1.6	—	[172]
$Zr_{47}Ti_{12.9}Nb_{2.8}Cu_{11}Ni_{9.6}Be_{16.7}$ + 25 vol% $Zr_{71}Ti_{16.3}Nb_{10}Cu_{1.8}Ni_{0.9}$	—	2	1.24	[151]
$Zr_{55}Cu_{30}Al_{10}Ni_5$	—	2	5	[176]

This model, first suggested by Gilbert et al. [145], predicts a Paris law exponent, m, of 2, even though the observed m values varied from 2 to 4. The reasons for the difference in the m values have been attributed to structural relaxation or damage of the near-tip material. But, Equation 8.26 has been shown to provide a reasonable description of the experimentally observed fatigue crack growth rates in BMGs with a proportionality factor of 0.01. Together with the presence of fatigue striations on fracture surfaces, this suggests a mechanism for cyclic crack advance in BMGs, which also involves repetitive blunting and resharpening of the crack tip. Hence, it could be concluded that the mechanism for crack growth is the same in both crystalline and BMG alloys.

Even though striations are shown to form during fatigue crack growth in both crystalline and BMG alloys, the spacing of the striations in BMGs appears to be larger than the value calculated from the growth rate, da/dN. In fact, it is noted that the observed striation spacing is an order of magnitude larger than the growth rate at high ΔK values, and two or more orders of magnitude larger at low ΔK values [177]. This disparity between the growth rate and the striation spacing shows that there is an accumulation of damage necessary prior to crack advance, similar to growth band formation in many polymers. It also implies that the applied load cycles are not efficient enough to produce crack extension, especially at low ΔK values. It is also noted that the striations do not extend over the width of the specimen; rather, they are broken up in many places along the crack front. Therefore, it can be inferred that the entire crack front does not extend with a single loading cycle, and that non-uniform extension of the crack front occurs during steady-state fatigue crack growth. Similar results were reported for BMG composites [151] and other BMG alloys [148,178–180].

8.7.3.1 Effect of Environment

Since materials generally show degradation in a corrosive environment, it is expected that the fatigue behavior of BMG alloys would be similar. Morrison et al. [168] investigated the fatigue behavior of $Zr_{52.5}Cu_{17.9}Ni_{14.6}Al_{10}Ti_5$ both in air and in 0.6 N NaCl electrolyte. They noted that at high stress levels, the corrosion-fatigue life was similar to the fatigue lives observed in air. But, the environment became increasingly detrimental with a decrease in stress. For example, the fatigue endurance limit of the BMG alloy in 0.6 N NaCl electrolyte was only about 50 MPa, approximately 10% of the value measured in air.

Ritchie and coworkers [167,172] studied the fatigue crack growth behavior of $Zr_{41.2}Ti_{13.8}Cu_{12.5}Ni_{10}Be_{22.5}$ at room temperature in different environments: air, deionized water, NaCl of different normalities, Na_2SO_4, $NaClO_4$, and some combinations of these. The authors reported that the crack growth rates were slightly higher in deionized water than in air. But, the crack growth rates increased dramatically, by as much as three orders of magnitude, when the BMG alloy was tested in 0.5 N NaCl solution compared with growth rates in air or deionized water. Additionally, they noted that fatigue crack growth in the aqueous solutions displayed an abrupt threshold, whereupon the growth rates increased by four to five orders of magnitude to reach a plateau (region II) regime where the crack growth was essentially independent of the stress intensity range. Fatigue crack growth rates in region II were found to be a function of the concentration of NaCl in solution. Specifically, the steady-state growth rates were decreased from $\sim 5 \times 10^{-7}$ to $\sim 1.5 \times 10^{-8}$ m per cycle with a decrease in the NaCl concentration from 0.5 to 0.005 N. However, the fatigue threshold value varied insignificantly over three orders of magnitude in NaCl concentration. Specifically, the K_{th} was constant at ~0.84 MPa $m^{0.5}$ when the NaCl concentration was 0.5, 0.05, or 0.005 N. Compared with a ΔK_{th} value of almost 3 MPa $m^{0.5}$ in air, its value was markedly reduced to ~0.9 MPa $m^{0.5}$ in the aqueous solution. This fatigue threshold in the aqueous solution was more than an order of magnitude lower than the fracture toughness of the alloy.

The fatigue crack growth rates were an order of magnitude slower in 0.5 N $NaClO_4$ and 0.5 N Na_2SO_4 solutions compared with their behavior in 0.5 N NaCl. Based on these observations, the authors concluded that the significant contribution of NaCl to increasing the fatigue crack growth rate could be attributed to a stress-corrosion crack growth mechanism, involving stress-assisted anodic dissolution at the crack tip.

8.7.3.2 Effect of Temperature

There has been only one study to investigate the effect of temperature on the crack growth rates during fatigue [181]. These authors studied the fatigue crack growth rates in a $Zr_{41.25}Ti_{13.75}Cu_{12.5}Ni_{10}Be_{22.5}$ BMG at different temperatures, from room temperature up to 220°C, which is lower than the glass transition temperature of 352°C. A sinusoidal loading waveform of 20 Hz was employed with a constant load ratio, R ($=K_{min}/K_{max}$) of 0.1. The effective stress intensity range $\Delta K_{th,eff}$, actually experienced by the crack tip, was defined as

$$\Delta K_{th,eff} = K_{max} - K_{cl} \tag{8.27}$$

where K_{cl} is the crack closure load determined from the initial deviation from linearity of the unloading load versus back-face strain data.

When the elevated temperature fatigue crack growth rates were measured, the authors noted a distinct mid-growth rate regime together with decreased growth rates in the near-threshold region. The mid-range growth rates were fitted to a Paris power law relationship, and the material constants C and m are listed in Table 8.6. Additionally, the effective threshold stress intensity factors, $\Delta K_{th,eff}$, are also listed, and are noted to be slightly less than ΔK_{th}.

The values of m are in the range 1.4–1.55, and are similar to the values reported earlier (see Table 8.5). Comparison of the fatigue data revealed that the mid-range crack growth rates were not significantly affected by the testing temperature. On the other hand, ΔK_{th} and $\Delta K_{th,eff}$ increased with increasing temperature (Figure 8.29). The $\Delta K_{th,eff}$ values, obtained after correcting for crack closure effects, also increased with increasing temperature, suggesting that removal of crack closure effects cannot account for the observed increase in ΔK_{th} with temperature. Fractured surfaces showed fatigue striations at growth rates above ~10^{-9} m per cycle for all temperatures, and at all growth rates for 220°C.

TABLE 8.6

Fatigue Crack Growth Rates of Vitreloy 1 at Different Temperatures

Test Temperature (°C)	C (m cycle^{-1}) (× 10^{-10} MPa $m^{0.5})^{-m}$	m	ΔK_{th} (MPa $m^{0.5}$)	$\Delta K_{th,eff}$ (MPa $m^{0.5}$)
25	6.6	1.4	1.25	—
100	7.1	1.4	1.07	1.07
140	5.9	1.4	1.14	1.11
180	7.4	1.45	1.24	1.21
220	6.8	1.55	1.40	1.38

Source: Hess, P.A. and Dauskardt, R.H., *Acta Mater.*, 52, 3525, 2004.

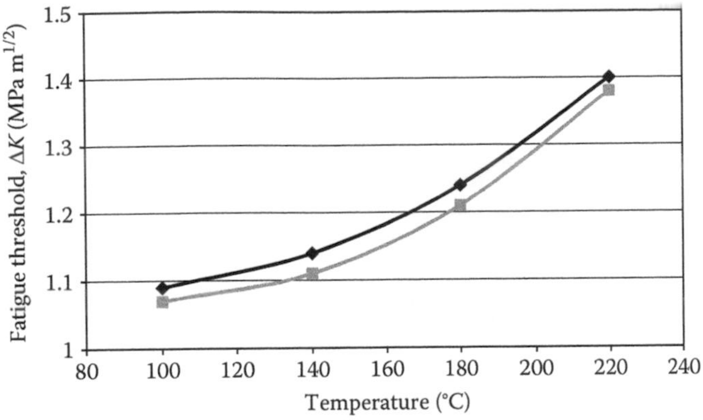

FIGURE 8.29
Variation of fatigue threshold, ΔK_{th}, as a function of testing temperature observed during fatigue testing of $Zr_{41.2}Ti_{13.8}Cu_{12.5}Ni_{10}Be_{22.5}$ BMG alloy at elevated temperatures. The upper curve is for ΔK_{th} and the lower curve is for $\Delta K_{th,eff}$, corrected for closure effects. Both curves show a similar trend.

8.7.4 Fatigue Fracture Surface Characterization

Scanning electron microscopy methods have been frequently used to characterize the surfaces of specimens that have been subjected to fatigue failure. The fatigue-fractured specimen surfaces generally show four main regions—a fatigue crack initiation site, a stable fatigue crack propagation region, an unstable final fast fracture region, and an apparent melting region.

In general, slip bands, deformation bands, twins, and grain boundaries are considered to be preferential sites for nucleation of fatigue cracks in ductile crystalline materials. But, such crystal defects are absent in BMG alloys and, therefore, fatigue crack initiation has to occur in a different way. In some cases, cracks are initiated from casting defects, such as porosity, cold shuts, polishing scratches, shallow pits, or oxide inclusions. In the case of bending tests, the cracks are always initiated on the tensile surface only and are associated with preexisting defects, such as gas pores or surface trenches. The size of the initiation site ranges from 5 to 10 μm.

Surrounding the initiation site is the stable crack propagation region, which has a thumbnail-shaped appearance. This region is basically perpendicular to the stress direction and contains finely spaced parallel marks (fatigue striations) oriented somewhat normal to the crack growth direction. These marks do not uniformly cover this region, and are especially evident in the outer portions of the bend test specimens, that is, maximum stress regions. These fatigue striations form via repetitive blunting and resharpening of the crack tip. During the early stages of crack propagation, these striations do not cover the entire crack front and are broken up in many places. Thus, the crack front probably does not grow uniformly during the early stages of crack growth.

The fast fracture region is characterized by a very rough surface and occupies most of the fracture surface.

The fatigue fracture surfaces of the $Zr_{50}Cu_{30}Al_{10}Ni_{10}$ specimen tested at σ_{max} = 1133 MPa and in air are shown in Figure 8.30 [157]. The fracture surface is basically perpendicular to the loading direction. In general, several crack initiation sites are found on the notched surface (Figure 8.30a). A fatigue crack originates from such sites and propagates

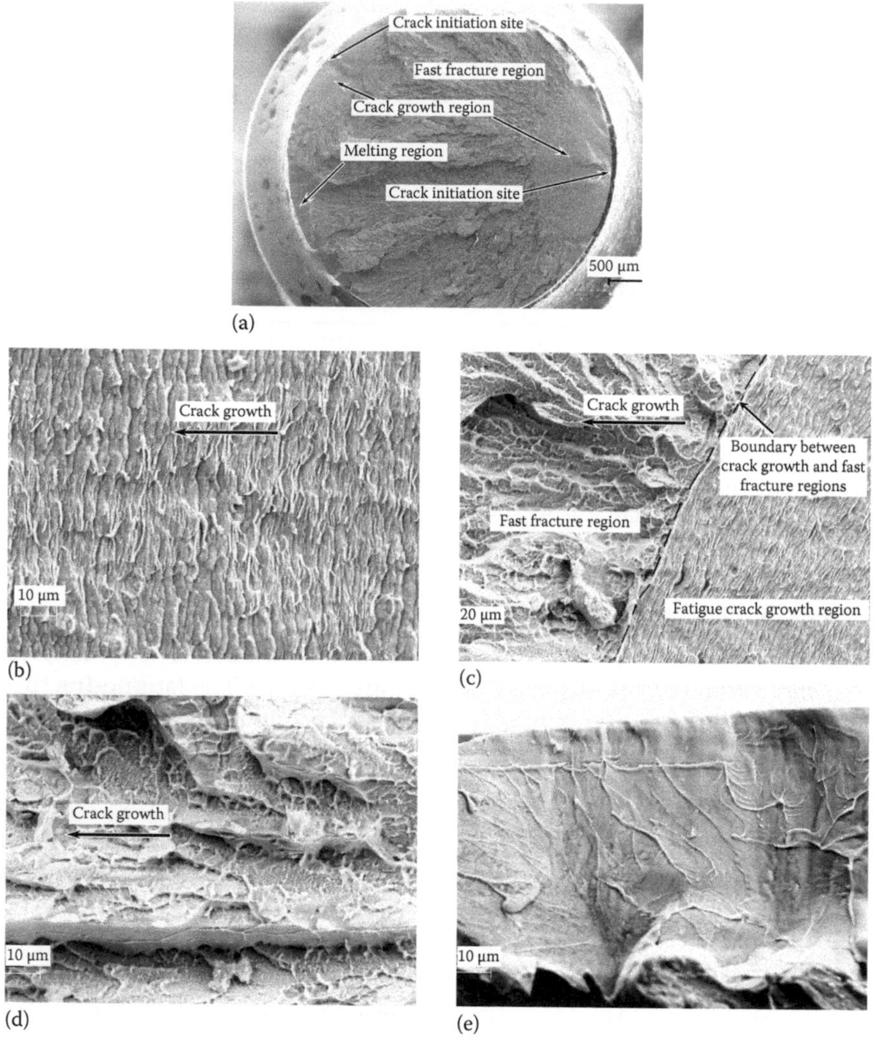

FIGURE 8.30
Fractographs of a $Zr_{50}Cu_{30}Al_{10}Ni_{10}$ BMG specimen tested at σ_{max} = 1133 MPa in air: (a) the overall fatigue fracture surface, (b) fatigue crack growth region, (c) boundary between the crack growth region and final fast fracture region, (d) final fast fracture region, and (e) local melting phenomenon. (Reprinted from Wang, G.Y. et al., *Intermetallics*, 12, 885, 2004. With permission.)

toward the interior of the specimen. The propagation region is of a thumbnail shape and relatively flat, and it exhibits a striation-type fracture (Figure 8.30b and c). The final fast fracture region is very rough and occupies most of the fracture surface. Further, the final fast fracture region shows a dimple-type morphology (Figure 8.30d). A clear boundary is also noticed between the crack propagation and fast fracture regions, which reveals that the fatigue and tensile fracture are probably controlled by different fracture mechanisms. In other words, the striation-type fracture mode was observed in the fatigue crack growth region, while the dimpled fracture was found in the fast fracture region. In addition to the three general features already described, another region is observed, which is close to the notched surface and relatively flat, generally with an inclined angle relative to the loading direction, in the melting area (Figure 8.30a). The distinct melting mark and vein pattern can be observed in the melting region at a high magnification by SEM (Figure 8.30e).

Very similar microstructural features of fatigue fracture are seen in many other BMG alloys subjected to fatigue. The main difference, of course, is the point of crack initiation, which is different in different types of specimens, depending on the way they have been cast and also on the type of surface defects. Further, it has also been reported that fatigue striations were not observed in the Fe-based BMG studied by Qiao et al. [142], and this is different from the Zr-based BMGs where striations are always observed. Comparing the fatigue fracture surfaces of monolithic and composite Zr-based BMG specimens, Wang et al. [155] noted some differences. For example, no fatigue striations were noticed in the BMG composite. It was also reported that the BMG composite showed a dimple morphology in the final fracture region that was different from the monolithic alloy.

Yang et al. [51] employed an IR camera system to capture the moment of fatigue fracture at high stress levels. Sparking occurred at the time of fracture due to an increase in temperature as a result of release of elastic energy. The sparking phenomenon was observed when the $Zr_{50}Cu_{30}Al_{10}Ni_{10}$ specimen was cyclically loaded at a stress level of 1500 MPa in air. But, the same phenomenon was not observed when specimens of $Zr_{50}Cu_{40}Al_{10}$ [157] and $Zr_{41.2}Ti_{13.8}Cu_{12.5}Ni_{10}Be_{22.5}$, or $Zr_{47}Ti_{12.9}Nb_{2.8}Ni_{9.6}Cu_{11}Be_{16.7}$ BMG composite [146] were tested. Even though sparking was not observed in these cases, the fracture section was very bright at the moment of fracture, suggesting that there was significant rise in temperature in this area. Due to the release of elastic energy of the BMG specimen in the final moment of fracture, the specimen temperature was estimated to be more than 900°C [45], a temperature sufficient to melt the BMG alloy. The fact that the BMG was molten was also inferred from the solidified droplet-like structure observed in the SEM on the fracture surface of the specimen.

8.7.5 Fatigue Fracture Mechanisms

Fatigue damage mechanisms have been thoroughly investigated and are well understood for ductile crystalline metals. Generally speaking, crystalline defects such as grain boundaries, slip bands, or deformation bands act as preferred sites for nucleation of fatigue cracks. These cracks then grow to a critical size during cyclic loading through a mechanism of crack blunting and resharpening. Once the crack reaches a critical size, catastrophic failure occurs and the specimen fractures.

8.7.5.1 Crack Initiation

BMG alloys are glassy in nature and they do not have these crystal defects. Therefore, the fatigue crack initiation mechanisms and the sites have to be different. Accordingly, it has been suggested (and also experimentally observed) that fatigue cracks initiate at surface irregularities, such as shallow pits; scratches introduced during polishing; and casting defects such as porosity, cold shuts, or oxide inclusions. In other studies, fatigue cracks were shown to initiate from shear bands [146,150]. Some mechanisms have also been proposed that do not require casting defects.

The excess quenched-in free volume in BMGs has been related to their fatigue behavior [147,148,164]. Cameron and Dauskardt [182] used a four-component amorphous

Lennard-Jones solid with atoms of different kinds to simulate a glassy metal using molecular dynamics methods. They found that cyclic loading in shear or tension increased the overall free volume with each cycle. It was also reported that localization of the free volume during cyclic loading could lead to eventual shear band formation. It was clearly demonstrated that the plastic deformation events occurred precisely in those areas where levels of free volume were higher. More importantly, it was shown that the accumulation of free volume and the resultant damage could happen at much lower cyclic loads than those required for the same processes to occur under monotonic loading conditions. Thus, the authors were able to rationalize the rapid initiation of fatigue damage and/or shear band formation in BMGs during fatigue.

Li et al. [183] conducted HRTEM investigations on a $Zr_{57}Ti_5Cu_{20}Ni_8Al_{10}$ glassy sample and observed a high concentration of nanometer-scale voids in shear bands, formed as a result of coalescence of excess free volume, once the applied stress was removed. By comparing the free energy of the shear band containing uniformly distributed free volume with that of the relaxed shear band with voids present, Li et al. [184] were able to show that coalescence was thermodynamically possible.

Using these observations, another mechanism was proposed to explain the crack initiation process in BMGs from a thermodynamic point of view. Since excess free volume in a shear band results in excess free energy relative to a relaxed glass with less free volume, Wright et al. [185] calculated the free energy in the shear band and correlated it with the free volume. It was shown that any free volume generated in the shear band during deformation was unstable, with the result that nanometer-scale voids formed spontaneously due to coalescence of the free volume. These voids then coarsen, and the size to which they grow is limited by growth kinetics. These voids can result in localized stress concentration during cyclic loading and a fatigue crack can initiate.

Void growth and linkage are facilitated by a tensile stress state, perhaps leading to fracture. On the other hand, a compressive stress state would hinder void growth, which may promote multiplication of shear bands, and this requires a larger strain to failure. Using this approach, the authors were able to rationalize the observation that failure under uniaxial tension occurs as the result of the propagation of a single shear band, whereas multiple shear bands can form under uniaxial compression without causing failure.

8.7.5.2 Crack Propagation

An important difference between crystalline materials and BMGs, as far as the fatigue behavior is concerned, is that nucleation of cracks seems to be relatively easier in BMGs and that the growth of the fatigue cracks largely accounts for the fatigue life. This is because once a crack is initiated on the surface of the BMG specimen (first through formation of a shear band and its growth to a characteristic length, in the order of 40–70 μm, when it transforms to a mode I form crack, oriented normal to the maximum stress), it can propagate without much difficulty due to the lack of microstructure. On the other hand, crack nucleation in crystalline alloys often accounts for 80% of the fatigue life [135].

As mentioned earlier, the fatigue crack growth data in BMG alloys have been fitted to the Paris power law equation, that is, the crack growth rate is related to the stress intensity factor range, ΔK, according to Equation 8.25. The exponent, m, has been found to be in the range 2–4, and this value is similar to what is observed in ductile crystalline metals. Again, similar to what is observed in ductile crystalline materials, fatigue striations develop in BMG alloys. These facts suggest that the mechanism for fatigue crack growth in BMGs may be similar to that in crystalline materials. That is, crack blunting and resharpening should also be occurring in BMGs.

Using this model for striation formation, the fatigue crack growth rate is directly proportional to the CTOD, *d*, according to the equation:

$$\frac{da}{dN} \propto \Delta d = \beta' \frac{(\Delta K)^2}{\sigma_y E'} \tag{8.28}$$

This equation is similar to Equation 8.26 except that, here, the proportionality constant has been shown as β'. This proportionality constant (β' is ~0.01–0.1 for mode I crack growth) is a function of the degree of slip reversibility and elastic-plastic properties of the material. Gilbert et al. [144] were able to show that a good correlation was obtained between the observed striation spacings and the calculated results with $\beta' = 0.01$.

It has been generally noted that the striation spacing observed in BMG alloys is much larger than what is indicated by the fatigue crack growth rate, by about one to two orders of magnitude [177]. This indicates that the crack does not extend uniformly through the width of the specimen and this has been verified experimentally by microstructural observations of the fracture surface. In the case of crystalline materials, it is assumed and also experimentally verified that one cycle during fatigue crack growth experiments forms one striation. From the observations of Hess et al. [177], it appears that this may not be true in BMGs. However, it is useful to note that two types of striations—coarse and fine—have been reported in the fatigue crack propagation region. The striations normally seen on the fracture surfaces of BMGs are coarse, and can be easily seen even at low magnifications. But, under higher magnifications, some fine striations are seen on the coarse striations [186]. It appears that the spacing of the fine striations matches the fatigue crack growth rate, *da*/*dN*.

8.7.6 Improvement of Fatigue Resistance

It is now accepted that fatigue crack initiation is the critical factor in determining the fatigue lives of BMGs. Once a fatigue crack forms, it can grow easily due to lack of microstructure. Therefore, the introduction of a crystalline phase by making it a composite might prevent the growth of small fatigue cracks and improve fatigue strength. Conflicting results are available in the literature on the fatigue behavior of BMG composites using this approach. Flores et al. [151] reported a more-than-doubled fatigue ratio for the composite in comparison to the monolithic glassy alloy. While the monolithic Vitreloy 1 ($Zr_{41.2}Ti_{13.8}Cu_{12.5}Ni_{10}Be_{22.5}$) had a fatigue ratio of 0.04 [144,145], the composite showed a fatigue ratio of 0.1. These authors explained the improvement on the basis of two possibilities. The first is that the second phase relieves the residual surface stresses or is able to accommodate any surface strains such that defects do not form. The other possibility is that the growth of defects is blocked by the second phase. Fujita et al. [153] also showed that a Ti-based BMG composite containing a dispersion of nanocrystals showed a very high fatigue ratio. The actual value for the monolithic alloy was not, however, reported.

Wang et al. [155], on the other hand, reported that the fatigue behavior of the composite was worse than that of the monolithic alloy. These authors reported a fatigue strength of

567 MPa for a $Zr_{41.2}Ti_{13.8}Cu_{12.5}Ni_{10}Be_{22.5}$ BMG alloy and 239 MPa for a BMG composite of the composition $Zr_{47}Ti_{12.9}Nb_{2.8}Ni_{9.6}Cu_{11}Be_{16.7}$, which contained a dendritic Zr–Ti–Nb crystalline phase dispersed in a glassy matrix, the composition of which is different from that of Vitreloy 1. Therefore, the comparison between the composite and the monolithic alloys may not be appropriate. A similar result of decreased fatigue strength was also reported in a $Cu_{47.5}Zr_{38}Hf_{9.5}Al_5$ BMG composite [159]. Here again, the strength was not compared with the appropriate monolithic alloy. It is also important to realize that the test methods and the test volumes are different in these two groups of investigations.

From these results, it is clear that the fatigue resistance of BMGs and their composites is, in general, poorer than those of crystalline materials. To overcome this problem, Launey et al. [143] proposed a microstructural design strategy matching the microstructural length scales (of the second crystalline phase) to mechanical crack-length scales.

A Zr-based alloy with the composition $Zr_{39.6}Ti_{33.9}Nb_{7.6}Cu_{6.4}Be_{12.5}$ was processed in the semisolid state (liquid + solid mushy region in the phase diagram) by holding it isothermally for several minutes and then quenching when the liquid had transformed into the glassy state. The dendritic phase was a ductile body-centered cubic (bcc) solid solution with the composition $Ti_{45}Zr_{40}Nb_{14}Cu_1$. The composition of the glassy matrix phase was determined to be $Zr_{34}Ti_{17}Nb_2Cu_9Be_{38}$. Through microstructural observations, it was determined that the volume fraction of the crystalline phase was about 67% and that the characteristic thickness of the glassy regions separating the dendritic arms or neighboring dendrites was about 2 μm, which is smaller than the critical crack size that leads to unstable crack propagation.

Fatigue testing of this composite was done using a four-point bend test at a frequency of 25 Hz and an R value of 0.1. The results showed that the fatigue behavior of the BMG composite was far superior to that of the monolithic BMG. The fatigue strength was reported to be 340 MPa and the fatigue ratio was 0.28, much higher than the value of 0.04 for Vitreloy 1. (The fatigue strength of the glassy alloy is not reported here, however, similar to the composition of the matrix). A similar increase in the fatigue strength of BMG composites was reported by Flores et al. [151]. But, in a Cu-based BMG composite, it was reported that the fatigue ratio of the composite was about half that of the monolithic alloy [140].

It has been suggested that the second-phase dendrites are the essential features leading to the enhancement of the fatigue resistance. Even though composites were tested earlier, their fatigue ratios were not as high as those reported for the $Zr_{39.6}Ti_{33.9}Nb_{7.6}Cu_{6.4}Be_{12.5}$ alloy. The important difference is in the characteristic dimensions of the crystalline phase compared with the pertinent mechanical length scales. The small scale of the second phase (preferably of nanometer dimensions) will improve the fatigue strength by a factor of two to three. The fatigue cracks have been shown to take a meandering path along the matrix–dendrite interfaces cutting through dendrite arms and along existing shear bands in the glassy phase separating the dendrites.

Conventional silicate glasses have been known to be strengthened up to four to five times by allowing compressive stresses to prevail at the surface through tempering during cooling. The compressive residual stresses suppress the cracking that leads to brittle fracture. Therefore, another approach to improve the fatigue strength of BMGs is to introduce compressive residual stresses at the surface. Zhang et al. [187] subjected the Vitreloy 1 to shot peening and noted that the plasticity of the alloy had increased.

Since the shot-peened BMG alloys have surface compressive stresses, it is expected that these alloys will have much better fatigue resistance than alloys without shot peening.

These compressive stresses should hinder the process by which surface shear bands transform to mode I cracks as well as the growth of the cracks themselves. Menzel and Dauskardt [137] reported that the fatigue limit of specimens loaded in compression–compression was much higher than of those that experienced tensile stresses at least for part of the loading cycle. This suggests that, for the material to be effective, the surface of the BMG should experience such a large compressive residual stress that it will never experience any tensile stress during testing. But, one difficulty of shot-peened BMG alloys for fatigue testing could be the unacceptable surface roughness.

Raghavan et al. [188] subjected a $Zr_{21.5}Ti_{42}Cu_{15.5}Ni_{14.5}Be_{3.5}Al_3$ BMG alloy to shot-peening treatment to investigate whether the fatigue life was improved. They did not find any improvement in the fatigue life. They observed cracks in the sub-surface regions, and this was attributed to the domination of the compressive residual stress field on the surface over deformation-induced plastic flow softening.

Qiao et al. [140] compared the fatigue behavior of BMGs in both the monolithic and composite conditions and noted that the fatigue endurance limit of the $Cu_{47.5}Zr_{38}Hf_{9.5}Al_5$ composite was 378 MPa, higher than the endurance limit of 224 MPa for the monolithic $Cu_{47.5}Zr_{47.5}Al_5$ BMG. They had, however, noted that the fatigue life of the composite was shorter than that of the monolithic, when the stress level was higher than the fatigue endurance limit. This was explained by them on the basis of the free volume theory. As was explained earlier, localization of free volume could lead to formation of shear bands. It is also known that the stress required to form shear bands under cyclic stress is much lower than the yield stress. Since a BMG composite is softer than the monolithic BMG, at a given stress, it is easier to form shear bands in the composite than in the monolithic alloy. Therefore, the fatigue life of the BMG composite is lower than that of the monolithic, when the stress level is higher than the fatigue endurance limit. On the other hand, at low stress levels, the presence of the second phase possibly relieves the residual surface stresses or allows accommodation of the surface strains. Further, the shear bands that had formed earlier could be blocked by the second phase. Thus, the fatigue endurance limit for the composite is higher than that of the monolithic BMG.

In an interesting study, Chiang et al. [189] sputter-coated a 316L stainless steel specimen with a 200 nm thick glass-forming alloy, $Zr_{47}Cu_{31}Al_{13}Ni_9$. It was reported that the fatigue life of the alloy with the coating was longer than that of the one without the coating. This difference was more pronounced at lower applied stress ranges. For example, at stress levels of 750 MPa or higher, there was no substantial effect of coating on the fatigue life. But, at a stress level of 700 MPa, the coated sample had a life of over 10^7 cycles, whereas the uncoated one had $\sim 3 \times 10^5$ cycles. Depending on the maximum applied stress, the fatigue life had increased 30-fold and the fatigue limit was enhanced by 30%, from 550 MPa of the uncoated sample to 700 MPa of the coated sample. The fatigue life improvement in the coated samples was mainly attributed to the good intrinsic mechanical properties, excellent adherence, good ductility, and the existence of compressive residual stresses of the coatings. Similar improvements were also achieved on coated Haynes C-2000 alloy [190].

It is known that the fatigue endurance limit of crystalline alloys increases with an increase in the yield strength of the alloy in the crystalline state. But, such a correlation is not apparent in BMG alloys. While the fatigue ratio, defined as the ratio of the fatigue endurance limit to the tensile strength, for crystalline alloys is usually between 0.3 and 0.5, that for BMGs is much less and goes down to a value as low as 0.04. However, a preliminary correlation between the fatigue ratio and Poisson's ratio has been reported with the fatigue ratio increasing with increasing values of Poisson's ratio [156].

8.8 Yield Behavior

8.8.1 Yield Criterion

It is well known that crystalline materials start yielding when the applied stress is above the critical shear stress. This critical shear stress is dependent on the Miller indices of the particular plane in a single crystal. Since materials are usually subjected to uniaxial tensile or compressive stresses, the shear stress on any particular plane and in a specific direction is calculated by knowing the orientation of the planes and directions with respect to the loading axis. This is known as the *resolved shear stress*. Yielding is expected to occur when the resolved shear stress reaches the critical value for yielding. The von Mises criterion has been shown to be valid in a number of metals and alloys [134]. Another yield criterion that is commonly used is the Tresca criterion, which is frequently used in modeling plastic flow because of its simplicity, but it is not, in general, an accurate description of the behavior of polycrystalline materials. According to these criteria, yielding is supposed to be symmetric, that is, the yield stress is of the same magnitude, whether testing is done in tension or compression. Consequently, the angle at which shear failure occurs is not different from 45° when tested in tension or compression. It has been noted in several metallic glasses, however, that they display asymmetry, that is, their yield stresses are different in tension and compression, and therefore it has been suggested that the yield criterion in metallic glasses should be different.

By measuring the yield strength of $Pd_{40}Ni_{40}P_{20}$ metallic glass undergoing low temperature, localized plastic flow in uniaxial compression, plane-strain compression, pure shear, and tension, Donovan [50] demonstrated that yielding follows the Mohr–Coulomb criterion, rather than the von Mises criterion. Yielding according to the Mohr–Coulomb criterion depends not only on the applied shear stress, τ, but also on the stress normal to the shear displacement, σ_n. Since BMGs have a very high strength (1–4 GPa), the normal stress applied on the shear plane is also very high (~1–2 GPa), and should therefore play an important role in the yielding process. Further, it has long been known that disordered materials experience dilatation due to plastic deformation. In terms of the flow strength-to-modulus ratio (σ_o/E), this value is of the order of 10^{-3} and is negligible in crystalline materials. But, in the case of BMGs, it can be about 0.02, and therefore pressure effects are not negligible. The Mohr–Coulomb yield criterion can be expressed as

$$\tau_y = \tau_o \pm \alpha\sigma_n \tag{8.29}$$

where:

- τ_y is the effective yield stress for shear on the shearing plane
- τ_o is the shear resistance of the glass (i.e., the stress at which yielding occurs in pure shear)
- α is a system-specific coefficient that controls the strength of the normal stress effect
- σ_n is the normal stress on the shear plane

One could also consider a term, σ_p, due to the hydrostatic pressure, used in the Tresca yield criterion. But, based on uniaxial yield stress alone, it is quite difficult to discern an appreciable difference between σ_n and σ_p, especially when their values are comparable. Some investigators have included these two terms separately (and called it the *three-parameter Tresca criterion*) [191]. Note that the compressive normal stress on the shear plane opposes dilatation, increasing the effective yield stress. This is why the sign will be negative. But, in tension, this sign will be positive. This equation was originally proposed to explain the flow of granular materials, where the σ_n term arises from the geometric rearrangement of sliding particles and the friction between them. Thus, α is effectively the coefficient of friction. Therefore, it has been postulated that Equation 8.29 may be applicable to metallic glasses, because the relative movement of randomly packed atoms in a metallic glass is analogous to that of randomly packed particles in a granular solid [192,193]. Based on the concept of the STZ, an elementary unit of deformation, Schuh and Lund [58] provided an atomic-level explanation for the pressure-dependent yielding of metallic glasses and confirmed that strength asymmetry exists in metallic glasses. This confirms that BMGs obey the Mohr–Coulomb criterion. Based on their model, they predicted the coefficient of friction to be $\alpha = 0.123 \pm 0.004$ for the $Cu_{50}Zr_{50}$ glass. This value is also in good agreement with the value of 0.11 ± 0.003 for the $Pd_{40}Ni_{40}P_{20}$ glass [40] and the value of 0.13 observed for the $Zr_{41.25}Ti_{13.75}Cu_{12.5}Ni_{10}Be_{22.5}$ glass [192]. Using the value of $\alpha = 0.123 \pm 0.004$, Equation 8.29 predicts a compressive shear angle of $\theta = 41.5 \pm 0.15$, a value which is in good agreement with the range observed in a variety of BMGs.

8.8.2 Fracture Angle

In the case of BMGs also, it was noted that failure occurs in a shear mode, at an angle close to 45°. But, results on the measurement of angles have not been consistent. Further, very often it also happens that investigators notice that fracture occurs at an angle of about 45° to the loading axis, but fail to measure and report the actual value. Table 8.7 lists some of the data available regarding the shear angles in some BMG monolithic alloys and composites. From this listing, it is clear that the angles are usually less than 45° (the actual range is 39.5°–43.7°) if failure occurs in compression, and more than 45° if it occurs in tension. Figure 8.31 shows the shear-off angles in a $Pd_{40}Ni_{40}P_{20}$ BMG specimen that had failed in compression and tension modes [197]. Note that the shear angle is 56° in tension and 42° in compression.

8.8.3 Effect of Strain Rate on Fracture Strength

Sufficient data are not available, and the available results are too conflicting, to draw any conclusions on the effect of strain rate on the fracture strength of BMGs. Maddin and Masumoto [3] reported that the room temperature fracture strength of $Pd_{80}Si_{20}$ filaments decreased with increasing strain rate in the range 10^{-4} to 10^{-2} s^{-1}. But, they reported that the yield strength was independent of the strain rate in this range. Kawamura et al. [212] investigated the deformation behavior of rapidly solidified $Zr_{65}Al_{10}Ni_{10}Cu_{15}$ ribbons in the strain rate range of 10^{-4} to 1 s^{-1} and in the temperature range from room temperature up to 653 K (close to the T_g of 652 K). They noted that the deformation was inhomogeneous at low temperatures and at high strain rates and homogeneous at higher temperatures and lower strain rates. It was reported that when the glassy alloy was tested at a strain rate of 5×10^{-4} s^{-1}, it deformed in an inhomogeneous mode up to a temperature of 533 K, and above this temperature the deformation was homogeneous. At a temperature of 653 K, the strength was found to increase with increasing strain rate from 5×10^{-4} to 5×10^{-2} s^{-1}. At still higher strain rates, the deformation mode was again inhomogeneous. From these results, the authors concluded that the strength was strongly dependent on strain rate in both the homogeneous and inhomogeneous modes. The room temperature strength

TABLE 8.7

Shear Angles at Which Fracture Occurs in Different BMG Alloys and Composites

Material	Compression (C)/ Tension (T)	Strain Rate (s^{-1})	Angle (°)	References
$Cu_{46}Zr_{47}Al_7$ (4 mm dia rod)	C	1×10^{-4}	43	[194]
$Cu_{46}Zr_{47}Al_7$ (2 mm dia rod)	C	1×10^{-4}	45	[194]
$Cu_{47}Ti_{34}Zr_{11}Ni_8$	C	1×10^{-4}	45	[195]
$Cu_{47}Ti_{34}Zr_{11}Ni_8$ + Graphite	C	1×10^{-4}	45	[195]
$Fe_{71}Nb_6B_{23}$	C	1×10^{-4}	44	[81]
$Mg_{65}Cu_{7.5}Ni_{7.5}Zn_5Ag_5Y_{10}$	C	1×10^{-4}	41.3	[88]
$Ni_{60.25}Nb_{39.75}$	C	4×10^{-4}	42	[196]
$Pd_{40}Ni_{40}P_{20}$	C	—	41.9 ± 1.2	[50]
$Pd_{40}Ni_{40}P_{20}$	C	1×10^{-3}	42	[197]
$Ti_{50}Cu_{20}Ni_{23}Sn_7$	C	—	Break or split	[198]
$Ti_{56}Cu_{16.8}Ni_{14.4}Sn_{4.8}Nb_8$	C	—	27	[198]
$Ti_{56}Cu_{16.8}Ni_{14.4}Sn_{4.8}Nb_8$	C	—	Split ($\approx$0)	[198]
$Zr_{40}Ti_{14}Cu_{12}Ni_{10}Be_{24}$	C	1×10^{-4}	42	[25]
$Zr_{41.25}Ti_{13.75}Cu_{12.5}Ni_{10}Be_{22.5}$	C	10^{-2} to 10^{-4}	45	[58]
$Zr_{41.25}Ti_{13.75}Cu_{12.5}Ni_{10}Be_{22.5}$	C	1×10^{-3}	42	[25]
$Zr_{41.25}Ti_{13.75}Cu_{12.5}Ni_{10}Be_{22.5}$	C	1×10^{-3}	42	[54]
$Zr_{41.25}Ti_{13.75}Cu_{12.5}Ni_{10}Be_{22.5}$	C	1×10^{-3}	90	[54]
$Zr_{41.25}Ti_{13.75}Cu_{12.5}Ni_{10}Be_{22.5}$	C	—	45	[47]
$Zr_{52.25}Cu_{28.5}Ni_{4.75}Al_{9.5}Ta_5$	C	1×10^{-4}	32 ± 1	[199]
$Zr_{52.5}Ni_{14.6}Cu_{17.9}Al_{10}Ti_5$	C	—	40–45	[200]
$Zr_{52.5}Ni_{14.6}Cu_{17.9}Al_{10}Ti_5$	C	—	42	[198]
$Zr_{54.5}Cu_{20}Al_{10}Ni_8Ti_{7.5}$	C	—	42	[198]
$Zr_{55}Cu_{30}Al_{10}Ni_5$	C	—	Break or split	[198]
$Zr_{55}Cu_{30}Al_{10}Ni_5$	C	—	40–43	[198]
$Zr_{55}Cu_{30}Al_{10}Ni_5$	C	—	45	[201]
$(Zr_{55}Cu_{30}Al_{10}Ni_5)_{95}Ta_5$	C	—	31–33	[198]
$Zr_{59}Cu_{20}Al_{10}Ni_8Ti_3$	C	—	43	[198]
$Zr_{62}Ti_{10}Ni_{10}Cu_{14.5}Be_{3.5}$	C	—	41.6 ± 2.1	[202]
$Co_{70}Fe_5Si_{15}B_{10}$	T	—	60	[203]
$Cu_{60}Zr_{30}Ti_{10}$	T	—	54	[204]
$Fe_{70}Ni_{10}B_{20}$	T	—	60	[64]
$Ni_{49}Fe_{29}P_{14}B_6Si_2$	T	—	53	[205]
$Ni_{78}Si_{10}B_{12}$	T	—	55	[206]
$Pd_{40}Ni_{40}P_{20}$	T	—	54.7	[4]
$Pd_{40}Ni_{40}P_{20}$	T	1×10^{-3}	56	[55,197]
$Pd_{77.5}Cu_6Si_{16.5}$	T	4×10^{-4}	51	[207]
$Pd_{80}Si_{20}$	T	—	48–50	[41]
$Zr_{40}Ti_{14}Cu_{12}Ni_{10}Be_{24}$	T	1×10^{-4}	56	[25]
$Zr_{41.25}Ti_{13.75}Cu_{12.5}Ni_{10}Be_{22.5}$	T	1×10^{-3}	90	[54]
$Zr_{41.25}Ti_{13.75}Cu_{12.5}Ni_{10}Be_{22.5}$	T	—	57	[54]
$Zr_{41.25}Ti_{13.75}Cu_{12.5}Ni_{10}Be_{22.5}$	T	1×10^{-1}	56	[208]
$Zr_{52.5}Ni_{14.6}Cu_{17.9}Al_{10}Ti_5$	T	—	55–65	[199]
$Zr_{52.5}Ni_{14.6}Cu_{17.9}Al_{10}Ti_5$	T	1×10^{-3}	56	[45]

(Continued)

TABLE 8.7 (CONTINUED)

Material	Compression (C)/ Tension (T)	Strain Rate (s^{-1})	Angle (°)	References
$Zr_{52.5}Ni_{14.6}Cu_{17.9}Al_{10}Ti_5$	T	—	53–60	[45]
$Zr_{52.5}Ni_{14.6}Cu_{17.9}Al_{10}Ti_5$	T	—	56	[198]
$Zr_{52.5}Ni_{14.6}Cu_{17.9}Al_{10}Ti_5$	T	—	≈90	[198]
$Zr_{59}Cu_{20}Al_{10}Ni_8Ti_3$	T	—	54	[209]
$Zr_{59}Cu_{20}Al_{10}Ni_8Ti_3$	T	—	≈90	[198]
$Zr_{60}Ti_{14.7}Nb_{5.3}Cu_{5.6}Ni_{4.4}Be_{10}$	T	—	54	[210]
$Zr_{62}Ti_{10}Ni_{10}Cu_{14.5}Be_{3.5}$	T	—	57 ± 3.7	[202]
$Zr_{65}Ni_{10}Al_{7.5}Cu_{7.5}Pd_{10}$	T	—	50	[211]

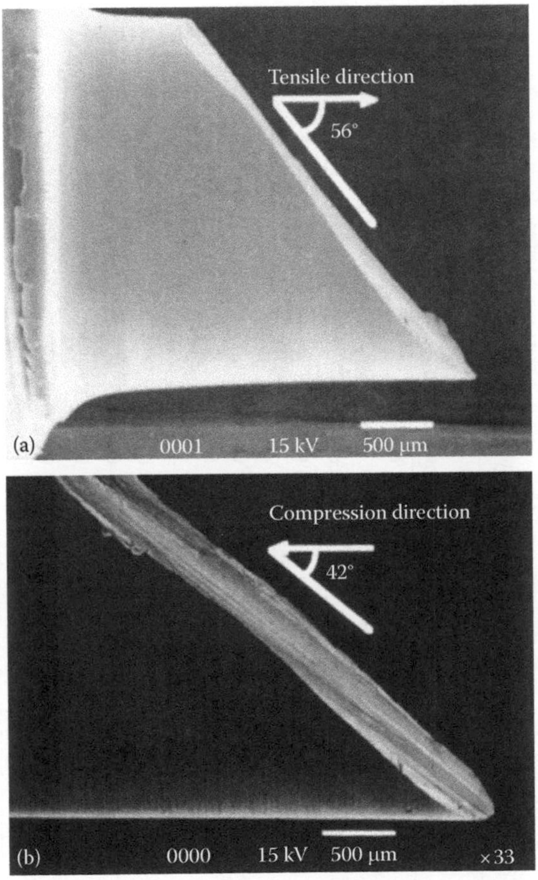

FIGURE 8.31
Appearance of the side surface of the $Pd_{40}Ni_{40}P_{20}$ BMG specimen fractured at the quasi-static strain rate of 1×10^{-3} s^{-1} (a) in tension and (b) in compression. Note that the shear angle is 56° in tension and 42° in compression. (Reprinted from Mukai, T. et al., *Intermetallics*, 10, 1071, 2002. With permission.)

of the alloy decreased with increasing strain rate in the strain rate range of 10^{-4} to 10 s^{-1}. Table 8.8 lists the data available on the effect of strain rate on the fracture strength. Data on melt-spun ribbon samples are also included here. It may be noted that the data are quite scattered, with the results showing that the strength increases in some cases, decreases in others, and sometimes does not change.

TABLE 8.8
Effect of Strain Rate on the Fracture Strength of Metallic Glass Alloys

Material	Specimen Shape	Loading Mode	Strain Rate Range (s^{-1})	(Effect on) Fracture Strength	References
$Ce_{60}Al_{15}Cu_{10}Ni_{15}$	2 mm dia rods	N	0.03–1 mN	Increases	[213]
Dy_3Al_2	20 μm thick ribbons	T	10^{-4} to 10^{-1}	Decreases	[214]
$Nd_{60}Fe_{20}Co_{10}Al_{10}$	5 mm dia rods	C	6×10^{-4} to 1×10^{3}	Increases	[215]
$Pd_{40}Ni_{40}P_{20}$	5 mm dia rods	C	3.3×10^{-5} to 2×10^{3}	Decreases	[197]
$Pd_{80}Si_{20}$	Ribbons	T	10^{-4} to 10^{-2}	No change	[3]
$Ti_{40}Zr_{25}Ni_8Cu_9Be_{18}$	1 and 3 mm dia rods	C	10^{-4} to 10^{3}	Increases	[216]
$Ti_{45}Zr_{16}Ni_9Cu_{10}Be_{20}$	1 mm dia rods	C	10^{-4} to 10^{-1}	Increases	[217]
$Zr_{41.25}Ti_{13.75}Cu_{12.5}Ni_{10}Be_{22.5}$ (Vit 1)	2.5 mm dia rods	—	10^{2} to 10^{4}	No change	[47]
$Zr_{41.25}Ti_{13.75}Cu_{12.5}Ni_{10}Be_{22.5}$ (Vit 1)	3.8–6.25 mm dia rods	C	10^{-3} to 10^{3}	No change	[218]
$Zr_{57}Ti_5Cu_{20}Ni8Al_{10}$	3 mm dia rods	C	1×10^{-4} to 3×10^{3}	Decreases	[57]
$Zr_{60}Ti_{14.7}Nb_{5.3}Cu_{5.6}Ni_{4.4}Be_{10}$	3 mm dia rods	C	2×10^{-4} to 3.7×10^{2}	Increases	[210]
$Zr_{65}Al_{10}Ni_{10}Cu_{15}$	20 μm thick ribbons	C	5×10^{-4} to 10^{0}	Increases	[212]
$Zr_{65}Al_{10}Ni_{10}Cu_{15}$	20 μm thick ribbons	T	1.67×10^{-4} to 5×10^{-1}	Increases	[219]

Note: N: nanoindentation; T: tension; C: compression.

In the case of melt-spun $Zr_{65}Al_{10}Ni_{10}Cu_{15}$ alloy ribbons, it was noted that the strength dropped gradually at low strain rates, but, it decreased very rapidly at higher strain rates of more than 10^{-1} s^{-1}. The sudden decrease in the fracture strength may be attributed to the possible presence of surface defects. The fracture strain and fracture stress of metallic glasses are more sensitive to internal flaws and cracks at high strain rates rather than at low strain rates. Under a high loading rate, the stress increased so rapidly that a large number of shear bands were initiated simultaneously at the defects. These multiple shear bands produce a rough fracture surface similar to that observed in tension (Figure 8.32) [197].

It has been shown that BMGs show serrations at low loading rates during nanoindentation [213,220] or at low strain rates during uniaxial loading [197]. One could use this to explain the reduced fracture strength of BMGs at high strain rates or even during dynamic loading. The emission of shear bands at high strain rates is not sufficiently fast to accommodate the applied strain rate and therefore causes an early fracture of the specimen.

Figure 8.33 shows the variation of fracture strength as a function of the applied strain rate for different BMG alloys; data for some melt-spun ribbons are also included [217].

8.8.4 Fracture Morphology

As mentioned earlier, metallic glasses undergo inhomogeneous deformation at low temperatures and high strain rates. This means that the deformation is mainly inhomogeneous at room temperature and that it occurs through initiation and propagation of shear bands. Therefore, when the surface of a glassy alloy specimen, fractured under uniaxial loading, is observed in an SEM, one typically observes a vein pattern, as shown in Figure 8.34. Such microstructures are observed whether the specimen fails in tension or compression and represent fracture characteristic of pure shear mode. It is suggested that a very high amount of shear is focused in the shear bands, as a result of which, a significant amount of energy is released when the specimen fractures, leading to a large temperature rise. This temperature rise can be so large that local softening or melting occurs. Consequently,

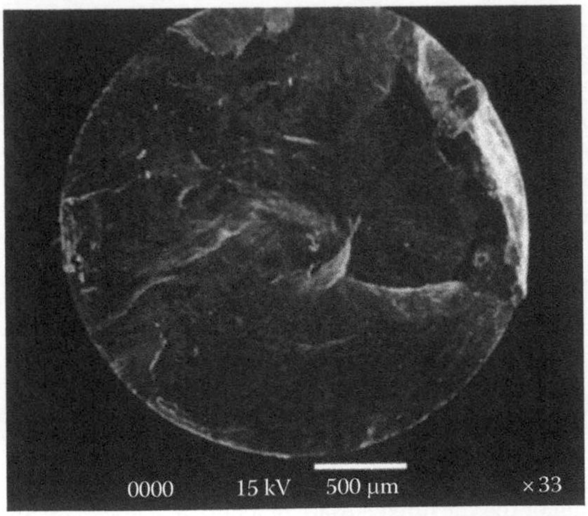

FIGURE 8.32
A top view of the fracture surface of the $Pd_{40}Ni_{40}P_{20}$ specimen deformed at a dynamic strain rate of 6×10^2 s^{-1}. It may be noted that the fracture surface is very rough as a result of the simultaneous operation of multiple shear bands at dynamic strain rates. (Reprinted from Mukai, T. et al., *Intermetallics*, 10, 1071, 2002. With permission.)

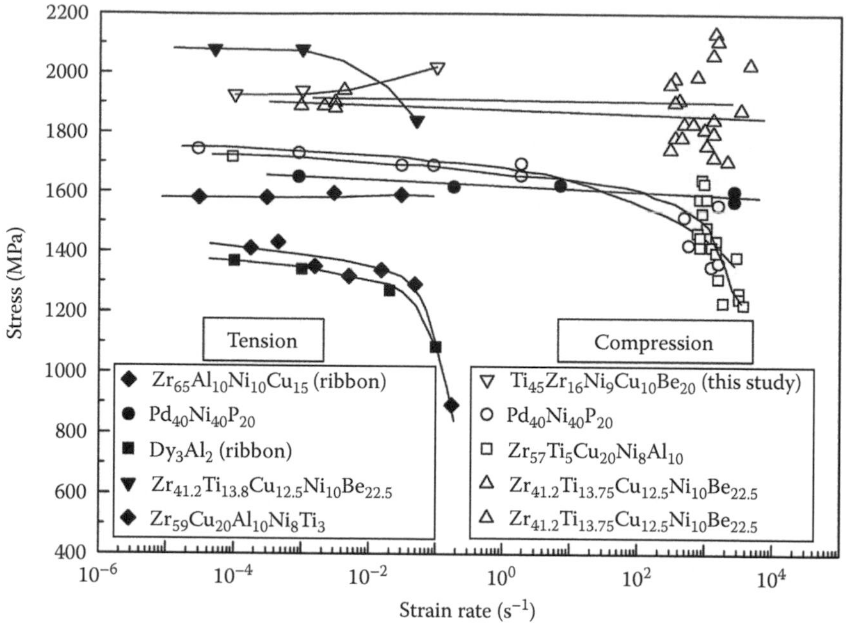

FIGURE 8.33
Variation of fracture strength as a function of applied strain rate in several BMG alloys tested in tension or compression. Data for some melt-spun glassy ribbons are also included. (Reprinted from Zhang, J. et al., *Mater. Sci. Eng. A*, 449–451, 290, 2007. With permission.)

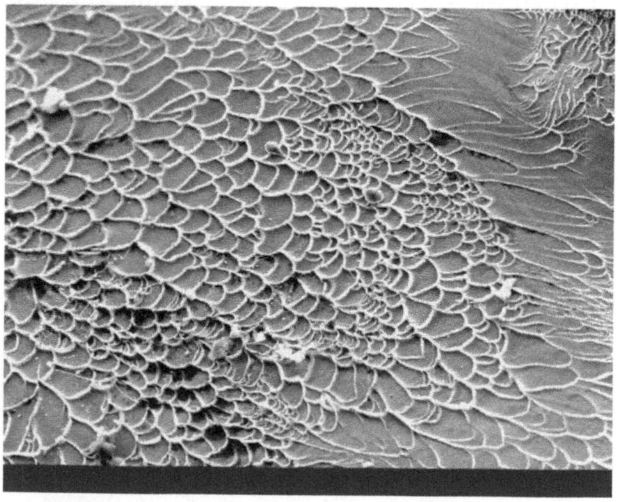

FIGURE 8.34
Scanning electron micrograph of the fractured surface of a bulk metallic glass alloy specimen. Note the vein pattern, which is typical of many metallic glasses that fracture along a shear band. Such microstructures are obtained both in tension and compression.

the viscosity of the material in the shear bands drops by many orders of magnitude. The mechanism for the significant drop in viscosity is still under debate. One proposal was that adiabatic heating occurs due to plastic flow. Another suggestion, which seems to have gained wider acceptance, is that there is an increase in the free volume and, consequently, the atomic mobility is considerably enhanced. Irrespective of the real reasons for the drop in viscosity, the shear bands are often compared to a fluid layer. The vein pattern observed on fractured surfaces can be visualized as the structure formed when two glass slides are pulled apart with a thin layer of a viscous fluid such as Vaseline in between them. Except for the ridges forming the veins, the fracture surface is flat.

The fracture behavior of samples failed in tension seems to be quite different from that on the compressive fracture surfaces. In addition to the vein-like structure, there are many round cores with different diameters on the tensile fracture surface. The veins seem to be radiating from these cores and propagating toward the outside. In the region of the cores, the fracture seems to take place in a normal fracture mode, rather than a pure shear mode as is typically observed under compression. From this morphology, it is suggested that the fracture of metallic glasses under tension should first originate from these cores induced by normal tensile stress on the plane, and then catastrophically propagate toward the outside of the cores in a shear mode driven by the shear stress. As a result, the tensile fracture surface of metallic glasses consists of a combined feature of cores and veins, which is quite different from the compressive fracture surface [209]. Figure 8.35 compares the typical features of $Zr_{59}Cu_{20}Al_{10}Ni_8Ti_3$ BMG fractures induced by compressive and tensile loading conditions.

Fracture occurs by a shear mode under both tensile and compressive loading conditions. But, an important difference noticed between the surface structures of specimens fractured in tension and compression is that the density of shear bands is much higher in the samples loaded under compression. Under tensile loading, one major shear band dominates the fracture process.

The fracture surfaces of BMG composites look different and are more complex mainly because of the presence of the crystalline phase. The appearance of the fracture surface depends on the volume fraction of the crystalline phase present in the composite. At large volume fractions, the ductility usually increases up to a point, but begins to drop at still

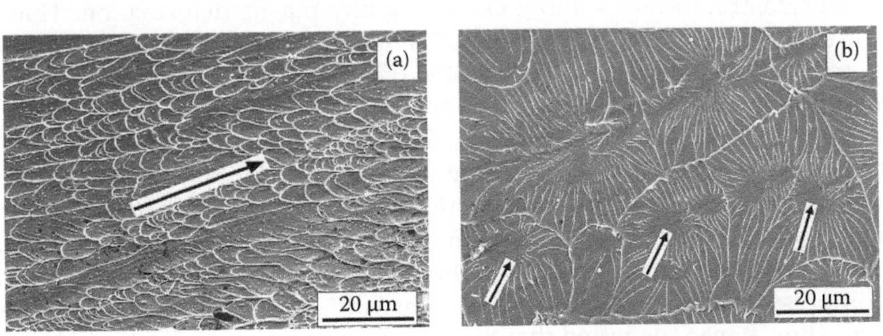

FIGURE 8.35
Comparison of the fracture surfaces of $Zr_{59}Cu_{20}Al_{10}Ni_8Ti_3$ BMG alloy that has failed under (a) compressive loading and (b) tensile loading. Notice that the specimen that has failed under compressive loading exhibits a vein-like pattern while the specimen that had failed in tension shows round cores with vein-like features radiating outward from their centers. The arrow in (a) shows the shear direction, while the arrows in (b) indicate the location of the round cores. (Reprinted from Zhang, Z.F. et al., *Acta Mater.*, 51, 1167, 2003. With permission.)

higher volume fractions of the crystalline phase. In such a case, in addition to the vein pattern, highly roughened surfaces with extensive local melting are observed in compression [199]. Further, with an increase in the volume fraction of the crystalline phase, the length scale and the volume fraction of the vein-like structure decrease. Dimples are also observed when the sample is loaded in tension [221]. These authors reported that with increasing volume fraction of dendrites in their $La_{86-y}Al_{14}(Cu, Ni)_y$ (y = 1–24) composites, the fracture surface transitioned to a more obvious dimpled surface characteristic of extensive plastic flow accompanied by microvoid formation and coalescence. This is typical of fracture in a crystalline material. Multiple shear bands were also clearly seen on the side surfaces of the deformed specimens and these were parallel to the direction of the fracture plane. The density of the shear bands increased with increasing volume fraction of the crystalline phase. Furthermore, secondary shear bands were also formed normal to the main shear bands, induced by the branching of the main shear bands.

In some cases, it has been reported that the vein-like pattern coexisted with river-like patterns and smooth areas under uniaxial compression [222]. In such cases, the fracture surface consists of a river-like pattern, occasional smooth featureless regions, and a vein pattern. The origin of the river-like patterns has been attributed to the normal stress, which acts locally on a shear plane. Occasionally, smooth regions form when a high rate of crack propagation is locally established after overcoming the threshold imposed by a crystalline particle.

8.9 BMG Composites

A critical requirement for a structural engineering material is that it should have an optimum combination of strength, ductility, toughness, and predictable and graceful (non-catastrophic) failure. As already discussed, BMGs are inherently brittle. They fracture immediately after yielding. Some of the BMG alloys show a limited amount of ductility during compression or bend testing. But, during a tensile test, the BMGs show yielding and then fail catastrophically without exhibiting any plastic deformation. That is, these materials are no different from a brittle material like a traditional glass or a ceramic. There has been some progress in recent years on the development of ductile/tough BMG composites and the available results have been reviewed recently [20].

Since it has been shown earlier that the mechanical behavior of conventional crystalline materials could be improved by introducing metallic glass ribbons into them [223,224], it was decided to explore whether the properties of metallic glasses could be improved by incorporating ductile metal fibers into them. Accordingly, Conner et al. [100] introduced tungsten and 1080 steel (music) wires into the $Zr_{41.25}Ti_{13.75}Cu_{12.5}Ni_{10}Be_{22.5}$ matrix and noted that the compressive strain to failure increased by 900% compared with that of the unreinforced BMG alloy. It was suggested that the compressive toughness came from restricting the propagation of shear bands, thus promoting the generation of multiple shear bands, which eventually delayed fracture. The tensile toughness increased as a result of ductile fiber delamination, fracture, and fiber pull-out. Subsequently, it was also demonstrated that the plasticity of BMGs could be enhanced by incorporating *in situ* formed ductile phases [225,226]. In this latter method, the alloy composition is designed in such a way that during solidification, a crystalline phase first precipitates out, usually in the form of a dendritic structure, and the remaining liquid solidifies into a glass. Solute partitioning occurs in this process and therefore the compositions of the glass and the crystalline phase are quite different. Huang et al. [226] produced a Zr–Ti–Nb–Cu–Ni–Be BMG composite starting from the alloy with the composition $Zr_{56}Ti_{14}Nb_5Cu_7Ni_6Be_{12}$ (the so-called LM002 alloy). The as-solidified microstructure contained dendrites dispersed in a glassy matrix and electron diffraction patterns confirmed that the crystalline phase had a bcc structure with a lattice parameter of 0.351 nm. Energy-filtered TEM images showed that the matrix was enriched with Be and Ni, while the dendrites were found to be depleted in Be and Ni. Ti and Zr were found to be distributed almost homogeneously in the composite. The concentration distribution of the different constituent elements was determined using EDS methods. Combining these results with those of the volume fraction of the dendrites determined by scanning electron microscopy, the composition of the dendrites was estimated to be $Zr_{70}Ti_{15}Nb_9Cu_4Ni_2$ and that of the glassy matrix to be $Zr_{49}Ti_{13}Nb_3Cu_9Ni_8Be_{18}$.

The second (crystalline) phase can be introduced into the BMG alloy in two different ways—*ex situ* and *in situ* methods. In the *ex situ* method, a crystalline phase is added separately to the melt prior to casting into the composite. Sometimes, the BMG melt is also infiltrated into a preform of crystalline particles or fibers, or one could also make glass/crystalline laminates. In these cases, the crystalline phase forms as discrete particles dispersed in a glassy matrix. On the other hand, in the *in situ* method, the alloy composition is designed in such a way that the second crystalline phase forms as dendrites during solidification. This is the more common practice used nowadays to produce BMG composites. In some cases, an external heat treatment of the BMG to partially crystallize it could also produce a composite.

It is generally expected that the crystalline phase does not react with the matrix material (although this is not a requirement). But, a reaction product is occasionally found at the interface between the glassy matrix and the reinforcement. Figure 8.36a shows a scanning electron micrograph of a BMG composite containing W as the reinforcement in a $Zr_{55}Cu_{30}Al_{10}Ni_5$ matrix. A bundle of the straight W wires was placed inside an evacuated quartz glass tube. A constriction was made about 1 cm above the place where the reinforcement was kept and the BMG alloy matrix ingots were placed above the neck. After melting the matrix alloy ingots, a slight positive pressure was applied to the melt that forced the liquid to infiltrate into the reinforcement bundle. It may be noted from Figure 8.36a

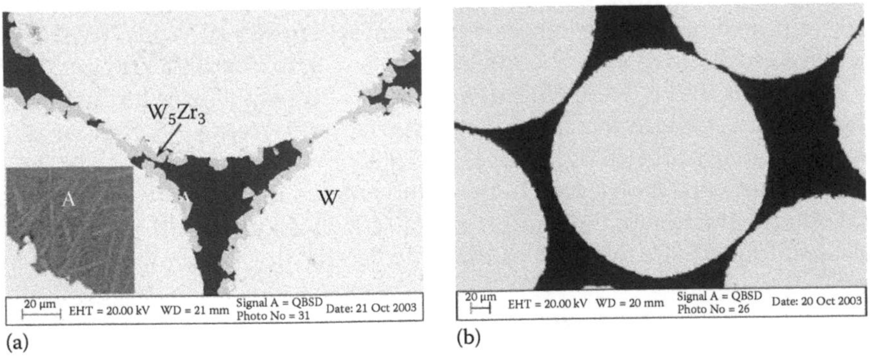

(a) (b)

FIGURE 8.36
Scanning electron micrographs of interfaces between the reinforcement and the Zr-based glassy matrix in composites prepared by the melt infiltration method. (a) W-fiber-reinforced composite showing the formation of a metastable W_5Zr_3 phase at the interface between the W fiber and $Zr_{55}Cu_{30}Al_{10}Ni_5$ BMG. (b) Absence of any reaction product at the interface in the W-fiber-reinforced composite when the BMG contained Nb in it ($Zr_{47}Ti_{13}Cu_{11}Ni_{10}Nb_3Be_{16}$). (Reprinted from Wang, M.L. et al., *Intermetallics*, 15, 1309, 2007. With permission.)

that Zr from the matrix and W reacted with each other to form a metastable W_5Zr_3 phase. Infiltration at higher temperatures and for longer periods of time resulted in the formation of the equilibrium W_2Zr phase instead of the metastable W_5Zr_3 phase. Such an interfacial reaction did not occur when the BMG alloy contained Nb ($Zr_{47}Ti_{13}Cu_{11}Ni_{10}Nb_3Be_{16}$), even though W was still used as the reinforcement (Figure 8.36b) [227].

It is also possible that the second (crystalline) phase could form *in situ* during subsequent processing of the BMG alloy, for example, during annealing, through partial crystallization of the glassy phase. In this case, the crystalline phase is usually of nanometric dimensions and is dispersed uniformly throughout the glassy matrix.

The properties of the BMG composite are determined by the nature of the second (crystalline) phase (usually referred to as *reinforcement* in the literature on composites), its size, volume fraction, and distribution in the matrix. Accordingly, a number of reinforcements of different types, sizes, and shapes have been tried out in Cu-, Fe-, Mg-, Ti-, and Zr-based BMG alloys. The type of reinforcements employed include metal wires, such as Cu, Mo, Nb, Ta, Ti, W, and steel; ceramic particles, such as CeO_2, Cr_2O_3, MgO, SiC, TiB, TiB_2, TiC, WC, Y_2O_3, ZrC, ZrN, and ZrO_2; and also carbon nanotubes, diamond, and graphite. Eckert et al. [228] and Qiao et al. [20] have recently published reviews on the mechanical properties of BMGs and their composites. The improvement in mechanical properties, especially in the toughness and ductility of the BMG alloys, has been quite impressive and Table 8.9 summarizes some of the typical combinations that have been tried out.

8.9.1 Mechanical Properties

BMG composites have been shown to have much better mechanical properties than their monolithic counterparts. Irrespective of the nature of the reinforcement, the fracture strength has been shown to increase with increasing volume fraction of the reinforcement. Associated with this increase in fracture strength is the plastic strain to fracture, which also has been shown to be higher with increasing volume fraction of the reinforcement. For example, the fracture strength (σ_f) and plastic strain to fracture (ε_p) for the monolithic $Cu_{60}Hf_{25}Ti_{15}$ BMG have been measured to be 2088 MPa and 1.6%, respectively. But, with the presence of 8 vol.% of a Nb-rich solid solution precipitate, these values increased to 2232 MPa and 14.1%, respectively [75]. Similarly, σ_f and ε_p for an $Mg_{65}Cu_{7.5}Ni_{7.5}Zn_5Ag_5Y_{10}$ BMG have been reported to be 490–650 MPa and 0%, respectively [241]. But, with the addition of TiB_2 reinforcement, these values were higher. For example with 10 vol.% TiB_2 addition σ_f increased to 992 MPa even though there was still no ductility. But, with an increase in TiB_2 to 20 vol.%, σ_f and ε_p were 1212 MPa and 3.2%, respectively. With further additions of TiB_2, the fracture strength increased, but ε_p started to decrease [92]. Similar trends have been reported in many other BMG composites.

The strengthening effect observed in the composites is associated with the nature, size, shape, and volume fraction of the reinforcement. The measured yield strength of the composite has been found to be the same as that calculated using the simple rule of mixtures:

$$\sigma_c = V_r \cdot \sigma_r + (1 - V_r) \cdot \sigma_m \tag{8.30}$$

where:

σ_c represents the yield strength of the composite
V_r is the volume fraction of the reinforcement (the crystalline phase here, irrespective of whether it is in the form of particles, fibers, dendrites, or other morphology)
σ_r and σ_m represent the yield strengths of the reinforcement and matrix, respectively

TABLE 8.9

Improvement in the Mechanical Properties of BMG Composites

Alloy	Reinforcement	Reinforcement Size (μm)	V_r (%)	σ_f (MPa)	ε_p (%)	References
$Cu_{60}Hf_{25}Ti_{15}$	Fully glassy	—	—	2088	1.6	[75]
$(Cu_{60}Hf_{25}Ti_{15})_{90}Nb_{10}$	Nb-rich SS dendrites	5–10	8	2232	14.1	[75]
$Cu_{50}Hf_{35}Ti_{10}Ag_5$	Fully glassy	—	—	2180	2.1[a]	[229]
$(Cu_{50}Hf_{35}Ti_{10}Ag_5)_{97}Ta_3$	Dendrites	30–40	4	2510	9.75[a]	[229]
$(Cu_{50}Hf_{35}Ti_{10}Ag_5)_{92}Ta_8$	Dendrites	18–23	17	2770	19.2[a]	[229]
$Cu_{60}Zr_{30}Ti_{10}$	Fully glassy	—	—	2080	3.3[a]	[230]
$(Cu_{60}Zr_{30}Ti_{10})_{95}Ta_5$	Ta(Ti, Zr) solid solution	3–20	9.2	2320	14.5[a]	[230]
$[(Fe_{0.5}Co_{0.5})_{0.75}B_{0.2}Si_{0.05}]_{96}Nb_4$	Fully glassy	—	—	3700	0	[231]
$\{[(Fe_{0.5}Co_{0.5})_{0.75}B_{0.2}Si_{0.05}]_{96}Nb_4\}_{99.75}Cu_{0.25}$	α-(Fe, Co) + $(Fe, Co)_{23}B_6$	3	13.6	4050	0.6	[231]
$La_{62}Al_{14}(Cu,Ni)_{24}$	—	—	—	561	0	[221]
$La_{72}Al_{14}(Cu,Ni)_{14}$	α-La dendrites	4–12	41	529	0.9	[221]
$La_{74}Al_{14}(Cu,Ni)_{12}$	α-La dendrites	4–12	50	521	4.1	[221]
$La_{85}Al_{14}(Cu,Ni)_1$	α-La dendrites	4–12	53	434	6	[221]
$Mg_{65}Cu_{20}Ag_5Gd_{10}$	Fully glassy	—	—	830	0	[232]
$Mg_{65}Cu_{20}Ag_5Gd_{10}$ + Nb	Nb	20–50	8	—	12.1	[232]
$Ni_{60.25}Nb_{39.75}$	Glass + nanocrystals	10–15	—	—	5.3	[196]
$Ti_{50}Cu_{25}Ni_{15}Sn_5Ta_5$	Nanocrystals	—	—	—		[233]
$Ti_{45}Zr_5Cu_{44}Ni_5Ta_1$	β-Ti phase + Ti_2Ni	—	—	—	3	[233]
$Zr_{50}Cu_{40}Al_{10}$	Fully glassy	—	—	1890	0.7	[234]
$(Zr_{50}Cu_{40}Al_{10})_{91}Ta_9$	Ta	—	—	2180	15.9	[234]
$Zr_{48}Cu_{36}Al_8Ag_8$	—	—	—	1885	—	[235]
$Zr_{48}Cu_{36}Al_8Ag_8$	Ta	40	10	2600	31[a]	[235]
$Zr_{57}Nb_5Al_{10}Cu_{15.4}Ni_{12.6}$	Mo wires	250	80	—	20	[236]
$Zr_{57}Nb_5Al_{10}Cu_{15.4}Ni_{12.6}$	Ta wires	250	80	—	27	[236]
$Zr_{52.5}Cu_{17.9}Ni_{14.6}Al_{10}Ti_5$	—	—	—	—	3	[237]
$Zr_{52.5}Cu_{17.9}Ni_{14.6}Al_{10}Ti_5$	Graphite	25–44	10	—	15	[237]
$Zr_{41.25}Ti_{13.75}Cu_{12.5}Ni_{10}Be_{22.5}$	—	—		—	<0.5%	[95]

(Continued)

TABLE 8.9 (CONTINUED)

Improvement in the Mechanical Properties of BMG Composites

Alloy	Reinforcement	Reinforcement Size (μm)	V_r (%)	σ_f (MPa)	ε_p (%)	References
$Zr_{41.25}Ti_{13.75}Cu_{12.5}Ni_{10}Be_{22.5}$	W	—	20	—	16%	[95]
$Zr_{56.2}Ti_{13.8}Nb_5Cu_{6.9}Ni_{5.6}Be_{12.5}$	β-Zr(Ti, Nb) dendrites	—	—	—	8.26[a]	[173]
$Zr_{56.2}Ti_{13.8}Nb_5Cu_{6.9}Ni_{5.6}Be_{12.5}$	β-Zr(Ti, Nb) dendrites	20–50 μm × 1–3	—	1757	8.82	[238]
$Zr_{56.2}Ti_{13.8}Nb_5Cu_{6.9}Ni_{5.6}Be_{12.5}$	β-Zr(Ti, Nb) spherical	18 μm	—	1800	12	[238]
$Zr_{38}Ti_{17}Cu_{10.5}Co_{12}Be_{22.5}$	Fully glassy	—	—	1942	2.4	[239]
$Zr_{38}Ti_{17}Cu_{10.5}Co_{12}Be_{22.5}$	W	As-cast	80	1852	82	[239]
$Zr_{38}Ti_{17}Cu_{10.5}Co_{12}Be_{22.5}$	W	As-extruded	80	2112	53	[239]
$Zr_{52.5}Cu_{17.9}Ni_{14.6}Al_{10}Ti_5$	Fully glassy	—	—	—	3	[237]
$Zr_{52.5}Cu_{17.9}Ni_{14.6}Al_{10}Ti_5$	Graphite	25–44	10	—	15	[237]
$Zr_{36.6}Ti_{31.4}Nb_7Cu_{5.9}Be_{19.1}$	—	—	42	—	7.6	[240]
$Zr_{38.3}Ti_{32.9}Nb_{7.3}Cu_{6.2}Be_{15.3}$	—	—	51	—	8.9	[240]
$Zr_{39.6}Ti_{33.9}Nb_{7.6}Cu_{6.4}Be_{12.5}$	—	—	67	—	11.5	[240]

Note: V_r: volume fraction of the reinforcement; σ_f: fracture strength; ε_p: plastic strain.

[a] Total strain to fracture.

Thus, the presence of hard reinforcements increases the strength to higher values than the softer reinforcements. The strength increases with increasing volume fraction of the reinforcement, as described. The increase in fracture strength of the BMG composite is also associated with an increase in plastic strain to failure. As in the case of the monolithic alloys, this increase in plasticity is associated with the multiplication of shear bands. The best combination of high strength and improved ductility is obtained when the reinforcement is in the form of ductile dendrites [229]. But, when the dendritic phase has been changed into a spherical shape, the properties appear to improve further [238].

Let us now look at the effect of the different properties of the reinforcement on the mechanical behavior of the BMG composites.

8.9.2 Effect of Particle Shape

Even though not many investigations have been reported so far on this aspect, the particle shape seems to have a significant effect on the ductility of composites. The structure and mechanical behavior of the $Zr_{56.2}Ti_{13.8}Nb_5Ni_{5.6}Be_{12.5}$ alloy were studied earlier by producing this alloy in the form of a composite in which the second phase existed as dendrites dispersed in a glassy matrix [173,225]. This was the first alloy composition to produce *in situ* BMG composites. This composite showed improvement in ductility in comparison with the monolithic BMG. To improve the ductility of the BMG further, and based on the analysis of shear band propagation and liquid–solid phase transformation, Sun et al. [238,242] developed an innovative processing technique to alter the microstructure and transform the dendritic morphology of the crystalline phase into spherical particles. The technique that these authors developed can be briefly described as follows. When the liquid alloy is brought down in temperature from the fully liquid state into the mushy zone (liquid + solid region), both the liquid and solid will attain equilibrium provided sufficient time is allowed at the given temperature. Further, to minimize the surface energy, the solid phase will assume a spherical shape. If the alloy is now quenched from this temperature, the solid phase will continue to have the spherical morphology and the liquid will transform into the glassy state. By maintaining the alloy with the above overall chemistry of $Zr_{56.2}Ti_{13.8}Nb_5Ni_{5.6}Be_{12.5}$ at 900°C for 5 min, the two-phase alloy was quenched into cylindrical molds of 8 mm diameter. Scanning electron microscopy studies confirmed that the second phase had a spherical morphology after this processing method, in contrast to the dendritic morphology when the alloy was quenched directly from the liquid state. Figure 8.37 shows the two morphologies obtained in this alloy. A similar processing technique was also employed to obtain the spherical morphology of the second phase in another Zr-based alloy with the composition $Zr_{60}Ti_{14.7}Nb_{5.3}Cu_{5.6}Ni_{4.4}Be_{10}$ [242]. Therefore, using this method, it should be possible to obtain different compositions and amounts of the crystalline phase (and also of the glassy phase) by holding the alloy in the two-phase liquid + solid region at different temperatures.

The ductility values obtained with these two morphologies for the second phase have been shown to be quite different. The alloy with the second phase having a spherical morphology showed much higher ductility than the one with the dendritic morphology, *viz.*, 12% plastic strain for the spherical particles instead of 8.8% for the dendritic morphology. Figure 8.38 shows the room temperature compressive stress–strain curves for the $Zr_{41.2}Ti_{13.8}Cu_{12.5}Ni_{10}Be_{22.5}$ alloy (Vit 1), $Zr_{56.2}Ti_{13.8}Nb_5Ni_{5.6}Be_{12.5}$ alloy with the second phase in a dendritic morphology (designated S1), and $Zr_{60}Ti_{14.7}Nb_{5.3}Cu_{5.6}Ni_{4.4}Be_{10}$ alloy, with the second phase as spherical particles (designated S2). It can be clearly seen that the ductility of the S1 alloy is higher than that of Vit 1, and the ductility of S2 is higher than that of even S1.

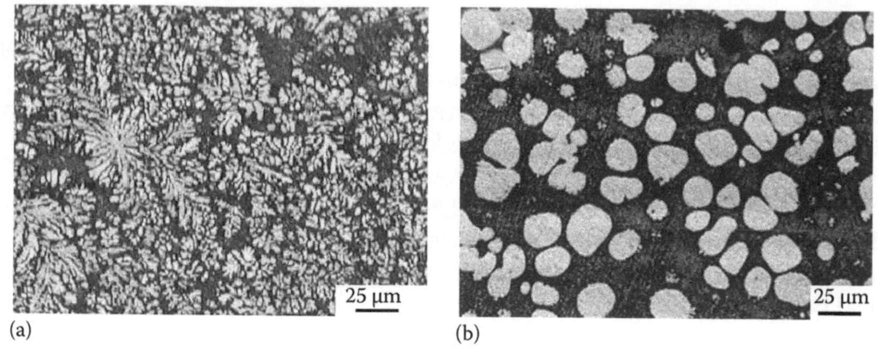

FIGURE 8.37
Typical microstructures of the $Zr_{56.2}Ti_{13.8}Nb_5Cu_{6.9}Ni_{5.6}Be_{12.5}$ alloy processed by two different techniques. (a) Optical micrograph obtained from an alloy produced by quenching directly from the liquid state. (b) Scanning electron micrograph from an alloy obtained by quenching the alloy first into the mushy (liquid + solid) state, holding there for some time, and then casting into the copper mold. Note that the second (crystalline) phase has a dendritic morphology in (a) and a spherical morphology in (b). (Reprinted from Sun, G.Y. et al., *Intermetallics*, 15, 632, 2007. With permission.)

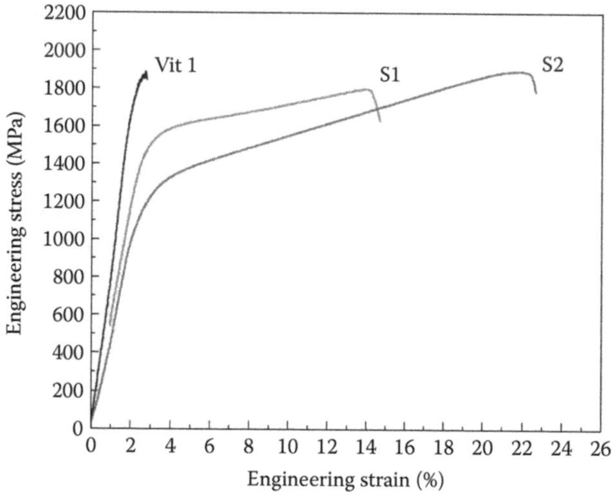

FIGURE 8.38
Room temperature compression stress–strain curves of a monolithic BMG alloy with the composition $Zr_{41.2}Ti_{13.8}Cu_{12.5}Ni_{10}Be_{22.5}$, S1 ($Zr_{56.2}Ti_{13.8}Nb_5Ni_{5.6}Be_{12.5}$ BMG composite with the second phase present as spherical particles), and S2 ($Zr_{60}Ti_{14.7}Nb_{5.3}Cu_{5.6}Ni_{4.4}Be_{10}$ alloy with the second phase as spherical particles) alloys. Note that the ductility of S2 is higher than that of S1, which is higher than that of Vit 1. (Reprinted from Sun, G.Y. et al., *Scr. Mater.*, 55, 375, 2006. With permission.)

Table 8.10 compares the mechanical properties of these three alloys with the second phase in different morphologies.

8.9.3 Effect of Volume Fraction

Schuh et al. [10] have analyzed the effect of the shape of the reinforcement (second phase) and its volume fraction in terms of the yield strength and plastic strain to fracture in both compression and tension. They calculated the yield strength in terms of the rule

TABLE 8.10

Comparison of the Mechanical Properties of BMG Composites with Different Morphologies of the Second (Crystalline) Phase

Alloy	Morphology of the Second Phase	V_r	σ_y (MPa)	σ_f (MPa)	ε_y (%)	ε_p (%)	References
$Zr_{41.2}Ti_{13.8}Cu_{12.5}Ni_{10}Be_{22.5}$	Fully glass	—	1900	1900	2	2	[95]
$Zr_{56.2}Ti_{13.8}Nb_5Ni_{5.6}Be_{12.5}$	Dendritic	25	1300	1700	1.2	6.8	[225]
$Zr_{56.2}Ti_{13.8}Nb_5Ni_{5.6}Be_{12.5}$ (S1)	Dendritic	30	1208	1757	1.78	8.82	[238]
$Zr_{56.2}Ti_{13.8}Nb_5Ni_{5.6}Be_{12.5}$ (S1)	Spherical	30	1350	1800	2.32	12	[238]
$Zr_{60}Ti_{14.7}Nb_{5.3}Cu_{5.6}Ni_{4.4}Be_{10}$ (S2)	Spherical	55	—	1900	—	22.3	[242]

Note: V_r: volume fraction of the reinforcement (second phase); σ_y: yield strength; σ_f: fracture strength; ε_y: strain at yielding; ε_p: plastic strain.

of mixtures (Equation 8.30). Even though this equation provides an upper bound on the strength of the composite and is applicable only in cases where the strain is the same in both the matrix and the reinforcement, the authors were able to observe some general trends and derive conclusions.

It was reported that at low volume fractions of the crystalline phase ($V_r < 0.3$), the yield stress of the composite was very close to that predicted by Equation 8.30. This observation suggests that the volume fraction, at low values, does not alter the macroscopic constitutive response of the glassy matrix. The plastic strain to failure will, of course, be larger, as shown in Table 8.9, for different composites. At values of $V_r > 0.3$, the composite response seems to depend on the morphology of the reinforcement. The yield strength of the composites with the fibers and large particles is the same as that calculated using the rule of mixtures. But, the yield strength for *in situ* processed composites with reinforcement as dendrites is higher than that of the rule of mixtures, and for nanocrystalline composites it is lower. From this, it appears that the dendritic morphology is more effective in inhibiting macroscopic plastic deformation of the matrix than any other, which may be related to the length scales associated with the plastic deformation of metallic glasses. But, the recent observation that spherical particles could further increase the strength needs to be looked into carefully.

As has been pointed out repeatedly, BMGs are very strong, but they suffer from lack of ductility, which seems to be an inherent characteristic noted during a tensile test. The reason for the development of BMG composites has essentially been to alleviate this problem. It has been observed that the ductility of the composites increases with increasing values of V_r [173,242]. But, the maximum value of the plastic strain to failure in compression reported so far is only about 20% (see Tables 8.9 and 8.10). The values are much lower in tension. Schuh et al. [10] have discussed in detail the reasons for this behavior in terms of shear band formation and shear band propagation.

8.9.4 Shear Band Initiation

Plastic flow is initiated in single-phase materials in some locally perturbed regions. Structural inhomogeneities and defects that introduce stress concentrations will therefore promote shear band initiation. In the case of composites, however, three different possibilities have been identified as stress concentrators.

First, due to the thermal mismatch strains that develop during cooling from the liquid state, significant residual stresses may be present in both the matrix and the reinforcement. Thermal residual stresses are generated in a composite due to the mismatch of coefficients of thermal expansion between the matrix and the reinforcement. Since these stresses can reach several hundred MPa, they can significantly affect the mechanical behavior of the composite. By conducting neutron powder diffraction studies and finite element modeling on a W-fiber-reinforced BMG ($Zr_{41.2}Ti_{13.8}Cu_{12.5}Ni_{10}Be_{22.5}$), Dragoi et al. [243] estimated that significant residual stresses existed in both the phases. They estimated that stresses of over 480 MPa were present in the matrix.

Second, under the action of an applied external load, both the phases will initially deform elastically. But, due to the differences in the elastic properties of the matrix and the reinforcement, some stress concentrations will occur in the matrix. Balch et al. [244] conducted high-energy synchrotron *in situ* x-ray diffraction experiments to measure strains in the crystalline reinforcing particles (W or Ta) in a $Zr_{57}Nb_5Al_{10}Cu_{15.4}Ni_{12.6}$ BMG. It was shown that the metallic particles yield and strain harden while transferring load to the largely elastic matrix. In the fully elastic regime, the stiffer reinforcement unloads the matrix; but, after the particles start plastically deforming, the stress supported by the elastic matrix increases. Even though the average level of "overloading" (difference between average matrix stress and the applied stress) at the highest applied stresses was only about 20 MPa, the localized stresses near the matrix–particle interface were considerably higher and thus initiated matrix shear banding and subsequent failure.

Lastly, at higher loads, if the crystalline phase yields before the matrix, an additional contribution to the stress concentration comes from the plastic misfit strain between the matrix and the reinforcement.

Shear bands are initiated once the stress concentration in the glassy matrix is sufficiently large that the appropriate yield criterion (von Mises or Mohr–Coulomb) is satisfied around the reinforcements. The stresses at which shear bands are initiated may be too low to support the propagation of a shear front into the matrix. In such a situation, plastic deformation is localized around the reinforcements. But, once the stress reaches a level high enough to drive the shear front, then large-scale plasticity sets in. The distinction between these two stresses (one to initiate a shear and the other to propagate the shear front) can be seen by carefully monitoring the elastic strain in the reinforcement.

8.9.5 Shear Band Propagation

Available experimental evidence that the increase in plastic strain to failure by the incorporation of a second crystalline phase is rather limited suggests that the crystalline particles are not effective barriers to the development of shear bands. It is reported that, in some cases, a crystalline particle can deflect a shear band that intersects it if the shear displacement is small [245] or in some cases even without intersecting it due to the interaction between the stress concentrations associated with the shear band and with the crystal. But, in many cases, the stress under which a shear band propagates has been found to be sufficient to either cleave or plastically deform the reinforcement, if it happens to be ductile [53] or fracture brittle particles [246]. These interactions were found to be much stronger during compressive rather than tensile stresses.

Since it has been noted that BMG composites with dendrites exhibit higher plasticity, it has been suggested that the stress concentration in the glassy matrix changes due to the dimensions of the dendrite features, which are typically about 10 μm, and thus influences the shear band formation. It is also possible that the newly formed shear bands are likely to intersect a dendrite arm before propagating very far, thus making the dendrite arms more effective barriers to shear front propagation. Using this concept of proximity of "obstacles" to shear band propagation, Hoffmann et al. [240] designed Zr-based BMG composites with different volume fractions of the crystalline phase and noted that the plastic strain to fracture increased with increasing volume fraction of the dendritic phase. But, it should be realized that the formation of dendrites is sensitive to cooling rate during solidification and therefore significant variations can be obtained in the microstructure. Further, the dendritic phase in the alloys investigated so far has been softer and, consequently, the strength is reduced, even though some ductility has been achieved. But, a side advantage is that the fracture toughness of such composites is quite high. This is not true if the crystalline phase formed *in situ* due to annealing is an intermetallic that is typically brittle. It is in this context that the mushy-zone processing suggested by Sun et al. [242] will be useful to produce equiaxed crystalline phases. In fact, Hoffmann et al. [240] also employed this technique to obtain a more uniform microstructure in their composites.

8.9.6 Superelasticity

One of the unique features of BMGs is their large elastic strain of about 2%, which is about four times larger than that (≃0.5%) of conventional crystalline alloys. On the other hand, crystalline alloys, e.g., special shape memory-type compounds, can exhibit a much larger elastic strain of about 5% by utilizing stress-induced reversible martensitic transformation. By synthesizing BMG composites containing shape memory intermetallic compounds, much effort has been spent on increasing the elastic strain while maintaining the high strength. Some results demonstrate that composites of glass and shape memory compounds exhibit superelastic strain of about 5%, but their yield strength is nearly the same as that of the intermetallic compound. Until now, there have been no reports of yield strengths above 2500 MPa and large elastic strain of about 15%. However, it is useful to discuss the superelastic behavior of some BMG composites including B2-type compounds, even though the yield strength is much lower than that (≃1600 MPa) for BMGs that have been used as practical structural materials.

In addition to the development of ductile Zr-based BMG composites by reinforcing the glassy phase with a dendritic structure [225,238,240], ductile Zr–Cu–Al composite alloys consisting of glassy plus metastable B2-type CuZr intermetallic compounds have also been developed [247–249]. The homogeneous dispersion of the CuZr compound in the glassy matrix causes a significant increase in plasticity. However, there is no evidence on the significant enhancement of superelastic properties for composite alloys containing the B2 CuZr compound.

Very recently, Tsarkov et al. [249] reported that 3 mm diameter rods of Ti–Ni–Cu–Zr–(Co–Y)–(B) alloy cast in a copper mold produced a unique mixed structure of TiNi intermetallic compound surrounded by a glassy phase. Figure 8.39 shows an SEM image of the cast TiNiCuZr alloy. It consists of TiNi compound with a grain size of about 10–20 μm surrounded by an amorphous phase with a width of about 2–3 μm. The composite structure, consisting of about 75–80 vol.% of the TiNi compound, is presumed to be formed through the process of primary solidification of the TiNi compound, followed by solidification of the amorphous phase from the remaining Cu–Zr rich alloy liquid. Thus, we can clearly notice a clear distinction between the TiNi and CuZr-rich regions. This interesting behavior seems to reflect the difference in attractive bonding forces among the constituent elements.

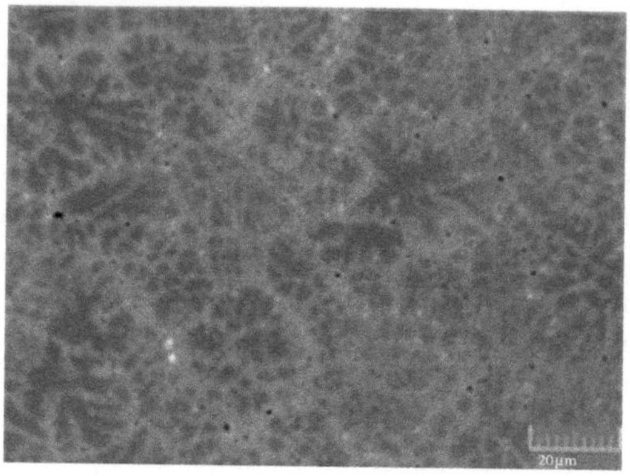

FIGURE 8.39
SEM image of the as-cast $Ti_{40}Ni_{39.5}Cu_8Zr_{10}Co_2Y_{0.5}$ BMG consisting of dark B2 phase crystals and bright interfacial amorphous phase. (Reprinted from Tsarkov, A.A. et al., *J. Alloys Compd.*, 658, 402, 2016. With permission.)

Figure 8.40a shows the compressive stress–strain curves of cast Ti-Ni-Cu-Zr-(Co-Y)-(B) alloys [249], where three clear stages can be noticed: A, elastic deformation; B, superelasticity caused by martensitic transformation; and C, plastic deformation. The first stage can be characterized as low yield stress of about 320 MPa and large elastic strain of about 2.5%. It is noticeable that the composite alloy exhibits the superelastic strain behavior resulting from the main NiTi intermetallic compound phase. In the third stage, one can see high flow stress of about 1800–2000 MPa, high fracture strength of about 2000–2300 MPa, and large plastic strain of about 10%–20%. Thus, the composite alloy exhibits high fracture strength and large plastic strain, in addition to the large superelastic strain. The fracture strength greatly exceeds that (195–690 MPa) of NiTi intermetallic compounds, and hence, the significant increase in the fracture strength is due to the existence of the glassy phase. In addition, the fracture strength of 2325 MPa is higher than those of binary Cu–Zr, ternary Cu–Zr–Ti, and quaternary Cu–Zr–Ti–Ni BMGs. The enhancement of fracture strength and fracture plastic strain is due to the increase in plastic deformability, presumably because of the combined effects of the easy generation of shear bands along the NiTi intermetallic compound/Cu–Zr glassy phase and the difficulty in propagating cracks from the glassy phase into the intermetallic compound. The effective combination is due to the narrow width and low volume fraction of the glassy phase.

Tsarkov et al. [249] also reported that replacement of Ti by 2% Co and 0.5% Y ($Ti_{40}Ni_{39.5}Cu_8Zr_{10}Co_2Y_{0.5}$) caused a significant increase in the yield strength up to 1955 MPa in the deformation stage in conjunction with the maintenance of superelastic strains, as shown in Figure 8.40b, resulting in the simultaneous achievement of high fracture strength of 2620 MPa and large elastic strain. These enhancements are due to the solid solution strengthening of Co and Nb in the NiTi intermetallic compound. All these data were obtained in a uniaxial compressive load condition. Considering the possible applications of these interesting properties, it is instructive to explore if similar high strength and large superelastic elongations can be achieved even in tensile and bending deformation modes.

8.9.7 High Strength-High Ductility BMGs

In the previous sections, the topics of strength and ductility were discussed independently. Considering the applications of BMGs, it is important to look at the simultaneous achievement of high strength and high ductility for BMGs, even though the two have a distinct

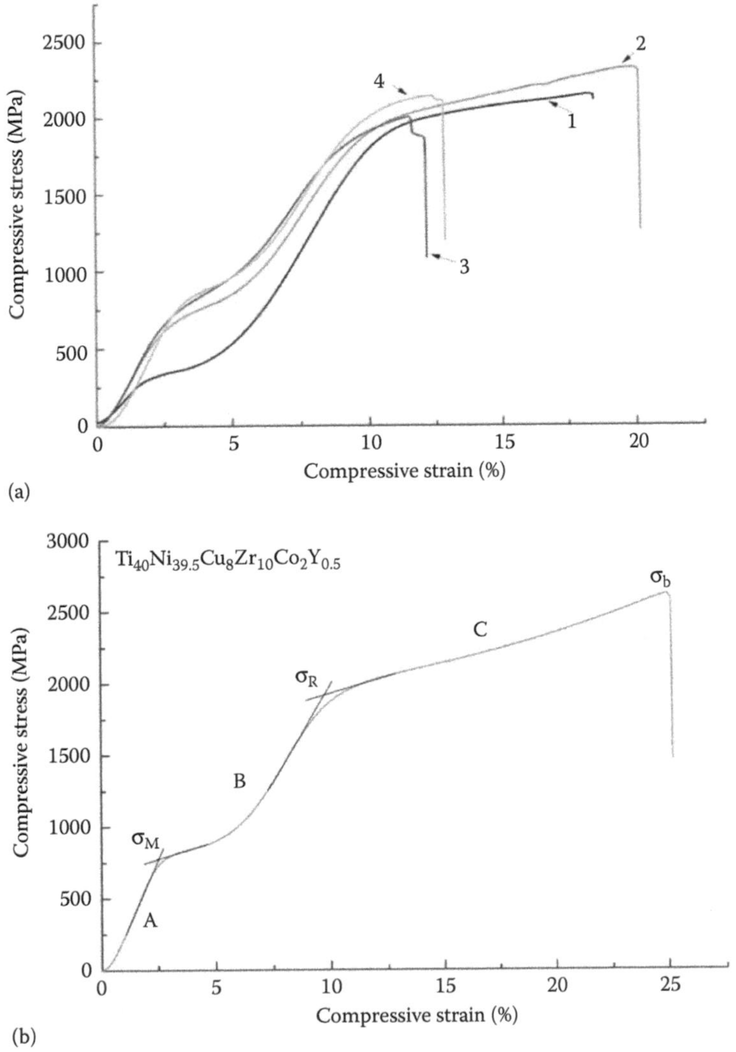

FIGURE 8.40
(a) Stress–strain curves of the as-cast (1) $Ti_{42}Ni_{39}Cu_9Zr_{10}$, (2) $Ti_{42}Ni_{39}Cu_7Zr_{10}Co_2$, (3) $Ti_{42}Ni_{38.5}Cu_7Zr_{10}Co_2Y_{0.5}$, and (4) $Ti_{42}Ni_{38}Cu_7Zr_{10}Co_2Y_{0.5}B_{0.5}$ samples. (b) Compressive stress–strain curve of the $Ti_{40}Ni_{39.5}Cu_8Zr_{10}Co_2Y_{0.5}$ composite material. (Reprinted from Tsarkov, A.A. et al., *J. Alloys Compd.*, 658, 402, 2016. With permission.)

trade-off relationship. Information from some successful investigations will provide us with an efficient way to develop a BMG with high strength and high ductility.

Figure 8.41 summarizes the relationship between yield strength and compressive plastic strain for various types of monolithic (consisting of a single glassy phase) and composite (with a mixed structure of glassy and crystalline phases) BMGs [250]. It is seen that the alloys can be mainly classified into the following three groups: (1) high strength and small plastic strain, (2) low strength and large plastic strain, and (3) high strength and large plastic strain. Group 3 consists of glass + dendritic bcc Ta- or Nb-rich solid solution phases in Cu–Ti–Hf–Ta, Cu–Ti–Hf–Nb, Cu–Ti–Hf–Ta–Ag, and Zr–Ni–Cu–Al alloy with high Zr content.

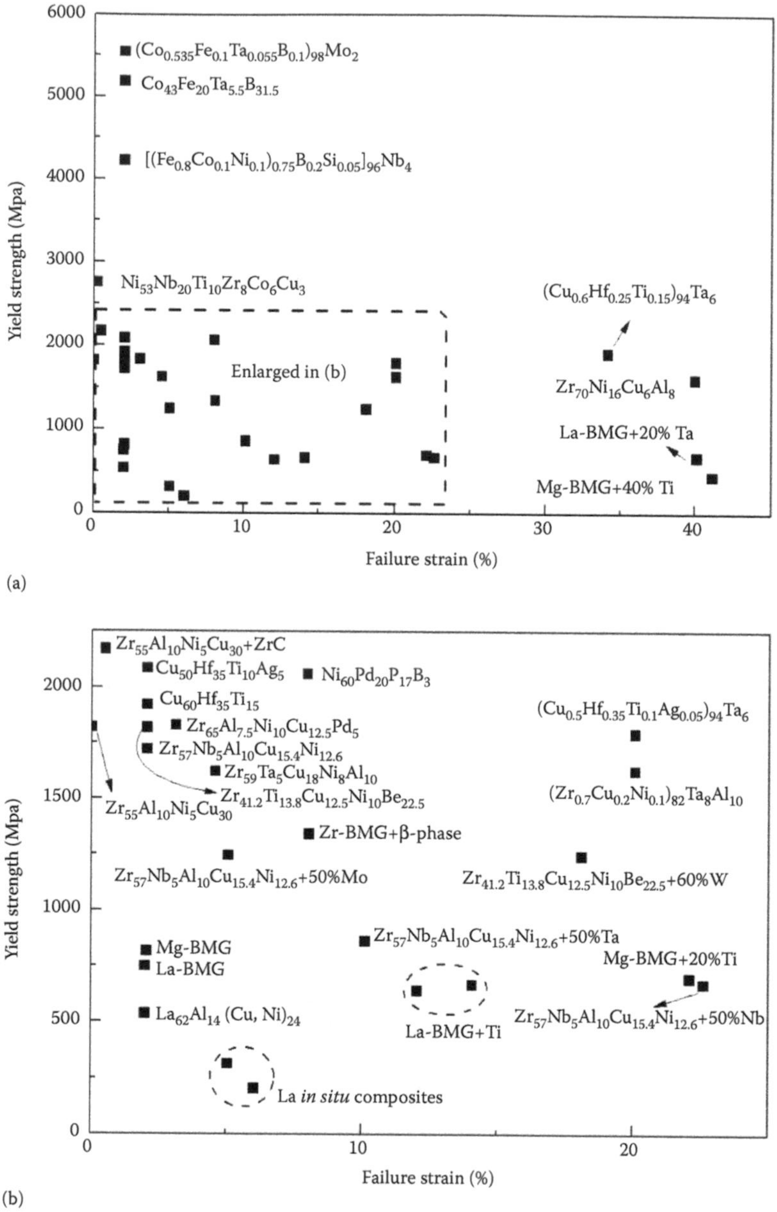

FIGURE 8.41
Plots comparing the yield strength and failure strain of different BMG composites. In general, the *in situ* composites show toughening, but with a large reduction in strength. The Cu-BMG + 6% Ta appears to possess the best combination of strength and plasticity of all Cu-based composites and even surpasses some of the $Zr_{57}Nb_5Al_{10}Cu_{15.4}Ni_{12.6}$-based *ex situ* composites.

As an example, Figure 8.42 shows the stress–strain curve of the Cu–Ti–Hf–Ta BMG, together with its microstructure [251]. The structure consists of bcc β-Ta(Ti,Hf) dendrite phase embedded in the glassy matrix. The dendrite phase is about 15 μm in length and 3.5 μm in width and its volume fraction is approximately 11%. The combined values of high strength of 2125 MPa and large plastic strain of 34% are believed to be the best

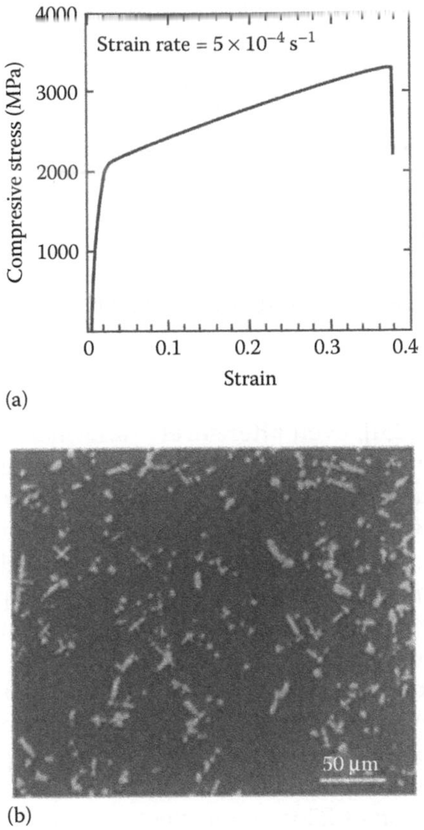

FIGURE 8.42
Nominal compressive stress–strain curve and SEM image of the central region in the transverse cross section of the cast 2 mm diameter rod of $(Cu_{0.6}Hf_{0.25}Ti_{0.15})_{94}Ta_6$ alloy with a mixed structure of glassy and dendritic bcc-β(Ta) phase. (Reprinted from Qin, C. et al., *Mater. Trans.*, 45, 2936, 2004. With permission.)

combined characteristics of all the metallic glass-based alloys reported to date. This best combination is presumably due to the synergistic effect of (1) rather high strength and good plasticity of the β-Ta(Ti,Hf) phase itself, (2) high strength and appreciable plasticity of the Cu-based glassy matrix, (3) easy initiation of shear deformation bands at the interface between β-Ta(Ti,Hf) and glassy phases, (4) suppression of the penetration of shear bands at the interface between β-Ta(Ti,Hf) and glassy matrix, and (5) homogeneous dispersion of dendritic β-Ta(Ti,Hf) phases with a suitable volume fraction. In particular, the dendritic morphology that is suitable for the initiation of shear band and the suppression of the penetration of shear band at the interface plays an important role in the extremely good combined characteristics. In addition, the dendritic morphology of the β-Ta(Ti,Hf) phase indicates that the Ta-based phase with high melting temperature was dissolved in the Cu-rich liquid and then precipitated during cooling from a high temperature. The dissolution of Ta-rich phase is attributed to the coexistence of Hf, an element with a rather high melting temperature. The dispersion of Ta-rich dendritic precipitates in glassy matrix is different from a number of previous data on the BMGs containing Ta.

When we look at the BMGs containing Ta, there are a number of papers showing the positive influence of Ta addition on the mechanical properties of BMGs such as

Zr–Al–Ni–Cu–Ta and Ti–Cu–Ni–Sn–Ta [252–254]. The morphology of the Ta-rich phase is nearly spherical, and the spherical particles are composed of nearly pure Ta metal because of its much higher melting temperature. Little is known about the formation of Ta-rich phase containing other solute elements with a dendritic morphology demonstrating the true dissolution of Ta into the liquid phase. The data shown in Ref. [251] are believed to be the first evidence for the precipitation of the dendritic Ta-rich phase containing other solute elements of Hf and Ti.

The dispersion of dendritic bcc β-(Ti,Zr) phase in Zr–Ti–Be–Ni–Cu metallic glasses causes a significant increase in plasticity in conjunction with a decrease in the yield strength [173], while the mixed phase Cu–Ti–Hf–Ta bulk alloy keeps nearly the same yield strength level as that of the base Cu–Hf–Ti bulk glassy alloy. The development of high-strength and high-ductility composite BMGs is possible by controlling the component and the size of the dendritic phase.

As shown in Figure 8.43 [251], even after severe deformation to about 34%, no distinct work-hardening phenomenon was observed for the present composite in spite of the existence of the dendritic crystalline phase. This is because the strength is dominated by the glassy matrix phase and the intense crossing among shear bands does not cause appreciable work hardening under a compressive applied load. This behavior is different from the work-hardening behavior of Zr–Ti–Be–Ni–Cu BMGs consisting of glass + β(Ti,Zr) phases. The difference is presumably due to the higher volume fraction of β-(Ti,Zr) phase, more distinctive dendrite morphology of the β-(Ti,Zr) phase and the lower yield strength of the β-(Ti,Zr) phase for the Be-containing Zr-based alloy [251]. The key factor in synthesizing a glassy alloy with high yield strength and significant work hardening seems to be to control the morphology and volume fraction of the second phase precipitates.

In addition to the dispersion of bcc-β (dendritic) phase in the glassy matrix of Zr- and Cu-based alloys, the dispersion of B2-type intermetallic compounds with a nearly spherical morphology has attracted increasing interest because the B2 phase has superelasticity as well as shape memory characteristics [255–258]. Table 8.11 summarizes the alloy systems, typical alloy compositions, as-cast structure, and mechanical properties of BMGs with mixed structures of glass + bcc-β phase and glass + B2 phase. As shown in Figure 8.44

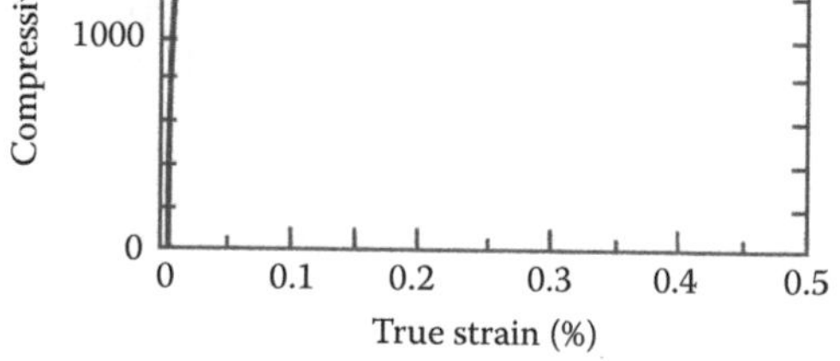

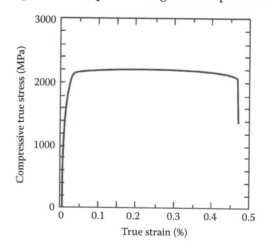

FIGURE 8.43
True compressive stress–strain curve of the cast 2 mm diameter rod of $(Cu_{0.6}Hf_{0.25}Ti_{0.15})_{94}Ta_6$ alloy with a mixed structure of glassy and dendritic bcc-β(Ta) phase. (Reprinted from Qin, C. et al., *Mater. Trans.*, 45, 2936, 2004. With permission.)

TABLE 8.11

Typical Alloy Compositions, As-Cast Structure, and Mechanical Properties of Bulk Glassy Alloys with Mixed Structures of Glass + I, Glass + bcc-β Phase, and Glass + B2 Phase

Alloy	As-Cast Structure	Young's Modulus (GPa)	Yield Strength (MPa)	Maximum Strength (MPa)	Fracture Strain (%)	References
$(Cu_{0.5}Hf_{0.35}Ti_{0.1}Ag_{0.05})_{92}Ta_8$	Glass + β	108	2770	—	19.2	[229]
$(Cu_{0.6}Hf_{0.25}Ti_{0.1})_{94}Ta_6$	Glass + β	—	2125	—	34	[251]
$(Cu_{0.6}Hf_{0.25}Ti_{0.1})_{96}Nb_4$	Glass + β	130	2405	—	2.8	[259]
$(Ti_{0.4}Zr_{0.1}Cu_{0.36}Pd_{0.14})_{99}Ta_1$	Glass + β	—	2060	2100	5	[260]
$(Ti_{0.4}Zr_{0.1}Cu_{0.36}Pd_{0.14})_{97}Nb_3$	Glass + β	80	2050	—	6.5	[261]
$Zr_{41.25}Ti_{13.75}Cu_{12.5}Ni_{10}Be_{22.5}$	Glass + β	96	1300	—	17.5	[100]
$Zr_{56.25}Ti_{13.75}Nb_5Be_{22.5}Cu_{1.25}Ni_1$	Glass + β	110	1300	—	1.2	[225]
$(Zr_{0.62}Cu_{0.154}Ni_{0.126}Al_{0.1})_{94}Ta_6$	Glass + β	—	1800	2090	7.5	[253]
$(Zr_{0.7}Cu_{0.2}Ni_{0.1})_{82}Ta_8Al_{10}$	Glass + β	—	1700	—	17	[245]
$Ni_{40}Cu_{10}Ti_{33}Zr_{16}Si_1$	Glass + B2	78	1200	1900	3.4	[255]
$Zr_{48.5}Cu_{46.5}Al_5$	Glass + B2	—	1894	—	7.7	[262]
$Zr_{65}Al_{7.5}Ni_{10}Cu_{12.5}Ag_5$	Glass + I	86	1900	—	2.2	[263,264]
$Zr_{65}Al_{7.5}Ni_{10}Cu_{12.5}Pd_5$	Glass + I	88	1780	1830	3.1	[265]

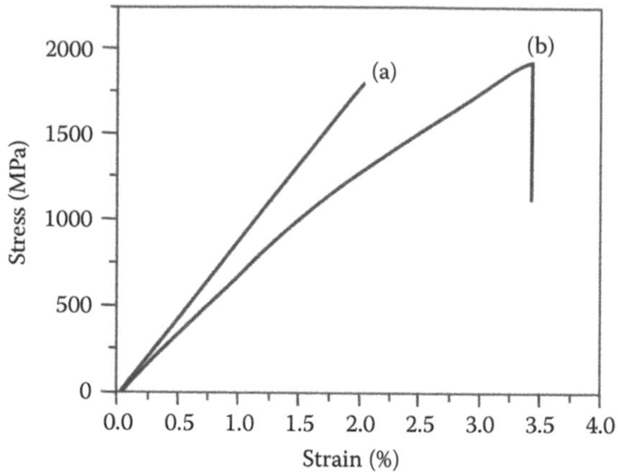

FIGURE 8.44
Compressive stress–strain curves for 3 mm diameter rod of $Ni_{40}Cu_{10}Ti_{33}Zr_{16}Si_1$ alloy: (a) fast cooled, (b) slowly cooled button. The slowly cooled button shows higher strength and better elongation. (Reprinted from Choi-Yim, H. et al., *Scr. Mater.*, 53, 1467, 2005. With permission.)

for $Ni_{40}Cu_{10}Ti_{33}Zr_{16}Si_1$, this alloy yields in compression at 1200 MPa, fails at 1900 MPa, and exhibits an elastic strain of 3.4%.

8.9.8 Icosahedral Particle-Dispersion Effect

It has generally been recognized that the high stability of the supercooled liquid in multicomponent alloys is due to the development of icosahedral-like medium-range ordered atomic configurations for metal–metal-type BMGs. Reflecting the unique atomic configuration, the primary precipitating phase from the supercooled liquid is an icosahedral phase. It is, therefore, valuable to present the influence of icosahedral precipitates on the mechanical properties for metal–metal-type BMGs in an independent section.

Zr-rich $Zr_{65}(Al,Ni,Cu)_{35}$ BMGs can be formed in the diameter range up to 16 mm [266] and exhibit the largest supercooled liquid region, of over 110 K, in Zr-based alloy systems [267]. The addition of noble metals, such as Pd, Pt, Ag, and Au, causes the precipitation of nanoscale icosahedral phase as the primary precipitate phase from the supercooled liquid, as shown in Figure 8.45 [268]. The resulting nanoquasicrystal-dispersed alloys exhibit enhancement of tensile and compressive strength as well as compressive ductility [269]. These results indicate that the nanoscale icosahedral precipitates are not always harmful as a structural modification method to develop high-ductility BMGs.

Recently, Yamada et al. [270] have reported that the 3% Au-containing Zr–Cu–Ni–Al BMG exhibits very large compressive strains of about 20%, as shown in Figure 8.46, and that the significant improvement of ductility was due to the deformation-induced precipitation. The precipitates were very fine (10 nm in size), had a small interparticle spacing of about 5 nm, and lay in the limited region just near the shear bands. However, no definite information on the crystal structure of the precipitates was obtained because of their fine sizes, as shown in Figure 8.47. Since the addition of small amounts of noble metals to $Zr_{60\text{-}65}Al_{10}Ni_{5\text{–}10}Cu_{20}$ BMG causes the primary precipitation of icosahedral phase from the supercooled liquid, the deformation-induced precipitates have been presumed to be

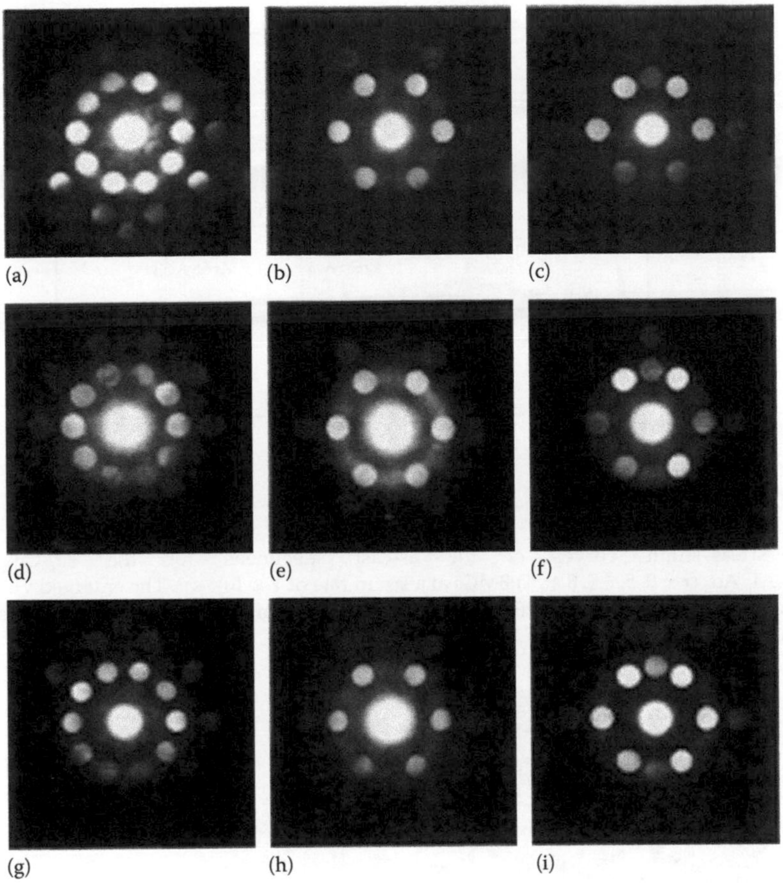

FIGURE 8.45
Nanobeam electron diffraction patterns revealing the five-, three-, and two-fold symmetries of the $Zr_{65}Al_{7.5}Ni_{10}Cu_{12.5}M_5$ (M = Pd, Au, or Pt) glassy alloys annealed for 120 s at 773 or 753 K, respectively. (a through c): M = Pd; (d through f): M = Au; and (g through i): M = Pt. (Reprinted from Inoue, A. et al., *Mater. Trans.*, 40, 1181, 1999. With permission.)

the icosahedral phase. The deformation-induced precipitates increase the viscosity of the shear band region and suppress the progress of localized shear deformation, resulting in the generation of new shear bands at different places, and leading to the increase in plasticity.

There have been a number of results on the improvement of plasticity related to the precipitation of an icosahedral phase [271–273]. Hereafter, it seems important to clarify the compositional criterion in relation to nanoscale icosahedral phase formation for obtaining the improvement of plasticity for the BMGs containing deformation-induced precipitates. In addition, among the noble metals only Au seems to be effective in increasing the plasticity, and its effectiveness is presumed to be due to the easy precipitation of icosahedral phase, because the precipitation temperature of icosahedral phase from the supercooled liquid region is the lowest for the Au-containing alloy [274]. However, Au is very expensive to be used for structural applications and hence the search for a much cheaper element with a similar effect is required.

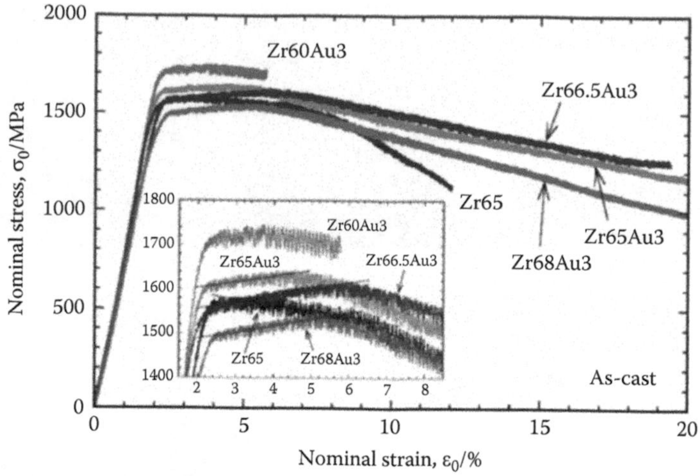

FIGURE 8.46
Compressive stress–strain curves of the as-cast specimens for the $Zr_{65}Cu_{20}Ni_5Al_{10}$ and $Zr_{60+x}Cu_{22-x}Ni_5Al_{10}Au_3$ (x = 0, 5, 6.5, 8 at.%) BMGs at a strain rate of 1×10^{-4} s^{-1}. The enlarged view of the strain range from 2% to 8% is also shown in the figure. (Reprinted from Yamada, M. et al., *J. Japan Inst. Met. Mater.*, 78, 449, 2014. With permission.)

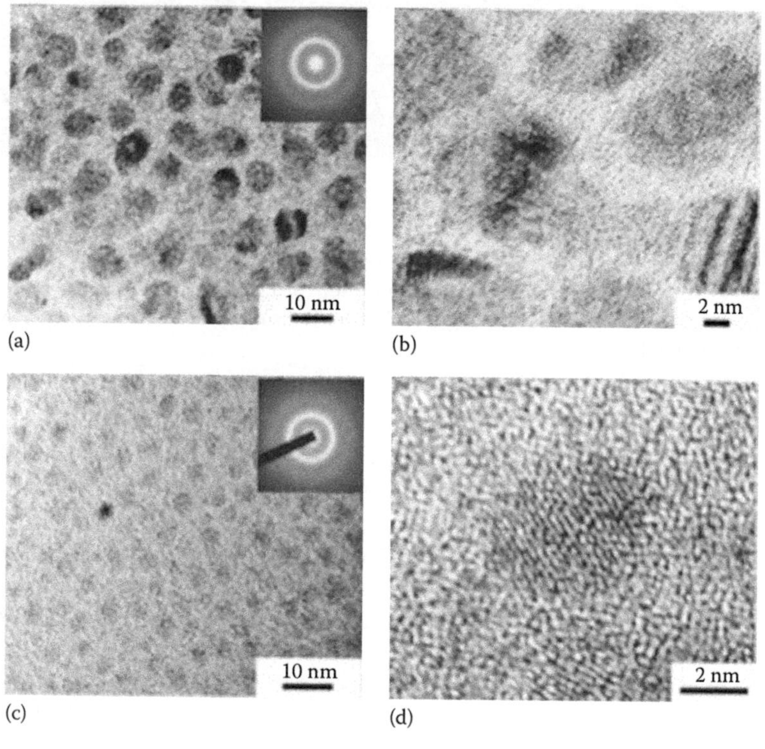

FIGURE 8.47
TEM images of the specimens near the principal shear bands after compression test for the as-cast (a, b) and annealed (c, d) specimens at (T_g-100)K for the $Zr_{66.5}Cu_{15.5}Ni_5Al_{10}Au_3$ BMG with selected area diffraction patterns. (a, c) Low-magnification images. (b, d) High-resolution images. (Reprinted from Yamada, M. et al., *J. Japan Inst. Met. Mater.*, 78, 449, 2014. With permission.)

8.10 Concluding Remarks

We have seen that the mechanical behavior of BMGs, due to their disordered nature, is substantially different from that of crystalline materials, which exhibit long-range order. In this chapter, we have discussed deformation mechanisms, formation of shear bands, temperature rise near shear bands, and strength and ductility of BMGs. The yielding behavior, fatigue, and fracture behavior of both monolithic and composite BMGs have also been discussed.

The inherent deformation mechanisms of BMGs have been shown to be different from their crystalline counterparts. The deformation of metallic glasses has been described in terms of the STZs, whose nature seems to be still unclear. BMGs have been shown to be much stronger than their crystalline counterparts, and their fracture strength approaches values as high as 5 GPa. There has been a lot of work to understand the ductility and plasticity of BMGs. But, the basic factors responsible for imparting ductility to these novel materials are still being investigated. It is clear that the BMG composites seem to hold much promise for commercial use in view of their increased ductility. The increased ductility appears to be determined not only by the Poisson's ratio, but also by the chemical homogeneity and the composition being close to an intermetallic in the crystalline state. Recent efforts have focused on simultaneously achieving high strength and high ductility in the same material. Significant differences exist between groups on the fatigue behavior of BMGs. The availability of reproducible and high-purity materials with controlled free volume can resolve some of the issues raised in the literature. Further research is required to obtain a clear understanding of the deformation behavior of BMGs.

References

1. Klement, Jr., W., R.H. Willens, and P. Duwez (1960). Non-crystalline structure in solidified gold–silicon alloys. *Nature* 187: 869–870.
2. Masumoto, T. and R. Maddin (1971). The mechanical properties of palladium–20 a/o silicon alloy quenched from the liquid state. *Acta Metall.* 19: 725–741.
3. Maddin, R. and T. Masumoto (1972). The deformation of amorphous palladium–20 at.% silicon. *Mater. Sci. Eng.* 9: 153–162.
4. Pampillo, C.A. (1975). Review: Flow and fracture in amorphous alloys. *J. Mater. Sci.* 10: 1194–1227.
5. Davis, L.A. (1978). Strength, ductility and toughness. In *Metallic Glasses,* eds. J.J. Gilman and H.J. Leamy, pp. 190–223. Materials Park, OH: ASM International.
6. Li, J.C.M. (1982). Mechanical properties of amorphous metals and alloys. In *Ultra-Rapid Quenching of Liquid Alloys,* ed. H. Herman. Vol. 20 of Treatise on Materials Science and Technology, pp. 325–389. New York: Academic Press.
7. Kimura, H. and T. Masumoto (1983). Strength, ductility and toughness: A study in model mechanics. In *Amorphous Metallic Alloys,* ed. F.E. Luborsky, pp. 187–230. London, UK: Butterworths.
8. Inoue, A., K. Kita, T. Zhang, and T. Masumoto (1989). An amorphous $La_{55}Al_{25}Ni_{20}$ alloy prepared by water quenching. *Mater. Trans., JIM* 30: 722–725.
9. Peker, A. and W.L. Johnson (1993). A highly processable metallic glass: $Zr_{41.2}Ti_{13.8}Cu_{12.5}Ni_{10.0}Be_{22.5}$. *Appl. Phys. Lett.* 63: 2342–2344.

10. Schuh, C.A., T.C. Hufnagel, and U. Ramamurty (2007). Mechanical behavior of amorphous alloys. *Acta Mater.* 55: 4067–4109.
11. Chen, M.W. (2008). Mechanical behavior of metallic glasses: Microscopic understanding of strength and ductility. *Ann. Rev. Mater. Res.* 38: 445–469.
12. Wang, G.Y., P.K. Liaw, and M.L. Morrison (2009). Progress in studying the fatigue behavior of Zr-based bulk metallic glasses and their composites. *Intermetallics* 17: 579–590.
13. Trexler, M.M. and N.N. Thadhani (2010). Mechanical properties of bulk metallic glasses. *Prog. Mater. Sci.* 55: 759–839.
14. Suryanarayana, C. (2012). Mechanical behavior of emerging materials. *Mater. Today* 15: 486–498.
15. Wang, W.H. (2012). The elastic properties, elastic models and elastic perspectives of metallic glasses. *Prog. Mater. Sci.* 57: 487–656.
16. Greer, A.L., Y.Q. Cheng, and E. Ma (2013). Shear bands in metallic glasses. *Mater. Sci. Eng. R* 74: 71–132.
17. Sun, B.A. and W.H. Wang (2015). The fracture of bulk metallic glasses. *Prog. Mater. Sci.* 74: 211–307.
18. Wang, W.H., Y. Yang, T.G. Nieh, and C.T. Liu (2015). On the source of plastic flow in metallic glasses: Concepts and models. *Intermetallics* 67: 81–86.
19. Huffnagel, T.C., C.A. Schuh, and M.L. Falk. (2016). Deformation of metallic glasses: Recent developments in theory, simulations, and experiments. *Acta Mater.* 109: 375–393.
20. Qiao, J.W., H.L. Jia, and P.K. Liaw. (2016) Metallic glass matrix composites. *Mater. Sci. Eng.* R100: 1–69.
21. Spaepen, F. (1977). A microscopic mechanism for steady state inhomogeneous flow in metallic glasses. *Acta Metall.* 25: 407–415.
22. Argon, A.S. (1979). Plastic deformation in metallic glasses. *Acta Metall.* 27: 47–58.
23. Steif, P.S., F. Spaepen, and J.W. Hutchinson (1982). Strain localization in amorphous metals. *Acta Metall.* 30: 447–455.
24. Leamy, H.J., H.S. Chen, and T.T. Wang (1972). Plastic flow and fracture of metallic glass. *Metall. Trans.* 3: 699–708.
25. Wright, W.J., R. Saha, and W.D. Nix (2001). Deformation mechanisms of the $Zr_{40}Ti_{14}Ni_{10}Cu_{12}Be_{24}$ bulk metallic glass. *Mater. Trans.* 42: 642–649.
26. Schuh, C.A. and T.G. Nieh (2003). A nanoindentation study of serrated flow in bulk metallic glasses. *Acta Mater.* 51: 87–99.
27. Wang, K., T. Fujita, Y.Q. Zeng, N. Nishiyama, A. Inoue, and M.W. Chen (2008). Micromechanisms of serrated flow in a $Ni_{50}Pd_{30}P_{20}$ bulk metallic glass with a large compression plasticity. *Acta Mater.* 56: 2834–2842.
28. Kawamura, T., T. Shibata, A. Inoue, and T. Masumoto (1998). Workability of the supercooled liquid in the $Zr_{65}Al_{10}Ni_{10}Cu_{15}$ bulk metallic glass. *Acta Mater.* 46: 253–263.
29. Wang, G., S.S. Fang, X.S. Xiao, Q. Hua, H.Z. Gu, and Y.D. Dong (2004). Microstructure and properties of $Zr_{65}Al_{10}Ni_{10}Cu_{15}$ amorphous plates rolled in the supercooled liquid region. *Mater. Sci. Eng. A* 373: 217–220.
30. Lewandowski, J.J., M. Shazly, and A. Shamimi Nouri (2006). Intrinsic and extrinsic toughening of metallic glasses. *Scr. Mater.* 54: 337–341.
31. Schroers, J. and N. Paton (2006). Amorphous metal alloys form like plastics. *Adv Mater. Proc.* 164: 61–63.
32. Kawamura, Y., T. Nakamura, and A. Inoue (1998). Superplasticity in $Pd_{40}Ni_{40}P_{20}$ metallic glass. *Scr. Mater.* 39: 301–306.
33. Kim, W.J., Y.K. Sa, J.B. Lee, and H.G. Jeong (2006). High-strain-rate superplasticity of $Zr_{65}Al_{10}Ni_{10}Cu_{15}$ sheet fabricated by squeeze casting method. *Intermetallics* 14: 377–381.
34. Wang, G., J. Shen, J.F. Sun, Y.J. Huang, J. Zou, Z.P. Lu, Z.H. Stachurski, and B.D. Zhou (2005). Superplasticity and superplastic forming ability of a Zr–Ti–Ni–Cu–Be bulk metallic glass in the supercooled liquid region. *J. Non-Cryst. Solids* 351: 209–217.
35. Nieh, T.G., J. Wadsworth, C.T. Liu, T. Ohkubo, and Y. Hirotsu (2001). Plasticity and structural instability in a bulk metallic glass deformed in the supercooled liquid region. *Acta Mater.* 49: 2887–2896.

36. Nieh, T.G., C.A. Schuh, J. Wadsworth, and Y. Li (2002). Strain rate dependent deformation in bulk metallic glasses. *Intermetallics* 10: 1177–1182.
37. Wang, W.K., H. Iwasaki, C. Suryanarayana, and T. Masumoto (1983). Crystallization characteristics of an amorphous $Ti_{80}Si_{20}$ alloy at high pressures. *J. Mater. Sci.* 18: 3765–3772.
38. Suryanarayana, C., W.K. Wang, H. Iwasaki, and T. Masumoto (1980). High-pressure synthesis of A15 Nb_3Si phase from amorphous Nb-Si alloys. *Solid State Commun.* 34: 861–863.
39. Ashby, M.F. (1972). A first report on deformation-mechanism maps. *Acta Metall.* 20: 887–897.
40. Frost, H.J. and M.F. Ashby (1982). *Deformation-Mechanism Maps: The Plasticity and Creep of Metals and Ceramics*. Oxford, UK: Pergamon Press.
41. Megusar, J., A.S. Argon, and N.J. Grant (1979). Plastic flow and fracture in $Pd_{80}Si_{20}$ near T_g. *Mater. Sci. Eng.* 38: 63–72.
42. Schuh, C.A., A.C. Lund, and T.G. Nieh (2004). New regime of homogeneous flow in the deformation map of metallic glasses: Elevated temperature nanoindentation experiments and mechanistic modeling. *Acta Mater.* 52: 5879–5891.
43. Furukawa, A. and H. Tanaka (2009). Inhomogeneous flow and fracture of glassy materials. *Nat. Mater.* 8: 601–609.
44. Flores K.M. and R.H. Dauskardt (1999). Local heating associated with crack tip plasticity in Zr–Ti–Ni–Cu–Be bulk amorphous metals. *J. Mater. Res.* 14: 638–643.
45. Liu, C.T., L. Heatherly, D.S. Easton, C.A. Carmichael, J.H. Schneibel, C.H. Chen, J.L. Wright, M.H. Yoo, J.A. Horton, and A. Inoue (1998). Test environments and mechanical properties of Zr-base bulk amorphous alloys. *Metall. Mater. Trans. A* 29 A: 1811–1820.
46. Masumoto T. and R. Maddin (1975). Structural stability and mechanical properties of amorphous metals. *Mater. Sci. Eng.* 19: 1–24.
47. Bruck H.A., A.J. Rosakis, and W.L. Johnson (1996). The dynamic compressive behavior beryllium bearing bulk metallic glasses. *J. Mater. Res.* 11: 503–511.
48. Kim, J.J., Y.Y. Choi, S. Suresh, and A.S. Argon (2002). Nanocrystallization during nanoindentation of a bulk amorphous metal alloy at room temperature. *Science* 295: 654–657.
49. Wright, W.J., R.B. Schwarz, and W.D. Nix (2001). Localized heating during serrated plastic flow in bulk metallic glasses. *Mater. Sci. Eng. A* 319–321: 229–232.
50. Donovan, P.E. (1989). A yield criterion for $Pd_{40}Ni_{40}P_{20}$ metallic glass. *Acta Metall.* 37: 445–456.
51. Yang, B., P.K. Liaw, G. Wang, M.L. Morrison, C.T. Liu, R.A. Buchanan, and Y. Yokoyama (2004). In-situ thermographic observation of mechanical damage in bulk-metallic glasses during fatigue and tensile experiments. *Intermetallics* 12: 1265–1274.
52. Donovan P.E. and W.M. Stobbs (1981). The structure of shear bands in metallic glasses. *Acta Metall.* 29: 1419–1426.
53. Pekarskaya, E., C.P. Kim, and W.L. Johnson (2001). In situ transmission electron microscopy studies of shear bands in a bulk metallic glass based composite. *J. Mater. Res.* 16: 2513–2518.
54. Lowhaphandu, P., S.L. Montgomery, and J.J. Lewandowski (1999). Effects of superimposed hydrostatic pressure on flow and fracture of a Zr–Ti–Ni–Cu–Be bulk amorphous alloy. *Scr. Mater.* 41: 19–24.
55. Mukai, T., T.G. Nieh, Y. Kawamura, A. Inoue, and K. Higashi (2002). Dynamic response of a $Pd_{40}Ni_{40}P_{20}$ bulk metallic glass in tension. *Scr. Mater.* 46: 43–47.
56. Zhang, Y. and A.L. Greer (2006). Thickness of shear bands in metallic glasses. *Appl. Phys. Lett.* 89: 071907-1–071907-3.
57. Hufnagel, T.C., T. Jiao, Y. Li, L.Q. Xing, and K.T. Ramesh (2002). Deformation and failure of $Zr_{57}Ti_5Cu_{20}Ni_8Al_{10}$ bulk metallic glass under quasi-static and dynamic compression. *J. Mater. Res.* 17: 1441–1445.
58. Schuh, C.A. and A.C. Lund (2003) Atomistic basis for the plastic yield criterion of metallic glass. *Nat. Mater.* 2: 449–452.
59. Bai, Y. and B. Dodd (1992). *Adiabatic Shear Localization: Occurrence, Theories, and Applications*. Oxford, UK: Pergamon Press.
60. Bever, M.B., D.L. Holt, and A.L. Titchner (1973). The stored energy of cold work. *Prog. Mater. Sci.* 17: 5–177.

61. Neuhauser, H. (1978). Rate of shear band formation in metallic glasses. *Scr. Metall.* 12: 471–474.
62. Yang, B., M.L. Morrison, P.K. Liaw, R.A. Buchanan, G.Y. Wang, C.T. Liu, and M. Denda (2005). Dynamic evolution of nanoscale shear bands in a bulk-metallic glass. *Appl. Phys. Lett.* 86: 141904-1–141904-3.
63. Gilbert, C.J., J.W. Ager III, V. Schroeder, R.O. Ritchie, J.P. Lloyd, and J.R. Graham (1999). Light emission during fracture of a Zr–Ti–Ni–Cu–Be bulk metallic glass. *Appl. Phys. Lett.* 74: 3809–3811.
64. Bengus, V.Z., E.D. Tabachnikova, S.E. Shumilin, Y.I. Golovin, M.V. Makarov, A.A. Shibkov, J. Miškuf, K. Csach, and V. Ocelik (1993). Some peculiarities of ductile shear failure of amorphous alloy ribbons. *Int. J. Rapid Solidif.* 8: 21–31.
65. Yang, B., C.T. Liu, T.G. Nieh, M.L. Morrison, P.K. Liaw, and R.A. Buchanan (2006). Localized heating and fracture criterion for bulk metallic glasses. *J. Mater. Res.* 21: 915–922.
66. Lewandowski, J.J. and A.L. Greer (2006). Temperature rise at shear bands in metallic glasses. *Nat. Mater.* 5: 15–18.
67. Chen, H., Y. He, G.J. Shiflet, and S.J. Poon (1994). Deformation-induced nanocrystal formation in shear bands of amorphous alloys. *Nature* 367: 541–543.
68. Jiang, W.H., F.E. Pinkerton, and M. Atzmon (2003). Deformation-induced nanocrystallization in an Al-based amorphous alloy at subambient temperature. *Scr. Mater.* 48: 1195–1200.
69. Jiang, W.H. and M. Atzmon (2006). Mechanically-assisted nanocrystallization and defects in amorphous alloys: A high-resolution transmission electron microscopy study. *Scr. Mater.* 54: 333–336.
70. Lee, M.H., D.H. Bae, D.H. Kim, W.T. Kim, D.J. Sordelet, K.B. Kim, and J. Eckert (2008). Nanocrystallization at shear bands in bulk metallic glass matrix composites. *Scr. Mater.* 58: 651–654.
71. Chen, M.W., A. Inoue, W. Zhang, and T. Sakurai (2006). Extraordinary plasticity of ductile bulk metallic glasses. *Phys. Rev. Lett.* 96: 245502-1–245502-4.
72. Inoue, A., W. Zhang, T. Tsurui, A.R. Yavari, and A.L. Greer (2005). Unusual room-temperature compressive plasticity in nanocrystal-toughened bulk copper-zirconium glass. *Philos. Mag. Lett.* 85: 221–229.
73. Inoue, A., B.L. Shen, H. Koshiba, H. Kato, and A.R. Yavari (2003). Cobalt-based bulk glassy alloy with ultrahigh strength and soft magnetic properties. *Nat. Mater.* 2: 661–663.
74. Inoue, A. and W. Zhang (2004). Formation, thermal stability and mechanical properties of Cu–Zr and Cu–Hf binary glassy alloy rods. *Mater. Trans.* 45: 584–587.
75. Qin, C.L., W. Zhang, K. Asami, H.M. Kimura, X.M. Wang, and A. Inoue (2006). A novel Cu-based BMG composite with high corrosion resistance and excellent mechanical properties. *Acta Mater.* 54: 3713–3719.
76. Qin, C.L., W. Zhang, K. Asami, N. Ohtsu, and A. Inoue (2005). Glass formation, corrosion behavior and mechanical properties of bulk glassy Cu–Hf–Ti–Nb alloys. *Acta Mater.* 53: 3903–3911.
77. Fu, H.M., H.F. Zhang, H. Wang, Q.S. Zhang, and Z.Q. Hu (2005). Synthesis and mechanical properties of Cu-based bulk metallic glass composites containing in-situ TiC particles. *Scr. Mater.* 52: 669–673.
78. Das, J., M.B. Tang, K.B. Kim, R. Theissmann, F. Baier, W.H. Wang, and J. Eckert (2005). "Work-hardenable" ductile bulk metallic glass. *Phys. Rev. Lett.* 94: 205501-1–205501-4.
79. Xu, D.H., B. Lohwongwatana, G. Duan, W.L. Johnson, and C. Garland (2004). Bulk metallic glass formation in binary Cu-rich alloy series – $Cu_{100-x}Zr_x$ ($x = 34, 36, 38.2, 40$ at.%) and mechanical properties of bulk $Cu_{64}Zr_{36}$ glass. *Acta Mater.* 52: 2621–2624.
80. Zhang, W., Q. Zhang, C. Qin, and A. Inoue (2008). Synthesis and properties of Cu–Zr–Ag–Al glassy alloys with high glass-forming ability. *Mater. Sci. Eng. B* 148: 92–96.
81. Zhang, Q., W. Zhang, G. Xie, and A. Inoue (2007). Glass-forming ability and mechanical properties of the ternary Cu–Zr–Al and quaternary Cu–Zr–Al–Ag bulk metallic glasses. *Mater. Trans.* 48: 1626–1630.
82. Gu, X.J., S.J. Poon, and G.J. Shiflet (2007). Mechanical properties of iron-based bulk metallic glasses. *J. Mater. Res.* 22: 344–351.

83. Yang, W., H. Liu, Y. Zhao, A. Inoue, K. Jiang, J. Huo, H. Ling, Q. Li, and B. Chen (2014). Mechanical properties and structural features of novel Fe-based bulk metallic glasses with unprecedented plasticity. *Sci. Rep.* 4: 6233.
84. Yao, J.H., J.Q. Wang, and Y. Li (2008). Ductile Fe–Nb–B bulk metallic glass with ultrahigh strength. *Appl. Phys. Lett.* 92: 251906-1–251906-3.
85. Amiya, K., A. Urata, N. Nishiyama, and A. Inoue (2004). Fe–B–Si–Nb bulk metallic glasses with high strength above 4000 MPa and distinct plastic elongation. *Mater. Trans.* 45: 1214–1218.
86. Liu, F.J., Q.W. Yang, S.J. Pang, C.L. Ma, and T. Zhang (2008). Ductile Fe-based BMGs with high glass forming ability and high strength. *Mater. Trans.* 49: 231–234.
87. Inoue, A., B.L. Shen, and C.T. Chang (2004). Super-high strength of over 4000 MPa for Fe-based bulk glassy alloys in $[(Fe_{1-x}Co_x)_{0.75}B_{0.2}Si_{0.05}]_{96}Nb_4$ system. *Acta Mater.* 52: 4093–4099.
88. Gu, X.J., S.J. Poon, and G.J. Shiflet (2007). Effects of carbon content on the mechanical properties of amorphous steel alloys. *Scr. Mater.* 57: 289–292.
89. Chen, D., A. Takeuchi, and A. Inoue (2007). Gd–Co–Al and Gd–Ni–Al bulk metallic glasses with high glass forming ability and good mechanical properties. *Mater. Sci. Eng. A* 457: 226–230.
90. Gu, X., Y.F. Zheng, S.P. Zhong, T.F. Xi, J.Q. Wang, and W.H. Wang (2010). Corrosion of, and cellular responses to Mg–Zn–Ca bulk metallic glasses. *Biomaterials* 31: 1093–1103.
91. Yuan, G.Y., K. Amiya, and A. Inoue (2005). Structural relaxation, glass-forming ability and mechanical properties of Mg–Cu–Ni–Gd alloys. *J. Non-Cryst. Solids* 351: 729–735.
92 . Xu, Y.K., H. Ma, J. Xu, and E. Ma (2005). Mg-based bulk metallic glass composites with plasticity and gigapascal strength. *Acta Mater.* 53: 1857–1866.
93. Zhang, T. and A. Inoue (2002). New bulk glassy Ni-based alloys with high strength of 3000 MPa. *Mater. Trans.* 43: 708–711.
94. Na, J.H., J.M. Park, K.H. Han, B.J. Park, W.T. Kim, and D.H. Kim (2006). The effect of Ta addition on the glass forming ability and mechanical properties of Ni–Zr–Nb–Al metallic glass alloys. *Mater. Sci. Eng. A* 431: 306–310.
95. Yao, K.F., Y.Q. Yang, and B. Chen (2007). Mechanical properties of Pd–Cu–Si bulk metallic glass. *Intermetallics* 15: 639–643.
96. Liu, L., A. Inoue, and T. Zhang (2005). Formation of bulk Pd–Cu–Si–P glass with good mechanical properties. *Mater. Trans.* 46: 376–378.
97. Schroers, J. and W.L. Johnson (2004). Ductile bulk metallic glass. *Phys. Rev. Lett.* 93: 255506-1–255506-4.
98. Ma, C.L., H. Soejima, S. Ishihara, K. Amiya, N. Nishiyama, and A. Inoue (2004). New Ti-based bulk glassy alloys with high glass-forming ability and superior mechanical properties. *Mater. Trans.* 45: 3223–3227.
99. Xie, G.Q., D.V. Louzguine-Luzgin, H.M. Kimura, and A. Inoue (2007). Ceramic particulate reinforced $Zr_{55}Cu_{30}Al_{10}Ni_5$ metallic glassy matrix composite fabricated by spark plasma sintering. *Mater. Trans.* 48: 1600–1604.
100. Conner, R.D., R.B. Dandliker, and W.L. Johnson (1998). Mechanical properties of tungsten and steel fiber reinforced $Zr_{41.25}Ti_{13.75}Cu_{12.5}Ni_{10}Be_{22.5}$ metallic glass matrix composites. *Acta Mater.* 46: 6089–6102.
101. Choi-Yim, H., R.D. Conner, F. Szuecs, and W.L. Johnson (2002). Processing, microstructure and properties of ductile metal particulate reinforced $Zr_{57}Nb_5Al_{10}Cu_{15.4}Ni_{12.6}$ bulk metallic glass composites. *Acta Mater.* 50: 2737–2745.
102. Johnson, W.L. and K. Samwer (2006). A universal criterion for plastic yielding of metallic glasses with a $(T/T_g)^{2/3}$ temperature dependence. *Phys. Rev. Lett.* 95: 195501-1–195501-4.
103. Chen, H.S. (1980). Glassy metals. *Rep. Prog. Phys.* 43: 353–432.
104. Yang, B., C.T. Liu, and T.G. Nieh (2006). Unified equation for the strength of bulk metallic glasses. *Appl. Phys. Lett.* 88: 221911-1–221911-3.
105. Liu, Y.H., C.T. Liu, W.H. Wang, A. Inoue, T. Sakurai, and M.W. Chen (2009). Thermodynamic origins of shear band formation and universal scaling law of metallic glass strength. *Phys. Rev. Lett.* 103: 065504-1–065504-4.

106. Inoue, A. (2000). Stabilization of metallic supercooled liquid and bulk amorphous alloys. *Acta Mater.* 48: 279–306.
107. Lewandowski, J.J., W.H. Wang, and A.L. Greer (2005). Intrinsic plasticity or brittleness of metallic glasses. *Philos. Mag. Lett.* 85: 77–87.
108. Pugh, S.F. (1954). Relations between the elastic moduli and plastic properties of polycrystalline pure metals. *Philos. Mag.* 45: 823–843.
109. Kelly, A., W.R. Tyson, and A.H. Cottrell (1967). Ductile and brittle crystals. *Philos. Mag.* 15: 567–586.
110. Cottrell, A.H. (1997). The art of simplification in materials science. *MRS Bull* 22: 15–19.
111. Chen H.S., J.T. Krause, and E. Coleman (1975). Elastic constants, hardness and their implications to flow properties of metallic glasses. *J. Non-Cryst. Solids* 18: 157–171.
112. Gu, X.J., A.-G. McDermott, S.J. Poon, and G.J. Shiflet (2006). Critical Poisson's ratio for plasticity in Fe–Mo–C–B–Ln bulk amorphous steel. *Appl. Phys. Lett.* 88: 211905-1–211905-3.
113. Liu, Y.H., G. Wang, R.J. Wang, D.Q. Zhao, M.X. Pan, and W.H. Wang (2007). Superplastic bulk metallic glasses at room temperature. *Science* 315: 1385–1388.
114. Yu, P. and H.Y. Bai (2008). Poisson's ratio and plasticity in CuZrAl bulk metallic glasses. *Mater. Sci. Eng. A* 485: 1–4.
115. Kawashima, A., Y.Q. Zeng, M. Fukuhara, H. Kurishita, N. Nishiyama, H. Miki, and A. Inoue (2008). Mechanical properties of a $Ni_{60}Pd_{20}P_{17}B_3$ bulk glassy alloy at cryogenic temperatures. *Mater. Sci. Eng. A* 498: 475–481.
116. Yokoyama, Y., K. Fujita, A.R. Yavari, and A. Inoue (2009). Malleable hypoeutectic Zr–Ni–Cu–Al bulk glassy alloys with tensile plastic elongation at room temperature. *Philos. Mag. Lett.* 89: 322–334.
117. Makino, A., X. Li, K. Yubuta, C.T. Chang, T. Kubota, and A. Inoue (2009). The effect of Cu on the plasticity of Fe–Si–B–P–based bulk metallic glass. *Scr. Mater.* 60: 277–280.
118. Méar, F.O., T. Wada, D.V. Louzguine-Luzgin, and A. Inoue (2009). Highly inhomogeneous compressive plasticity in nanocrystal-toughened Zr–Cu–Ni–Al bulk metallic glass. *Philos. Mag. Lett.* 89: 276–281.
119. Zhang, T., A.P. Tsai, A. Inoue, and T. Masumoto (1991). Production of amorphous alloy balloon. *Boundary* 7: 39–43.
120. Wu, F.F., Z.F. Zhang, S.X. Mao, and J. Eckert (2009). Effect of sample size on ductility of metallic glass. *Philos. Mag. Lett.* 89: 178–184.
121. Madge, S.V., D.V. Louzguine-Luzgin, J.J. Lewandowski, and A.L. Greer (2012). Toughness, extrinsic effects and Poisson's ratio of bulk metallic glasses. *Acta Mater.* 60: 4800–4809.
122. Yokoyama, Y., K. Fujita, A.R. Yavari, and A. Inoue (2009). Malleable hypoeutectic Zr-Ni-Cu-Al bulk glassy alloys with tensile plastic elongation at room temperature. *Philos. Mag. Lett.* 89: 322–334.
123. Inoue, A. and A. Takeuchi (2010). Recent development and applications of bulk glassy alloys. *Int. J. Appl. Glass Sci.* 1: 273–295.
124. Yoshida, N., K. Fujita, Y. Yokoyama, H. Kimura, and A. Inoue (2007). Effect of Zr composition and heat-treatment on fracture toughness in Zr-Cu-Al bulk metallic glass. *J. Jpn. Inst. Met.* 71: 730–735.
125. Kawashima, A., Y. Yokoyama, and A. Inoue (2010). Zr-based bulk glassy alloy with improved resistance to stress corrosion cracking in sodium chloride solutions. *Corros. Sci.* 52: 2950–2957.
126. Kumar, G., S. Prades-Rodel, A. Blatter, and J. Schroers (2011). Unusual brittle behavior of Pd-based bulk metallic glass. *Scr. Mater.* 65: 585–587.
127. Zhao, Y.Y., A. Inoue, C.T. Chang, J. Liu, B.L. Shen, X.M. Wang, and R.W. Li (2014). Composition effect on intrinsic plasticity or brittleness in metallic glasses. *Sci. Rep.* 4: 5733.
128. Mattern, N., A. Shariq, B. Schwarz, U. Vainio, and J. Eckert (2012). Structural and magnetic nanoclusters in $Cu_{50}Zr_{50-x}Gd_x$ (x = 5 at.%) metallic glasses. *Acta Mater.* 60: 1946–1956.
129. Inoue, A. and A. Takeuchi (2011). Recent development and application products of bulk glassy alloys. *Acta Mater.* 59: 2243–2267.

130. Yang, W., H. Liu, Y. Zhao, A. Inoue, K. Jiang, J. Huo, H. Ling, Q. Li, and B. Shen (2014). Mechanical properties and structural features of novel Fe-based bulk metallic glasses with unprecedented plasticity. *Sci. Rep.* 4: 06233.
131. Kong, F.L., C.T. Chang, A. Inoue, E. Shalaan, and F. Al-Marzouki (2014). Fe-based amorphous soft magnetic alloys with high saturation magnetization and good bending ductility. *J. Alloys Compd.* 615: 163–166.
132. Han, Y., F. Kong, C. Chang, S. Zhu, A. Inoue, E. -S. Shalaan, and F. Al-Marzouki (2015). Syntheses and corrosion behaviors of Fe-based amorphous soft magnetic alloys with high-saturation magnetization near 1.7 T. *J. Mater. Res.* 30: 547–555.
133. Han, Y., F.L. Kong, F.F. Han, A. Inoue, S.L. Zhu, E. Shalaan, and F. Al-Marzouki (2016). New Fe-based soft magnetic amorphous alloys with high saturation magnetization and good corrosion resistance for dust core application. *Intermetallics* 76 18–25.
134. Courtney, T.H. (2000). *Mechanical Behavior of Materials*, 2nd edn. New York: McGraw-Hill.
135. Suresh, S. (1998). *Fatigue of Materials*, 2nd edn. Cambridge, UK: Cambridge University Press.
136. Peter W.H., R.A. Buchanan, C.T. Liu, and P.K. Liaw (2003). The fatigue behavior of a zirconium-based bulk metallic glass in vacuum and air. *J. Non-Cryst. Solids* 317: 187–192.
137. Menzel, B.C. and R.H. Dauskardt (2006). The fatigue endurance limit of a Zr-based bulk metallic glass. *Scr. Mater.* 55: 601–604.
138. Wang, G.Y., J.D. Landes, A. Peker, and P.K. Liaw (2007). Comments on "The fatigue endurance limit of a Zr-based bulk metallic glass." *Scr. Mater.* 57: 65–68.
139. Menzel, B.C. and R.H. Dauskardt (2007). Response to comments on "The fatigue endurance limit of a Zr-based bulk metallic glass." *Scr. Mater.* 57: 69–71.
140. Qiao, D.C., G.J. Fan, P.K. Liaw, and H. Choo (2007). Fatigue behaviors of the $Cu_{47.5}Zr_{47.5}Al_5$ bulk-metallic glass (BMG) and $Cu_{47.5}Zr_{38}Hf_{9.5}Al_5$ BMG composite. *Int. J. Fatigue* 29: 2149–2154.
141. Freels, M., P.K. Liaw, G.Y. Wang, Q.S. Zhang, and Z.Q. Hu (2007). Stress-life fatigue behavior and fracture-surface morphology of a Cu-based bulk metallic glass. *J. Mater. Res.* 22: 374–381.
142. Qiao, D.C., G.Y. Wang, P.K. Liaw, V. Ponnambalam, S.J. Poon, and G.J. Shiflet (2007). Fatigue behavior of an $Fe_{48}Cr_{15}Mo_{14}Er_2C_{15}B_6$ amorphous steel. *J. Mater. Res.* 22: 544–550.
143. Launey, M.E., D.C. Hoffmann, W.L. Johnson, and R.O. Ritchie (2009). Solution to the problem of the poor cyclic fatigue resistance of bulk metallic glasses. *Proc. Natl. Acad. Sci. USA* 106: 4986–4991.
144. Gilbert, C.J., J.M. Lippmann, and R.O. Ritchie (1998). Fatigue of a Zr–Ti–Cu–Ni–Be bulk amorphous metal: Stress/life and crack-growth behavior. *Scr. Mater.* 38: 537–542.
145. Gilbert, C.J., V. Schroeder, and R.O. Ritchie (1999). Mechanisms for fracture and fatigue-crack propagation in a bulk metallic glass. *Metall. Mater. Trans. A* 30 A: 1739–1753.
146. Menzel, B.C. and R.H. Dauskardt (2006). Stress-life fatigue behavior of a Zr-based bulk metallic glass. *Acta Mater.* 54: 935–943.
147. Launey, M.E., R. Busch, and J.J. Kruzic (2006). Influence of structural relaxation on the fatigue behavior of a $Zr_{41.25}Ti_{13.75}Ni_{10}Cu_{12.5}Be_{22.5}$ bulk amorphous alloy. *Scr. Mater.* 54: 483–487.
148. Launey, M.E., R. Busch, and J.J. Kruzic (2008). Effects of free volume changes and residual stresses on the fatigue and fracture behavior of a Zr–Ti–Ni–Cu–Be bulk metallic glass. *Acta Mater.* 56: 500–510.
149. Qiao, D.C., P.K. Liaw, C. Fan, Y.H. Lin, G.Y. Wang, H. Choo, and R.A. Buchanan (2006). Fatigue and fracture behavior of $(Zr_{58}Ni_{13.6}Cu_{18}Al_{10.4})_{99}Nb_1$ bulk-amorphous alloy. *Intermetallics* 14: 1043–1050.
150. Morrison, M.L., R.A. Buchanan, P.K. Liaw, B.A. Green, G.Y. Wang, C.T. Liu, and J.A. Horton (2007). Four-point-bending-fatigue behavior of the Zr-based Vitreloy 105 bulk metallic glass. *Mater. Sci. Eng. A* 467: 190–197.
151. Flores, K.M., W.L. Johnson, and R.H. Dauskardt (2003). Fracture and fatigue behavior of a Zr–Ti–Nb ductile phase reinforced bulk metallic glass matrix composite. *Scr. Mater.* 49: 1181–1187.
152. Nakai, Y. and S. Hosomi (2007). Fatigue crack initiation and small-crack propagation in Zr-based bulk metallic glass. *Mater. Trans.* 48: 1770–1773.

153. Fujita, K., T. Hashimoto, W. Zhang, N. Nishiyama, C. Ma, H.M. Kimura, and A. Inoue (2008). Ultrahigh fatigue strength in Ti-based bulk metallic glass. *Rev. Adv. Mater. Sci.* 18: 137–139.
154. Wang, G.Y., P.K. Liaw, A. Peker, B. Yang, M.L. Benson, W. Yuan, W.H. Peter et al. (2005). Fatigue behavior of Zr–Ti–Ni–Cu–Be bulk metallic glass. *Intermetallics* 13: 429–435.
155. Wang, G.Y., P.K. Liaw, A. Peker, M. Freels, W.H. Peter, R.A. Buchanan, and C.R. Brooks (2006). Comparison of fatigue behavior of a bulk metallic glass and its composite. *Intermetallics* 14: 1091–1097.
156. Wang, G.Y., P.K. Liaw, Y. Yokoyama, A. Peker, W.H. Peter, B. Yang, M. Freels et al. (2007). Studying fatigue behavior and Poisson's ratio of bulk metallic glasses. *Intermetallics* 15: 663–667.
157. Wang, G.Y., P.K. Liaw, W.H. Peter, B. Yang, Y. Yokoyama, M.L. Benson, B.A. Green et al. (2004). Fatigue behavior of bulk metallic glasses. *Intermetallics* 12: 885–892.
158. Qiao, D.C., G.Y. Wang, W.H. Jiang, Y. Yokoyama, P.K. Liaw, and H. Choo (2007). Compression-compression fatigue and fracture behaviors of $Zr_{50}Al_{10}Cu_{37}Pd_3$ bulk metallic glass. *Mater. Trans.* 48: 1828–1833.
159. Keryvin, V., Y. Nadot, and Y. Yokoyama (2007). Fatigue pre-cracking and toughness of the $Zr_{55}Cu_{30}Al_{10}Ni_5$ bulk metallic glass for two oxygen levels. *Scr. Mater.* 57: 145–148.
160. Zhang, Z.F., J. Eckert, and L. Schultz (2003). Tensile and fatigue fracture mechanisms of a Zr-based bulk metallic glass. *J. Mater. Res.* 18: 456–465.
161. Peter, W.H., P.K. Liaw, R.A. Buchanan, C.T. Liu, C.R. Brooks, J.A. Horton, C.A. Carmichael, and J.L. Wright (2002). Fatigue behavior of $Zr_{52.5}Al_{10}Ti_5Cu_{17.9}Ni_{14.6}$ bulk metallic glass. *Intermetallics* 10: 1125–1129.
162. Yokoyama, Y., N. Nishiyama, K. Fukaura, H. Sunada, and A. Inoue (1999). Rotating-beam fatigue strength of $Pd_{40}Cu_{30}Ni_{10}P_{20}$ bulk amorphous alloy. *Mater. Trans., JIM* 40: 696–699.
163. Yokoyama, Y., K. Fukaura, and A. Inoue (2004). Effect of Ni addition on fatigue properties of bulk glassy $Zr_{50}Cu_{40}Al_{10}$ alloys. *Mater. Trans.* 45: 1672–1678.
164. Yokoyama, Y., P.K. Liaw, M. Nishijima, K. Hiraga, R.A. Buchanan, and A. Inoue (2006). Fatigue strength enhancement of cast $Zr_{50}Cu_{40}Al_{10}$ glassy alloys. *Mater. Trans.* 47: 1286–1293.
165. Brandes, E.A. and G.B. Brook, eds. (1992). *Smithells Metals Reference Book*, 7th edn. London, UK: Butterworth-Heinemann.
166. Wang, G.Y., P.K. Liaw, Y. Yokoyama, W.H. Peter, B. Yang, M. Freels, R.A. Buchanan, C.T. Liu, and C.R. Brooks (2007). Influence of air and vacuum environment on fatigue behavior of Zr-based bulk metallic glasses. *J. Alloys Compd.* 434–435: 68–70.
167. Schroeder, V. and R.O. Ritchie (2006). Stress-corrosion fatigue-crack growth in a Zr-based bulk amorphous metal. *Acta Mater.* 54: 1785–1794.
168. Morrison, M.L., R.A. Buchanan, P.K. Liaw, B.A. Green, G.Y. Wang, C.T. Liu, and J.A. Horton (2007). Corrosion-fatigue studies of the Zr-based Vitreloy 105 bulk metallic glass. *Mater. Sci. Eng. A* 467: 198–206.
169. Maruyama, N., K. Nakazawa, and T. Hanawa (2002). Fatigue properties of Zr-based bulk amorphous alloy in phosphate buffered saline solution. *Mater. Trans.* 43: 3118–3121.
170. Schroeder, V., C.J. Gilbert, and R.O. Ritchie (1999). Effect of aqueous environment on fatigue-crack propagation behavior in a Zr-based bulk amorphous metal. *Scr. Mater.* 40: 1057–1061.
171. Kawashima, A., H. Kurishita, H.M. Kimura, and A. Inoue (2007). Effect of chloride ion concentration on the fatigue crack growth rate of a $Zr_{55}Al_{10}Ni_5Cu_{30}$ bulk metallic glass. *Mater. Trans.* 48: 1969–1972.
172. Ritchie, R.O., V. Schroeder, and C.J. Gilbert (2000). Fracture, fatigue and environmentally-assisted failure of a Zr-based bulk amorphous metal. *Intermetallics* 8: 469–475.
173. Szuecs, F., C.P. Kim, and W.L. Johnson (2001). Mechanical properties of $Zr_{56.2}Ti_{13.8}Nb_{5.0}Cu_{6.9}Ni_{5.6}Be_{12.5}$ ductile phase reinforced bulk metallic glass composite. *Acta Mater.* 49: 1507–1513.
174. Murali, P. and U. Ramamurty (2005). Embrittlement of a bulk metallic glass due to sub-T_g annealing. *Acta Mater.* 53: 1467–1478.
175. Gilbert, C.J., R.O. Ritchie, and W.L. Johnson (1997). Fracture toughness and fatigue-crack propagation in a Zr–Ti–Cu–Ni–Be bulk metallic glass. *Appl. Phys. Lett.* 71: 476–478.

176. Yokoyama, Y., K. Fukaura, and H. Sunada (2000). Fatigue properties and microstructures of $Zr_{55}Cu_{30}Al_{10}Ni_5$ bulk glassy alloys. *Mater. Trans., JIM* 41: 675–680.
177. Hess, P.A., B.C. Menzel, and R.H. Dauskardt (2006). Fatigue damage in bulk metallic glass II: Experiments. *Scr. Mater.* 54: 355–361.
178. Fujita, K., A. Inoue, and T. Zhang (2000). Fatigue crack propagation in a nanocrystalline Zr-based bulk metallic glass. *Mater. Trans.* 41: 1448–1453.
179. Zhang, H., Z.G. Wang, K.Q. Qiu, Q.S. Zang, and H.F. Zhang (2003). Cyclic deformation and fatigue crack propagation of a Zr-based bulk amorphous metal. *Mater. Sci. Eng. A* 356: 173–180.
180. Schroeder, V., C.J. Gilbert, and R.O. Ritchie (2001). A comparison of the mechanisms of fatigue-crack propagation behavior in a Zr-based bulk amorphous metal in air and an aqueous chloride solution. *Mater. Sci. Eng. A* 317: 145–152.
181. Hess, P.A. and R.H. Dauskardt (2004). Mechanism of elevated temperature fatigue crack growth in Zr–Ti–Cu–Ni–Be bulk metallic glass. *Acta Mater.* 52: 3525–3533.
182. Cameron, K.K. and R.H. Dauskardt (2006). Fatigue damage in bulk metallic glass I: Simulation. *Scr. Mater.* 54: 349–353.
183. Li, J., Z.L. Wang, and T.C. Hufnagel (2002). Characterization of nanometer-scale defects in metallic glasses by quantitative high-resolution transmission electron microscopy. *Phys. Rev. B* 65: 144201-1–144201-6.
184. Li, J., F. Spaepen, and T.C. Hufnagel (2002). Nanometre-scale defects in shear bands in a metallic glass. *Philos. Mag. A* 82: 2623–2630.
185. Wright, W.J., T.C. Hufnagel, and W.D. Nix (2003). Free volume coalescence and void formation in shear bands in metallic glass. *J. Appl. Phys.* 93: 1432–1437.
186. Wang, G.Y., P.K. Liaw, Y. Yokoyama, A. Inoue, and C.T. Liu (2008). Fatigue behavior of Zr-based bulk metallic glasses. *Mater. Sci. Eng. A* 494: 314–323.
187. Zhang, Y., W.H. Wang, and A.L. Greer (2006). Making metallic glasses plastic by control of residual stress. *Nat. Mater.* 5: 857–860.
188. Raghavan, R., R. Ayer, H.W. Jin, C.N. Marzinsky, and U. Ramamurty (2008). Effect of shot peening on the fatigue life of a Zr-based bulk metallic glass. *Scr. Mater.* 59: 167–170.
189. Chiang, C.L., J.P. Chu, F.X. Liu, P.K. Liaw, and R.A. Buchanan (2006). A 200 nm thick glass-forming metallic film for fatigue-property enhancements. *Appl. Phys. Lett.* 88: 131902-1–131902-3.
190. Liu, F.X., P.K. Liaw, W.H. Jiang, C.L. Chiang, Y.F. Gao, Y.F. Guan, J.P. Chu, and P.D. Rack (2007). Fatigue-resistance enhancements by glass-forming metallic films. *Mater. Sci. Eng. A* 468–470: 246–252.
191. Li, J.C.M. and J.B.C. Wu (1976). Pressure and normal stress effects in shear yielding. *J. Mater. Sci.* 11: 445–457.
192. Vaidyanathan, R., M. Dao, G. Ravichandran, and S. Suresh (2001). Study of mechanical deformation in bulk metallic glass through instrumented indentation. *Acta Mater.* 49: 3781–3789.
193. Lewandowski, J.J. and P. Lowhaphandu (2002). Effect of hydrostatic pressure on the flow and fracture of a bulk amorphous metal. *Philos. Mag. A* 82: 3427–3441.
194. Jiang, F., D.H. Zhang, L.C. Zhang, Z.B. Zhang, L. He, J. Sun, and Z.F. Zhang (2007). Microstructure evolution and mechanical properties of $Cu_{46}Zr_{47}Al_7$ bulk metallic glass composite containing CuZr crystallizing phases. *Mater. Sci. Eng. A* 467: 139–145.
195. Sun, Y.F., C.H. Shek, S.K. Guan, B.C. Wei, and J.Y. Geng (2006). Formation, thermal stability and deformation behavior of graphite-flakes reinforced Cu-based bulk metallic glass matrix composites. *Mater. Sci. Eng. A* 435–436: 132–138.
196. Chen, L.Y., Z.D. Fu, W. Zeng, G.Q. Zhang, Y.W. Zeng, G.L. Xu, S.L. Zhang, and J.Z. Jiang (2007). Ultrahigh strength binary Ni– Nb bulk glassy alloy composite with good ductility. *J. Alloys Compd.* 443: 105–108.
197. Mukai, T., T.G. Nieh, Y. Kawamura, A. Inoue, and K. Higashi (2002). Effect of strain rate on compressive behavior of a $Pd_{40}Ni_{40}P_{20}$ bulk metallic glass. *Intermetallics* 10: 1071–1077.
198. Zhang, Z.F., G. He, J. Eckert, and L. Schultz (2003). Fracture mechanism in bulk metallic glassy materials. *Phys. Rev. Lett.* 91: 045505-1–045505-4.

199. He, G., Z.F. Zhang, W. Löser, J. Eckert, and L. Schultz (2003). Effect of Ta on glass formation, thermal stability and mechanical properties of a $Zr_{52.25}Cu_{28.5}Ni_{4.75}Al_{9.5}Ta_5$ bulk metallic glass. *Acta Mater.* 51: 2383–2395.
200. He, G., J. Lu, Z. Bian, D.J. Chen, G.L. Chen, G.C. Tu, and G.J. Chen (2001). Fracture morphology and quenched-in precipitates induced embrittlement in a Zr-base bulk glass. *Mater. Trans.* 42: 356–364.
201. Hasegawa, M., D. Nagata, T. Wada, and A. Inoue (2006). Preparation and mechanical properties of dispersed-ZrN glassy composite alloys containing pores. *Acta Mater.* 54: 3221–3226.
202. Lowhaphandu, P., L.A. Ludrosky, S.L. Montgomery, and J.J. Lewandowski (2000). Deformation and fracture toughness of a bulk amorphous Zr–Ti–Ni–Cu–Be alloy. *Intermetallics* 8: 487–492.
203. Noskova, N.I., N.F. Vildanova, Yu.I. Filippov, and A.P. Potapov (1985). Strength, plasticity, and fracture of ribbons of $Fe_5Co_{70}Si_{15}B_{10}$ amorphous alloy. *Phys. Stat. Sol. (a)* 87: 549–557.
204. Inoue, A., W. Zhang, T. Zhang, and K. Kurosaka (2001). High-strength Cu-based bulk glassy alloys in Cu–Zr–Ti and Cu–Hf–Ti ternary systems. *Acta Mater.* 49: 2645–2652.
205. Davis, L.A. and T.T. Yeow (1980). Flow and fracture of a Ni-Fe metallic glass. *J. Mater. Sci.* 15: 230–236.
206. Alpas, A.T., J. Edwards, and C.N. Reid (1989). Fracture and fatigue crack propagation in a nickel-base metallic glass. *Metall. Trans. A* 20: 1395–1409.
207. Takayama, S. (1979). Serrated plastic flow in metallic glasses. *Scr. Metall.* 13: 463–467.
208. Sergueeva, A.V., N.A. Mara, J.D. Kuntz, D.J. Branagan, and A.K. Mukherjee (2004). Shear band formation and ductility of metallic glasses. *Mater. Sci. Eng. A* 383: 219–223.
209. Zhang, Z.F., J. Eckert, and L. Schultz (2003). Differences in compressive and tensile fracture mechanisms of $Zr_{59}Cu_{20}Al_{10}Ni_8Ti_3$ bulk metallic glass. *Acta Mater* 51: 1167–1179.
210. Qiao, J.W., Y. Zhang, P. Feng, Q.M. Zhang, and G.L. Chen (2009). Strain rate response of mechanical behaviors for a Zr-based bulk metallic glass matrix composite. *Mater. Sci. Eng. A* 515: 141–145.
211. Inoue, A., H.M. Kimura, and T. Zhang (2000). High-strength aluminum- and zirconium-based alloys containing nanoquasicrystalline particles. *Mater. Sci. Eng. A* 294–296: 727–735.
212. Kawamura, T., T. Shibata, A. Inoue, and T. Masumoto (1996). Deformation behavior of $Zr_{65}Al_{10}Ni_{10}Cu_{15}$ glassy alloy with wide supercooled liquid region. *Appl. Phys. Lett.* 69: 1208–1210.
213. Wei, B.C., L.C. Zhang, T.H. Zhang, D.M. Xing, J. Das, and J. Eckert (2007). Strain rate dependence of plastic flow in Ce-based bulk metallic glass during nanoindentation. *J. Mater. Res.* 22: 258–263.
214. Sergueeva, A.V., N.A. Mara, D.J. Branaga, and A.K. Mukherjee (2004). Strain rate effect on metallic glass ductility. *Scr. Mater.* 50: 1303–1307.
215. Liu, L.F., L.H. Dai, Y.L. Bai, B.C. Wei, and G.S. Yu (2005). Strain rate-dependent compressive deformation behavior of Nd-based bulk metallic glass. *Intermetallics* 13: 827–832.
216. Ma, W.F., H.C. Kou, J.S. Li, H. Chang, and L. Zhou (2009). Effect of strain rate on compressive behavior of Ti-based bulk metallic glass at room temperature. *J. Alloys Compd.* 472: 214–218.
217. Zhang, J., J.M. Park, D.H. Kim, and H.S. Kim (2007). Effect of strain rate on compressive behavior of $Ti_{45}Zr_{16}Ni_9Cu_{10}Be_{20}$ bulk metallic glass. *Mater. Sci. Eng. A* 449–451: 290–294.
218. Ghatu, S., R.J. Dowding, and L.J. Kecskes (2002). Characterization of uniaxial compressive response of bulk amorphous Zr–Ti–Cu–Ni–Be alloy. *Mater. Sci. Eng. A* 334: 33–40.
219. Kawamura, T., T. Shibata, A. Inoue, and T. Masumoto (1997). Superplastic deformation of $Zr_{65}Al_{10}Ni_{10}Cu_{15}$ metallic glass. *Scr. Mater.* 37: 431–436.
220. Schuh, C.A., T.G. Nieh, and Y. Kawamura (2002). Rate dependence of serrated flow during nanoindentation of a bulk metallic glass. *J. Mater. Res.* 17: 1651–1654.
221. Lee, M.L., Y. Li, and C.A. Schuh (2004). Effect of a controlled volume fraction of dendritic phases on tensile and compressive ductility in La-based metallic glass matrix composites. *Acta Mater.* 52: 4121–4131.
222. Kusy, M., U. Kühn, A. Concustell, A. Gebert, J. Das, J. Eckert, L. Schultz, and M.D. Baro (2006). Fracture surface morphology of compressed bulk metallic glass-matrix-composites and bulk metallic glass. *Intermetallics* 14: 982–986.
223. Cytron, S.J. (1982). A metallic glass-metal matrix composite. *J. Mater. Sci. Lett.* 1: 211–213.

224. Long, Y. and T.H. Courtney (1990). Fracture behavior of laminated metal metallic glass composites. *Metall. Mater. Trans. A* 21: 2159–2168.
225. Hays, C.C., C.P. Kim, and W.L. Johnson (2000). Microstructure controlled shear band pattern formation and enhanced plasticity of bulk metallic glasses containing *in situ* formed ductile phase dendrite dispersions. *Phys. Rev. Lett.* 84: 2901–2904.
226. Huang, Y.L., A. Bracchi., T. Niermann, M. Seibt, D. Danilov, B. Nestler, and S. Schneider (2005). Dendritic microstructure in the metallic glass matrix composite $Zr_{56}Ti_{14}Nb5Cu_7Ni_6Be_{12}$. *Scr. Mater.* 53: 93–97.
227. Wang, M.L., G.L. Chen, X. Hui, Y. Zhang, and Z.Y. Bai (2007). Optimized interface and mechanical properties of W fiber/Zr-based bulk metallic glass composites by minor Nb addition. *Intermetallics* 15: 1309–1315.
228. Eckert, J., J. Das, S. Pauly, and C. Duhamel (2007). Mechanical properties of bulk metallic glasses and composites. *J. Mater. Res.* 22: 285–301.
229. Bian, Z., H. Kato, C.L. Qin, W. Zhang, and A. Inoue (2005). Cu–Hf–Ti–Ag–Ta bulk metallic glass composites and their properties. *Acta Mater.* 53: 2037–2048.
230. Kim, Y.C., D.H. Kim, and J.C. Lee (2003). Formation of ductile Cu-based bulk metallic glass matrix composite by Ta addition. *Mater. Trans.* 44: 2224–2227.
231. Shen, B.L., H. Men, and A. Inoue (2006). Fe-based bulk glassy alloy composite containing *in-situ* formed α-(Fe,Co) and $(Fe,Co)_{23}B_6$ microcrystalline grains. *Appl. Phys. Lett.* 89: 101915-1–101915-3.
232. Pan, D.G., H.F. Zhang, A.M. Wang, and Z.Q. Hu (2006). Enhanced plasticity in Mg-based bulk metallic glass composite reinforced with ductile Nb particles. *Appl. Phys. Lett.* 89: 261904-1–261904-3.
233. Yamamoto, T., H. Ito, M. Hasegawa, and A. Inoue (2007). Mechanical properties and microstructures of composites of Ti-based metallic glass and β-Ti. *Mater. Trans.* 48: 1812–1815.
234. Okazaki, K., W. Zhang, and A. Inoue (2006). Microstructure and mechanical properties of $(Zr_{0.5}Cu_{0.4}Al_{0.1})_{100-x}Ta_x$ bulk metallic glass composites. *Mater. Trans.* 47: 2571–2575.
235. Zhang, Q.S., W. Zhang, G.Q. Xie, and A. Inoue (2007). Unusual plasticity of the particulate-reinforced Cu–Zr-based bulk metallic glass composites. *Mater. Trans.* 48: 2542–2544.
236. Choi-Yim, H., S.B. Lee, and R.D. Conner (2008). Mechanical behavior of Mo and Ta wire-reinforced bulk metallic glass composites. *Scr. Mater.* 58: 763–766.
237. Siegrist, M.E. and J.F. Löffler (2007). Bulk metallic glass-graphite composites. *Scr. Mater.* 56: 1079–1082.
238. Sun, G.Y., G. Chen, and G.L. Chen (2007). Comparison of microstructures and properties of Zr-based bulk metallic glass composites with dendritic and spherical bcc phase precipitates. *Intermetallics* 15: 632–634.
239. Xue, Y.F., H.N. Cai, L. Wang, F.C. Wang, and H.F. Zhang (2007). Strength-improved Zr-based metallic glass/porous tungsten phase composite by hydrostatic extrusion. *Appl. Phys. Lett.* 90: 081901-1–081901-3.
240. Hoffmann, D.C., J.Y. Suh, A. Wiest, G. Duan, M.L. Lind, M.D. Demetriou, and W.L. Johnson (2008). Designing metallic glass matrix composites with high toughness and tensile ductility. *Nature* 451: 1085–1090.
241. Ma, H., E. Ma, and J. Xu (2003). A new $Mg_{65}Cu_{7.5}Ni_{7.5}Zn_5Ag_5Y_{10}$ bulk metallic glass with strong glass-forming ability. *J. Mater. Res.* 18: 2288–2291.
242. Sun, G.Y., G. Chen, C.T. Liu, and G.L. Chen (2006). Innovative processing and property improvement of metallic glass based composites. *Scr. Mater.* 55: 375–378.
243. Dragoi, D., E. Üstündag, B. Clausen, and M.A.M. Bourke (2001). Investigation of thermal residual stresses in tungsten-fiber/bulk metallic glass matrix composites. *Scr. Mater* 45: 245–252.
244. Balch, D.K., E. Üstündag, and D.C. Dunand (2003). Elasto-plastic load transfer in bulk metallic glass composites containing ductile particles. *Metall. Mater. Trans. A* 34A: 1787–1797.
245. Hufnagel, T.C., C. Fan, R.T. Ott, J. Li, and S. Brennan (2002). Controlling shear band behavior in metallic glasses through microstructural design. *Intermetallics* 10: 1163–1166.
246. Donovan, P.E. and W.M. Stobbs (1983). Shear band interactions with crystals in partially crystallized metallic glasses. *J. Non-Cryst. Solids* 55: 61–76.

247. Liu, Z., R. Li, G. Liu, K. Song, S. Pauly, T. Zhang, and J. Eckert (2012). Pronounced ductility in CuZrAl ternary bulk metallic glass composites with optimized microstructure through melt adjustment, *AIP Advances* 2: 032176.
248. Song, K.K., S. Pauly, Y. Zhang, R. Li, S. Gorantla, N. Narayanan, U. Kühn, T. Gemming, and J. Eckert (2012). Triple yielding and deformation mechanisms in metastable $Cu_{47.5}Zr_{47.5}Al_5$ composites. *Acta Mater.* 60: 6000–6012.
249. Tsarkov, A.A., A.Y. Churyumov, V.Y. Zadorozhnyy, and D.V. Louzguine-Luzgin (2016). High-strength and ductile (Ti–Ni)-(Cu–Zr) crystalline/amorphous composite materials with superelasticity and TRIP effect. *J. Alloys Compd.* 658: 402–407.
250. Madge, S.V., D.V. Louzguine-Luzgin, A. Inoue, and A.L. Greer (2012). Large compressive plasticity in a La-based glass-crystal composite. *Metals* 3: 41–48.
251. Qin, C., W. Zhang, H. Kimura, and A. Inoue (2004). Excellent mechanical properties of Cu-Hf-Ti-Ta bulk glassy alloys containing in-situ dendrite Ta-based bcc phase. *Mater. Trans.* 45: 2936–2940.
252. Saida, J. and A. Inoue (2002). Icosahedral quasicrystalline phase formation in Zr–Al–Ni–Cu glassy alloys by the addition of V, Nb and Ta. *J. Non-Cryst. Solids* 312–314: 502–507.
253. Dong, W., H. Zhang, W. Sun, A. Wang, H. Li, and Z. Hu (2006). Zr–Cu–Ni–Al–Ta glassy matrix composites with enhanced plasticity. *J. Mater. Res.* 21: 1490–1499.
254. He, G., J. Eckert, W. Loser, and L. Schultz (2003). Novel Ti-base nanostructure-dendrite composite with enhanced plasticity. *Nat. Mater.* 2: 33–37.
255. Choi-Yim, H., R.D. Conner, and W.L. Johnson (2005). In situ composite formation in the Ni–(Cu)–Ti–Zr–Si system. *Scr. Mater.* 53: 1467–1470.
256. Wu, F.-F., K.C. Chan, S.-S. Jiang, S.-H. Chen, and G. Wang (2014). Bulk metallic glass composite with good tensile ductility, high strength and large elastic strain limit. *Sci. Rep.* 4: 5302.
257. Kozachkov, H., J. Kolodziejska, W.L. Johnson, and D.C. Hofmann (2013). Effect of cooling rate on the volume fraction of B2 phases in a CuZrAlCo metallic glass matrix composite. *Intermetallics* 39: 89–93.
258. Kim, K.B., S. Yi, H. Choi-Yim, J. Das, W.L. Johnson, and J. Eckert (2005). Interfacial instability-driven amorphization/nanocrystallization in a bulk $Ni_{45}Cu_5Ti_{33}Zr_{16}Si_1$ alloy during solidification. *Phys. Rev. B* 72: 092102-1–092102-4.
259. Qin, C., W. Zhang, K. Asami, N. Ohtsu, and A. Inoue (2005). Glass formation, corrosion behavior and mechanical properties of bulk glassy Cu–Hf–Ti–Nb alloys. *Acta Mater.* 53: 3903–3911.
260. Qin, F., X. Wang, and A. Inoue (2007). Effects of Ta on microstructure and mechanical property of Ti–Zr–Cu–Pd–Ta alloys. *Mater. Trans.* 48: 2390–2394.
261. Qin, F.X., X.M. Wang, G.Q. Xie, and A. Inoue (2008). Distinct plastic strain of Ni-free Ti–Zr–Cu–Pd–Nb bulk metallic glasses with potential for biomedical applications, *Intermetallics* 16: 1026–1030.
262. Sun, Y.F., B.C. Wei, Y.R. Wang, W.H. Li, T.L. Cheung, and C.H. Shek (2005). Plasticity-improved Zr–Cu–Al bulk metallic glass matrix composites containing martensite phase, *Appl. Phys. Lett.* 87: 051905-1--051905-3.
263. Inoue, A., T. Zhang, M.W. Chen, T. Sakurai, J. Saida, and M. Matsushita (2000). Formation and properties of Zr-based bulk quasicrystalline alloys with high strength and good ductility. *J. Mater. Res.* 15: 2195–2208.
264. Inoue, A., T. Zhang, M.W. Chen, and T. Sakurai (1999). Mechanical properties of bulk amorphous Zr–Al–Cu–Ni–Ag alloys containing nanoscale quasicrystalline particles. *Mater. Trans., JIM* 40: 1382–1389.
265. Inoue, A., T. Zhang, J. Saida, M. Matsushita, M.W. Chen, and T. Sakurai (1999). High strength and good ductility of bulk quasicrystalline base alloys in $Zr_{65}Al_{7.5}Ni_{10}Cu_{17.5-x}Pd_x$ system. *Mater. Trans., JIM* 40: 1137–1143.
266. Inoue, A., T. Zhang, N. Nishiyama, K. Ohba, and T. Masumoto (1993). Preparation of 16 mm diameter rod of amorphous $Zr_{65}Al_{7.5}Ni_{10}Cu_{17.5}$ alloy. *Mater. Trans., JIM* 34: 1234–1237.
267. Inoue, A. (1995). High strength bulk amorphous alloys with low critical cooling rates. *Mater. Trans., JIM* 36: 866–875.

268. Inoue, A., T. Zhang, J. Saida, M. Matsushita, M.W. Chen, and T. Sakurai (1999). Formation of icosahedral quasicrystalline phase in Zr–Al–Ni–Cu–M (M = Ag, Pd, Au or Pt) systems. *Mater. Trans., JIM* 40: 1181–1184.
269. Inoue, A., T. Zhang, J. Saida, and M. Matsushita (2000). Enhancement of strength and ductility in Zr-based bulk amorphous alloys by precipitation of quasicrystalline phase. *Mater. Trans., JIM* 41: 1511–1520.
270. Yamada, M., T. Yamasaki, K. Fujita, Y. Yokoyama, and D.H. Kim (2014). Effects of Au-addition on plastic deformation ability of Zr–Cu–Ni–Al bulk metallic glasses. *J. Japan Inst. Met. Mater.* 78: 449–458.
271. Kato, H., J. Saida, and A. Inoue (2004). Influence of hydrostatic pressure during casting on as cast structure and mechanical properties in $Zr_{65}Al_{7.5}Ni_{10}Cu_{17.5-3x}Pd_x$ (x = 0, 17.5) alloys. *Scr. Mater.* 51: 1063–1068.
272. Park, E., and D. Kim (2005). Design of bulk metallic glasses with high glass forming ability and enhancement of plasticity in metallic glass matrix composites: A review. *Met. Mater. Internat.* 11: 19–27.
273. Saida, J., H. Kato, A.D.H. Setyawan, K. Yoshimi, and A. Inoue (2007). Deformation-induced nanoscale dynamic transformation studies in Zr–Al–Ni–Pd and Zr–Al–Ni–Cu bulk metallic glasses. *Mater. Trans.* 48: 1327–1335.
274. Yamasaki, T., M. Yamada, K. Fujita, H. Kato, and D. Hyang Kim (2016). Effects of noble metal additions on plastic deformation of Zr–Cu–Ni–Al based bulk metallic glasses. *J. Japan Soc. Powder and Powder Metall.* 63: 230–238.

9

Magnetic Properties

9.1 Introduction

The physical and mechanical properties and the chemical behavior of bulk metallic glasses (BMGs) have been described in the previous chapters. It has been shown that these BMGs have excellent combinations of properties and, based on these, some applications have been suggested. These will be described in Chapter 10. But, the most important application to which the melt-spun glassy ribbons have been put is in transformer core laminations, based on the excellent soft magnetic properties of these alloys. Therefore, a significant amount of effort has also been spent on investigating the magnetic properties of BMG alloys.

Magnetic properties are of fundamental importance for several applications in the electrical and electronic industries. Since the basic phenomena and terminology have been described in several physics textbooks, these will not be repeated here. The interested reader could refer to Cullity and Graham [1] for a good introduction to the different aspects of magnetic materials.

Inoue and Gook synthesized the first Fe-based BMG alloy (in the form of melt-spun ribbons) in 1995 in the $Fe_{72}Al_5Ga_2P_{11}C_6B_4$ system [2]. This alloy had a wide supercooled liquid region of 61 K. This discovery was immediately followed by the synthesis of another Fe-based BMG, *viz.*, the $Fe_{72}Al_5Ge_2P_{11}C_6B_4$ alloy [3]. Many other Fe-based BMGs were subsequently synthesized both by researchers in Inoue's group and others [4]. Currently, the largest diameter of an Fe-based BMG is 16 mm in an $Fe_{41}Co_7Cr_{15}Mo_{14}C_{15}B_6Y_2$ alloy produced by the drop casting technique [5].

The magnetic properties of BMGs have been investigated mostly in Fe-based BMGs, and a few investigations have also been reported on Co-based BMGs. The hard magnetic properties of Nd- and Pr-based alloys have also been studied.

The nature of magnetic investigations in BMG alloys has followed trends very similar to those in the case of melt-spun glassy ribbons. In fact, even for BMG compositions, several researchers have been studying the magnetic behavior using melt-spun ribbons. Some minor differences were noted in the magnetic properties of melt-spun ribbons and bulk rods, especially in those properties that are affected by structural relaxation, for example, magnetostriction and coercivity. It has been shown that the saturation magnetization of the alloys is not any different whether measured on the melt-spun ribbon form or powder [6] or between the melt-spun ribbon and bulk rods of different diameters [7]. This is illustrated in the hysteresis loops presented in Figure 9.1 for $Fe_{65}Co_{10}Ga_5P_{12}C_4B_4$ and $Fe_{62.8}Co_{10}B_{13.5}Si_{10}Nb_3Cu_{0.7}$ alloys, respectively.

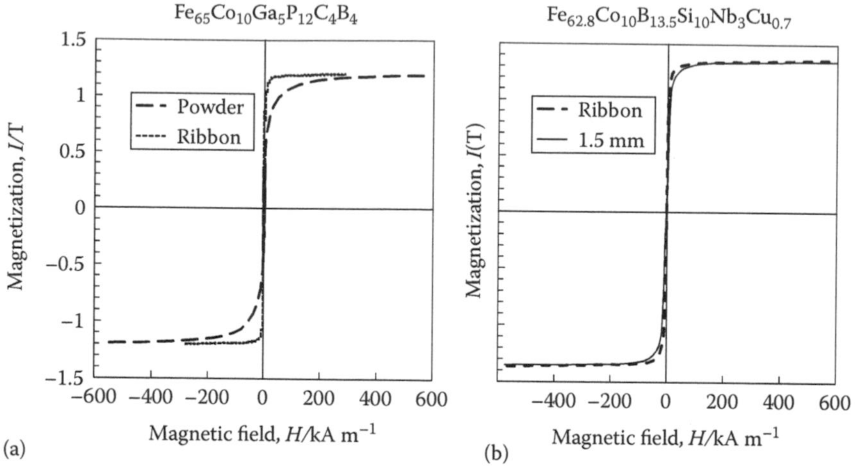

FIGURE 9.1
Hysteresis loops for the glassy (a) $Fe_{65}Co_{10}Ga_5P_{12}C_4B_4$ alloy in the melt-spun ribbon and gas-atomized powder conditions. (Reprinted from Shen, B.L. and Inoue, A., *J. Mater. Res.*, 18, 2115, 2003. With permission.) (b) $Fe_{62.8}Co_{10}B_{13.5}Si_{10}Nb_3Cu_{0.7}$ alloy in the melt-spun ribbon of 20 μm thickness and bulk rod of 1.5 mm diameter. (Reprinted from Shen, B.L. and Inoue, A., *J. Mater. Res.*, 19, 2549, 2004. With permission.)

9.2 Soft Magnetic Materials

The most desirable soft magnetic properties of the material include high saturation magnetization, low coercivity, and high electrical resistivity. Coupled with these, good corrosion resistance and high mechanical strength will be useful. Since we are dealing with BMGs, the glass-forming ability (GFA) of the alloy should also be high. But, it will be realized that it is not always easy to achieve a good combination of all the desired features in one alloy composition, and therefore some sacrifices will have to be made. For example, alloying additions may be made to improve the GFA, but the saturation magnetization will come down. On the other hand, if we wish to attain the highest saturation magnetization, then the alloying elements will have to be maintained at a low level, and in this case, glass formation may not always be easy or, in some cases, only partial amorphization may occur.

The magnetic behavior of BMGs has been studied mostly in Fe-based BMGs. A very large number of studies have also been conducted on Fe-based melt-spun ribbons starting from the pioneering investigation of Duwez and Lin [8] on the Fe–C–P system in 1967. However, an important difference between the investigations on melt-spun ribbons and BMGs is that while both metal–metalloid- and metal–metal-type alloys have been investigated in the thin film category, only the metal–metalloid type of alloys have been studied in the BMG group. Studies on the magnetic properties of BMG alloys in the metal–metal-type category are conspicuous by their absence.

9.2.1 Effect of Alloying Elements

Different alloying elements have been tried to improve the GFA of alloys so that they could be fabricated into large-diameter rods. Co is one such element added from 0 to 20 at.% to Fe-based $Fe_{70-x}Co_xHf_5Mo_7B_{15}Y_3$ BMG alloys prepared from commercially pure raw materials. Three millimeter diameter rods of alloys of all compositions were cast by the copper mold casting technique. It was shown that while the Co-free alloy contained mostly the glassy phase, it also contained a small volume fraction of the $Fe_{23}B_6$ crystalline phase. But, the GFA of the alloy increased with the addition of Co and a fully glassy phase was obtained up to 12 at.% Co additions. However, when the Co addition was increased to 20 at.%, a crystalline phase (most probably the α-(Fe, Co) solid solution) appeared. From these observations, the authors [9] concluded that the addition of Co increased the GFA of the alloy in a limited composition range. The magnetic properties of the alloys, however, showed a mixed trend. In comparison to the glass+crystal composite, the fully glassy alloys showed a low coercivity (H_c) value in the range of 1–4 Oe. The saturation magnetization (I_s) decreased initially with the addition of Co, up to about 8 at.%, and then started to increase on further additions up to 20 at.%. Thus, the glass+crystal composites showed higher values of both saturation magnetization and coercivity, suggesting that fully glassy alloys show good soft magnetic properties, albeit with a slight reduction in H_c.

The addition of Co to Fe-based magnetic alloys is expected to decrease the I_s value due to the smaller magnetic moment of Co. This is why the I_s value decreased initially with Co addition. But, the I_s value increased at higher Co contents, presumably because of the chemical short-range order (CSRO) in the glassy phase [10]. With increasing Co addition, the CSRO was expected to increase and at 20 at.% Co, the I_s was the highest because of the precipitation of the α-(Fe, Co) solid solution phase. The highest I_s value observed in the Co-free alloy and the alloy with 20 at.% Co were explained on the basis that both the alloys contained strongly ferromagnetic phases, *viz.*, $Fe_{23}B_6$ and α-(Fe, Co), respectively. A similar phenomenon was also reported in the FeCoNiZrMoB system [11].

It was reported earlier that the addition of Nb to Fe- and Co-based alloys improves their GFA [12,13]. Since Fe–B–Si alloys have been known to exhibit good soft magnetic properties, about 4 at.% Nb was added to the $(Fe_{1-x}Ni_x)_{0.75}B_{0.2}Si_{0.05}$ system [14]. While the GFA of the alloy certainly improved, the saturation magnetization, I_s, came down from 1.1 to 0.8 T. But, the coercivity, H_c, decreased from 2.3 to 1.2 A m^{-1} and the permeability, μ_e, increased from 16,000 to 24,000 as the Ni content increased from $x=0.1$–0.4. This was attributed to the high degree of amorphicity and structural homogeneity in the glassy alloy due to the improved GFA by the addition of Ni.

On the other hand, the addition of Fe to Co-based BMG alloys was shown to improve not only the GFA (as evidenced by the increase in the critical diameter for glass formation from 2 to 4 mm), but also the magnetic properties. With increasing Fe content from $x=0.1$–0.4 in the $[(Co_{1-x}Fe_x)_{0.75}B_{0.2}Si_{0.05}]_{96}Nb_4$ alloys, the I_s increased from 0.71 to 0.97 T. The H_c value increased slightly from 0.7 to 1.8 A m^{-1} and the μ_e value decreased from 32,500 to 14,800. The magnetostriction, λ_s, increased from 0.55×10^{-6} to 5.76×10^{-6}. Similar improvements in GFA with a slight effect on the soft magnetic properties have also been reported on the addition of Mo to $Fe_{79}P_{10}C_4B_4Si_3$ [15] and Cr to $Fe_{76}Mo_2Ga_2P_{10}C_4B_4Si_2$ [16]. It has been suggested that the large fractions of metalloid and nonmagnetic refractory elements present could also be responsible for the low saturation magnetization observed in the multicomponent alloys [17].

Thus, from this description it is clear that alloying additions could be so chosen as to improve the GFA of the alloy such that large bulk rods could be obtained. But, the nature and amount of the alloying addition will decide the magnitude of change in the magnetic properties. The maximum possible amount of Fe is required to achieve the highest saturation magnetization value.

Table 9.1 lists the magnetic properties of some select Co- and Fe-based glassy alloys including both melt-spun ribbons and BMG rods.

TABLE 9.1

Magnetic Properties of Selected BMG Alloys

Alloy	Form	Thickness/Size (mm)	I_s (T)	H_c (A m^{-1})	μ_e	λ_s (10^{-6})	T_c (K)	References
$Co_{43}Fe_{20}Ta_{5.5}B_{31.5}$	Ribbon	0.02	0.51	0.9	40,000	—	—	[18]
$Co_{43}Fe_{20}Ta_{5.5}B_{31.5}$	Rod	2	0.49	—	—	—	—	[18]
$Co_{43}Fe_{20}Ta_{5.5}B_{31.5}$	Rod	3	0.49	0.25	550,000	—	—	[19]
$(Co_{0.705}Fe_{0.045}Si_{0.1}B_{0.15})_{96}Nb_4$	Rod	1	0.59	<3	—	—	—	[20]
$[(Co_{0.6}Fe_{0.4})_{0.75}B_{0.2}Si_{0.05}]_{96}Nb_4$	Ribbon	0.02	—	—	—	—	—	[20]
$Fe_{30}Co_{30}Ni_{15}Si_8B_{17}$	Rod	1.2	0.92	3	—	—	—	[21]
$Fe_{37.5}Ni_{22.5}Cr_5Co_{10}B_{15}Si_{10}$	Ribbon	20–25	0.89	3720	—	7.6	—	[22]
$Fe_{40}Co_{40}Cu_{0.5}Al_2Zr_9Si_4B_{4.5}$	Ribbon	—	1.18	3	—	—	736	[23]
$Fe_{52}Co_{9.5}Nd_3Dy_{0.5}B_{35}$	Ribbon	—	1.26	1	—	15.7	—	[24]
$Fe_{56}Co_7Ni_7Zr_{10}B_{20}$	Ribbon	—	0.92	—	5,100	—	567	[25]
$Fe_{56}Co_7Ni_7Zr_{10}B_{20}$	Ribbon	0.015	0.92	5.2	—	—	—	[26]
$Fe_{62}Nb_8B_{30}$	Ribbon	0.035	0.75	—	—	9.8	516	[27]
$Fe_{62}Co_{9.5}Gd_{3.5}Si_{10}B_{15}$	Ribbon		0.98	5	—	—	596	[23]
$Fe_{62}Co_{9.5}Nd_3Dy_{0.5}B_{25}$	Ribbon		1.37	4	—	17.9	—	[24]
$(Fe_{0.75}B_{0.15}Si_{0.1})_{96}Nb_4$	Rod	1.5	1.2	3.7	9,600	—	—	[28]
$[(Fe_{0.8}Co_{0.1}Ni_{0.1})_{0.75}B_{0.2}Si_{0.05}]_{96}Nb_4$	Rod	2.5	1.1	3	16,000	—	—	[28]
$[(Fe_{0.6}Co_{0.1}Ni_{0.3})_{0.75}B_{0.2}Si_{0.05}]_{96}Nb_4$	Rod	3	0.8	2.5	19,000	—	—	[28]
$[(Fe_{0.6}Co_{0.2}Ni_{0.2})_{0.75}B_{0.2}Si_{0.05}]_{96}Nb_4$	Rod	4	0.86	2.5	19,000	—	—	[28]
$[(Fe_{0.6}Co_{0.3}Ni_{0.1})_{0.75}B_{0.2}Si_{0.05}]_{96}Nb_4$	Rod	4	0.9	2	21,000	—	—	[28]
$[(Fe_{0.8}Co_{0.1}Ni_{0.1})_{0.75}B_{0.2}Si_{0.05}]_{96}Nb_4$	Rod	2.5	1.1	3	18,000	—	—	[29]
$Fe_{76}Al_4P_{12}B_4Si_4$	Ribbon	<0.025	1.38	3	—	—	623	[30]
$Fe_{76}Si_9B_{10}P_5$	Rod	2.5	1.51	1	—	—	—	[21]
$Fe_{76}Mo_2Ga_2P_{10}C_4B_4Si2$	Rod	2	1.32	3	—	—	—	[21]
$Fe_{76}P_5(Si_{0.3}B_{0.5}C_{0.2})_{19}$	Rod	3	1.44	1	17,000	—	680	[31]
$(Fe_{0.85}Co_{0.15})_{77}Ga_2P_{10}C_5B_{3.5}Si_{2.5}$	Rod	3	1.4	5	—	—	694	[32]
$Fe_{78}Ga_2P_{12}C_4B_4$	Ribbon	0.02	1.34	2	—	—	—	[33]
$Fe_{78}Ga_2P_{9.5}C_4B_4Si_{2.5}$	Rod	2	1.4	3	—	—	625	[34]
$Fe_{70}Al_5Ga_2P_{9.65}C_{5.75}B_{4.6}Si_3$	Ring	1	1.19	2.2	110,000	21	—	[35,36]
$Fe_{77}Al_{2.14}Ga_{0.86}P_{8.4}C_5B_4Si_{2.6}$	Ribbon	0.022	1.5	3	4,400	—	—	[37]
$Fe_{40}Co_{20}Ni_{15}P_{10}C_{10}B_5$	Rod	2.5	1.13	3.8	—	—	—	[38]
$Fe_{71.4}Si_4B_{20}Nb_4Cu_{0.6}$	Rod	1.5	1.24	1.4	24,700	—	—	[39]
$(Fe_{0.3}Co_{0.6}Ni_{0.1})_{68}(B_{0.811}Si_{0.189})_{27}Nb_5$	Rod	5	0.65	0.7	22,400	—	—	[40]

9.2.2 Effect of Annealing

Annealing of the fully glassy alloy, especially in the temperature range between the Curie temperature, T_c, and the crystallization temperature, T_x, has been shown to significantly enhance the soft magnetic properties of the alloys. In fact, the development of the FINEMET alloys in the late 1980s was based on this concept [41]. Annealing of the Fe–Si–B–M glassy alloys containing small amounts of up to 1.5 at.% of M (=Cu, Nb, Mo, W, Ta, etc.) was shown to result in the precipitation of fine (10 nm in size) α-Fe(Si, B) crystalline particles. Since Cu and Fe have a positive heat of mixing, they segregate to form Fe-rich, Cu-rich, and Nb-rich regions. Fe-rich regions become the nuclei for the α-Fe solid solution. Since the Cu-rich and Nb-rich regions have a higher crystallization temperature, they do not crystallize. Therefore, the microstructure consists of fine α-Fe grains dispersed in a glassy matrix. It has been found that these FINEMET alloys show low coercivity, low saturation magnetostriction, and low core losses in comparison to the fully glassy alloys.

Annealing of the fully glassy BMG alloys has also been carried out to enhance the magnetic properties of these alloys. Since the FINEMET alloys contain the α-Fe solid solution phase dispersed in the glassy matrix, it is important that the BMG alloys exhibit at least two stages of crystallization to achieve the desired microstructure. This means that BMGs showing a single stage of crystallization cannot be annealed to produce such a microstructure. For example, Inoue et al. [42] added a small amount of Cu and slightly reduced the B content in the Fe–Si–B–Nb alloy to a composition of $Fe_{72.5}Si_{10}B_{12.5}Nb_4Cu_1$ to achieve a change in the crystallization mode. They observed that while the Cu-free alloy showed essentially one crystallization peak, the Cu-containing alloy showed multiple crystallization peaks. Consequently, on annealing the glassy alloy at a temperature beyond the first crystallization peak, the bcc α-Fe solid solution precipitated out from the glassy matrix. The magnetic properties of such magnetic alloys are much better than those of the fully glassy alloy, and they are also comparable to those of FINEMET and NANOPERM alloys. These results are presented in Table 9.2.

Similar types of investigations have been reported in different Fe-based alloys. Chang et al. [14] reported that crystallization in the $(Fe_{0.75}B_{0.2}Si_{0.05})_{96}Nb_4$ glassy alloy took place in one stage (Figure 9.2a). Shen and Inoue [7] reported that on adding 0.7 at.% Cu to an Fe–Co–Si–B–Nb alloy, crystallization took place in two stages (Figure 9.2b). Such a situation is useful in obtaining a fine dispersion of the α-Fe solid solution in a glassy phase on annealing the glassy alloy at a temperature above the first exothermic peak. Figure 9.3 shows a bright-field transmission electron micrograph (TEM) of the 1.5 mm diameter rod of the $Fe_{62.8}Co_{10}B_{13.5}Si_{10}Nb_3Cu_{0.7}$ glassy alloy annealed at 873 K for 5 min. It may be seen that the bcc α-(Fe, Co) grains with a size of between 10 and 15 nm are dispersed homogeneously in the glassy matrix phase.

The addition of Nb to Fe–B–Si alloys has been clearly shown to improve their GFA [43]. For example, a glassy alloy phase could be obtained when the $Fe_{75}B_{15}Si_{10}$ alloy was melt-spun. But a glassy alloy could not be obtained by copper mold casting. With the addition of Nb, though, the GFA increased and rods of 0.5 mm diameter could be obtained on adding 1 at.% Nb. The maximum-sized rod of 1.5 mm diameter could be obtained on adding 4 at.% Nb. This investigation was subsequently extended to the $[(Fe,Co,Ni)_{0.75}B_{0.20}Si_{0.05}]_{96}Nb_4$ system and the glass-forming composition range was mapped out (Figure 9.4a). BMG rods of 5 mm diameter could be identified in the Fe–Co–rich composition of $[(Fe_{0.6}Co_{0.4})_{0.75}B_{0.20}Si_{0.05}]_{96}Nb_4$. The maximum diameter could be further increased in this alloy to 7.7 mm by using the fluxing treatment [45].

TABLE 9.2

Effect of Annealing on the Magnetic Properties of Glassy Alloys

Alloy	Sample	Annealing Conditions	Structure	I_s (T)	H_c (A m^{-1})	μ_e	References
$(Fe_{0.75}Si_{0.1}B_{0.15})_{96}Nb_4$	1.5 mm rod	—	Glassy	1.47	2.9	17,000	[43]
$Fe_{72.5}Si_{10}B_{12.5}Nb_4Cu_1$	40 μm ribbon	883 K/5 min	α-Fe in glass	1.23	0.7	80,000	[42]
$Fe_{72.5}Si_{10}B_{12.5}Nb_4Cu_1$	0.5 mm rod	883 K/5 min	α-Fe in glass	1.21	1.8	32,000	[42]
$Fe_{73.5}Si_{13.5}B_9Nb_3Cu_1$ (FINEMET)	18 μm ribbon	823 K/1 h	α-Fe in glass	1.24	0.53	100,000	[41]
$Fe_{86}Zr_7B_6Cu_1$ (NANOPERM)	21 μm ribbon	873 K/1 h	α-Fe in glass	1.52	3.2	41,000	[44]

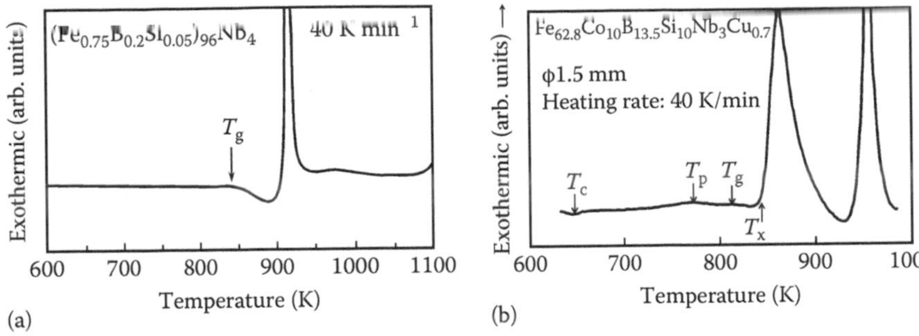

FIGURE 9.2
(a) DSC curve of a melt-spun $(Fe_{0.75}B_{0.2}Si_{0.05})_{96}Nb_4$ alloy showing that crystallization takes place in one stage. (Reprinted from Chang, C.T. et al., *Appl. Phys. Lett.*, 89, 051912-1, 2006. With permission.) (b) DSC curve of $Fe_{62.8}Co_{10}B_{13.5}Si_{10}Nb_3Cu_{0.7}$ alloy showing that crystallization takes place in two stages. (Reprinted from Shen, B.L. and Inoue, A., *J. Mater. Res.*, 19, 2549, 2004. With permission.) This situation helps in obtaining the precipitation of a crystalline bcc α-Fe solid solution phase in a glassy matrix on annealing the alloy above the first crystallization peak.

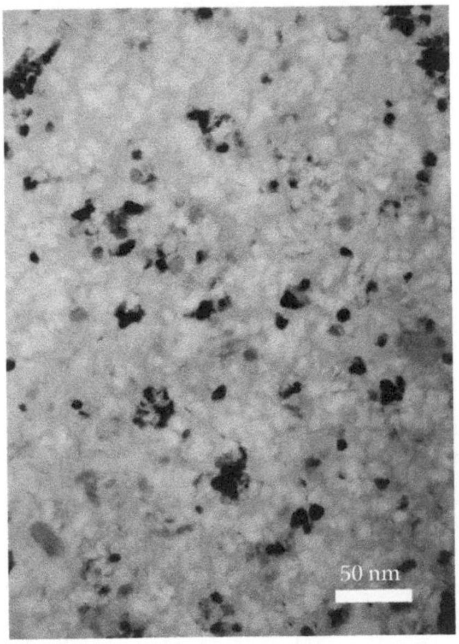

FIGURE 9.3
Bright-field TEM of the $Fe_{62.8}Co_{10}B_{13.5}Si_{10}Nb_3Cu_{0.7}$ glassy alloy annealed at 873 K for 5 min. Note that 10–15 nm-size grains of α-(Fe, Co) phase are uniformly dispersed in the glassy matrix. (Reprinted from Shen, B.L. and Inoue, A., *J. Mater. Res.*, 19, 2549, 2004. With permission.)

The magnetic properties of these multicomponent BMG alloys in the $[(Fe,Co,Ni)_{0.75}B_{0.20}Si_{0.05}]_{96}Nb_4$ system were investigated by measuring the I_s and H_c as a function of composition on melt-spun ribbons annealed for 5 min at a temperature 50 K below the glass transition temperature. Figure 9.4b and c show the variation of I_s and H_c as a function of composition. The saturation magnetization showed high values of over 1.3 T in the Fe-rich Fe–Co–B–Si–Nb alloys and it decreased with increasing Co and Ni contents. However, almost all the alloys showed a low value of coercivity ($<2.5\ A\ m^{-1}$) that decreased gradually with increasing Co content. The lowest values of $\leq 1\ A\ m^{-1}$ were obtained in the most Co-rich alloys.

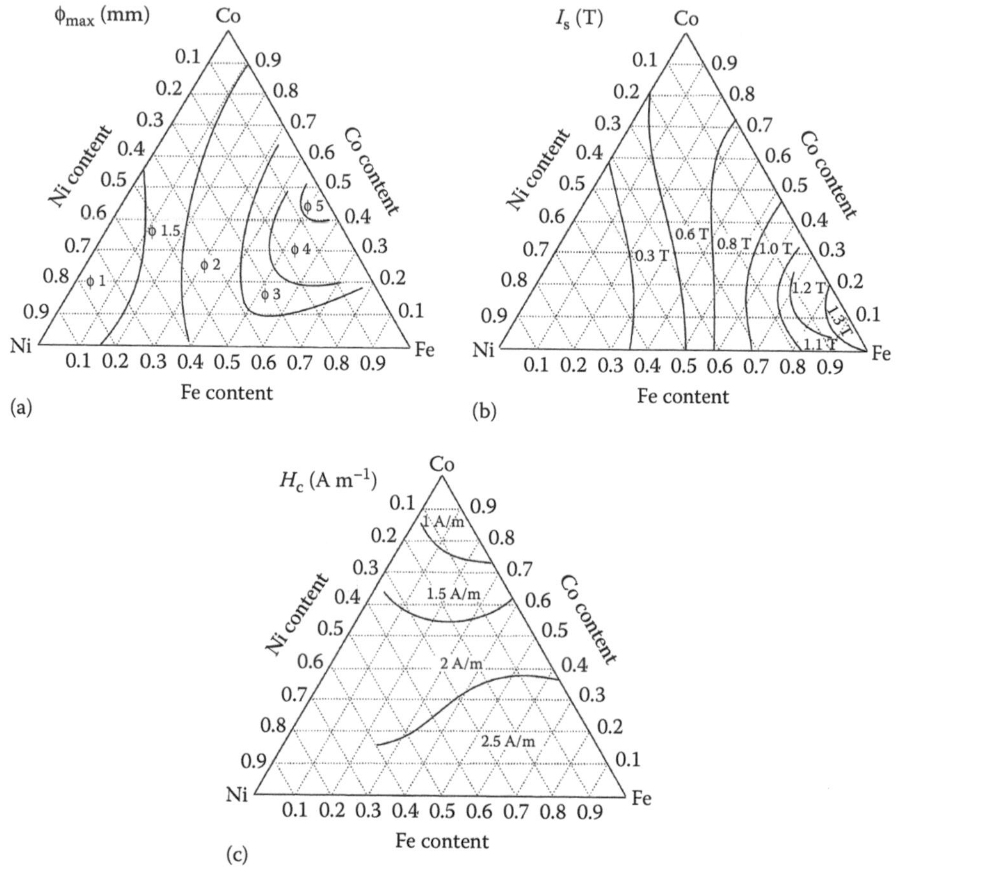

FIGURE 9.4
Plots showing (a) the maximum diameter of the glass formed, ϕ; (b) saturation magnetization, I_s; and (c) coercivity, H_c, as a function of composition in the $[(Fe_{1-x-y}Co_xNi_y)_{0.75}B_{0.20}Si_{0.05}]_{96}Nb_4$ alloys.

Irrespective of whether the bcc α-Fe solid solution phase or some other crystalline phase was forming, the magnetic properties of glassy alloys were generally improved on annealing. Table 9.3 lists several examples. In every case, the coercivity decreased and the permeability increased on annealing. There was not much change in the saturation magnetization values. Additionally, such samples showed the lowest core losses [48].

9.3 Nanocrystalline Alloys

The observations of the preceding paragraphs make it very clear that in order to achieve a high saturation magnetization, it is necessary to have as high an Fe content as possible and decrease the other metal and metalloid contents to the minimum value that is necessary to obtain the glassy phase. Investigating on these lines, Makino et al. [54] selected the $Fe_{94-x}Nb_6B_x$ alloy system to arrive at the optimum composition to achieve the best soft magnetic properties. They noted that a fully glassy phase was obtained only when the Nb content was a minimum of 10.5 at.%. Saturation magnetization of higher than 1.6 T was obtained in

TABLE 9.3

Comparison of the Magnetic Properties of Melt-spun Glassy Alloys in the As-Quenched Glassy Condition and After Annealing

Alloy	Annealing Conditions	I_s (T)	H_c (A m^{-1})	μ_e	λ_s (10^{-6})	$(BH)_{max}$ (kJ m^{-3})	References
$Fe_{56}Co_7Ni_7Zr_{10}B_{20}$	Glassy	0.92	—	5,100	—	—	[25]
$Fe_{56}Co_7Ni_7Zr_{10}B_{20}$	750 K/10 min	0.96	2.41	17,700	—	—	[25]
$Fe_{73}Al_5Ga_2P_{11}C_5B_4$	Glassy	1.24	4	—	23	—	[46]
$Fe_{73}Al_5Ga_2P_{11}C_5B_4$	583 K/2 h	1.27	3	—	25	—	[46]
$Fe_{73}Al_5Ga_2P_{11}C_5B_4$	713 K/210 min	1.26	1	—	25	—	[46]
$Fe_{74}Al_4Ga_2P_{12}B_4Si_4$	Glassy	1.03	19.1	1,900	21	—	[47]
$Fe_{74}Al_4Ga_2P_{12}B_4Si_4$	723 K/10 min	1.14	6.4	19,000	—	—	[47]
$Fe_{76}Al_4P_{12}B_4Si_4$	Glassy	1.08	12.7	2,600	30	—	[47]
$Fe_{76}Al_4P_{12}B_4Si_4$	723 K/10 min	1.24	2.6	21,000	—	—	[47]
$Fe_{77}Al_{2.14}Ga_{0.86}P_{8.4}C_5B_4Si_{2.6}$	Glassy	1.39	9	—	32	10	[46]
$Fe_{77}Al_{2.14}Ga_{0.86}P_{8.4}C_5B_4Si_{2.6}$	623 K/2 h	1.42	5	—	35	7.8	[46]
$Fe_{77}Al_{2.14}Ga_{0.86}P_{8.4}C_5B_4Si_{2.6}$	713 K/10 min	1.41	4	—	33	7	[46]
$Fe_{78}Si_{13}B_9$	Glassy	1.55	9	—	27	8.5	[46]
$Fe_{78}Si_{13}B_9$	653 K/2 h	1.6	5	—	30	7.4	[46]
$Fe_{80}B_{20}$	Glassy	1.65	12	—	31	7.4	[46]
$Fe_{80}B_{20}$	593 K/2 h	1.68	5	—	33	7.1	[46]
$Fe_{80}Zr_4Ti_3Cu_1B_{12}$	Glassy	0.94	48.5	7,100	—	—	[48]
$Fe_{80}Zr_4Ti_3Cu_1B_{12}$	673 K/1 h	1.45	4.5	14,800	—	—	[48]
$Fe_{80}Zr_4Ti_3Cu_1B_{12}$	723 K/1 h	1.46	5.0	15,700	—	—	[48]
$Fe_{80}Zr_4Ti_3Cu_1B_{12}$	773 K/1 h	1.44	5.5	15,600	—	—	[48]
$Fe_{80}P_{12}B_4Si_4$	Glassy	1.13	1.3	5,800	31	—	[47]
$Fe_{80}P_{12}B_4Si_4$	723 K/10 min	1.34	1.1	22,000	—	—	[47]
$Fe_{81}B_{10}Si_{5.5}B_{3.5}$	660 K/10 min	1.65	3.3	13,000	—	—	[49]
$Fe_{82}B_{10}Si_5P_3$	647 K/10 min	1.68	3.9	15,000	—	—	[49]
$Fe_{83}B_9Si_5P_3$	631 K/10 min	1.68	5.4	8,500	—	—	[49]
$Fe_{84}B_{8.5}Si_{4.5}P_3$	602 K/10 min	1.70	6.2	8,300	—	—	[49]
$Fe_{85}B_8Si_4P_3$	588 K/10 min	1.65	17.4	7,300	—	—	[49]
$Fe_{82}B_{10}P_4Si_3C_1$	Glassy	1.62	2.2	12,100	—	—	[50]

(Continued)

TABLE 9.3 (CONTINUED)

Alloy	Annealing Conditions	I_s (T)	H_c (A m^{-1})	μ_e	λ_s (10^{-6})	$(BH)_{max}$ (kJ m^{-3})	References
$Fe_{82}Cr_2B_8P_4Si_3C_1$	Glassy	1.49	4.6	13,900	—	—	[50]
$Fe_{82}Nb_1B_9P_4S_{i3}C_1$	Glassy	1.57	4.6	15,400	—	—	[50]
$Fe_{82}Cr_2Nb_1B_8P_4Si_2C_1$	Glassy	1.34	4.4	13,500	—	—	[50]
$Fe_{67}Co_{20}B_{13}$	Glassy	1.62	69	—	—	—	[51]
$Fe_{67}Co_{20}B_{13}$	566 K/10 min	1.76	35	—	—	—	[51]
$Fe_{66}Co_{20}B_{14}$	Glassy	1.48	44	—	—	—	[51]
$Fe_{66}Co_{20}B_{14}$	585 K/10 min	1.71	22	—	—	—	[51]
$Fe_{65}Co_{20}B_{15}$	Glassy	1.45	46	—	—	—	[51]
$Fe_{65}Co_{20}B_{15}$	610 K/10 min	1.69	24	—	—	—	[51]
$Fe_{84}Mo_2B_8Si_2P_4$	593 K/10 min	1.53	7.3	—	—	—	[52]
$Fe_{83}Mo_2B_8Si_2P_4C_1$	612 K/10 min	1.51	4.6	—	—	—	[52]
$Fe_{84}Cr_2B_8Si_2P_4$	589 K/10 min	1.50	9.1	—	—	—	[52]
$Fe_{83}Cr_2B_8Si_2P_4C_1$	600 K/10 min	1.45	6.9	—	—	—	[52]
$(Fe_{0.8}Co_{0.2})_{83}B_9Si_5P_3$	663 K/10 min	1.72	3.5	—	—	—	[53]
$(Fe_{0.8}Co_{0.2})_{84}B_{8.5}Si_{4.5}P_3$	641 K/10 min	1.74	4.9	—	—	—	[53]
$(Fe_{0.8}Co_{0.2})_{85}B_8Si_4P_3$	625 K/10 min	1.76	5.8	—	—	—	[53]
$(Fe_{0.8}Co_{0.2})_{86}B_{7.5}Si_{53.5}P_3$	609 K/10 min	1.69	15.2	—	—	—	[53]
$(Fe_{0.8}Co_{0.2})_{87}B_{97}Si_3P_3$	610 K/10 min	1.82	26.9	—	—	—	[53]

an alloy with $x \leq 9$, where the alloys have a mixture of the glassy + α-Fe composite structure. The highest permeability, μ_e, of 28,000 and the smallest grain size of 10.5 nm were obtained at $x = 11$. Thus, a single composition was not going to offer the best soft magnetic properties. Therefore, they limited the Nb content in the alloy to 6 at.% and small amounts of Cu and P were added to this alloy to optimize the composition at $Fe_{84.9}Nb_6B_8P_1Cu_{0.1}$. They were also able to fabricate this alloy by melt spinning in air. This alloy contained the α-Fe phase with an average grain size of 10 nm dispersed in a glassy matrix and showed excellent soft magnetic properties with a saturation magnetization of 1.61 T, coercivity of 4.7 A m^{-1}, and permeability of 41,000. The core losses for this alloy were very low at 0.11 W kg^{-1}.

The best soft magnetic properties were achieved by crystallizing the glassy phase to produce a uniform nanostructure. In FINEMET and other alloys, the volume fraction of the nanocrystalline phase is very large, reaching a value of almost 90%. Since a glassy precursor is a prerequisite to achieve this microstructure, one needs to have a minimum amount of different metals and/or metalloids to first produce the glassy phase. Further, to achieve the large volume fraction of the nanocrystalline phase, the presence of metals such as Cu, Nb, Mo, W, and Ta is required. The presence of these nonmagnetic metals can significantly reduce the saturation magnetization. Further, these metallic elements are quite expensive. To overcome these difficulties and the requirement of metallic elements and also to achieve a very high saturation magnetization, Makino et al. [55,56] developed novel Fe-based alloys that contain only metalloids. The generic composition of their alloys was $Fe_{83.3-84.3}Si_4B_8P_{3-4}Cu_{0.7}$, and was based on the $Fe_{82}Si_9B_9$ alloy that proved most promising among the Fe-based melt-spun magnetic alloys. In the modified composition, P substitutes for B and Cu for Fe. (The presence of a small amount of Cu is necessary to achieve nanocrystallization.) The large concentration of Fe provides a high saturation magnetization, and the absence of the metallic elements (other than Fe) ensures that the alloys are not expensive.

Figure 9.5a shows the x-ray diffraction (XRD) patterns of the melt-spun ribbons of $Fe_{82}Si_9B_9$ and $Fe_{81.7}Si_9B_7P_2Cu_{0.3}$ alloys. The XRD patterns show a single peak centered at $2\theta = 45°$, identified as the 110 diffraction peak of α-Fe for the ternary alloy. The grain size of this phase, as determined from the peak width, was 93 nm. The grain size is expected to decrease with the addition of P and/or Cu. The Cu- and P-containing alloy appears to be glassy. Figure 9.5b and c show a high-resolution TEM image and the diffraction pattern, respectively, from the $Fe_{81.7}Si_9B_7P_2Cu_{0.3}$ alloy. While the XRD pattern and the electron diffraction pattern showed only a diffuse and broad peak/ring, indicative of the glassy nature of the alloy, the high-resolution TEM image did not reveal it to be fully glassy. Instead, it showed the presence of an extremely small crystalline-like phase of about 3 nm or smaller in diameter. This phase appeared to be randomly dispersed in the glassy matrix. From the lattice fringes in the micrographs, it was concluded that they represented the {1 1 0} planes of α-Fe. From these observations, the authors concluded that the 93 nm-sized grains present in the ternary alloy transformed to very fine grains of about 3 nm through the simultaneous addition of 2–4 at.% P and 0.3 at.% Cu. On annealing these alloys at temperatures in between the two crystallization peaks, a uniform nanocrystalline structure with a grain size of about 10 nm was produced.

Table 9.4 lists the soft magnetic properties of $Fe_{83.3}Si_4B_8P_4Cu_{0.7}$ nanocrystalline alloy and compares them with some of the popular soft magnetic alloys, including the industrially used Fe-3.5 wt.% Si alloy. It is worth noting that the magnetic properties of the newly developed nanocrystalline alloys are better than those of the recently developed alloys (glassy, composite, or nanocrystalline) and also those of the commercially important crystalline alloys.

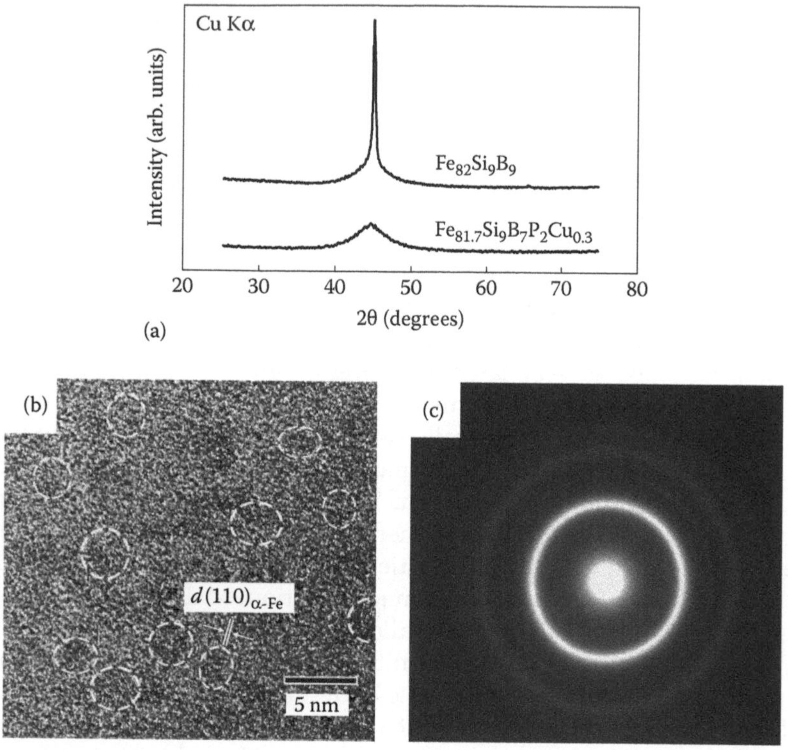

FIGURE 9.5
(a) X-ray diffraction patterns of the as-quenched $Fe_{82}Si_9B_9$ and $Fe_{81.7}Si_9B_7P_2Cu_{0.3}$ alloys. The $(1\ 1\ 0)_{\alpha\text{-Fe}}$ peak is reasonably sharp for the Fe–Si–B alloy and much wider for the Cu- and P-containing alloy. (b) High-resolution TEM image showing medium-range order and (c) the corresponding electron diffraction pattern from the $Fe_{81.7}Si_9B_7P_2Cu_{0.3}$ alloy. (Reprinted from Makino, A. et al., *Mater. Trans.*, 50, 204, 2009. With permission.)

9.4 Hard Magnetic Materials

There has been very little work done on the hard magnetic properties of the BMG alloys. Even though Nd–Fe [80] and Pr–Fe alloys [81] were produced in the glassy condition by melt-spinning methods, their section thickness was limited to a few tens of micrometers, typically <30 μm. The addition of Al was found to significantly increase the GFA of these alloys. As an example, Figure 9.6 shows the structure of the Nd–Fe–Al alloys in the as-quenched condition. Depending on the alloy composition, the as-spun ribbon contained either a fully glassy phase, or a crystalline phase, or a mixture of both. But, it is important to note that the glass-forming region in the system was very wide. A fully glassy phase was obtained in alloys containing 0–90 at.% Fe and 0–93 at.% Al. [82]. It was also subsequently reported [83] that very large rods of up to 10 mm diameter could be produced in alloys with 20–40 at.% Fe and 10–30 at.% Al. The largest diameter of 12 mm could be obtained in the $Nd_{70}Al_{10}Fe_{20}$ composition, and the maximum diameter of the glassy rod decreased with deviation from this composition. Therefore, most of the investigations were focused on $Nd_{90-x}Al_{10}Fe_x$ compositions with $x=20$, 25, or 30. In some cases, Fe is partially substituted by Co [84].

An important difference between these Nd–Al–Fe and other BMG alloys is that a clear glass transition, T_g, was not observed in these glassy alloys. Instead, one directly

TABLE 9.4

Grain Size and Soft Magnetic Properties of the Newly Developed Nanocrystalline Alloys and Their Comparison with Previously Reported Nanocrystalline, Glassy, and Crystalline Alloys

Material	Grain Size (nm)	I_s (T)	H_c (A m^{-1})	μ_e (at 1 kHz)	λ_s (10^{-6})	References
$Fe_{84.3}Si_4B_8P_3Cu_{0.7}$	17	1.94	10	16,000	3	[55]
$Fe_{83.3}Si_4B_8P_4Cu_{0.7}$	10	1.88	7	25,000	2	[55]
$Fe_{73.5}Si_{13.5}B_9Nb_3Cu_1$	20	1.24	0.5	150,000	2.1	[41]
$Fe_{90}Zr_7B_3$	13	1.7	5.8	30,000	−1.1	[57]
$Fe_{85.5}Zr_2Nb_4B_{8.5}$	11	1.64	3.0	60,000	−0.1	[58]
$Fe_{82.7}Si_2B_{14}Cu_{1.3}$	22	1.85	6.5	—	—	[59]
$(Fe_{0.7}Co_{0.3})_{88}Hf_7B_4Cu_1$	10	1.77	200	240	—	[60]
$(Fe_{0.85}B_{0.15})_{98.5}Cu_{1.5}$		1.83	6.9	60,000	—	[61]
$Fe_{80.5}Si_4B_{14}Cu_{1.5}$		1.80	5.7	—	—	[62]
$Fe_{80.6}Si_5B_{13}Cu_{1.4}$		1.80	5.7	—	—	[63]
$Fe_{82.5}Si_2B_{14}Cu_{1.5}$		1.84	6.5	—	—	[64]
$Fe_{82.65}Si_2B_{14}Cu_{1.35}$		1.84	6.5	—	—	[65]
$Fe_{82.7}Si_2B_{14}Cu_{1.3}$		1.85	6.5	—	—	[63]
$Fe_{82}Si_4B_{12}Nb_1Cu_1$		1.78	3.2	—	—	[66]
$Fe_{82.75}Si_4B_8P_4Cu_{1.25}$		1.83	2.1	31,600	—	[67]
$Fe_{83.3}Si_4B_8P_4Cu_{0.7}$		1.88	7	25,000	—	[68]
$Fe_{84.3}Si_4B_8P_3Cu_{0.7}$		1.94	10	16,000	—	[69]
$Fe_{83}Si_4B_8P_4Cu_1$		1.82	2.3	30,500	—	[67]
$Fe_{82.65}Si_2B_9P_5Cu_{1.35}$		1.8	1.1	37,000	—	[70]
$Fe_{83.5}B_{10}C_6Cu_{0.5}$		1.74	8	—	—	[71]
$Fe_{85.5}B_7C_7Cu_{0.5}$		1.81	30	—	—	[72]
$Fe_{82.7}B_{10}C_6Cu_{1.3}$		1.83	15	—	—	[71]
$Fe_{83}B_{10}C_6Cu_1$		1.78	5	—	—	[71]
$Fe_{85.5}B_7C_6Si_1Cu_{0.5}$		1.8	10	—	—	[72]
$Fe_{83}B_{10}C_4Si_2Cu_1$		1.78	4	13,600	—	[73]
$Fe_{67}Co_{20}B_{13}$		1.76	35	—	—	[74]
$Fe_{66}Co_{20}B_{14}$		1.71	22	—	—	[74]
$Fe_{65}Co_{20}B_{15}$		1.69	24	—	—	[74]
$(Fe_{0.83}P_{0.16}Cu_{0.01})_{98.5}Al_{1.5}$		1.54	5.4	—	—	[75]
$Fe_{82.75}P_9C_7Cu_{0.75}Nb_{0.5}$		1.66	6.8	29,000	—	[76]
$(Fe_{0.65}Co_{0.35})_{84}Si_4B_8P_3Cu_1$		1.68	5.4	—	29,600	[77]
$Fe_{81.2}Co_4Si_{0.5}B_{9.5}P_4C_{0.8}$		1.84	7	—	—	[78]
$Fe_{78}Si_9B_{13}$	Glassy	1.58	1.6	10,000	27	[79]
Fe-3.5 wt.% Si	Crystalline	1.97	41	770	6.8	[44]

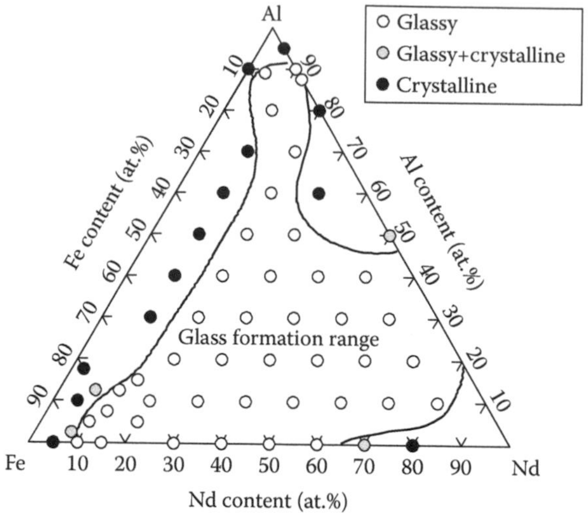

FIGURE 9.6
Structure of melt-spun Nd–Fe–Al alloys. Depending on the composition, the as-quenched alloy contains either a fully glassy phase, or a crystalline phase, or a combination of both. Note that the composition range in which the fully glassy phase forms is very wide. (Reprinted from Inoue, A. et al., *Mater. Trans., JIM*, 37, 99, 1996. With permission.)

observed only the crystallization temperature and no supercooled liquid region was noted. The absence of T_g in these alloys was thought to be due to the fact that T_g is higher than the crystallization temperature, T_x [85]. But, it is now believed that it is due to the chemical inhomogeneity of the glassy phase [86]. A clear T_g was later observed in a 3 mm diameter rod of $Nd_{60}Al_{20}Fe_{20}$ BMG alloy on reheating the glassy alloy at a heating rate of 20 K min^{-1} [87].

Inoue et al. [82] reported that $Nd_{70}Al_{10}Fe_{20}$ and $Nd_{60}Al_{10}Fe_{30}$ alloys in the as-cast glassy condition exhibited hard magnetic properties at room temperature. It was also reported that this property disappeared in the fully crystallized condition [83]. Another important observation was that while the BMG alloys were reported to exhibit the hard magnetic properties, the melt-spun ribbons exhibited only the soft magnetic properties [80]. These interesting features of the Nd–Fe–Al alloys created a lot of interest in the magnetics community, and detailed studies were conducted on the origin of the hard magnetic behavior of these alloys. Table 9.5 lists the hard magnetic properties of the Nd- and Pr-based BMG alloys.

Figure 9.7 presents the hysteresis curves for the glassy $Nd_{70}Al_{10}Fe_{20}$ glassy rods of different diameters. The hysteresis loop is significantly different for samples of different diameters. Even though no distinct difference in B_r was clearly seen, one could notice a tendency for the H_c value to increase with increasing sample diameter from 1 to 5 mm. Accordingly, for the melt-spun ribbon of 30 μm thickness, the H_c value was significantly lower than it was for the bulk rods. Whereas the H_c value was 209 kA m^{-1} for the 1 mm diameter rod sample, it was only 5 kA m^{-1} for the thin ribbon sample. This may be contrasted with the situation in soft magnetic materials, where no significant differences were noted either between the ribbon and bulk samples or between the ribbon and powder forms (Figure 9.1).

Another important observation reported was that the hard magnetic properties completely disappeared in the fully crystallized alloys. Figure 9.8 shows the variation of

TABLE 9.5

Hard Magnetic Properties of Different Nd- and Pr-Based BMG Alloys

Alloy	Sample Dimensions	I_s (T)	B_r (T)	H_c (kA m^{-1})	T_c (K)	$(BH)_{max}$ (kJ m^{-3})	References
$Nd_{60}Al_{20}Fe_{20}$	3 mm rod	12.5	8	160	—	—	[87]
$Nd_{60}Al_{10}Fe_{30}$	3 mm rod	16	11	230	—	—	[87]
$Nd_{60}Al_{10}Fe_{30}$	5 mm rod	—	0.112	288	—	19	[82]
$Nd_{60}Al_{10}Fe_{30}$	12 mm rod	—	0.086	321	—	16	[83]
$Nd_{60}Al_{10}Fe_{20}Co_{10}$	5 mm rod	10.82 (A m^2 kg^{-1})	7.22	326.3	465	—	[84]
$Nd_{65}Al_{10}Fe_{25}$	5 mm rod	11.79	7.53	271.4	446	—	[84]
$Nd_{65}Al_{10}Fe_{20}Co_5$	5 mm rod	12.18	8.05	302.4	452	—	[84]
$Nd_{65}Al_{10}Fe_{15}Co_{10}$	5 mm rod	9.91	6.38	301.2	466	—	[84]
$Nd_{65}Al_{10}Fe_{25}$	12 mm rod	—	0.076	315	—	15	[83]
$Nd_{70}Al_{10}Fe_{20}$	30 µm thick ribbon	—	0.010	5	—	—	[82]
$Nd_{70}Al_{10}Fe_{20}$	5 mm rod	—	0.059	209	—	5	[82]
$Nd_{70}Al_{10}Fe_{20}$	12 mm rod	—	0.076	292	—	13	[83]
$Pr_{60}Al_{10}Fe_{30}$	1 mm rod	—	0.09	321	—	13	[85]

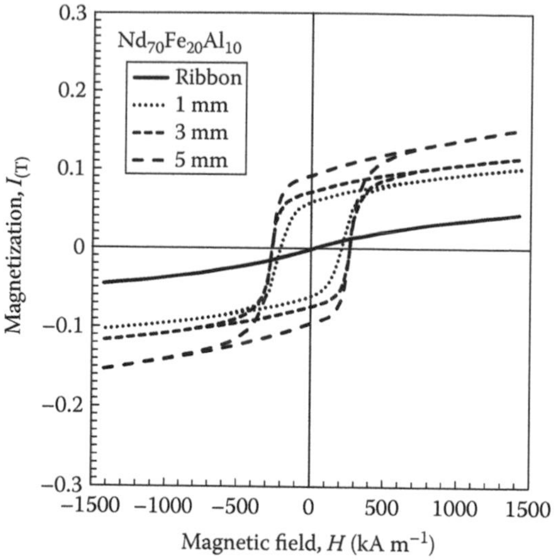

FIGURE 9.7
Hysteresis loops of the as-cast glassy $Nd_{70}Al_{10}Fe_{20}$ alloy cylinders with diameters of 1, 3, and 5 mm. The data for a melt-spun glassy ribbon of 30 μm thickness are also included for comparison. (Reprinted from Inoue, A. et al., *Mater. Trans., JIM*, 37, 99, 1996. With permission.)

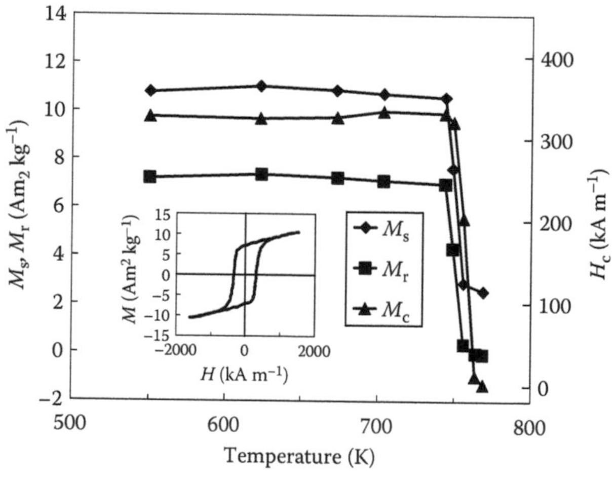

FIGURE 9.8
Variation of coercivity, H_c, saturation magnetization, M_s, and remanence magnetization, M_r, for the $Nd_{60}Al_{10}Fe_{20}Co_{10}$ BMG alloy isothermally annealed for 30 min at different temperatures. The inset shows the hysteresis loop obtained at room temperature. Note that the room temperature values remain almost unchanged on annealing up to about 740 K and completely disappear in the fully crystallized condition at 760 K. (Reprinted from Wei, B.C. et al., *Phys. Rev. B*, 64, 012406-1, 2001. With permission.)

coercivity, saturation magnetization, and remanence with the isothermal annealing temperature for the $Nd_{60}Al_{10}Fe_{20}Co_{10}$ BMG alloy. The inset shows the hysteresis loop at room temperature. It may be noted that the hard magnetic properties exhibited by the alloy at room temperature were retained almost unchanged on isothermal annealing up to about 740 K and completely disappeared after full crystallization of the BMG alloy above about 760 K [88].

The most characteristic property of the hard magnetic alloys is their coercivity. It has been suggested that the high coercivity in these alloys is related to the presence of crystalline clusters in the glassy matrix. These nanocrystalline magnetic clusters have a random anisotropy and are dispersed in the glass. The strong magnetic exchange coupling between these clusters was expected to lead to the high coercivity [80]. High-resolution TEM studies on these alloys have clearly shown the presence of clusters. The size of these clusters has been varyingly reported to be from about 0.5 nm [85] to about 5 nm [84]. From the typical morphology of the fine particles dispersed in the glassy matrix, it was suggested that they could be formed due to phase separation either in the melt or in the slightly undercooled liquid into two metastable phases [89]. Other reasons have also been suggested [90].

9.5 Magnetocaloric Properties

The magnetocaloric effect (MCE) utilizes the magnetothermodynamic phenomenon, in which the heating or cooling of magnetic materials is achieved by varying the magnetic field [91,92]. Magnetic refrigeration based on MCE has been regarded as a potential alternative to replace the conventional gas compression/expansion refrigeration due to its merits such as environmental friendliness and relatively high efficiency. The two main motivations for studying glassy materials for possible magnetocaloric applications are (1) their ability to change their magnetization reversibly in very low magnetic fields and (2) the flatter top of the temperature dependence of the magnetic entropy change [93].

Glassy materials have additional characteristics that are desirable for magnetic refrigerants: low cost, higher electrical resistivity than crystalline materials (which minimizes eddy current losses), high corrosion resistance, tunability of the transition temperature by alloying, good mechanical properties (especially in the case of BMGs), and negligible magnetic hysteresis. Moreover, the fact that these materials display negligible magnetic anisotropy fundamentally simplifies the study of their magnetic transition, making them a good testing ground for analyzing the physics behind the MCE and for developing thermodynamic models to represent their response [94]. Table 9.6 presents the data for a few glassy alloys.

9.6 Concluding Remarks

The magnetic properties of BMGs and melt-spun ribbons have been investigated in multicomponent alloy systems. Both soft and hard magnetic alloys have been investigated. It was noted that the number and type of investigations are far greater in the soft magnetic materials category. The magnetic properties have been studied both in the as-quenched and annealed conditions.

There appears to be a large scope for enhancing the soft magnetic properties of materials. Whereas the highest saturation magnetization in a fully glassy alloy is only about 1.5

TABLE 9.6

Critical Diameter and Magnetocaloric Properties under Applied Fields of 1.5 and 5 T for the Typical Ternary Fe-based Glassy Alloys

Compositions (at.%)	Critical diameter	T_C (K)	$-\Delta S_M$ (J kg^{-1} K)		RC_{FWHM} (J kg^{-1})	References
			1.5 T	5 T	1.5 T	
$Fe_{80}P_{13}C_7$	Rods (2 mm)	579	2.20	5.05	125.6	[94]
$(Fe_{0.76}B_{0.24})_{96}Nb_4$	Ribbons (D_C<0.5 mm)	559	1.51	3.86	120.8	
$Fe_{88}Zr_8B_4$		285	—	3.30	—	[95]
$Fe_{91}Zr_7B_2$		230	—	2.80	—	
$Fe_{80}Cr_8B_{12}$		328	1.00	2.59	115	[96]
$Fe_{77}Cr_8B_{15}$		375	1.18	2.98	61	
$Fe_{84}Nb_7B_9$		299	1.44	—	—	[97]
$Fe_{79}Nb_7B_{14}$		372	1.07	—	—	
$Fe_{73}Nb_7B_{20}$		419	0.97	—	—	
$Fe_{86}Y_5Zr_9$		284	0.89	—	—	[98]
$Fe_{81}Y_{10}Zr_9$		470	1.12	—	—	
$Fe_{70}Mn_{10}B_{20}$		438	1.00	—	117	[99]
$Fe_{65}Mn_{15}B_{20}$		340	0.87	—	98	
$Fe_{60}Mn_{20}B_{20}$		210	0.60	—	83	
$Fe_{56}Mn_{24}B_{20}$		162	0.50	—	68	
$Fe_{85}Zr_{10}B_5$		300	1.39	—	—	[100]
$Fe_{82}Mn_8Zr_{10}$		210	—	2.78	—	[101]
$Fe_{84}Mn_6Zr_{10}$		218	—	2.29	—	
$Fe_{86}Mn_4Zr_{10}$		228	—	2.51	—	
$Fe_{80}Mn_{10}Zr_{10}$		195	—	2.33	—	
$Fe_{88}Sn_2Zr_{10}$		290	—	4.10	—	[102]
$Fe_{86}Sn_4Zr_{10}$		300	—	3.30	—	

Note: RC_{FWHM}, refrigerant capacity at full width half maximum.

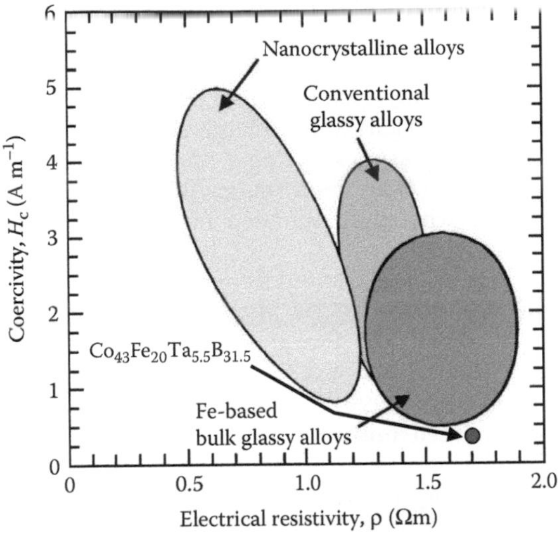

FIGURE 9.9
Plot relating the coercivity, H_c, and electrical resistivity, ρ, of different types of magnetic alloys. Note that both the Co-based and Fe-based BMG alloys have the desirable combination of low coercivity and high electrical resistivity.

T, that in the FINEMET-type alloys (in which a large volume fraction of the nanocrystalline α-Fe phase is uniformly dispersed in the glassy matrix) is only about 1.3 T. These low values have been essentially due to the presence of a reasonably large percentage of nonmagnetic metals in the alloy. But, it has been shown recently that a fully nanocrystalline alloy without any metallic elements (other than Fe) can exhibit a very high saturation magnetization of 1.9 T. The other soft magnetic properties of these alloys can be tailored by suitable annealing treatments. As mentioned in the introduction, a good soft magnetic alloy should have high saturation magnetization, low coercivity, and high electrical resistivity. Figure 9.9 shows the plot of coercivity versus electrical resistivity for a number of magnetic alloys. Values for the nanocrystalline alloys, melt-spun ribbons, and Co-based and Fe-based BMG alloys are shown. It may be noted that both Fe-based and Co-based BMG alloys show the desirable combination of low coercivity and high electrical resistivity.

The number of studies on hard magnetic materials is limited. But, it is interesting to note that hard magnetism was detected only in the BMG condition and not in the melt-spun ribbons. Further, the desired value of coercivity could be obtained by controlling the diameter of the BMG rods.

References

1. Cullity, B.D. and C.D. Graham (2009). *Introduction to Magnetic Materials*, 2nd edn. Hoboken, NJ: John Wiley & Sons.
2. Inoue, A. and J.S. Gook (1995). Fe-based ferromagnetic glassy alloys with wide supercooled liquid region. *Mater. Trans., JIM* 36: 1180–1183.
3. Inoue, A. and J.S. Gook (1995). Multicomponent Fe-based glassy alloys with wide supercooled liquid region before crystallization. *Mater. Trans., JIM* 36: 1282–1285.

4. Suryanarayana, C. and A. Inoue (2013). Iron-based bulk metallic glasses. *Internat. Mater. Rev.* 58: 131–166.
5. Shen, J., Q.J. Chen, J.F. Sun, H.B. Fan, and G. Wang (2005). Exceptionally high glass-forming ability of an FeCoCrMoCBY alloy. *Appl. Phys. Lett.* 86: 151907-1–151807-3.
6. Shen, B.L. and A. Inoue (2003). Fabrication of large-size Fe-based glassy cores with good soft magnetic properties by spark plasma sintering. *J. Mater. Res.* 18: 2115–2121.
7. Shen, B.L. and A. Inoue (2004). Soft magnetic properties of bulk nanocrystalline Fe–Co–B–Si–Nb–Cu alloy with high saturated magnetization of 1.35 T. *J. Mater. Res.* 19: 2549–2552.
8. Duwez, P. and S.C.H. Lin (1967). Amorphous ferromagnetic phase in iron–carbon–phosphorus alloys. *J. Appl. Phys.* 38: 4096–4097.
9. Guo, S.F., Z.Y. Wy, and L. Liu (2009). Preparation and magnetic properties of FeCoHfMoBY bulk metallic glasses. *J. Alloys Compd.* 468: 54–57.
10. Kraus, L., V. Haslar, and P. Duhaj (1994). Correlation between saturation magnetization, magnetostriction and creep-induced anisotropy in amorphous $(FeCo)_{85}B_{15}$ alloys. *IEEE Trans. Mag.* 30: 530–532.
11. Liu, D.Y., W.S. Sun, H.F. Zhang, and Z.Q. Hu (2004). Preparation, thermal stability and magnetic properties of Fe–Co–Ni–Zr–Mo–B bulk metallic glass. *Intermetallics* 12: 1149–1152.
12. Inoue, A., B.L. Shen, and C.T. Chang (2004). Super-high strength of over 4000 MPa for Fe-based bulk glassy alloys in $[(Fe_{1-x}Co_x)_{0.75}B_{0.2}Si_{0.05}]_{96}Nb_4$ system. *Acta Mater.* 52: 4093–4099.
13. Chang, C.T., B.L. Shen, and A. Inoue (2006). Co–Fe–B–Si–Nb bulk glassy alloys with superhigh strength and extremely low magnetostriction. *Appl. Phys. Lett.* 88: 011901-1–011901-3.
14. Chang, C.T., B.L. Shen, and A. Inoue (2006). FeNi-based bulk glassy alloys with superhigh mechanical strength and excellent soft-magnetic properties. *Appl. Phys. Lett.* 89: 051912-1–051912-3.
15. Shen, B.L., M. Akiba, and A. Inoue (2006). Excellent soft-ferromagnetic bulk glassy alloys with high saturation magnetization. *Appl. Phys. Lett.* 88: 131907-1–131907-3.
16. Shen, B.L., M. Akiba, and A. Inoue (2006). Effect of Cr addition on the glass-forming ability, magnetic properties, and corrosion resistance of FeMoGaPCBSi bulk glassy alloys. *J. Appl. Phys.* 100: 043523-1–043523-5.
17. Pawlik, P. and H.A. Davies (2003). The bulk glass forming abilities and mechanical and magnetic properties of Fe–Co–Zr–Mo–W–B alloys. *J. Non-Cryst. Solids* 329: 17–21.
18. Shen, B.L., H. Koshiba, A. Inoue, H.M. Kimura, and T. Mizushima (2001). Bulk glassy $Co_{43}Fe_{20}Ta_{5.5}B_{31.5}$ alloy with high glass-forming ability and good soft magnetic properties. *Mater. Trans.* 42: 2136–2139.
19. Inoue, A., B.L. Shen, H. Koshiba, H. Kato, and A.R. Yavari (2004). Ultra-high strength above 5000 MPa and soft magnetic properties of Co–Fe–Ta–B bulk glassy alloys. *Acta Mater.* 52: 1631–1637.
20. Inoue, A. and B.L. Shen (2002). Formation and soft magnetic properties of Co–Fe–Si–B–Nb bulk glassy alloys. *Mater. Trans.* 43: 1230–1234.
21. Makino, A., T. Kubota, C.T. Chang, M. Makabe, and A. Inoue (2007). FeSiBP bulk metallic glasses with unusual combination of high magnetization and high glass-forming ability. *Mater. Trans.* 48: 3024–3027.
22. Sulitanu, N., F. Brînză, and F.M. Tufescu (2005). Effect of Co substitution for Ni on the microstructure and magnetic properties of (Fe, Ni)-based amorphous alloys produced by melt spinning. *J. Non-Cryst. Solids* 351: 418–425.
23. Mitra, A., H.-Y. Kim, B.L. Shen, N. Nishiyama, and A. Inoue (2003). Crystallization and magnetic properties of $Fe_{40}Co_{40}Cu_{0.5}Al_2Zr_9Si_4B_{4.5}$ and $Fe_{62}Co_{9.5}Gd_{3.5}Si_{10}B_{15}$ amorphous alloys. *Mater. Trans.* 44: 1562–1565.
24. Zhang, W., Y. Long, M. Imafuku, and A. Inoue (2002). Thermal stability and soft magnetic properties of (Fe, Co)–(Nd, Dy)–B glassy alloys with high boron concentrations. *Mater. Trans.* 43: 1974–1978.

25. Inoue, A., T. Zhang, T. Itoi, and A. Takeuchi (1997). New Fe–Co–Ni–Zr–B amorphous alloys with wide supercooled liquid regions and good soft magnetic properties. *Mater. Trans., JIM* 38: 359–362.
26. Inoue, A., H. Koshiba, T. Zhang, and A. Makino (1997). Thermal and magnetic properties of $Fe_{56}Co_7Ni_7Zr_{10-x}Nb_xB_{20}$ amorphous alloys with wide supercooled liquid range. *Mater. Trans., JIM* 38: 577–582.
27. Gercsi, ZS F. Mazaleyrat, S.N. Kane, and L.K. Varga (2004). Magnetic and structural study of $(Fe_{1-x}Co_x)_{62}$ Nb_8B_{30} bulk amorphous alloys. *Mater. Sci. Eng. A* 375–377: 1048–1052.
28. Shen, B.L., C.T. Chang, and A. Inoue (2007). Formation, ductile deformation behavior and soft-magnetic properties of (Fe,Co,Ni)–B–Si–Nb bulk glassy alloys. *Intermetallics* 15: 9–16.
29. Inoue, A. and B.L. Shen (2004). A new Fe-based bulk glassy alloy with outstanding mechanical properties. *Adv. Mater.* 16: 2189–2192.
30. Pawlik, P., H.A. Davies, and M.R.J. Gibbs (2004). The glass forming abilities and magnetic properties of Fe–Al–Ga–P–B–Si and Fe–Al–Ga–P–B–C alloys. *Mater. Sci. Eng. A* 375–377: 372–376.
31. Chang, C.T., T. Kubota, A. Makino, and A. Inoue (2009). Synthesis of ferromagnetic Fe-based bulk glassy alloys in the Fe–Si–B–P–C system. *J. Alloys Compd.* 473: 368–372.
32. Amiya, K., A. Urata, N. Nishiyama, and A. Inoue (2007). Thermal stability and magnetic properties of (Fe, Co)–Ga–(P, C, B, Si) bulk glassy alloys. *Mater. Sci. Eng. A* 449–451: 356–359.
33. Shen, B.L., H.M. Kimura, A. Inoue, and T. Mizushima (2001). Bulk glassy $Fe_{78-x}Co_xGa_2P_{12}C_4B_4$ alloys with high saturation magnetization and good soft magnetic properties. *Mater. Trans.* 42: 1052–1055.
34. Shen, B.L. and A. Inoue (2002). Bulk glassy Fe–Ga–P–C–B–Si alloys with high glass-forming ability, high saturation magnetization and good soft magnetic properties. *Mater. Trans.* 43: 1235–1239.
35. Mizushima, T., K. Ikarashi, S. Yoshida, A. Makino, and A. Inoue (1999). Soft magnetic properties of ring shape bulk glassy Fe–Al–Ga–P–C–B–Si alloy prepared by copper mold casting. *Mater. Trans., JIM* 40: 1019–1022.
36. Makino, A., A. Inoue, and T. Mizushima (2000). Soft magnetic properties of Fe-based bulk amorphous alloys. *Mater. Trans., JIM* 41: 1471–1477.
37. Mizushima, T., A. Makino, S. Yoshida, and A. Inoue (1999). Low core losses and soft magnetic properties of Fe–Al–Ga–P–C–B–Si glassy alloy ribbons with large thicknesses. *J. Appl. Phys.* 85: 4418–4420.
38. Miao, H., C. Chang, Y. Li, Y. Wang, X. Jia, and W. Zhang (2015). Fabrication and properties of soft magnetic Fe–Co–Ni–P–C–B bulk metallic glasses with high glass-forming ability. *J. Non-Cryst. Solids* 421: 24–29.
39. Liu, Q., J. Mo, H. Liu, L. Xue, L. Hou, W. Yang, L. Dou, B. Shen, and L. Dou (2016). Effects of Cu substitution for Nb on magnetic properties of Fe-based bulk metallic glasses. *J. Non-Cryst. Solids* 443: 108–111.
40. Dong, Y., Q. Man, C. Chang, B. Shen, X.-M. Wang, and R.-W. Li (2015). Preparation and magnetic properties of $(Co_{0.6}Fe_{0.3}Ni_{0.1})_{70-x}$ $(B_{0.811}Si_{0.189})_{25+x}$ Nb_5 bulk glassy alloys. *J. Mater. Sci.: Materials in Electronics* 26: 7006–7012.
41. Yoshizawa, Y., S. Oguma, and K. Yamauchi (1988). New Fe-based soft magnetic alloys composed of ultrafine grain structure. *J. Appl. Phys.* 64: 6044–6046.
42. Inoue, A., B.L. Shen, and T. Ohsuna (2002). Soft magnetic properties of nanocrystalline Fe–Si–B–Nb–Cu rod alloys obtained by crystallization of cast amorphous phase. *Mater. Trans.* 43: 2337–2341.
43. Inoue, A. and B.L. Shen (2002). Soft magnetic bulk glassy Fe–B–Si–Nb alloys with high saturation magnetization above 1.5 T. *Mater. Trans.* 43: 766–769.
44. Makino, A., K. Suzuki, A. Inoue, and T. Masumoto (1991). Low core loss of a bcc $Fe_{86}Zr_7B_6Cu_1$ alloy with nanoscale grain size. *Mater. Trans., JIM* 32: 551–556.
45. Bitoh, T., A. Makino, A. Inoue, and A.L. Greer (2006). Large bulk soft magnetic $[(Fe_{0.5}Co_{0.5})_{0.75}B_{0.20}Si_{0.05}]_{96}Nb_4$ glassy alloy prepared by B_2O_3 flux melting and water quenching. *Appl. Phys. Lett.* 88: 182510-1–182510-3.

46. Bitoh, T., A. Makino, and A. Inoue (2004). Magnetization process and coercivity of Fe–(Al, Ga) – (P, C, B, Si) soft magnetic glassy alloys. *Mater. Trans.* 45: 1219–1227.
47. Inoue, A. and R.E. Park (1996). Soft magnetic properties and wide supercooled liquid region of Fe-P-B-Si Base amorphous alloys. *Mater. Trans., JIM* 37: 1715–1721.
48. Miglierini, M., R. Vittek, and M. Hasiak (2007). Impact of annealing temperature on magnetic microstructure of $Fe_{80}Zr_4Ti_3Cu_1B_{12}$ rapidly quenched alloy. *Mater. Sci. Eng. A* 449–451: 419–422.
49. Kong, F.L., C.T. Chang, A. Inoue, E. Shalaan, and F. Al-Marzouki (2014). Fe-based amorphous soft magnetic alloys with high saturation magnetization and good bending ductility. *J. Alloys Compd.* 615: 163–166.
50. Han, Y., C.T. Chang, S.L. Zhu, A. Inoue, D.V. Louzguine-Luzgin, E. Shalaan, and F. Al-Marzouki (2014). Fe-based soft magnetic amorphous alloys with high saturation magnetization above 1.5 T and high corrosion resistance. *Intermetallics* 54: 169–175.
51. Han, Y., A. Inoue, F.L. Kong, C.T. Chang, S.L. Shu, E. Shalaan, and F. Al-Marzouki (2016). Softening and good ductility for nanocrystal-dispersed amorphous Fe–Co–B alloys with high saturation magnetization above 1.7 T. *J. Alloys Compd.* 657: 237–245.
52. Han, Y., F.L. Kong, F.F. Han, A. Inoue, S.L. Zhu, E. Shalaan, and F. Al-Marzouki (2016). New Fe-based soft magnetic amorphous alloys with high saturation magnetization and good corrosion resistance for dust core application. *Intermetallics* 76: 18–25.
53. Han, Y., J. Ding, F.L. Kong, A. Inoue, S.L. Zhu, Z. Wang, E. Shalaan, and F. Al-Marzouki (2017). FeCo-based soft magnetic alloys with high B_s approaching 1.75 T and good bending ductility. *J. Alloys Compd.* 691: 364–368.
54. Makino, A., T. Bitoh, A. Inoue, and T. Masumoto (2003). Nb-poor Fe–Nb–B nanocrystalline soft magnetic alloys with small amount of P and Cu prepared by melt spinning in air. *Scr. Mater.* 48: 869–874.
55. Makino, A., H. Men, T. Kubota, K. Yubuta, and A. Inoue (2009). FeSiBPCu nanocrystalline soft magnetic alloys with high B_s of 1.9 Tesla produced by crystallizing hetero-amorphous phase. *Mater. Trans.* 50: 204–209.
56. Makino, A., H. Men, T. Kubota, K. Yubuta, and A. Inoue (2009). New Fe-metalloids based nanocrystalline alloys with high B_s of 1.9 T and excellent magnetic softness. *J. Appl. Phys.* 105: 07A308-1–07A308-3.
57. Suzuki, K., A. Makino, A. Inoue, N. Kataoka, and T. Masumoto (1991). High saturation magnetization and soft magnetic properties of bcc Fe–Zr–B and Fe–Zr–B–M (M=Transition Metal) alloys with nanoscale grain size. *Mater. Trans., JIM* 32: 93–102.
58. Suzuki, K., M. Kikuchi, A. Makino, A. Inoue, and T. Masumoto (1991). Changes in microstructure and soft magnetic properties of an $Fe_{86}Zr_7B_6Cu_1$ amorphous alloy upon crystallization. *Mater. Trans., JIM* 32: 961–968.
59. Ohta, M. and Y. Yoshizawa (2007). Magnetic properties of nanocrystalline $Fe_{82.65}Cu_{1.35}Si_xB_{16-x}$ alloys ($x=0$–7). *Appl. Phys. Lett.* 91: 062517-1–062517-3.
60. Iwanabe, H., B. Lu, M.E. McHenry, and D.E. Laughlin (1991). Thermal stability of the nanocrystalline Fe–Co–Hf–B–Cu alloy. *J. Appl. Phys.* 85: 4424–4426.
61. Ohta, M. and Y. Yoshizawa (2007). Improvement of soft magnetic properties in $(Fe_{0.85}B_{0.15})_{100-x}Cu_x$ melt-spun alloys. *Mater. Trans.* 48: 2378–2380.
62. Ohta, M. and Y. Yoshizawa (2008). Magnetic properties of high B_s Fe–Cu–Si–B nanocrystalline soft magnetic alloys. *J. Mag. Mag. Mater.* 320: E750–E753.
63. Ohta, M. and Y. Yoshizawa (2007). New high B_s Fe-based nanocrystalline soft magnetic alloys. *Jpn. J. Appl. Phys.* 46: L477–L479.
64. Ohta, M. and Y. Yoshizawa (2008). Cu addition effect on soft magnetic properties in Fe–Si–B alloy system. *J. Appl. Phys.* 103: 07E722-1–07E722-3.
65. Ohta, M. and Y. Yoshizawa (2007). Magnetic properties of nanocrystalline $Fe_{82.65}Cu_{1.35}Si_xB_{16-x}$ alloys (x=0–7). *Appl. Phys. Lett.* 91: 062517–062517-3.
66. Ohta, M. and Y. Yoshizawa (2009). High B_s nanocrystalline $Fe_{84-x-y}Cu_xNb_ySi_4B_{12}$ alloys (x=0.0–1.4, y=0.0–2.5). *J. Mag. Mag. Mater.* 321: 2220–2224.

67. Kong, F., A. Wang, X.D. Fan, H. Men, B.L. Shen, G.Q. Xie, A. Makino, and A. Inoue (2011). High B_s $Fe_{84-x}Si_4B_8P_4Cu_x$ (x = 0–1.5) nanocrystalline alloys with excellent magnetic softness. *J. Appl. Phys.* 109: 07A303-1–07A303-3.
68. Makino, A., H. Men, K. Yubuta, and T. Kubota (2009). Soft magnetic FeSiBPCu heteroamorphous alloys with high Fe content. *J. Appl. Phys.* 105: 013922-1–13922-4.
69. Makino, A., H. Men, T. Kubota, K. Yubuta, and A. Inoue (2009). New excellent soft magnetic FeSiBPCu nanocrystallized alloys with high B_s of 1.9 T from nanohetero-amorphous phase. *IEEE Trans. Mag.* 45: 4302–4305.
70. Kong, F.L., H. Men, T.C. Liu, and B.L. Shen (2012). Effect of P to B concentration ratio on soft magnetic properties in FeSiBPCu nanocrystalline alloys. *J. Appl. Phys.* 111: 07A311-1–07A311-3.
71. Fan, X.D., H. Men, A.B. Ma, and B.L. Shen (2013). Soft magnetic properties in $Fe_{84-x}B_{10}C_6Cu_x$ nanocrystalline alloys. *J. Mag. Mag. Mater.* 326: 22–27.
72. Fan, X.D. and B.L. Shen (2015). Crystallization behavior and magnetic properties in high Fe content FeBCSiCu alloy system. *J. Mag. Mag. Mater.* 385: 277–281.
73. Fan, X., H. Men, A. Ma, and B. Shen (2012). The influence of Si substitution on soft magnetic properties and crystallization behavior in $Fe_{83}B_{10}C_{6-x}Si_xCu_1$ alloy system. *Science China Technological Sciences*, 55: 2416–2419.
74. Han, Y., A. Inoue, F.L. Kong, C.T. Chang, S.L. Shu, E. Shalaan, and F. Al-Marzouki (2016). Softening and good ductility for nanocrystal-dispersed amorphous Fe–Co–B alloys with high saturation magnetization above 1.7 T. *J. Alloys Compd.* 657: 237–245.
75. Zhu, J.S. and Y.G. Wang (2015). Effect of Al addition on the glass forming ability, thermal stability and soft magnetic properties of $(Fe_{0.83}P_{0.16}Cu_{0.01})_{100-x}Al_x$ nanocrystalline alloys. *J Alloys Compd.* 652: 220–224.
76. Xiang, R., S.X. Zhou, B.S. Dong, and Y.G. Wang (2015). Effect of Nb addition on the magnetic properties and microstructure of FePCCu nanocrystalline alloy. *J. Mater. Sci.: Materials in Electronics* 26: 4091–4096.
77. Xiang, R., S.X. Zhou, B.S. Dong, G.Q. Zhang, Z.Z. Li, and Y.G. Wang (2015). Role of Co on microstructure, crystallization behavior and soft magnetic properties of $(Fe_{1-x}Co_x)_{84}Si_4B_8P_3Cu_1$ nanocrystalline alloys. *J. Mater. Sci.: Materials in Electronics* 26: 2076–2081.
78. Takenaka, K., A.D. Setyawan, Y. Zhang, P. Sharma, N. Nishiyama, and A. Makino (2015). Production of nanocrystalline (Fe, Co)–Si–B–P–Cu alloy with excellent soft magnetic properties for commercial applications, *Mater. Trans.* 56: 372–376.
79. Nathasingh, D.M. and H.H. Liebermann (1987). Transformer applications of amorphous alloys in power distribution systems. *IEEE Trans. Power Delivery*, PWRD-2: 843–850.
80. Nagayama, K., H. Ino, N. Saito, Y. Nakagawa, E. Kita, and K. Siratori (1990). Magnetic properties of amorphous Fe–Nd alloys. *J. Phys. Soc. Jpn.* 59: 2483–2495.
81. Croat, J.J. (1981). Magnetic properties of melt-spun Pr-Fe alloys. *J. Appl. Phys.* 52: 2509–2511.
82. Inoue, A., T. Zhang, W. Zhang, and A. Takeuchi (1996). Bulk Nd–Fe–Al amorphous alloys with hard magnetic properties. *Mater. Trans., JIM* 37: 99–108.
83. Inoue, A., T. Zhang, A. Takeuchi, and W. Zhang (1996). Hard magnetic bulk amorphous Nd–Fe–Al alloys of 12 mm diameter made by suction casting. *Mater. Trans., JIM* 37: 636–640.
84. Pan, M.X., B.C. Wei, L. Xia, W.H. Wang, D.Q. Zhao, Z. Zhang, and B.S. Han (2002). Magnetic properties and microstructural characteristics of bulk Nd–Al–Fe–Co glassy alloys. *Intermetallics* 10: 1215–1219.
85. Inoue, A., A. Takeuchi, and T. Zhang (1998). Ferromagnetic bulk amorphous alloys. *Metall. Mater. Trans. A* 29A: 1779–1793.
86. Wei, B.C., W. Löser, L. Xia, S. Roth, M.X. Pan, W.H. Wang, and J. Eckert (2002). Anomalous thermal stability of Nd–Fe–Co–Al bulk metallic glass. *Acta Mater.* 50: 4357–4367.
87. Xia, L., S.S. Fang, C.L. Jo, and Y.D. Dong (2006). Glass forming ability and microstructure of hard magnetic $Nd_{60}Al_{20}Fe_{20}$ glass forming alloy. *Intermetallics* 14: 1098–1101.
88. Wei, B.C., W.H. Wang, M.X. Pan, B.S. Han, Z.R. Zhang, and W.R. Hu (2001). Domain structure of a $Nd_{60}Al_{10}Fe_{20}Co_{10}$ bulk metallic glass. *Phys. Rev. B* 64: 012406-1–012406-4.

89. Schneider, S., A. Bracchi, K. Samwer, M. Seibt, and P. Thiagarajan (2002). Microstructure-controlled magnetic properties of the bulk glass-forming alloy $Nd_{60}Fe_{30}Al_{10}$. *Appl. Phys. Lett.* 80: 1749–1751.
90. Frankwicz, P.S., S. Ram, and H.-J. Fecht (1996). Observation of a metastable intermediate phase in water quenched $Zr_{65.0}Al_{7.5}Ni_{10.0}Cu_{17.5}$ cylinders. *Mater. Lett.* 28: 77–82.
91. Pecharsky, V.K. and K.A. Gschneidner Jr. (1999). Magnetocaloric effect and magnetic refrigeration. *J. Mag. Mag. Mater.* 200: 44–56.
92. Waske, A., B. Schwarz, N. Mattern, and J. Eckert (2013). Magnetocaloric (Fe–B)-based amorphous alloys. *J. Mag. Mag. Mater.* 329: 101–104.
93. Brück, E. (2007). Magnetocaloric refrigeration at ambient temperature. In: K.H.J. Buschow (ed.), *Handbook of Magnetic Materials*, pp. 235–291. Amsterdam: Elsevier.
94. Yang, W., J. Huo, H. Liu, J. Li, L. Song, Q. Li, L. Xue, B. Shen, and A. Inoue (2016). Extraordinary magnetocaloric effect of Fe-based bulk glassy rods by combining fluxing treatment and J quenching technique. *J. Alloys Compd.* 684: 29–33.
95. Álvarez, P., P. Gorria, J.S. Marcos, L.F. Barquín, and J.A. Blanco (2010). The role of boron on the magneto-caloric effect of FeZrB metallic glasses. *Intermetallics* 18: 2464–2467.
96. Franco, V., A. Conde, and L.F. Kiss (2008). Magnetocaloric response of FeCrB amorphous alloys: Predicting the magnetic entropy change from the Arrott-Noakes equation of state. *J. Appl. Phys.* 104: 033903-1–033903-5.
97. Min, S.G., K.S. Kim, S.C. Yu, and K.W. Lee (2007). The magnetization behavior and magnetocaloric effect in amorphous Fe-Nb-B ribbons. *Mater. Sci. Eng. A* 449–451: 423–425.
98. Kim, K.S., S.G. Min, S.C. Yu, S.K. Oh, Y.C. Kim, and K.Y. Kim (2006). The large magnetocaloric effect in amorphous $Fe_{91-x}Y_xZr_9$ ($x = 0, 5, 10$) alloys. *J. Mag. Mag. Mater.* 304: e642–e644.
99. Caballero-Flores, R., V. Franco, A. Conde, and L.F. Kiss (2010). Influence of Mn on the magnetocaloric effect of nanoperm-type alloys. *J. Appl. Phys.* 108: 073921-1–073921-5.
100. Wang, Y.Y. and X.F. Bi (2009). The role of Zr and B in room temperature magnetic entropy change of FeZrB amorphous alloys. *Appl. Phys. Lett.* 95: 262501-1–262501-3.
101. Min, S.G., K.S. Kim, S.C. Yu, H.S. Suh, and S.W. Lee (2005). Analysis of magnetization and magnetocaloric effect in amorphous FeZr Mn ribbons. *J. Appl. Phys.* 97: 10M310-1–10M310-3.
102. Phan, T.L., N.H. Dan, T.D. Thanh, N.T. Mai, T.A. Ho, S.C. Yu, A.T. Le, and M.H. Phan (2015). Magnetic properties and magnetocaloric effect in $Fe_{90-x}Sn_xZr_{10}$ alloy ribbons. *J. Korean Phys. Soc.* 66: 1247–1252.

10

Applications

10.1 Introduction

The previous chapters have described the criteria for bulk metallic glass (BMG) formation, the different methods for their preparation, their crystallization behavior, and a description of their physical, chemical, and magnetic properties. From these descriptions, it is clear that BMGs have an interesting combination of properties. They exhibit very high strength (both in tension and compression), large elastic elongation limit, and in some cases both high strength and large ductility, very high hardness, excellent corrosion resistance, and a good combination of soft magnetic properties. Even though the melt-spun glassy ribbons also have somewhat similar properties, BMGs exhibit some special characteristics.

10.2 Special Characteristics of Bulk Metallic Glasses

One of the greatest advantages of BMGs is that they could be cast into large section thicknesses. As mentioned earlier, because of the presence of a large number of components in the alloy system and its composition being close to deep eutectics, the critical cooling rate required to form a glassy phase is very low, typically 1–10 K s^{-1}. Consequently, it becomes possible to produce large section thicknesses in the fully glassy state. Glassy rods with diameters of >10 mm have been produced in alloy systems based on Ca, Ce, Co, Cu, Fe, La, Mg, Nd, Ni, Pd, Pt, Ti, Y, and Zr. Currently, the largest diameter of a rod that could be cast into a fully glassy state has increased from 72 mm in a $Pd_{40}Cu_{30}Ni_{10}P_{20}$ alloy [1] to 80 mm in a $Pd_{42.5}Cu_{30}Ni_{7.5}P_{20}$ alloy [2]. But, there does not appear to be any theoretical limit on the maximum section thickness that can be achieved for a fully glassy structure, as long as one can design an appropriate composition, and solidify the melt at a rate higher than the critical cooling rate for glass formation. The advantages of a large section thickness for industrial applications are limitless. Parts could be produced not only for electronic applications or microelectromechanical systems (MEMS) devices but also for large components.

BMGs are also characterized by a large supercooled liquid region, defined as the temperature interval between the glass transition, T_g, and crystallization temperatures, T_x, that is, $\Delta T_x = T_x - T_g$. This value can be very large and the highest value reported so far is 131 K in a $Pd_{43}Ni_{10}Cu_{27}P_{20}$ alloy [3]. The viscosity of the BMG is very low in this temperature regime and, therefore, it is very easy for the BMG to be conveniently fabricated into complex shapes. In fact, superplastic forming (SPF) of BMGs in this temperature interval has been achieved [4–8]. It has been frequently pointed out that the SPF of BMGs is very similar to the processing of plastics [5]. Since SPF is an established commercial practice, this should prove extremely beneficial and economical in the processing of BMGs. The only point that needs to be kept in mind during SPF of BMGs is that the time available for processing the BMGs is limited, since they tend to crystallize. The time for crystallization is short at high temperatures and is reasonably long at lower temperatures. Of course, the actual values are different for different types of BMG alloys.

In the case of SPF operations, the time for processing has to be shorter than the time required for crystallization. Since the time for crystallization depends on the pressure and temperature also, these parameters also need to be optimized.

But, if filling of a die is involved [7], then the formability of a BMG in the supercooled liquid region can be quantified using the Hagen–Poiseuille equation:

$$p = 16v\eta \frac{L}{d^2} \tag{10.1}$$

with p as the required pressure to move a liquid with viscosity, η, at a velocity, v, through a channel of thickness, d, and length, L. The maximum time available for the forming process is given by the time to reach crystallization, t_{cryst}. Substituting $v = L/t_{\mathrm{cryst}}$ in Equation 10.1, we get the maximum length that can be filled as

$$L = \sqrt{\frac{p t_{\mathrm{cryst}} d^2}{16\eta}} \tag{10.2}$$

This filling length can be used to quantify the formability of the BMG and to determine the optimum processing conditions.

Another very important characteristic of BMGs is their large elastic strain limit. Unlike crystalline alloys, where it is much less (about 0.5%), most BMG alloys exhibit an elastic strain limit of about 2% at room temperature (see Chapter 8). The most powerful combination of low Young's modulus and large elastic strain limit can be useful in a variety of commercial and large-volume applications.

Yet another special feature of BMGs is their high yield (or fracture) strength and hardness. Because of this, BMGs could be used in applications where scratch and wear resistance becomes important.

The last feature that is characteristic of BMGs is the ability to achieve a very good surface finish. Since the BMGs are glassy and so do not have microstructural features such as grains and grain boundaries, the surface of BMGs is very smooth, and they can also be polished to a very high surface finish, if required. In fact, such good surface finishes were earlier obtained only in oxide glasses and perhaps in polymeric materials.

As different applications require different forms of the glassy material, BMG alloy compositions are produced in the form of rods, sheets, plates, spheres, pipes, and others. Figure 10.1 shows photographs of the different forms and shapes in which BMGs are synthesized.

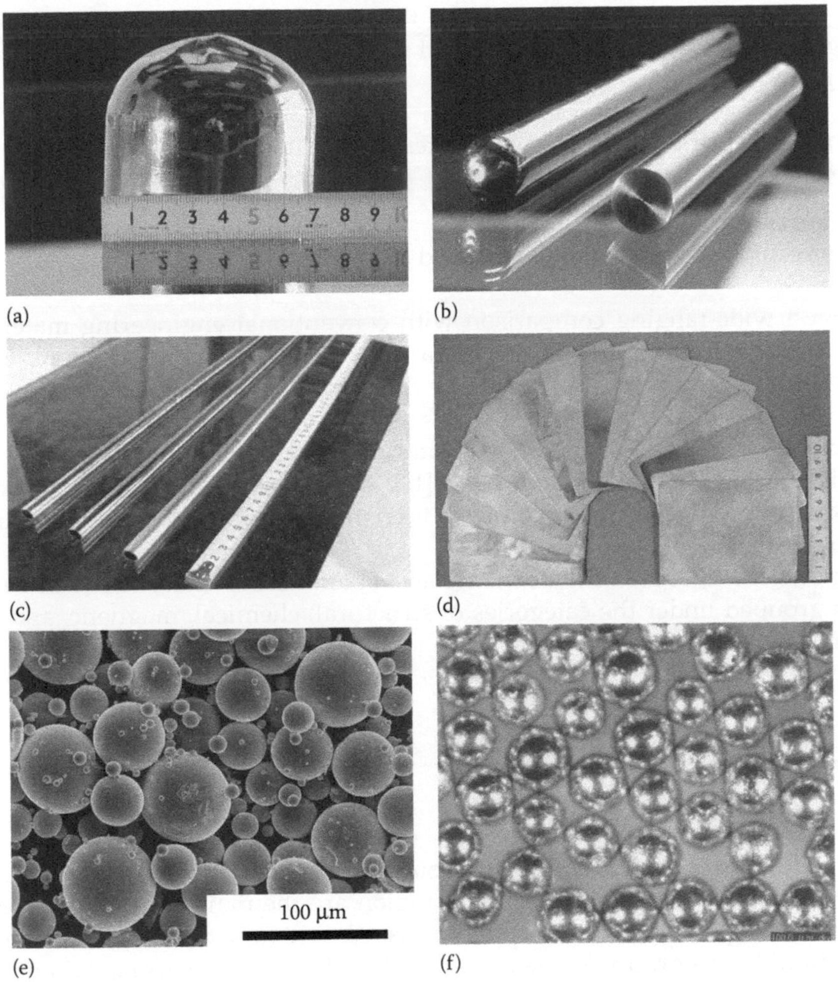

FIGURE 10.1
Different forms in which BMGs have been produced. (a) Cast cylinder, (b) rods, (c) pipes, (d) sheets, (e) powder, and (f) spheres.

It is probably not just one single property that is likely to be important in selecting a material (whether crystalline or glassy) for any specific application; more often than not it is a combination of a few properties. Added to this, the continuous availability and constant supply of the material, and its compatibility with the environment and the human body (if it is to be used as a replacement in the body) are also important. But, the most important aspect that a manufacturer looks for in any material is the cost involved (the cost of the material and the cost of fabrication). Unless it is less expensive than the existing competing material, it is unlikely to replace it, irrespective of the enhanced properties and performance. The only exception to this could be in some critical applications where a material is not available or an application where the cost does not play an important role, for example, in military, space, or life-saving medical applications. Succinct summaries of some of the potential applications of BMGs have recently been presented [9,10].

Ashby and Greer [11] have summarized the attractive and unattractive attributes of metallic glasses and suggested that a useful starting point [12] to search for applications for these materials could be to

1. Identify the attributes of the new material that are better than those of existing materials.
2. Identify the attributes that are worse.
3. Explore applications that exploit (1) and are insensitive to (2).

Based on a wide-ranging comparison with conventional engineering materials, they have shown that, currently, metallic glasses are restricted to niche applications. But, due to their outstanding properties, there could be many more future applications awaiting, for example, in MEMS devices.

Liquidmetal Technologies in the United States (http://www.liquidmetal.com) has already commercialized some of the BMG products [13]. Another company is BMG Corporation in Japan (http://www.bmg-japan.co.jp), which is currently manufacturing BMG samples for testing purposes.

Let us now look at some of the existing and potential applications of BMG materials. These are grouped under the categories of structural, chemical, magnetic, and miscellaneous applications.

10.3 Structural Applications

The high yield (or fracture) strength, low Young's modulus, large elastic strain limit, and easy formability in the supercooled liquid region are the main attributes of BMGs that make them attractive for structural applications. We will now discuss the different possible applications in this group. As mentioned earlier, the large supercooled liquid region in BMG alloys offers an excellent opportunity to form complex shapes easily. This is mainly because the plastic flow of the material in this temperature regime is Newtonian in nature (i.e., the strain rate is proportional to the applied stress). This attribute of BMG alloys has been extensively exploited to produce different types of parts with complex shapes, such as gears, coiled springs, and others. The sizes of these parts are much smaller than have been achieved using conventional crystalline alloys.

10.3.1 Sporting Goods

BMGs first found widespread application in sporting goods due to their desirable mechanical properties, *viz.*, high strength and large elastic elongation limit. The excellent mechanical properties of Zr-based BMGs were exploited commercially in golf clubs, followed by tennis rackets, baseball and softball bats, skis and snowboards, bicycle parts, scuba gear, fishing equipment, and marine applications [13].

The BMGs are generally produced in the form of rods or plates. But, for their application to golf clubs, it is necessary to make near-net moldings that are heavy and have a large area. Therefore, Kakiuchi et al. [14] have developed the "mold-clamp casting method" to produce 250 × 220 × 3 mm flat sheets of a Zr-based BMG alloy, $Zr_{55}Cu_{30}Al_{10}Ni_5$. The copper mold used in this method has two parts. In the first stage, the master alloy is arc-melted on the lower copper mold, which is cooled by water. After the alloy is molten, the lower mold is pushed into the press stage. In this stage, an inclined upper copper mold presses down onto the molten alloy, spreading it uniformly. Due to the efficient heat extraction, the molten alloy solidifies into the glassy state. The mechanical properties of such mold-clamp cast BMGs were found to be superior to those produced by the regular copper mold casting method. The contour of the produced glassy alloy face was properly arranged by machining, and this was then bonded to the concavity provided in the face of the head body of golf clubs made of Ti-6Al-4V.

To "fly" is one of the most critical performance requirements for both clubs and balls in a golf game. The main factor that causes a club to let a ball fly is a property of repulsion between them. To improve the repulsion, efforts have been made in the past mainly on modifying the ball materials, because the repulsion is largely governed by the energy loss arising from the deformation of the ball. In the present case, it has been possible to increase the repulsion efficiency using the Zr-based BMG golf club head. For example, the repulsive efficiency (defined as the ratio of ball velocity/club head velocity) was found to be 1.43 for the BMG alloy face, whereas it is only 1.405 for the Ti-alloy face. The overall flying distance was 225 m for the BMG alloy face, whereas it is only 213 m for the Ti-alloy face.

All these improvements are essentially due to the mechanical properties of the BMG alloy. The BMG has a very high yield strength coupled with a large elastic strain limit, typically about 2%, which is more than twice that in a crystalline material. Hence, the modulus of resilience, *U*, calculated as the area under the elastic portion of the stress–strain curve works out to be

$$U = \tfrac{1}{2}\sigma_y \cdot \varepsilon_y = \tfrac{1}{2}E\varepsilon_y^2 \tag{10.3}$$

where:

σ_y and ε_y are the yield stress and elastic strain limit, respectively
E is Young's modulus

Since the ε_y value for BMGs is at least twice that in a crystalline material, the modulus of resilience is at least four times that of a crystalline material. It is suggested that 99% of the impact energy from a BMG head is transferred to the ball. This value should be compared with 70% for a titanium head. Figure 10.2 shows the outer shapes of the golf club heads made of a Zr-based BMG manufactured by SRI Sports Ltd. (Kobe, Japan), a subsidiary of Dunlop Corporation in Japan.

Liquidmetal Technologies calls this high-energy transfer "pure energy transfer." This company has started marketing baseball bats, tennis rackets, and other sporting goods (Figure 10.3). The HEAD Radical Liquidmetal tennis rackets seem to offer large sweet spots, plenty of control, and impressive feel, with very little vibration.

10.3.2 Precision Gears

Forming complicated structures from BMGs in their supercooled liquid region offers many advantages. Firstly, BMGs are ideal materials for small geometries since they are highly homogeneous even on a nanoscale, due to the absence of crystalline features such as grains in the glassy material. Secondly, the solidification shrinkage during SPF of BMGs in the supercooled liquid region is about an order of magnitude less than what

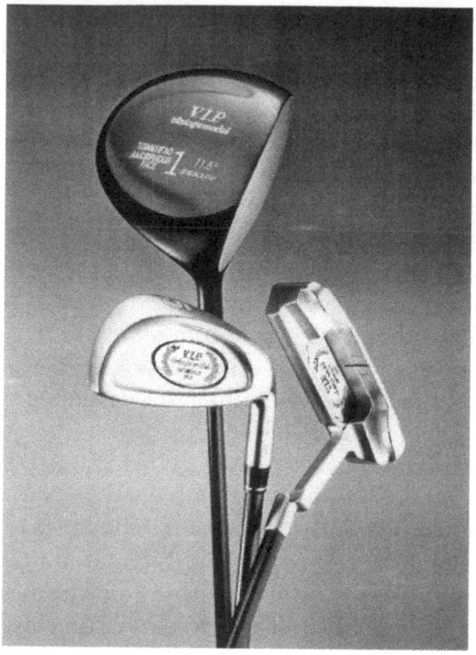

FIGURE 10.2
Outer shapes of commercial golf club heads in wood-, iron-, and putter-type forms where the face materials are made of Zr-based BMG alloy. (Reprinted from Kakiuchi, H. et al., *Mater. Trans.*, 42, 678, 2001. With permission.)

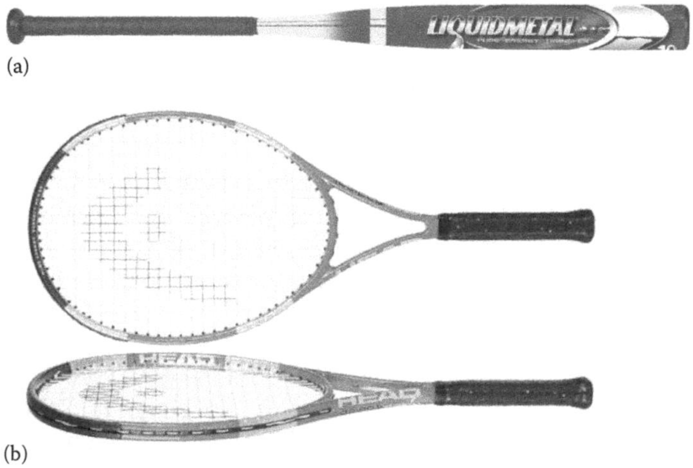

FIGURE 10.3
(a) Baseball bat and (b) tennis racket made of Liquidmetal (BMG) alloys.

it is in typical casting alloys. This is due to the absence of first-order phase transformations (liquid-to-solid), which introduce significant shrinkage into the casting. Thirdly, the degree of porosity in the formed part is very significantly reduced. It is estimated that if a feedstock material that contains 2% porosity with pores of 10 μm or larger is formed under a pressure of 50 MPa, the porosity after SPF is reduced to 0.004% [7]. Lastly, the superior surface imprintability and net-shape forming capability have been found to be very attractive in the field of micro-machines [15].

Because of the excellent filling characteristics of BMG alloys, it has been possible to produce extremely small parts of complex design using BMGs. Let us now look at the fabrication of a micro-gear, referred to in the literature as a *sun-carrier.* A conventional sun-gear is usually produced by assembling five individual machined parts. But, using the BMG alloys, it is possible to produce the sun-gear in just one step [16].

Figure 10.4a shows a schematic illustration of the two dies used. The cavity in the upper die (die A) consists of two concentric cylindrical holes with diameters of 1.7 and 0.6 mm and heights of 0.3 and 0.4 mm, respectively. Figure 10.4b shows a schematic illustration

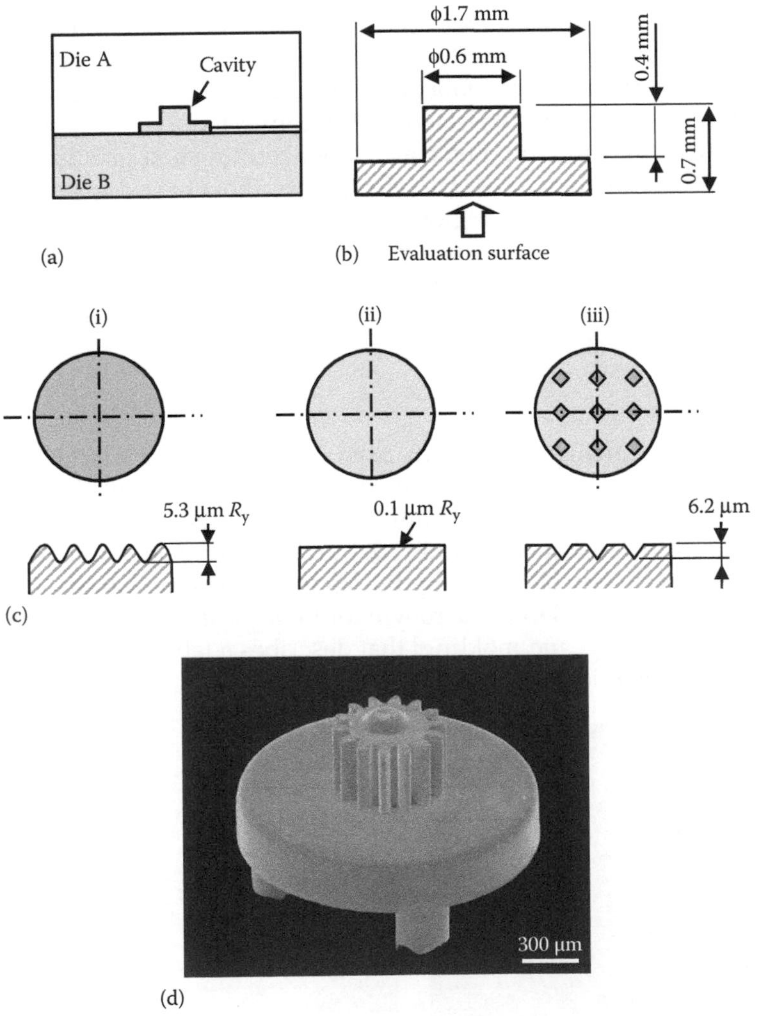

FIGURE 10.4
(a) Schematic illustration of the die assembly; (b) schematic of the specimen with the dimensions indicated; (c) schematic illustrations of the top surface of Die B prepared by (i) electro-discharge machining, (ii) polishing, and (iii) Vickers indentation. (d) External appearance of the Ni-based BMG sun-carrier fabricated by the precision die casting technique from an electro-discharge machined mold. (Reprinted from Ishida, M. et al., *Mater. Trans.*, 45, 1239, 2004. With permission.)

of the specimen. The cavity has the simplified geometry of the micro-gear. The filling characteristics of the lower die were investigated by preparing the top plane of the lower die (die B) in three different finishes—electro discharge machining, polishing, and Vickers indentation—as shown in Figure 10.4c. The $Ni_{53}Nb_{20}Zr_8Ti_{10}Co_6Cu_3$ BMG and a conventional Al–Si–Cu die-casting alloy (ADC 12) were used to study the filling characteristics. It was shown that, in comparison to the Al die casting alloy, the BMG had superior filling characteristics. It was also shown that the BMG could faithfully reproduce the surface features of the die, thus ensuring that the surface imprintability of the BMGs was good. The die cavity is filled using a precision die casting technique in the temperature range of 1300–1690 K. The actual sun-carrier made with an $Ni_{53}Nb_{20}Zr_8Ti_{10}Co_6Cu_3$ BMG alloy is shown in Figure 10.4d. It consists of a micro-gear, carrier plate, and three pins. The outer diameter of the gear is 0.65 mm, it has 14 teeth with a module of 0.04, and the micro-gear is seated on a carrier plate 1.7 mm in diameter. Three pins, with a diameter of 0.30 mm and a length of 0.45 mm, are located at the bottom of the carrier plate for rotating planetary gears.

Similar complex gears were also fabricated by Schroers et al. [7] by microreplication from <100> Si molds. Since the fabricated part is connected to a large reservoir, it needs to be separated from the reservoir. But, conventional cutting and grinding techniques are incapable of holding tight tolerances. Therefore, the authors have processed the BMG and the mold in the supercooled liquid region and used a scraper to hot separate the reservoir from the part. This method resulted in a plane surface on which further layers could be fabricated for MEMS devices. The parts could also be etched out to obtain 3-D micro parts. Figure 10.5 shows a complex gear and a coil-shaped spring made by such a process from a $Zr_{44}Ti_{11}Cu_{10}Ni_{10}Be_{25}$ BMG alloy.

10.3.3 Motors

Micro-geared motors with high torque have been used in different engineering fields. With advancing technology and the need for miniaturization, the size of the motors has been constantly decreasing. For example, the minimum size of the motors was 12 mm in 1985, and it was reduced to 7 mm in 2000 and to 2.4 mm in 2005. Generally, micro-gears with complicated 3-D shapes are fabricated by lithography, micromachining, chemical or ion etching, or LIGA. (LIGA is a German acronym for *LIthographie, Galvanoformung, Abformung* [lithography, electroplating, and molding] that describes a fabrication technology used to create high-aspect-ratio microstructures.)

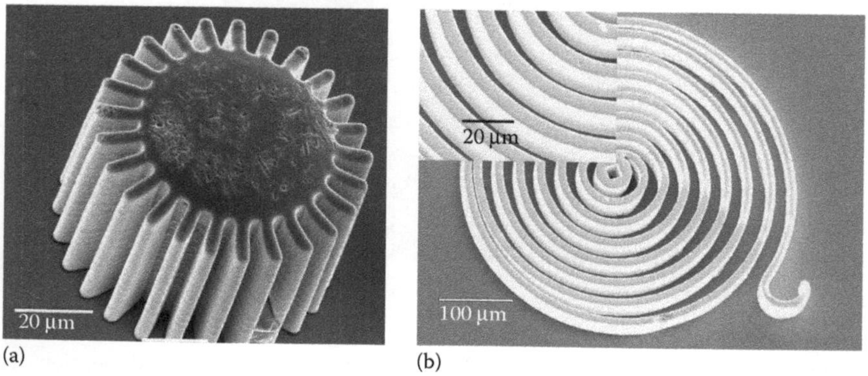

FIGURE 10.5
(a) A complex micro-gear and (b) coil shape spring made from a $Zr_{44}Ti_{11}Cu_{10}Ni_{10}Be_{25}$ BMG alloy. (Reprinted from Schroers, J. et al., *Mater. Sci. Eng. A*, 449–451, 898, 2007. With permission.)

The wear resistance behavior of a 2.4 mm diameter gear was evaluated using sliding-wear and rolling-wear tests [17]. It was reported that the wear loss (volume) of the Ni-based BMG was larger than that of the carbon steel under sliding-wear conditions, but smaller under rolling-wear conditions. Since the Ni-based BMG has a high hardness, this result suggests that factors other than the hardness may be playing an important role in determining the wear loss. The wear loss was, however, different depending on the type of material used for the gears.

Figure 10.6a shows the appearance of the gear made of the Ni-based BMG after 2500 h of durability testing (corresponding to 1875 million revolutions), and Figure 10.6b shows the appearance of the gear made of the carbon steel after 8 h (corresponding to 6 million revolutions). The gear teeth of the carbon steel are worn off and heavily damaged even after just 8 h of use, while the teeth of the BMG alloy are in very good condition even after 2500 h of use [17].

Figure 10.7 shows the comparative wear behavior of gears made with different types of materials in a 2.4 mm diameter geared motor. The gear made with the carbon steel is used as a reference. The wear life of the gear was 1.6 times higher than an all-steel gear when a BMG alloy gear replaced one of the gears. However, the wear life increased by seven times when more gears were replaced with BMG alloys. However, in the case of all BMG alloy gears, the wear life was 313 times higher than all steel gears [18].

Encouraged by this, Inoue et al. [19] have fabricated the world's smallest micro-geared motor (1.5 mm in diameter and 9.4 mm in length) using a high-strength Ni-based BMG alloy. The components of this geared motor cannot be made by any mechanical machining methods. Figure 10.8 shows the different components of this motor. It consists of a sun-carrier, an output shaft, and six pieces of planetary gear, all made out of the Ni-based BMG with the composition $Ni_{53}Nb_{20}Zr_8Ti_{10}Co_6Cu_3$. It was confirmed that this micro-geared motor had high rotating torques of 0.1 m Nm at two stages of a stacked gear-ratio reduction system and 0.6 m Nm at three stages, which were 6–20 times higher than the vibration force for a conventional geared motor with a diameter of 4.5 mm in mobile telephones. These micro-geared motors are expected to be used in advanced medical equipment such as endoscopes, micropumps, rotablators, and catheters for thrombus removal, precision optics, micro-industries, micro-factories, and so on (Figure 10.9). The micro-geared motor diameter has been further reduced to 0.9 mm, which is more suitable for biomedical application [10].

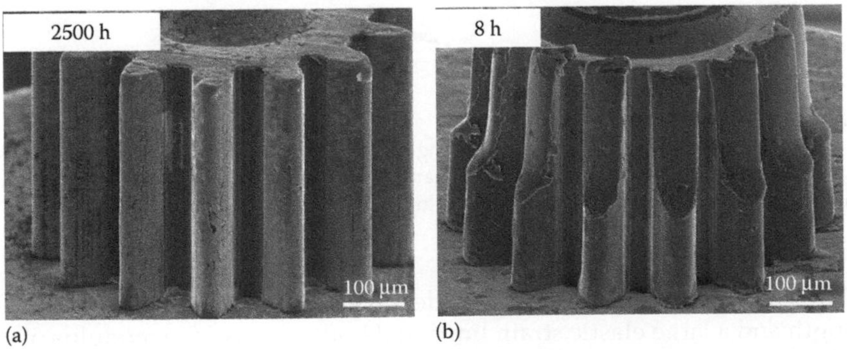

FIGURE 10.6
SEM images of the gears after the durability tests. (a) The gear made out of the Ni-based BMG alloy after 2500 h of use (1875 million revolutions) and (b) the carbon steel gear after 8 h of use (6 million revolutions). Notice the serious damage in the carbon steel gear even after just 8 h of use, while the BMG gear is intact even after 2500 h of use. (Reprinted from Ishida, M. et al., *Mater. Sci. Eng. A*, 449–451, 149, 2007. With permission.)

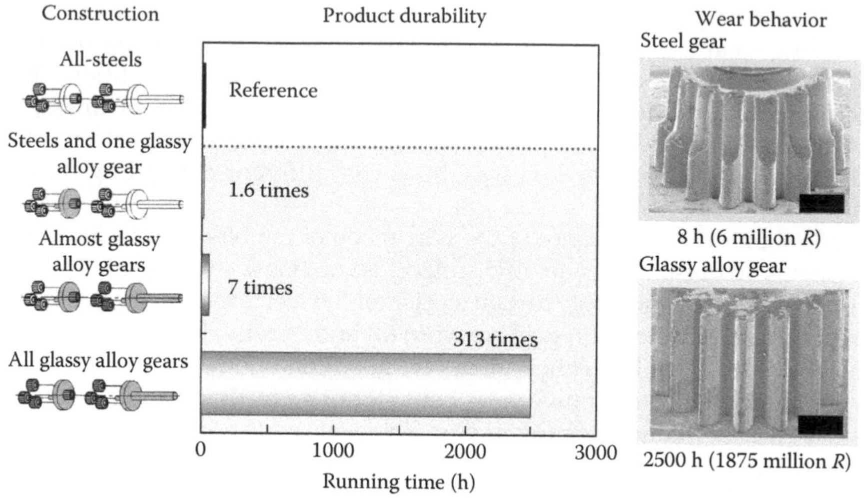

FIGURE 10.7
Comparative wear resistance behavior of gears made with different materials in a 2.4 mm diameter geared motor. (Reprinted from Inoue, A. et al., *Mater. Sci. Eng. A*, 441, 18, 2006. With permission.)

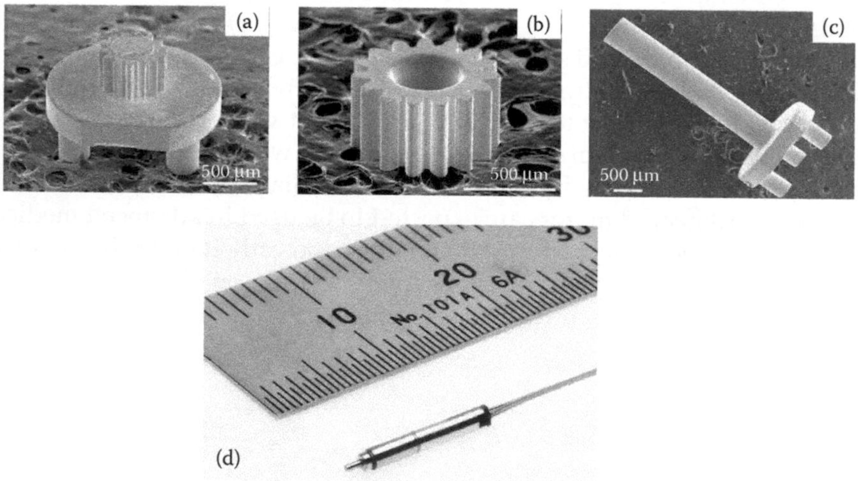

FIGURE 10.8
Precision micro-gear parts produced by injection casting of an $Ni_{53}Nb_{20}Zr_8Ti_{10}Co_6Cu_3$ BMG alloy: (a) sun-carrier, (b) planetary gear, and (c) an output shaft. (d) Micro-geared motor with a diameter of 1.5 mm and a length of 9.4 mm fabricated from the $Ni_{53}Nb_{20}Zr_8Ti_{10}Co_6Cu_3$ BMG alloy.

10.3.4 Automobile Valve Springs

The mechanical properties of BMGs are ideal for spring materials. They exhibit high yield strength and a large elastic strain limit, and both increase the modulus of resilience. Further, the combination of high strength and low Young's modulus allow the spring wires to be slimmer and the springs themselves to be shorter. Additionally, by using the appropriate BMG alloy compositions in automobile valve springs, it should be possible to decrease the weight of the engines by reducing the cylinder head sizes and consequently the fuel consumption. It was estimated that if the conventional valve springs made of

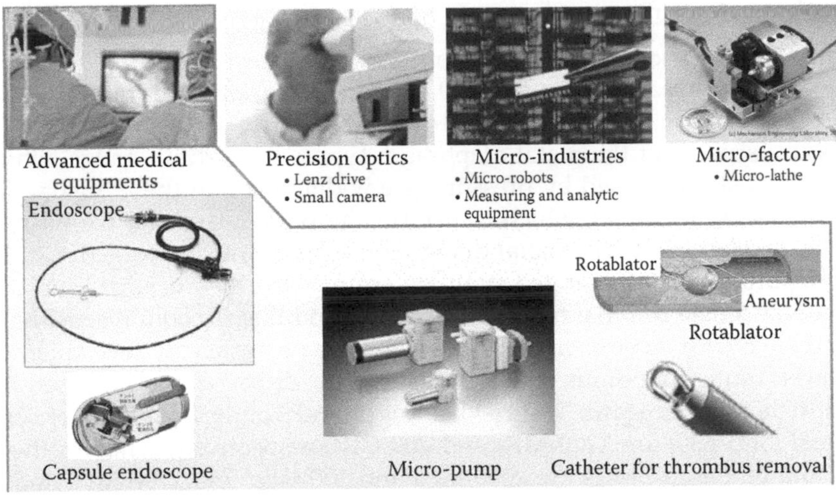

FIGURE 10.9
Projected application areas for micro-geared motors.

oil-tempered and shot-peened Si–Cr steel were replaced with Zr- or Ti-based BMGs, the overall weight of the engine would come down by 4 kg (about 10 lb).

Son et al. [20] have developed a method to produce wires by the rotating disk quenching method. They were able to produce wires with diameters ranging from 1.4 to 5 mm. In this process, the molten alloy was injected into a semicircular groove on a rotating copper disk of 500 mm diameter. By varying the process parameters, such as nozzle diameter (500–900 μm), injection pressure (30–50 kPa), and the rotation speed of the copper disk (0.47–0.58 m s^{-1}), the authors were able to optimize the conditions to achieve a wire with a uniform cross section and good surface finish. The metallic wires were then wound over a metallic mold heated to 683–700 K (which is about 50 K above the glass transition temperature of the alloy investigated) to obtain helical springs. To avoid crystallization, coiling was completed within 20–30 min, followed by cooling to room temperature. Figure 10.10 shows the helical springs of $Zr_{55}Cu_{30}Al_{10}Ni_5$ alloy produced by the coiling of glass wires of 1 and 2 mm diameter.

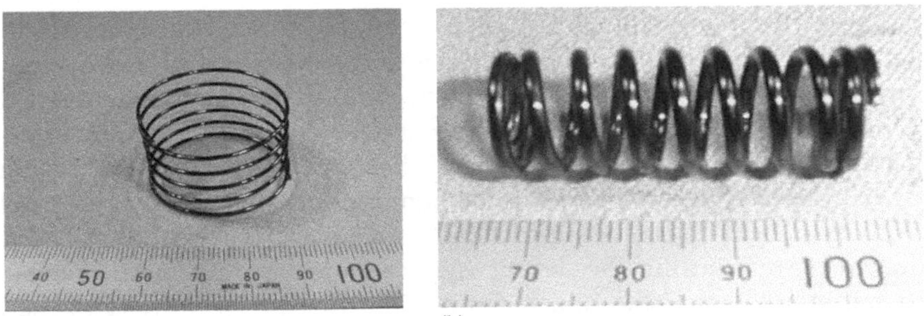

(a) (b)

FIGURE 10.10
Helical springs of $Zr_{55}Cu_{30}Al_{10}Ni_5$ BMG alloy produced by the coiling of wires of (a) 1 mm and (b) 2 mm in diameter. (Reprinted from Son, K. et al., *Mater. Sci. Eng. A*, 449–451, 248, 2007. With permission.)

10.3.5 Diaphragms for Pressure Sensors

Pressure sensors are used for fuel-injection control to lower the fuel consumption of automobiles and thus reduce toxic and CO_2 exhaust emissions. Miniaturized and high-sensitivity pressure sensors for braking oil-pressure control are also desirable for active safety. Pressure sensors of the metal-diaphragm type with a strain gauge, which detect the magnitude of strain caused by pressure, are widely used in the industry. Generally, the sensors act as transducers, which transform the parameters that we wish to detect into strain or displacement. The metal diaphragm is made of cold-forged stainless steel. Since the magnitude of the generated strain determines the measurement sensitivity, it is useful to employ a material that has a low Young's modulus. To obtain sensitivity higher than that possible with a commercial diaphragm, it is necessary to develop a new material with lower Young's modulus and higher strength. Zr-based and Ni-based BMGs are very appropriate in this regard. The yield strength and Young's modulus of commercial stainless steel (SUS 630) are 1200 MPa and 200 GPa, respectively, whereas these values for the Zr- and Ni-based BMGs are 2000 MPa and 100 GPa, and 2700 MPa and 135 GPa, respectively. Thus, it is clear that the Zr-based BMGs with a lower Young's modulus are appropriate for a high-sensitivity diaphragm, and the Ni-based BMG with a high strength is suitable for a high-pressure resistance diaphragm. In other words, since the Young's modulus of the Zr-based BMG is about half that of the stainless steel, the sensitivity can be correspondingly doubled simply by replacing stainless steel with the Zr-based BMG.

The strain, ε, in the bottom plane of the diaphragm can be expressed as

$$\varepsilon = \alpha \frac{Pa^2}{Et^2} \tag{10.4}$$

where:

- α is a constant
- P is the applied pressure
- a is the effective diameter of the diaphragm
- E is Young's modulus
- t is the thickness of the diaphragm

In such situations, BMGs offer a large potential to produce miniaturized diaphragms with high sensitivity and high-pressure capacity because of their low Young's modulus and high strength.

For pressure sensing, a strain gauge with a fine pattern needs to be deposited on the bottom plane of the diaphragm. Since the deposition substrate is heated by conventional chemical vapor deposition (CVD) methods, which could lead to crystallization of the glassy material, a low-temperature deposition process was developed (Cat-CVD and excimer-laser-annealing methods). Figure 10.11 shows the Zr-based $Zr_{55}Cu_{30}Al_{10}Ni_5$ BMG diaphragm, complete with a strain gauge made by this low-temperature deposition process [21].

The sensitivity of both the commercial stainless steel and Zr-based BMG alloy sensors were evaluated by monitoring the output voltage under a given applied pressure. The results are summarized in Table 10.1. It can be seen that the sensitivity of the Zr-based BMG diaphragm was about four times that of a conventional diaphragm [22].

Figure 10.12 shows the different components in an automobile where pressure sensors could be used. It is expected that the market for the sensors in the automobile sector

FIGURE 10.11
Zr-based BMG diaphragm with a strain gauge deposited at low temperatures. (Reprinted from Nishiyama, N. et al., *Mater. Sci. Eng. A*, 449–451, 79, 2007. With permission.)

TABLE 10.1

Output Voltage for Pressure Sensors at a Testing Pressure of 20 MPa Using Commercial Stainless Steel (SUS 630) and Zr-Based $Zr_{55}Cu_{30}Al_{10}Ni_5$ BMG with Strain Gauges Deposited under Identical Conditions

	Output Voltage (mV) Using	
Process	**Stainless Steel (SUS 630)**	**Zr-Based $Zr_{55}Cu_{30}Al_{10}Ni_5$ BMG**
Plasma CVD	60	—
Cat-CVD	50	100
Excimer laser annealing	110	230

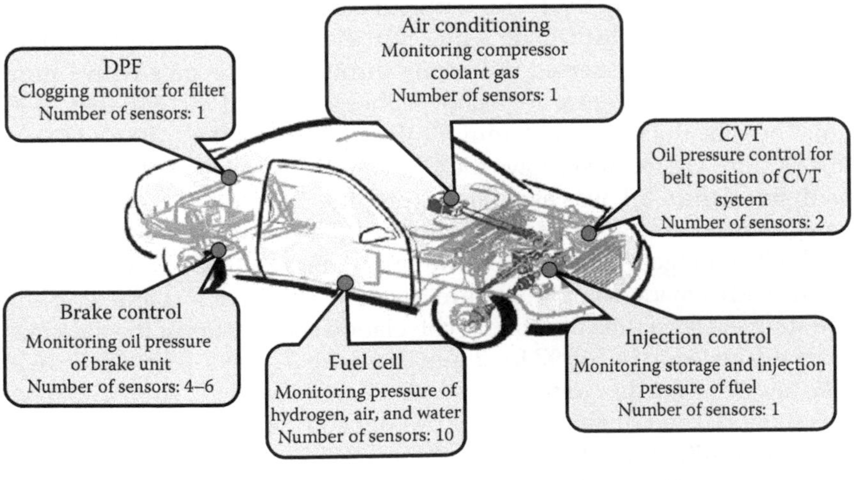

FIGURE 10.12
Expected market for pressure sensors to be used in different parts of an automobile.

alone will be about $100 million. Nagano Keiki Co. Ltd. in Japan has already constructed their mass production facility with a capacity of 5 million sensor pieces a year.

10.3.6 Pipes for a Coriolis Mass Flowmeter

The Coriolis mass flowmeter (CMF) is a unique flowmeter used to directly measure the mass of a fluid. The value obtained using this instrument is independent of the environmental temperature as well as the structure of the measured fluid, a feature that is appealing for applications in different industries, including petroleum, coal or natural gas, semiconductor, biomedical, and advanced chemical industries. This flowmeter, commercialized in 1979, consists of a sensing pipe through which the fluid whose mass is to be measured flows, and a vibration generator that gives a primary flexural vibration to the sensing pipe. These vibrational amplitudes are proportional to the mass of the fluid. By measuring the maximum elastic deformation, δ_{max}, in the pipe, one can calculate the Coriolis force, F_C (the force exerted by the fluid against the wall of the pipe), and from that the mass of the fluid. The δ_{max} is given as

$$\delta_{max} = \frac{4F_C l^3}{3\pi E\left(d_2^4 - d_1^4\right)} \tag{10.5}$$

where:

l, d_1, and d_2 are the length and inner and outer diameters of the pipe
E is Young's modulus

Thus, one can see that δ_{max} is inversely proportional to Young's modulus and therefore a higher sensitivity can be obtained by using a material that has a low E value.

Commercial CMFs are fabricated using stainless steel or Hastelloy for their high corrosion resistance. BMGs have the advantages of high corrosion resistance coupled with high yield strength and low Young's modulus. Because of the high strength, BMG pipes with thinner wall widths can be used. The low Young's modulus leads to a large elastic deformation for the same applied force. From these synergistic effects, the sensitivity when using a BMG pipe is several times higher than with a commercial stainless steel pipe.

Glassy alloy pipes of $Ti_{50}Cu_{25}Ni_{15}Zr_5Sn_5$ with 2 and 6 mm diameters were produced by the suction-casting method. While the pipe with 2 mm diameter was glassy with nanoparticles of about 5 nm in diameter dispersed uniformly throughout, the pipe with 6 mm diameter was mostly crystalline [22]. Figure 10.13 shows photographs of the Ti-based BMG pipes of 2 and 6 mm. The tube thickness is 0.8 mm for the 6 mm diameter tube and 0.2 mm for the 2 mm diameter tube. Figure 10.14 shows the CMF fabricated using the Ti-based BMG pipes.

The sensitivity of the CMF was evaluated using the Ti-based BMG and compared with an electromagnetic flowmeter as a *de facto* standard. The measured values indicate similar linearity in the flow range from 0.1 to 1000 mL min^{-1}. Figure 10.15 shows the obtained sensitivities of the CMFs made using the 6 mm diameter and 2 mm diameter Ti-BMG tubes and also the stainless steel tube. The linear correlation coefficients of the data measured by each CMF are evaluated to be 0.9997 for 2 mm diameter Ti-BMG, 0.995 for the 6 mm diameter Ti-BMG, and 0.997 for the stainless steel. The 2 mm diameter Ti-BMG-based CMF has a sensitivity about 30 times higher than the present commercial units using stainless steel pipes [23]. These results suggest that the Ti-based BMG pipes with low Young's modulus and high strength enhance the performance of the CMFs.

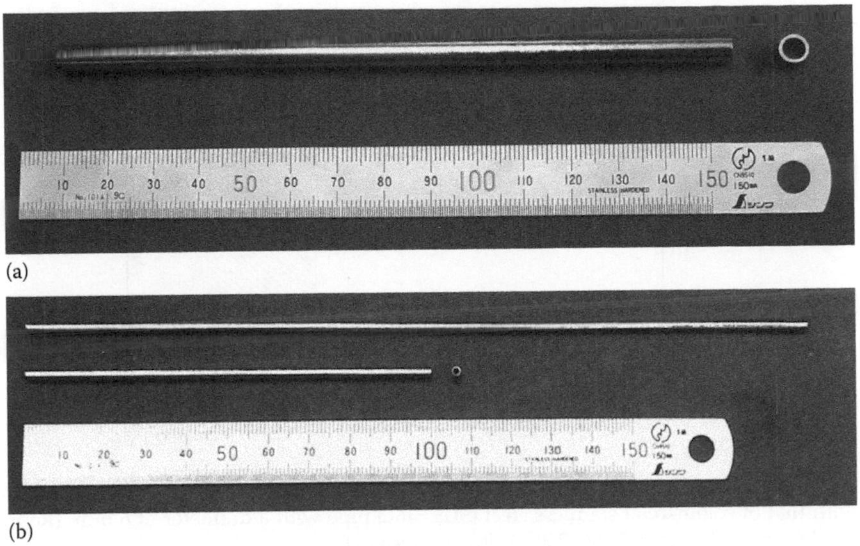

FIGURE 10.13
Ti-based BMG pipes with outer diameters of (a) 6 and (b) 2 mm. (Reprinted from Nishiyama, N. et al., *J. Non-Cryst. Solids*, 353, 3615, 2007. With permission.)

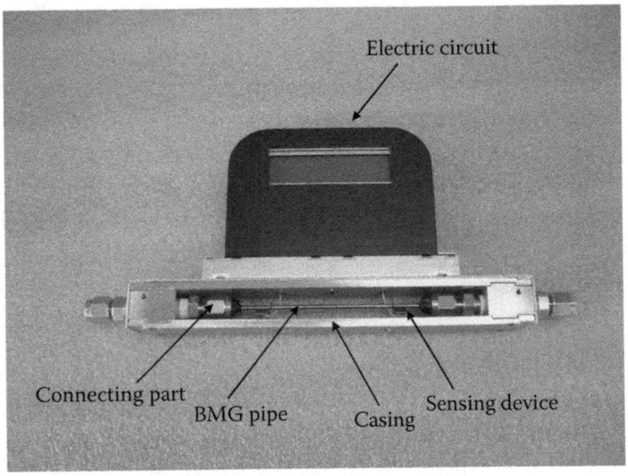

FIGURE 10.14
CMF developed using the Ti-based BMG pipes.

10.3.7 Optical Mirror Devices

The main requirements for a mirror are smooth surface finish and high reflectivity. These requirements are easily met in the case of noncrystalline (or glassy) materials since they do not have any grain structure and therefore no grain boundaries to scatter the light. Both polymeric and oxide glasses have been used for such purposes since they are also transparent.

Metallic glasses (including BMGs) are noncrystalline and they can be easily formed into complex shapes in the supercooled liquid condition. Further, they also exhibit metallic luster. Therefore, BMGs have been explored for the reflective parts of optical devices. Hata

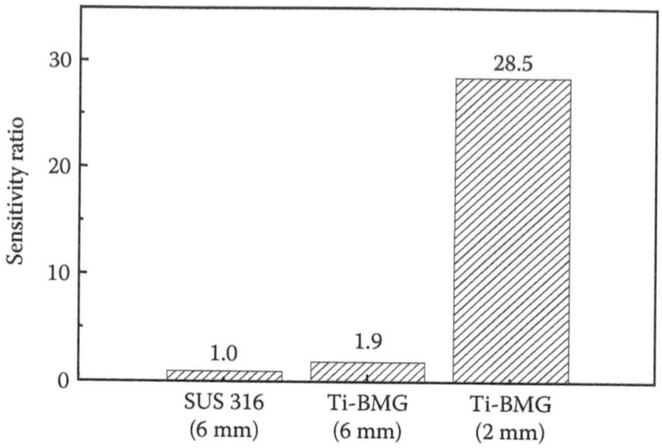

FIGURE 10.15
Sensitivity obtained by the CMF using the Ti-based BMG with diameters of 2 and 6 mm. Their sensitivity is compared with that of commercial stainless steel (SUS 316L) pipe with a diameter of 6 mm. (Reprinted from Ma, C.L. et al., *Mater. Sci. Eng. A*, 407, 201, 2005. With permission.)

et al. [24] presented an example of an optical mirror produced by viscous flow forming of a Zr-based BMG with the composition $Zr_{55}Cu_{30}Al_{10}Ni_5$. This alloy composition was chosen because it has a wide supercooled liquid region of 86 K. By investigating the relationship between strain rate sensitivity and temperature, it was shown that the strain rate sensitivity exponent, *m*, was almost 1.0 at 700 and 720 K, suggesting that the BMG could be considered an ideal Newtonian liquid at these temperatures. Based on this viscous workability in the supercooled liquid state, a concave spherical mirror was prepared at a forming temperature of 730 K under an applied pressure of 30 MPa for under 8 min. The Zr-based BMG mirror thus formed had a smooth surface and exhibited good reflectivity. The surface profile obtained by laser interferometry showed a maximum surface roughness of 90 nm and a deviation from sphericity of 500 nm. These close dimensional tolerances suggest that this route using BMGs can form optical mirrors with high dimensional accuracy.

10.3.8 Structural Parts for Aircraft

BMGs are being considered for aircraft parts, specifically for the slat-track cover surrounding a set of guide rails at the front of the wings (Figure 10.16a). Such parts are conventionally fabricated by machining, although in recent times, SPF of crystalline Ti alloys has also been employed. In contrast, BMGs can be formed at lower temperatures and at lower stresses in the supercooled liquid region. Generally, high corrosion resistance is required for the inner face of the cover, because of exposure to ambient atmosphere moistened by rain or seawater. Zero leakage is required for the outer face of the cover because it acts as part of the fuel tank. In addition, reduced weight and volume are desirable to reduce fuel consumption.

Using high-strength BMGs, thinner and lighter parts can be prepared and, additionally, low-cost production and high throughput can be expected. The $Zr_{55}Cu_{30}Al_{10}Ni_5$ BMG alloy was fabricated into the slat track cover through viscous flow forming at a gas pressure of 3.5 MPa at a temperature of 10 K above the glass transition temperature of this alloy (T_g = 680 K). The desired shape was obtained (Figure 10.16b), showing that crystalline parts fabricated by SPF methods can be conveniently obtained by viscous flow forming of the BMGs, at low applied stresses and lower temperatures [25].

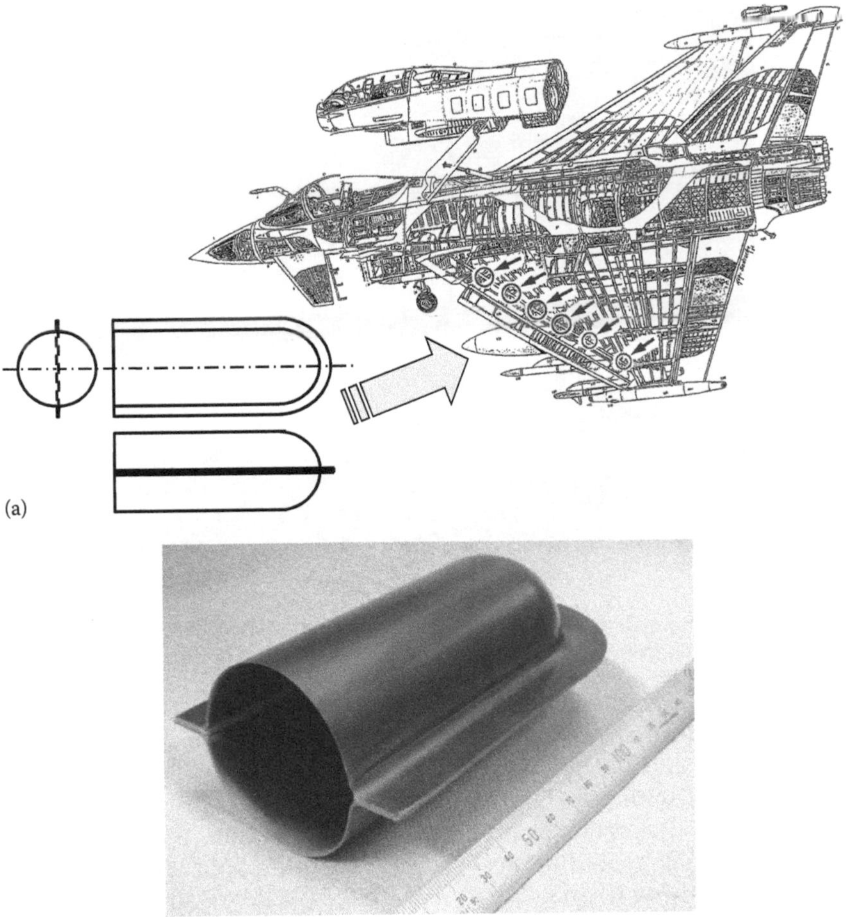

(a)

(b)

FIGURE 10.16
(a) Some areas (see small arrows) of an aircraft in which BMGs could be exploited. (b) Slat track cover fabricated by joining two identical parts obtained by viscous flow forming of a $Zr_{55}Cu_{30}Al_{10}Ni_5$ BMG plate.

10.3.9 Shot Peening Balls

Shot peening is a commercial process to induce residual compressive stresses on the surface of a specimen to improve its fatigue resistance. The shot peening medium used is steel balls since they are cheap. The requirements of the material to be used as shots are high strength, good ductility, high endurance against cyclic bombardment load, and high corrosion resistance. Further, since these are used in large volume, their cost must be low. That is why traditionally high-carbon steels are used for this purpose.

It has been possible to produce Fe-based alloy compositions in a fully glassy state of large section thicknesses of up to 16 mm in an Fe-based BMG alloy with the composition $Fe_{41}Co_7Cr_{15}Mo_{14}C_{15}B_6Y_2$ [26]. This alloy cannot be used for shot peening purposes, since it is brittle. Further, for peening purposes, the BMG needs to be in the form of shots. Therefore, a high-speed atomization process was developed for producing glassy alloy spheres in the range of 0.1–2 mm. The production capacity was 240 t $year^{-1}$.

Figure 10.17 shows the outer shape and surface finish of the commercialized $Fe_{44}Co_5Ni_{24}Mo_2B_{17}Si_8$ glassy alloy shots with a diameter of about 80 μm. These were

FIGURE 10.17
Size and surface finish of commercial $Fe_{44}Co_5Ni_{24}Mo_2B_{17}Si_8$ glassy alloy shots of 80 μm diameter produced by water atomization. (Reprinted from Inoue, A. et al., *Mater. Trans.*, 44, 2391, 2003. With permission.)

produced by water atomization and sieved to the required powder size distribution [27]. The shot peening performance of this steel was compared with two conventional crystalline steels: a high-speed steel with the composition Fe–1.15C–4Cr–5Mo–2.5V-6.5W–8Co (wt.%) and a cast steel with the composition Fe–1C–0.9Si–0.7Mn (wt.%). Table 10.2 compares the mechanical properties of these three steels. Results were obtained on bombarding two commercial crystalline steel sheets having the approximate compositions of Fe–0.83C–0.25Si–0.5Mn (wt.%) (JIS-G4801) and Fe–1.5C–12Cr–1Mo–0.35V (wt.%) (SKD11) with the Fe-based glassy alloy shots and the cast steel (crystalline) shots. The Fe-based BMG performed much better than the other two crystalline steels in all aspects. Table 10.3 summarizes the observations.

The crystalline steel bombarded with the glassy shots had a very homogeneous crater-like surface pattern. The average crater size was measured to be about 20 μm for the glassy alloy shots and about 7 μm for the cast steel shots. Close observation of the crater pattern suggests that the bombarded area generated by one bombardment was much larger for the glassy alloy shots. The average height of the crater edges was measured to be 15 μm for the glassy alloy shots and 5 μm for the cast steel shots. Further, on bombarding for 40 s, the depth of the affected region was 100 μm for the glassy alloy shots and about 45 μm for the cast steel shots. Thus, it is clear that, under identical conditions of shot peening, the glassy alloy shot could produce a much larger affected area than the cast steel shots. More importantly, the surface compressive stresses were higher, about 1600 MPa, with the BMG shots, while they were only about 1470 MPa with the cast steel shots. The depth up to which the compressive surface stresses were retained was higher when bombarded with the BMG shots. A compressive stress of 500 MPa was retained up to a depth of 27 μm in the case of BMG shots and only 18 μm in the case of the cast steel shot. Most importantly, the endurance lifetime of the BMG shot was an order of magnitude longer than that of the cast steels.

TABLE 10.2

Mechanical Properties of the Fe-Based BMG and Two Other Cast Steel Shots Used in the Investigation

Property	Fe-Based BMG $Fe_{44}Co_5Ni_{24}Mo_2B_{17}Si_8$	High-Speed Steel Fe–1.15C–4Cr–5Mo–2.5V–6.5W–8Co (wt.%)	Cast Steel Fe–1C–0.9Si–0.7Mn (wt.%)
Young's modulus (GPa)	80	215	210
Fracture strength (MPa)	3200	2100	1100
Vickers hardness	930	815	810
Density (g cm^{-3})	7.4	7.7	7.55

TABLE 10.3

Effects of Bombardment with Fe-Based BMG Alloy Shots and Cast Steel Shots for 40 s on Two Commercial Steel Sheets

Attribute	Fe-Based BMG	Cast Steel
Crater size (μm)	20	7
Average crater height (μm)	15	5
Depth of affected region (μm)	100	45
Surface Vickers hardness	510	480
Maximum compressive stress on the surface (MPa)	1600	1470
Depth (μm) of the region at which the compressive stress was 500 MPa	27	18
Endurance time needed to final rupture of the peening shots (h)	28	—

10.3.10 Metallic Glassy Screw Bolts

It is known that bulk glassy alloy sheets in the Zr-based alloy system can be cold rolled to about 60%–70% reduction in thickness and the resulting crossing of shear bands can increase the plastic strain in conjunction with a slight decrease in Young's modulus and yield strength, and an increase in fracture strength [28]. However, little is known about the production of the final products for BMGs by cold working, though there are data on the influence of cold working on the structure, thermal stability, mechanical properties, and corrosion behavior of BMGs [10]. Thus, it is useful to explore the possibility of producing the BMG in the final shape by a single cold working process within a short time; this can accelerate the commercialization of BMGs.

Recently, Yamanaka et al. [29] have reported that a screw glassy alloy bolt with a net final shape could be formed in only 6 s, through a single-pass step in a cold thread rolling process (Figure 10.18) of $Zr_{55}Al_{10}Ni_5Cu_{30}$ BMG rod. As shown in Figure 10.19, the bolt prepared by the thread rolling process has a size of 8 mm in shank length, 3 mm in outer diameter, and 0.5 mm in screw pitch. The thread rolling deformation has been confirmed to occur through shear sliding deformation mode on the basis of detailed observation of surface deformation mode in some intermediate thread-rolled stages. The severely deformed regions at the screw top edge have been confirmed to maintain the glassy phase without any appreciable crystalline phase by the focused beam x-ray diffraction method [29].

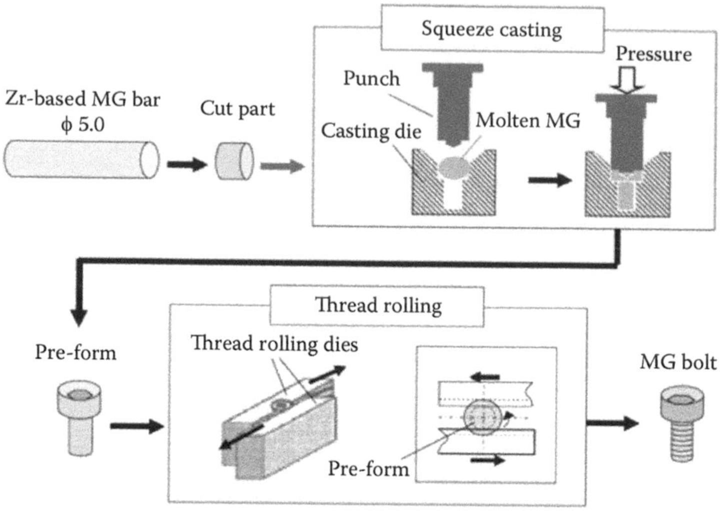

FIGURE 10.18
Experimental procedure for making Zr-based bulk metallic glass (MG) bolt. (Reprinted from Yamanaka, S. et al., *Mater. Trans.*, 52, 243, 2011. With permission.)

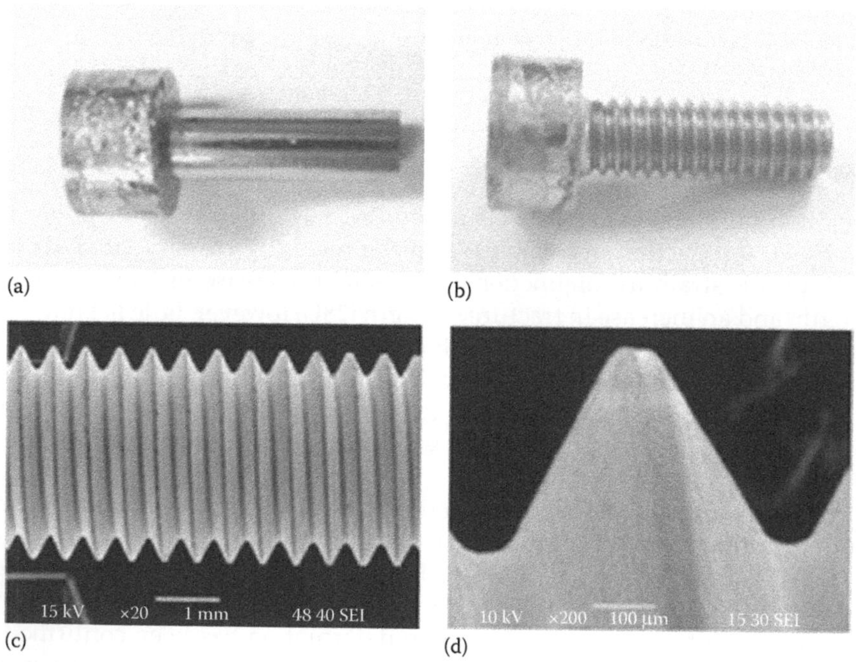

FIGURE 10.19
(a) Pre-form for BMG bolt processed by squeeze casting; (b) thread-rolled BMG bolt at room temperature (293 K); (c) scanning electron micrograph of the screw thread of the thread-rolled BMG bolt (b); (d) magnified image of ridge, groove, and flank of the screw thread (c). (Reprinted from Yamanaka, S. et al., *Mater. Trans.*, 52, 243, 2011. With permission.)

Figure 10.20 shows the tensile strength–deformation curves of the Zr–Al–Ni–Cu glassy alloy screw bolt prepared by thread rolling, together with the data for glassy alloy bolt prepared only by mechanical machining, and for conventional Cr–Mo steel bolt with the same size [30]. The thread-rolled glassy alloy bolt exhibited higher yield strength than those for the mechanically machined bolt and Cr–Mo steel bolt. It is worth noting that the thread rolling caused the distinct enhancement of yield strength, tensile fracture strength, and plasticity as compared with the bolt prepared by mechanical machining. The tensile fracture of the Zr-based thread-rolled glassy bolt occurred along the nearly maximum shear stress plane and the fracture surface consisted mainly of a well-developed vein pattern, as shown in Figure 10.21 [30]. This tendency is consistent with the enhancement of the mechanical properties of BMGs subjected to cold rolling [31] and amorphous alloy wires subjected to cold drawing [32]. Their enhancements have also been demonstrated to be related to the existence of compressive residual stresses [30].

The resulting screw bolts have been tested for several years and there is no deterioration in the endurance property obtained to date. The endurance property is also related to corrosion resistance and stress-corrosion cracking sensitivity. Although the $Zr_{55}Al_{10}Ni_5Cu_{30}$ BMG alloy did not exhibit good stress-corrosion cracking resistance, the increase in Zr content up to 70 at.% provided high resistance to stress corrosion cracking in conjunction with an increase in

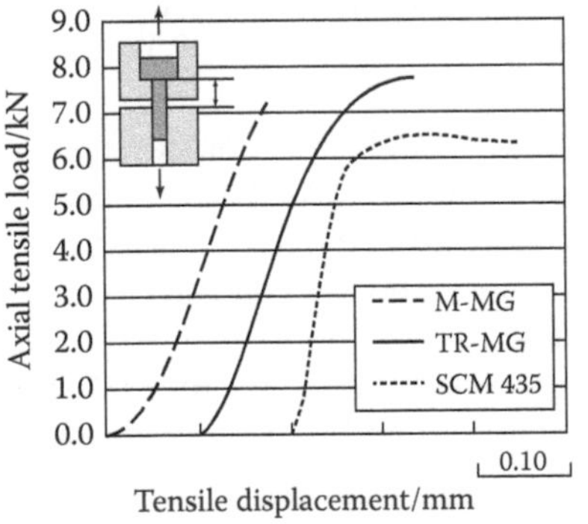

FIGURE 10.20
Experimental results of the tensile load-displacement curves of machined bulk metallic glass bolt (M-MG), thread-rolled MG bolt (TR-MG), and SCM435 steel bolt. (Reprinted from Yamanaka, S. et al., *J. Mater. Proc. Technol.*, 214, 2593, 2014. With permission.)

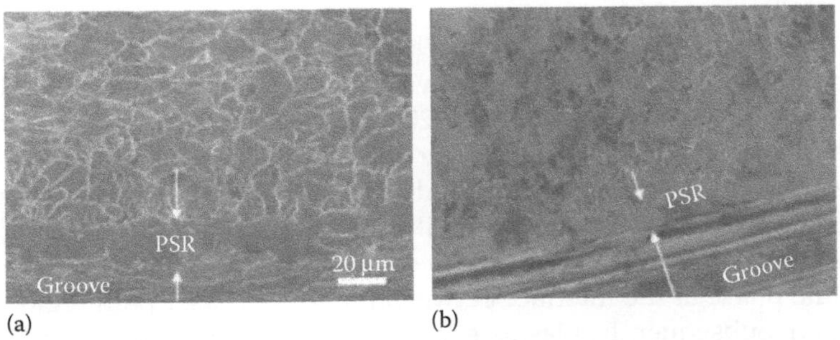

FIGURE 10.21
SEM images of fracture surfaces near the grooves of (a) thread-rolled and (b) machined bulk metallic glass bolts after tensile test. (Reprinted from Yamanaka, S. et al., *J. Mater. Proc. Technol.*, 214, 2593, 2014. With permission.)

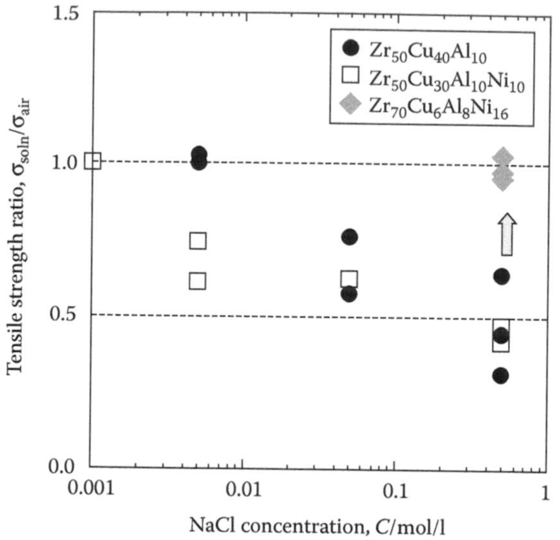

FIGURE 10.22
Slow strain rate technique (SSRT) test results obtained for $Zr_{70}Cu_6Al_8Ni_{16}$ BMG in 0.5 M NaCl at an initial strain rate of 5×10^{-6} s^{-1}. The results for $Zr_{50}Cu_{40}Al_{10}$ and $Zr_{50}Cu_{30}Al_{10}Ni_{10}$ BMGs are also shown for comparison. (Reprinted from Kawashima, A. et al., *Corros. Sci.*, 52, 2950, 2010. With permission.)

compressive plasticity and fracture toughness. For example, Figure 10.22 shows the change in the yield stress of $Zr_{70}Cu_6Al_8Ni_{16}$ and $Zr_{50}Al_{10}Ni_{10}Cu_{30}$ BMGs tested in 0.5 M NaCl solution at room temperature by the slow strain rate tensile testing method [33]. Although the 50% Zr BMG alloy showed significant degradation of yield strength in NaCl solution, no appreciable reduction of yield stress was detected for the 70% Zr bulk glassy alloy, while also indicating high resistance to stress-corrosion cracking. Thus, the 70% Zr glassy alloy rod is expected to exhibit much longer endurance time than the 50% Zr glassy alloy bolt.

10.3.11 Applications to Surface Coatings

Much attention has been paid in recent years to the formation, properties, and application of BMG surface coatings by utilizing their high glass-forming ability. This is because the development of Fe-based glassy alloy surface coatings on various kinds of substrate materials is promising for significant extension of BMGs as structural materials without material size limitation.

Before the synthesis of Fe-based BMGs in 1995 [34], the focus was on glassy alloy coatings using ordinary glassy alloys, which required very high cooling rates, by using lasers [35–37]. In 1987, Yoshioka et al. [38] tried to deposit a 20 μm thick glassy surface layer of Ni–Cr–P–B alloy by a laser processing method, consisting of remelting the melt-spun glassy alloy sheet using laser heating, followed by the reinforced flow of liquid nitrogen gas, which led to high resolidification rates. Others also tried a similar approach, but it was difficult to produce a homogeneous amorphous phase without any crystalline phase because of the precipitation of the crystalline phase at the interface between the former laser path region and the next laser path region. Subsequently, plasma spray [39] and explosive welding [40] techniques were also tried, with similar difficulties. Thus, when glassy alloys with the requirement of high cooling rates were used, it was difficult to obtain a fully glassy coating in thicknesses larger than about 0.1 mm and consequently there were no actual applications.

Since the synthesis of BMGs, it has become possible to form glassy coatings with a thickness >0.2 mm and some application examples have been presented, particularly for Fe–Cr–Mo–B–C BMGs, for example, by the high-velocity oxy fuel (HVOF) technique. This simple process is very attractive because of the simultaneous achievement of high deposition rates, high cooling rates, and wide deposition areas [41]. Several highly dense glassy Fe–Cr–Mo–P–C, Fe–Cr–Mo–(P,B,C), and Fe–Cr–Mo–C–B BMG compositions have been deposited using this method [36,41,42].

Figure 10.23a shows the cross-sectional structure of the $Fe_{43}Cr_{16}Mo_{16}C_{15}B_{10}$ glassy alloy layer coating on an Al-based crystalline alloy [42]. The HVOF spray coating conditions were such that a dense coating of 200–250 μm thickness was obtained with true bonding achieved by using an oxygen gas velocity of 1800 standard cubic feet per hour (SCFPH). X-ray diffraction confirmed that the resulting spray-coated layer was fully glassy (Figure 10.23b). Cu-, Ni-, and Zr-based glassy coatings have also been deposited.

Based on this success, intensive efforts have been made to use four types of glassy coatings for practical applications. These are (1) Fe–Cr–Mo–C–B alloys, which are in practical use in Japan; (2) Fe–Cr–Mo–C–B–Y alloys; (3) Fe–Cr– B–Nb alloys; and (4) Fe–Cr–B–Si–P and Fe–P–B–Si–C alloys. Alloy types (1) to (3) have been developed for structural and anti-irradiation materials, while type (4) is suitable for magnetic sensor applications. Similar alloy components have also been used for surface coatings in China, Germany, and Korea.

Among the three structural Fe-based alloys mentioned in the preceding paragraphs, because of their inexpensive nature, Fe–Cr–Mo–C–B alloys and their glassy coatings have been used as practical materials with the features of high hardness (of about 1000–1100, higher than that of hard chromium-coated plate), high wear resistance, high corrosion resistance (3 orders of magnitude higher than that of 304 stainless steel), and high oxidation resistance at elevated temperatures.

In contrast with the hard chromium phase with the same thickness and subjected to the same adhesion strength (bending) test, neither abrasion nor swelling were seen in the severely bent region of the glassy coating. In addition, the wear loss weight was 0.01 g for the glass coating, 0.41 g for the hard chromium plating, and 0.11 g for the CrN coating, indicating that the glassy alloy coating has much better wear resistance [43].

Figure 10.24 shows the surface appearance of the Fe–Cr–Mo–C–B glass coating that was formed on an Al-based alloy pipe by the HVOF method [44]. As a practical example, Figure 10.25a shows the surface appearance of the vessel coated with the Fe-based

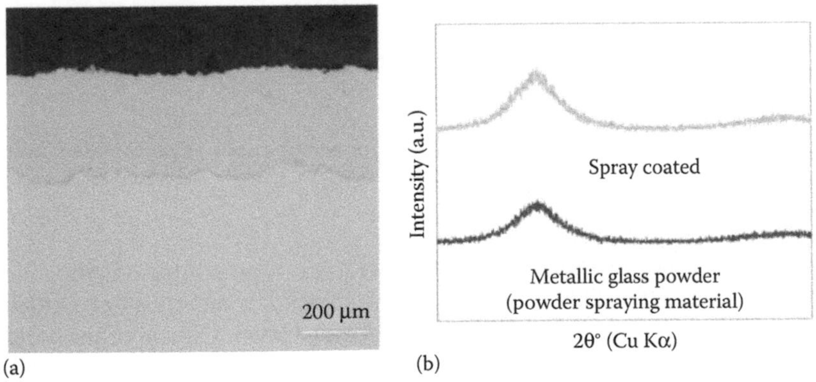

FIGURE 10.23
(a) Optical micrograph of the cross section of the $Fe_{43}Cr_{16}Mo_{16}C_{15}B_{10}$ coating and (b) XRD patterns of the powder and its sprayed coating. (Reprinted from Kim, H.G. et al., *Mater. Sci. Forum*, 580–582, 467, 2008. With permission.)

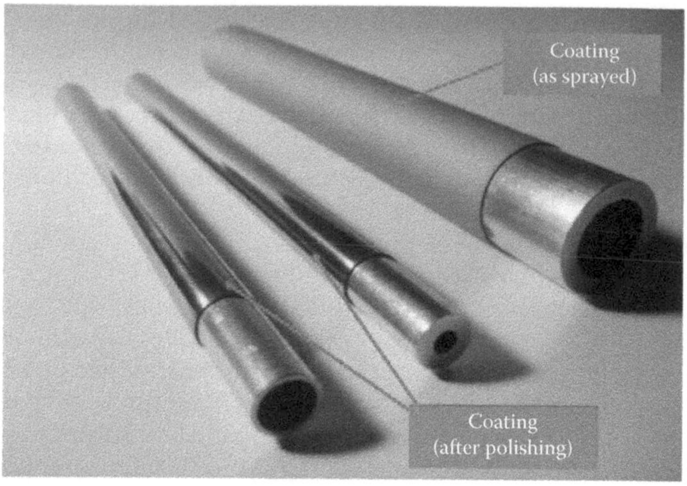

FIGURE 10.24
Appearance of metallic glass coating sprayed on an aluminum-based alloy pipe.

(a) (b)

FIGURE 10.25
Surface appearance of vessels for holding Pb-free high-temperature-type soldering alloy liquid (a) for the Fe-based BMG-coated vessel in the as-prepared state and after 2000 h use in the liquid state, and (b) for the 304 stainless steel after 500 h use.

BMG and used for holding Pb-free high-temperature-type soldering alloy liquid in the as-prepared state and after 2000 h use [44]. No distinct damage due to corrosion and erosion attacks was seen on the inner glass-coated surface layer. On the other hand, as shown in Figure 10.25b, a similar vessel made of 304 stainless steel and used in the same condition was severely attacked and holes formed even after a short use time of about 500 h. Thus, the Fe–Cr–Mo–C–B glassy alloy-coated layer exhibits extremely high corrosion and erosion resistance [44].

In addition to the structural application of the Fe-based glass coating, the Fe–Cr–B–Si–P and Fe–B–C–P–Si glassy alloy coatings exhibit rather good soft magnetic properties with high saturation magnetization even in the spray-coated state and have been commercialized as a highly sensitive magnetic sensor to detect torque, surface strain, rotation speed, and so on. [43]. The Fe–Cr–Mo–C–B–Y and Fe–Cr–Mo–C–B–Tm coatings have been used to protect against nuclear irradiation damage by utilizing the high absorption capacity of the glassy structure for irradiation. The Fe–Cr–Mo–C–B–Nb coatings have also been applied to the edge part of the boring bar, which requires high hardness, high corrosion resistance, and high erosion wear resistance in a highly corrosive atmosphere in the oil production industry in Brazil. These application fields of Fe–Cr–Mo–metalloid base BMGs as structural/functional materials are expected to be extended significantly in the near future due to an excellent combination of properties that cannot be obtained in the Fe-based crystalline alloys. The common characteristics of the Fe-based coatings include high hardness (1100–1300), high T_g (841–884 K), high T_x (893–921 K), very low corrosion losses (about 0.01 mm year^{-1} in HCl solution), and low wear losses (0.01 g h^{-1}).

10.4 Chemical Applications

The superior corrosion resistance of BMGs over their crystalline counterparts plays a major role in this group of applications. In addition to this, the high strength and easy formability also assume importance. BMGs have been specifically considered as the most appropriate materials for fuel cell separators.

Fuel cell systems are known to have a higher efficiency in comparison to internal combustion engines by directly converting the chemical energy of fuels to electrical energy. The use of fuel cells is expected to reduce the consumption of fossil fuels and CO_2 emissions that lead to global warming. There has been recent progress in the development of proton exchange membrane fuel cells (PEMFCs) by utilizing the superior corrosion resistance and viscous deformability of BMGs.

Because of its high output current density and low-temperature operation, the PEMFC is desirable for household use and for automobile applications. However, the main constituent parts, *viz.*, the membrane, catalyst, and separator, still present problems that need to be solved. In particular, the functions of the separator are to transfer the fuel and the oxidizer to the reaction site, contain the reaction products, accumulate the electricity generated, and mechanically support the cell. The separator, having such significant roles, accounts for more than 60% of the weight of the fuel cell and more than 30% of the production cost. Therefore, the mass, volume, and cost of the fuel cell can be significantly brought down by substituting the present separator substrate, fine graphite, with another suitable material. Even though stainless steel is an attractive substitute, it was reported that a drastic drop in output voltage was expected to occur due to the formation of a surface passive layer. Therefore, BMGs are being examined as a replacement material. BMGs should be ideal for fuel cell separators because of their high strength, superior corrosion resistance, and excellent formability in the supercooled liquid region. Fe-based BMGs are inexpensive, and also have the required corrosion resistance. But, they have been discounted for application because of their brittleness. On the other hand, the corrosion rates of Ni-based glassy alloys in dilute H_2SO_4 solution (simulating the actual PEMFC conditions around 368 K) are one order of magnitude lower than those of stainless steel. Jayaraj et al. [45] have also shown that the corrosion resistance of Ni-based glassy alloys is superior to that of Fe-based glassy alloys under cathodic conditions. Consequently, attention has focused on Ni-based BMGs.

A series of Ni-based Ni–Nb–Ti–Zr BMG alloys were investigated, and preliminary corrosion studies were conducted on these alloys in boiling H_2SO_4. Based on these results, it was shown that the corrosion rate decreased with increasing Nb content and that there was no benefit beyond about 15 at.% of Nb. Therefore, the alloy with the composition $Ni_{60}Nb_{15}Ti_{15}Zr_{10}$ was chosen for further study [46].

A prototype fuel cell separator with grooves 0.9 mm wide and 0.6 mm deep with a pitch of 2.7 mm was fabricated. Figure 10.26a and b shows a photograph and the cross-section profile after the groove was formed, respectively. It may be noted that the cross-section profile is extremely good. Tests using the prototype show that the Ni-based BMG separator generates a high voltage at the actual operation current, higher than that generated by a stainless steel separator. In addition to this, the durability test at a current density of 0.5 A m^{-2} shows no degradation at 350 h (Figure 10.26c). Hence, it was concluded that the Ni-based BMG is a promising material for the next-generation separators in fuel cells.

It has also been shown that the glassy Ni–Nb–Pt–Sn alloy can be an excellent anode material for application in fuel cells since it has a higher catalytic activity as well as a longer lifetime than the polycrystalline platinum that is being presently used [47].

$Ni_{65}Cr_{15}P_{16}B_4$ [48] and $Zr_{55}Al_{10}Ni_5Cu_{30}$ [49] coatings were also formed on Al-based substrates by the HVOF spray coating method and gas tunnel type plasma spray method, respectively. The Ni-based glassy alloy coating was tested with the aim of applying it to fuel cell separator plates. Figure 10.27 shows a cross-sectional microstructure of the Ni-based glassy alloy coating on an Al alloy. The film was about 200 μm in thickness and no large pinholes connecting the surface and the substrate were seen in the glassy film. It is also worth noting that the boundary between the film and the substrate was not flat because the abrasive blasting on the Al substrate was conducted before the spray coating to increase the adhesion of the film to the substrate.

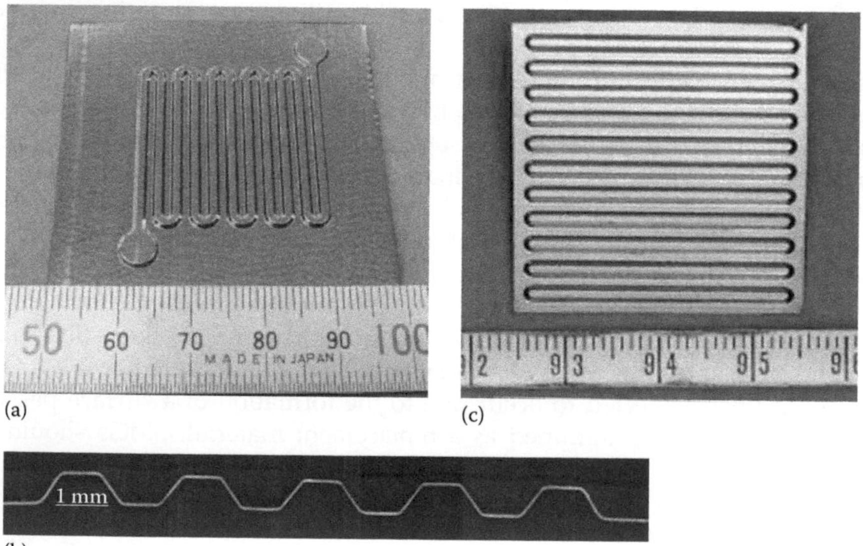

(b)

FIGURE 10.26
(a) Prototype fuel cell separator using an Ni-based BMG sheet, (b) cross-sectional morphology of the groove-formed specimen, and (c) appearance of the BMG separator after power generation for 350 h. (Reprinted from Inoue, A. et al., *Mater. Trans.*, 46, 1706, 2005. With permission.)

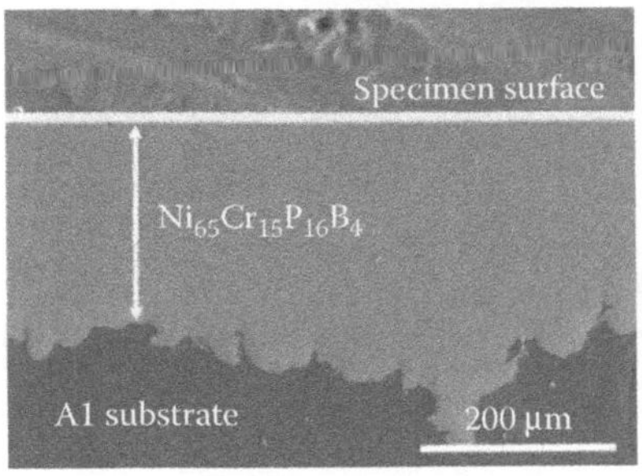

FIGURE 10.27
A cross section of the $Ni_{65}Cr_{15}P_{16}B_4$ BMG surface coating deposited on an Al alloy substrate. (Reprinted from Kim, S-C. et al., *Mater. Trans.*, 51, 1609, 2010. With permission.)

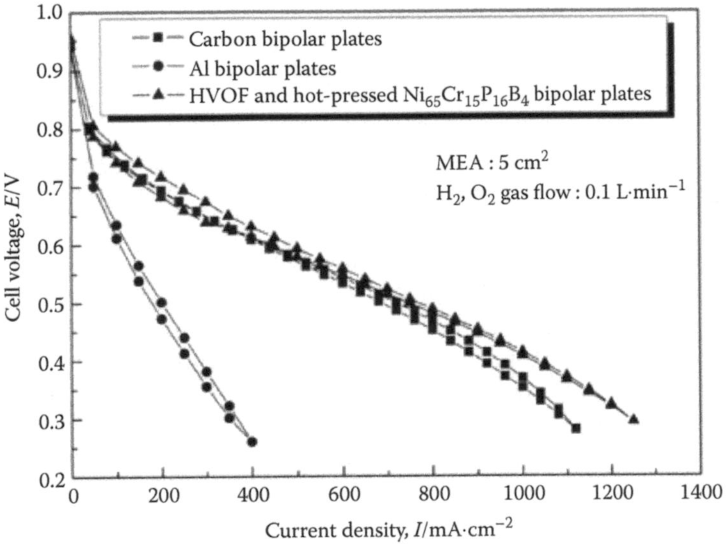

FIGURE 10.28
I-V curves of the single cell with carbon, Al, and $Ni_{65}Cr_{15}P_{16}B_4$ BMG bipolar plates measured after 50 cycle repetitions. (Reprinted from Kim, S-C. et al., *Mater. Trans.*, 51, 1609, 2010. With permission.)

Figure 10.28 shows the I-V curves of the single fuel cell with the Ni-based BMG alloy bipolar plates together with the data for conventional graphite and Al. The measurement was made at the cell temperature of 353 K and the results obtained after 50 repetitions to age the membranes. At the cell voltage of 0.5 V, the single cell employing the Ni-based BMG bipolar plates generated a current density of 750 mA cm^{-2}, which is higher than the 700 mA cm^{-2} for graphite and 200 mA cm^{-2} for Al. Thus, the single cell with the Ni-based BMG-coated bipolar plates shows better I-V performance than that with Al bipolar plates because of much better corrosion resistance, which can suppress the increase in contact electrical resistance. Besides, it is noticed that the I-V performance of the single cell with the Ni-based BMG-coated bipolar plates is as high as that with the graphite bipolar plates, implying that the corrosion resistance of the Ni-based glass is so high that the coating can be applied to bipolar plates in practice.

Long-time durability tests conducted at a constant current density of 200 mA cm^{-2} at 353 K with the single cell showed that the cell voltage measured with the BMG bipolar plates did not drop during the test, even though the voltage had dropped slightly

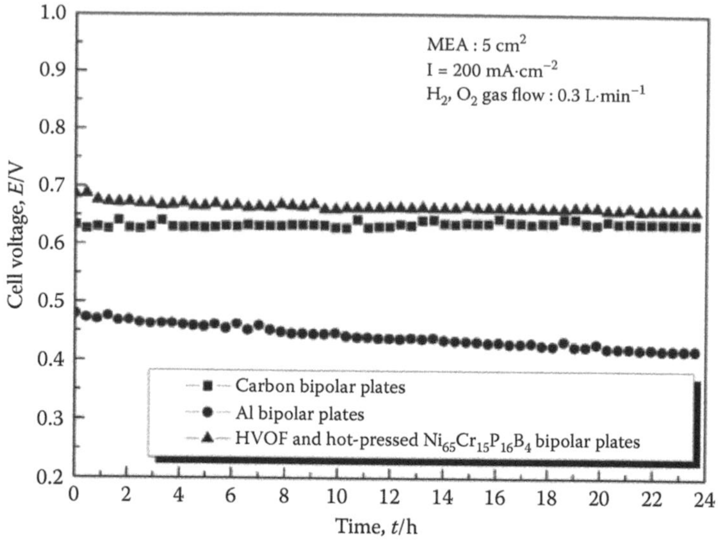

FIGURE 10.29
Long-time durability tests conducted with a single cell having graphite, Al, and $Ni_{65}Cr_{15}P_{16}B_4$ BMG bipolar plates. (Reprinted from Kim, S-C. et al., *Mater. Trans.*, 51, 1609, 2010. With permission.)

in the initial stage of the measurement due to the formation of the passive surface film (Figure 10.29). Yamaura et al. [50–52] earlier used Ni-based glassy alloys for bipolar plates by hot-pressing the melt-spun alloy ribbons to sheets of 50 μm thickness and 50–100 mm in width at around T_g. It was reported that the cell performance using those glassy alloy bipolar plates was lower than that with graphite bipolar plates. This may be because it was difficult to mount such a thin bipolar plate on the graphite frame without gas leak and also without an increase in the contact resistance. However, in the study using the HVOF technique, glass-coated bipolar plates were produced in a bulk form, and they were easy to mount in a single fuel cell.

10.5 Magnetic Applications

The magnetic properties of melt-spun glassy ribbons have been exploited for a variety of applications. The outstanding soft magnetic properties of the Fe-based melt-spun ribbons have been exploited in power distribution transformers and several other applications [53–55]. Present-day electronic devices have been going down in size, and miniaturization is reaching micrometer and even nanometer levels. MEMS and nano-electromechanical systems (NEMS) are the operative words of today. In addition to this, multifunctionality has been another requirement. All of these will lead to faster processing and further integration, supported by semiconductor integrated-circuit technologies. The required dimensional control for production is from 50 to 5 nm. Thus, the development of nanometer-accurate linear actuators is highly beneficial. Further applications would be in such devices as X–Y stages for accurate positioning of cell-operation manipulators in the biomedical industry. Soft-magnetic BMGs with high permeability and low coercivity appear to be the most appropriate materials for the magnetic yokes of such linear actuators.

10.5.1 Magnetic Actuators

Nishiyama et al. [56] have produced 30 mm long, 20 mm wide, and 1 mm thick $Fe_{73}Ga_4P_{11}C_5B_4Si_3$ BMG plates by the squeeze casting (a combined casting and forging) process. This alloy has excellent soft magnetic properties (I_s = 1.32 T; μ_{max} = 110,000; H_c = 33 A m^{-1}) and these are improved by annealing (μ_{max} = 160,000 and H_c = 2.7 A m^{-1}). Using this material, a yoke and a prototype linear actuator were fabricated, and these are shown in Figure 10.30. Tests show that the force generated by the actuator using the Fe-based BMG yoke is higher, in the frequency range of 20–45 Hz, than when Sr-ferrite is used as a permanent magnet. The large Lorentz force suggests that linear actuators using Fe-based BMG yokes can show good acceleration and deceleration. In addition, the high relative permeability (μ) of the BMG permits the actuators to be driven with high-frequency current pulses, leading to fine pitch control through a feedback system.

Based on the high μ and low H_c of Fe-based BMGs, it has been suggested that they are appropriate for low-loss magnetic cores for choke coils [57]. Cores using Fe-based BMGs exhibit a constant μ of ~110 for frequencies up to 10 MHz, comparable to commercial cores. The $Fe_{70}Al_5Ga_2P_{9.65}C_{5.75}B_{4.6}Si_3$ BMG showed a low core loss of 610 kW m^{-3} at 100 kHz in a magnetic field of 0.1 T. Using these excellent high-frequency magnetic properties, as well as viscous formability, magnetic shielding sheets for laptop computers are being developed (Figure 10.31).

10.5.2 Hyperthermia Glassy Alloys

Submicron-size glassy alloy magnetic powders are required for hyperthermia applications because they can be easily targeted to damaged body organs by means of external magnetic fields. In order to enable such an application, it is essential to develop soft magnetic materials with Curie temperatures around human body temperature (310 K), good corrosion resistance in the human body, and that do not include any toxic elements. Until now, Fe-oxide submicron powders have been developed for this purpose [58]. Fe-based glassy alloy powders also seem to be good candidates as hyperthermia materials because

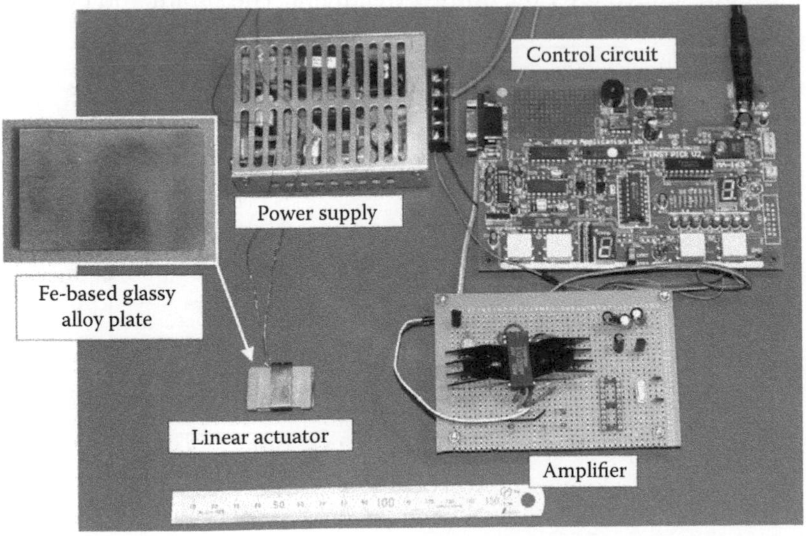

FIGURE 10.30
Magnetic yoke made of an $Fe_{73}Ga_4P_{11}C_5B_4Si_3$ BMG plate for a prototype linear actuator.

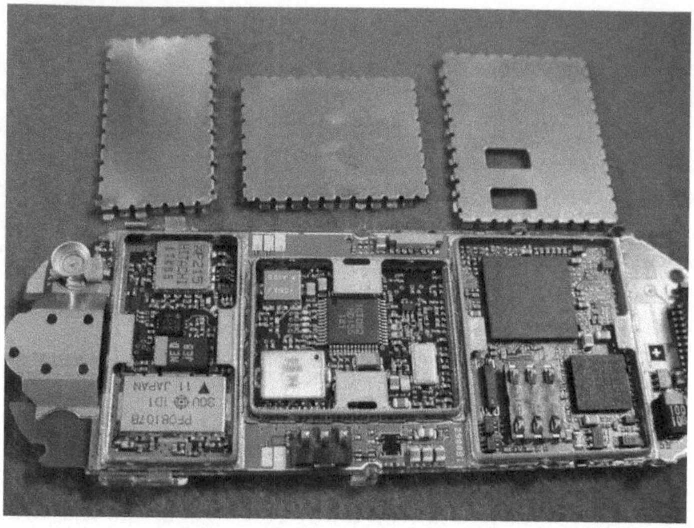

FIGURE 10.31
BMG magnetic shielding sheets for laptop computers.

of their wide alloy composition flexibility and continuous changes in various functional properties over wide ranges.

Recently, Chiriac et al. [59,60] have reported that $Fe_{79.7-x}Nb_{0.3}Cr_xB_{20}$ (x = 11–13.5 at.%) glassy submicron powders with low Curie temperatures are suitable for hyperthermia applications. The glassy alloy submicron powders were produced by melt spinning, followed by embrittlement treatment by annealing, high-energy ball-milling in a wet environment including surfactant, washing to remove the excess surfactant, and then drying. The resulting glassy alloy submicron powders exhibit T_c of 290–330 K (Figure 10.32) and the powder size is in the range 25–40 nm after milling for 4 h in air (Figure 10.33) [59]. These powders have the advantage that the glassy structure of the submicron powders remains almost unchanged after the milling treatment. The submicron powders had good

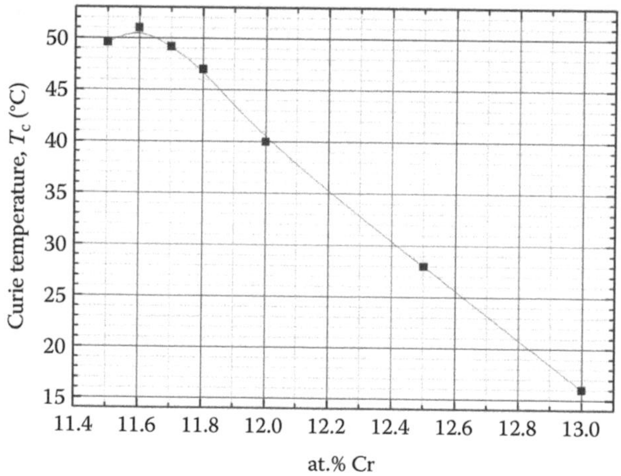

FIGURE 10.32
Change in Curie temperature (T_C) with Cr content for glassy alloy $Fe_{79.7-x}Cr_xNb_{0.3}B_{20}$ (x = 11.5 and 13 at.%) ribbons. (Reprinted from Chiriac, H. et al., *J. Appl. Phys.*, 115, 17B520, 2014. With permission.)

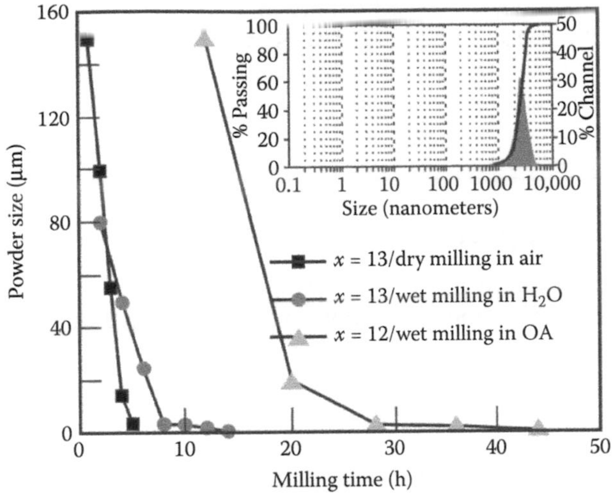

FIGURE 10.33
Milling time dependence of powder size for $Fe_{79.7-x}Cr_xNb_{0.3}B_{20}$ (x = 12 and 13 at.%) powders prepared by high-energy ball milling from glassy alloy ribbon precursors. Inset shows the size distribution of $Fe_{67.7}Cr_{12}Nb_{0.3}B_{20}$ powders, milled in oleic acid for 44 h, and determined by dynamic light scattering. (Reprinted from Chiriac, H. et al., *J. Appl. Phys.*, 115, 17B520, 2014. With permission.)

heating efficiency that could reach about 10°C within the first 500 s in the AC field of 350 mT (f = 153 kHz) and then showed a nearly constant value, being independent of the heating time, as shown in Figure 10.34. The equilibrium temperature agrees with the Curie temperature of the glassy alloy powder. It is noticed that the Curie temperature does not show appreciable change even after the milling treatment to submicron powders from the melt-spun ribbon. The change in Cr content in the range from 12.5 to 11.5 at.% can

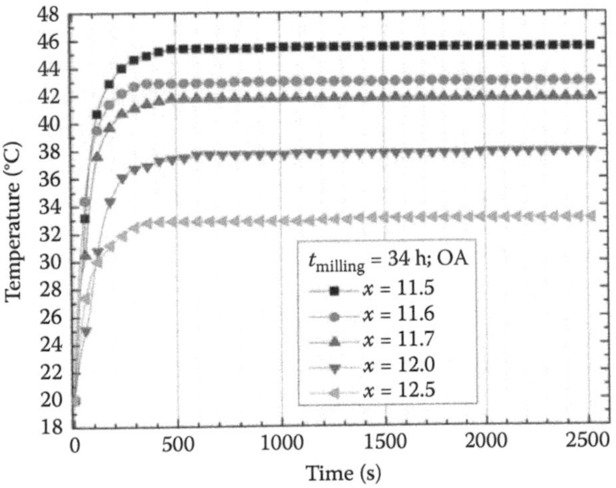

FIGURE 10.34
Change in the heating efficiency curves of $Fe_{79.7-x}Cr_xNb_{0.3}B_{20}$ (x = 11.5 and 12.5 at.%) MA powders with milling time and Cr content. The tests were carried out in an AC field of 350 mT (f = 153 kHz) created by a home-made magnetic-induction hyperthermia unit. (Reprinted from Chiriac, H. et al., *J. Appl. Phys.*, 115, 17B520, 2014. With permission.)

tune the heating efficiently between 33°C and 46°C. Being able to control the equilibrium temperature makes it possible to heat the diseased organ to the desired temperature using the powders. The ability to achieve this effect, which has been shown in animals, may be possible in the human body in the near future.

10.6 Miscellaneous Applications

The high strength and wear resistance of BMGs, along with their biocompatibility, smooth surface finish, and aesthetic appearance, determine the type of material to be used and the appropriate applications.

10.6.1 Jewelry

Jewelry has fascinated humankind for several hundreds of years. The two important material characteristics that are important for jewelry are their aesthetic appearance and their ability to retain their luster and brightness, or inertness to the environment. Gold is one metal that has been extensively used for jewelry because of its special attributes. It is estimated that about two-thirds of today's newly mined gold goes to the jewelry industry, and the remaining one-third is for the electronics and dental industries, coinage and bullion, aerospace, glass making, and decorative gilding. A smooth surface finish and an aesthetic appearance are considered important in this group of applications.

Gold is very soft and ductile and so it can be fabricated into complex designs. In fact, it is so soft (its Vickers hardness is 8) that it has to be alloyed with Ag and Cu to make it easy to handle and impart some scratch and wear resistance. It can have a very good surface finish and can be aesthetically pleasing due to its bright yellow color. Lastly, since it is a noble metal, it will not corrode and, therefore, its surface luster and bright appearance can be retained.

Due to the special attributes of BMGs based on Au and Pt, Schroers et al. [61] have developed these alloys for jewelry applications. The authors chose one composition for each of the alloys: $Au_{49}Ag_{5.5}Pd_{2.3}Cu_{26.9}Si_{16.3}$ (referred to as LM18kAu) and $Pt_{57.5}Cu_{14.7}Ni_{5.3}P_{22.5}$ (referred to as LM850Plat). Both these alloys have very good glass-forming ability, as evidenced by the production of 5 mm thick plates for the Au-based alloy and 16 mm thick plates for the Pt-based alloy. Conventional processing of Pt-based alloys is difficult because of their high liquidus temperature (over 1800°C), reaction with the crucible material, tarnishing and oxidation, and solidification shrinkage, due to thermal contraction and phase transformation (liquid-to-solid) shrinkage. Table 10.4 lists some selected mechanical properties of these BMGs and their approximate crystalline counterparts. It may be noted that the BMG alloys are harder and stronger, show higher elastic limits, and are also lighter. Additionally, the S index, a measure of the formability of the alloy, is defined as

$$S = \frac{T_x - T_g}{T_\ell - T_g} = \frac{\Delta T_x}{T_\ell - T_g} \tag{10.6}$$

where T_x, T_g, and T_ℓ represent the crystallization, glass transition, and liquidus temperatures, respectively, The S values are 0.24 and 0.34 for the Au-based and Pt-based alloys, respectively, and these are much higher than those for $Zr_{44}Ti_{11}Cu_{10}Ni_{10}Be_{25}$ and $Pt_{42.5}Cu_{27}Ni_{9.5}P_{21}$, which are considered very good glass formers. These BMG alloys can be superplastically

TABLE 10.4
Selected Properties of Au-Based and Pt-Based BMGs and Their Approximate Crystalline Counterparts

Material	Density (g cm^{-3})	Yield Strength (MPa)	Elastic Elongation (%)	Hardness	$S = \Delta T_x/(T_\ell - T_g)$
$Au_{49}Ag_{5.5}Pd_{2.3}Cu_{26.9}Si_{16.3}$ (LM18kAu)	11	1200	1.5	360	0.24
Au–Ag–Cu (18k)	15.4	350	<0.5	150	—
$Pt_{57.5}Cu_{14.7}Ni_{5.3}P_{22.5}$ (LM850Plat)	15.3	1400	1.3	402	0.34
Pt/Ir850/150	21.5	420	<0.5	160	—

formed in the supercooled liquid region at 150°C for 200 s under a pressure of 100 MPa for the Au-based alloy, and at 270°C for 100 s under a pressure of 28 MPa for the Pt-based alloy. Figure 10.35 shows the net-shape formed Au-based and Pt-based alloys. The smooth surface finish is worth noticing.

Because of the application of pressure during the SPF operation, the shrinkage porosity in the processed material is significantly reduced, as is the porosity in the formed product. Additionally, due to the high strength of the alloys, they should be more scratch and wear resistant than the conventional crystalline alloys. Further, it has been suggested that the large elastic strain limit, combined with the high strength, of the BMG alloys should open up opportunities for new jewelry designs. Liquidmetal Technologies has been commercializing some of these jewelry items. The additional jewelry items commercialized by Liquidmetal Technologies include watch casings (to replace Ni and other metals, which can cause allergic reactions), fountain pen nibs, and finger rings.

The Pd- and Pt-based BMG alloys have been known to possess very high glass-forming ability in a wide composition range. Using this characteristic, Tanaka Noblemetal Co. Ltd. have produced some ornamental objects, such as wind bells, necklaces, tiepins, and ordinary and wedding rings, from the $Pt_{48.75}Pd_{9.75}Cu_{19.5}P_{22}$ (at.%) alloy that does not contain Ni, as shown in Figure 10.36.

(a) (b)

FIGURE 10.35
Net shape formability using Au-based and Pt-based BMG alloys. (a) The Au-based alloy was formed at 150°C for 200 s under a pressure of 100 MPa. (b) The Pt-based alloy formed at 270°C for 100 s under a pressure of 28 MPa using pellets as feedstock material. (Reprinted from Schroers, J. et al., *Mater. Sci. Eng. A*, 449–451, 235, 2007. With permission.)

FIGURE 10.36
Ornaments made of metallic glasses.

It is important here to note that these glassy ornaments have been produced by the sand mold casting method, which is similar to the one used to produce conventional castings, for example, cast iron. This is believed to be the first time that BMGs with a maximum thickness of several millimeters were produced by the sand casting process. This method can lead to a significant increase in the freedom of size, outer shape, and morphology of cast materials that are suitable for the mass production of various ornamental materials with complicated outer morphology. Even though the cooling rate during sand casting is considerably lower than in copper mold casting, it is notable that the rate exceeds the critical cooling rate (about 0.1–1 K s^{-1}) for Pd–Pt–Cu–P alloys, thus producing glassy alloys. Even at present, Ishizuka Co. Ltd. in Japan is selling various kinds of ornamental articles prepared by the sand or copper mold casting methods. However, one should recognize that the surface of the sand cast articles is not always smooth and does not show a good shiny luster.

10.6.2 Biomedical Applications

Many metallic materials have been used as implants in the human body. But, for a material to be used as an implant, the metallic material should be biocompatible and inert (it should have a high corrosion resistance), and have mechanical properties that are very similar to the part that is being replaced. Apart from the toxicity issue, the mismatch of the modulus of elasticity (Young's modulus) between the metallic implant and the human part is a serious concern, since this could develop undesirable stresses (stress shielding) and consequent damage. Hence, a biomedical replacement material should have

- Reasonably low density
- Little or no cytotoxic metal in its composition
- High strength and long fatigue life
- Low Young's modulus, comparable with the part that is being replaced
- Large room temperature plasticity so that it can be easily formed
- Good casting properties so that it can be easily cast into defect-free materials

These requirements could be different depending on the part to be replaced.

Different metallic materials, including stainless steel, vitallium and other Co-based alloys, and Ti alloys have been used for biomedical applications. However, the mechanical properties of some of these alloys have been found to be inadequate, as has their reasonably high Young's modulus. BMGs are highly suitable for this purpose since they have very high yield strength, large elastic elongation limits of about 2%, and low Young's modulus, and they can be easily cast into shapes. Further, Ti-based alloys have low density, excellent biocompatibility, and corrosion resistance. Therefore, Ti-based BMGs appear to be ideal materials for biomedical applications.

Many of the Ti-based BMGs contain toxic elements such as Al, Ni, and Be. Therefore, Zhu et al. [62] developed a Ti-based BMG that did not contain any of the toxic elements. The alloy series had a composition of $Ti_{40}Zr_{10}Cu_{40-x}Pd_{10+x}$ (with $x = 0, 2, 4, 6, 8$, and 10) and it was possible to obtain these alloy compositions in a bulk glassy form with diameters of 4–7 mm. For example, 7 mm diameter glassy rods could be obtained in $Ti_{40}Zr_{10}Cu_{36}Pd_{14}$ and $Ti_{40}Zr_{10}Cu_{34}Pd_{16}$ alloy compositions, and the $Ti_{40}Zr_{10}Cu_{36}Pd_{14}$ BMG alloy was chosen for further study. This alloy exhibited a compressive strength of 1950 MPa, Young's modulus of 82 GPa, and an elastic elongation of 2.3% [62].

Hydroxyapatite (HAp) is the main mineral constituent of teeth and bone, and it shows excellent biocompatibility with hard tissue, skin, and muscle tissue. Further, HAp does not exhibit any cytotoxic effects and can directly bond to the bone. Therefore, HAp has been frequently used as either a coating or a reinforcement in different materials. Coating with HAp to improve the surface bioactivity of Ti alloys was not effective, since it flaked off due to the poor ceramic/metal interface bonding. To overcome this problem, Zhu et al. [63] produced $Ti_{40}Zr_{10}Cu_{36}Pd_{14}$ BMG alloy composites containing 6 vol.% HAp through solidification methods. The compressive fracture strength, Young's modulus, and elastic elongation limit were almost the same for both the monolithic and the composite alloy, suggesting that the addition of HAp did not adversely affect the mechanical properties. Since wetting of HAp particles occurred spontaneously with the molten metal, it can be assumed that the interfacial bonding between the glassy matrix and the HAp phases was good.

A bone-like apatite film formed on the Ti-based BMG alloy surface after electrochemical and chemical treatments. It was noted during a scratch test that the HAp film started peeling off at loads of >25 g with a hydrothermal treatment, whereas it required only 5–10 g without the hydrothermal treatment.

The monolithic $Ti_{40}Zr_{10}Cu_{36}Pd_{14}$ BMG alloy was found to be a good replacement for teeth and in plates and screws for fixing bones (Figure 10.37).

10.6.3 Medical Devices

As mentioned in Section 10.6.2, many BMGs have no cytotoxic effects, they are corrosion resistant, and they are biocompatible. Further, they have a high tensile strength and are therefore wear resistant. In addition, the easy formability and control of the surface finish offer excellent opportunities for these materials to be used in the medical industry.

These features are taken into consideration in using BMGs to develop products for reconstructive devices, fracture fixations, spinal implants, ophthalmic surgery, and cataract surgery. The BMGs have also been shown to yield surgical blades that are sharper and longer lasting than steel, less expensive than diamond, and more consistently manufacturable (since they are produced from a single mold) and ready for use. Other edged tool applications include knives and razor blades.

Vitreloys (trade name for Zr-based BMG alloys manufactured by Liquidmetal Technologies) are also being used for materials in knee-replacement devices, pacemaker casings, and other implants.

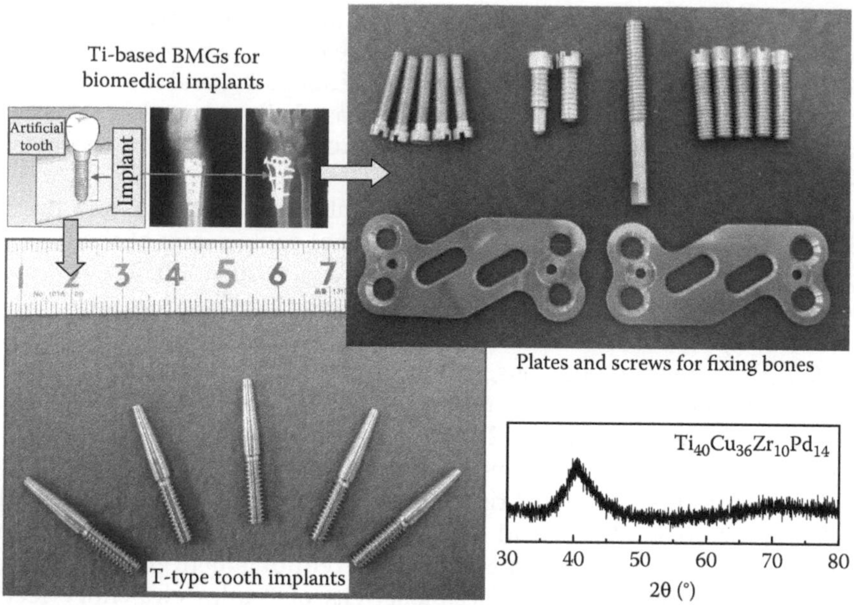

FIGURE 10.37
$Ti_{40}Zr_{10}Cu_{36}Pd_{14}$ BMG alloy implants for T-type teeth and plates and screws for fixing bones.

In addition to the Zr–Al–Ni–Cu-based BMGs without toxic and allergic elements, some Ni- and Be-free Ti-based BMGs have also been developed, typical examples being the Ti–Zr–Cu–Pd [62] and Ti–Zr–Cu–Pd–Sn [64] systems. The addition of 1%–5% Nb or Ta to the Ti-based BMGs results in the formation of finely dispersed sub-nanoscale bcc β-phase clusters dispersed in the glassy matrix. Just like the nanoscale crystalline, quasicrystalline, or dendritic dispersions, these clusters also increase the yield strength and fracture strain significantly. That is, the clusters also play a role in the propagation of shear bands to the whole specimen as well as the initiation/generation of shear bands along the interface with glassy matrix.

These Ti-based BMGs with glassy or glass and cluster phases also exhibit such high corrosion resistance in Hanks' solution to be good enough for use in clinical instruments [65]. The combination of electrochemical treatment in an aqueous solution of 5 M NaOH as an electrolyte and then the immersion treatment in Hanks' solution causes the formation of a highly dense HAp phase on the outer surface of the Ti-based BMGs [66], indicating that the Ti-based BMG also has good biocompatibility.

The Ti-based BMGs also have good viscous flow ability and nano-imprintability in the supercooled liquid region that enable nanoscale forming to a fine well-controlled material shape, which is favorable for clinical instrument applications. Figure 10.38 shows the outer surface appearance of Ti–Zr–Cu–Pd–Sn glassy alloy tweezers produced by the sequential process of casting, viscous flow deformation, nano-imprinting, and then laser welding. The edge part in the tweezers and the remaining supporting part are made of the Ti-based BMG and the conventional Ti–Al–V alloy, respectively. It is notable that the inside part of the tweezers has a very fine surface ruggedness pattern which enables the picking up of small objects. This ruggedness pattern was produced by the nano-imprinting process in the supercooled liquid region. The commercial value of the glassy alloy tweezers is currently being evaluated.

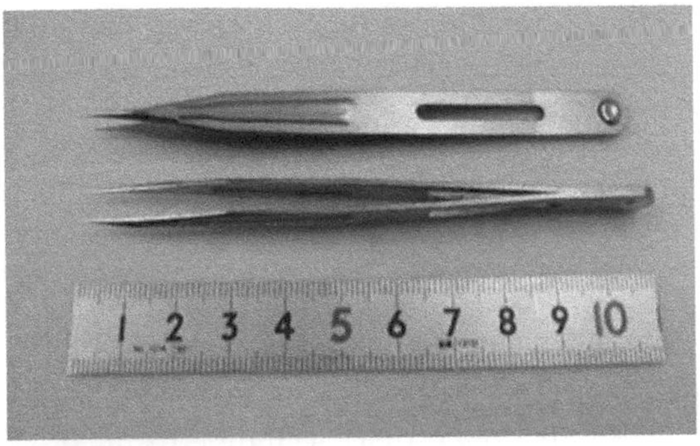

FIGURE 10.38
Outer surface appearance of Ti–Zr–Cu–Pd–Sn glassy alloy tweezers.

10.6.4 Others

Since BMGs can be stronger and lighter at the same time, and can be easily molded into different shapes in the supercooled liquid region, they have been used to produce components for liquid crystal display casings on cell phones, the "uncrushable" SanDisk U3 Cruzer titanium Flash Drive, and to create a Vitreloy-encased laptop that rolls up like a piece of paper.

Liquidmetal Technologies has recently developed glassy alloys for use in the field of hinge components for electronic housings in the mobile phone industry. With the increased complexity in handset circuitry, a greater demand is placed on the performance of internal structural components. Consequently, the traditional materials have been found to be inadequate in meeting the new performance benchmarks.

Currently used materials, such as zinc, magnesium, stainless steel, and even titanium, have inherent design and performance limitations. Other lower-strength metals are also not suitable since they have less resilience to impact from dropping or the daily wear-and-tear of repeated flip-phone actions. The high strength and large elasticity enable thinner profiles to be made.

The kinetic energy penetrator (KEP), the key component of the highly effective armor piercing ammunition system, currently utilizes depleted uranium (DU) because of its high density and self-sharpening behavior. Ballistic tests have shown that the Vitreloy + W composite KEP exhibits self-sharpening behavior, similar to the DU-KEP. It has been mentioned that examples of new military applications for BMGs include composite armor; lightweight casings for ordnance; MEMS casings and components; thin-walled casings and components for electronics; casings for night sights and optical devices; missile components such as fins, nosecones, gimbals, and bodies; aircraft fasteners; and electromagnetic pulse (EMP) and electromagnetic interference (EMI) shielding.

10.7 Concluding Remarks

BMGs have an interesting combination of properties. They exhibit very high strength, large elastic elongations, low Young's modulus, good corrosion resistance, and the ability to be easily formed in the supercooled liquid state. Coupled with the fact that these materials do not have a crystal structure and consequently no grains and grain boundaries, the surface finish is extremely smooth. Such a fine texture has earlier been possible only in polymeric materials and oxide glasses. Thus, the potential applications for these materials seem to be unlimited. In fact, as described in the preceding paragraphs, several different existing and potential applications for metallic glasses, with special reference to the BMGs, have been reviewed.

The prospect of BMGs being used as engineering materials is still not fully clear. However, the present results imply that metallic glass exhibits a new feature, that is, "smaller is safer." Therefore, metallic glasses can be potential candidates as materials in MEMS [67].

Even though some specific applications have been described, and parts have gone into production, this production is not currently being pursued. The reasons may be different for different products. But, it has been explained by John Kang, chairman and CEO of Liquidmetal Technologies, Pyeongtaek, South Korea, that "Manufacturing process limitations, higher-than-expected production costs, unpredictable customer adoption cycles, short product shelf-life, and intense pricing pressures have made it difficult to compete profitably in this commodity-driven market." He goes on to say that "Processes are not yet refined to the point that we can cost-effectively manufacture price-sensitive, commodity products" [68].

The excellent combination of the mechanical properties of BMGs has been touted as very special. While it is true that they have very high strength and hardness, their wear resistance does not appear to be satisfactory. One reason for this unsatisfactory performance is that wear resistance of a material may not depend just on the hardness of the material; other factors could also be playing a role. It has been occasionally mentioned that the wear resistance of a BMG/crystalline part combination is worse than that of the crystalline/crystalline combination. One could think of the following reasons for this decreased wear resistance of the BMG/crystalline combination. Firstly, it could be because the surface roughness of the crystalline and BMG parts is different and, therefore, the BMG may be wearing out faster. Secondly, the wear debris gets deposited at the interface, and this could cause some stress concentration, and therefore the components may be failing sooner. Lastly, it is possible that the deformation/wear mechanism is different in the crystalline materials and the BMGs. But, if instead of the BMG/crystalline combination, one uses a BMG/BMG combination, the wear resistance appears to be much better than the crystalline/crystalline or BMG/crystalline combinations.

The low Young's modulus and high elastic elongation could be used advantageously. This is because, for a given stress, the elastic elongation is large and, therefore, the BMG materials offer a high sensitivity for detecting elastic deformation. Further, if this deformation was transformed into other measurable units, such as electric signals or flow rate, for example, this would be extremely useful for sensor applications. Some of these have been described in Section 10.3.6.

Another important attribute of BMGs is that they can be fabricated into complex shapes in the supercooled liquid region. Further, they can be fabricated into net shapes. While it is true that one can produce parts with complex design, the time available for forming is determined by the time for crystallization of the glassy material. This, of course, can be determined and the process is carried out without allowing the crystallization process to occur. But, it has been mentioned that the actual forming of the BMG parts in the supercooled liquid region is more expensive than forming the crystalline parts. Therefore, if the properties and performance of the crystalline part are satisfactory for the intended application, the use of BMGs may not be advantageous.

The supercooled liquid region, $\Delta T_x = T_x - T_g$, where T_g and T_x represent the glass transition and crystallization temperature, respectively, is reasonably large for BMG alloys. But, more than just the ΔT_x value, the ratio $\Delta T_x/T_m$, where T_m is the melting point of the alloy, seems to be more important in determining the thermal stability of the glassy alloy.

The soft magnetic properties of BMGs are very good. They have a low coercivity and very high permeability at high frequencies. The saturation magnetization can also be suitably adjusted by optimizing the chemical composition. But, in comparison to the melt-spun ribbons, the composition range in which the BMGs could be obtained seems to be limited, and they require a greater number of components. Consequently, the saturation magnetization is low. In other words, there appear to be many possibilities to obtain metallic glasses using the melt-spinning technique; however, BMGs also have some specific advantages, such as their high glass-forming ability and ease of forming operations in the supercooled liquid state.

From this discussion, it is clear that while BMGs have some unique features, there are also difficulties and unattractive features associated with commercializing them. The specific advantages of obtaining the glassy alloys in the BMG condition over the melt-spun ribbon condition for commercial applications are still not very clear. It is becoming more and more apparent that the special attributes of BMGs can be better exploited when the volume of the product produced using the BMG is small, that is, for small-sized products. Processing costs could be significantly reduced in this case. Thus, one could think of imprinting, micro-gears, MEMS parts, ID tags, and so on, as potential applications. Alternatively, the BMGs could also be used in applications where cost is not the main consideration; for example, in military applications, space applications, surgery, and security areas. Additionally, it is worthwhile using the basic scientific knowledge gained in the study of BMGs to develop new *crystalline* materials (maybe nanostructured materials) at a low cost.

It is most important to remember that for successful commercial exploitation of BMGs, it is essential that we have a low-cost production process for mass production and commercialization. The BMGs amply satisfy these requirements, because forming them in the supercooled liquid region could be rapid and also less expensive. The current high cost of the BMGs can be a big deterrent for their use in consumer goods. Extensive usage of the BMGs in several different applications could increase their volume of production and consequently bring down their cost.

Finally, as encouraging information on the application of BMGs, there have been recent reports that some companies in China have been applying net-shape cast Zr–Al–Ni–Cu base BMG alloys as structural parts, which include a SIM card tray, a pin-type spring and case, and so on, in smartphones. These applications simultaneously utilize high strength, large elastic strain, high corrosion resistance, high wear resistance, surface smoothness, and good net castability [69]. This new trend seems to be promising for significant extension of BMG applications in the future.

References

1. Inoue, A., N. Nishiyama, and H.M. Kimura (1997). Preparation and thermal stability of bulk amorphous $Pd_{40}Cu_{30}Ni_{10}P_{20}$ alloy cylinder of 72 mm in diameter. *Mater. Trans., JIM* 38: 179–183.
2. Nishiyama, N., K. Takenaka, H. Miura, N. Saisoh, Y. Zeng, and A. Inoue (2012). The world's biggest glassy alloy ever made. *Intermetallics* 30: 19–24.
3. Lu, I.-R., G. Wilde, G.P. Görler, and R. Willnecker (1999). Thermodynamic properties of Pd-based glass-forming alloys. *J. Non-Cryst. Solids* 250–252: 577–581.
4. Kawamura, Y., T. Shibata, A. Inoue, and T. Masumoto (1997). Superplastic deformation of $Zr_{65}Al_{10}Ni_{10}Cu_{15}$ metallic glass. *Scr. Mater.* 37: 431–436.

5. Schroers, J. (2005). The superplastic forming of bulk metallic glasses. *JOM* 57(5): 35–39.
6. Kim, W.J., J.B. Lee, and H.G. Jeong (2006). Superplastic gas pressure forming of $Zr_{65}Al_{10}Ni_{10}Cu_{15}$ metallic glass sheets fabricated by squeeze mold casting. *Mater. Sci. Eng. A* 428: 205–210.
7. Schroers, J., T. Nguyen, S. O'Keeffe, and A. Desai (2007). Thermoplastic forming of bulk metallic glass—applications for MEMS and microstructure fabrication. *Mater. Sci. Eng. A* 449–451: 898–902.
8. Liu, Y.H., G. Wang, R.J. Wang, D.Q. Zhao, M.X. Pan, and W.H. Wang (2007). Superplastic bulk metallic glasses at room temperature. *Science* 315: 1385–1388.
9. Inoue, A. and N. Nishiyama (2007). New bulk metallic glasses for applications as magnetic-sensing, chemical, and structural materials. *MRS Bull.* 32(8): 651–658.
10. Inoue, A. and A. Takeuchi (2011). Recent development and application products of bulk glassy alloys. *Acta Mater.* 59: 2243–2267.
11. Ashby, M.F. and A.L. Greer (2006). Metallic glasses as structural materials. *Scr. Mater.* 54: 321–326.
12. Salimon, A., Y. Bréchet, M.F. Ashby, and A.L. Greer (2004). Selection of applications for a material. *Adv. Eng. Mater.* 6: 249–265.
13. Liquidmetal Technologies homepage. http://www.liquidmetal.com.
14. Kakiuchi, H., A. Inoue, M. Onuki, Y. Takano, and T. Yamaguchi (2001). Application of Zr-based bulk glassy alloys to golf clubs. *Mater. Trans.* 42: 678–681.
15. Sharma, P., W. Zhang, K. Amiya, H.M. Kimura, and A. Inoue (2005). Nanoscale patterning of Zr–Al–Cu–Ni metallic glass thin films deposited by magnetron sputtering. *J. Nanosci. Nanotech.* 5: 416–420.
16. Ishida, M., H. Takeda, D. Watanabe, K. Amiya, N. Nishiyama, K. Kita, Y. Saotome, and A. Inoue (2004). Fillability and imprintability of high-strength Ni-based bulk metallic glass prepared by the precision die-casting technique. *Mater. Trans.* 45: 1239–1244.
17. Ishida, M., H. Takeda, N. Nishiyama, K. Kita, Y. Shimizu, Y. Saotome, and A. Inoue (2007). Wear resistivity of super-precision microgear made of Ni-based metallic glass. *Mater. Sci. Eng. A* 449–451: 149–154.
18. Inoue, A., B.L. Shen, and A. Takeuchi (2006). Fabrication, properties and applications of bulk glassy alloys in late transition metal-based systems. *Mater. Sci. Eng. A* 441: 18–25.
19. Inoue, A., B.L. Shen, and A. Takeuchi (2006). Developments and applications of bulk glassy alloys in late transition metal base system. *Mater. Trans.* 47: 1275–1285.
20. Son, K., H. Soejima, N. Nishiyama, X.M. Wang, and A. Inoue (2007). Process development of metallic glass wires by a groove quenching technique for production of coil springs. *Mater. Sci. Eng. A* 449–451: 248–252.
21. Nishiyama, N., K. Amiya, and A. Inoue (2007). Recent progress of bulk metallic glasses for strain-sensing devices. *Mater. Sci. Eng. A* 449–451: 79–83.
22. Nishiyama, N., K. Amiya, and A. Inoue (2007). Novel applications of bulk metallic glass for industrial products. *J. Non-Cryst. Solids* 353: 3615–3621.
23. Ma, C.L., N. Nishiyama, and A. Inoue (2005). Fabrication and characterization of Coriolis mass flowmeter made form Ti-based glass tubes. *Mater. Sci. Eng. A* 407: 201–206.
24. Hata, S., N. Yamada, Y. Saotome, A. Inoue, and A. Shimokohbe (1998). Precision and micromachining of metallic glasses—formability of Zr-based metallic glasses in the supercooled liquid state. In *Proceedings of China-Japan Bilateral Conference on Advanced Manufacturing Engineering*, Huangshan City, China, 1998, pp. 81–86.
25. Soejima, H., N. Nishiyama, H. Takehisa, M. Shimanuki, and A. Inoue (2005). Viscous flow forming of Zr-based bulk metallic glasses for industrial products. *J. Metastable Nanocryst. Mater.* 24–25: 531–534.
26. Shen, J., Q. Chen, J. Sun, H. Han, and G. Wang (2005). Exceptionally high glass-forming ability of an FeCoCrMoCBY alloy. *Appl. Phys. Lett.* 86: 151907-1–151907-3.
27. Inoue, A., I. Yoshii, H.M. Kimura, K. Okumura, and J. Kurosaki (2003). Enhanced shot peening effect for steels by using Fe-based glassy alloy shots. *Mater. Trans.* 44: 2391–2395.

28. Inoue, A. (2000). Stabilization of metallic supercooled liquid and bulk amorphous alloys. *Acta Mater.* 48: 279–306.
29. Yamanaka, S., K. Amiya, Y. Saotome, and A. Inoue (2011). Plastic working of metallic glass bolts by cold thread rolling. *Mater. Trans.* 52: 243–249.
30. Yamanaka, S., K. Amiya, and Y. Saotome (2014). Effects of residual stress on elastic plastic behavior of metallic glass bolts formed by cold thread rolling. *J. Mater. Process. Technol.* 214: 2593–2599.
31. Yokoyama, Y., K. Yamano, K. Fukaura, H. Sunada, and A. Inoue (2001). Enhancement of ductility and plasticity of $Zr_{55}Cu_{30}Al_{10}Ni_5$ bulk glassy alloy by cold rolling. *Mater. Trans.* 42: 623–632.
32. Masumoto, T., I. Ohnaka, A. Inoue, and M. Hagiwara (1981). Production of Pd–Cu–Si amorphous wires by melt spinning method using rotating water. *Scr. Metall.* 15: 293–296.
33. Kawashima, A., Y. Yokoyama, and A. Inoue (2010). Zr-based bulk glassy alloy with improved resistance to stress corrosion cracking in sodium chloride solutions. *Corros. Sci.* 52: 2950–2957.
34. Inoue, A., Y. Shinohara, and J.S. Gook (1995). Thermal and magnetic properties of bulk Fe-based glassy alloys prepared by copper mold casting. *Mater. Trans., JIM* 36: 1427–1433.
35. Otsubo, F., H. Era, and K. Kishitake (2000). Formation of amorphous Fe–Cr–Mo–8P–2C coatings by the high velocity oxy-fuel process. *J. Thermal Spray Techn.* 9: 494–498.
36. Otsubo, F. and K. Kishitake (2005). Corrosion resistance of Fe–16%Cr–30%Mo–(C,B,P) amorphous coatings sprayed by HVOF and APS processes. *Mater. Trans.* 46: 80–83.
37. Kobayashi, A., S. Yano, H. Kimura, and A. Inoue (2008). Mechanical property of Fe-base metallic glass coating formed by gas tunnel type plasma spraying. *Surf. Coat. Technol.* 202: 2513–2518.
38. Yoshioka, H., K. Asami, A. Kawashima, and K. Hashimoto (1987). Laser-processed corrosion-resistant amorphous Ni-Cr-P-B surface alloys on a mild steel. *Corros. Sci.* 27: 981–995.
39. Kobayashi, A., S. Yano, H. Kimura, and A. Inoue (2008). Fe-based metallic glass coatings produced by smart plasma spraying process. *Mater. Sci. Eng. B* 148: 110–113.
40. Liu, W.D., K.X. Liu, Q.Y. Chen, J.T. Wang, H.H. Yan. and X.J. Li (2009). Metallic glass coating on metal plate by adjusted explosive welding technique. *Appl. Sur. Scif.* 255: 9343–9347.
41. Kim, H.G., K. Nakata, T. Tsumura, M. Sugiyama, T. Igarashi, M. Fukumoto, H. Kimura, and A. Inoue (2006). Properties of metallic glass coating on an aluminum alloy substrate produced using a HVOF spraying process. In *Characterization and Control of Interfaces for High Quality Advanced Materials II*, eds. K. Ewsuk, K. Nogi, R. Waesche, Y. Umakoshi, T. Hinklin, K. Uematsu, T. Tomsia, H. Abe, H. Kamiya, and M. Naito. Hoboken, NJ: John Wiley & Sons, Inc.; *Ceramic Trans.* 198: 69–77.
42. Kim, H.G., K. Nakata, T. Tsumura, M. Sugiyama, T. Igarashi, M. Fukumoto, H. Kimura, and A. Inoue (2008). Effect of particle size distribution of the feedstock powder on the microstructure of bulk metallic glass sprayed coating by HVOF on aluminum alloy substrate. *Mater. Sci. Forum* 580–582: 467–470.
43. Igarashi, T. (2012). Characteristics and practical application of thermal spraying metallic glasses coating (GALOA) (in Japanese), *Thermal Spray. Tech.* 32(2): 93–96.
44. Sugiyama, M., T. Igarashi, T. Okano, H. Kimura, and A. Inoue (2007). Development of the glassy metal coating by thermal spray and application for the Pb-free solder erosion resistant. *Mater. Japan* 46(1): 31–33.
45. Jayaraj, J., Y.C. Kim, H.K. Seok, K.B. Kim, and E. Fleury (2007). Development of metallic glasses for bipolar plate applications. *Mater. Sci. Eng. A* 449–451: 30–33.
46. Inoue, A., T. Shimizu, S. Yamaura, Y. Fujita, S. Takagi, and H.M. Kimura (2005). Development of glassy alloy separators for a proton exchange membrane fuel cell (PEMFC). *Mater. Trans.* 46: 1706–1710.
47. Sistiaga, M. and A.R. Pierna (2003). Application of amorphous materials for fuel cells. *J. Non-Cryst. Solids* 329: 184–187.
48. Kim, S., S. Yamaura, Y. Shimizu, K. Nakashima, T. Igarashi, A. Makino, and A. Inoue (2010). Production of $Ni_{65}Cr_{15}P_{16}B_4$ metallic glass-coated bipolar plate for fuel cell by high velocity oxy-fuel (HVOF) spray coating method. *Mater. Trans.* 51: 1609–1613.

49. Kobayashi, A., T. Kuroda, H. Kimura and A. Inoue (2010). Effect of spraying condition on property of Zr-based metallic glass coating by gas tunnel type spraying. *Mater. Sci. Eng.* B 173: 122–125.
50. Kim, S.C., S. Yamaura, T. Igarashi, Y. Shimizu, K. Nakashima, A. Makino, and A. Inoue (2011). Production of Ni–Cr–P–B alloy-coated bipolar Plates for PEMFC by HVOF spray-coating and their surface analysis by XPS. *J. Japan Inst. Met.* 75: 122–130.
51. Kim, S.C., S. Yamaura, A. Makino, and A. Inoue (2011). Production of Ni–P amorphous alloy-coated bipolar plate for PEM fuel cell by electroless plating. *J. Japan Inst. Met.* 75: 557–561.
52. Yamaura, S., K. Katsumata, M. Hattori, and T. Yogo (2013). Production of Ni-based glassy alloy-coated bipolar plate with hydrophilic surface for PEMFC and its evaluation by electrochemical impedance spectroscopy. *Mater. Trans.* 54: 1324–1329.
53. Smith, C.H. (1993). Applications of rapidly solidified soft magnetic alloys. In *Rapidly Solidified Alloys: Processes, Structures, Properties, Applications*, ed. H.H. Liebermann, pp. 617–663. New York: Marcel Dekker.
54. Hasegawa, R. (2001). Applications of amorphous magnetic alloys in electronic devices. *J. Non-Cryst. Solids* 287: 405–412.
55. Hasegawa, R. (2004). Applications of amorphous magnetic alloys. *Mater. Sci. Eng. A* 375–377: 90–97.
56. Nishiyama, N., K. Amiya, and A. Inoue (2004). Bulk metallic glasses for industrial products: New structural and functional applications. In *Amorphous and Nanocrystalline Metals*, Symposium Proceedings, eds. R. Busch, T.C. Hufnagel, J. Eckert, A. Inoue, W.L. Johnson, and A.R. Yavari, Vol. 806, pp. 387–392. Warrendale, PA: Materials Research Society.
57. Yoshida, S., T. Mizushima, T. Hatanai, and A. Inoue (2000). Preparation of new amorphous powder cores using Fe-based glassy alloy. *IEEE Trans. Magn.* 36: 3424–3429.
58. Pankhurst, Q.A., N.T.K. Thanh, S.K. Jones, and J. Dobson (2009). Progress in applications of magnetic nanoparticles in biomedicine. *J. Phys. D: Appl. Phys.* 42: 224001 (15 pages).
59. Chiriac, H., N. Lupu, M. Lostun, G. Ababei, M. Grigoraş, and C. Dănceanu (2014). Low T_C Fe–Cr–Nb–B glassy submicron powders for hyperthermia applications. *J. Appl. Phys.* 115: 17B520 (3 pages).
60. Chiriac, H., L. Whitmore, M. Grigoras, G. Ababei, G. Stoian, and N. Lupu (2015). Influence of Cr on the nanoclusters formation and superferromagnetic behavior of Fe–Cr–Nb–B glassy alloys. *J. Appl. Phys.* 117: 17B522.
61. Schroers, J., B. Lohwongwatana, W.L. Johnson, and A. Peker (2007). Precious bulk metallic glasses for jewelry applications. *Mater. Sci. Eng. A* 449–451: 235–238.
62. Zhu, S.L., X.M. Wang, F.X. Qin, and A. Inoue (2007). A new Ti-based bulk glassy alloy with potential for biomedical application. *Mater. Sci. Eng. A* 459: 233–237.
63. Zhu, S.L., X.M. Wang, G.Q. Xie, F.X. Qin, M. Yoshimura, and A. Inoue (2008). Formation of Ti-based bulk glassy alloy/hydroxyapatite composite. *Scr. Mater.* 58: 287–290.
64. Zhu, S.L., X.M. Wang, and A. Inoue (2008). Glass-forming ability and mechanical properties of Ti-based bulk glassy alloys with large diameters of up to 1 cm. *Intermetallics* 16: 1031–1035.
65. Qin, F., G. Xie, X. Wang, T. Wada, M. Song, K. Furuya, M. Yoshimura, M. Tsukamoto, and A. Inoue (2009). Microstructure and electrochemical properties of PVD TiN, (Ti,Al) N-coated Ti-based bulk metallic glasses. *Mater. Trans.* 50: 1313–1317.
66. Onoki, T., X. Wang, S. Zhu, N. Sugiyama, Y. Hoshikawa, M. Akao, N. Matsushita, A. Nakahira, E. Yasuda, M. Yoshimura, and A. Inoue (2009). Effects of growing integrated layer [GIL] formation on bonding behavior between hydroxyapatite ceramics and Ti-based bulk metallic glasses via hydrothermal hot-pressing. *Mater. Sci. Eng. B* 161: 27–30.
67. Wu, F.F., Z.F. Zhang, S.X. Mao, and J. Eckert (2009). Effect of sample size on ductility of metallic glass. *Philos. Mag. Lett.* 89: 178–184.
68. Telford, M. (2004). The case for bulk metallic glass. *Mater. Today* 7: 36–43.
69. Second Conference on Structural and Functional Glassy Alloys, organized by B.L. Shen, Xuzhou, China, May 19–20, 2017.

11

Epilogue

11.1 Introduction

We have so far surveyed the different growth stages of bulk metallic glasses (BMGs), starting from their inception in 1989, to their present-day status of becoming a mature field. We had looked at the formative stages, when BMGs were produced by water quenching in an La-based alloy, and then a major breakthrough, when a 14 mm diameter BMG rod was produced in a Zr-based alloy. It is estimated that about 8000 research publications are available in the open literature on the different aspects of BMGs, and international conferences are periodically organized in different parts of the world.

The properties of BMGs do not appear to be significantly different from those of melt-spun metallic glass ribbons. But there are differences between BMGs and melt-spun glassy ribbons. It has been shown that BMGs offer excellent opportunities to fabricate complex-shaped products because of their excellent forming characteristics in the supercooled liquid region, i.e., the temperature interval between the glass transition and crystallization temperatures. Further, the supercooled liquid region has provided a most convenient way to study the physical properties (especially the transport properties) in this otherwise inaccessible region. In comparison with their crystalline counterparts, BMGs have shown impressive property enhancements. The question on everyone's mind is: where do we go from here and in which direction?

It is always difficult to predict the future. It is more so if one has to predict the future trends of an adolescent technology such as BMGs, since it is known that it takes about 15–20 years for any invention to be successfully marketed for commercial applications. Since BMGs were first reported only in 1989, we can consider that BMGs are just round the corner for serious consideration as suitable materials to replace the existing materials.

Before we venture into predicting the future of BMGs, let us see the current status of BMGs. Figure 11.1 shows the best properties that have been achieved so far. BMGs have been prepared with a section thickness of a few millimeters to a few centimeters, with the maximum diameter achieved being 80 mm in a Pd-based alloy [1]. Large supercooled liquid regions of 131 K were obtained in a Pd-based BMG alloy [2]. Extremely high fracture strengths of over 5 GPa were obtained in a Co-based BMG alloy [3]. Plastic elongations of nearly 20% were achieved in a Pt-based BMG alloy [4]. Interesting soft magnetic properties of saturation magnetization of 1.5 T, coercivity of 1 A m^{-1}, and magnetostriction of 2×10^{-6} were obtained in an Fe-based BMG alloy [5]. A reasonably high magnetic energy product of 19 kJ m^{-3} was obtained in an Nd-based BMG alloy [6]. BMG composites containing a large volume fraction of the reinforcements and with good ductility have been synthesized [7]. All these facts suggest that significant improvements have been achieved in the synthesis and properties of these novel materials.

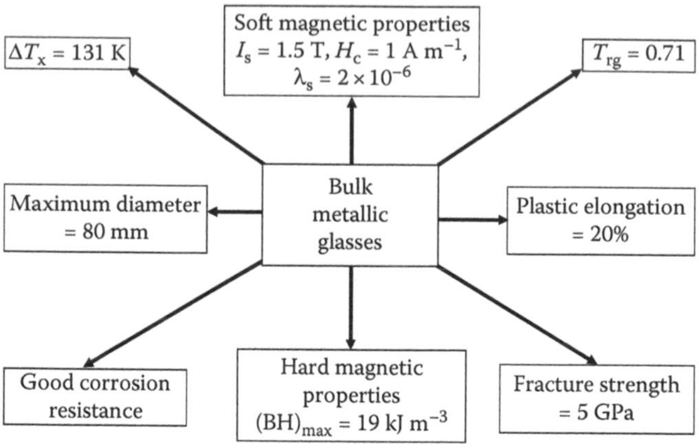

FIGURE 11.1
Current achievements of BMGs.

11.2 Size and Shape

There has been a constant drive to increase the diameter of the BMG rod/section thickness of BMGs in different alloy systems. This has been achieved through innovative chemistries and design principles based on an approach to searching 3-D compositional space to improve the glass-forming ability (GFA) [8]. Figure 11.2 shows the increases in the critical diameters of Cu-based and Mg-based BMGs. It is interesting to note that the maximum diameter for a fully glassy rod has increased up to 25 mm in a Cu-based alloy [9] and up to 27 mm in an Mg-based alloy [10]. Further, the largest diameter of 72 mm for a fully glassy rod established in 1997 for a Pd-based alloy [11] has been surpassed only recently [1].

It will be most desirable to obtain even larger critical diameters in different alloy systems. For example, the potential for applications will certainly increase if one is able to synthesize fully glassy rods with a critical diameter of over 10 cm. But, it would be even better if this could be achieved in the most useful commercial alloy systems, e.g., those based on Al and Fe. The present scientific understanding of the GFA of alloys has not been able to produce rods larger than 16 mm in the Fe-based alloy systems, and it has not been possible to obtain even a 1 mm diameter rod in the Al alloy system. Therefore, it would be extremely useful if BMG rods could be obtained with diameters of about 30 and 5 mm in the Fe- and Al-based alloy systems, respectively. Precise and reliable computer simulation methods to identify the components of an alloy system possessing very good GFA could be a valuable asset in this direction.

It is, however, interesting to note that, most recently, 50 mm diameter Ti [12] and 73 mm diameter Zr [13] BMGs were produced. Addition of rare earths has been known to improve the GFA of alloys. Other additions, such as Ag, also have been shown to increase the GFA. Thus, one can possibly use the well-established principles to design suitable compositions to produce large-diameter glasses.

Current reports indicate that it has been possible to produce BMG alloys in different forms, such as rods, sheets, tubes, spheres, and others. The dimensions of these products have been limited. The likelihood that these products will be used in applications will significantly increase when sheets with a reasonably large surface area (1 × 1 m), longer rods or cylinders with a length of over 2 m, and long tubes with a small wall thickness of about

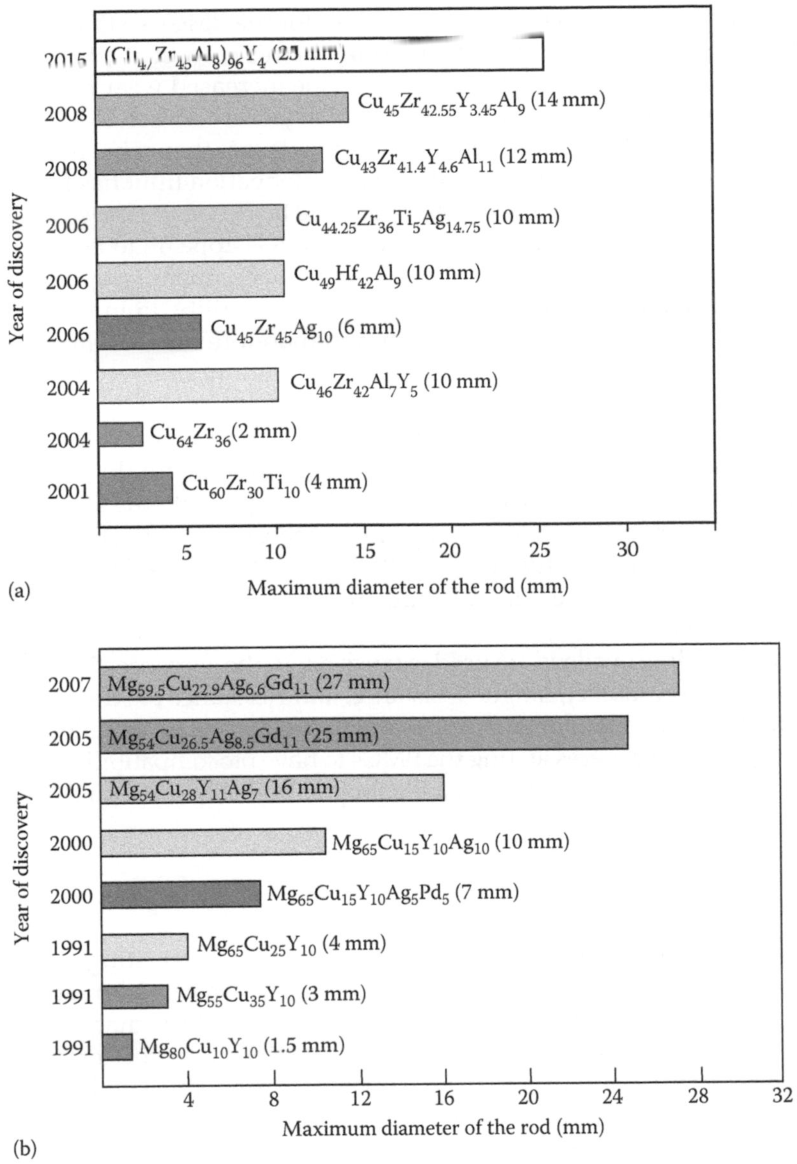

FIGURE 11.2
Increase in the critical diameter of BMG alloys based on (a) copper and (b) magnesium.

0.1 mm can be produced. All these are doable, and the existing manufacturing methods need to be modified to achieve these goals.

11.3 Mechanical Properties

Ultra-high strength has been reported in a large number of BMG alloys, with the strength exceeding about 3 GPa; the highest strength reported was over 5 GPa. Since it is well known that high strength (and hardness) and wear resistance go together, it is only natural to hope to obtain higher and higher strength values for the BMGs so that they can be used in applications where wear resistance is important. In this respect, it will be useful to have even higher strength, say exceeding 6 GPa, to provide increased wear resistance.

Associated with the high strength, the elastic elongation limit of BMG alloys has been reported to be about 2%. This value is about four times larger than what has been traditionally achieved in a crystalline alloy. This high elastic elongation limit has found interesting applications in springs and sporting goods. But, it will be beneficial if we can achieve a still higher value. The reduced modulus of elasticity of BMGs appears to be a very important property that has already helped in finding a number of commercial applications.

The Achilles' heel of BMG alloys has been their low ductility. In most cases, the BMGs experience only elastic deformation in tension, and therefore the ductility has been virtually zero. But, through innovative approaches, some amount of ductility has been introduced into the BMG alloys—mostly through the introduction of a ductile reinforcement. Some success has been achieved in developing superelastic alloys [14]. But, if a reasonable amount of ductility, say about 5%, could be achieved in a monolithic BMG alloy (without sacrificing the strength by introducing a ductile component), the fabrication of BMGs could be achieved more easily. Additionally, if the fracture toughness of the BMGs could also be increased to over 120 MPa $m^{1/2}$, the BMGs could be more forgiving and tolerant of the presence of cracks in the alloys. Fatigue resistance and stress-corrosion resistance could also be improved.

Aluminum and titanium alloys are used extensively in the aerospace industry. High specific strength, good formability, fatigue resistance, and resistance to stress-corrosion cracking are the properties needed for such applications. Further, titanium alloys are also used for biological applications, necessitating the BMGs to have biocompatibility. Improved BMG alloys based on Al and Ti with these desirable properties will be welcome developments.

11.4 Magnetic Properties

Metallic glasses in general, and BMGs in particular, exhibit interesting magnetic properties. They show both soft and hard magnetic behavior. If we can learn any lessons from the experience of melt-spun metallic glasses, it is possible that the magnetic properties of BMG alloys will be the most important ones in exploiting them for commercial applications. But, due to the large number of alloying elements needed to synthesize BMG alloys, the saturation magnetization of BMGs is low. The presence of rare earth elements in the BMGs is also not desirable. Therefore, if BMGs can be synthesized without the rare earth elements and with the minimum of alloying elements, the saturation magnetization can be high, say above about 1.8 T. Further, such BMGs will also be more affordable, since the expensive rare earth elements are not being used. Additionally, even though the magnetostriction of the BMGs is relatively low, a zero value has not been achieved so far. Hence, the development of alloy compositions with zero magnetostriction and ultra-high permeability is an urgent need.

11.5 Fundamental Properties

A vast scientific base has been developed with the synthesis and characterization of BMGs. A very large number of BMG compositions have been synthesized in the glassy state. In spite of this, we do not seem to have a clear idea of the GFA of alloys and which compositions will produce fully glassy alloys with a large section thickness. The mechanism for the high stability of the supercooled liquid is still not clear. Further, it is still a matter of speculation regarding the width of the supercooled liquid region, ΔT_x. For example, is it possible to obtain a ΔT_x value of over 200 K in any alloy system, and if so, how do we achieve it? Can we also predict the glass transition temperature? Can it be a really very low value, say 300 K? Can we identify alloys with a high reduced glass transition temperature, T_{rg}? All these are still open questions without any clear answers.

It has been possible to study the rheological behavior of metallic glasses in the supercooled liquid region. Interesting observations have been made on the diffusion behavior of BMGs, and many other investigations have been conducted to obtain basic information about the structure and properties of BMGs. While the results obtained have produced useful information, they have also raised additional questions. Thus, there appears to be a lot of scope to continue to delve deeper into the subject and generate new ideas.

11.6 Concluding Remarks

BMGs have been shown to exhibit an interesting combination of physical, mechanical, chemical, and magnetic properties. New scientific information has been generated during the last 25 years, and the rate at which progress is being made during the last 5–10 years is phenomenal. A number of applications have been suggested for these materials. There are still many unsolved problems, and the chances of these novel materials being put to industrial uses are very high once solutions are found for these problems.

As mentioned in Chapter 10, the adoption of a new material into the existing systems is besieged by many hurdles. Among the many issues to be resolved, the most important is the cost. Therefore, if we are able to produce these BMG alloy parts at low cost and with high reliability, chances are that these will be adopted. In addition, we also need to look into the recycling of the used parts. Further, a very profitable avenue to be pursued will be to impart novel functionalities to BMGs through forming composites and multiphase BMGs, and this is likely to open up new areas of research and increase their application potential. The fact that one has to consider the full cycle of the material from cradle to grave cannot be overemphasized.

References

1. Nishiyama, N., K. Takenaka, H. Miura, N. Saidoh, Y. Zeng, and A. Inoue (2012). The world's biggest glassy alloy ever made. *Intermetallics* 30: 19–24. (doi:10.1016/j.intermet.2012.03.020)
2. Lu, I.-R., G. Wilde, G.P. Görler, and R. Willnecker (1999). Thermodynamic properties of Pd-based glass-forming alloys. *J. Non-Cryst. Solids* 250–252: 577–581.
3. Inoue, A., B.L. Shen, H. Koshiba, H. Kato, and A.R. Yavari (2003). Cobalt-based bulk glassy alloy with ultrahigh strength and soft magnetic properties. *Nat. Mater.* 2: 661–663. (doi:10.1038/nmat982)
4. Schroers, J. and W.L. Johnson (2004). Ductile bulk metallic glass. *Phys. Rev. Lett.* 93: 255506-1–255506-4.

5. Makino, A., T. Kubota, C.T. Chang, M. Makabe, and A. Inoue (2007). FeSiBP bulk metallic glasses with unusual combination of high magnetization and high glass-forming ability. *Mater. Trans.* 48: 3024–3027. (doi:10.2320/matertrans.MRP2007198)
6. Inoue, A., T. Zhang, W. Zhang, and A. Takeuchi (1996). Bulk Nd–Fe–Al amorphous alloys with hard magnetic properties. *Mater. Trans. JIM* 37: 99–108. (doi:10.2320/matertrans1989.37.99)
7. Hoffmann, D.C., J.Y. Suh, A. Wiest, G. Duan, M.L. Lind, M.D. Demetriou, and W.L. Johnson (2008). Designing metallic glass matrix composites with high toughness and tensile ductility. *Nature* 451: 1085–1090. (doi:10.1038/nature06598)
8. Li, Y., S.J. Poon, G.J. Shiflet, J. Xu, D.H. Kim, and J.F. Löffler (2007). Formation of bulk metallic glasses and their composites. *MRS Bull.* 32: 624–628. (doi:10.1557/mrs2007.123)
9. Deng, L., B. Zhou, H.S. Yang, X. Jiang, B. Jiang, and X.G. Zhang (2015). Roles of minor rare-earth elements addition in formation and properties of Cu–Zr–Al bulk metallic glasses. *J. Alloys Compd.* 632: 429–434. (doi:10.1016/j.jallcom.2015.01.036)
10. Zheng, Q., J. Xu, and E. Ma (2007). High glass-forming ability correlated with fragility of Mg–Cu(Ag)–Gd alloys. *J. Appl. Phys.* 102: 113519-1–113519-5.
11. Inoue, A., N. Nishiyama, and H.M. Kimura (1997). Preparation and thermal stability of bulk amorphous $Pd_{40}Cu_{30}Ni_{10}P_{20}$ alloy cylinder of 72 mm diameter. *Mater. Trans. JIM* 38: 179–183. (doi:10.2320/matertrans1989.38.179)
12. Zhang, L., M.Q. Tang, Z.W. Zhu, H.M. Fu, H.W. Zhang, A.M. Wang, H. Li, H.F. Zhang, and Z.Q. Hu (2015). Compressive plastic metallic glasses with exceptional glass forming ability in the Ti–Zr–Cu–Fe–Be alloy system. *J. Alloys Compd.* 638: 349–355. (doi:10.1016/j.jallcom.2015.03.120)
13. Lou, H.B., X.D. Wang, F. Xu, S.Q. Ding, Q.P. Cao, K. Hono, and J.Z. Jiang (2011). 73 mm-diameter bulk metallic glass rod by copper mould casting. *Appl. Phys. Lett.* 99: 051910-1–051910-3.
14. Tsarkov, A.A., A.Yu. Churyumov, V.Yu. Zadorozhnyy, and D.V. Louizguine-Luzgin (2016). High-strength and ductile (Ti–Ni)–(Cu–Zr) crystalline/amorphous composite materials with superelasticity and TRIP effect. *J. Alloys Compd.* 658: 402–407.

Index

G

H

I

Y

Z

Milton Keynes UK
Ingram Content Group UK Ltd.
UKHW051904071024
449327UK00025B/2088

9 780367 657505